STUDENT SOLUTIONS

LAUREL TECHNICAL S

INTERMEDIATE ALGEBRA

SECOND EDITION

K. ELAYN MARTIN-GAY

PRENTICE HALL
Upper Saddle River, NJ 07458

Acquisitions Editor: ***Melissa Acuña/Ann Marie Jones***
Production Editor: ***James Buckley***
Production Supervisor: ***Barbara Murray***
Art Director: ***Amy Rosen***
Buyer: ***Alan Fischer***
Supplements Editor: ***April Thrower***

Simon & Schuster/A Viacom Company
Upper Saddle River, NJ 07458

Printed in the United States of America

10 9 8 7 6 5 4 3 2

ISBN 0-13-258096-9

Prentice-Hall International (UK) Limited, *London*
Prentice-Hall of Australia Pty. Limited, *Sydney*
Prentice-Hall Canada, Inc., *Toronto*
Prentice-Hall Hispanoamericana, S.A., *Mexico*
Prentice-Hall of India Private Limited, *New Delhi*
Prentice-Hall of Japan, Inc., *Tokyo*
Simon & Schuster Asia Pte. Ltd., *Singapore*
Editora Prentice-Hall do Brasil, Ltda., *Rio de Janeiro*

Table of Contents

Study for Success

Consider the following tips for becoming more successful in your mathematics course.

- Attend each class. If you must miss a class, borrow notes from another student and make a copy. Call and get the homework assignment so that you do not get behind. Consider viewing the videotapes that accompany this textbook if they are available at your school's library or media resource center.

- In your mathematics classroom or lab, be sure you have a good view of the board or overhead projector screen and that you can hear your instructor or lab leader.

- Come to class prepared. Use a notebook just for this class, preferably consisting of a section for notes and a section for homework. Bring the appropriate writing utensils.

- You may be forced to take notes quickly during class. Set up your own abbreviation system. Take notes as thoroughly as possible. After class you can rewrite your notes more legibly and fill in any gaps.

- It is important to keep good notes in your mathematics class because these will become yet another resource when you do your homework, study for exams, and prepare for later courses. If you can't remember how to solve a problem when doing your homework, check your notes to see how your instructor suggested to do the problem. You can also refer to the examples in the textbook or this Study Guide.

- Be sure you understand what has been assigned for homework. You can always double-check the assignment with another student in the class. Do your homework after class as soon as possible. This way the information is still fresh in your mind.

- You may find it helpful to find one or more study partners. Consider working together after class to review the material covered during class and to be available as resources to one another while completing homework assignments. You may also find that explaining concepts to a study partner helps to clarify the concepts in your own mind. When preparing for tests and exams, you and your study partners may find it helpful to quiz each other on major concepts.

- Practice is what improves your skills in every area, especially mathematics. Practice is what will make you become comfortable with problems. If there are certain assigned problems that you are having difficulty with, do more of this type until you can do them easily. Remember that this Study Guide provides you with lots of additional problems for practice.

- As you do homework problems and read the textbook, write down your questions. Don't rely on your memory, you may have forgotten an important part of your question by the time you reach your class.

- As you read your textbook, don't just read the examples. The entire section is important in your understanding of the material. Don't forget that this Study Guide provides additional examples if you require further clarification. The numerous additional exercises in this Study Guide will also give you extra practice on any topics for which you need it.

- Don't be afraid to ask questions. If something is not clear, ask for further explanation. Usually there are other students with the same questions.

- Be sure to hand in assignments on time. Don't lose points needlessly.

- Take advantage of all extra-credit opportunities if they exist.

- When doing homework or taking a test, give yourself plenty of paper. Show every step of a problem in an organized manner. This will help reduce the number of careless mistakes.

- When you are studying for a test or exam, you may find the practice chapter tests in this Study Guide a helpful resource of review and self-assessment. There are also two different practice final exams included in this Study Guide.

- It is important to study your mathematics as many days as you can possibly fit into your schedule. Even if it is just for 10 or 15 minutes, this is beneficial. Do not wait until just before a test to do a crash study session. The material should be understood and learned, not memorized. The exercises in this Study Guide provide you with additional resources for practicing your skills and keeping them current. Most likely you will take another mathematics course after this one. You are gaining tools in this class that you will use in the next class.

- Many times students can do their homework assignments but then "freeze up" on a test. Their minds go blank, they panic, and, hence, they do not do well on their test. If this happens to you, it can be overcome by practicing taking tests. When studying for a test, make up an actual test for yourself or use one if the practice tests in this Study Guide if appropriate. Then find a quiet spot and pretend you are in class. Take this test just as if you were in class. The more you practice taking tests, the more comfortable you will be in class during the actual event.

Consider the following guidelines for working cooperatively in groups:

- Be sure to have each group member introduce himself or herself to the rest of the group.
- Agree on what your group must do and how you will get it done.
- Be courteous and listen carefully to other group members.
- Create an atmosphere that allows group members to be comfortable in asking for help when needed.
- Remember that not everyone works at the same pace. Be patient. Offer your help. Receive help graciously.
- Keep in mind that all contributions made by group members are valuable.
- Ask for your instructor's help only when all other sources of assistance have been exhausted.
- If you finish early, double-check your work. As appropriate, reflect on your work. Can you make a general rule about the solution or describe a real-world use for what you learned?

Chapter 1

Exercise Set 1.1

1. $5x = (5)(7) = 35$

3. $9.8z = (9.8)(3.1) = 30.38$

5. $xy = (17)(11) = 187$

7. $ab = \left(\frac{1}{2}\right)\left(\frac{3}{4}\right) = \frac{3}{8}$

9. $3x + y = (3)(6) + 4 = 18 + 4 = 22$

11. $qr - s = (1)(14) - 3 = 14 - 3 = 11$

13. $400t = (400)(5) = 2000$ miles

15. $lw = (5.1)(4) = 20.4$ square feet

17. $7098t = (7098)(5.2) = \$36,909.60$

19. $\{1, 2, 3, 4, 5\}$

21. $\{11, 12, 13, 14, 15, 16\}$

23. $\{0\}$

25. $\{0, 2, 4, 6, 8\}$

27. $\{0, 2, 4, 6\}$

−1 0 2 4 6

29. $\left\{\frac{1}{2}, \frac{2}{3}\right\}$

−1 $-\frac{1}{2}$ 0 $\frac{1}{2}$ $\frac{2}{3}$ 1

31. $\{-2, -6, -10\}$

−10 −8 −6 −4 −2

33. Answers may vary.

35. $\{3, 0, \sqrt{36}\}$

37. $\{3, \sqrt{36}\}$

39. $\{\sqrt{7}\}$

41. $-11 \in \{x|x$ is an integer$\}$

43. $0 \notin \{x|x$ is a positive integer$\}$

45. $12 \notin \{1, 3, 5, \ldots\}$

47. $0 \notin \{1, 2, 3, \ldots\}$

49. True

51. True

53. False

55. False

57. True

59. False

61. Answers may vary.

63. $2x$

65. $2x + 5$

67. $x - 10$

69. $x + 2$

71. $\frac{x}{11}$

73. $3x + 12$

75. $x - 17$

77. $2(x + 3)$

79. $\frac{5}{4 - x}$

81. 1991 ⇒ \$355,000,000
1992 ⇒ \$1,392,000,000
1993 ⇒ \$1,025,000,000
1994 ⇒ \$1,526,000,000

Exercise Set 1.2

1. $4c = 7$

3. $3(x + 1) = 7$

5. $\frac{n}{5} = 4n$

7. $z - 2 = 2z$

9. $0 > -2$

11. $\frac{12}{3} = \frac{8}{2}$

13. $-7.9 < -7.09$

15. $7x \leq -21$

17. $-2 + x \neq 10$

19. $2(x - 6) > \frac{1}{11}$

21. 6.2

23. $-\frac{4}{7}$

25. $\frac{2}{3}$

27. 0

29. $\frac{1}{5}$

31. $-\frac{1}{8}$

33. -4

35. Undefined

37. $\frac{8}{7}$

39. Zero. For every real number x, $0 \cdot x \neq 1$, so 0 has no reciprocal. It is the only real number that has no reciprocal because if $x \neq 0$, then $x \cdot \frac{1}{x} = 1$ by definition.

41. $7x + y = y + 7x$

43. $z \cdot w = w \cdot z$

45. $\frac{1}{3} \cdot \frac{x}{5} = \frac{x}{5} \cdot \frac{1}{3}$

47. No

49. $5 \cdot (7x) = (7 \cdot 5) \cdot x$

51. $(x + 1.2) + y = x + (1.2 + y)$

53. $(14z) \cdot y = 14(z \cdot y)$

55. $12 - (5 - 3) = 10$; $(12 - 5) - 3 = 4$; Subtraction is not associative.

57. $3 \cdot x + 3 \cdot 5 = 3x + 15$

59. $8 \cdot 2a + 8 \cdot b = 16a + 8b$

61. $2(6x + 5y + 2z) = 12x + 10y + 4z$

63. $y - 7 = 6$

65. $3y < -17$

67. $2(x - 6) = -27$

69. $x - 4 \geq 3x$

71. $2y - 6 = \frac{1}{8}$

73. $\frac{n+5}{2} > 2n$

75. $6 + 3x$

77. 0

79. 7

81. $(10 \cdot 2)y$

83. $a(b + c) = ab + ac$

Exercise Set 1.3

1. $-(2) = -2$

3. $-(-4) = 4$

5. 0

7. $-[-(-3)] = -3$

9. Answers may vary.

11. $-3 + 8 = 5$

13. $-14 + (-10) = -24$

15. $-4.3 - 6.7 = -11$

17. $13 - 17 = -4$

19. $\frac{11}{15} - \left(-\frac{3}{5}\right) = \frac{11}{15} + \frac{9}{15} = \frac{20}{15} = \frac{4}{3}$

21. $19 - 10 - 11 = 9 - 11 = -2$

23. $(-5)(12) = -60$

25. $(-8)(-10) = 80$

27. $\frac{-12}{-4} = 3$

29. $\frac{0}{-2} = 0$

31. $(-4)(-2)(-1) = 8(-1) = -8$

33. $-\frac{6}{7} \div 2 = \frac{-3}{7} = -\frac{3}{7}$

35. $\left(-\frac{2}{7}\right)\left(-\frac{1}{6}\right) = \frac{2}{42} = \frac{1}{21}$

37. $-7^2 = -49$

39. $(-6)^2 = (-6)(-6) = 36$

41. $(-2)^3 = (-2)(-2)(-2) = 4(-2) = -8$

43. Answers may vary.

45. 7

47. $\frac{1}{3}$

49. 4

51. 3

53. $-4 + 7 = 3$

55. $-9 + (-3) = -12$

57. $-4 - (-19) = -4 + 19 = 15$

59. $6.3 - 18.5 = -12.2$

61. $(-4)(-7)(0) = 0$

63. $\left(-\frac{2}{3}\right)\left(\frac{6}{4}\right) = -\frac{12}{12} = -1$

65. $-14 - 7 = -21$

67. $-\frac{4}{5} - \left(-\frac{3}{10}\right) = -\frac{8}{10} + \frac{3}{10} = -\frac{5}{10} = -\frac{1}{2}$

69. $8 - 14 = -6$

71. $-\frac{34}{2} = -17$

73. $16 - 8 - 9 = 8 - 9 = -1$

75. $\sqrt[3]{8} = 2$

77. $\sqrt{100} = 10$

79. $-\frac{1}{6} \div \frac{9}{10} = -\frac{1}{6} \cdot \frac{10}{9} = -\frac{10}{54} = -\frac{5}{27}$

81. $1 - \frac{1}{5} - \frac{3}{7} = \frac{35}{35} - \frac{7}{35} - \frac{15}{35} = \frac{13}{35}$

83. $10{,}203 - 5998 = 4205$ meters

85. b

87. d

89. Yes. Two players have 6 points each, (third player has 0 points), or two players have 5 points each, (third has 2 points).

91. $\sqrt{273} = 16.5227$

93. $\sqrt{19.6} = 4.4272$

95. 13.2%

97. $12.9\% - 2.1\% = 10.8\%$

Exercise Set 1.4

1. $3(5-7)^4 = 3(-2)^4 = 3(16) = 48$

3. $-3^2 + 2^3 = -9 + 8 = -1$

5. $\frac{3-(-12)}{-5} = \frac{3+12}{-5} = \frac{15}{-5} = -3$

7. $|3.6 - 7.2| + |3.6 + 7.2|$
$= |-3.6| + |10.8|$
$= 3.6 + 10.8 = 14.4$

9. $\frac{(3-\sqrt{9})-(-5-1.3)}{-3}$
$= \frac{(3-3)-(-6.3)}{-3}$
$= \frac{0+6.3}{-3} = \frac{6.3}{-3} = -2.1$

11. $\frac{|3-9|-|-5|}{-3} = \frac{6-5}{-3} = \frac{1}{-3} = -\frac{1}{3}$

13. $(-3)^2 + 2^3 = 9 + 8 = 17$

15. $\frac{3(-2+1)}{5} - \frac{-7(2-4)}{1-(-2)}$
$= \frac{3(-1)}{5} - \frac{-7(-2)}{1+2}$
$= \frac{-3}{5} - \frac{14}{3}$
$= \frac{-9}{15} - \frac{70}{15}$
$= \frac{-79}{15}$

17. $\frac{\left(-\frac{3}{10}\right)}{\left(\frac{42}{50}\right)} = -\frac{3}{10} \cdot \frac{50}{42} = -\frac{5}{14}$

19. $\frac{-1.682 - 17.895}{(-7.102)(-4.691)} = -\frac{19.577}{33.3155} = -0.588$

21. $x^2 + z^2 = (-2)^2 + 3^2 = 4 + 9 = 13$

23. $-5(-x+3y)$
$= -5(-(-2)+3(-5))$
$= -5(2-15)$
$= -5(-13) = 65$

25. $\frac{3z-y}{2x-z}$
$= \frac{3(3)-(-5)}{2(-2)-3}$
$= \frac{9+5}{-4-3} = \frac{14}{-7} = -2$

27. $\frac{y_2 - y_1}{x_2 - x_1} = \frac{2-(-3)}{4-2} = \frac{2+3}{2} = \frac{5}{2}$

29. a.

y	5	7	10	100
$8+2y$	18	22	28	208

b. Increase

31. a.

x	10	100	1000
$\frac{100x+5000}{x}$	600	150	105

b. Decrease

33. $0.05n$

35. $0.05n + 0.1d$

37. $25 - x$

39. $180 - x$

41. $6.49x$

43. $x + 2$

45. $\$0.0825p$

47. $\$31 + 0.29x$

49. $-9x+4x+18-10x$
$=-9x+4x-10x+18$
$=(-9+4-10)x+18$
$=-15x+18$

51. $5k-(3k-10)$
$=5k-3k+10$
$=(5-3)k+10=2k+10$

53. $(3x+4)-(6x-1)$
$=3x+4-6x+1$
$=3x-6x+4+1$
$=(3-6)x+5=-3x+5$

55. $3(x-2)+x+15$
$=3x-6+x+15$
$=3x+x-6+15$
$=(3+1)x+9=4x+9$

57. $-(n+5)+(5n-3)$
$=-n-5+5n-3$
$=-n+5n-5-3$
$=(-1+5)n-8=4n-8$

59. $4(6n-3)-3(8n+4)$
$=24n-12-24n-12$
$=24n-24n-12-12$
$=(24-24)n-24=0n-24=-24$

61. $3x-2(x-5)+x$
$=3x-2x+10+x$
$=3x-2x+x+10$
$=(3-2+1)x+10=2x+10$

63. $-1.2(5.7x-3.6)+8.75x$
$=-6.84x+4.32+8.75x$
$=-6.84x+8.75x+4.32$
$=(-6.84+8.75)x+4.32=1.91x+4.32$

65. $8.1z+7.3(z+5.2)-6.85$
$=8.1z+7.3z+37.96-6.85$
$=(8.1+7.3)z+31.11=15.4z+31.11$

67. No; Cylinder 1 volume = 63.2 in.^3
Cylinder 2 volume = 81.8 in.^3

69. 10 million

71. 42 million

73. Increasing

Chapter 1 - Review

1. $7x=7(3)=21$

2. $st=(1.6)(5)=8$

3. $90t=90(3600)=324{,}000$

4. $\{-1, 1, 3\}$

5. $\{-2, 0, 2, 4, 6\}$

6. $\{\ \}$

7. $\{\ \}$

8. $\{6, 7, 8, \ldots\}$

9. $\{\ldots, -1, 0, 1, 2\}$

10. True

11. False

12. True, since $\sqrt{169}=13$.

13. True, since zero is not an element of the empty set.

14. False, since π is irrational.

15. True, since π is a real number.

16. False, since $\sqrt{4}=2$.

17. True, since -9 is a rational number.

18. True

19. True, since C is not a subset of B.

20. False, since all integers are rational numbers.

21. True, since the empty set is a subset of all sets.

22. True, since every set is a subset of itself.

23. True, since every element of D is also an element of C.

24. True, since every integer is real.

25. True, since every irrational number is also a real number.

26. False, since B does not contain the set $\{5\}$.

27. True, since $\{5\}$ is a subset of B.

28. $\left\{5,\ \frac{8}{2},\ \sqrt{9}\right\}$

29. $\left\{5,\ \frac{8}{2},\ \sqrt{9}\right\}$

30. $\left\{5,\ -\frac{2}{3},\ \frac{8}{2},\ \sqrt{9},\ 0.3,\ 1\frac{5}{8},\ -1\right\}$

31. $\left\{\sqrt{7},\ \pi\right\}$

32. $\left\{5, -\frac{2}{3}, \frac{8}{2}, \sqrt{9}, 0.3, \sqrt{7}, 1\frac{5}{8}, -1, \pi\right\}$

33. $\left\{5, \frac{8}{2}, \sqrt{9}, -1\right\}$

34. $12 = -4x$

35. $n + 2n = -15$

36. $4(y + 3) = -1$

37. $6(t - 5) = 4$

38. $z - 7 = 6$

39. $9x - 10 = 5$

40. $x - 5 \geq 12$

41. $-4 < 7y$

42. $\frac{2}{3} \neq 2\left(n + \frac{1}{4}\right)$

43. $t + 6 \leq -12$

44. Associative Property of Addition

45. Distributive Property

46. Additive Inverse Property

47. Commutative Property of Addition

48. Associative and Commutative Properties of Multiplication. To see this: $XYZ = (XY)Z = X(YZ) = (YZ)X = YZX$

49. Multiplicative Inverse Property

50. Multiplication Property of Zero

51. Associative Property of Multiplication

52. Additive Identity Property

53. Multiplicative Identity Property

54. $-\left(-\frac{3}{4}\right) = \frac{3}{4}$

55. -0.6

56. $-0 = 0$

57. -1

58. $\frac{1}{\left(-\frac{3}{4}\right)} = -\frac{4}{3}$

59. $\frac{1}{0.6}$

60. Undefined

61. 1

62. $5x - 15z = 5(x - 3z)$

63. $(3 + x) + (7 + y)$

64. $0 = 2 + (-2)$, for example

65. $1 = 2 \cdot \frac{1}{2}$, for example

66. $[(3.4)(0.7)]5 = (3.4)[(0.7)5]$

67. $7 + 0$

68. $-9 > -12$

69. $0 > -6$

70. $-3 < -1$

71. $7 = |-7|$

72. $-5 < -(-5)$, where $-(-5) = 5$

73. $-(-2) > -2$, where $-(-2) = 2$

74. $-7 + 3 = -4$

75. $-10 + (-25) = -35$

76. $5(-0.4) = -2$

77. $(-3.1)(-0.1) = 0.31$

78. $-7 - (-15) = -7 + 15 = 8$

79. $9 - (-4.3) = 13.3$

80. $(-6)(-4)(0)(-3) = 0$

81. $(-12)(0)(-1)(-5) = 0$

82. $(-24) \div 0 =$ undefined

83. $0 \div (-45) = 0$

84. $(-36) \div (-9) = 4$

85. $(60) \div (-12) = -5$

86. $\left(-\frac{4}{5}\right) - \left(-\frac{2}{3}\right) = -\frac{12}{15} + \frac{10}{15} = -\frac{2}{15}$

87. $\left(\frac{5}{4}\right) - \left(-2\frac{3}{4}\right) = \frac{5}{4} + \frac{11}{4} = \frac{16}{4} = 4$

88. $1 - \frac{1}{4} - \frac{1}{3} = \frac{12}{12} - \frac{3}{12} - \frac{4}{12} = \frac{5}{12}$

89. $-5 + 7 - 3 - (-10)$
$= 2 - 3 + 10 = -1 + 10 = 9$

90. $8 - (-3) + (-4) + 6 = 8 + 3 - 4 + 6 = 13$

91. $3(4-5)^4 = 3(-1)^4 = 3 \cdot 1 = 3$

92. $6(7-10)^2 = 6(-3)^2 = 6(9) = 54$

93. $\left(-\frac{8}{15}\right) \cdot \left(-\frac{2}{3}\right)^2 = -\frac{8}{15} \cdot \frac{4}{9} = -\frac{32}{135}$

94. $\left(-\frac{3}{4}\right)^2 \cdot \left(-\frac{10}{21}\right) = \left(\frac{9}{16}\right)\left(-\frac{10}{21}\right) = -\frac{15}{56}$

95. $\frac{-\frac{6}{15}}{\frac{8}{25}} = -\frac{6}{15} \cdot \frac{25}{8} = -\frac{3}{3} \cdot \frac{5}{4} = -\frac{5}{4}$

96. $\frac{\frac{4}{9}}{-\frac{8}{45}} = \frac{4}{9} \cdot \left(-\frac{45}{8}\right) = -\frac{5}{2}$

97. $-\frac{3}{8} + 3(2) \div 6$
$= -\frac{3}{8} + 6 \div 6$
$= -\frac{3}{8} + 1 = -\frac{3}{8} + \frac{8}{8} = \frac{5}{8}$

98. $5(-2) - (-3) - \frac{1}{6} + \frac{2}{3}$
$= -10 + 3 - \frac{1}{6} + \frac{2}{3}$
$= -7 - \frac{1}{6} + \frac{2}{3}$
$= -\frac{42}{6} - \frac{1}{6} + \frac{4}{6} = -\frac{39}{6} = -6\frac{1}{2}$

99. $\left|2^3 - 3^2\right| - |5-7|$
$= |8-9| - |-2| = |-1| - 2 = 1 - 2 = -1$

100. $\left|5^2 - 2^2\right| + |9 \div (-3)|$
$= |25-4| + |-3| = |21| + 3 = 21 + 3 = 24$

101. $(2^3 - 3^2) - (5-7)$
$= (8-9) - (-2) = -1 + 2 = 1$

102. $(5^2 - 2^4) + [9 \div (-3)] = 25 - 16 + (-3) = 6$

103. $\frac{(8-10)^3 - (-4)^2}{2 + 8(2) \div 4}$
$= \frac{(-2)^3 - 16}{2 + 16 \div 4} = \frac{-8-16}{2+4} = -\frac{24}{6} = -4$

104. $\frac{(2+4)^2+(-1)^5}{12 \div 2 \cdot 3 - 3}$
$= \frac{6^2+(-1)}{6 \cdot 3 - 3} = \frac{36-1}{18-3} = \frac{35}{15} = \frac{7}{3}$

105. $\frac{(4-9)+4-9}{10-12 \div 4 \cdot 8}$
$= \frac{-5+4-9}{10-3 \cdot 8} = -\frac{-1-9}{10-24} = \frac{-10}{-14} = \frac{5}{7}$

106. $\frac{3-7-(7-3)}{15+30 \div 6 \cdot 2}$
$= \frac{-4-(4)}{15+5 \cdot 2} = \frac{-8}{15+10} = -\frac{8}{25}$

107. $\frac{\sqrt{25}}{4+3 \cdot 7} = \frac{5}{4+21} = \frac{5}{25} = \frac{1}{5}$

108. $\frac{\sqrt{64}}{24-8 \cdot 2} = \frac{8}{24-16} = \frac{8}{8} = 1$

109. $x^2 - y^2 + z^2$
$= 0^2 - 3^2 + (-2)^2 = 0 - 9 + 4 = -5$

110. $\frac{5x+z}{2y} = \frac{5(0)+(-2)}{2(3)} = \frac{0-2}{6} = \frac{-2}{6} = -\frac{1}{3}$

111. $\frac{-7y-3z}{-3}$
$= \frac{-7(3)-3(-2)}{-3}$
$= \frac{-21+6}{-3} = \frac{-15}{-3} = 5$

112. $(x-y+z)^2$
$= [0-3+(-2)]^2 = (-5)^2 = 25$

113. **a.**

r	1	10	100
$2\pi r$	6.28	62.8	628

b. Increase

Chapter 1 - Test

1. True, –2.3 lies to the right of –2.33 on the number line.

2. False, $-6^2 = -36$, while $(-6)^2 = 36$.

3. False, $-5 - 8 = -13$, while $-(5-8) = -(-3) = 3$.

4. False, $(-2)(-3)(0) = 0$, while $\frac{(-4)}{0}$ is undefined.

5. True, natural numbers are synonymous with positive integers.

6. False, for example, $\frac{1}{2}$ is a rational number which is not an integer.

7. $5 - 12 \div 3(2) = 5 - 4 \cdot 2 = 5 - 8 = -3$

8. $|4-6|^3 - (1-6^2)$
$= |-2|^3 - (1-36)$
$= 2^3 - (-35) = 8 + 35 = 43$

9. $(4-9)^3 - |-4-6|^2$
$= (-5)^3 - |-10|^2$
$= -125 - 100 = -225$

10. $\left[3|4-5|^5 - (-9)\right] \div (-6)$
$= \left[3|-1|^5 + 9\right] \div (-6)$
$= (3 \cdot 1^5 + 9) \div (-6)$
$= (3+9) \div (-6) = 12 \div (-6) = -2$

11. $\frac{6(7-9)^3+(-2)}{(-2)(-5)(-5)}$
$= \frac{6(-2)^3-2}{10(-5)}$
$= \frac{6(-8)-2}{-50} = \frac{-48-2}{-50} = \frac{-50}{-50} = 1$

12. $q^2 - r^2 = 4^2 - (-2)^2 = 16 - 4 = 12$

13. $\frac{5t-3q}{3r-1}=\frac{5(1)-3(4)}{3(-2)-1}=\frac{5-12}{-6-1}=\frac{-7}{-7}=1$

14. a.

x	1	3	10	20
$5.75x$	5.75	17.25	57.50	115.00

b. Increase

15. $2|x+5|=30$

16. $\frac{(6-y)^2}{7}<-2$

17. $\frac{9z}{|-12|}\neq 10$

18. $3\left(\frac{n}{5}\right)=-n$

19. $20=2x-6$

20. $-2=\frac{x}{x+5}$

21. Distributive Property

22. Associative Property of Addition

23. Additive Inverse Property

24. Multiplication Property of Zero

25. $0.05n+0.1d$

Chapter 2

Section 2.1

Mental Math

1. $3x + 5x + 6 + 15 = 8x + 21$

3. $5n + n + 3 - 10 = 6n - 7$

5. $8x - 12x + 5 - 6 = -4x - 1$

Exercise Set 2.1

1. $x + 2.8 = 1.9$
$x + 2.8 - 2.8 = 1.9 - 2.8$
$x = -0.9$
$\{-0.9\}$

3. $5x - 4 = 26$
$5x = 30$
$x = 6$
$\{6\}$

5. $-4.1 - 7z = 3.6$
$-7z = 7$
$z = -1.1$
$\{-1.1\}$

7. $5y + 12 = 2y - 3$
$3y = -15$
$y = -5$
$\{-5\}$

9. $8x - 5x + 3 = x - 7 + 10$
$3x + 3 = x + 3$
$2x = 0$
$x = 0$
$\{0\}$

11. $5x + 12 = 2(2x + 7)$
$5x + 12 = 4x + 14$
$x = 2$
$\{2\}$

13. $3(x - 6) = 5x$
$3x - 18 = 5x$
$-18 = 2x$
$x = -9$
$\{-9\}$

15. $-2(5y - 1) - y = -4(y - 3)$
$-10y + 2 - y = -4y + 12$
$-11y + 2 = -4y + 12$
$-7y = 10$
$y = -\frac{10}{7}$
$\left\{-\frac{10}{7}\right\}$

17. a. $4(x + 1) + 1 = 4x + 4 + 1 = 4x + 5$

b. $4(x + 1) + 1 = -7$
$4x + 5 = -7$
$4x = -12$
$x = -3$
$\{-3\}$

c. Answers may vary.

19. $\frac{x}{2} + \frac{2}{3} = \frac{3}{4}$
$6x + 8 = 9$
$6x = 1$
$x = \frac{1}{6}$
$\left\{\frac{1}{6}\right\}$

21. $\frac{3t}{4} - \frac{t}{2} = 1$
$3t - 2t = 4$
$t = 4$
$\{4\}$

23. $\frac{n-3}{4} + \frac{n+5}{7} = \frac{5}{14}$
$7(n - 3) + 4(n + 5) = 10$
$7n - 21 + 4n + 20 = 10$
$11n - 1 = 10$
$11n = 11$
$n = 1$
$\{1\}$

25. $\frac{3x-1}{9}+x=\frac{3x+1}{3}+4$
$(3x-1)+9x=3(3x+1)+36$
$3x-1+9x=9x+3+36$
$12x-1=9x+39$
$3x=40$
$x=\frac{40}{3}$
$\left\{\frac{40}{3}\right\}$

27. $4(n+3)=2(6+2n)$
$4n+12=12+4n$ or
$4n+12=4n+12$, an identity
$\{n|n \text{ is a real number}\}$
Therefore, all real numbers are solutions.

29. $3(x-1)+5=3x+2$
$3x-3+5=3x+2$
$3x+2=3x+2$
Therefore, all real numbers are solutions.
$\{x|x \text{ is a real number}\}$

31. Answers may vary.

33. $-5x+1.5=-19.5$
$-5x=-21$
$x=4.2$
$\{4.2\}$

35. $x-10=-6x+4$
$7x=14$
$x=2$
$\{2\}$

37. $3x-4-5x=x+4+x$
$-2x-4=2x+4$
$-4x=8$
$x=-2$
$\{-2\}$

39. $5(y+4)=4(y+5)$
$5y+20=4y+20$
$y=0$
$\{0\}$

41. $0.6x-10=1.4x-14$
$4=0.8x$
$x=5$
$\{5\}$

43. $6x-2(x-3)=4(x+1)+4$
$6x-2x+6=4x+4+4$
$4x+6=4x+8$
$0=2$
Therefore, no solution exists.
$\{\ \}$

45. $\frac{3}{8}+\frac{b}{3}=\frac{5}{12}$
$9+8b=10$
$8b=1$
$b=\frac{1}{8}$
$\left\{\frac{1}{8}\right\}$

47. $z+3(2+4z)=6(z+1)+5z$
$z+6+12z=6z+6+5z$
$13z+6=11z+6$
$2z=0$
$z=0$
$\{0\}$

49. $\frac{3t+1}{8}=\frac{5+2t}{7}+2$
$7(3t+1)=8(5+2t)+112$
$21t+7=40+16t+112$
$21t+7=16t+152$
$5t=145$
$t=29$
$\{29\}$

51. $\frac{m-4}{3}-\frac{3m-1}{5}=1$
$5(m-4)-3(3m-1)=15$
$5m-20-9m+3=15$
$-4m-17=15$
$-4m=32$
$m=-8$
$\{-8\}$

53. $\frac{x}{5}-\frac{x}{4}=\frac{1}{2}(x-2)$
$4x-5x=10(x-2)$
$-x=10x-20$
$-11x=-20$
$x=\frac{20}{11}$
$\left\{\frac{20}{11}\right\}$

55. $5(x-2)+2x=7(x+4)$
$5x-10+2x=7x+28$
$7x-10=7x+28$
$0=38$
Therefore, no solution exists.
$\{\ \}$

57. $y+0.2=0.6(y+3)$
$y+0.2=0.6y+1.8$
$0.4y=1.6$
$y=4$
$\{4\}$

59. $2y+5(y-4)=4y-2(y-10)$
$2y+5y-20=4y-2y+20$
$7y-20=2y+20$
$5y=40$
$y=8$
$\{8\}$

61. $2(x-8)+x=3(x-6)+2$
$2x-16+x=3x-18+2$
$3x-16=3x-16$
Therefore, all real numbers are solutions.
$\{x|x \text{ is a real number}\}$

63. $\frac{5x-1}{6}-3x=\frac{1}{3}+\frac{4x+3}{9}$
$3(5x-1)-54x=6+2(4x+3)$
$15x-3-54x=6+8x+6$
$-39x-3=8x+12$
$-47x=15$
$x=-\frac{15}{47}$
$\left\{-\frac{15}{47}\right\}$

65. $-2(b-4)-(3b-1)=5b+3$
$-2b+8-3b+1=5b+3$
$-5b+9=5b+3$
$-10b=-6$
$b=\frac{6}{10}=\frac{3}{5}$
$\left\{\frac{3}{5}\right\}$

67. $1.5(4-x)=1.3(2-x)$
$6-1.5x=2.6-1.3x$
$-0.2x=-3.4$
$x=17$
$\{17\}$

69. $\frac{1}{4}(a+2)=\frac{1}{6}(5-a)$
$3(a+2)=2(5-a)$
$3a+6=10-2a$
$5a=4$
$a=\frac{4}{5}$
$\left\{\frac{4}{5}\right\}$

71. $y+y+y+y=4y$

73. $z+(z+1)+(z+2)=3z+3$

75. $0.05x+0.10(x+3)$
$=0.05x+0.10x+0.30$
$=0.15x+0.30$

77. $4x+3(2x+1)=4x+6x+3=10x+3$

79. $3.2x+4=5.4x-7$
$3.2x+4-4=5.4x-7-4$
$3.2x=5.4x-11$
$k=-11$

81. $\frac{x}{6}+4=\frac{x}{3}$
$6\left(\frac{x}{6}+4\right)=6\left(\frac{x}{3}\right)$
$x+24=2x$
$k=24$

83. $2.569x=-12.48534$
$\frac{2.569x}{2.569}=\frac{-12.48534}{2.569}$
$x=-4.86$
$\{-4.86\}$

85. $2.86z-7.95=-3.57$
$2.86z-7.95+7.95=-3.57+7.95$
$2.86z=4.38$
$\frac{2.86z}{2.86}=\frac{4.38}{2.86}$
$z=1.5315$
$\{1.5315\}$

87. Not a fair game

89. $y_2-y_1=4-(-3)=4+3=7$

91. $\frac{y_3}{3y_1}=\frac{0}{3(-3)}=\frac{0}{-9}=0$

93. $y_1 - y_2 = -3 - 4 = -7$

95. $7x^2 + 2x - 3 = 6x(x+4) + x^2$
$7x^2 + 2x - 3 = 6x^2 + 24x + x^2$
$7x^2 + 2x - 3 = 7x^2 + 24x$
$2x - 3 = 24x$
$-3 = 22x$
$-\frac{3}{22} = x$
$\left\{-\frac{3}{22}\right\}$

97. $x(x+1) + 16 = x(x+5)$
$x^2 + x + 16 = x^2 + 5x$
$x + 16 = 5x$
$16 = 4x$
$4 = x$
$\{4\}$

Exercise Set 2.2

1. $4(x-2) = 6x + 2$
$4x - 8 = 6x + 2$
$-10 = 2x$
$-5 = x$

3. Let x = one number
Then $270 - x$ = other number
$x = 5(270 - x)$
$x = 1350 - 5x$
$6x = 1350$
$x = 225$; $270 - 225 = 45$
225 and 45

5. $260(0.30) = 78$

7. $16(0.12) = 1.92$

9. $1{,}943{,}000{,}000(0.21) = 408{,}030{,}000$
Approximately 408 million acres

11. $110{,}000(0.47) = 51{,}700$

13. $100 - 12 - 39 - 8 - 21 = 20\%$

15. $300(0.39) = 117$

17. Let x = seats in B737-200
Then $2x$ = seats in B767-300
$x + 2x = 336$
$3x = 336$
$x = 112$
$2x = 224$
B737-200 has 112 seats
B767-300 has 224 seats

19. Let x = price before taxes
$x + 0.08x = 464.40$
$1.08x = 464.40$
$x = 430$
\$430

21. Let x = length of side of square
$4x = 3(x+6)$
$4x = 3x + 18$
$x = 18$, $x + 6 = 24$
square's side is 18 cm
triangle's side is 24 cm

23. Let x = width of room
Then $2x + 2$ = length of room
$2x + 2(2x+2) = 40$
$2x + 4x + 4 = 40$
$6x + 4 = 40$
$6x = 36$
$x = 6$
$2x + 2 = 14$
width is 6 cm; length is 14 cm

25. $20 - 20(0.75) = 20 - 15 = 5$ years

27. Let x = width
Then $5(x+1)$ = length
$x + 5(x+1) = 55.4$
$x + 5x + 5 = 55.4$
$6x + 5 = 55.4$
$6x = 50.4$
$x = 8.4$
$5(x+1) = 47$
width is 8.4 meters; length is 47 meters

29. Let x = measure of second angle
Then $2x$ = measure of first angle
$3x - 12$ = measure of third angle
$x + 2x + 3x - 12 = 180$
$6x - 12 = 180$
$6x = 192$
$x = 32$
$2x = 64$
$3x - 12 = 84$
64°, 32°, 84°

31. (1.5 billion)(0.44) = 660 million hours

33. Let x = 1st integer
Then $x + 2$ = 2nd integer
$x + 4$ = 3rd integer
$2x + x + 4 = 268{,}222$
$3x = 268{,}218$
$x = 89{,}406$
$x + 2 = 89{,}408,\ x + 4 = 89{,}410$

35. Let x = 1st integer
$x + 1$ = 2nd integer
$x + 2$ = 3rd integer
$x + x + 1 + x + 2 = 3(x + 1)$
$3x + 3 = 3x + 3$
Identity, any three consecutive integers will work.

37. Let x = base angle
Then $3x - 10$ = third angle
$2x + 3x - 10 = 180$
$5x - 10 = 180$
$5x = 190$
$x = 38$
$3x - 10 = 104$
38°, 38°, 104°

39. Let x = operators in 1995
$x - 0.233x = 46{,}000$
$0.767x = 46{,}000$
$x = 59{,}974$

41. Greater use of calling cards and voice recognition technology, for example

43. $R = 24x$
$C = 100 + 20x$
$24x = 100 + 20x$
$4x = 100$
$x = 25$
25 skateboards

45. $R = 7.50x$
$C = 4.50x + 2400$
$7.50x = 4.50x + 2400$
$3x = 2400$
$x = 800$
800 books

47. It loses money.

49. $2a + b - c$
$= 2(5) + (-1) - 3$
$= 10 - 1 - 3 = 6$

51. $4ab - 3bc$
$= 4(-5)(-8) - 3(-8)(2)$
$= 160 + 48 = 208$

53. $n^2 - m^2 = (-3)^2 - (-8)^2 = 9 - 64 = -55$

55. $P + Prt = 3000 + 3000(0.0325)(2) = 3195$

57. $\frac{1}{3}Bh = \frac{1}{3}(53.04)(6.89) = 121.8152$

Section 2.3

Mental Math

1. $2x + y = 5$
$y = -2x + 5$

3. $a - 5b = 8$
$a = 5b + 8$

5. $5j + k - h = 6$
$k = h - 5j + 6$

Exercise Set 2.3

1. $D = rt$
$\frac{D}{r} = \frac{rt}{r}$
$\frac{D}{r} = t$

3. $I = PRT$
$\frac{I}{PT} = \frac{PRT}{PT}$
$\frac{I}{PT} = R$

5. $9x - 4y = 16$
$9x - 4y - 9x = 16 - 9x$
$-4y = 16 - 9x$
$\frac{-4y}{-4} = \frac{16 - 9x}{-4}$
$y = \frac{9x - 16}{4}$

7. $P = 2L + 2W$
$P - 2L = 2L + 2W - 2L$
$P - 2L = 2W$
$\frac{P-2L}{2} = \frac{2W}{2}$
$\frac{P-2L}{2} = W$

9. $J = AC - 3$
$J + 3 = AC - 3 + 3$
$J + 3 = AC$
$\frac{J+3}{C} = \frac{AC}{C}$
$\frac{J+3}{C} = A$

11. $W = gh - 3gt^2$
$W = g(h - 3t^2)$
$\frac{W}{h-3t^2} = \frac{g(h-3t^2)}{h-3t^2}$
$\frac{W}{h-3t^2} = g$

13. $T = C(2 + AB)$
$T = 2C + ABC$
$T - 2C = 2C + ABC - 2C$
$T - 2C = ABC$
$\frac{T-2C}{AC} = \frac{ABC}{AC}$
$\frac{T-2C}{AC} = B$

15. $C = 2\pi r$
$\frac{C}{2\pi} = \frac{2\pi r}{2\pi}$
$\frac{C}{2\pi} = r$

17. $E = I(r + R)$
$E = Ir + IR$
$E - IR = Ir + IR - IR$
$E - IR = Ir$
$\frac{E-IR}{I} = \frac{Ir}{I}$
$\frac{E}{I} - R = r$

19. $s = \frac{n}{2}(a + L)$
$2 \cdot s = 2 \cdot \frac{n}{2}(a + L)$
$2s = n(a + L)$
$2s = an + Ln$
$2s - an = an + Ln - an$
$2s - an = Ln$
$\frac{2s-an}{n} = \frac{Ln}{n}$
$\frac{2s}{n} - a = L$

21. $\frac{1}{u} - \frac{1}{v} = \frac{1}{w}$
$\frac{v-u}{uv} = \frac{1}{w}$
$w(v - u) = uv$
$\frac{w(v-u)}{v-u} = \frac{uv}{v-u}$
$w = \frac{uv}{v-u}$

23. $N = 3st^4 - 5sv$
$N - 3st^4 = 3st^4 - 5sv - 3st^4$
$N - 3st^4 = -5sv$
$\frac{N-3st^4}{-5s} = \frac{-5sv}{-5s}$
$\frac{3st^4 - N}{5s} = v$

25. $S = \frac{a}{1-r}$
$S(1-r) = \frac{a}{1-r}(1-r)$
$S - Sr = a$
$S - Sr - S = a - S$
$-Sr = a - S$
$\frac{-Sr}{-S} = \frac{a-S}{-S}$
$r = \frac{S-a}{S}$

27. $S = 2LW + 2LH + 2WH$
$S - 2LW = 2LW - 2LW + H(2L + 2W)$
$S - 2LW = H(2L + 2W)$
$\frac{S-2LW}{2L+2W} = \frac{H(2L+2W)}{2L+2W}$
$\frac{S-2LW}{2L+2W} = H$

29. $A = P\left(1+\frac{r}{n}\right)^{nt} = 3500\left(1+\frac{0.03}{n}\right)^{10n}$

n	1	2	4	12	365
A	\$4703.71	\$4713.99	\$4719.22	\$4722.74	\$4724.45

31. 12 times a year; you earn more interest

33. $A = P\left(1+\frac{r}{n}\right)^{nt} = 6000\left(1+\frac{0.04}{n}\right)^{5n}$

a. $A = 6000\left(1+\frac{0.04}{2}\right)^{5\cdot 2} = \$7313.97$

b. $A = 6000\left(1+\frac{0.04}{4}\right)^{5\cdot 4} = \$7321.14$

c. $A = 6000\left(1+\frac{0.04}{12}\right)^{5\cdot 12} = \$7325.98$

35. $C = \frac{5}{9}(F-32)$
$C = \frac{5}{9}(104-32)$
$C = \frac{5}{9}(72)$
$C = 40°$

37. $\frac{25}{0.01} = 2500$
$\sqrt{2500} = 50$ cm

39. $d = rt$ or $t = \frac{d}{r}$
where $d = 2(90) = 180$ mi and
$r = 50$ mi/hr, so $t = \frac{180}{50} = 3.6$ hr
Thus, she takes 3.6 hours or 3 hours, 36 minutes to make the round trip.

41. $A = s^2$ where $s = 64$ ft, so
$A = 64^2 = 4096$ ft^2
Thus, $\frac{4096}{24} \approx 171$ packages of tiles should be bought.

43. $V = \pi r^2 h$
1 mile = 5280 feet
1.3 miles = 6864 feet
$3800 = \pi r^2(6864)$
$0.42 = r$
0.42 feet

45. $\frac{168 \text{ mi}}{1 \text{ hr}} \cdot \frac{5280 \text{ ft}}{1 \text{ mi}} \cdot \frac{1 \text{ hr}}{60 \text{ min}} \cdot \frac{1 \text{ min}}{60 \text{ sec}}$
= 246.4 ft/sec
$d = rt$
60.5 feet = 246.4 ft/sec · t
.25 sec = t

47. To receive from and transmit to fixed locations on the earth, for example

49. $A = P\left(1+\frac{r}{n}\right)^{nt}$
$A = 10,000\left(1+\frac{0.085}{4}\right)^{4\cdot 2}$
$= 10,000(1+0.02125)^8$
= 10,000(1.183196) = \$11,831.96
\$11,831.96 – \$10,000 = \$1831.96

51. The area of one pair of walls is $2 \cdot 14 \cdot 8 = 224$ ft^2 and the area of the other walls is $2 \cdot 16 \cdot 8 = 256$ ft^2 for a total of 480 ft^2. Multiplying by 2, the number of coats, yields 960 ft^2. Dividing this by 500 yields 1.92. Thus, 2 gallons should be purchased.

53. Volume = Volume of sphere + Volume of cylinder
$V = \frac{4}{3}\pi r^3 + \pi r^2 h$
$V = \frac{4}{3}\pi(4.2)^3 + \pi(4.2)^2(21.2)$
$V = 1485.19$ m^3

55. $d = rt$ or $t = \frac{d}{r}$ where $d = 135$ mi and $r = 60$ mi/hr so $t = \frac{135}{60} = 2.25$ hr
Thus, it takes him 2.25 hours or 2 hours and 15 minutes.

57. $V = \pi r^2 h$
$V = \pi(2.3)^2(18.3)$
$V = 304.12816\ \text{m}^3$

59. $\frac{1}{8}$

61. $\frac{1}{8}$

63. $\frac{1}{8} + \frac{3}{8} = \frac{4}{8} = \frac{1}{2}$

65. $\frac{1}{8} + \frac{1}{8} + \frac{1}{8} = \frac{3}{8}$

67. 0

69. 0

71. $\{-3, -2, -1\}$

73. $\{-3, -2, -1, 0, 1\}$

75. Answers may vary.

Exercise Set 2.4

1. $(-\infty, -3)$
−6 −5 −4 −3 −2 −1 0 1 2

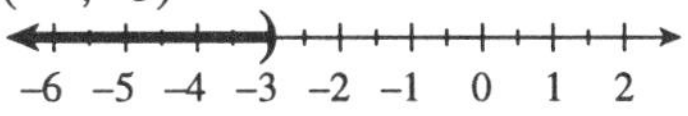

3. $[0.3, \infty)$
−0.2 −0.1 0.0 0.1 0.2 0.3 0.4 0.5 0.6

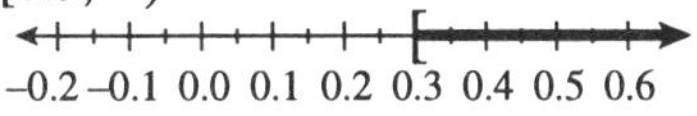

5. $(5, \infty)$
−2 −1 0 1 2 3 4 5 6

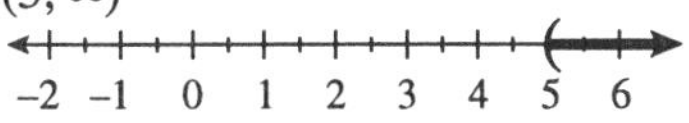

7. $(-2, 5)$
−2 −1 0 1 2 3 4 5 6

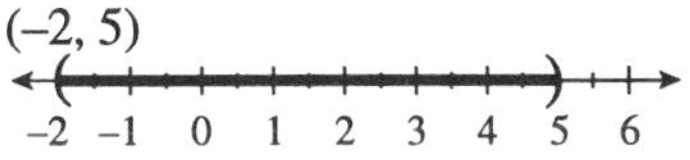

9. $(-1, 5)$
−2 −1 0 1 2 3 4 5 6

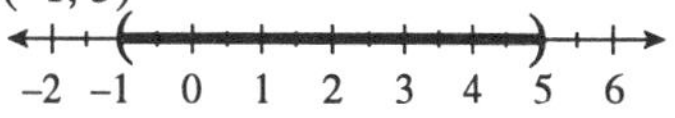

11. Use a parenthesis when the end-value is not included; use a bracket when the end-value is included.

13. $5x + 3 > 2 + 4x$
$x > -1$
$(-1, \infty)$
−4 −3 −2 −1 0 1 2 3 4

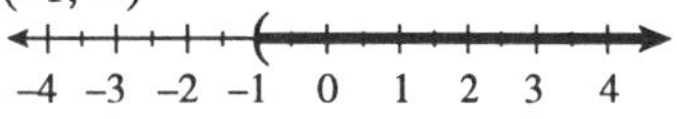

15. $8x - 7 \le 7x - 5$
$x \le 2$
$(-\infty, 2]$
−2 −1 0 1 2 3 4 5 6

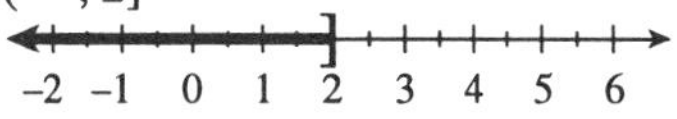

17. $5x < -23.5$
$x < -4.7$
$(-\infty, -4.7)$
−8 −4.7 0 8

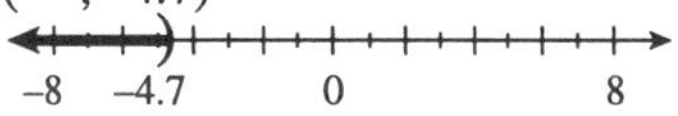

19. $-3x \ge \frac{1}{2}$
$x \le -\frac{1}{6}$
$\left(-\infty, -\frac{1}{6}\right]$
−1 $-\frac{1}{6}$ 0 1

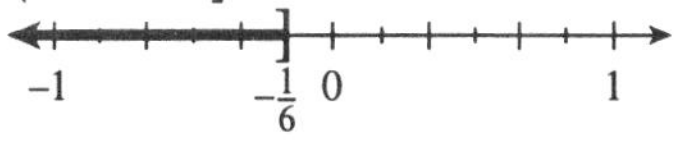

21. $15 + 2x \ge 4x - 7$
$22 \ge 2x$
$11 \ge x$
$x \le 11$
$(-\infty, 11]$
−3 −1 1 3 5 7 9 11 13

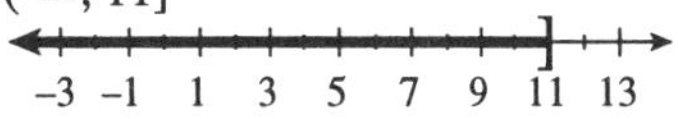

23. $\frac{3x}{4} \ge 2$
$x \ge \frac{8}{3}$
$\left[\frac{8}{3}, \infty\right)$
2 $\frac{8}{3}$ 3

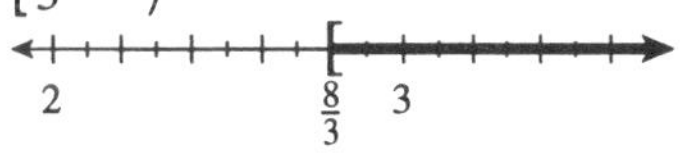

25. $3(x-5) < 2(2x-1)$
$3x - 15 < 4x - 2$
$-13 < x$
$x > -13$
$(-13, \infty)$

−17 −15 −13 −11 −9 −7 −5 −3 −1

27. $\frac{1}{2} + \frac{2}{3} \geq \frac{x}{6}$
$\frac{7}{6} \geq \frac{x}{6}$
$7 \geq x$
$x \leq 7$
$(-\infty, 7]$

0 1 2 3 4 5 6 7 8

29. $4(x-1) \geq 4x - 8$
$4x - 4 \geq 4x - 8$
$-4 \geq -8$
All real numbers.
$(-\infty, \infty)$

−8 −6 −4 −2 0 2 4 6 8

31. $7x < 7(x-2)$
$7x < 7x - 14$
$0 < -14$
No solution exists.
{ }

−8 −6 −4 −2 0 2 4 6 8

33. Answers may vary.

35. $7.3x > 43.8$
$x > 6$
$(6, \infty)$

0 1 2 3 4 5 6 7 8

37. $-4x \leq \frac{2}{5}$
$x \geq -\frac{1}{10}$
$\left[-\frac{1}{10}, \infty\right)$

$-\frac{4}{10}$ $-\frac{3}{10}$ $-\frac{2}{10}$ $-\frac{1}{10}$ 0 $\frac{1}{10}$ $\frac{2}{10}$ $\frac{3}{10}$ $\frac{4}{10}$

39. $-2x + 7 \geq 9$
$-2x \geq 2$
$x \leq -1$

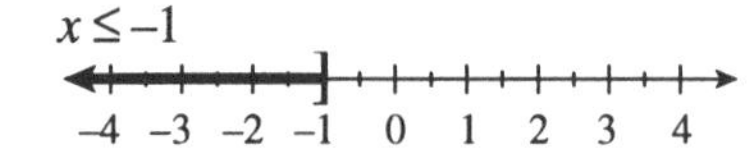

41. $4(2x+1) > 4$
$2x + 1 > 1$
$2x > 0$
$x > 0$
$(0, \infty)$

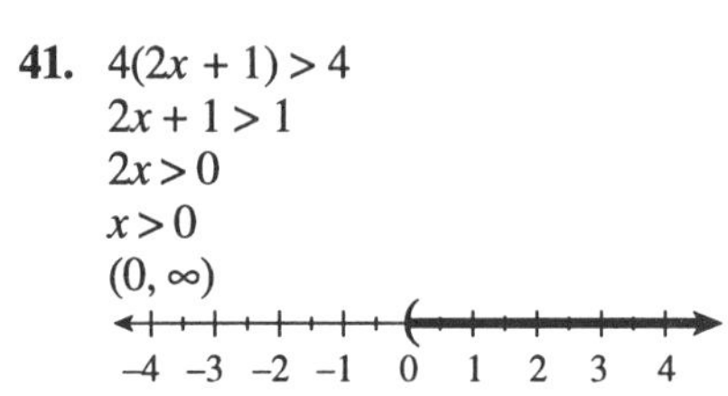

43. $\frac{x+7}{5} > 1$
$x + 7 > 5$
$x > -2$
$(-2, \infty)$

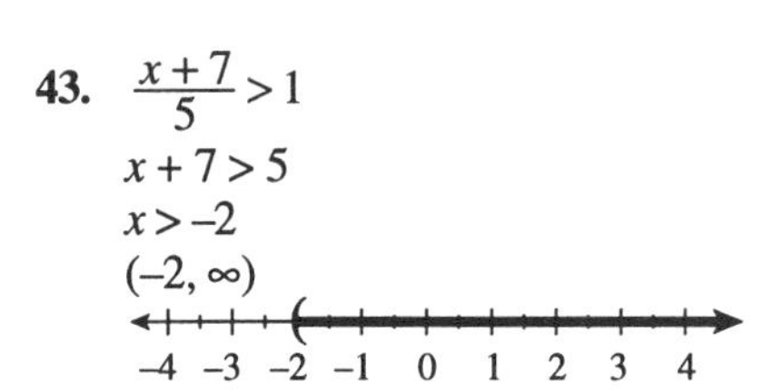

45. $\frac{-5x+11}{2} \leq 7$
$-5x + 11 \leq 14$
$-5x \leq 3$
$x \geq -\frac{3}{5}$
$\left[-\frac{3}{5}, \infty\right)$

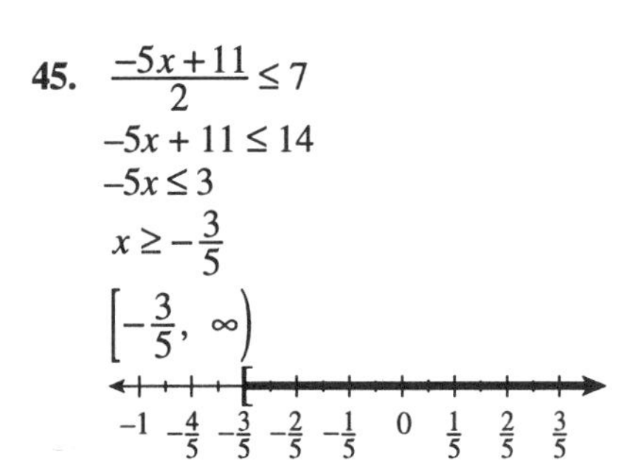

47. $8x - 16.4 \leq 10x + 2.8$
$-19.2 \leq 2x$
$-9.6 \leq x$
$x \geq -9.6$
$[-9.6, \infty)$

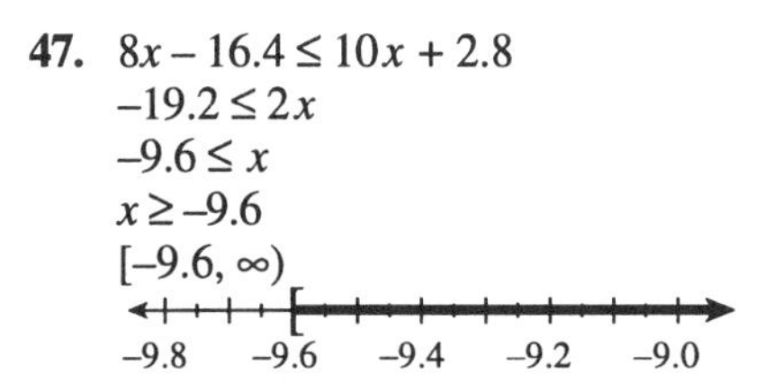

49. $2(x-3) > 70$
$x - 3 > 35$
$x > 38$
$(38, \infty)$

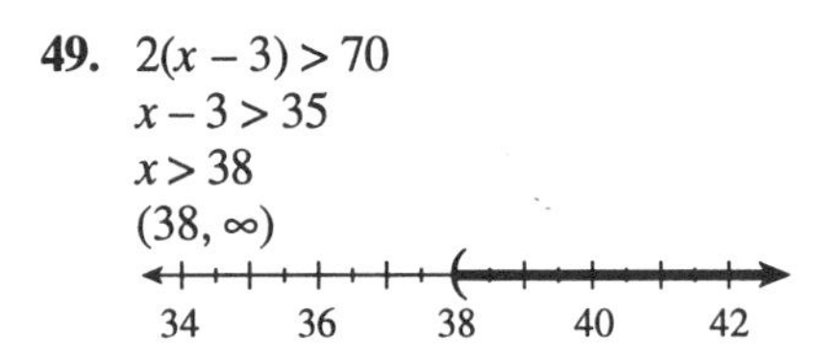

51. $-5x + 4 \leq -4(x-1)$
$-5x + 4 \leq -4x + 4$
$0 \leq x$
$x \geq 0$
$[0, \infty)$

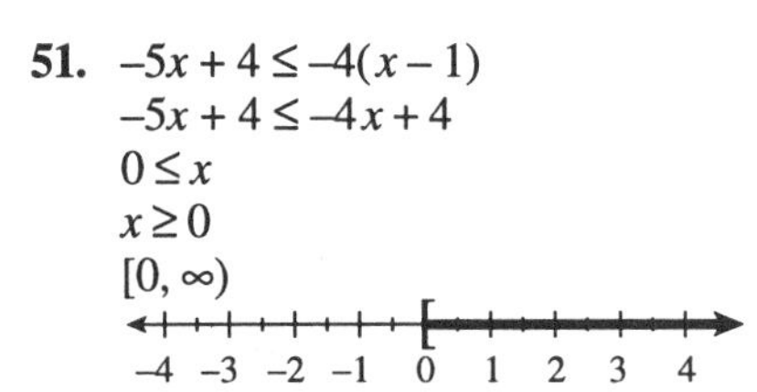

53. $\frac{1}{4}(x-7) \geq x+2$
$x-7 \geq 4x+8$
$-15 \geq 3x$
$-5 \geq x$
$x \leq -5$
$(-\infty, -5]$

−6 −5 −4 −3 −2 −1 0 1 2

55. $\frac{2}{3}(x+2) < \frac{1}{5}(2x+7)$
$10(x+2) < 3(2x+7)$
$10x+20 < 6x+21$
$4x < 1$
$x < \frac{1}{4}$
$\left(-\infty, \frac{1}{4}\right)$

$-1\ -\frac{3}{4}\ -\frac{1}{2}\ -\frac{1}{4}\ 0\ \frac{1}{4}\ \frac{1}{2}\ \frac{3}{4}\ 1$

57. $4(x-6)+2x-4 \geq 3(x-7)+10x$
$4x-24+2x-4 \geq 3x-21+10x$
$6x-28 \geq 13x-21$
$-7 \geq 7x$
$-1 \geq x$
$x \leq -1$
$(-\infty, -1]$

−4 −3 −2 −1 0 1 2 3 4

59. $\frac{5x+1}{7} - \frac{2x-6}{4} \geq -4$
$4(5x+1)-7(2x-6) \geq -112$
$20x+4-14x+42 \geq -112$
$6x+46 \geq -112$
$6x \geq -158$
$x \geq -\frac{79}{3}$
$\left[-\frac{79}{3}, \infty\right)$

$-27 \quad -\frac{79}{3} \quad -26$

61. $\frac{-x+2}{2} - \frac{1-5x}{8} < -1$
$4(-x+2)-(1-5x) < -8$
$-4x+8-1+5x < -8$
$x+7 < -8$
$x < -15$
$(-\infty, -15)$

−17 −16 −15 −14 −13

63. $0.8x+0.6x \geq 4.2$
$1.4x \geq 4.2$
$x \geq 3$
$[3, \infty)$

−2 −1 0 1 2 3 4 5 6

65. $\frac{x+5}{5} - \frac{3+x}{8} \geq -\frac{3}{10}$
$8(x+5)-5(3+x) \geq 4(-3)$
$8x+40-15-5x \geq -12$
$3x+25 \geq -12$
$3x \geq -37$
$x \geq -\frac{37}{3}$

$-13 \quad -\frac{37}{3} \quad -12$

67. $\frac{x+3}{12} + \frac{x-5}{15} < \frac{2}{3}$
$5(x+3)+4(x-5) < 20(2)$
$5x+15+4x-20 < 40$
$9x-5 < 40$
$9x < 45$
$x < 5$
$(-\infty, 5)$

0 1 2 3 4 5 6 7 8

69. Let x = her score on the final. Then
$\frac{72+67+82+79+2x}{6} \geq 60$
$300+2x \geq 360$
$2x \geq 60$
$x \geq 30$
Therefore, she must score at least 30 on the final exam.

71. Let x = the weight of the luggage and cargo. Then
$6(160)+x \leq 2000$
$960+x \leq 2000$
$x \leq 1040$
The plane can carry a maximum of 1040 lbs of luggage and cargo.

73. Let x = the weight of the envelope. Then
$0.32 + 0.23(x - 1) \le 4.00$
$0.32 + 0.23x - 0.23 \le 4.00$
$0.23x + 0.09 \le 4.00$
$0.23x \le 3.91$
$x \le 17.00$
Thus, at most 17 ounces can be mailed for \$4.00.

75. Let n = the number of calls made in a given month. Then solve
$25 \le 13 + 0.06n$
$12 \le 0.06n$
$200 \le n$
$n \ge 200$
Therefore, Plan 1 is more economical than Plan 2 if at least 200 calls are made.

77. $F \ge \frac{9}{5}C + 32$
$F \ge \frac{9}{5}(500) + 32$
$F \ge 932°$

79. $s = 651.2t + 27{,}821$
$t = \frac{s - 27{,}821}{651.2}$

a. $t = \frac{30{,}000 - 27{,}821}{651.2}$
$t = 3.346 \approx 4$ years
$1989 + 4 = 1993$

b. Answers may vary.

81. Decreasing

83. $w = -3.26t + 87.79$
$w = -3.26(10) + 87.79$
$w = 55.19$
55.19 lb/person/year

85. $w = -3.26t + 87.79$
$t = \frac{87.79 - w}{3.26}$
$t = \frac{87.79 - 50}{3.26}$
$t = 11.59 \approx 12$ years
$1990 + 12 = 2002$

87. The line graph representing consumption of whole milk is above that representing consumption of skim milk.

89. $w = -3.26t + 87.79$
$s = 1.25t + 22.75$
$w = s$
$-3.26t + 87.79 = 1.25t + 22.75$
$87.79 - 22.75 = 1.25t + 3.26t$
$65.04 = 4.51t$
$14.42 = t$
Approximately 14 years
$1990 + 14 = 2004$

91. 0, 1, 2, 3, 4, 5, 6, 7

93. –6, –7, –8, ...

95. $(-7, 1]$

–7 –6 –5 –4 –3 –2 –1 0 1

97. $[-2.5, 5.3)$

–2.5 0 5.3

Exercise Set 2.5

1. $C \cup D = \{2, 3, 4, 5, 6, 7\}$

3. $A \cap D = \{4, 6\}$

5. $A \cup B = \{..., -2, -1, 0, 1, ...\}$

7. $B \cap D = \{5, 7\}$

9. $x < 5$ and $x > -2$
$-2 < x < 5$
$(-2, 5)$

–2 –1 0 1 2 3 4 5 6

11. $x + 1 \ge 7$ and $3x - 1 \ge 5$
$x \ge 6$ and $3x \ge 6$
$x \ge 2$
$x \ge 6$
$[6, \infty)$

0 1 2 3 4 5 6 7 8

13. $4x + 2 \le -10$
$4x \le -12$
$x \le -3$
$(-\infty, -3]$

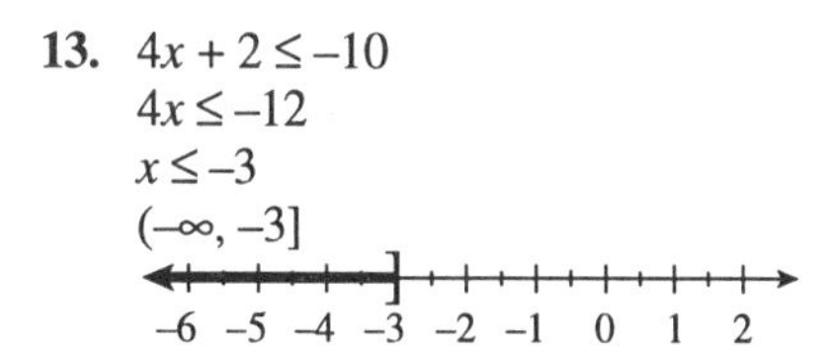

15. $5 < x - 6 < 11$
$11 < x < 17$
$(11, 17)$

17. $-2 \le 3x - 5 \le 7$
$3 \le 3x \le 12$
$1 \le x \le 4$
$[1, 4]$

19. $1 \le \frac{2}{3}x + 3 \le 4$

$-2 \le \frac{2}{3}x \le 1$

$-3 \le x \le \frac{3}{2}$

$\left[-3, \frac{3}{2}\right]$

21. $-5 \le \frac{x+1}{4} \le -2$
$-20 \le x + 1 \le -8$
$-21 \le x \le -9$
$[-21, -9]$

23. $x < -1$ or $x > 0$
$(-\infty, -1) \cup (0, \infty)$

25. $-2x \le -4$ or $5x - 20 \ge 5$
$x \ge 2$ or $5x \ge 25$
$x \ge 5$
$x \ge 2$
$[2, \infty)$

27. $3(x - 1) < 12$ or $x + 7 > 10$
$x - 1 < 4$ or $x > 3$
$x < 5$
All real numbers.
$(-\infty, \infty)$

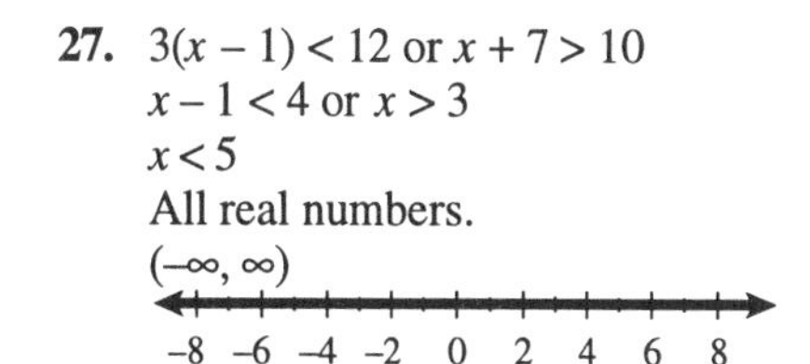

29. Answers may vary.

31. $x < 2$ and $x > -1$
$-1 < x < 2$
$(-1, 2)$

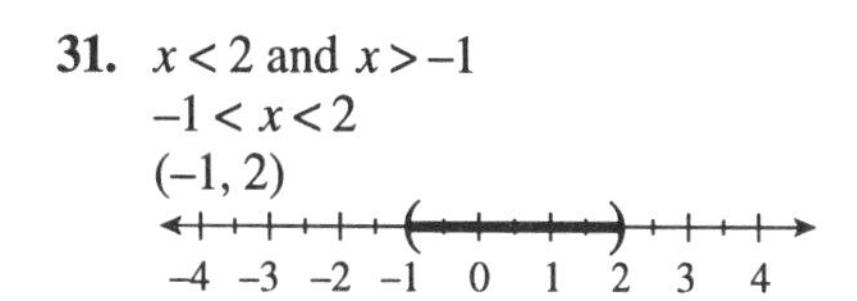

33. $x < 2$ or $x > -1$
All real numbers.
$(-\infty, \infty)$

35. $x \ge -5$ and $x \ge -1$
$x \ge -1$
$[-1, \infty)$

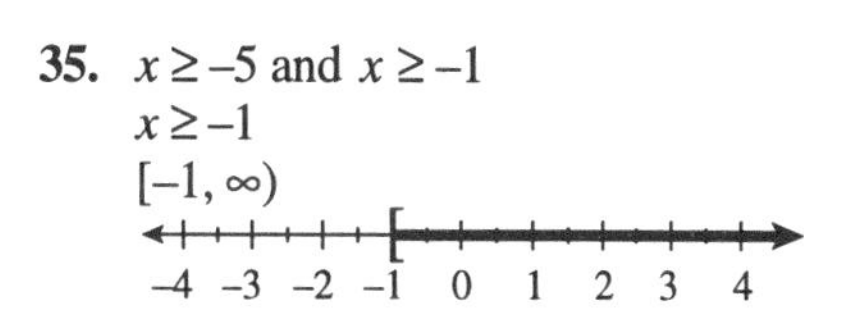

37. $x \ge -5$ or $x \ge -1$
$x \ge -5$
$[-5, \infty)$

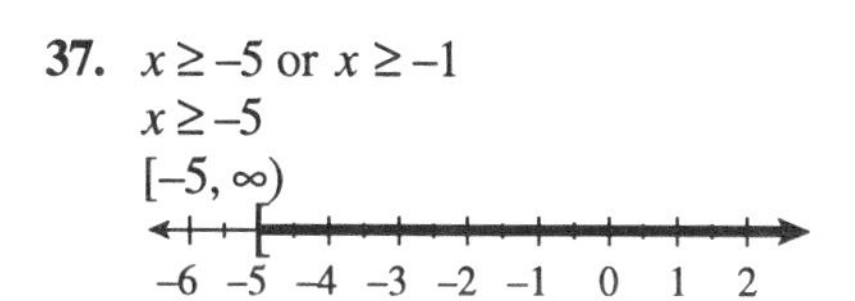

39. $0 \le 2x - 3 \le 9$
$3 \le 2x \le 12$

$\frac{3}{2} \le x \le 6$

$\left[\frac{3}{2}, 6\right]$

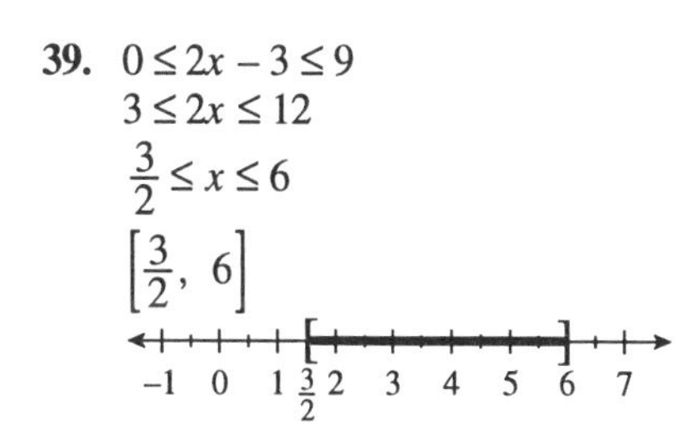

41. $\frac{1}{2} < x - \frac{3}{4} < 2$

$\frac{5}{4} < x < \frac{11}{4}$

$\left(\frac{5}{4}, \frac{11}{4}\right)$

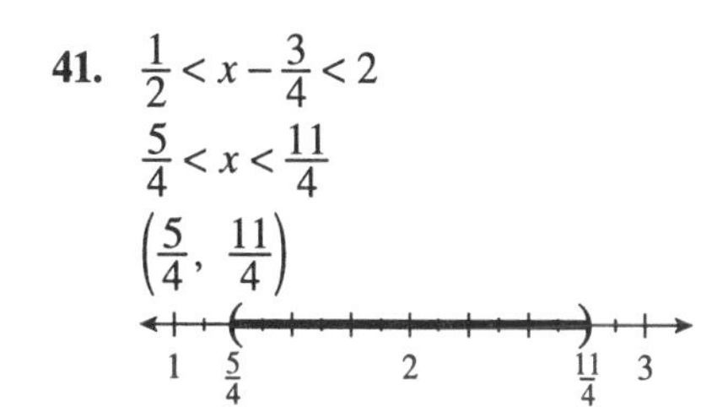

43. $x+3 \geq 3$ and $x+3 \leq 2$
$x \geq 0$ and $x \leq -1$
No real number
{ }

–4 –3 –2 –1 0 1 2 3 4

45. $3x \geq 5$ or $-x-6<1$
$x \geq \frac{5}{3}$ or $-x<7$
$x>-7$
$(-7, \infty)$

–8 –6 –4 –2 0 2 4 6 8

47. $0<\frac{5-2x}{3}<5$
$0<5-2x<15$
$-5<-2x<10$
$-5<x<\frac{5}{2}$
$\left(-5, \frac{5}{2}\right)$

–5 –4 –3 –2 –1 0 1 2 $\frac{5}{2}$ 3

49. $-6<3(x-2) \leq 8$
$-2<x-2 \leq \frac{8}{3}$
$0<x \leq \frac{14}{3}$
$\left(0, \frac{14}{3}\right]$

–2 –1 0 1 2 3 4 $\frac{14}{3}$ 6

51. $-x+5>6$ and $1+2x \leq -5$
$-x>1$ and $2x \leq -6$
$x<-1$ and $x \leq -3$
$x \leq -3$
$(-\infty, -3]$

–6 –5 –4 –3 –2 –1 0 1 2

53. $3x+2 \leq 5$ or $7x>29$
$3x \leq 3$ or $x>\frac{29}{7}$
$x \leq 1$
$(-\infty, 1] \cup \left(\frac{29}{7}, \infty\right)$

–1 0 1 2 3 $\frac{29}{7}$ 5 6 7

55. $5-x>7$ and $2x+3 \geq 13$
No solution. { }

–4 –3 –2 –1 0 1 2 3 4

57. $-\frac{1}{2} \leq \frac{4x-1}{6}<\frac{5}{6}$
$-3 \leq 4x-1<5$
$-2 \leq 4x<6$
$-\frac{1}{2} \leq x<\frac{3}{2}$
$\left[-\frac{1}{2}, \frac{3}{2}\right)$

$-\frac{1}{2}$ $-\frac{1}{4}$ 0 $\frac{1}{4}$ $\frac{1}{2}$ $\frac{3}{4}$ 1 $\frac{5}{4}$ $\frac{3}{2}$

59. $\frac{1}{15}<\frac{8-3x}{15}<\frac{4}{5}$
$1<8-3x<12$
$7<-3x<4$
$-\frac{4}{3}<x<\frac{7}{3}$
$\left(-\frac{4}{3}, \frac{7}{3}\right)$

$-\frac{4}{3}$ 0 $\frac{7}{3}$

61. $0.3<0.2x-0.9<1.5$
$1.2<0.2x<2.4$
$6<x<12$
$(6, 12)$

5 6 7 8 9 10 11 12 13

63. $-29 \leq C \leq 35$
$-29 \leq \frac{5}{9}(F-32) \leq 35$
$-52.2 \leq F-32 \leq 63$
$-20.2 \leq F \leq 95$
$-20.2° \leq F \leq 95°$

65. $70 \leq \frac{68+65+75+78+2x}{6} \leq 79$
$420<286+2x \leq 474$
$134 \leq 2x \leq 188$
$67 \leq x \leq 94$

67. 1992 and 1993

69. $|-7|-|19|=7-19=-12$

71. $-(-6)-|-10| = 6-10 = -4$

73. $|x| = 7$
$x = -7, 7$

75. $|x| = 0$
$x = 0$

77. $2x-3 < 3x+1 < 4x-5$
$2x-3 < 3x+1$ and $3x+1 < 4x-5$
$-4 < x$ and $6 < x$
$x > 6$

0 1 2 3 4 5 6 7 8

79. $-3(x-2) \le 3-2x \le 10-3x$
$-3x+6 \le 3-2x$ and $3-2x \le 10-3x$
$3 \le x$ and $x \le 7$
$3 \le x \le 7$

0 1 2 3 4 5 6 7 8

81. $5x-8 < 2(2+x) < -2(1+2x)$
$5x-8 < 4+2x$ and $4+2x < -2-4x$
$3x < 12$ and $6 < -6x$
$x < 4$ and $-1 > x$
$x < -1$

−2 −1 0 1 2 3 4 5 6

Section 2.6

Mental Math

1. $|-7| = -(-7) = 7$

3. $-|5| = -5$

5. $-|-6| = -(-(-6)) = -6$

7. $|-3|+|-2|+|-7|$
$= -(-3) + -(-2) + -(-7)$
$= 3+2+7 = 12$

Exercise Set 2.6

1. $|x| = 7$
$x = 7$ or $x = -7$
$\{7, -7\}$

3. $|3x| = 12.6$
$3x = 12.6$ or $3x = -12.6$
$x = 4.2$ or $x = -4.2$
$\{4.2, -4.2\}$

5. $|2x-5| = 9$
$2x-5 = 9$ or $2x-5 = -9$
$2x = 14$ or $2x = -4$
$x = 7$ or $x = -2$
$\{7, -2\}$

7. $\left|\frac{x}{2}-3\right| = 1$
$\frac{x}{2}-3 = 1$ or $\frac{x}{2}-3 = -1$
$\frac{x}{2} = 4$ or $\frac{x}{2} = 2$
$x = 8$ or $x = 4$
$\{8, 4\}$

9. $|z|+4 = 9$
$|z| = 5$
$z = 5$ or $z = -5$
$\{5, -5\}$

11. $|3x|+5 = 14$
$|3x| = 9$
$3x = 9$ or $3x = -9$
$x = 3$ or $x = -3$
$\{3, -3\}$

13. $|2x| = 0$
$2x = 0$
$x = 0$
$\{0\}$

15. $|4n+1|+10 = 4$
$|4n+1| = -6$ which is impossible.
$\{\ \}$

17. $|5x-1| = 0$
$5x-1 = 0$
$5x = 1$
$x = \frac{1}{5}$
$\left\{\frac{1}{5}\right\}$

19. $|x| = 5$

21. $|5x-7|=|3x+11|$
$5x-7=3x+11$
or $5x-7=-(3x+11)$
$2x=18$ or $5x-7=-3x-11$
$x=9$ or $8x=-4$
$x=9$ or $x=-\frac{1}{2}$
$\left\{9, -\frac{1}{2}\right\}$

23. $|z+8|=|z-3|$
$z+8=z-3$ or $z+8=-(z-3)$
$8=-3$ or $z+8=-z+3$
no solution or $2z=-5$
$z=-\frac{5}{2}$
$\left\{-\frac{5}{2}\right\}$

25. Answers may vary.

27. $|x|=4$
$x=4$ or $x=-4$
$\{4, -4\}$

29. $|y|=0$
$y=0$
$\{0\}$

31. $|z|=-2$ is impossible.
{ }

33. $|7-3x|=7$
$7-3x=7$ or $7-3x=-7$
$-3x=0$ or $-3x=-14$
$x=0$ or $x=\frac{14}{3}$
$\left\{0, \frac{14}{3}\right\}$

35. $|6x|-1=11$
$|6x|=12$
$6x=12$ or $6x=-12$
$x=2$ or $x=-2$
$\{2, -2\}$

37. $|4p|=-8$ is impossible.
{ }

39. $|x-3|+3=7$
$|x-3|=4$
$x-3=4$ or $x-3=-4$
$x=7$ or $x=-1$
$\{7, -1\}$

41. $\left|\frac{z}{4}+5\right|=-7$ is impossible.
{ }

43. $|9v-3|=-8$ is impossible.
{ }

45. $|8n+1|=0$
$8n+1=0$
$8n=-1$
$n=-\frac{1}{8}$
$\left\{-\frac{1}{8}\right\}$

47. $|1+6c|-7=-3$
$|1+6c|=4$
$1+6c=4$ or $1+6c=-4$
$6c=3$ or $6c=-5$
$c=\frac{1}{2}$ or $c=-\frac{5}{6}$
$\left\{\frac{1}{2}, -\frac{5}{6}\right\}$

49. $|5x+1|=11$
$5x+1=11$ or $5x+1=-11$
$5x=10$ or $5x=-12$
$x=2$ or $x=-\frac{12}{5}$
$\left\{2, -\frac{12}{5}\right\}$

51. $|4x-2|=|-10|$ or $|4x-2|=10$
$4x-2=10$ or $4x-2=-10$
$4x=12$ or $4x=-8$
$x=3$ or $x=-2$
$\{3, -2\}$

53. $|5x+1| = |4x-7|$
$5x + 1 = 4x - 7$ or $5x + 1 = -(4x - 7)$
$x = -8$ or $5x + 1 = -4x + 7$
$x = -8$ or $9x = 6$
$x = -8$ or $x = \frac{2}{3}$
$\left\{-8, \frac{2}{3}\right\}$

55. $|6+2x| = -|-7|$ or $|6+2x| = -7$, which is impossible.
{ }

57. $|2x-6| = |10-2x|$
$2x - 6 = 10 - 2x$ or $2x - 6 = -(10 - 2x)$
$4x = 16$ or $2x - 6 = -10 + 2x$
$x = 4$ or $-6 = -10$ which is impossible
{4}

59. $\left|\frac{2x-5}{3}\right| = 7$
$\frac{2x-5}{3} = 7$ or $\frac{2x-5}{3} = -7$
$2x - 5 = 21$ or $2x - 5 = -21$
$2x = 26$ or $2x = -16$
$x = 13$ or $x = -8$
{13, −8}

61. $2 + |5n| = 17$
$|5n| = 15$
$5n = 15$ or $5n = -15$
$n = 3$ or $n = -3$
{3, −3}

63. $\left|\frac{2x-1}{3}\right| = |-5|$
$\left|\frac{2x-1}{3}\right| = 5$
$\frac{2x-1}{3} = 5$ or $\frac{2x-1}{3} = -5$
$2x - 1 = 15$ or $2x - 1 = -15$
$2x = 16$ or $2x = -14$
$x = 8$ or $x = -7$
{8, −7}

65. $|2y-3| = |9-4y|$
$2y - 3 = 9 - 4y$ or $2y - 3 = -(9 - 4y)$
$6y = 12$ or $2y - 3 = -9 + 4y$
$y = 2$ or $6 = 2y$
$y = 2$ or $3 = y$
{2, 3}

67. $\left|\frac{3n+2}{8}\right| = |-1|$ or $\left|\frac{3n+2}{8}\right| = 1$
$\frac{3n+2}{8} = 1$ or $\frac{3n+2}{8} = -1$
$3n + 2 = 8$ or $3n + 2 = -8$
$3n = 6$ or $3n = -10$
$n = 2$ or $n = -\frac{10}{3}$
$\left\{2, -\frac{10}{3}\right\}$

69. $|x+4| = |7-x|$
$x + 4 = 7 - x$ or $x + 4 = -(7 - x)$
$2x = 3$ or $x + 4 = -7 + x$
$x = \frac{3}{2}$ or $4 = -7$, which is impossible.
$\left\{\frac{3}{2}\right\}$

71. $\left|\frac{8c-7}{3}\right| = -|-5|$ or $\left|\frac{8c-7}{3}\right| = -5$ which is impossible.
{ }

73. Answers may vary.

75. $100 - 20 - 45 = 35\%$

77. (9.75 billion)(0.45) = 4.39 billion

79. $|x| \geq -2$
−2, −1, 0, 1, 2, for example

81. $|y| < 0$
No solution

Exercise Set 2.7

1. $|x| \leq 4$
$-4 \leq x \leq 4$
[−4, 4]

−8 −6 −4 −2 0 2 4 6 8

3. $|x-3|<2$
$-2<x-3<2$
$1<x<5$
$(1, 5)$

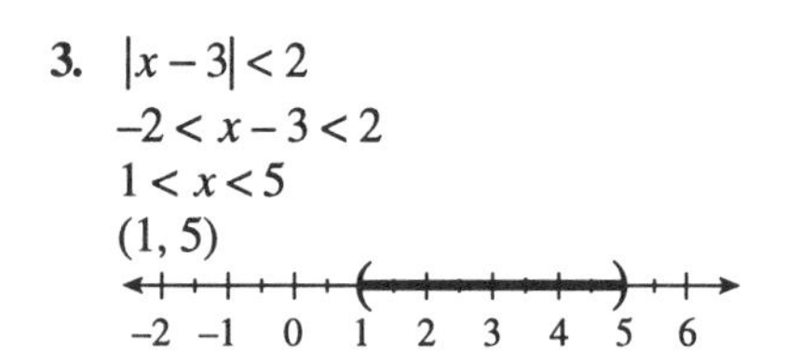

5. $|x+3|<2$
$-2<x+3<2$
$-5<x<-1$
$(-5, -1)$

7. $|2x+7|\le 13$
$-13\le 2x+7\le 13$
$-20\le 2x\le 6$
$-10\le x\le 3$
$[-10, 3]$

9. $|x|+7\le 12$
$|x|\le 5$
$-5\le x\le 5$
$[-5, 5]$

11. $|3x-1|<-5$
No real solutions.
$\{\ \}$

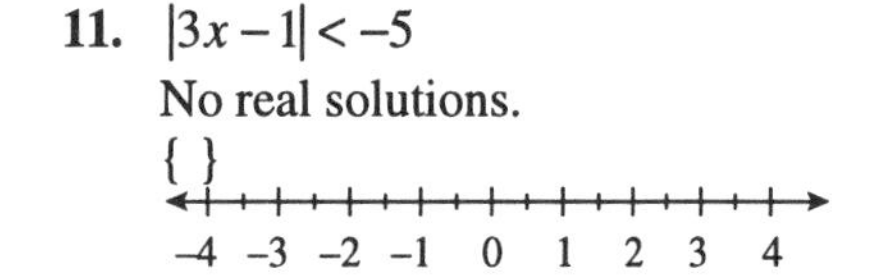

13. $|x-6|-7\le -1$
$|x-6|\le 6$
$-6\le x-6\le 6$
$0\le x\le 12$
$[0, 12]$

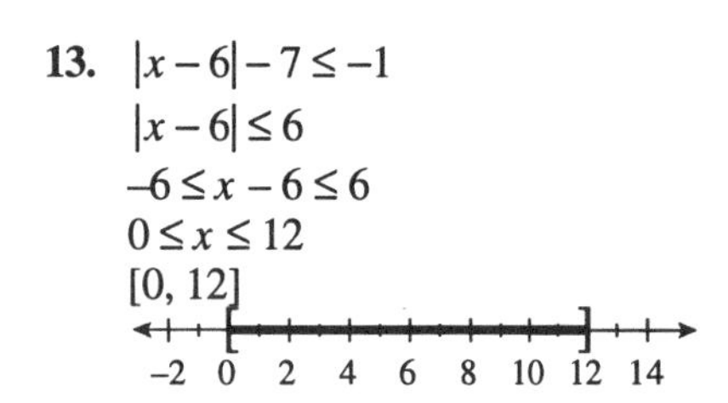

15. $|x|>3$
$x<-3$ or $x>3$
$(-\infty, -3)\cup(3, \infty)$

17. $|x+10|\ge 14$
$x+10\le -14$ or $x+10\ge 14$
$x\le -24$ or $x\ge 4$
$(-\infty, -24]\cup[4, \infty)$

19. $|x|+2>6$
$|x|>4$
$x<-4$ or $x>4$
$(-\infty, -4)\cup(4, \infty)$

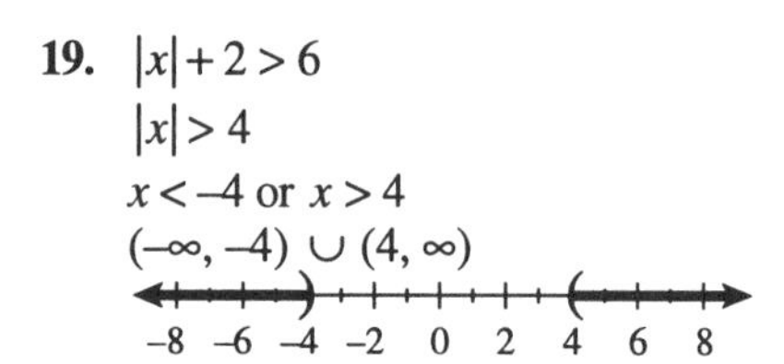

21. $|5x|>-4$
All real numbers. $(-\infty, \infty)$

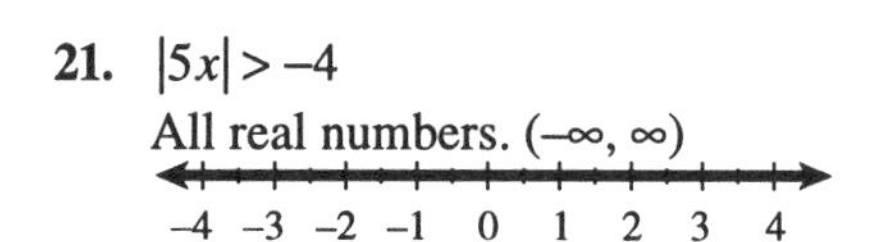

23. $|6x-8|+3>7$
$|6x-8|>4$
$6x-8<-4$ or $6x-8>4$
$6x<4$ or $6x>12$
$x<\frac{2}{3}$ or $x>2$
$\left(-\infty, \frac{2}{3}\right)\cup(2, \infty)$

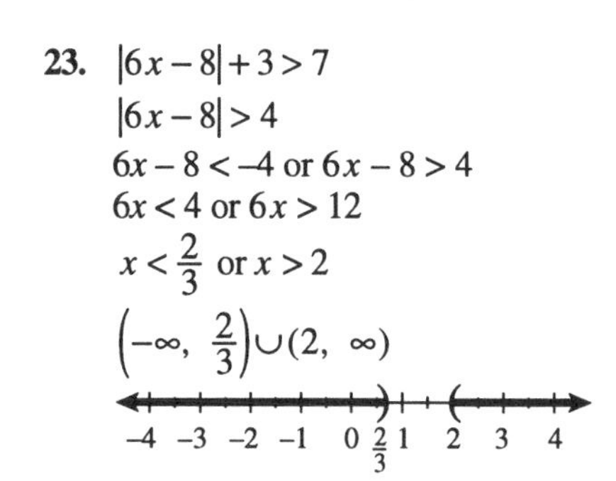

25. $|x|\le 0$
$|x|=0$
$x=0$
$\{0\}$

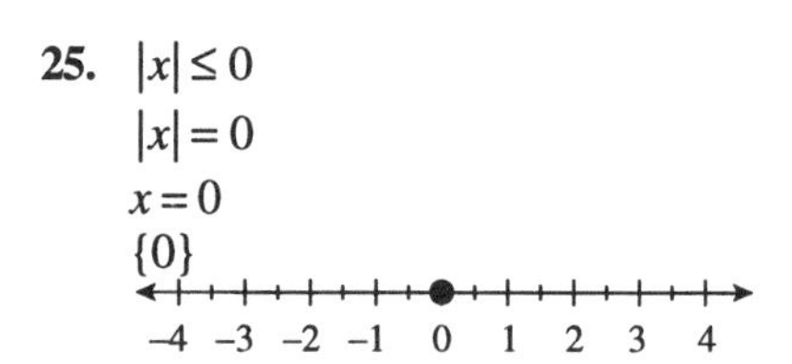

27. $|8x+3|>0$ only excludes $|8x+3|=0$
$8x+3=0$
$8x=-3$
$x=-\frac{3}{8}$
All reals except $-\frac{3}{8}$.
$\left(\infty, -\frac{3}{8}\right)\cup\left(-\frac{3}{8}, \infty\right)$

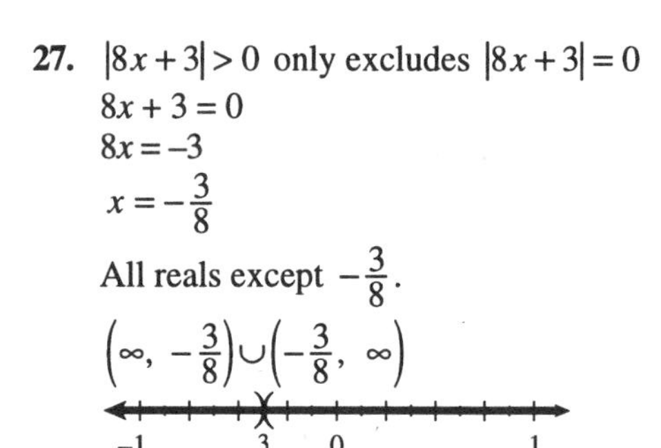

29. $|x| < 7$

31. $|x| \le 5$

33. $|x| \le 2$
$-2 \le x \le 2$
$[-2, 2]$
−4 −3 −2 −1 0 1 2 3 4

35. $|y| > 1$
$y < -1$ or $y > 1$
$(-\infty, -1) \cup (1, \infty)$
−4 −3 −2 −1 0 1 2 3 4

37. $|x-3| < 8$
$-8 < x-3 < 8$
$-5 < x < 11$
$(-5, 11)$
−5 −3 −1 1 3 5 7 9 11

39. $|6x-8| > 4$
$6x-8 < -4$ or $6x-8 > 4$
$6x < 4$ or $6x > 12$
$x < \frac{2}{3}$ or $x > 2$
$\left(-\infty, \frac{2}{3}\right) \cup (2, \infty)$
−4 −3 −2 −1 0 $\frac{2}{3}$ 1 2 3 4

41. $5 + |x| \le 2$
$|x| \le -3$
No real solution.
{ }
−4 −3 −2 −1 0 1 2 3 4

43. $|x| > -4$
All real numbers.
$(-\infty, \infty)$
−4 −3 −2 −1 0 1 2 3 4

45. $|2x-7| \le 11$
$-11 \le 2x-7 \le 11$
$-4 \le 2x \le 18$
$-2 \le x \le 9$
$[-2, 9]$

47. $|x+5| + 2 \ge 8$
$|x+5| \ge 6$
$x+5 \le -6$ or $x+5 \ge 6$
$x \le -11$ or $x \ge 1$
$(-\infty, -11] \cup [1, \infty)$
−13 −11 −9 −7 −5 −3 −1 1 3

49. $|x| > 0$ excludes only $|x| = 0$, or $x = 0$
All reals except $x = 0$
$(-\infty, 0) \cup (0, \infty)$
−4 −3 −2 −1 0 1 2 3 4

51. $9 + |x| > 7$
$|x| > -2$
All real numbers.
$(-\infty, \infty)$
−4 −3 −2 −1 0 1 2 3 4

53. $6 + |4x-1| \le 9$
$|4x-1| \le 3$
$-3 \le 4x-1 \le 3$
$-2 \le 4x \le 4$
$-\frac{1}{2} \le x \le 1$
$\left[-\frac{1}{2}, 1\right]$
−1 $-\frac{3}{4}$ $-\frac{1}{2}$ $-\frac{1}{4}$ 0 $\frac{1}{4}$ $\frac{1}{2}$ $\frac{3}{4}$ 1

55. $\left|\frac{2}{3}x + 1\right| > 1$
$\frac{2}{3}x + 1 < -1$ or $\frac{2}{3}x + 1 > 1$
$\frac{2}{3}x < -2$ or $\frac{2}{3}x > 0$
$x < -3$ or $x > 0$
$(-\infty, -3) \cup (0, \infty)$
−6 −5 −4 −3 −2 −1 0 1 2

57. $|5x+3| < -6$

No real numbers.

$\{\ \}$

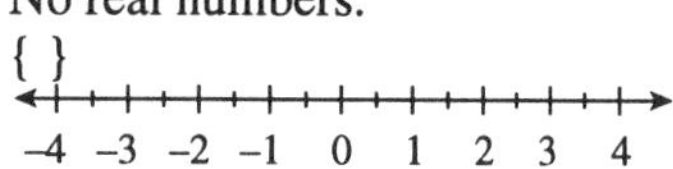

59. $|8x+3| \geq 0$

All real numbers.

$(-\infty, \infty)$

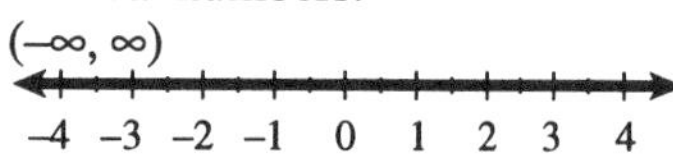

61. $|1+3x|+4 < 5$

$|1+3x| < 1$

$-1 < 1+3x < 1$

$-2 < 3x < 0$

$-\frac{2}{3} < x < 0$

$\left(-\frac{2}{3}, 0\right)$

–1 $-\frac{2}{3}$ $-\frac{1}{3}$ 0 $\frac{1}{3}$ $\frac{2}{3}$ 1

63. $|x|-3 \geq -3$

$|x| \geq 0$

All real numbers.

$(-\infty, \infty)$

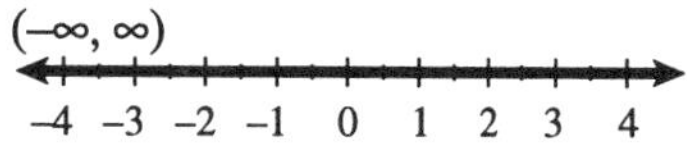

65. $|8x|-10 > -2$

$|8x| > 8$

$8x < -8$ or $8x > 8$

$x < -1$ or $x > 1$

$(-\infty, -1) \cup (1, \infty)$

–4 –3 –2 –1 0 1 2 3 4

67. $\left|\frac{x+6}{3}\right| > 2$

$\frac{x+6}{3} < -2$ or $\frac{x+6}{3} > 2$

$x+6 < -6$ or $x+6 > 6$

$x < -12$ or $x > 0$

$(-\infty, -12) \cup (0, \infty)$

–14 –12 –10 –8 –6 –4 –2 0 2

69. $|2(3+x)| > 6$

$2|3+x| > 6$

$|3+x| > 3$

$3+x < -3$ or $3+x > 3$

$x < -6$ or $x > 0$

$(-\infty, -6) \cup (0, \infty)$

–10 –8 –6 –4 –2 0 2 4 6

71. $\left|\frac{5(x+2)}{3}\right| < 7$

$\frac{5}{3}|x+2| < 7$

$|x+2| < \frac{21}{5}$

$-\frac{21}{5} < x+2 < \frac{21}{5}$

$-\frac{31}{5} < x < \frac{11}{5}$

$\left(-\frac{31}{5}, \frac{11}{5}\right)$

$-\frac{31}{5}$ 0 $\frac{11}{5}$

73. $-15+|2x-7| \leq -6$

$|2x-7| \leq 9$

$-9 \leq 2x-7 \leq 9$

$-2 \leq 2x \leq 16$

$-1 \leq x \leq 8$

$[-1, 8]$

–1 8

75. $\left|2x+\frac{3}{4}\right|-7 \leq -2$

$\left|2x+\frac{3}{4}\right| \leq 5$

$-5 \leq 2x+\frac{3}{4} \leq 5$

$-\frac{23}{4} \leq 2x \leq \frac{17}{4}$

$-\frac{23}{8} \leq x \leq \frac{17}{8}$

$\left[-\frac{23}{8}, \frac{17}{8}\right]$

–4 $-\frac{23}{8}$ 0 $\frac{17}{8}$ 4

77. $|2x-3|<7$
$-7<2x-3<7$
$-4<2x<10$
$-2<x<5$
$(-2, 5)$

79. $|2x-3|=7$
$2x-3=7$ or $2x-3=-7$
$2x=10$ or $2x=-4$
$x=5$ or $x=-2$
$\{5, -2\}$

81. $|x-5|\geq 12$
$x-5\leq -12$ or $x-5\geq 12$
$x\leq -7$ or $x\geq 17$
$(-\infty, -7] \cup [17, \infty)$

83. $|9+4x|=0$
$9+4x=0$
$4x=-9$
$x=-\frac{9}{4}$
$\left\{-\frac{9}{4}\right\}$

85. $|2x+1|+4<7$
$|2x+1|<3$
$-3<2x+1<3$
$-4<2x<2$
$-2<x<1$
$(-2, 1)$

87. $|3x-5|+4=5$
$|3x-5|=1$
$3x-5=-1$ or $3x-5=1$
$3x=4$ or $3x=6$
$x=\frac{4}{3}$ or $x=2$
$\left\{\frac{4}{3}, 2\right\}$

89. $|x+11|=-1$
No real solution.
{ }

91. $\left|\frac{2x-1}{3}\right|=6$
$\frac{2x-1}{3}=-6$ or $\frac{2x-1}{3}=6$
$2x-1=-18$ or $2x-1=18$
$2x=-17$ or $2x=19$
$x=-\frac{17}{2}$ or $x=\frac{19}{2}$
$\left\{-\frac{17}{2}, \frac{19}{2}\right\}$

93. $\left|\frac{3x-5}{6}\right|>5$
$\frac{3x-5}{6}<-5$ or $\frac{3x-5}{6}>5$
$3x-5<-30$ or $3x-5>30$
$3x<-25$ or $3x>35$
$x<-\frac{25}{3}$ or $x>\frac{35}{3}$
$\left(-\infty, -\frac{25}{3}\right)\cup\left(\frac{35}{3}, \infty\right)$

95. $3.5-0.05<x<3.5+0.05$
$3.45 \text{ m} < x < 3.55 \text{ m}$

97. $\frac{1}{6}$

99. 0

101. $\frac{1}{6}+\frac{1}{6}=\frac{2}{6}=\frac{1}{3}$

103. $3x-4y=12$
$3(2)-4y=12$
$6-4y=12$
$-4y=6$
$y=-\frac{6}{4}$
$y=-\frac{3}{2}=-1.5$

105. $3x-4y=12$
$3x-4(-3)=12$
$3x+12=12$
$3x=0$
$x=0$

Chapter 2 - Review

1. $4(x-5)=2x-14$
$4x-20=2x-14$
$2x=6$
$x=3$
$\{3\}$

2. $x + 7 = -2(x + 8)$
$x + 7 = -2x - 16$
$3x = -23$
$x = -\frac{23}{3}$
$\left\{-\frac{23}{3}\right\}$

3. $3(2y - 1) = -8(6 + y)$
$6y - 3 = -48 - 8y$
$14y = -45$
$y = -\frac{45}{14}$
$\left\{-\frac{45}{14}\right\}$

4. $-(z + 12) = 5(2z - 1)$
$-z - 12 = 10z - 5$
$-7 = 11z$
$-\frac{7}{11} = z$
$\left\{-\frac{7}{11}\right\}$

5. $n - (8 + 4n) = 2(3n - 4)$
$n - 8 - 4n = 6n - 8$
$-3n = 6n$
$0 = 9n$
$n = 0$
$\{0\}$

6. $4(9v + 2) = 6(1 + 6v) - 10$
$36v + 8 = 6 + 36v - 10$
$36v + 8 = 36v - 4$
$8 = -4$
No real number.
$\{\ \}$

7. $0.3(x - 2) = 1.2$
$x - 2 = 4$
$x = 6$
$\{6\}$

8. $1.5 = 0.2(c - 0.3)$
$1.5 = 0.2c - 0.06$
$1.56 = 0.2c$
$7.8 = c$
$\{7.8\}$

9. $-4(2 - 3h) = 2(3h - 4) + 6h$
$-8 + 12h = 6h - 8 + 6h$
$-8 + 12h = 12h - 8$
All real numbers.
$\{h|h \text{ is a real number}\}$

10. $6(m - 1) + 3(2 - m) = 0$
$6m - 6 + 6 - 3m = 0$
$3m = 0$
$m = 0$
$\{0\}$

11. $6 - 3(2g + 4) - 4g = 5(1 - 2g)$
$6 - 6g - 12 - 4g = 5 - 10g$
$-6 - 10g = 5 - 10g$
$-6 = 5$
No real number.
$\{\ \}$

12. $20 - 5(p + 1) + 3p = -(2p - 15)$
$20 - 5p - 5 + 3p = -2p + 15$
$15 - 2p = -2p + 15$
All real numbers.
$\{p|p \text{ is a real number}\}$

13. $\frac{x}{3} - 4 = x - 2$
$x - 12 = 3x - 6$
$-6 = 2x$
$x = -3$
$\{-3\}$

14. $\frac{9}{4}y = \frac{2}{3}y$
$\frac{27}{12}y - \frac{8}{12}y = 0$
$\frac{19}{12}y = 0$
$y = 0$
$\{0\}$

15. $\frac{3n}{8} - 1 = 3 + \frac{n}{6}$
$9n - 24 = 72 + 4n$
$5n = 96$
$n = \frac{96}{5}$
$\left\{\frac{96}{5}\right\}$

16. $\frac{z}{6}+1=\frac{z}{2}+2$
$z+6=3z+12$
$-6=2z$
$-3=z$
$\{-3\}$

17. $\frac{y}{4}-\frac{y}{2}=-8$
$y-2y=-32$
$-y=-32$
$y=32$
$\{32\}$

18. $\frac{2x}{3}-\frac{8}{3}=x$
$2x-8=3x$
$-8=x$
$\{-8\}$

19. $\frac{b-2}{3}=\frac{b+2}{5}$
$5b-10=3b+6$
$2b=16$
$b=8$
$\{8\}$

20. $\frac{2t-1}{3}=\frac{3t+2}{15}$
$10t-5=3t+2$
$7t=7$
$t=1$
$\{1\}$

21. $\frac{2(t+1)}{3}=\frac{2(t-1)}{3}$
$t+1=t-1$
$1=-1$
No real number.
$\{\ \}$

22. $\frac{3a-3}{6}=\frac{4a+1}{15}+2$
$5(3a-3)=2(4a+1)+30\cdot 2$
$15a-15=8a+2+60$
$15a-15=8a+62$
$7a=77$
$a=11$
$\{11\}$

23. $\frac{x-2}{5}+\frac{x+2}{2}=\frac{x+4}{3}$
$6(x-2)+15(x+2)=10(x+4)$
$6x-12+15x+30=10x+40$
$21x+18=10x+40$
$11x=22$
$x=2$
$\{2\}$

24. $\frac{2z-3}{4}-\frac{4-z}{2}=\frac{z+1}{3}$
$3(2z-3)-6(4-z)=4(z+1)$
$6z-9-24+6z=4z+4$
$8z=37$
$z=\frac{37}{8}$
$\left\{\frac{37}{8}\right\}$

25. $2(x-3)=3x+1$
$2x-6=3x+1$
$-7=x$

26. Let x = smaller number
Then $x+5$ = larger number
$x+x+5=285$
$2x=280$
$x=140$
$x+5=145$
140 and 145

27. $130(0.40)=52$

28. $8(0.015)=0.12$

29. Let x = salary for high school graduate
$x+0.85x=37{,}000$
$1.85x=37{,}000$
$x=\$20{,}000$

30. Let n = the first integer,
$n+1$ = the second integer
$n+2$ = the third integer, and
$n+4$ = the fourth integer.
$(n+1)+(n+2)+(n+3)-2n=16$
$n+6=16$
$n=10$
Therefore, the integers are 10, 11, 12, and 13.

31. Let x = smaller odd integer
Then $x + 2$ = larger odd integer
$5x = 3(x+2) + 54$
$5x = 3x + 6 + 54$
$2x = 60$
$x = 30$
Since this is not odd, no such consecutive integers exist.

32. Let w = the width of the playing field and $2w - 5$ = the length of the playing field. Then
$2w + 2(2w - 5) = 230$
$2w + 4w - 10 = 230$
$6w - 10 = 230$
$6w = 240$
$w = 40$
$2w - 5 = 2(40) - 5 = 80 - 5 = 75$
Therefore, the field is 75 meters long and 40 meters wide.

33. Let n = the number of miles driven. Then
$2(19.95) + 0.12(n - 200) = 46.86$
$39.9 + 0.12n - 24 = 46.86$
$0.12n + 15.9 = 46.86$
$0.12n = 30.96$
$n = 258$
He drove 258 miles.

34. Solve $R = C$
$16.50x = 4.50x + 3000$
$12x = 3000$
$x = 250$
Thus, 250 calculators must be produced and sold in order to break even.

35. $R = 40x$
$C = 20x + 100$
$40x = 20x + 100$
$20x = 100$
$x = 5$
$R = 40 \cdot 5 = 200$
5 plants, \$200

36. $V = LWH$
$W = \frac{V}{LH}$

37. $C = 2\pi r$
$\frac{C}{2\pi} = r$

38. $5x - 4y = -12$
$5x + 12 = 4y$
$y = \frac{5x+12}{4}$

39. $5x - 4y = -12$
$5x = 4y - 12$
$x = \frac{4y-12}{5}$

40. $y - y_1 = m(x - x_1)$
$m = \frac{y - y_1}{x - x_1}$

41. $y - y_1 = m(x - x_1)$
$y - y_1 = mx - mx_1$
$y - y_1 + mx_1 = mx$
$\frac{y - y_1 + mx_1}{m} = x$

42. $E = I(R + r)$
$\frac{E}{I} = R + r$
$r = \frac{E}{I} - R$

43. $S = vt + gt^2$
$S - vt = gt^2$
$\frac{S - vt}{t^2} = g$

44. $T = gr + gvt$
$T = g(r + vt)$
$g = \frac{T}{r + vt}$

45. $I = Prt + P$
$I = P(rt + 1)$
$\frac{I}{rt+1} = P$

46. $A = \frac{h}{2}(B + b)$
$\frac{2A}{h} = B + b$
$B = \frac{2A}{h} - b$

47. $V = \frac{1}{3}\pi r^2 h$

$\frac{3V}{\pi r^2} = h$

48. $R = \frac{r_1 + r_2}{2}$

$2R = r_1 + r_2$

$r_1 = 2R - r_2$

49. $\frac{V_1}{T_1} = \frac{V_2}{T_2}$

$T_2 V_1 = T_1 V_2$

$T_2 = \frac{T_1 V_2}{V_1}$

50. $\frac{1}{a} + \frac{1}{b} = \frac{1}{c}$

$\frac{1}{b} = \frac{1}{c} - \frac{1}{a}$

$\frac{1}{b} = \frac{a-c}{ca}$

$b = \frac{ca}{a-c}$

51. $\frac{2}{x} - \frac{3}{y} = \frac{1}{z}$

$2yz - 3xz = xy$

$2yz - xy = 3xz$

$y(2z - x) = 3xz$

$y = \frac{3xz}{2z - x}$

52. $R = \frac{R_1 R_2}{R_1 + R_2}$

$R(R_1 + R_2) = R_1 R_2$

$RR_1 + RR_2 = R_1 R_2$

$RR_1 = R_1 R_2 - RR_2$

$RR_1 = (R_1 - R)R_2$

$R_2 = \frac{RR_1}{R_1 - R}$

53. $C = \frac{2AB}{A-B}$

$CA - CB = 2AB$

$CA - 2AB = CB$

$A(C - 2B) = CB$

$A = \frac{CB}{C - 2B}$

54. $\frac{x-y}{5} + \frac{y}{4} = \frac{2x}{3}$

$12(x - y) + 15y = 20(2x)$

$12x - 12y + 15y = 40x$

$12x + 3y = 40x$

$3y = 28x$

$y = \frac{28}{3}x$

55. $\frac{b+c}{d} - \frac{b}{c} = \frac{5}{c}$

$c(b+c) - db = 5d$

$c(b+c) = 5d + db$

$c(b+c) = d(5+b)$

$\frac{c(b+c)}{5+b} = d$

56. $A = P\left(1 + \frac{r}{n}\right)^{nt} = 3000\left(1 + \frac{0.03}{n}\right)^{7n}$

a. $A = 3000\left(1 + \frac{0.03}{2}\right)^{14} = \$3695.27$

b. $A = 3000\left(1 + \frac{0.03}{52}\right)^{364} = \$3700.81$

57. $F = \frac{9}{5}C + 32$

$90 = \frac{9}{5}C + 32$

$58 = \frac{9}{5}C$

$\frac{290}{9} = C$

$32.2° \approx C$

58. Let x = width

Then $x + 2$ = length

$(x + 4)(x + 2 + 4) = x(x + 2) + 88$

$(x + 4)(x + 6) = x^2 + 2x + 88$

$x^2 + 10x + 24 = x^2 + 2x + 88$

$8x = 64$

$x = 8$

$x + 2 = 10$

Width is 8 in.; Length is 10 in.

59. Area $= 18 \times 21 = 378$ ft^2

Packages $= \frac{378}{24} = 15.75$

16 packages

60. $V_{\text{box}} = lwh = 8(5)(3) = 120 \text{ in.}^3$, while
$V_{\text{cyl}} = \pi r^2 h = \pi \cdot 3^2 \cdot 6 = 54\pi \approx 170 \text{ in.}^3$
Therefore, the cylinder holds more ice cream.

61. $d = rt$
$130 = r(2.25)$
$57.8 = r$
58 mph

62. $3(x-5) > -(x+3)$
$3x - 15 > -x - 3$
$4x > 12$
$x > 3$
$(3, \infty)$

63. $-2(x+7) \geq 3(x+2)$
$-2x - 14 \geq 3x + 6$
$-20 \geq 5x$
$-4 \geq x$
$(-\infty, -4]$

64. $4x - (5 + 2x) < 3x - 1$
$4x - 5 - 2x < 3x - 1$
$2x - 5 < 3x - 1$
$-4 < x$
$(-4, \infty)$

65. $3(x-8) < 7x + 2(5-x)$
$3x - 24 < 7x + 20 - 2x$
$-34 < 2x$
$-17 < x$
$(-17, \infty)$

66. $24 \geq 6x - 2(3x-5) + 2x$
$24 \geq 6x - 6x + 10 + 2x$
$24 \geq 10 + 2x$
$14 \geq 2x$
$7 \geq x$
$x \leq 7$
$(-\infty, 7]$

67. $48 + x \geq 5(2x+4) - 2x$
$48 + x \geq 10x + 20 - 2x$
$28 \geq 7x$
$4 \geq x$
$(-\infty, 4]$

68. $\frac{x}{3} + \frac{1}{2} > \frac{2}{3}$
$\frac{x}{3} > \frac{1}{6}$
$x > \frac{1}{2}$
$\left(\frac{1}{2}, \infty\right)$

69. $x + \frac{3}{4} < -\frac{x}{2} + \frac{9}{4}$
$4x + 3 < -2x + 9$
$6x < 6$
$x < 1$
$(-\infty, 1)$

70. $\frac{x-5}{2} \leq \frac{3}{8}(2x+6)$
$4(x-5) \leq 3(2x+6)$
$4x - 20 \leq 6x + 18$
$-38 \leq 2x$
$-19 \leq x$ or $x \geq -19$
$[-19, \infty)$

71. $\frac{3(x-2)}{5} > \frac{-5(x-2)}{3}$
$9x - 18 > -25x + 50$
$34x > 68$
$x > 2$
$(2, \infty)$

72. Let n = the number of lbs. of laundry. Then solve
$15 < 0.50(10) + 0.40(n-10)$
$15 < 5 + 0.4n - 4$
$15 < 1 + 0.4n$
$14 < 0.4n$
$35 < n$ or $n > 35$
Therefore, it is more economical to use the housekeeper for more than 35 lbs. of laundry.

73. $500 \leq F \leq 1000$
$500 \leq \frac{9}{5}C + 32 \leq 1000$
$468 \leq \frac{9}{5}C \leq 968$
$260 \leq C \leq 537.8$
260° to 538° C

74. Let x = the minimum score that the last judge can give. Then

$\frac{9.5+9.7+9.9+9.7+9.7+9.6+9.5+x}{8} \geq 9.65$

$67.6 + x \geq 77.2$

$x \geq 9.6$

Therefore, the last judge must give Nana at least a 9.6 for her to win a silver medal.

75. $4000 \leq 2x + 500 \leq 8000$

$3500 \leq 2x \leq 7500$

$1750 \leq x \leq 3750$

\$1750 to \$3750

76. $1 \leq 4x - 7 \leq 3$

$8 \leq 4x \leq 10$

$2 \leq x \leq 2.5$

$[2, 2.5]$

77. $-2 \leq 8 + 5x < -1$

$-10 \leq 5x < -9$

$-2 \leq x < -\frac{9}{5}$

$\left[-2, -\frac{9}{5}\right)$

78. $-3 < 4(2x - 1) < 12$

$-\frac{3}{4} < 2x - 1 < 3$

$\frac{1}{4} < 2x < 4$

$\frac{1}{8} < x < 2$

$\left(\frac{1}{8}, 2\right)$

79. $-6 < x - (3 - 4x) < -3$

$-6 < x - 3 + 4x < -3$

$-3 < 5x < 0$

$-\frac{3}{5} < x < 0$

$\left(-\frac{3}{5}, 0\right)$

80. $\frac{1}{6} < \frac{4x-3}{3} \leq \frac{4}{5}$

$5 < 10(4x - 3) \leq 24$

$5 < 40x - 30 \leq 24$

$35 < 40x \leq 54$

$\frac{7}{8} < x \leq \frac{27}{20}$

$\left(\frac{7}{8}, \frac{27}{20}\right]$

81. $0 \leq \frac{2(3x+4)}{5} \leq 3$

$0 \leq 6x + 8 \leq 15$

$-8 \leq 6x \leq 7$

$-\frac{4}{3} \leq x \leq \frac{7}{6}$

$\left[-\frac{4}{3}, \frac{7}{6}\right]$

82. $x \leq 2$ and $x > -5$

$-5 < x \leq 2$

$(-5, 2]$

83. $x \leq 2$ or $x > -5$

$(-\infty, \infty)$

84. $3x - 5 > 6$ or $-x < -5$

$3x > 11$ or $x > 5$

$x > \frac{11}{3}$

$\left(\frac{11}{3}, \infty\right)$

85. $-2x \leq 6$ and $-2x + 3 < -7$

$x \geq -3$ and $-2x < -10$

$x > 5$

$(5, \infty)$

86. $|x - 7| = 9$

$x - 7 = 9$ or $x - 7 = -9$

$x = 16$ or $x = -2$

$\{16, -2\}$

87. $|8 - x| = 3$

$8 - x = 3$ or $8 - x = -3$

$x = 5$ or $x = 11$

$\{5, 11\}$

88. $|2x + 9| = 9$

$2x + 9 = 9$ or $2x + 9 = -9$

$2x = 0$ or $2x = -18$

$x = 0$ or $x = -9$

$\{0, -9\}$

89. $|-3x+4|=7$
$-3x+4=7$ or $-3x+4=-7$
$-3x=3$ or $-3x=-11$
$x=-1$ or $x=\frac{11}{3}$
$\left\{-1, \frac{11}{3}\right\}$

90. $|3x-2|+6=10$
$|3x-2|=4$
$3x-2=4$ or $3x-2=-4$
$3x=6$ or $3x=-2$
$x=2$ or $x=-\frac{2}{3}$
$\left\{2, -\frac{2}{3}\right\}$

91. $5+|6x+1|=5$
$|6x+1|=0$
$6x+1=0$
$6x=-1$
$x=-\frac{1}{6}$
$\left\{-\frac{1}{6}\right\}$

92. $-5=|4x-3|$
No real solution.
{ }

93. $|5-6x|+8=3$
$|5-6x|=-5$
Not possible
{ }

94. $|7x|-26=-5$
$|7x|=21$
$7x=21$ or $7x=-21$
$x=3$ or $x=-3$
$\{3,-3\}$

95. $-8=|x-3|-10$
$2=|x-3|$
$x-3=2$ or $x-3=-2$
$x=5$ or $x=1$
$\{1,5\}$

96. $\left|\frac{3x-7}{4}\right|=2$
$\frac{3x-7}{4}=2$ or $\frac{3x-7}{4}=-2$
$3x-7=8$ or $3x-7=-8$
$3x=15$ or $3x=-1$
$x=5$ or $x=-\frac{1}{3}$
$\left\{5, -\frac{1}{3}\right\}$

97. $\left|\frac{9-2x}{5}\right|=-3$
Not possible
{ }

98. $|6x+1|=|15+4x|$
$6x+1=15+4x$ or $6x+1=-(15+4x)$
$2x=14$ or $6x+1=-15-4x$
$x=7$ or $10x=-16$
$x=7$ or $x=-\frac{16}{10}=-\frac{8}{5}$
$\left\{7, -\frac{8}{5}\right\}$

99. $|x-3|=|7+2x|$
$x-3=7+2x$ or $x-3=-(7+2x)$
$-10=x$ or $x-3=-7-2x$
$x=-10$ or $3x=-4$
$x=-10$ or $x=-\frac{4}{3}$
$\left\{-10, -\frac{4}{3}\right\}$

100. $|5x-1|<9$
$-9<5x-1<9$
$-8<5x<10$
$-\frac{8}{5}<x<2$
$\left(-\frac{8}{5}, 2\right)$

101. $|6+4x|\geq 10$
$6+4x\leq -10$ or $6+4x\geq 10$
$4x\leq -16$ or $4x\geq 4$
$x\leq -4$ or $x\geq 1$
$(-\infty, -4] \cup [1, \infty)$

102. $|3x|-8>1$
$|3x|>9$
$3x<-9$ or $3x>9$
$x<-3$ or $x>3$
$(-\infty, -3) \cup (3, \infty)$

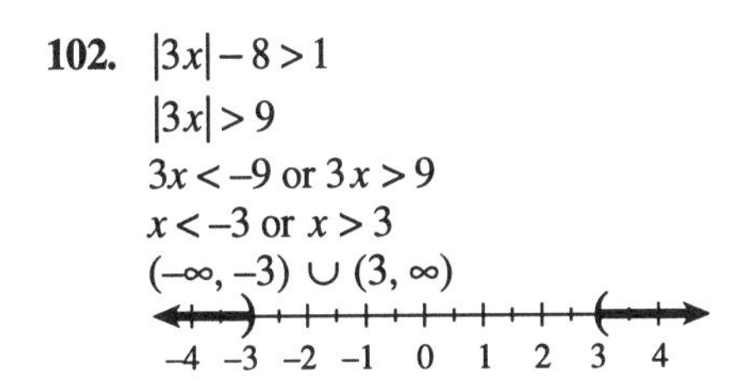

103. $9+|5x|<24$
$|5x|<15$
$-15<5x<15$
$-3<x<3$
$(-3, 3)$

104. $|6x-5|\le -1$
No real solution.
{ }

105. $|6x-5|\ge -1$
Since $|6x-5|$ is positive for all real numbers x, the solution is $(-\infty, \infty)$.

106. $\left|3x+\frac{2}{5}\right|\ge 4$
$3x+\frac{2}{5}\le -4$ or $3x+\frac{2}{5}\ge 4$
$3x\le -\frac{22}{5}$ or $3x\ge \frac{18}{5}$
$x\le -\frac{22}{15}$ or $x\ge \frac{6}{5}$
$\left(-\infty, -\frac{22}{15}\right]\cup\left[\frac{6}{5}, \infty\right)$

107. $\left|\frac{4x-3}{5}\right|<1$
$-1<\frac{4x-3}{5}<1$
$-5<4x-3<5$
$-2<4x<8$
$-\frac{1}{2}<x<2$
$\left(-\frac{1}{2}, 2\right)$

108. $\left|\frac{x}{3}+6\right|-8>-5$
$\left|\frac{x}{3}+6\right|>3$
$\frac{x}{3}+6<-3$ or $\frac{x}{3}+6>3$
$\frac{x}{3}<-9$ or $\frac{x}{3}>-3$
$x<-27$ or $x>-9$
$(-\infty, -27) \cup (-9, \infty)$

109. $\left|\frac{4(x-1)}{7}\right|+10<2$
$\left|\frac{4(x-1)}{7}\right|<-8$
Not possible
{ }

Chapter 2 - Test

1. $8x+14=5x+44$
$3x=30$
$x=10$
$\{10\}$

2. $3(x+2)=11-2(2-x)$
$3x+6=11-4+2x$
$3x+6=7+2x$
$x=1$
$\{1\}$

3. $3(y-4)+y=2(6+2y)$
$3y-12+y=12+4y$
$4y-12=12+4y$
$-12=12$
No real solution.
{ }

4. $7n-6+n=2(4n-3)$
$8n-6=8n-6$
All real numbers.
$(-\infty, \infty)$

5. $\frac{z}{2}+\frac{z}{3}=10$
$\frac{5z}{6}=10$
$z=\frac{60}{5}=12$
$\{12\}$

6. $\frac{7w}{4}+5=\frac{3w}{10}+1$
$35w+100=6w+20$
$29w=-80$
$w=-\frac{80}{29}$
$\left\{-\frac{80}{29}\right\}$

7. $|6x-5|=1$
$6x-5=-1$ or $6x-5=1$
$6x=4$ or $6x=6$
$x=\frac{2}{3}$ or $x=1$
$\left\{\frac{2}{3}, 1\right\}$

8. $|8-2t|=-6$
No real number.
$\{\ \}$

9. $3x-4y=8$
$3x-8=4y$
$y=\frac{3x-8}{4}$

10. $4(2n-3m)-3(5n-7m)=0$
$8n-12m-15n+21m=0$
$9m-7n=0$
$9m=7n$
$n=\frac{9m}{7}$

11. $S=gt^2+gvt$
$S=g(t^2+vt)$
$g=\frac{S}{t^2+vt}$

12. $F=\frac{9}{5}C+32$
$F-32=\frac{9}{5}C$
$C=\frac{5}{9}(F-32)$

13. $3(2x-7)-4x>-(x+6)$
$6x-21-4x>-x-6$
$2x-21>-x-6$
$3x>15$
$x>5$
$(5, \infty)$

14. $8-\frac{x}{2}\le 7$
$1\le\frac{x}{2}$
$2\le x$
$[2, \infty)$

15. $-3<2(x-3)\le 4$
$-3<2x-6\le 4$
$3<2x\le 10$
$\frac{3}{2}<x\le 5$
$\left(\frac{3}{2}, 5\right]$

16. $|3x+1|>5$
$3x+1<-5$ or $3x+1>5$
$3x<-6$ or $3x>4$
$x<-2$ or $x>\frac{4}{3}$
$(-\infty, -2)\cup\left(\frac{4}{3}, \infty\right)$

17. $x\ge 5$ and $x\ge 4$
$x\ge 5$
$[5, \infty)$

18. $x\ge 5$ or $x\ge 4$
$x\ge 4$
$[4, \infty)$

19. $-x>1$ and $3x+3\ge x-3$
$x<-1$ and $2x\ge -6$
$x\ge -3$
$-3\le x<-1$
$[-3, -1)$

102. $|3x| - 8 > 1$
$|3x| > 9$
$3x < -9$ or $3x > 9$
$x < -3$ or $x > 3$
$(-\infty, -3) \cup (3, \infty)$

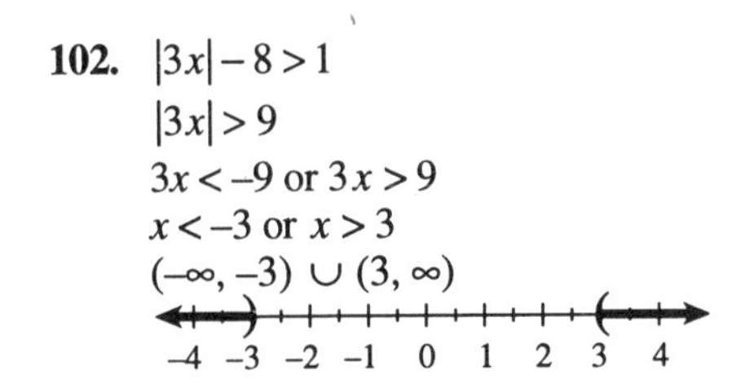

103. $9 + |5x| < 24$
$|5x| < 15$
$-15 < 5x < 15$
$-3 < x < 3$
$(-3, 3)$

104. $|6x - 5| \le -1$
No real solution.
$\{\ \}$

105. $|6x - 5| \ge -1$
Since $|6x - 5|$ is positive for all real numbers x, the solution is $(-\infty, \infty)$.

106. $\left|3x + \frac{2}{5}\right| \ge 4$
$3x + \frac{2}{5} \le -4$ or $3x + \frac{2}{5} \ge 4$
$3x \le -\frac{22}{5}$ or $3x \ge \frac{18}{5}$
$x \le -\frac{22}{15}$ or $x \ge \frac{6}{5}$
$\left(-\infty, -\frac{22}{15}\right] \cup \left[\frac{6}{5}, \infty\right)$

107. $\left|\frac{4x-3}{5}\right| < 1$
$-1 < \frac{4x-3}{5} < 1$
$-5 < 4x - 3 < 5$
$-2 < 4x < 8$
$-\frac{1}{2} < x < 2$
$\left(-\frac{1}{2}, 2\right)$

108. $\left|\frac{x}{3} + 6\right| - 8 > -5$
$\left|\frac{x}{3} + 6\right| > 3$
$\frac{x}{3} + 6 < -3$ or $\frac{x}{3} + 6 > 3$
$\frac{x}{3} < -9$ or $\frac{x}{3} > -3$
$x < -27$ or $x > -9$
$(-\infty, -27) \cup (-9, \infty)$

109. $\left|\frac{4(x-1)}{7}\right| + 10 < 2$
$\left|\frac{4(x-1)}{7}\right| < -8$
Not possible
$\{\ \}$

Chapter 2 - Test

1. $8x + 14 = 5x + 44$
$3x = 30$
$x = 10$
$\{10\}$

2. $3(x + 2) = 11 - 2(2 - x)$
$3x + 6 = 11 - 4 + 2x$
$3x + 6 = 7 + 2x$
$x = 1$
$\{1\}$

3. $3(y - 4) + y = 2(6 + 2y)$
$3y - 12 + y = 12 + 4y$
$4y - 12 = 12 + 4y$
$-12 = 12$
No real solution.
$\{\ \}$

4. $7n - 6 + n = 2(4n - 3)$
$8n - 6 = 8n - 6$
All real numbers.
$(-\infty, \infty)$

5. $\frac{z}{2}+\frac{z}{3}=10$
$\frac{5z}{6}=10$
$z=\frac{60}{5}=12$
$\{12\}$

6. $\frac{7w}{4}+5=\frac{3w}{10}+1$
$35w+100=6w+20$
$29w=-80$
$w=-\frac{80}{29}$
$\left\{-\frac{80}{29}\right\}$

7. $|6x-5|=1$
$6x-5=-1$ or $6x-5=1$
$6x=4$ or $6x=6$
$x=\frac{2}{3}$ or $x=1$
$\left\{\frac{2}{3}, 1\right\}$

8. $|8-2t|=-6$
No real number.
{ }

9. $3x-4y=8$
$3x-8=4y$
$y=\frac{3x-8}{4}$

10. $4(2n-3m)-3(5n-7m)=0$
$8n-12m-15n+21m=0$
$9m-7n=0$
$9m=7n$
$n=\frac{9m}{7}$

11. $S=gt^2+gvt$
$S=g(t^2+vt)$
$g=\frac{S}{t^2+vt}$

12. $F=\frac{9}{5}C+32$
$F-32=\frac{9}{5}C$
$C=\frac{5}{9}(F-32)$

13. $3(2x-7)-4x>-(x+6)$
$6x-21-4x>-x-6$
$2x-21>-x-6$
$3x>15$
$x>5$
$(5, \infty)$

14. $8-\frac{x}{2}\le 7$
$1\le\frac{x}{2}$
$2\le x$
$[2, \infty)$

15. $-3<2(x-3)\le 4$
$-3<2x-6\le 4$
$3<2x\le 10$
$\frac{3}{2}<x\le 5$
$\left(\frac{3}{2}, 5\right]$

16. $|3x+1|>5$
$3x+1<-5$ or $3x+1>5$
$3x<-6$ or $3x>4$
$x<-2$ or $x>\frac{4}{3}$
$(-\infty, -2)\cup\left(\frac{4}{3}, \infty\right)$

17. $x\ge 5$ and $x\ge 4$
$x\ge 5$
$[5, \infty)$

18. $x\ge 5$ or $x\ge 4$
$x\ge 4$
$[4, \infty)$

19. $-x>1$ and $3x+3\ge x-3$
$x<-1$ and $2x\ge -6$
$x\ge -3$
$-3\le x<-1$
$[-3, -1)$

20. $6x + 1 > 5x + 4$ or $1 - x > -4$
$x > 3$ or $5 > x$
$3 < x$ or $x < 5$
$(-\infty, \infty)$

21. $80(0.12) = 9.6$

22. Let x = Prudential sales
$x + 1.21x = 939{,}250{,}000$
$2.21x = 939{,}250{,}000$
$x = 425$ million dollars

23. Recall that $C = 2\pi r$ where $C = 78.5$. Then
$78.5 = 2\pi r$
$r = \frac{78.5}{2\pi} = \frac{39.25}{\pi}$
Also, recall that $A = \pi r^2$. So
$A = \pi\left(\frac{39.25}{\pi}\right)^2 \approx \frac{39.25^2}{3.14} \approx 490.63$
Dividing this by 60 yields approximately 8.18. Therefore, 8 hunting dogs could safely be kept in the pen.

24. Solve $R > C$
$7.4x > 3{,}910 + 2.8x$
$4.6x > 3910$
$x > 850$
Therefore, 850 sunglasses must be produced and sold in order for them to yield a profit.

25. $A = P\left(1 + \frac{r}{n}\right)^{nt}$

$A = 2500\left(1 + \frac{0.035}{4}\right)^{4\cdot 10} = \$3542.27$

Chapter 2 - Cumulative Review

1. **a.** $-3 + (-11) = -14$

b. $3 + (-7) = -4$

c. $-10 + 15 = 5$

d. $-8.3 + (-1.9) = -10.2$

e. $-\frac{1}{4} + \frac{1}{2} = -\frac{1}{4} + \frac{2}{4} = \frac{1}{4}$

f. $-\frac{2}{3} + \frac{3}{7} = -\frac{14}{21} + \frac{9}{21} = -\frac{5}{21}$

2. **a.** $\frac{20}{-4} = -5$

b. $\frac{-9}{-3} = 3$

c. $-\frac{3}{8} \div 3 = -\frac{3}{8} \cdot \frac{1}{3} = -\frac{1}{8}$

d. $\frac{-40}{10} = -4$

e. $\frac{-1}{10} \div \frac{-2}{5} = \frac{-1}{10} \cdot \frac{5}{-2} = \frac{1}{4}$

f. $\frac{8}{0}$ is undefined

3. **a.** $\sqrt{9} = 3$

b. $\sqrt{25} = 5$

c. $\sqrt{\frac{1}{4}} = \frac{1}{2}$

4. **a.** $3 + 2 \cdot 10 = 3 + 20 = 23$

b. $2(1-4)^2 = 2(-3)^2 = 2 \cdot 9 = 18$

c. $\frac{|-2|^3 + 1}{-7 - \sqrt{4}} = \frac{2^3 + 1}{-7 - 2} = \frac{8 + 1}{-9} = \frac{9}{-9} = -1$

d. $\frac{(6+2)-(-4)}{2-(-3)} = \frac{8+4}{2+3} = \frac{12}{5}$

5. **a.** $0.25x$

b. $2x$

c. $156x$

d. $0.09x$

6. **a.** $3xy - 2xy + 5 - 7 + xy = 2xy - 2$

b. $7x^2 + 3 - 5(x^2 - 4)$
$= 7x^2 + 3 - 5x^2 + 20 = 2x^2 + 23$

c. $(2.1x - 5.6) - (-x - 5.3)$
$= 2.1x - 5.6 + x + 5.3 = 3.1x - 0.3$

7. $2x+5=9$
$2x=4$
$x=2$
$\{2\}$

8. $6x-4=2+6(x-1)$
$6x-4=2+6x-6$
$6x-4=6x-4$
Identity
$\{x|x \text{ is a real number}\}$

9. Let x = original price
$x-0.08x=2162$
$0.92x=2162$
$x=2350$
\$2350

10. $V=lwh$
$\frac{V}{lw}=h$

11. $A=\frac{1}{2}(B+b)h$
$2A=Bh+bh$
$2A-Bh=bh$
$\frac{2A-Bh}{h}=b$

12. Let A = area
Let x = side of square
$0.01A=16$
$A=1600$
$A=x^2$
$1600=x^2$
$40=x$
The square has a side length of 40 cm.

13. a. $\frac{1}{4}x\leq\frac{3}{8}$
$x\leq\frac{3}{2}$
$\left(-\infty, \frac{3}{2}\right]$

b. $-2.3x<6.9$
$x>-3$
$(-3, \infty)$

14. $2(x+3)>2x+1$
$2x+6>2x+1$
$6>1$
True for all x
$(-\infty, \infty)$

15. $x-7<2$ and $2x+1<9$
$x<9$ and $2x<8$
$x<4$
$(-\infty, 4)$

16. $5x-3\leq 10$ or $x+1\geq 5$
$5x\leq 13$ or $x\geq 4$
$x\leq\frac{13}{5}$
$\left(-\infty, \frac{13}{5}\right]\cup[4, \infty)$

17. $|p|=2$
$p=-2$ or $p=2$
$\{-2, 2\}$

18. $|3x+2|=|5x-8|$
$3x+2=5x-8$ or $3x+2=-(5x-8)$
$10=2x$ or $3x+2=-5x+8$
$5=x$ or $8x=6$
$x=\frac{3}{4}$
$\left\{\frac{3}{4}, 5\right\}$

19. $|m-6|<2$
$-2<m-6<2$
$4<m<8$
$(4, 8)$

20. $|2x+9|+5>3$
$|2x+9|>-2$
Since an absolute value is always positive, this is true for all x.
$(-\infty, \infty)$

Chapter 3

Section 3.1 Graphing Calculator Box

1, $y = -3.2x + 7.9$

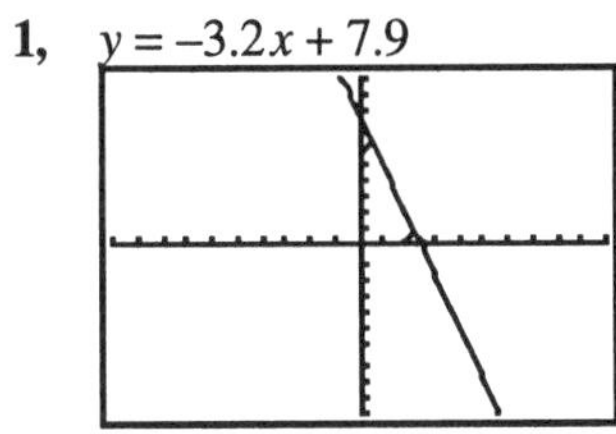

3. $y = \frac{1}{4}x - \frac{2}{3}$

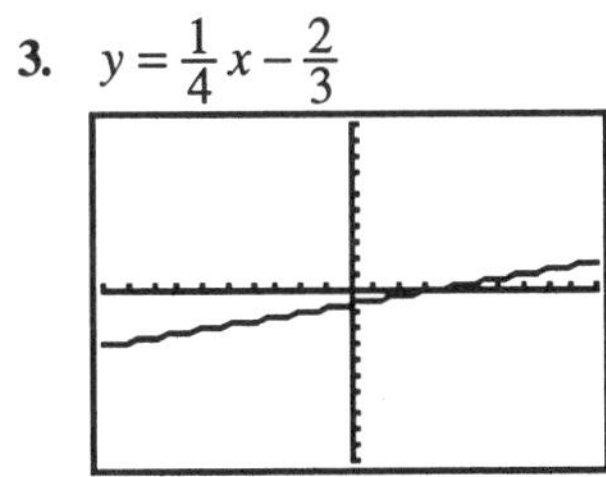

5. $y = |x - 3| + 2$

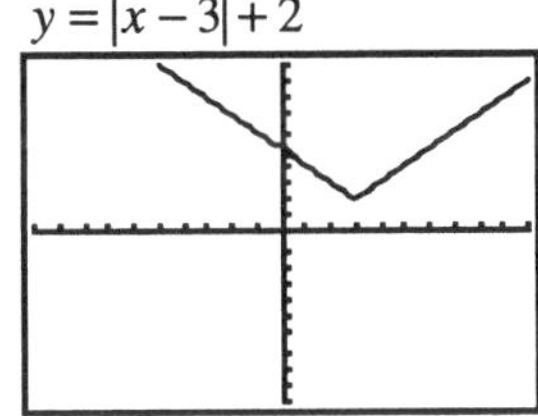

7. $y = x^2 + 3$

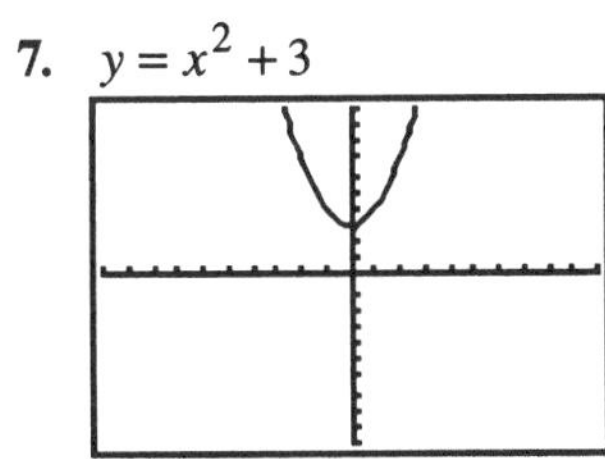

Section 3.1

Mental Math

1. $A(5, 2)$
3. $C(3, -1)$
5. $E(-5, -2)$
7. $G(-1, 0)$

Exercise Set 3.1

1. (3, 2) in quadrant I

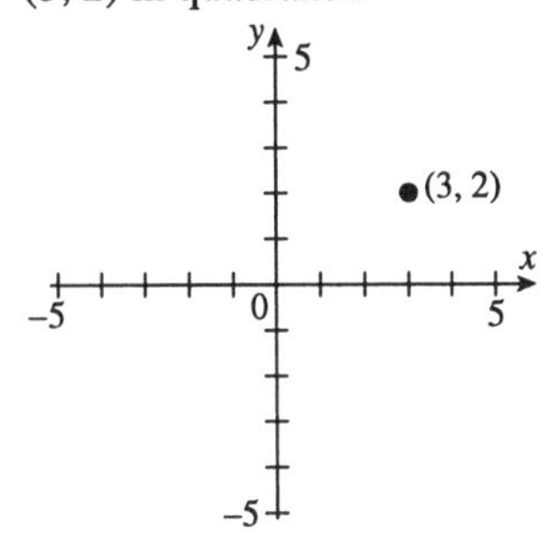

3. (–5, 3) in quadrant II

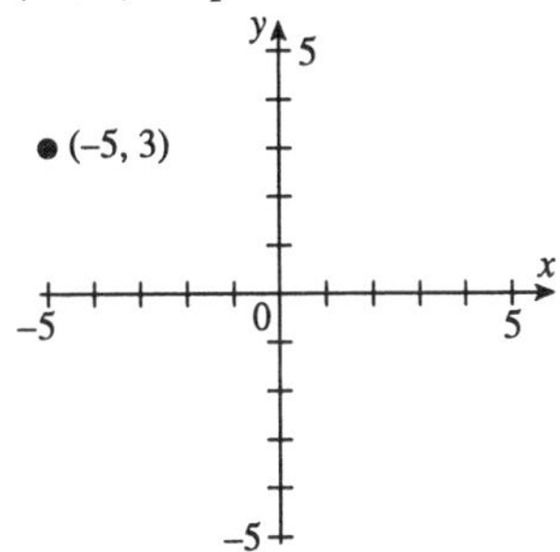

5. $\left(5\frac{1}{2}, -4\right)$ in quadrant IV

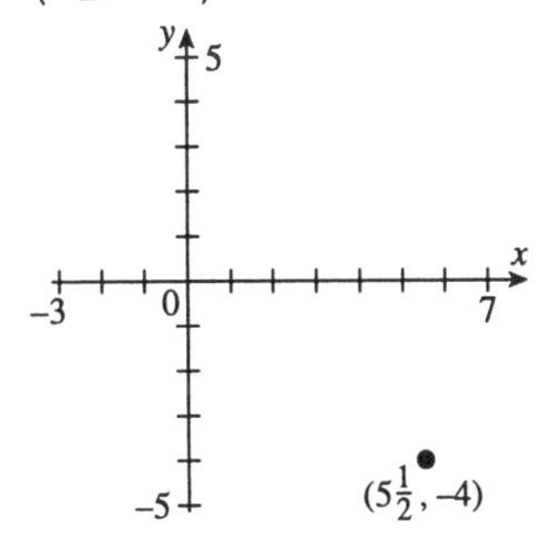

7. (0, 3.5) on y-axis

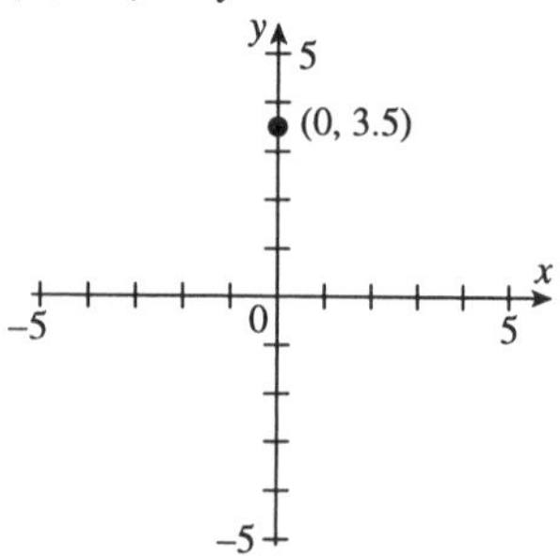

9. (−2, −4) in quadrant III

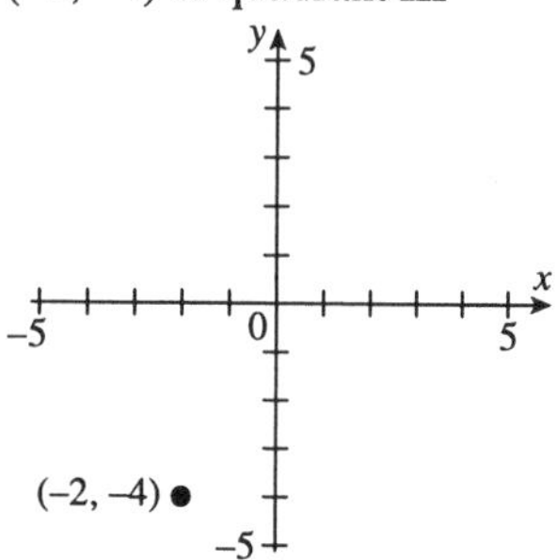

11. quadrant IV

13. x-axis

15. quadrant III

17. Let $x = 0,\ y = 5$
$y = 3x - 5$
$5 = 3 \cdot 0 - 5$
$5 = -5$
False; No
Let $x = -1,\ y = -8$
$y = 3x - 5$
$-8 = 3(-1) - 5$
$-8 = -3 - 5$
$-8 = -8$
True; Yes

19. Let $x = 1,\ y = 0$
$-6x + 5y = -6$
$-6(1) + 5(0) = -6$
$-6 + 0 = -6$
$-6 = -6$
True; Yes
Let $x = 2,\ y = \frac{6}{5}$
$-6x + 5y = -6$
$-6(2) + 5\left(\frac{6}{5}\right) = -6$
$-12 + 6 = -6$
$-6 = -6$
True; Yes

21. Let $x = 1,\ y = 2$
$y = 2x^2$
$2 = 2(1)^2$
$2 = 2$
True; Yes
Let $x = 3,\ y = 18$
$y = 2x^2$
$18 = 2(3)^2$
$18 = 2(9)$
$18 = 18$
True; Yes

23. Let $x = 2,\ y = 8$
$y = x^3$
$8 = 2^3$
$8 = 8$
True; Yes
Let $x = 3,\ y = 9$
$y = x^3$
$9 = 3^3$
$9 = 27$
False; No

25. Let $x = 1,\ y = 3$
$y = \sqrt{x} + 2$
$3 = \sqrt{1} + 2$
$3 = 1 + 2$
$3 = 3$
True; Yes
Let $x = 4,\ y = 4$
$y = \sqrt{x} + 2$
$4 = \sqrt{4} + 2$
$4 = 2 + 2$
$4 = 4$
True; Yes

27. Linear

29. Linear

31. Linear

33. Not linear

35. Linear

37. Not linear

39. Not linear

41. Linear

43. Linear

45. Not linear

47. Not linear

49. Not linear

51. Linear

53. $x + y = 3$
Find three ordered pair solutions.

x	y
0	3
3	0
1	2

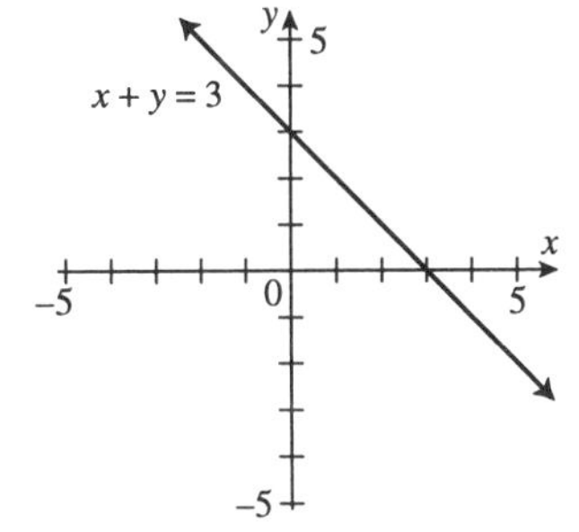

55. $y = 4x$
Find three ordered pair solutions.

x	y
0	0
1	4
2	8

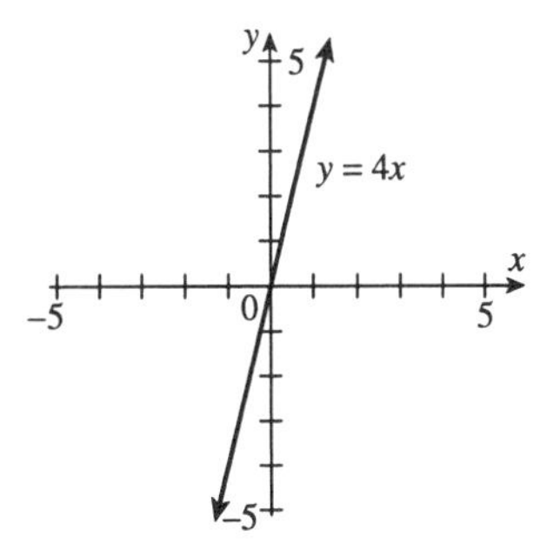

57. $y = 4x - 2$
Find three ordered pair solutions.

x	y
0	–2
1	2
$\frac{1}{2}$	0

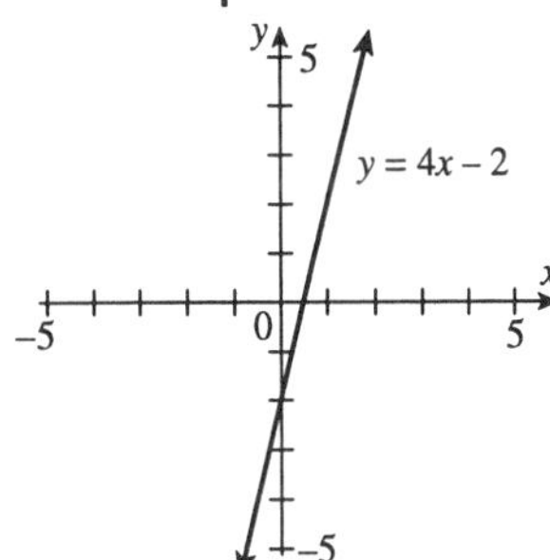

59. $y = |x| + 3$

x	y
–3	6
–2	5
–1	4
0	3
1	4
2	5
3	6

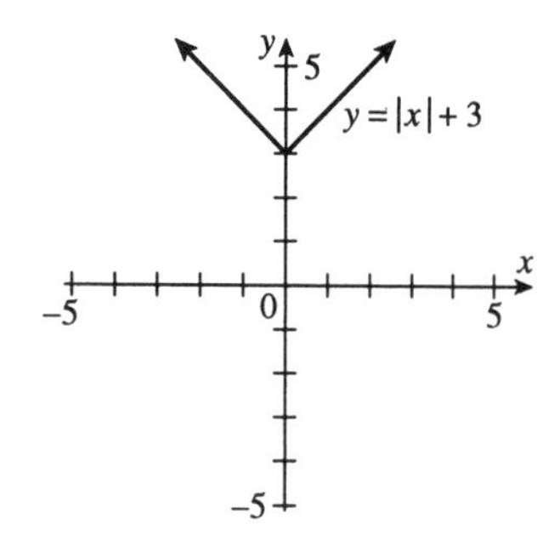

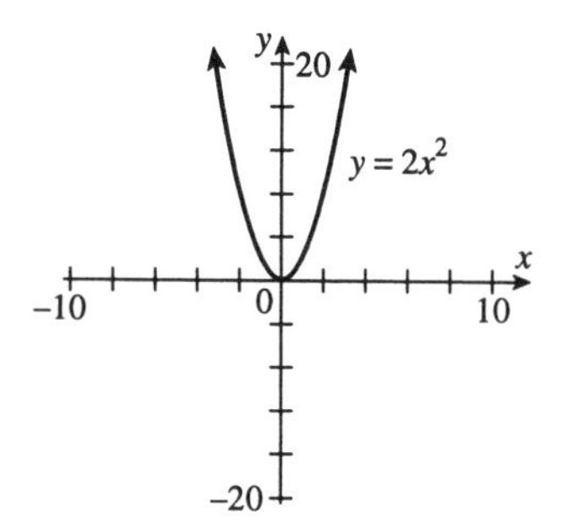

61. $2x - y = 5$
Find three ordered pair solutions.

x	y
0	−5
1	−3
2	−1

63. $y = 2x^2$

x	y
−3	18
−2	8
−1	2
0	0
1	2
2	8
3	18

65. $y = x^2 - 3$

x	y
−3	6
−2	1
−1	−2
0	−3
1	−2
2	1
3	6

67. $y = -2x$
Find three ordered pair solutions.

x	y
0	0
1	−2
−1	2

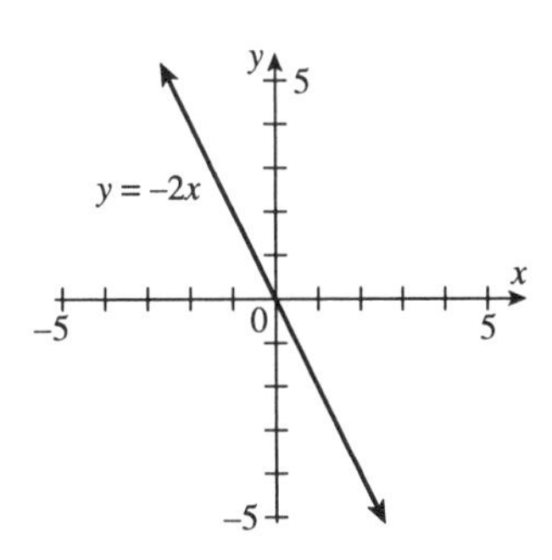

69. $y = -2x + 3$
Find three ordered pair solutions.

x	y
0	3
1	1
2	−1

71. $y = |x + 2|$

x	y
−4	2
−3	1
−2	0
−1	1
0	2
1	3

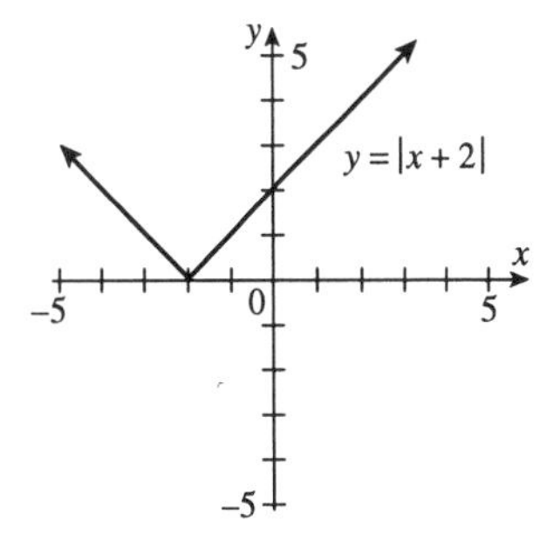

73. $y = x^3$

x	y
−3	−27
−2	−8
−1	−1
0	0
1	1
2	8

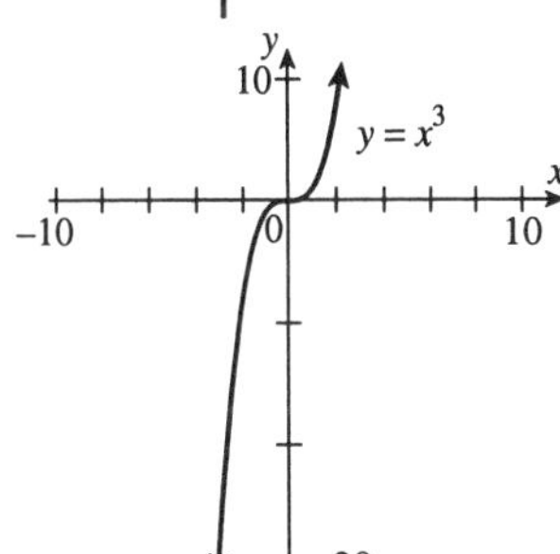

75. $y = -|x|$

x	y
−3	−3
−2	−2
−1	−1
0	0
1	−1
2	−2
3	−3

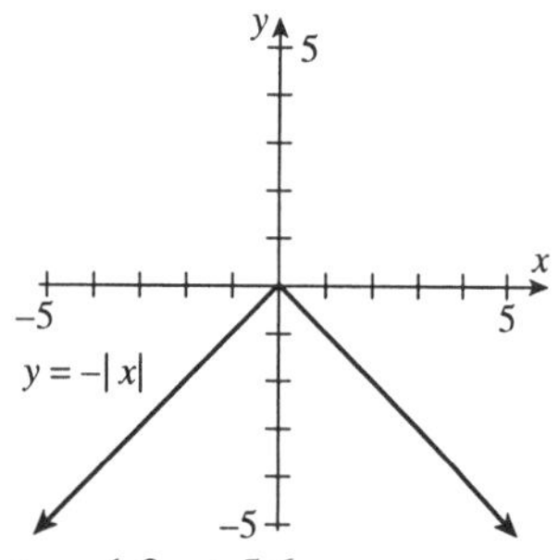

77. $y = -1.3x + 5.6$
Find three ordered pair solutions.

x	y
0	5.6
1	4.3
−1	6.9

79. $y = x^2 - 4x + 7$

x	y
0	7
1	4
2	3
3	4
4	7

81. a.

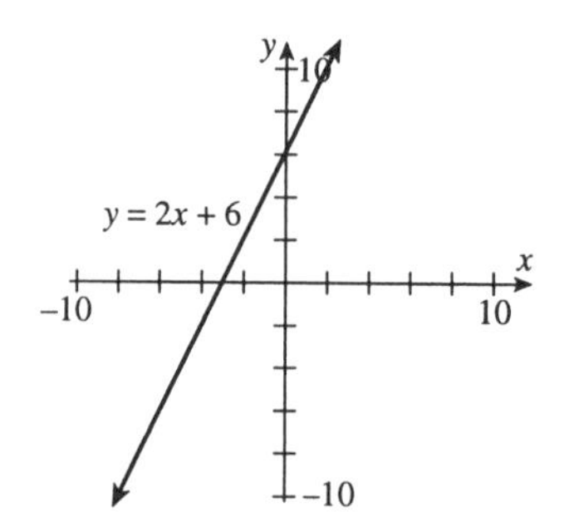

b. 14 inches

83. \$7000

85. 7000 − 6500 = \$500

87. Answers may vary.

89. They are parallel.

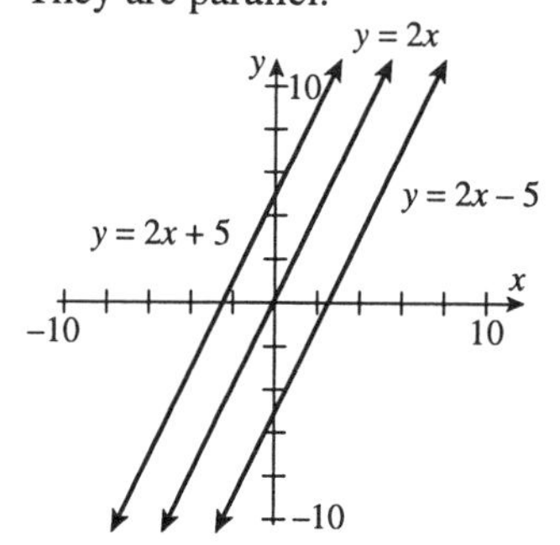

91. Answers may vary.

93. $y = -3 - 2x$

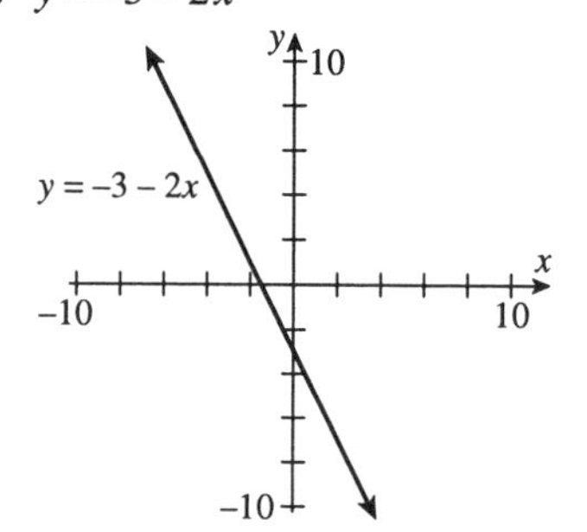

95. $y = 5 - x^2$

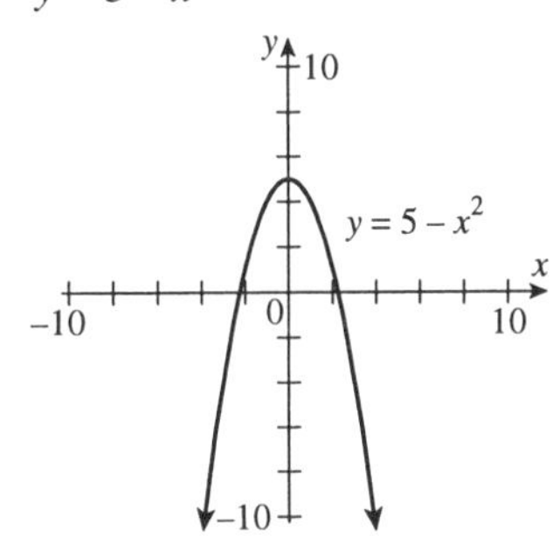

101. $5 + 7(x + 1) = 12 + 10x$
$5 + 7x + 7 = 12 + 10x$
$7x + 12 = 12 + 10x$
$7x = 10x$
$0 = 3x$
$0 = x$
$\{0\}$

103. $\frac{1}{6} + 2x = \frac{2}{3}$
$1 + 12x = 4$
$12x = 3$
$x = \frac{1}{4}$
$\left\{\frac{1}{4}\right\}$

105. $-3x > 18$
$x > -6$
$(-\infty, -6)$

107. $9x + 8 \le 6x - 4$
$3x \le -12$
$x \le -4$
$(-\infty, -4]$

Section 3.2 Graphing Calculator Box

1. $f(x) = |x|$
$g(x) = |x| + 1$

3. $f(x) = x$
$H(x) = x - 6$

5. $f(x) = -x^2$
$F(x) = -x^2 + 7$

Exercise Set 3.2

1. Domain = {−1, 0, −2, 5}
Range = {7, 6, 2}
The relation is a function.

3. Domain = {−2, 6, −7}
Range = {4, −3, −8}
The relation is not a function since −2 is paired with both 4 and −3.

5. Domain = {1}
Range = {1, 2, 3, 4}
The relation is not a function since 1 is paired with both 1 and 2 for example.

7. Domain = $\left\{\frac{3}{2}, 0\right\}$

Range = $\left\{\frac{1}{2}, -7, \frac{4}{5}\right\}$

The relation is not a function since $\frac{3}{2}$ is paired with both $\frac{1}{2}$ and −7.

9. Domain = {−3, 0, 3}
Range = {−3, 0, 3}
The relation is a function.

11. Domain = {−1, 1, 2, 3}
Range = {2, 1}
The relation is a function.

13. Domain = {Colorado, Alaska, Delaware, Illinois, Connecticut, Texas}
Range = {6, 1, 20, 30}
The relation is a function.

15. Domain = {32°, 104°, 212°, 50°}
Range = {0°, 40°, 10°, 100°}
The relation is a function.

17. Domain = {2, –1, 5, 100}
Range = {0}
The relation is a function.

19. The relation is a function.

21. Answers may vary.

23. Function

25. Not a function

27. Function

29. Domain = $[0, \infty)$
Range = All reals
The relation is not a function since it fails the vertical line test (try $x = 1$).

31. Domain = $[-1, 1]$
Range = All reals
The relation is not a function since it fails the vertical line test (try $x = 0$).

33. Domain = $(-\infty, \infty)$
Range = $(-\infty, -3] \cup [3, \infty)$
The relation is not a function since it fails the vertical line test (try $x = 3$).

35. Domain = $[2, 7]$
Range = $[1, 6]$
The relation is not a function since it fails the vertical line test.

37. Domain = $\{-2\}$
Range = $(-\infty, \infty)$
The relation is not a function since it fails the vertical line test.

39. Domain = $(-\infty, \infty)$
Range = $(-\infty, 3]$
The relation is a function.

41. Answers may vary.

43. Yes

45. No

47. Yes

49. Yes

51. Yes

53. No

55. $f(x) = 3x + 3$
$f(4) = 3(4) + 3 = 12 + 3 = 15$

57. $h(x) = 5x^2 - 7$
$h(-3) = 5(-3)^2 - 7 = 5(9) - 7$
$= 45 - 7 = 38$

59. $g(x) = 4x^2 - 6x + 3$
$g(2) = 4(2)^2 - 6(2) + 3$
$= 4(4) - 12 + 3$
$= 16 - 12 + 3$
$= 7$

61. $g(x) = 4x^2 - 6x + 3$
$g(0) = 4(0)^2 - 6(0) + 3$
$= 4(0) - 0 + 3$
$= 0 - 0 + 3 = 3$

63. $f(x) = \frac{1}{2}x$

a. $f(0) = \frac{1}{2}(0) = 0$

b. $f(2) = \frac{1}{2}(2) = 1$

c. $f(-2) = \frac{1}{2}(-2) = -1$

65. $g(x) = 2x^2 + 4$

a. $g(-11) = 2(-11)^2 + 4$
$= 2(121) + 4 = 242 + 4 = 246$

b. $g(-1) = 2(-1)^2 + 4$
$= 2(1) + 4 = 2 + 4 = 6$

c. $g\left(\frac{1}{2}\right) = 2\left(\frac{1}{2}\right)^2 + 4$
$= 2\left(\frac{1}{4}\right) + 4 = \frac{1}{2} + \frac{8}{2} = \frac{9}{2}$

67. $f(x) = -5$

a. $f(2) = -5$

b. $f(0) = -5$

c. $f(606) = -5$

69. $f(x) = 1.3x^2 - 2.6x + 5.1$

a. $f(2) = 1.3(2)^2 - 2.6(2) + 5.1$
$= 1.3(4) - 5.2 + 5.1$
$= 5.2 - 5.2 + 5.1 = 5.1$

b. $f(-2) = 1.3(-2)^2 - 2.6(-2) + 5.1$
$= 1.3(4) + 5.2 + 5.1$
$= 5.2 + 5.2 + 5.1 = 15.5$

c. $f(3.1) = 1.3(3.1)^2 - 2.6(3.1) + 5.1$
$= 1.3(9.61) - 8.06 + 5.1$
$= 12.493 - 8.06 + 5.1 = 9.533$

71. a. \$6.5 billion

b. $f(x) = 1.087x + 3.44$
$f(3) = 1.087(3) + 3.44$
$= 3.261 + 3.44 = 6.701$
\$6.701 billion

73. $f(x) = 1.087x + 3.44$
$f(20) = 1.087(20) + 3.44$
$= 21.74 + 3.44 = 25.18$
\$25.18 billion

75. $f(x) = x + 7$

77. $A(r) = \pi r^2$
$A(5) = \pi(5)^2 = 25\pi$
25π square cm

79. $V = x^3$
$V = (14)^3 = 2744$
2744 cubic in.

81. $H(f) = 2.59f + 47.24$
$H(46) = 2.59(46) + 47.24$
$= 119.14 + 47.24 = 166.38$
166.38 cm

83. $D(x) = \frac{136}{25}x$
$D(30) = \frac{136}{25}(30) = 163.2$
163.2 mg

85. $x - y = -5$

x	0	−5	1
y	5	0	6

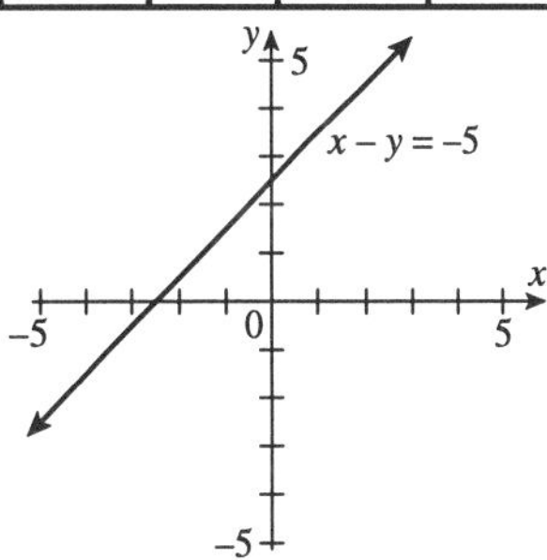

87. $7x + 4y = 8$

x	0	$\frac{8}{7}$	$\frac{12}{7}$
y	2	0	−1

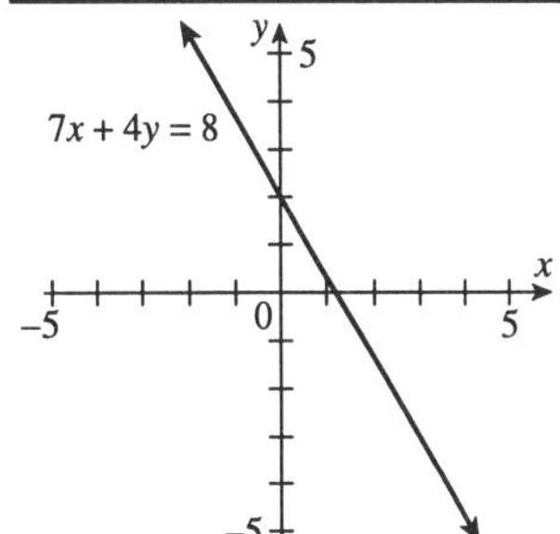

89. $y = 6x$

x	0	0	–1
y	0	0	–6

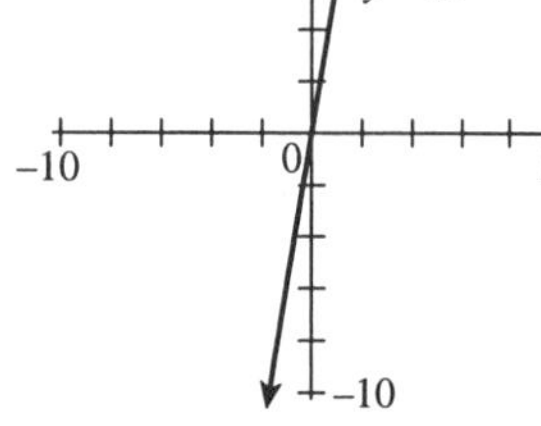

91. Yes; The two horizontal lengths on the bottom of the figure add to 45 m. The two vertical lengths not labeled on the figure add to 40 m.
$45 + 45 + 40 + 40 = 170$ m

93. $g(x) = -3x + 12$

a. $g(s) = -3s + 12$

b. $g(r) = -3r + 12$

95. $f(x) = x^2 - 12$

a. $f(12) = (12)^2 - 12 = 144 - 12 = 132$

b. $f(a) = a^2 - 12$

Section 3.3 Graphing Calculator Explorations

1. $x = 3.5y$
$y = \frac{x}{3.5}$

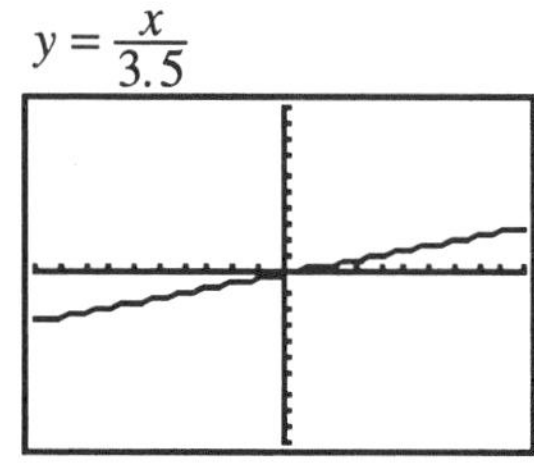

3. $5.78x + 2.31y = 10.98$
$2.31y = -5.78x + 10.98$
$y = \frac{-5.78}{2.31}x + \frac{10.98}{2.31}$

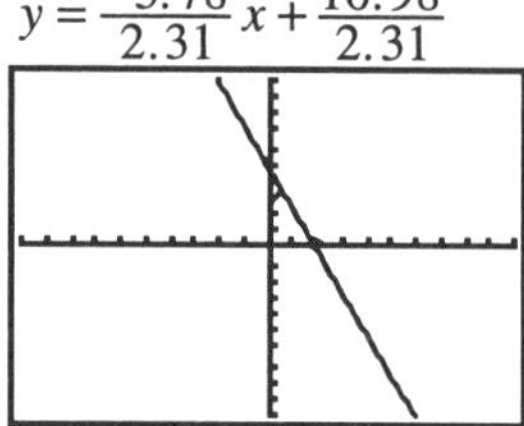

5. $y - |x| = 3.78$
$y = |x| + 3.78$

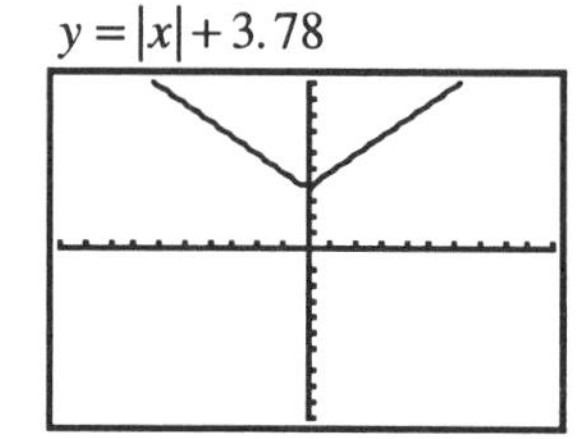

7. $y - 5.6x^2 = 7.7x + 1.5$
$y = 5.6x^2 + 7.7x + 1.5$

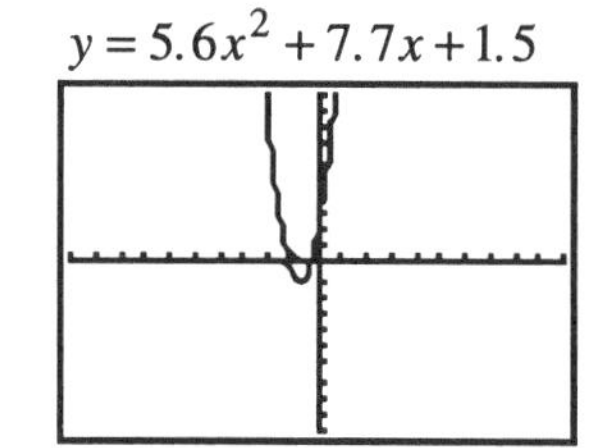

Exercise Set 3.3

1. $f(x) = -2x$
Find ordered pair solutions or use $m = -2$, $b = 0$.

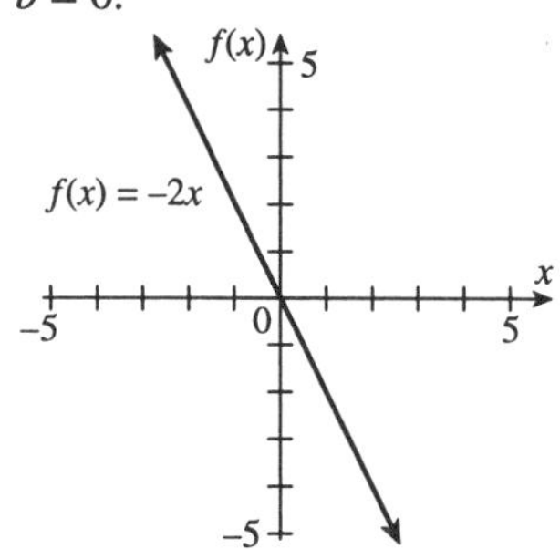

3. $f(x) = -2x + 3$
Find ordered pair solutions or use $m = -2$, $b = 3$.

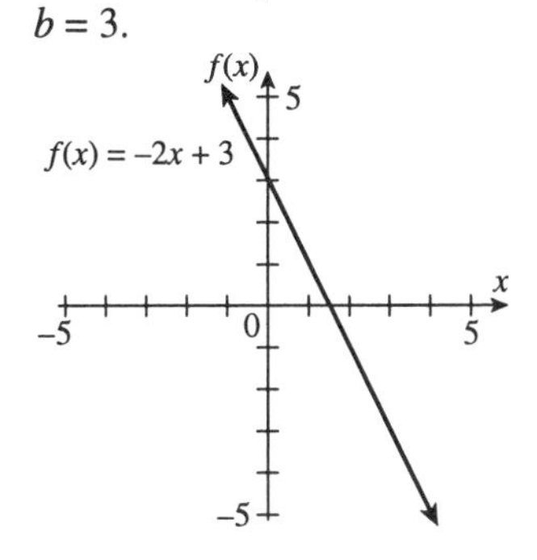

5. $f(x) = \frac{1}{2}x$

Find ordered pair solutions or use $m = \frac{1}{2}$, $b = 0$.

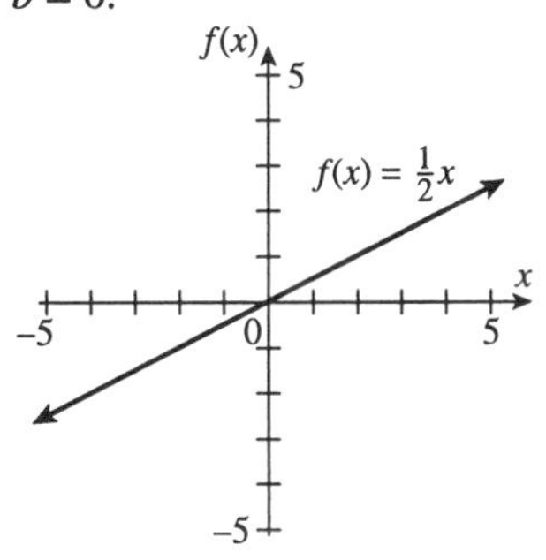

7. $f(x) = \frac{1}{2}x - 4$

Find ordered pair solutions or use $m = \frac{1}{2}$, $b = -4$.

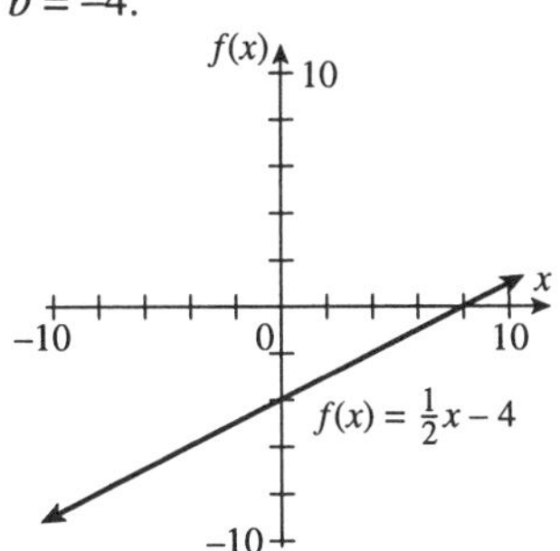

9. C

11. D

13. $x - y = 3$

Let $x = 0$	Let $y = 0$	Let $x = 2$
$0 - y = 3$	$x - 0 = 3$	$2 - y = 3$
$y = -3$	$x = 3$	$y = -1$

x	0	3	2
y	−3	0	−1

15. $x = 5y$

Let $x = 0$	Let $x = 5$	Let $x = -5$
$0 = 5y$	$5 = 5y$	$-5 = 5y$
$y = 0$	$y = 1$	$y = -1$

x	0	5	−5
y	0	1	−1

17. $-x + 2y = 6$

Let $x = 0$	Let $y = 0$	Let $x = -4$
$-0 + 2y = 6$	$-x + 2(0) = 6$	$-(-4) + 2y = 6$
$y = 3$	$x = -6$	$y = 1$

x	0	−6	−4
y	3	0	1

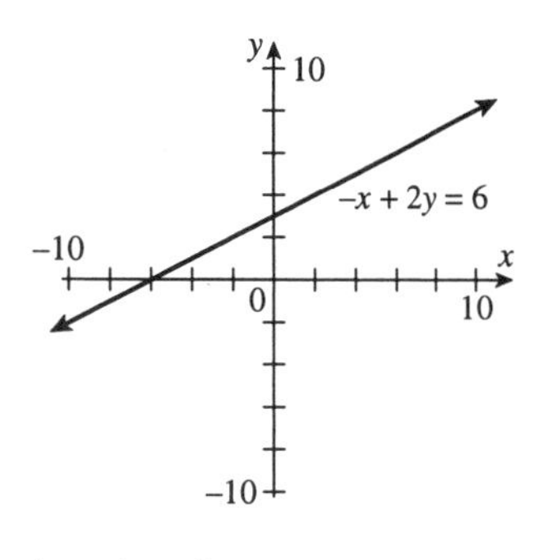

19. $2x - 4y = 8$

Let $x = 0$ Let $y = 0$ Let $x = 2$

$2(0) - 4y = 8$ $2x - 4(0) = 8$ $2(2) - 4y = 8$

$y = -2$ $x = 4$ $y = -1$

x	0	4	2
y	−2	0	−1

21. Answers may vary.

23. $x = -1$

Vertical line with x-intercept at −1

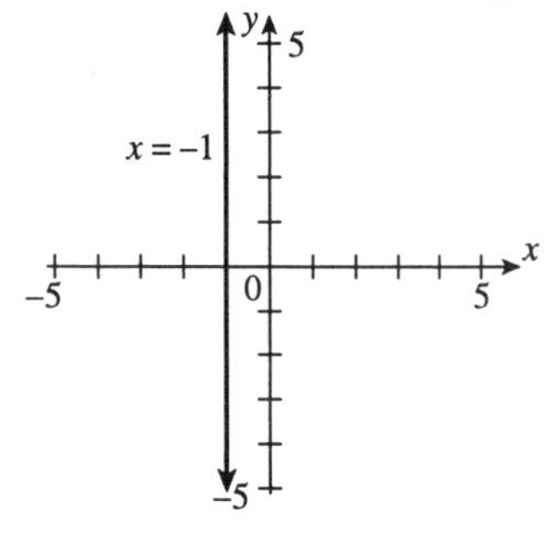

25. $y = 0$

Horizontal line with y-intercept at 0

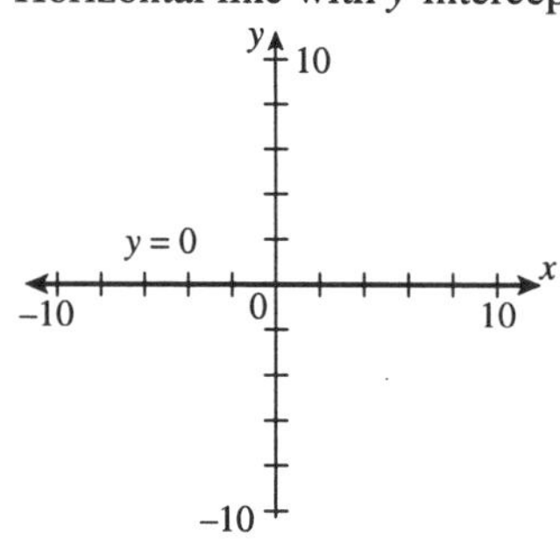

27. $y + 7 = 0$

$y = -7$

Horizontal line with y-intercept at −7

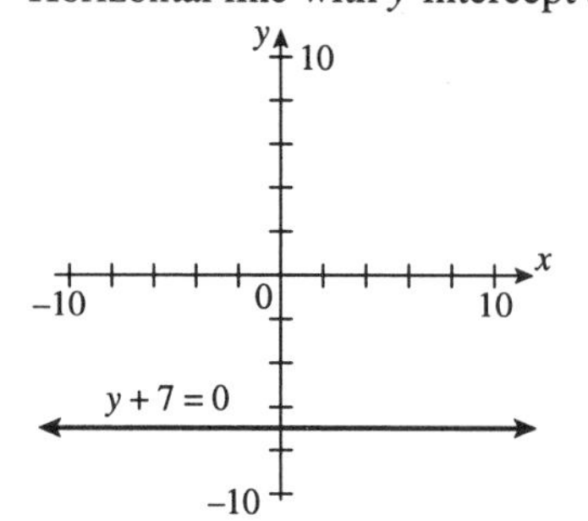

29. C

31. A

33. Answers may vary.

For exercises 35–59, find ordered pair solutions, or find the x-and y-intercepts, or find the slope and y-intercept with the equation in the form $y = mx + b$.

35. $x + 2y = 8$

37. $f(x) = \frac{3}{4}x + 2$ or $y = \frac{3}{4}x + 2$

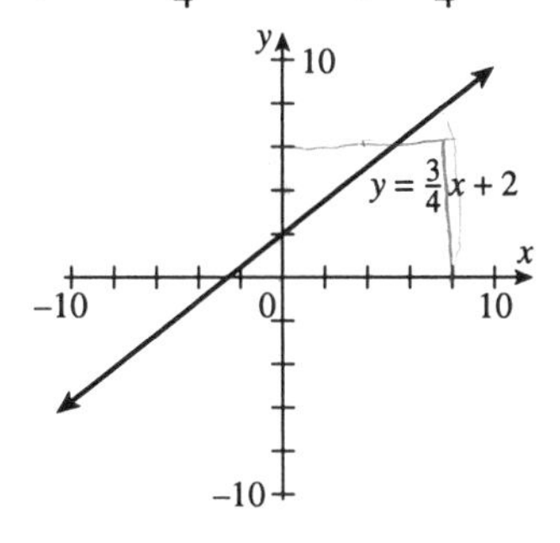

39. $x = -3$

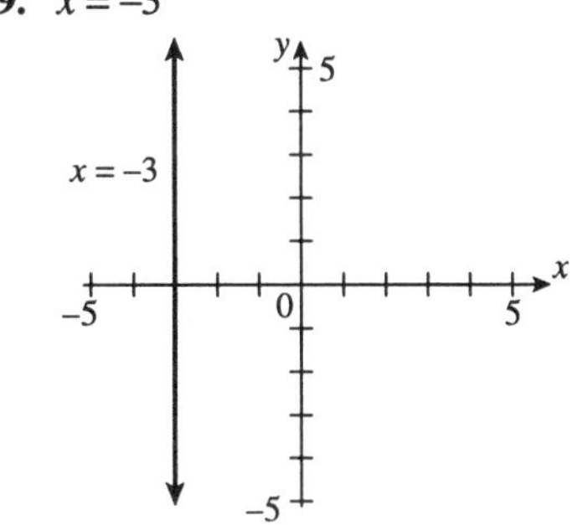

41. $3x + 5y = 7$

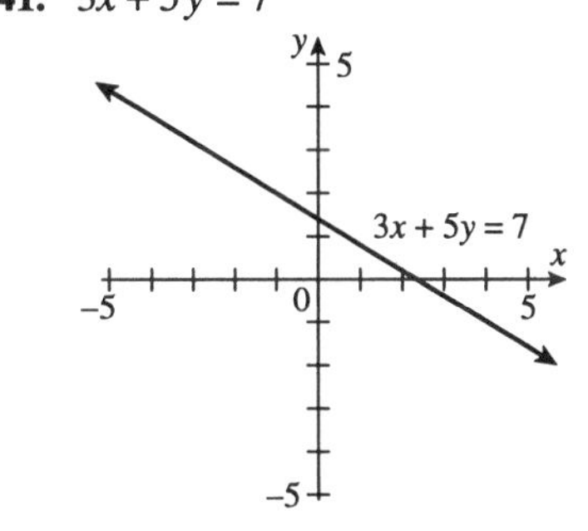

43. $f(x) = x$ or $y = x$

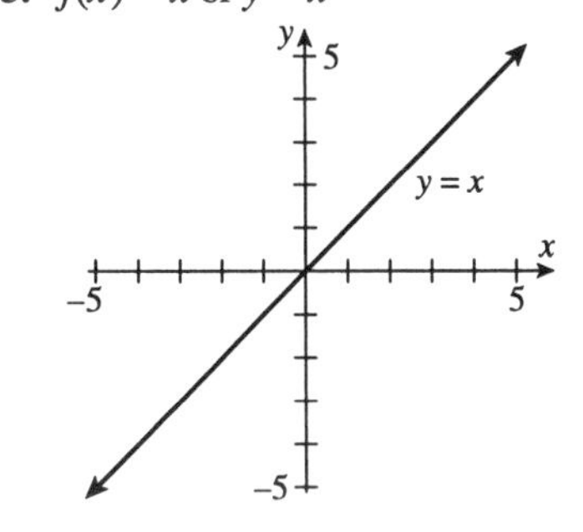

45. $x + 8y = 8$

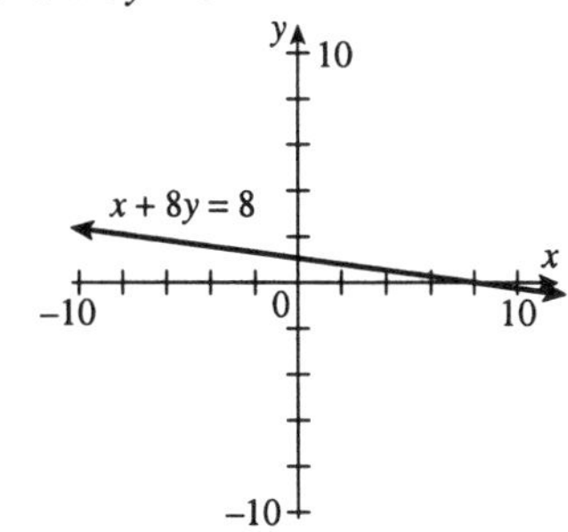

47. $5 = 6x - y$

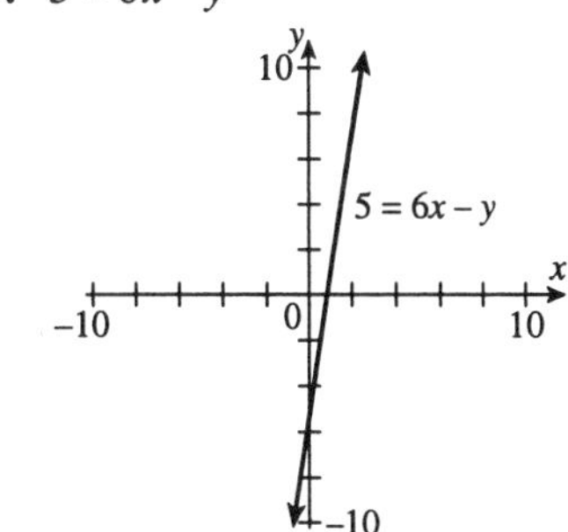

49. $-x + 10y = 11$

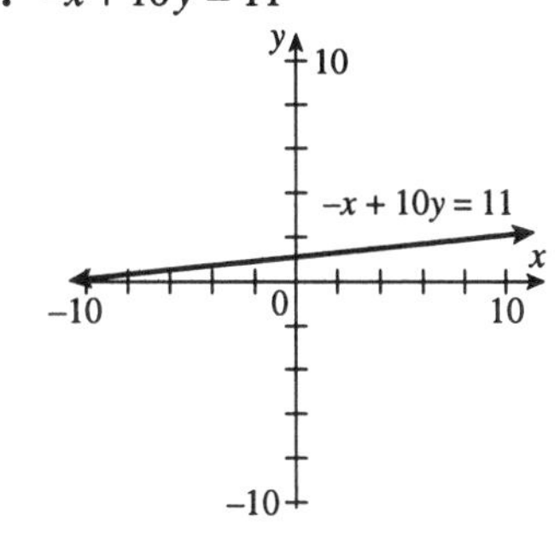

51. $y = 1$

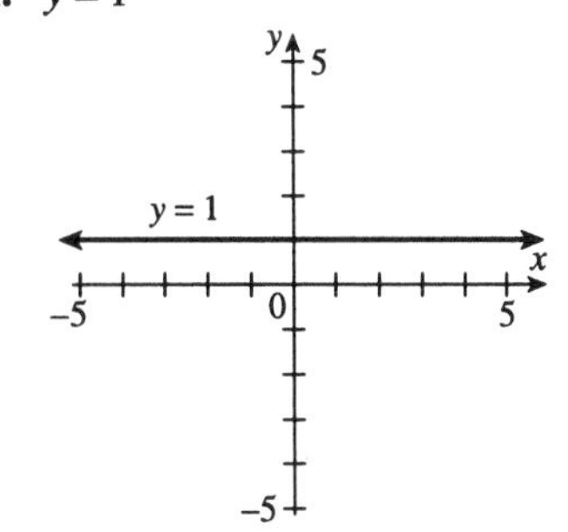

53. $f(x) = \frac{1}{2}x$ or $y = \frac{1}{2}x$

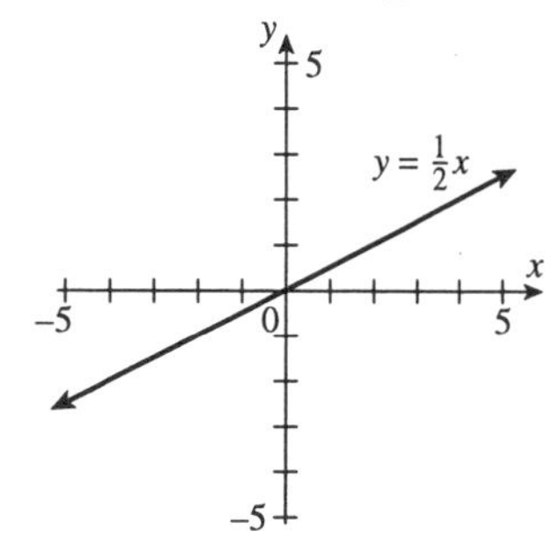

55. $x + 3 = 0$ or $x = -3$

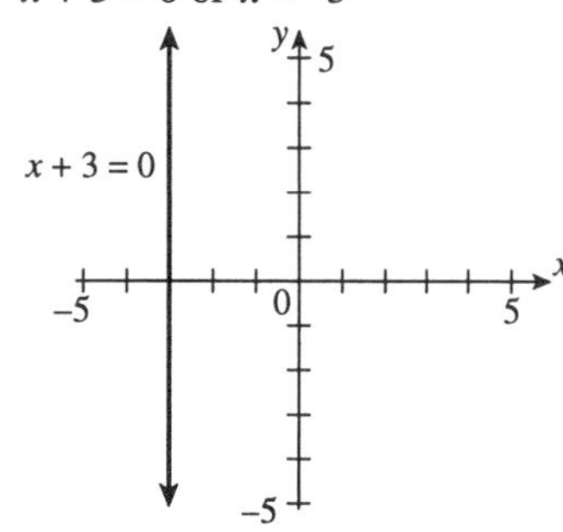

57. $f(x) = 4x - \frac{1}{3}$ or $y = 4x - \frac{1}{3}$

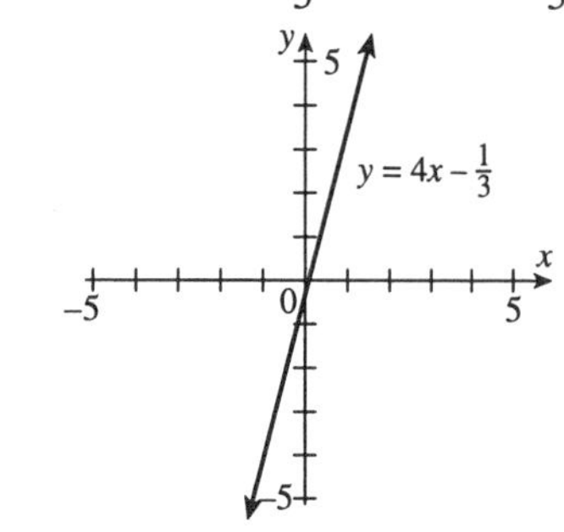

59. $2x + 3y = 6$

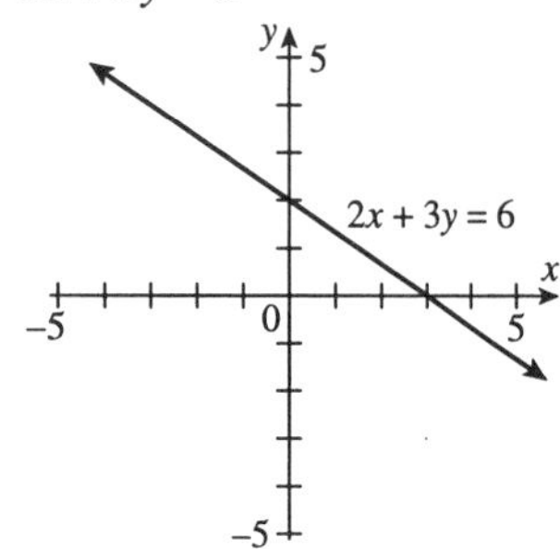

61. $2x + 3y = 1500$

a. $2(0) + 3y = 1500$
$3y = 1500$
$y = 500$
(0, 500); If no tables are produced, 500 chairs can be produced.

b. $2x + 3(0) = 1500$
$2x = 1500$
$x = 750$
(750, 0); If no chairs are produced, 750 tables can be produced.

c. $2(50) + 3y = 1500$
$100 + 3y = 1500$
$3y = 1400$
$y = 466.7$
466 chairs

63. $C(x) = 0.2x + 24$

a. $C(200) = 0.2(200) + 24$
$= 40 + 24 = 64$
$64

b.

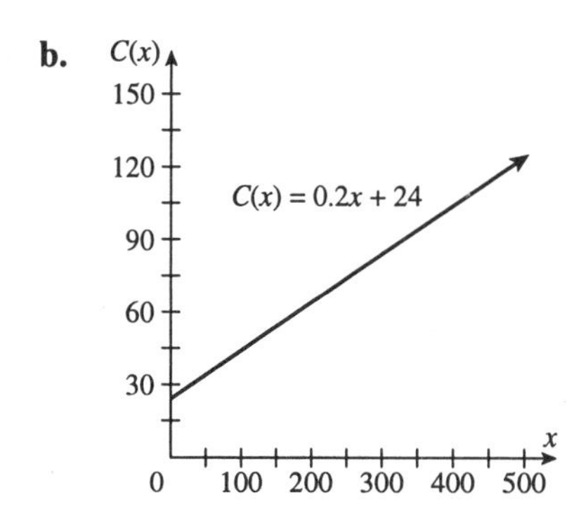

c. The line has a positive slope.

65. $f(t) = 91.7t + 747.8$

a. $f(20) = 91.7(20) + 747.8$
$= 1834 + 747.8 = 2581.8$
$2581.80

b. $2000 = 91.7t + 747.8$
$1252.2 = 91.7t$
$13.66 = t$
$1990 + 14 = 2004$

c. Answers may vary.

71. $|x-3|=6$
$x-3=6$ or $x-3=-6$
$x=9$ or $x=-3$
$\{9, -3\}$

73. $|2x+5|>3$
$2x+5<-3$ or $2x+5>3$
$2x<-8$ or $2x>-2$
$x<-4$ or $x>-1$
$(-\infty, -4)\cup(-1, \infty)$

75. $|3x-4|\le 2$
$-2\le 3x-4\le 2$
$2\le 3x\le 6$
$\frac{2}{3}\le x\le 2$
$\left[\frac{2}{3}, 2\right]$

77. $\frac{-6-3}{2-8}=\frac{-9}{-6}=\frac{3}{2}$

79. $\frac{-8-(-2)}{-3-(-2)}=\frac{-8+2}{-3+2}=\frac{-6}{-1}=6$

81. $\frac{0-6}{5-0}=-\frac{6}{5}$

Section 3.4 Graphing Calculator Explorations

1. $y=2.3x+6.7$
$x=5.1$, $y=18.4$

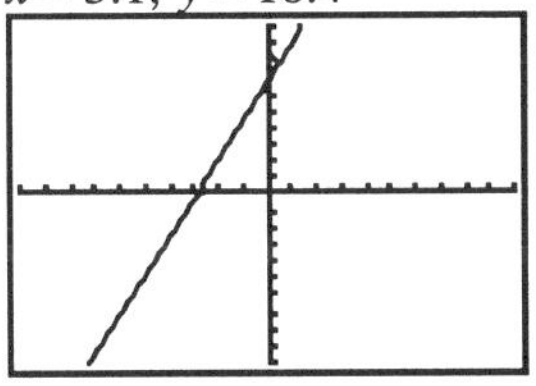

3. $y=-5.9x-1.6$
$x=-1.5$, $y=7.2$

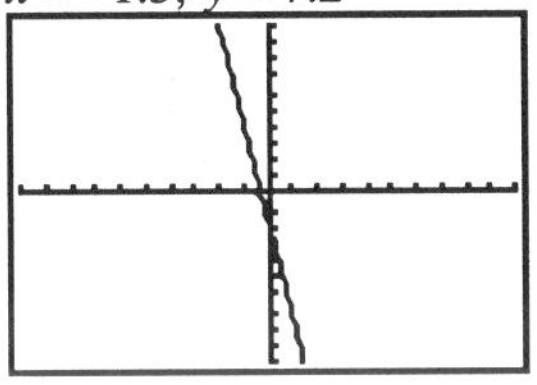

5. $y=x^2+5.2x-3.3$
$x=2.3$, $y=14.0$
$x=4.2, -9.4$, $y=36$

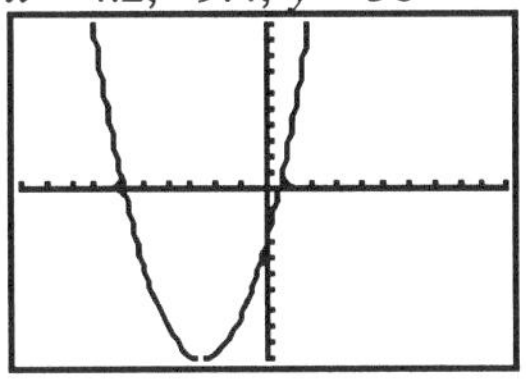

Section 3.4

Mental Math

1. Upward

3. Horizontal

Exercise Set 3.4

1. $m=\frac{11-2}{8-3}=\frac{9}{5}$

3. $m=\frac{8-1}{1-3}=\frac{7}{-2}=-\frac{7}{2}$

5. $m=\frac{3-8}{4-(-2)}=-\frac{5}{6}$

7. $m=\frac{-4-(-6)}{4-(-2)}=\frac{2}{6}=\frac{1}{3}$

9. $m=\frac{11-(-1)}{-12-(-3)}=\frac{12}{-9}=-\frac{4}{3}$

11. $m=\frac{5-5}{3-(-2)}=\frac{0}{5}=0$

13. $m=\frac{-5-1}{-1-(-1)}=-\frac{6}{0}$
= undefined slope

15. $m=\frac{0-6}{-3-0}=\frac{-6}{-3}=2$

17. $m=\frac{4-2}{-3-(-1)}=\frac{2}{-3+1}=\frac{2}{-2}=-1$

19. ℓ_2

21. ℓ_2

23. ℓ_2

25. **a.** $m \text{ for } \ell_1 = \frac{-2-4}{2-(-1)} = \frac{-6}{3} = -2$

$m \text{ for } \ell_2 = \frac{2-6}{-4-(-8)} = \frac{-4}{4} = -1$

$m \text{ for } \ell_3 = \frac{-4-0}{0-(-6)} = \frac{-4}{6} = -\frac{2}{3}$

b. Lesser

27. $f(x) = -2x + 6$
$m = -2,\ b = 6$

29. $-5x + y = 10$
$y = 5x + 10$
$m = 5,\ b = 10$

31. $-3x - 4y = 6$
$4y = -3x - 6$
$y = -\frac{3}{4}x - \frac{3}{2}$
$m = -\frac{3}{4},\ b = -\frac{3}{2}$

33. $f(x) = -\frac{1}{4}x$
$m = -\frac{1}{4},\ b = 0$

35. $f(x) = 2x - 3$
$m = 2,\ b = -3$
D

37. $f(x) = -2x - 3$
$m = -2,\ b = -3$
C

39. $y = -2$
$m = 0$

41. $x = 4$
m is undefined

43. $y - 7 = 0$
$y = 7$
$m = 0$

45. Answers may vary.

47. $f(x) = x + 2$
$m = 1,\ b = 2$

49. $4x - 7y = 28$
$7y = 4x - 28$
$y = \frac{4}{7}x - 4$
$m = \frac{4}{7},\ b = -4$

51. $2y - 7 = x$
$2y = x + 7$
$y = \frac{1}{2}x + \frac{7}{2}$
$m = \frac{1}{2},\ b = \frac{7}{2}$

53. $x = 7$
Slope is undefined.
There is no y-intercept.

55. $f(x) = \frac{1}{7}x$
$m = \frac{1}{7},\ b = 0$

57. $x - 7 = 0$
$x = 7$
Slope is undefined.
There is no y-intercept.

59. $2y + 4 = -7$
$2y = -11$
$y = -\frac{11}{2}$
$m = 0,\ b = -\frac{11}{2}$

61. $f(x) = 5x - 6$ $\quad g(x) = 5x + 2$
$m = 5$ $\quad m = 5$
$b = -6$ $\quad b = 2$
Parallel, since they have the same slope but different y-intercepts.

63. $2x - y = -10$ $\quad 2x + 4y = 2$
$y = 2x + 10$ $\quad y = -\frac{1}{2}x + \frac{1}{2}$
$m = 2$ $\quad m = -\frac{1}{2}$
Perpendicular, since the product of their slopes is -1.

65. $x + 4y = 7$ $\quad 2x - 5y = 0$
$y = -\frac{1}{4}x + \frac{7}{4}$ $\quad y = \frac{2}{5}x$
$m = -\frac{1}{4}$ $\quad m = \frac{2}{5}$
Neither, since their slopes are not equal, nor does their product equal –1.

67. Answers may vary.

69. Two points on line: (1, 0), (0, 3)
$m = \frac{3-0}{0-1} = \frac{3}{-1} = -3$

71. Two points on line: (–3, –1), (2, 4)
$m = \frac{4-(-1)}{2-(-3)} = \frac{4+1}{2+3} = \frac{5}{5} = 1$

73. $m = \frac{3}{25}$

75. $m = \frac{15}{100} = \frac{3}{20}$

77. $f(x) = x$
$m = 1$
The slope of a parallel line is 1.

79. $f(x) = x$
$m = 1$
The slope of a perpendicular line is –1.

81. $-3x + 4y = 0$
$y = \frac{3}{4}x$
$m = \frac{3}{4}$
The slope of a parallel line is $\frac{3}{4}$.

83. **a.** (6, 20)

b. (10, 13)

c. $m = \frac{13-20}{10-6} = -\frac{7}{4}$, or –1.75 yards per second

85. $-4x + 2y = 5$
$2x - y = 7$

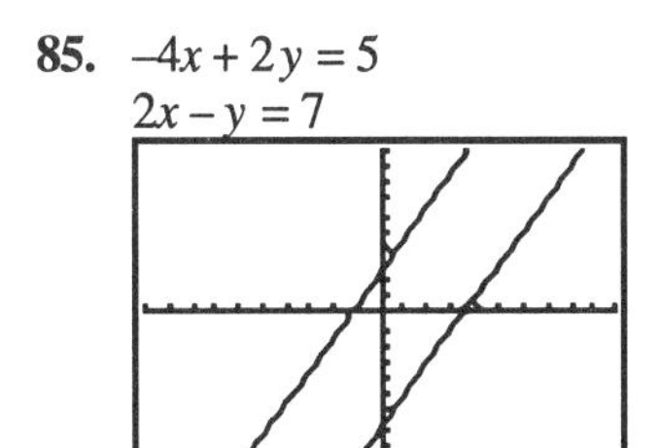

87. **a.** $y = \frac{1}{2}x + 1$
$y = x + 1$
$y = 2x + 1$

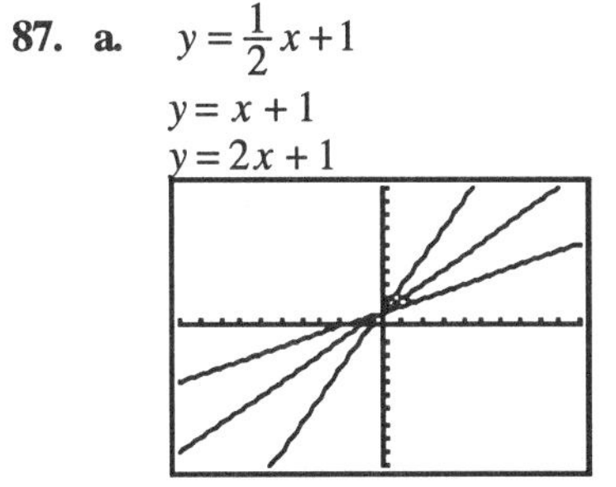

b. $y = -\frac{1}{2}x + 1$
$y = -x + 1$
$y = -2x + 1$

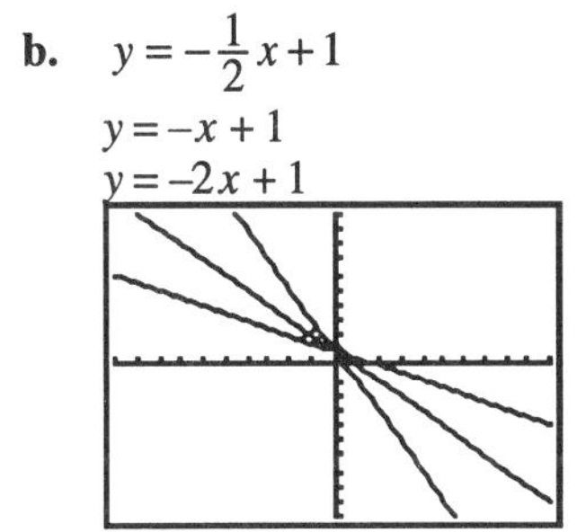

c. True

89. $P(B) = \frac{2}{11}$

91. $P(I \text{ or } T) = \frac{3}{11}$

93. $P(\text{vowel}) = \frac{4}{11}$

95. $y - 0 = -3[x - (-10)]$
$y = -3(x + 10)$
$y = -3x - 30$

97. $y - 9 = -8[x - (-4)]$
$y - 9 = -8(x + 4)$
$y - 9 = -8x - 32$
$y = -8x - 23$

Section 3.5

Mental Math

1. $m = -4,\ b = 12$

3. $m = 5,\ b = 0$

5. $m = \frac{1}{2},\ b = 6$

7. $y = 12x + 6$ $\quad$ $y = 12x - 2$
   $m = 12$ $\quad$ $m = 12$
   $b = 6$ $\quad$ $b = -2$
   Parallel, since the slopes are equal and their y-intercepts are different.

9. $y = -9x + 3$ $\quad$ $y = \frac{3}{2}x - 7$
   $m = -9$ $\quad$ $m = \frac{3}{2}$
   Neither, since the slopes are not equal and their product is not -1.

Exercise Set 3.5

1. $y = -x + 1$

3. $y = 2x + \frac{3}{4}$

5. $y = \frac{2}{7}x$

7. Point: (1, 3)
   Slope: $\frac{3}{2}$

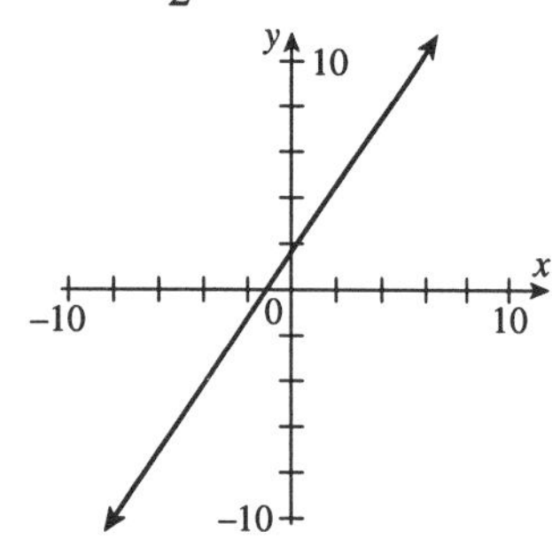

9. Point: (0, 0)
   Slope: 5

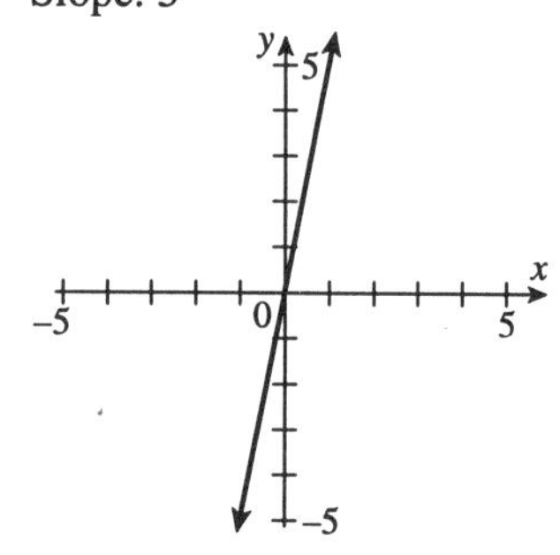

11. Point: (0, 7)
    Slope: –1

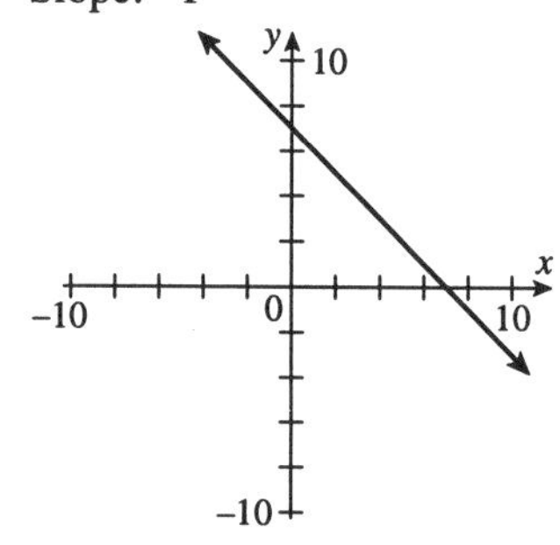

13. $y - 2 = 3(x - 1)$
    $y - 2 = 3x - 3$
    $3x - y = 1$

15. $y - (-3) = -2(x - 1)$
    $y + 3 = -2x + 2$
    $2x + y = -1$

17. $y - 2 = \frac{1}{2}(x - (-6))$
    $2(y - 2) = x + 6$
    $2y - 4 = x + 6$
    $x - 2y = -10$

19. $y - 0 = -\frac{9}{10}(x - (-3))$
    $10y = -9(x + 3)$
    $10y = -9x - 27$
    $9x + 10y = -27$

21. (0, 3), (1, 1)
$m = \frac{1-3}{1-0} = -\frac{2}{1} = -2$
$b = 3$
$y = -2x + 3$
$2x + y = 3$

23. (−2, 1), (4, 5)
$m = \frac{5-1}{4-(-2)} = \frac{4}{6} = \frac{2}{3}$
$y - 1 = \frac{2}{3}(x+2)$
$3y - 3 = 2x + 4$
$2x - 3y = -7$

25. $m = \frac{6-0}{4-2} = \frac{6}{2} = 3$
$y - 0 = 3(x - 2)$
$y = 3x - 6$
$f(x) = 3x - 6$

27. $m = \frac{13-5}{-6-(-2)} = \frac{8}{-4} = -2$
$y - 5 = -2(x - (-2))$
$y - 5 = -2(x + 2)$
$y - 5 = -2x - 4$
$y = -2x + 1$
$f(x) = -2x + 1$

29. $m = \frac{-3-(-4)}{-4-(-2)} = \frac{1}{-2} = -\frac{1}{2}$
$y - (-4) = -\frac{1}{2}(x - (-2))$
$2(y + 4) = -(x + 2)$
$2y + 8 = -x - 2$
$y = -\frac{1}{2}x - \frac{10}{2}$
$f(x) = -\frac{1}{2}x - 5$

31. $m = \frac{-9-(-8)}{-6-(-3)} = \frac{-1}{-3} = \frac{1}{3}$
$y - (-8) = \frac{1}{3}(x - (-3))$
$3(y + 8) = x + 3$
$3y + 24 = x + 3$
$3y = x - 21$
$f(x) = \frac{1}{3}x - 7$

33. Answers may vary.

35. $f(0) = -2$

37. $f(2) = 2$

39. $f(x) = -6$
$f(-2) = -6$
$x = -2$

41. $y = mx + b$
$-4 = 0(-2) + b$
$-4 = b$
$y = -4$

43. Every vertical line is in the form $x = c$. Since the line passes through the point (4, 7), its equation is $x = 4$.

45. Every horizontal line is in the form $y = c$. Since the line passes through the point (0, 5), its equation is $y = 5$.

47. $y = 4x - 2$ so $m = 4$
$y - 8 = 4(x - 3)$
$y - 8 = 4x - 12$
$y = 4x - 4$
$f(x) = 4x - 4$

49. $3y = x - 6$ or $y = \frac{1}{3}x - 2$
So $m_{\perp} = \frac{1}{3}$ and $m = -3$.
Now, $y - (-5) = -3(x - 2)$
$y + 5 = -3x + 6$
$y = -3x + 1$
$f(x) = -3x + 1$

51. $3x + 2y = 5$
$2y = -3x + 5$
$y = -\frac{3}{2}x + \frac{5}{2}$ so $m = \frac{-3}{2}$
Now, $y - (-3) = -\frac{3}{2}(x - (-2))$
$2(y + 3) = -3(x + 2)$
$2y + 6 = -3x - 6$
$y = -\frac{3}{2}x - 6$
$f(x) = -\frac{3}{2}x - 6$

53. $y - 3 = 2(x - (-2))$
$y - 3 = 2(x + 2)$
$y - 3 = 2x + 4$
$2x - y = -7$

55. $m = \frac{2-6}{5-1} = \frac{-4}{4} = -1$
$y - 6 = -1(x - 1)$
$y - 6 = -x + 1$
$y = -x + 7$
$f(x) = -x + 7$

57. $y = -\frac{1}{2}x + 11$
$2y = -x + 22$
$x + 2y = 22$

59. $m = \frac{-6-(-4)}{0-(-7)} = \frac{-2}{7}$
$y = -\frac{2}{7}x - 6$
$7y = -2x - 42$
$2x + 7y = -42$

61. $y - 0 = -\frac{4}{3}(x - (-5))$
$3y = -4(x + 5)$
$3y = -4x - 20$
$4x + 3y = -20$

63. $x = -2$

65. $2x + 4y = 8$
$4y = -2x + 8$
$y = -\frac{1}{2}x + 2$, so $m = -\frac{1}{2}$. Now,
$y - (-2) = -\frac{1}{2}(x - 6)$
$2(y + 2) = -x + 6$
$2y + 4 = -x + 6$
$x + 2y = 2$

67. $y = 12$

69. $8x - y = 9$
$8x - 9 = y$ or $y = 8x - 9$
so $m = 8$
$y - 1 = 8(x - 6)$
$y - 1 = 8x - 48$
$47 = 8x - y$ or $8x - y = 47$

71. $x = 5$

73. $m = \frac{-5-(-8)}{-6-2} = \frac{3}{-8} = -\frac{3}{8}$
$y - (-8) = -\frac{3}{8}(x - 2)$
$8(y + 8) = -3(x - 2)$
$8y + 64 = -3x + 6$
$y = -\frac{3}{8}x - \frac{29}{4}$
$f(x) = -\frac{3}{8}x - \frac{29}{4}$

75. a. (1, 32), (3, 96)
$m = \frac{96-32}{3-1} = \frac{64}{2} = 32$
$y - 32 = 32(x - 1)$
$y = 32x - 32 + 32$
$y = 32x$
$R(x) = 32x$

b. $R(4) = 32(4) = 128$ ft per second

77. a. (0, 97,500), (4, 109,800)
$m = \frac{109{,}800 - 97{,}500}{4-0} = 3075$
$y - 97500 = 3075(x - 0)$
$y = 3075x + 97500$
$P(x) = 3075x + 97{,}500$

b. $P(9) = 3075(9) + 97{,}500 = \$125{,}175$

79. a. (1, 30,000), (4, 66,000)
$m = \frac{66{,}000 - 30{,}000}{4-1} = 12{,}000$
$y - 30{,}000 = 12{,}000(x - 1)$
$y = 12{,}000x + 18{,}000$
$P(x) = 12{,}000x + 18{,}000$

b. $P(7) = 12{,}000(7) + 18{,}000$
$= \$102{,}000$

c. $126{,}000 = 12{,}000x + 18{,}000$
$x = \frac{126{,}000 - 18{,}000}{12{,}000}$
$x = 9$ years

81. a. (3, 10,000), (5, 8000)
$m = \frac{8000 - 10{,}000}{5-3} = -1000$
$y - 10{,}000 = -1000(x - 3)$
$y = -1000x + 13{,}000$

b. $y = -1000(3.5) + 13{,}000$
$y = 9500$
9500 Fun Noodles

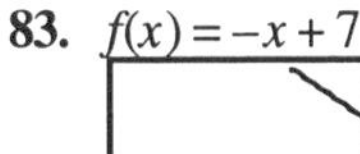

83. $f(x) = -x + 7$

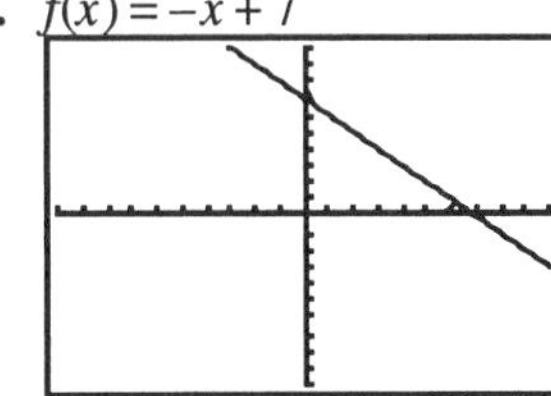

85. $4x + 3y = -20$

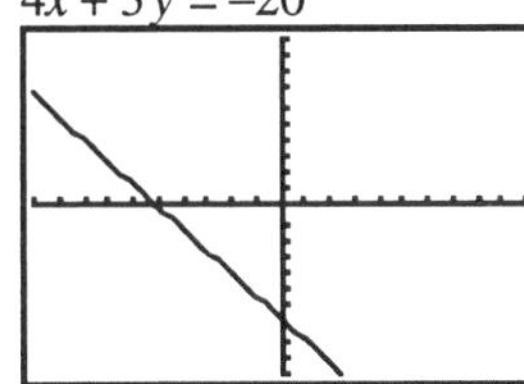

87. $2x - 7 \le 21$
$2x \le 28$
$x \le 14$
$(-\infty, 14]$

4 6 8 10 12 14 16 18 20

89. $5(x - 2) \ge 3(x - 1)$
$5x - 10 \ge 3x - 3$
$2x \ge 7$
$x \ge \frac{7}{2}$
$\left[\frac{7}{2}, \infty\right)$

−3 −2 −1 0 1 2 $\frac{7}{2}$ 5

91. $\frac{x}{2} + \frac{1}{4} < \frac{1}{8}$
$4x + 2 < 1$
$4x < -1$
$x < -\frac{1}{4}$
$\left(-\infty, -\frac{1}{4}\right)$

−1 $-\frac{3}{4}$ $-\frac{1}{2}$ $-\frac{1}{4}$ 0 $\frac{1}{4}$ $\frac{1}{2}$ $\frac{3}{4}$ 1

93. $m_{\perp} = \frac{1-(-1)}{-5-3} = \frac{2}{-8} = -\frac{1}{4}$, so $m = 4$
$m((3, -1), (-5, 1)) = \left(\frac{3-5}{2}, \frac{-1+1}{2}\right)$
$= (-1, 0)$
so $y - 0 = 4(x - (-1))$
$y = 4(x + 1)$
$y = 4x + 4$
$-4x + y = 4$

95. $m_{\perp} = \frac{-4-6}{-22-(-2)} = \frac{-10}{-20} = \frac{1}{2}$
so $m = -2$
$m((-2, 6), (-22, -4)) = \left(\frac{-2-22}{2}, \frac{6-4}{2}\right)$
$= (-12, 1)$
Now, $y - 1 = -2(x - (-12)) = -2(x + 12)$
so $y - 1 = -2x - 24$ or $2x + y = -23$

97. $m_{\perp} = \frac{7-3}{-4-2} = \frac{4}{-6} = \frac{-2}{3}$, so $m = \frac{3}{2}$
$m((2, 3), (-4, 7)) = \left(\frac{2-4}{2}, \frac{3+7}{2}\right)$
$= (-1, 5)$
Now, $y - 5 = \frac{3}{2}(x - (-1))$
$2y - 10 = 3(x + 1) = 3x + 3$
or $3x - 2y = -13$

Exercise Set 3.6

1. $x < 2$

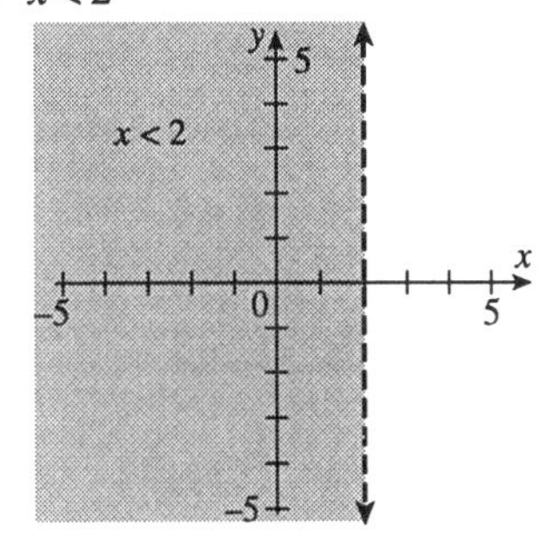

3. $x - y \ge 7$
$y \le x - 7$

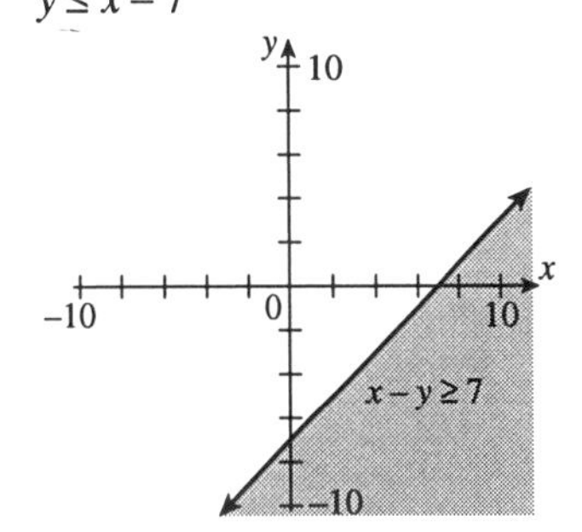

5. $3x + y > 6$
$y > -3x + 6$

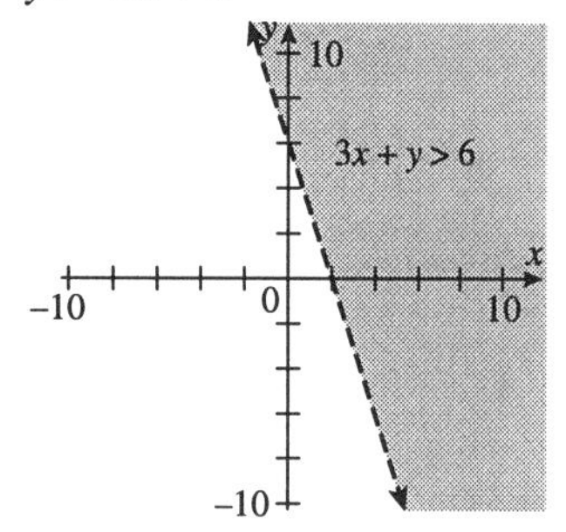

7. $y \le -2x$

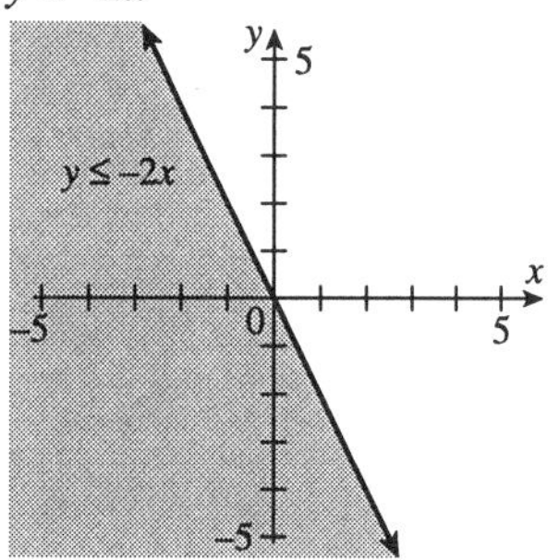

9. $2x + 4y \ge 8$
$4y \ge -2x + 8$
$y > -\frac{1}{2}x + 2$

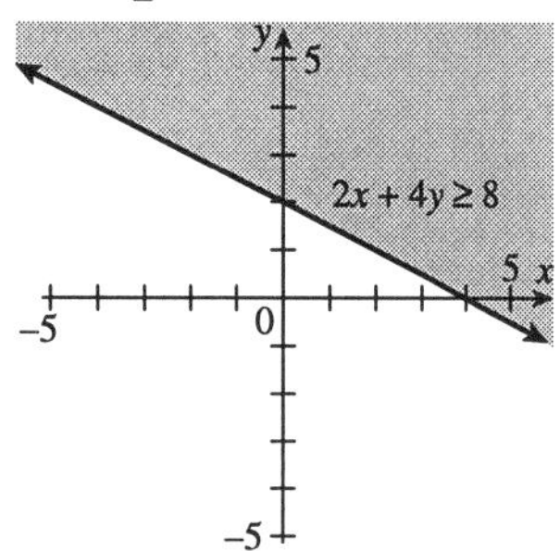

11. $5x + 3y > -15$
$3y > -5x - 15$
$y > -\frac{5}{3}x - 5$

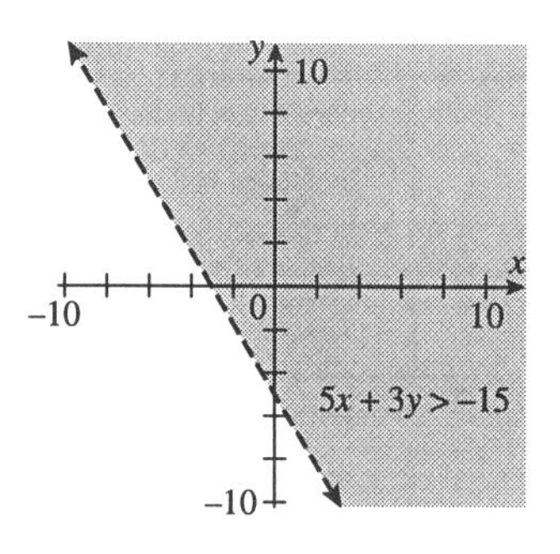

13. Use a dashed boundary line with < or >.

15. $x \ge 3$ and $y \le -2$

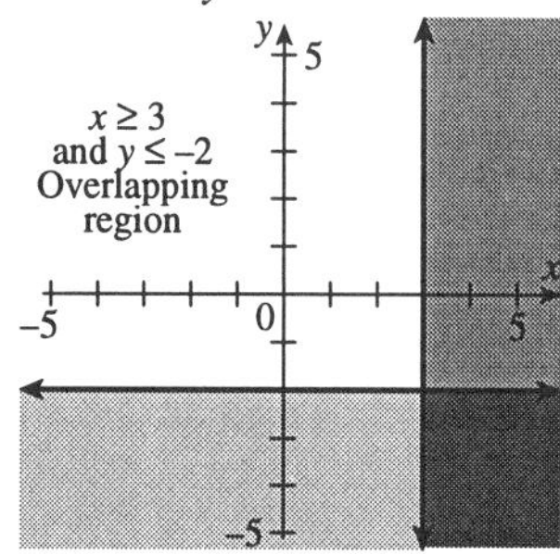

17. $x \le -2$ or $y \ge 4$

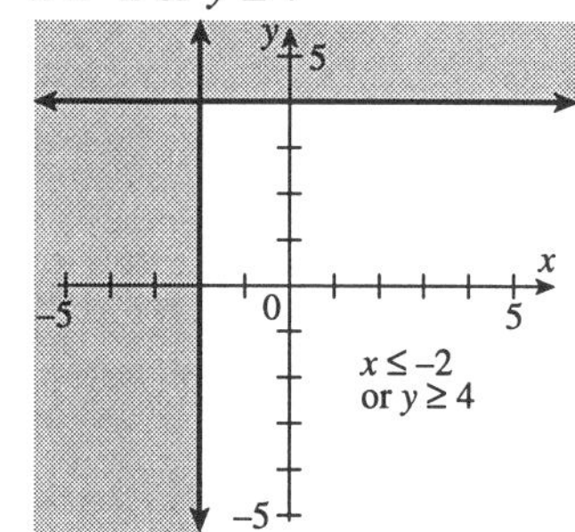

19. $x - y < 3$ and $x > 4$
$y > x - 3$ and $x > 4$

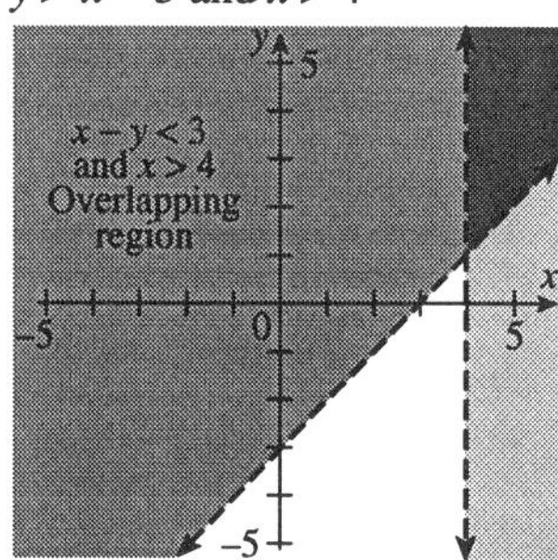

21. $x + y \le 3$ or $x - y \ge 5$
$y \le -x + 3$ or $y \le x - 5$

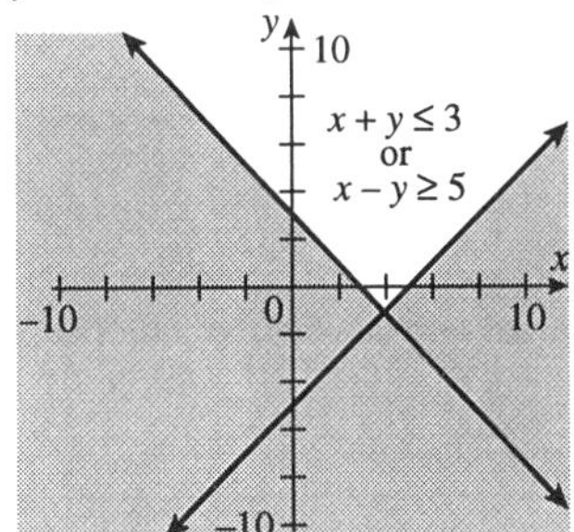

23. $y \ge -2$

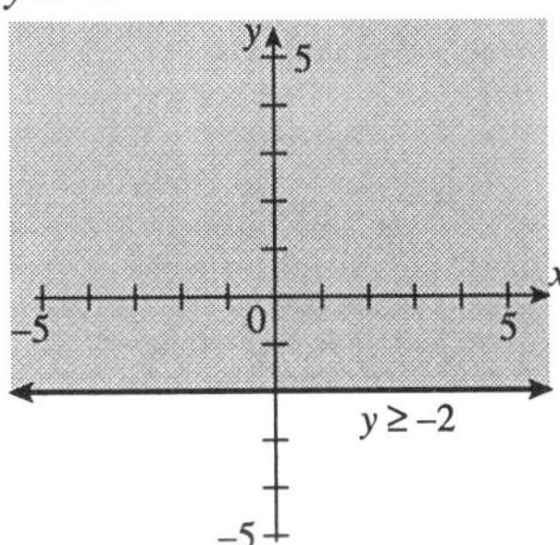

25. $x - 6y < 12$
$x - 12 < 6y$
$y > \frac{1}{6}x - 2$

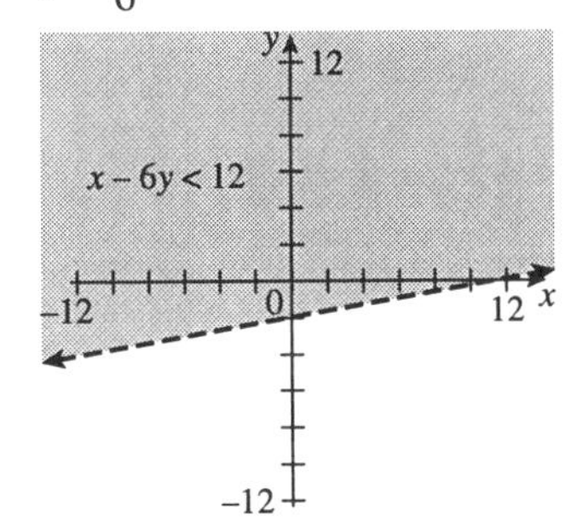

27. $x > 5$

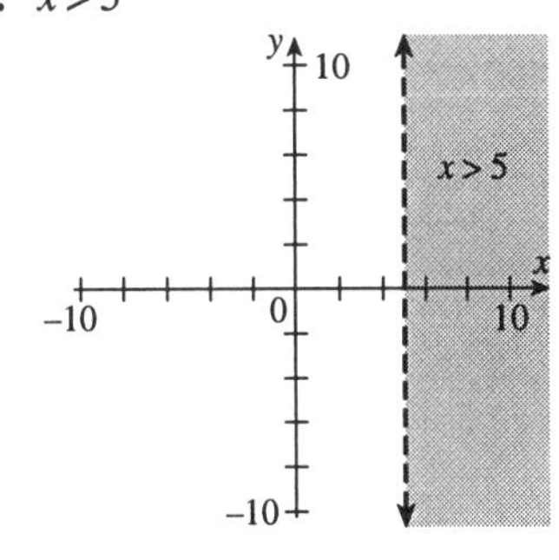

29. $-2x + y \le 4$
$y \le 2x + 4$

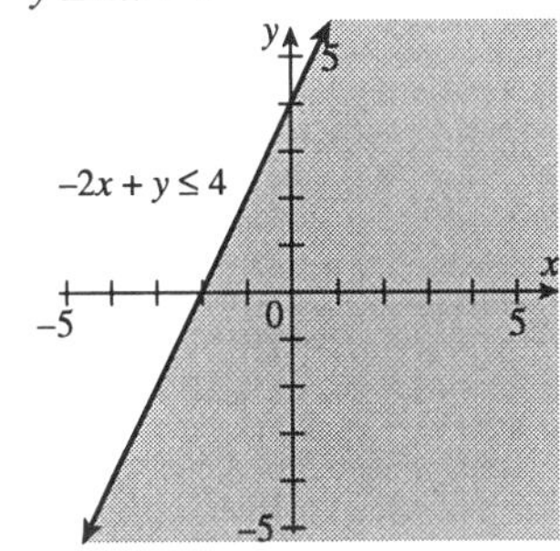

31. $x - 3y < 0$
$x < 3y$
$y > \frac{x}{3}$

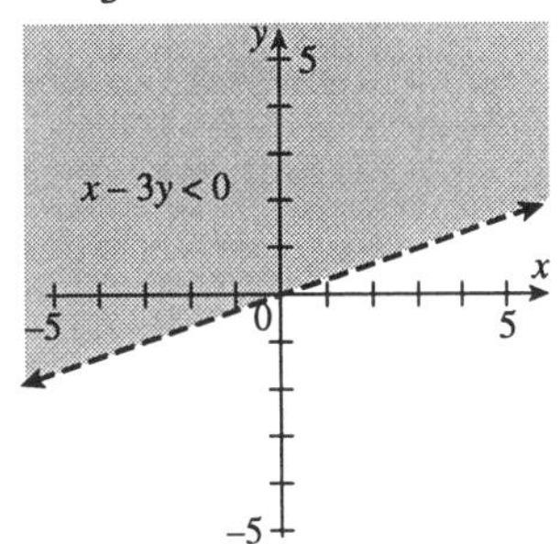

33. $3x - 2y \le 12$
$3x - 12 \le 2y$
$y \ge \frac{3}{2}x - 6$

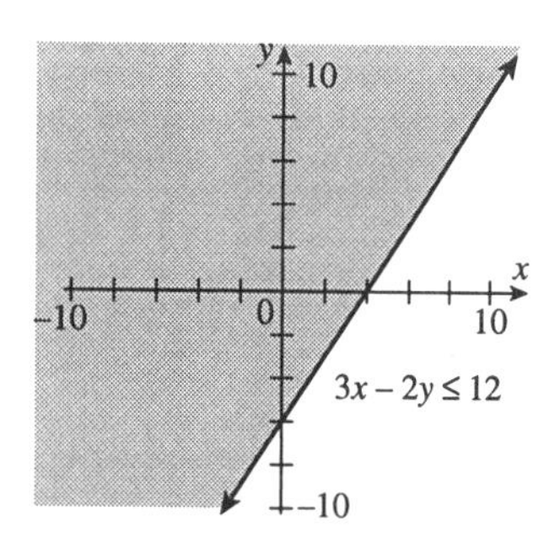

35. $x - y \geq 2$ or $y < 5$
$y \leq x - 2$ or $y < 5$

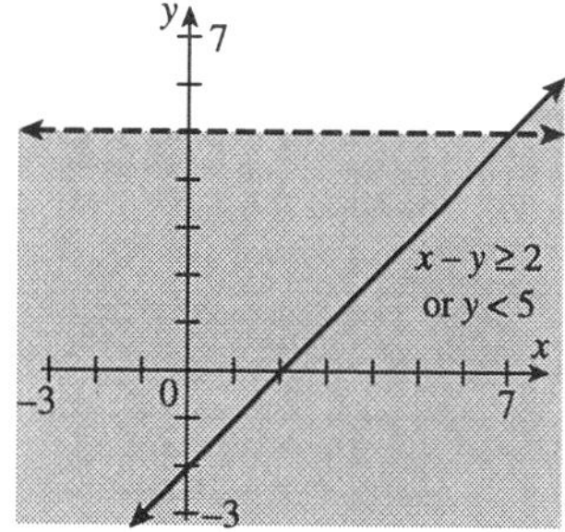

37. $x + y \leq 1$ and $y \leq -1$
$y \leq -x + 1$ and $y \leq -1$

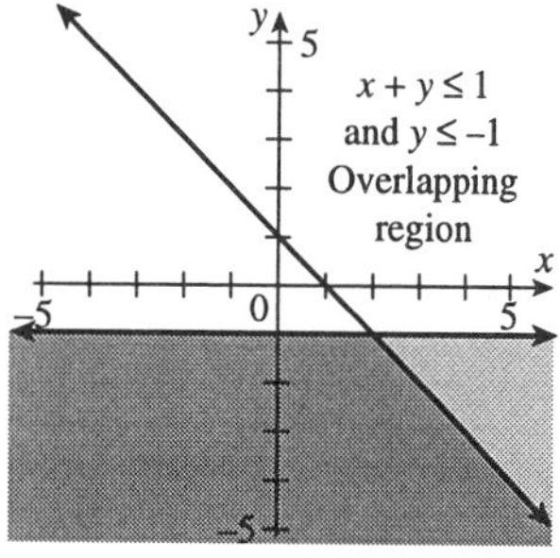

39. $2x + y > 4$ or $x \geq 1$
$y > -2x + 4$ or $x \geq 1$

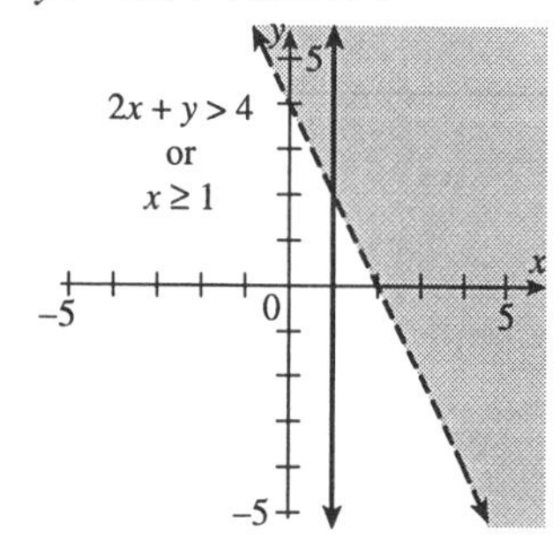

41. $-2 \leq x \leq 1$

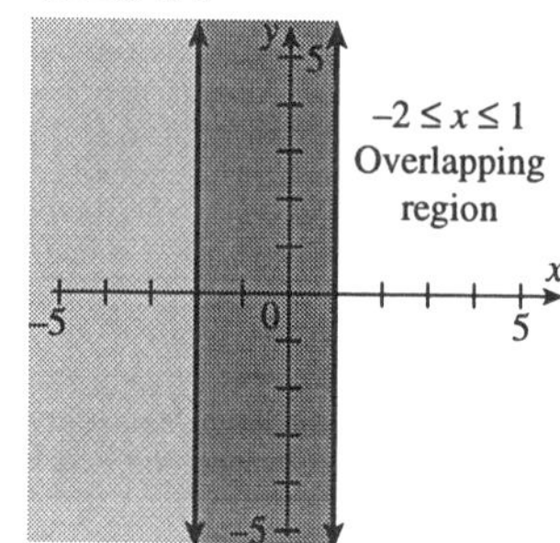

43. $x + y \leq 0$ or $3x - 6y \geq 12$
$y \leq -x$ or $3x - 12 \geq 6y$
$y \leq -x$ or $y \leq \frac{1}{2}x - 2$

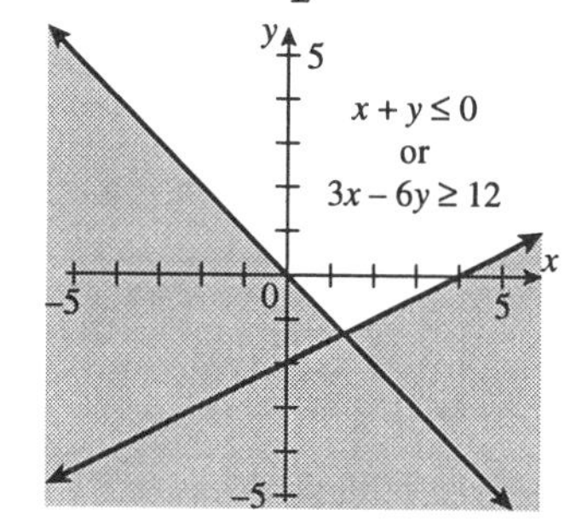

45. $2x - y > 3$ and $x \geq 0$
$y < 2x - 3$ and $x \geq 0$

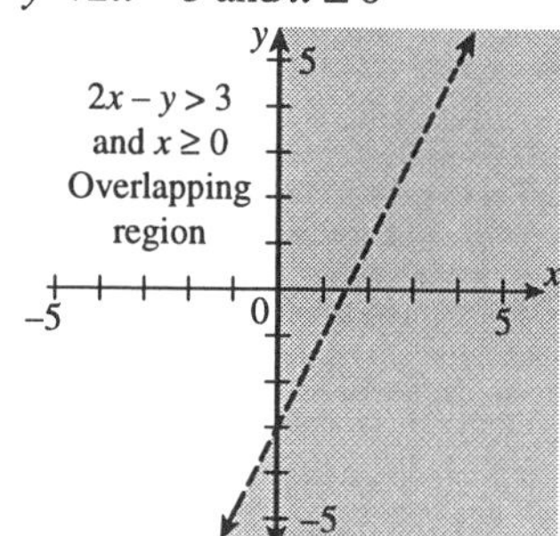

47. D

49. A

51. $x \geq 2$

53. $y \leq -3$

55. $y > 4$

57. $x < 1$

59. $x \le 20$ and $y \ge 10$

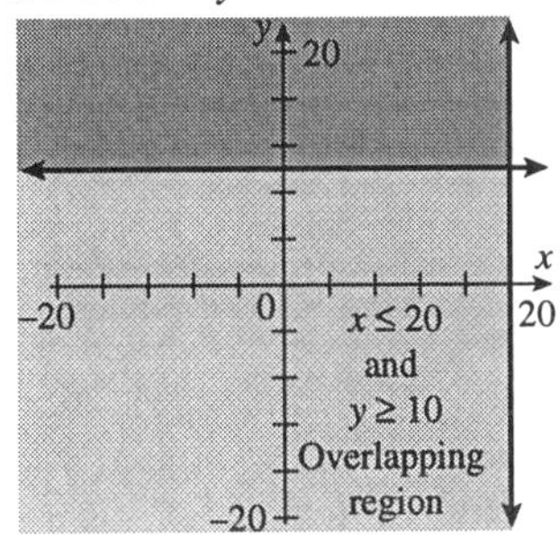

61. $\begin{cases} x \ge 0 \\ y \ge 0 \\ 2x + 4y \le 40 \end{cases}$

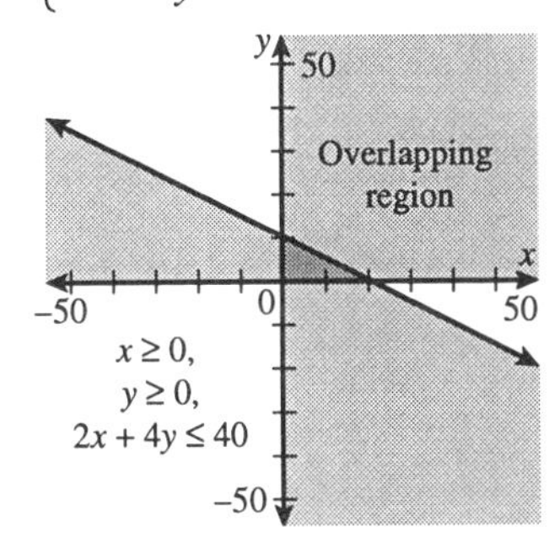

63. $3^2 = 3 \cdot 3 = 9$

65. $(-5)^2 = (-5)(-5) = 25$

67. $-2^4 = -(2)(2)(2)(2) = -16$

69. $\left(\frac{2}{7}\right)^2 = \frac{2^2}{7^2} = \frac{4}{49}$

71. Domain: $(-\infty, 2] \cup [2, \infty)$
Range: $(-\infty, \infty)$
This relation is not a function.

Chapter 3 - Review

1. $A(2, -1)$, quadrant IV
$B(-2, 1)$, quadrant II
$C(0, 3)$, y-axis
$D(-3, -5)$, quadrant III

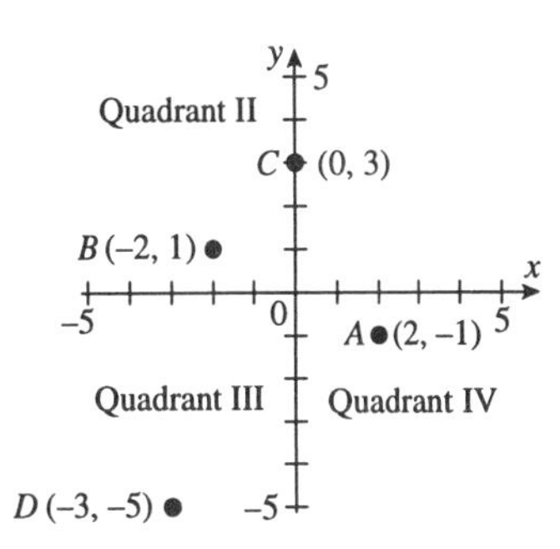

2. $A(-3, 4)$, quadrant II
$B(4, -3)$, quadrant IV
$C(-2, 0)$, x-axis
$D(-4, 1)$, quadrant II

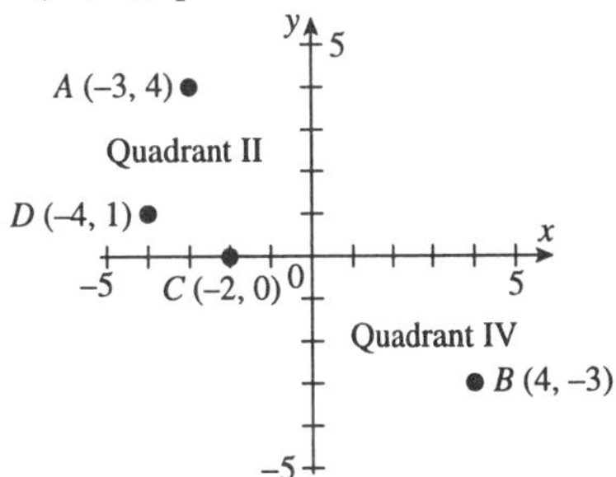

3. $7x - 8y = 56$
$(0, 56)$; No
$7(0) - 8(56) = 56$
$-448 = 56$, False
$(8, 0)$; Yes
$7(8) - 8(0) = 56$
$56 = 56$, True

4. $-2x + 5y = 10$
$(-5, 0)$; Yes
$-2(-5) + 5(0) = 10$
$10 = 10$, True
$(1, 1)$; No
$-2(1) + 5(1) = 10$
$-2 + 5 = 10$
$3 = 10$, False

5. $x = 13$
$(13, 5)$; Yes
$13 = 13$, True
$(13, 13)$; Yes
$13 = 13$, True

6. $y = 2$
(7, 2); Yes
$2 = 2$, True
(2, 7); No
$7 = 2$, False

7. $-2 + y = 6x$
$-2 + y = 6(7)$
$-2 + y = 42$
$y = 44$
(7, 44)

8. $y = 3x + 5$
$-8 = 3x + 5$
$-13 = 3x$
$-\frac{13}{3} = x$
$\left(-\frac{13}{3}, -8\right)$

9. $9 = -3x + 4y$

(9, 9)
(1, 3)
(−3, 0)
10
−10
0
10
−10

a. $9 = -3x + 4(0)$
$9 = -3x$
$-3 = x$
(−3, 0)

b. $9 = -3x + 4(3)$
$9 = -3x + 12$
$-3 = -3x$
$1 = x$
(1, 3)

c. $9 = -3(9) + 4y$
$9 = -27 + 4y$
$36 = 4y$
$9 = y$
(9, 9)

10. $y = -2x$

(−7, 14)
14
(0, 0)
−14
14
−14
(7, −14)

a. $y = -2(7)$
$y = -14$
(7, −14)

b. $y = -2(-7)$
$y = 14$
(−7, 14)

c. $y = -2(0)$
$y = 0$
(0, 0)

11. $x = 2y$

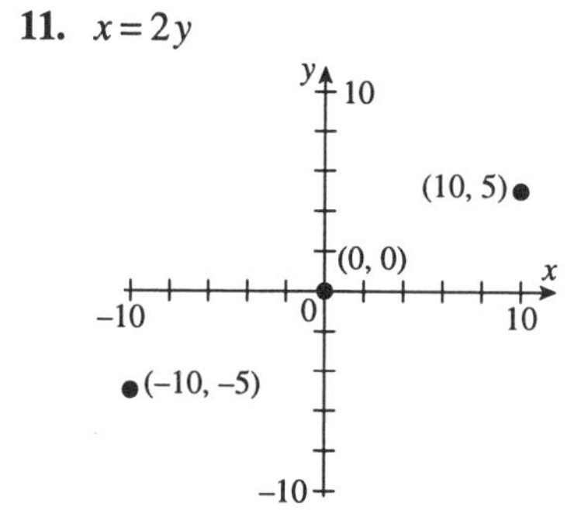

a. $x = 2(0)$
$x = 0$
(0, 0)

b. $x = 2(5)$
$x = 10$
(10, 5)

c. $x = 2(-5)$
$x = -10$
(−10, −5)

12. Linear

13. Linear

14. Not linear

15. Not linear

16. Linear

17. Linear

18. $3x - y = 4$
Find three ordered pair solutions, or find x- and y-intercepts, or find m and b.

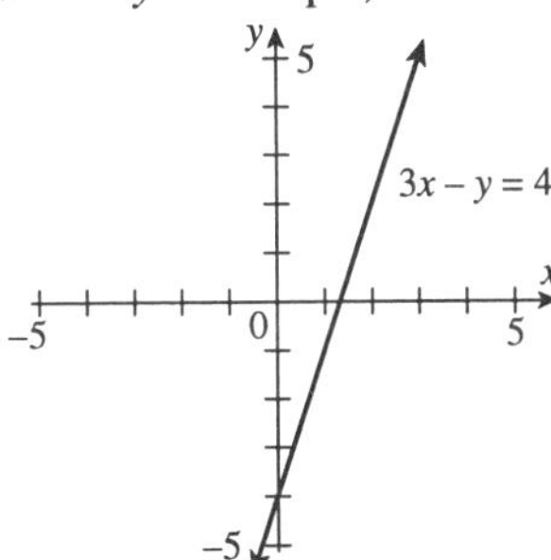

19. $x - 3y = 2$
Find three ordered pair solutions, or find x- and y-intercepts, or find m and b.

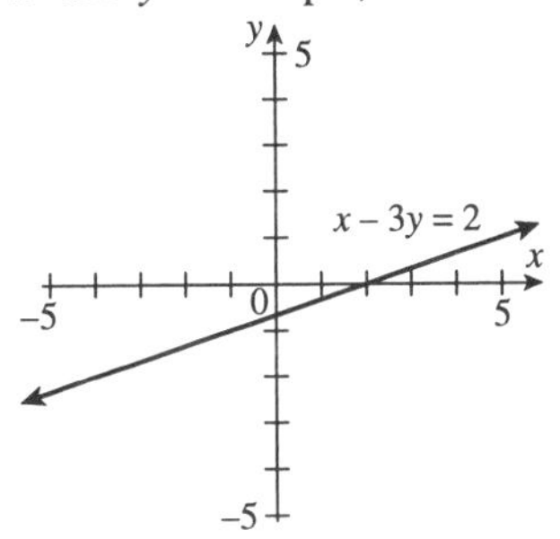

20. $y = |x| + 4$

x	y
−3	7
−2	6
−1	5
0	4
1	5
2	6
3	7

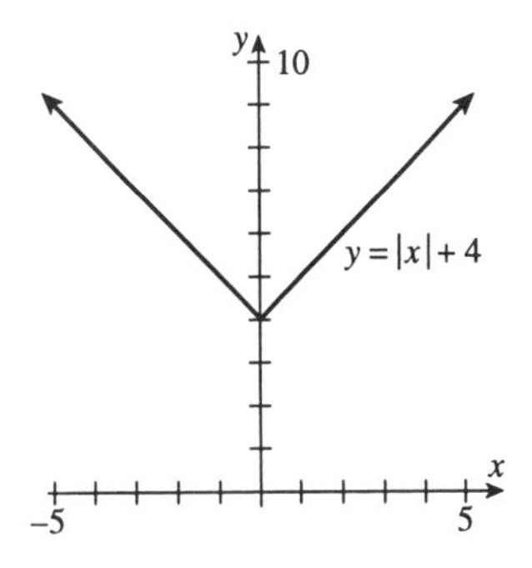

21. $y = x^2 + 4$

x	y
−3	13
−2	8
−1	5
0	4
1	5
2	8
3	13

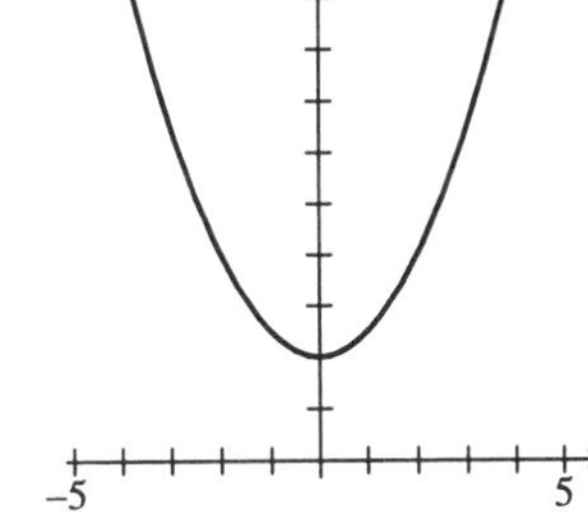

22. $y = -\frac{1}{2}x + 2$
Find three ordered pair solutions, or find x- and y-intercepts, or find m and b.

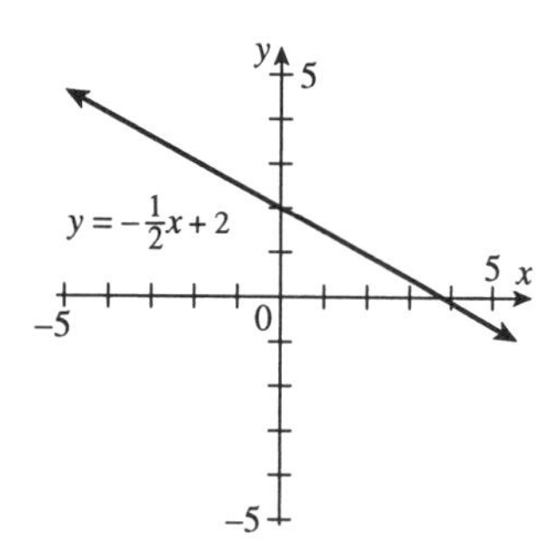

23. $y=-x+5$
Find three ordered pair solutions, or find x- and y-intercepts, or find m and b.

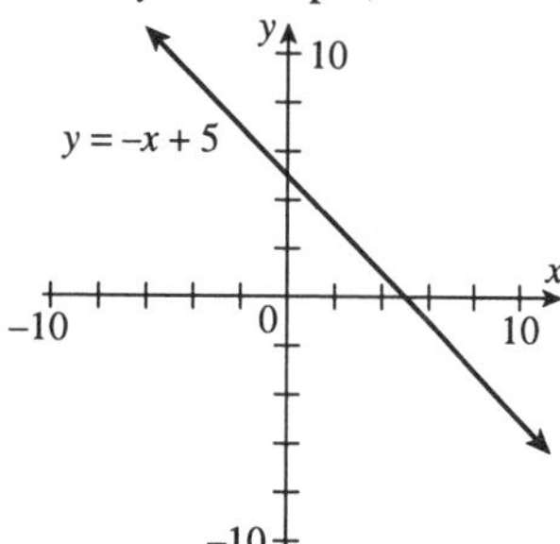

24. $y=-1.36x$
Find three ordered pair solutions, or find x- and y-intercepts, or find m and b.

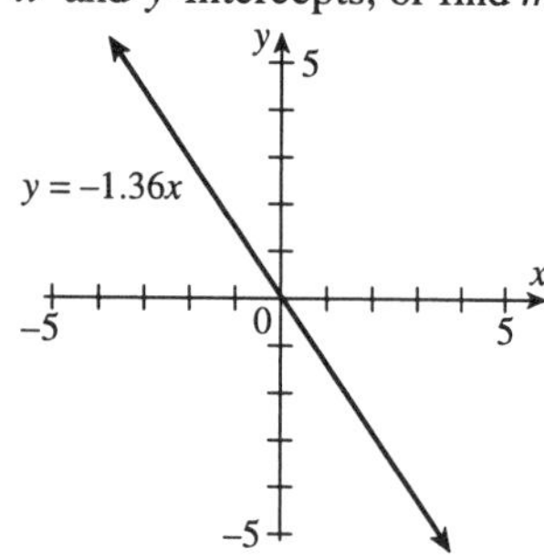

25. $y=2.1x+5.9$
Find three ordered pair solutions, or find x- and y-intercepts, or find m and b.

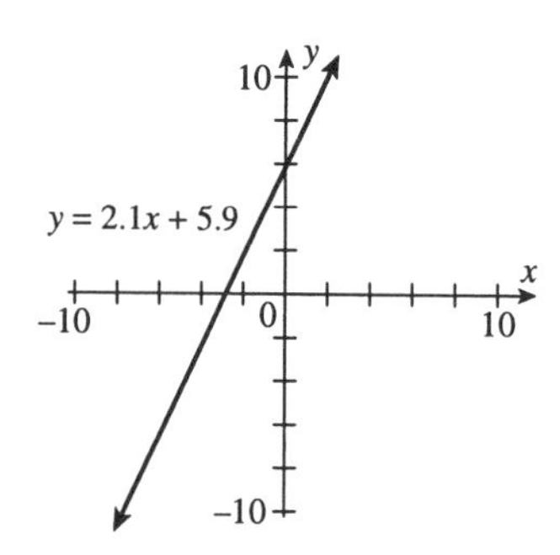

26. Domain: $\left\{-\frac{1}{2},\ 6,\ 0,\ 25\right\}$
Range: $\left\{\frac{3}{4},\ -12,\ 25\right\}$
Function

27. Domain: $\left\{\frac{3}{4},\ -12,\ 25\right\}$
Range: $\left\{-\frac{1}{2},\ 6,\ 0,\ 25\right\}$
Not a function

28. Domain: {2, 4, 6, 8}
Range: {2, 4, 5, 6}
Not a function

29. Domain: {Triangle, Square, Rectangle, Parallelogram}
Range: {3, 4}
Function

30. Domain: $(-\infty, \infty)$
Range: $(-\infty, -1] \cup [1, \infty)$
Not a function

31. Domain: $\{-3\}$
Range: $(-\infty, \infty)$
Not a function

32. Domain: $(-\infty, \infty)$
Range: $\{4\}$
Function

33. Domain: $[-1, 1]$
Range: $[-1, 1]$
Not a function

34. $f(x)=x-5$
$f(2)=2-5=-3$

35. $g(x) = -3x$
$g(0) = -3(0) = 0$

36. $g(x) = -3x$
$g(-6) = -3(-6) = 18$

37. $h(x) = 2x^2 - 6x + 1$
$h(-1) = 2(-1)^2 - 6(-1) + 1$
$= 2(1) + 6 + 1 = 9$

38. $h(x) = 2x^2 - 6x + 1$
$h(1) = 2(1)^2 - 6(1) + 1$
$= 2 - 6 + 1 = -3$

39. $f(x) = x - 5$
$f(5) = 5 - 5 = 0$

40. $J(x) = 2.54x$
$J(150) = 2.54(150) = 381$ pounds

41. $J(x) = 2.54x$
$J(2000) = 2.54(2000) = 5080$ pounds

42. $f(x) = x$ or $y = x$
$m = 1,\ b = 0$

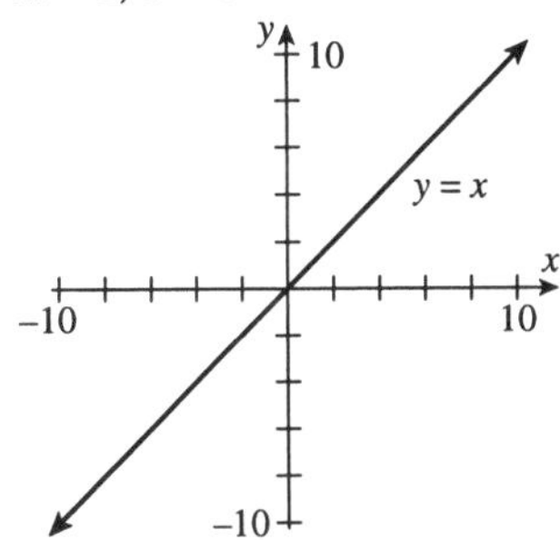

43. $f(x) = -\frac{1}{3}x$ or $y = -\frac{1}{3}x$
$m = -\frac{1}{3},\ b = 0$

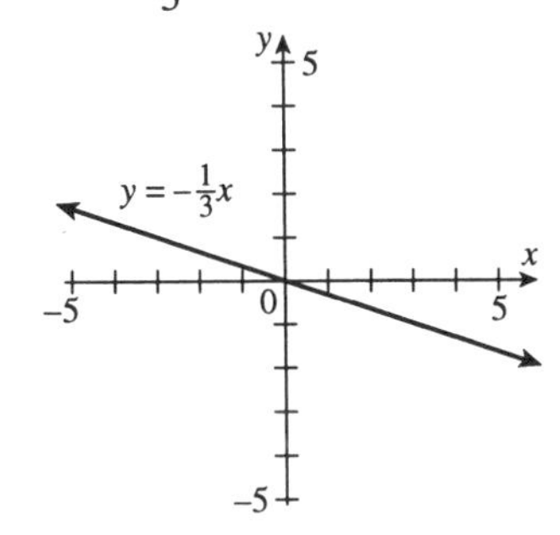

44. $g(x) = 4x - 1$ or $y = 4x - 1$
$m = 4,\ b = -1$

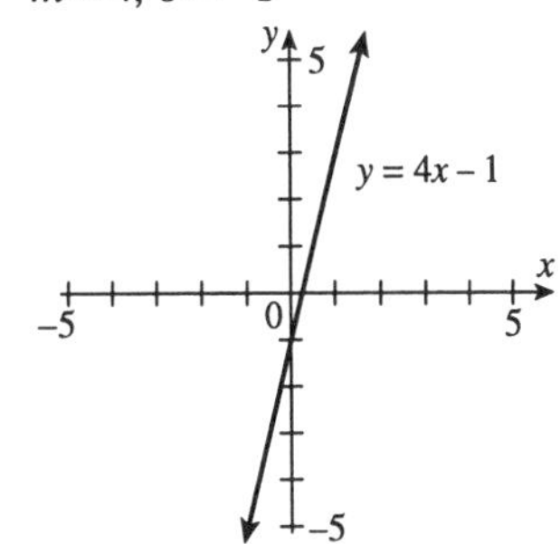

45. C

46. A

47. B

48. D

49. $4x + 5y = 20$

Let $x = 0$	Let $y = 0$
$4(0) + 5y = 20$	$4x + 5(0) = 20$
$y = 4$	$x = 5$
$(0, 4)$	$(5, 0)$

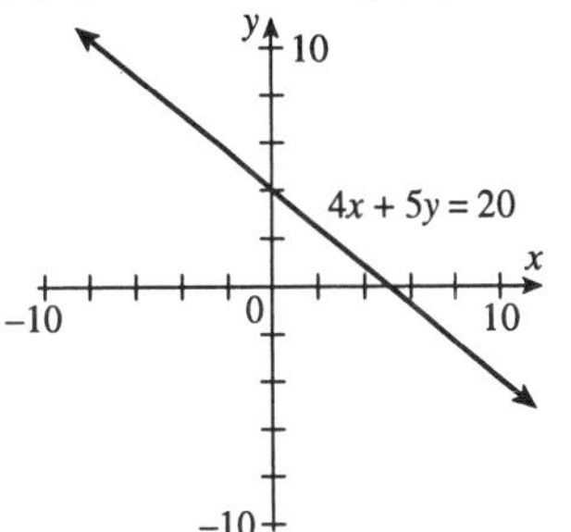

50. $3x - 2y = -9$

Let $x = 0$	Let $y = 0$
$3(0) - 2y = -9$	$3x - 2(0) = -9$
$y = \frac{9}{2}$	$x = -3$
$\left(0,\ \frac{9}{2}\right)$	$(-3, 0)$

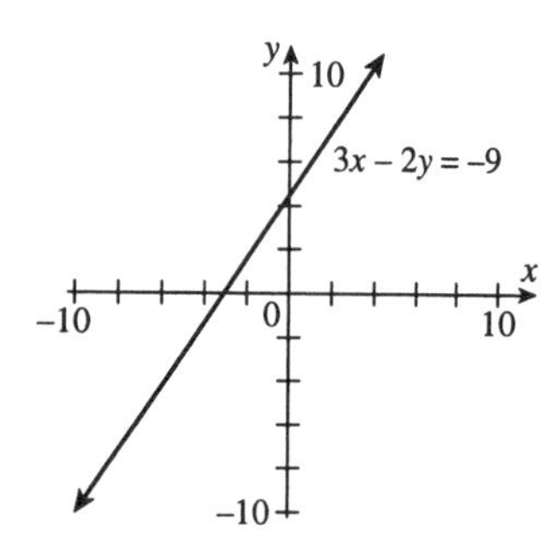

51. $4x - y = 3$

Let $x = 0$ Let $y = 0$

$4(0) - y = 3$ $4x - 0 = 3$

$y = -3$ $x = \frac{3}{4}$

$(0, -3)$ $\left(\frac{3}{4}, 0\right)$

52. $2x + 6y = 9$

Let $x = 0$ Let $y = 0$

$2(0) + 6y = 9$ $2x + 6(0) = 9$

$y = \frac{3}{2}$ $x = \frac{9}{2}$

$\left(0, \frac{3}{2}\right)$ $\left(\frac{9}{2}, 0\right)$

53. $y = 5$

Horizontal line with y-intercept 5.

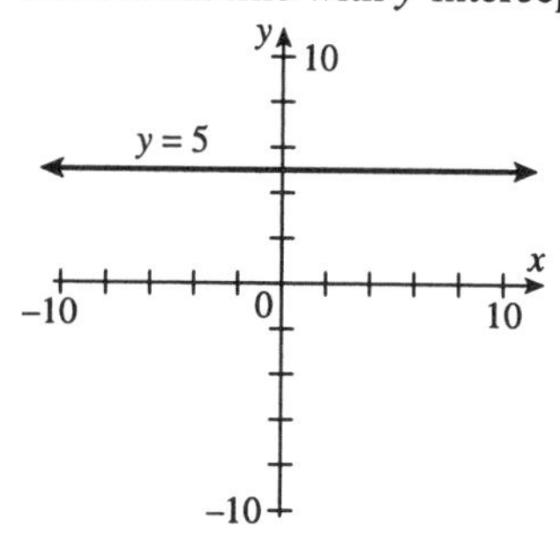

54. $x = -2$

Vertical line with x-intercept −2.

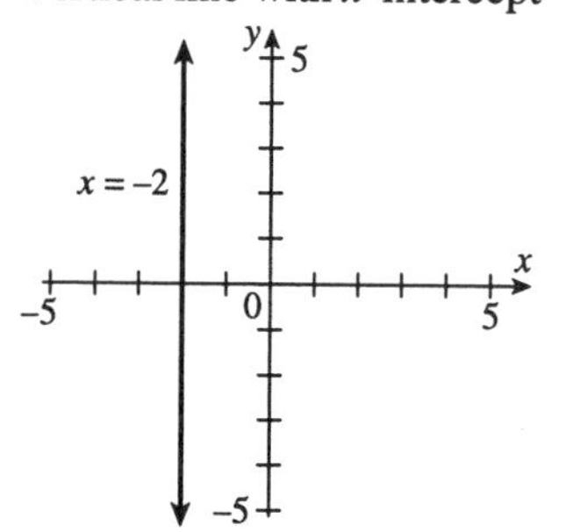

55. $C(x) = 0.3x + 42$

a. $C(150) = 0.3(150) + 42$
$= 45 + 42 = 87$
\$87

b. $m = 0.3$, $b = 42$

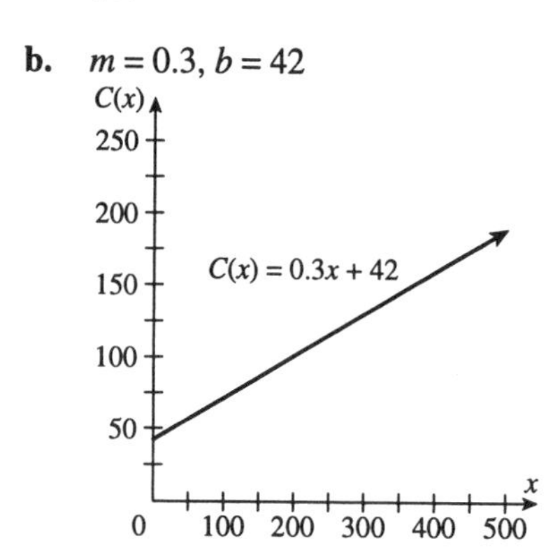

56. $m = \frac{-4-8}{6-2} = \frac{-12}{4} = -3$

57. $m = \frac{13-9}{5-(-3)} = \frac{4}{8} = \frac{1}{2}$

58. $m = \frac{6-(-4)}{-3-(-7)} = \frac{10}{4} = \frac{5}{2}$

59. $m = \frac{7-(-2)}{-5-7} = \frac{9}{-12} = -\frac{3}{4}$

60. $6x - 15y = 20$
$6x - 20 = 15y$
$y = \frac{2}{5}x - \frac{4}{3}$
$m = \frac{2}{5}$ and $b = -\frac{4}{3}$

61. $4x + 14y = 21$
$14y = -4x + 21$
$y = -\frac{2}{7}x + \frac{3}{2}$
$m = -\frac{2}{7},\ b = \frac{3}{2}$

62. $y - 3 = 0$
$y = 3$
Slope = 0

63. $x = -5$
Vertical line
Slope is undefined.

64. ℓ_2

65. ℓ_2

66. ℓ_2

67. ℓ_1

68. $f(x) = -2x + 6$ $g(x) = 2x - 1$
$m = -2$ $m = 2$
Neither; The slopes are not the same and their product is not –1.

69. $-x + 3y = 2$ $6x - 18y = 3$
$y = \frac{1}{3}x + \frac{2}{3}$ $y = \frac{1}{3}x - \frac{1}{6}$
$m = \frac{1}{3},\ b = \frac{2}{3}$ $m = \frac{1}{3},\ b = -\frac{1}{6}$
Parallel, since their slopes are equal and their y-intercepts are different.

70. Horizontal lines have slope = 0.
The y-intercept is –1.
$y = -1$

71. Vertical lines have undefined slope.
The x-intercept is –2.
$x = -2$

72. The slope is undefined.
The x-intercept is –4.
$x = -4$

73. Horizontal line with y-intercept = 5
$y = 5$

74. $y - 5 = 3(x - (-3))$
$y - 5 = 3(x + 3)$
$y - 5 = 3x + 9$
$3x - y = -14$

75. $y - (-2) = 2(x - 5)$
$y + 2 = 2x - 10$
$2x - y = 12$

76. $m = \frac{-2-(-1)}{-4-(-6)} = -\frac{1}{2}$
Now, $y - (-1) = -\frac{1}{2}(x - (-6))$
$2(y + 1) = -(x + 6)$
$2y + 2 = -x - 6$
$x + 2y = -8$

77. $m = \frac{-8-3}{-4-(-5)} = \frac{-11}{1} = -11$
$y - 3 = -11(x - (-5))$
$y - 3 = -11x - 55$
$11x + y = -52$

78. $x = 4$ has undefined slope
A line perpendicular to $x = 4$ has slope = 0. y-intercept at 3
$y = 3$

79. Slope = 0, y-intercept = –5
$y = -5$

80. $y = -\frac{2}{3}x + 4$
$f(x) = -\frac{2}{3}x + 4$

81. $y = -x - 2$
$f(x) = -x - 2$

82. $6x + 3y = 5$
$3y = -6x + 5$
$y = -2x + \frac{5}{3}$, so $m = -2$. Now,
$y - (-6) = -2(x - 2)$
$y + 6 = -2x + 4$
$y = -2x - 2$
$f(x) = -2x - 2$

83. $3x + 2y = 8$
$2y = -3x + 8$
$y = -\frac{3}{2}x + 4$
$m = -\frac{3}{2}$ Now,
$y - (-2) = -\frac{3}{2}(x - (-4))$
$y + 2 = -\frac{3}{2}x - 6$
$y = -\frac{3}{2}x - 8$
$f(x) = -\frac{3}{2}x - 8$

84. $4x + 3y = 5$
$3y = -4x + 5$
$y = -\frac{4}{3}x + \frac{5}{3}$, so $m_{\perp} = -\frac{4}{3}$ and $m = \frac{3}{4}$
Now, $y - (-1) = \frac{3}{4}(x - (-6))$
$4(y + 1) = 3(x + 6)$
$4y + 4 = 3x + 18$
$y = \frac{3}{4}x + \frac{7}{2}$
$f(x) = \frac{3}{4}x + \frac{7}{2}$

85. $2x - 3y = 6$
$3y = 2x - 6$
$y = \frac{2}{3}x - 2$
$m = \frac{2}{3}$, so the slope of the perpendicular line is $-\frac{3}{2}$.
$y - 5 = -\frac{3}{2}(x - (-4))$
$y - 5 = -\frac{3}{2}x - 6$
$y = -\frac{3}{2}x - 1$
$f(x) = -\frac{3}{2}x - 1$

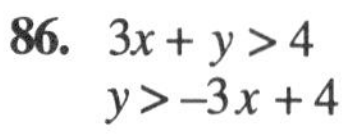

86. $3x + y > 4$
$y > -3x + 4$

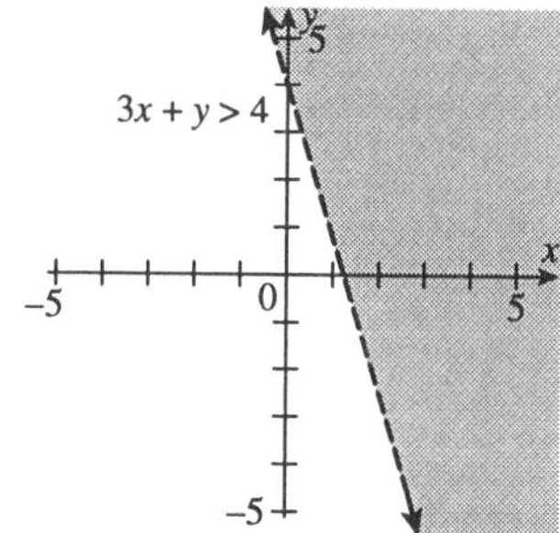

87. $\frac{1}{2}x - y < 2$
$-y < -\frac{1}{2}x + 2$
$y > \frac{1}{2}x - 2$

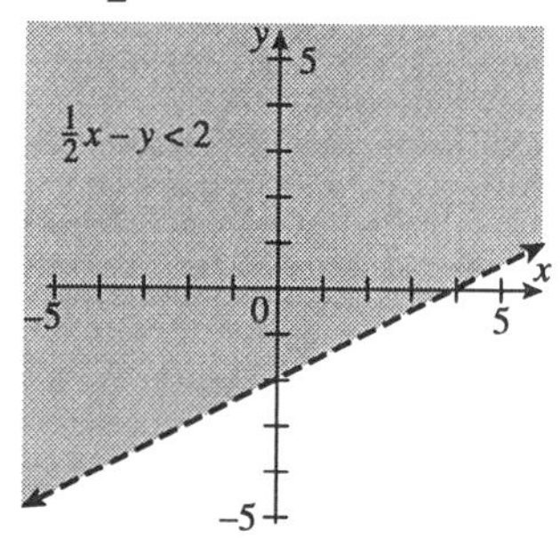

88. $5x - 2y \le 9$
$5x - 9 \le 2y$
$y \ge \frac{5}{2}x - \frac{9}{2}$

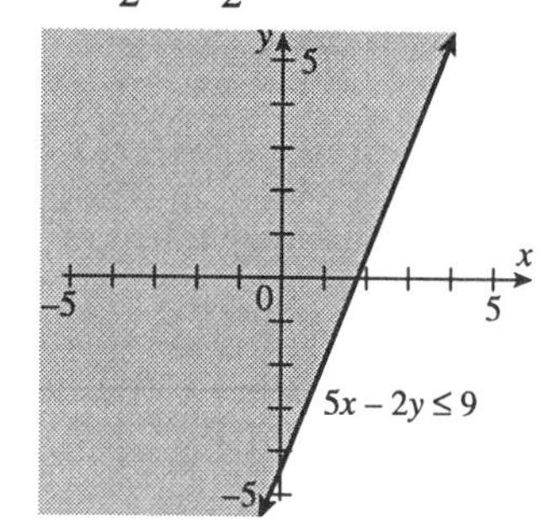

89. $3y \ge x$

$y \ge \frac{1}{3}x$

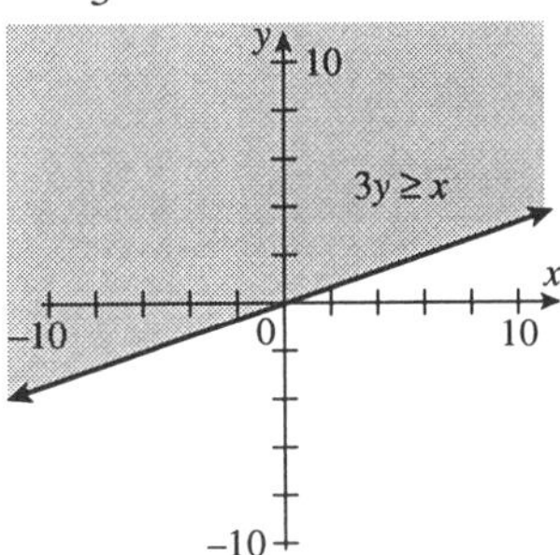

90. $y < 1$

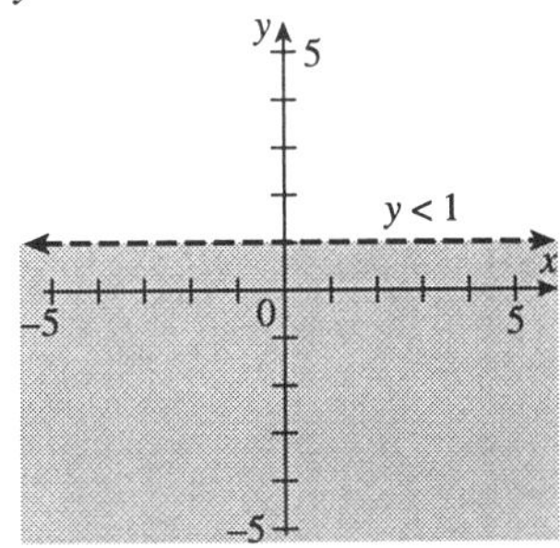

91. $x > -2$

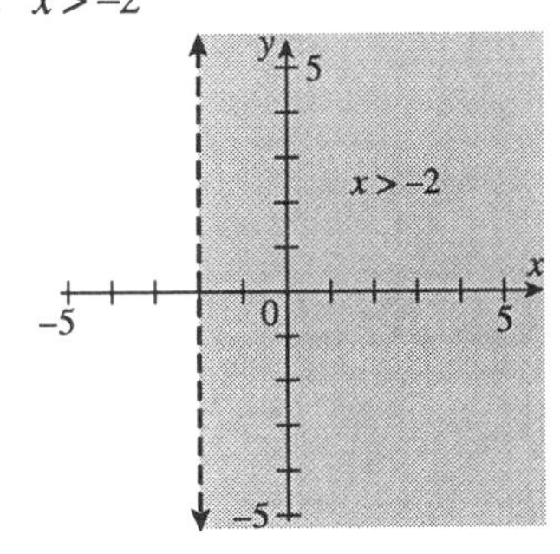

92. $y > 2x + 3$ or $x \le -3$

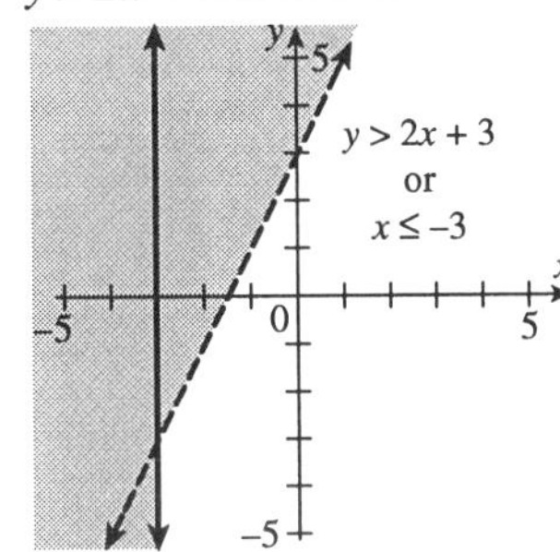

93. $2x < 3y + 8$ and $y \ge -2$

$y > \frac{2}{3}x - \frac{8}{3}$ and $y \ge -2$

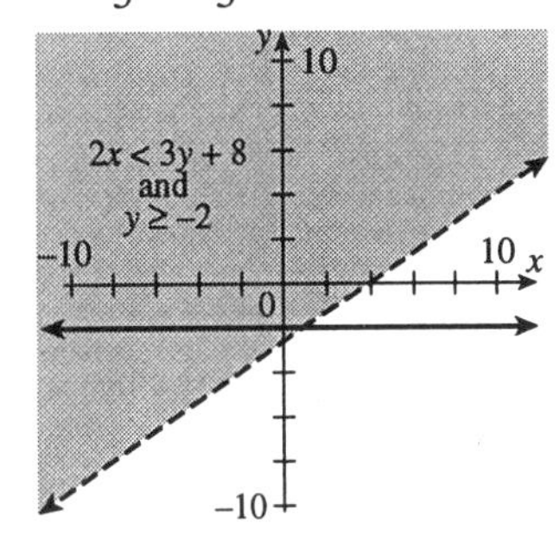

Chapter 3 - Test

1. $A(6, -2)$ in quadrant IV

$B(4, 0)$ on the x-axis

$C(-1, 6)$ in quadrant II

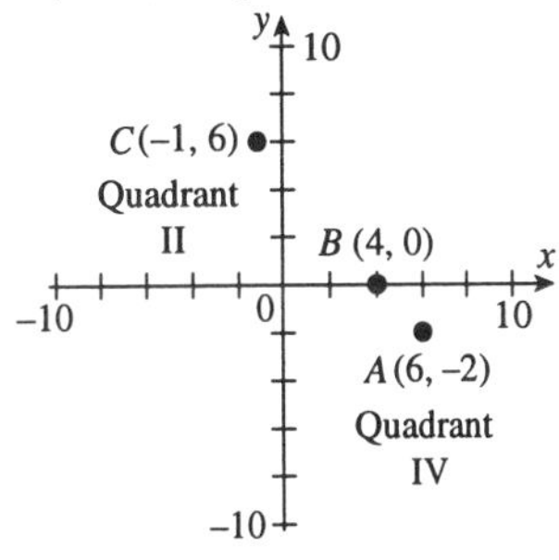

2. $2y - 3x = 12$

$x = -6$: $2y - 3(-6) = 12$

$2y + 18 = 12$

$2y = -6$

$y = -3$

$(-6, -3)$

3. $2x - 3y = -6$

$3y = 2x + 6$

$y = \frac{2}{3}x + 2$

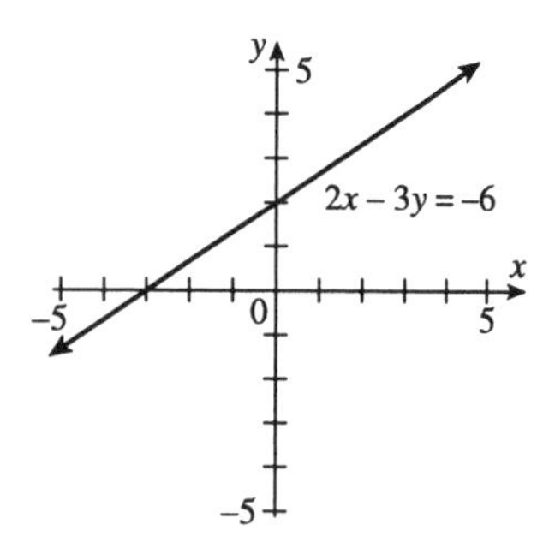

4. $4x+6y=7$
$6y=-4x+7$
$y=-\frac{2}{3}x+\frac{7}{6}$

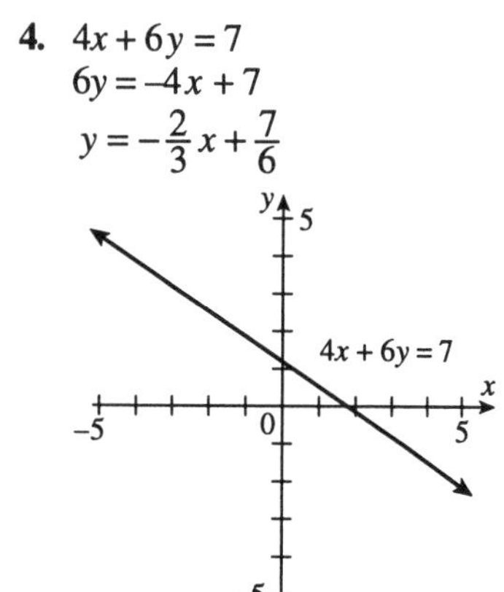

5. $y=\frac{2}{3}x$

6. $y=-3$

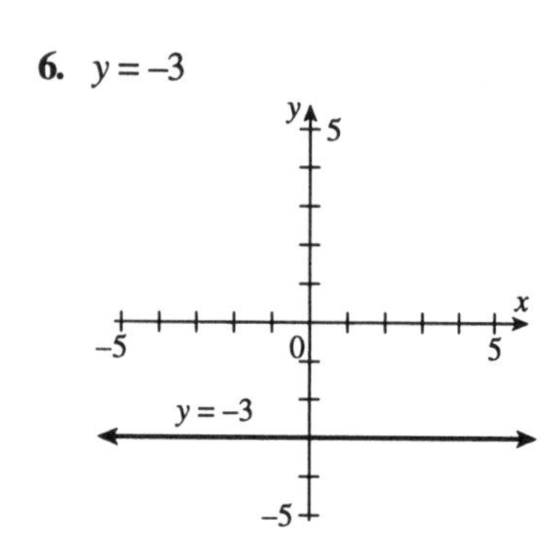

7. $m=\frac{10-(-8)}{-7-5}=\frac{18}{-12}=-\frac{3}{2}$

8. $3x+12y=8$
$12y=-3x+8$
$y=-\frac{1}{4}x+\frac{2}{3}$
$m=-\frac{1}{4}$ and $b=\frac{2}{3}$

9. $f(x)=(x-1)^2$

x	$f(x)$
−2	9
−1	4
0	1
1	0
2	1
3	4
4	9

10. $g(x)=|x|+2$

x	$g(x)$
−3	5
−2	4
−1	3
0	2
1	3
2	4
3	5

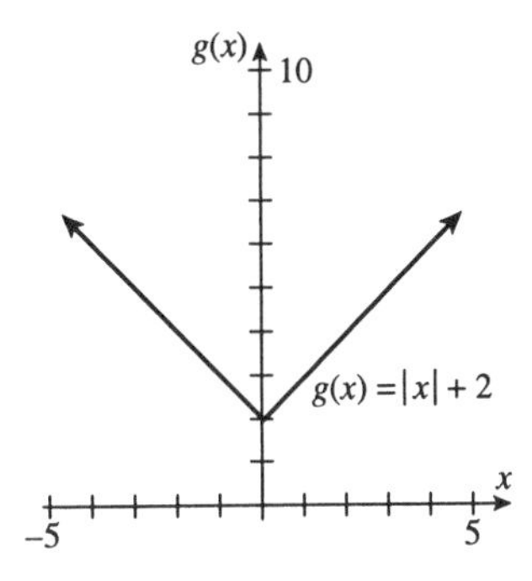

11. $y = -8$

12. $x = -4$

13. $y = -2$

14. $y - (-1) = -3(x - 4)$
$y + 1 = -3x + 12$
$3x + y = 11$

15. $y = 5x + (-2)$
$5x - y = 2$

16. $m = \frac{-3 - (-2)}{6 - 4} = -\frac{1}{2}$
Now, $y - (-2) = -\frac{1}{2}(x - 4)$
$2(y + 2) = -(x - 4)$
$2y + 4 = -x + 4$
$y = -\frac{1}{2}x$
$f(x) = -\frac{1}{2}x$

17. $3x - y = 4$
$y = 3x - 4$
so $m_{\perp} = 3$ and $m = -\frac{1}{3}$
Now, $y - 2 = -\frac{1}{3}(x - (-1))$
$3(y - 2) = -(x + 1)$
$3y - 6 = -x - 1$
$y = -\frac{1}{3}x + \frac{5}{3}$
$f(x) = -\frac{1}{3}x + \frac{5}{3}$

18. $2y + x = 3$
$2y = -x + 3$
$y = -\frac{1}{2}x + \frac{3}{2}$, so $m = -\frac{1}{2}$
Now, $y - (-2) = -\frac{1}{2}(x - 3)$
$2(y + 2) = -(x - 3)$
$2y + 4 = -x + 3$
$y = -\frac{1}{2}x - \frac{1}{2}$
$f(x) = -\frac{1}{2}x - \frac{1}{2}$

19. $2x - 5y = 8$
$5y = 2x - 8$
$y = \frac{2}{5}x - \frac{8}{5}$, so $m_1 = \frac{2}{5}$,
$m_2 = \frac{-1 - 4}{-1 - 1} = \frac{-5}{-2} = \frac{5}{2}$
Therefore, lines L_1 and L_2 are neither parallel nor perpendicular.

20. $x \leq -4$

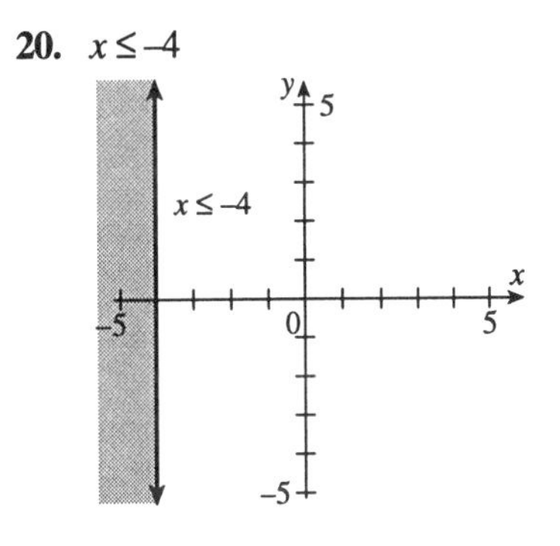

21. $y > -2$

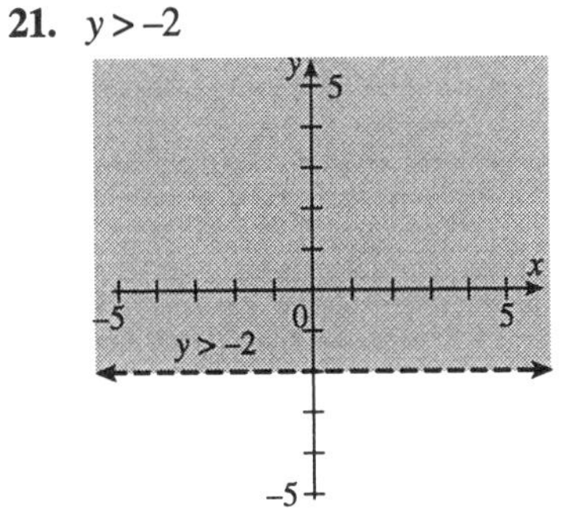

22. $2x - y > 5$
$y < 2x - 5$

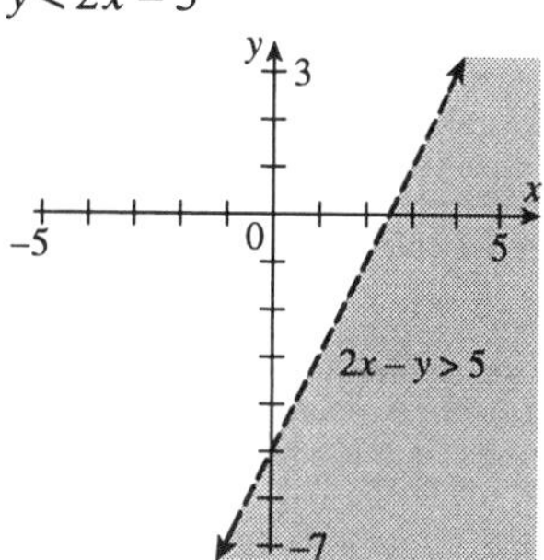

23. $2x + 4y < 6$ and $y \le -4$
$4y < -2x + 6$ and $y \le -4$
$y < -\frac{1}{2}x + \frac{3}{2}$ and $y \le -4$

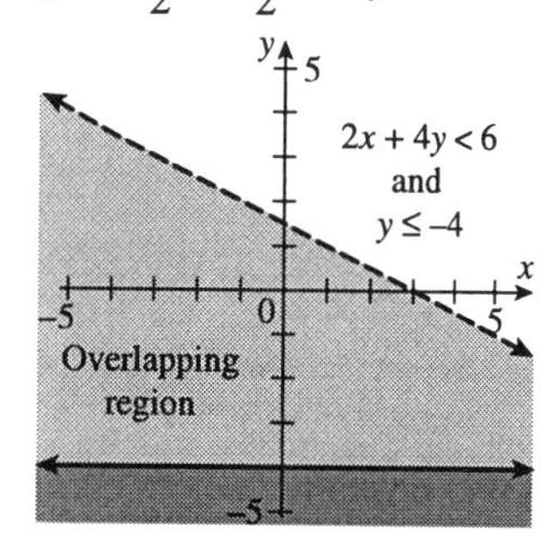

24. Domain: $(-\infty, \infty)$
Range: $\{5\}$
Function

25. Domain: $\{-2\}$
Range: $(-\infty, \infty)$
Not a function

26. Domain: $(-\infty, \infty)$
Range: $[0, \infty)$
Function

27. Domain: $(-\infty, \infty)$
Range: $(-\infty, \infty)$
Function

28. $f(x) = 708x + 13{,}570$

a. $f(5) = 708(5) + 13{,}570 = \$17{,}110$

b. $f(15) = 708(15) + 13{,}570 = \$24{,}190$

c. $30{,}000 = 708x + 13{,}570$
$x = \frac{30{,}000 - 13{,}570}{708}$
$x = 23.21$
$1985 + 24 = 2009$

Chapter 3 - Cumulative Review

1. a. $11 + 2 - 7 = 6$

b. $-5 - 4 + 2 = -7$

2. $-6x - 1 + 5x = 3$
$-x = 4$
$x = -4$
$\{-4\}$

3. Let x = smaller number
Then $72 - x$ = larger number
$72 - x = 2x + 3$
$69 = 3x$
$23 = x$
$72 - 23 = 49$
23 and 49

4. $3y - 2x = 7$
$3y = 2x + 7$
$y = \frac{2}{3}x + \frac{7}{3}$

5. Let y = income per month
x = amount of sales
$y = 0.2x + 600$
$1500 = 0.2x + 600$
$900 = 0.2x$
$4500 = x$
$4500

6. $-1 \le \frac{2x}{3} + 5 \le 2$
$-6 \le \frac{2}{3}x \le -3$
$-18 \le 2x \le -9$
$-9 \le x \le -\frac{9}{2}$
$\left[-9, -\frac{9}{2}\right]$

7. $\left|\frac{x}{2}-1\right|=11$

$\frac{x}{2}-1=11$ or $\frac{x}{2}-1=-11$

$\frac{x}{2}=12$ or $\frac{x}{2}=-10$

$x=24$ or $x=-20$

$\{-20, 24\}$

8. $|y-3|>7$

$y-3>7$ or $y-3<-7$

$y>10$ or $y<-4$

$(-\infty, 4) \cup (10, \infty)$

9. $\left|\frac{2(x+1)}{3}\right| \le 0$

An absolute value cannot be negative.

$\frac{2(x+1)}{3}=0$

$x+1=0$

$x=-1$

$\{-1\}$

10.

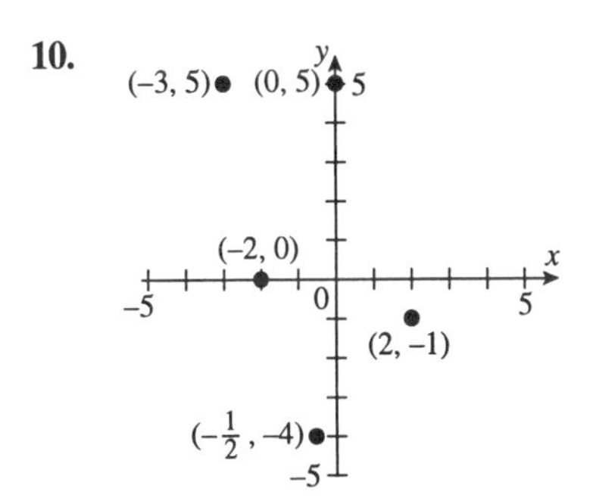

a. (2, −1) lies in quadrant IV

b. (0, 5) is not in any quadrant

c. (−3, 5) lies in quadrant II

d. (−2, 0) is not in any quadrant

e. $\left(-\frac{1}{2}, -4\right)$ lies in quadrant III

11. $y=|x|$

x	y
−3	3
−2	2
−1	1
0	0
1	1
2	2
3	3

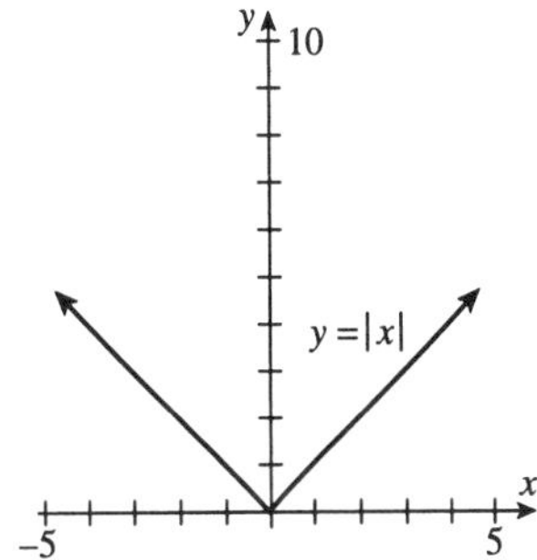

12. a. Domain: {2, 0, 3}
Range: {3, 4, −1}

b. Domain: {−4, −3, −2, −1, 0, 1, 2, 3}
Range: {1}

c. Domain: {Eric, Miami, Escondido, Gary, Waco}
Range: {109, 359, 104, 117}

d. Domain: [−3, 5]
Range: [−2, 4]

13. a. Yes

b. Yes

c. No

d. Yes

e. No

f. No

14. $x - 3y = 6$

Let $x = 0$	Let $y = 0$
$0 - 3y = 6$	$x - 3(0) = 6$
$y = -2$	$x = 6$
$(0, -2)$	$(6, 0)$

15. $y = -3$
Horizontal line with y-intercept -3.

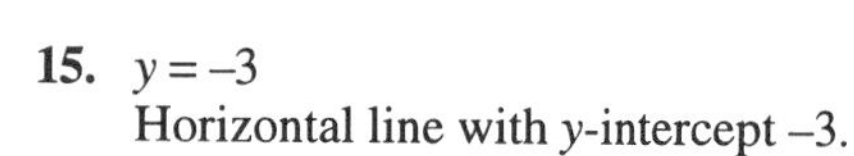

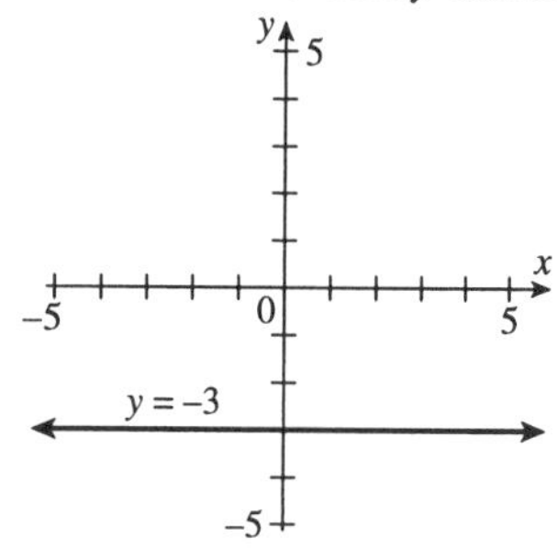

16. $m = \dfrac{6-(-7)}{-3-5} = \dfrac{13}{-8} = -\dfrac{13}{8}$

17. $3x - 4y = 4$
$-4y = -3x + 4$
$y = \dfrac{3}{4}x - 1$
$m = \dfrac{3}{4},\ b = -1$

18. $y - (-5) = -3(x - 1)$
$y + 5 = -3x + 3$
$3x + y = -2$

19. Slope = 0
y-intercept = 3
$y = 3$

20. $3x \geq y$

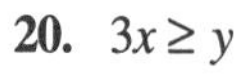

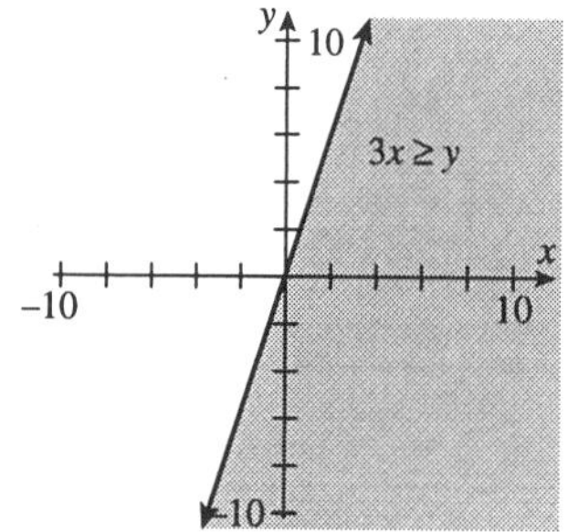

21. $x + \frac{1}{2}y \geq -4$ or $y \leq -2$

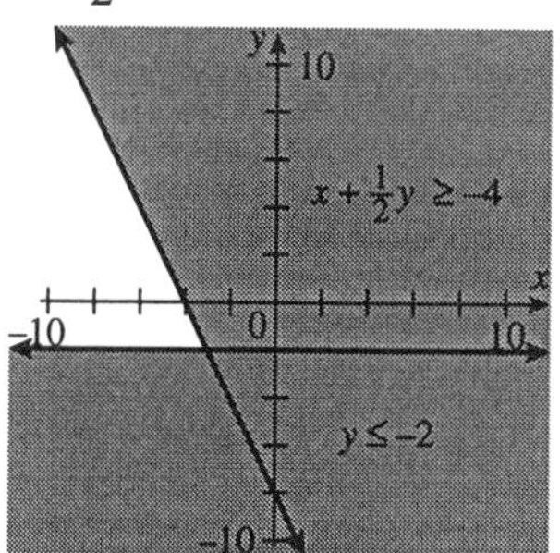

Chapter 4

Section 4.1 Graphing Calculator Explorations

1. $y = -1.65x + 3.65$
$y = 4.56x - 9.44$
$\{(2.11, .17)\}$

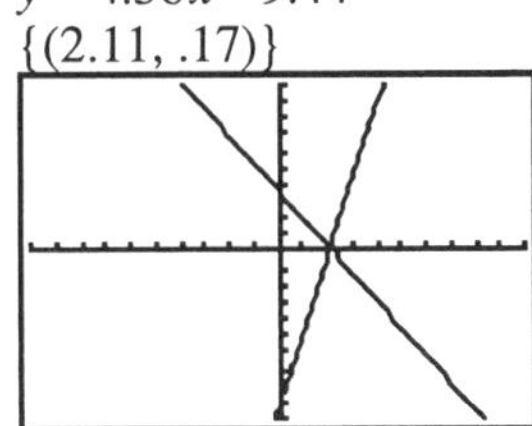

3. $2.33x - 4.72y = 10.61$
$5.86x + 6.22y = -8.89$
or
$y = \frac{2.33x - 10.61}{4.72}$
$y = \frac{-5.86x - 8.89}{6.22}$
$\{(0.57, -1.97)\}$

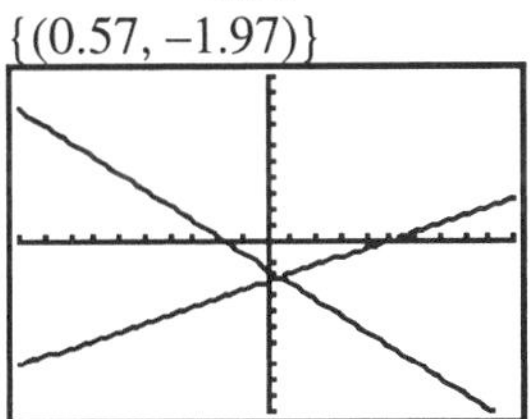

Exercise Set 4.1

1. $\begin{cases} x + y = 1 \\ x - 2y = 4 \end{cases}$
$\{(2, -1)\}$

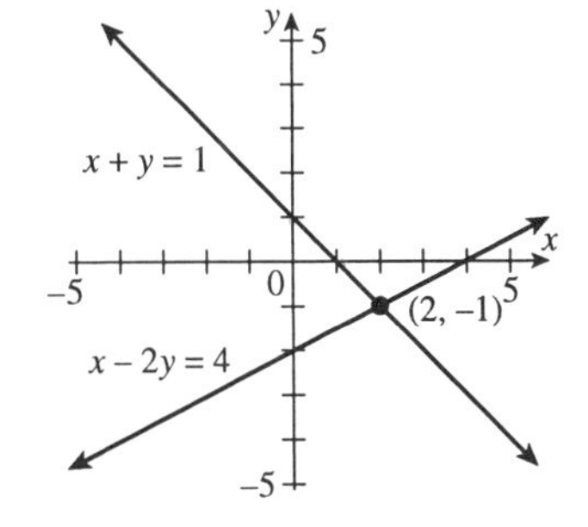

3. $\begin{cases} 2y - 4 = 0 \\ x + 2y = 5 \end{cases}$
$\{(1, 2)\}$

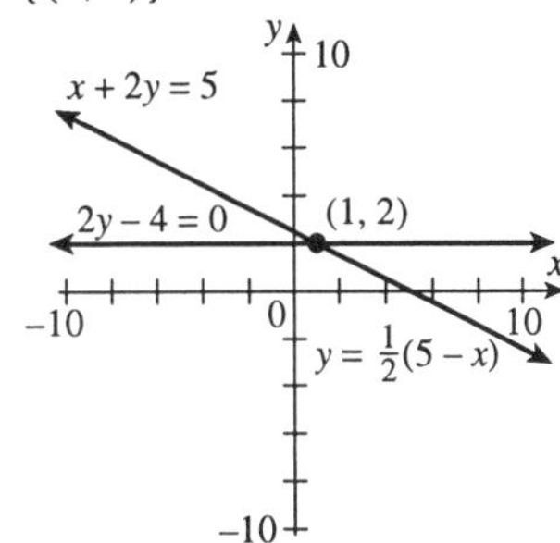

5. $\begin{cases} 3x - y = 4 \\ 6x - 2y = 4 \end{cases}$
$\{\ \}$

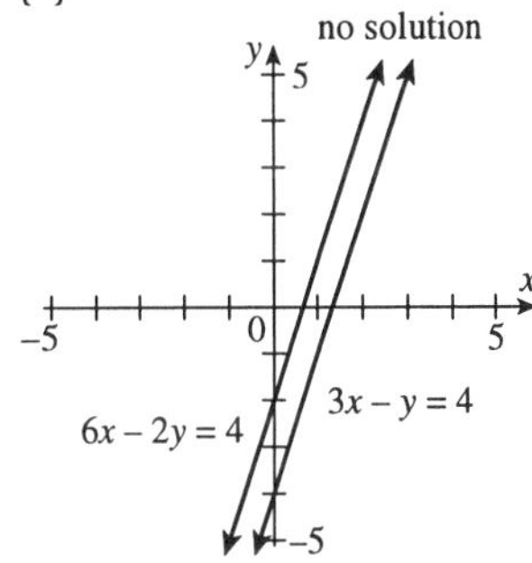

7. No

9. $\begin{cases} x + y = 10 \\ y = 4x \end{cases}$
Substituting we get:
$x + 4x = 10$
$5x = 10$
$x = 2$
Substituting back we get:
$y = 4(2)$
$y = 8$
$\{(2, 8)\}$

11. $\begin{cases} 4x - y = 9 \\ 2x + 3y = -27 \end{cases}$

Solve equation 1 for y:
$4x - 9 = y$
$y = 4x - 9$
Now substitute.
$2x + 3(4x - 9) = -27$
$2x + 12x - 27 = -27$
$14x = 0$
$x = 0$
Substitute back.
$y = 4(0) - 9 = -9$
$\{(0, -9)\}$

13. $\begin{cases} \frac{1}{2}x + \frac{3}{4}y = -\frac{1}{4} \\ \frac{3}{4}x - \frac{1}{4}y = 1 \end{cases}$

$\to \begin{cases} 2x + 3y = -1 \\ 3x - y = 4 \end{cases}$

Now solve equation 2 for y.
$3x - 4 = y$
$y = 3x - 4$
Substitute.
$2x + 3(3x - 4) = -1$
$2x + 9x - 12 = -1$
$11x = 11$
$x = 1$
Substitute back.
$y = 3(1) - 4$
$y = -1$
$\{(1, -1)\}$

15. $\begin{cases} \frac{x}{3} + y = \frac{4}{3} \\ -x + 2y = 11 \end{cases}$

$\to \begin{cases} x + 3y = 4 \\ x + 2y = 11 \end{cases}$

Solve equation 2 for x.
$2y - 11 = x$
$x = 2y - 11$
Substitute.
$2y - 11 + 3y = 4$
$5y = 15$
$y = 3$
Substitute back.
$x = 2(3) - 11$
$x = -5$
$\{(-5, 3)\}$

17. $\begin{cases} 2x - 4y = 0 \\ x + 2y = 5 \end{cases}$

$\to \begin{cases} x - 2y = 0 \\ x + 2y = 5 \end{cases}$

Add.
$2x = 5$
$x = \frac{5}{2}$
Substitute back.
$\frac{5}{2} - 2y = 0$
$\frac{5}{2} = 2y$
$y = \frac{5}{4}$
$\left\{\left(\frac{5}{2}, \frac{5}{4}\right)\right\}$

19. $\begin{cases} 5x + 2y = 1 \\ x - 3y = 7 \end{cases}$

$\to \begin{cases} 5x + 2y = 1 \\ -5x + 15y = -35 \end{cases}$

Add.
$17y = -34$
$y = -2$
Substitute back.
$x - 3(-2) = 7$
$x + 6 = 7$
$x = 1$
$\{(1, -2)\}$

21. $\begin{cases} 5x - 2y = 27 \\ -3x + 5y = 18 \end{cases}$

$\to \begin{cases} 15x - 6y = 81 \\ -15x + 25y = 90 \end{cases}$

Add.
$19y = 171$
$y = 9$
Substitute back.
$5x - 2(9) = 27$
$5x - 18 = 27$
$5x = 45$
$x = 9$
$\{(9, 9)\}$

23. $\begin{cases} 3x-5y=11 \\ 2x-6y=2 \end{cases}$

$\rightarrow \begin{cases} 3x-5y=11 \\ x-3y=1 \end{cases}$

$\rightarrow \begin{cases} 3x-5y=11 \\ -3x+9y=-3 \end{cases}$

Add.
$4y=8$
$y=2$
Substitute back.
$x-3(2)=1$
$x-6=1$
$x=7$
$\{(7, 2)\}$

25. $\begin{cases} x-2y=4 \\ 2x-4y=4 \end{cases}$

$\rightarrow \begin{cases} 2x-4y=8 \\ 2x-4y=4 \end{cases}$

Inconsistent system.
$\{\ \}$

27. $\begin{cases} 3x+y=1 \\ 2y=2-6x \end{cases}$

$\rightarrow \begin{cases} 3x+\ y=1 \\ 6x+2y=2 \end{cases}$

$\rightarrow \begin{cases} 6x+2y=2 \\ 6x+2y=2 \end{cases}$

Dependent system.
$\{(x, y)|3x+y=1\}$

29. Answers may vary.

31. $\begin{cases} 2x+5y=8 \\ 6x+\ y=10 \end{cases}$

$\rightarrow \begin{cases} -6x-15y=-24 \\ 6x+\ y=\ 10 \end{cases}$

Add.
$-14y=-14$
$y=1$
Substitute back.
$6x+1=10$
$6x=9$
$x=\frac{9}{6}=\frac{3}{2}$
$\left\{\left(\frac{3}{2}, 1\right)\right\}$

33. $\begin{cases} x+\ y=1 \\ x-2y=4 \end{cases}$

Subtract.
$3y=-3$
$y=-1$
Substitute back.
$x+(-1)=1$
$x=2$
$\{(2, -1)\}$

35. $\begin{cases} \frac{1}{3}x+\ y=\frac{4}{3} \\ -\frac{1}{4}x-\frac{1}{2}y=-\frac{1}{4} \end{cases}$

$\rightarrow \begin{cases} x+3y=4 \\ -x-2y=-1 \end{cases}$

Add.
$y=3$
Substitute back.
$x+3(3)=4$
$x+9=4$
$x=-5$
$\{(-5, 3)\}$

37. $\begin{cases} 2x+6y=8 \\ 3x+9y=12 \end{cases}$

$\rightarrow \begin{cases} x+3y=4 \\ x+3y=4 \end{cases}$

Dependent system.
$\{(x, y)|x+3y=4\}$

39. $\begin{cases} 4x+2y=5 \\ 2x+\ y=-1 \end{cases}$

$\rightarrow \begin{cases} 4x+2y=5 \\ 4x+2y=-2 \end{cases}$

Inconsistent system.
$\{\ \}$

41. $\begin{cases} 10y - 2x = 1 \\ 5y = 4 - 6x \end{cases}$

$\begin{cases} 10y = 2x + 1 \\ 10y = -12x + 8 \end{cases}$

$2x + 1 = -12x + 8$
$14x = 7$
$x = \frac{7}{14} = \frac{1}{2}$
Substitute back.
$10y - 2\left(\frac{1}{2}\right) = 1$
$10y - 1 = 1$
$10y = 2$
$y = \frac{2}{10} = \frac{1}{5}$
$\left\{\left(\frac{1}{2}, \frac{1}{5}\right)\right\}$

43. $\begin{cases} \frac{3}{4}x + \frac{5}{2}y = 11 \\ \frac{1}{16}x - \frac{3}{4}y = -1 \end{cases}$

$\rightarrow \begin{cases} 3x + 10y = 44 \\ x - 12y = -16 \end{cases}$

$\rightarrow \begin{cases} 3x + 10y = 44 \\ -3x + 36y = 48 \end{cases}$

Add.
$46y = 92$
$y = 2$
$x - 12(2) = -16$
$x - 24 = -16$
$x = 8$
$\{(8, 2)\}$

45. $\begin{cases} x = 3y + 2 \\ 5x - 15y = 10 \end{cases}$

Substitute.
$5(3y + 2) - 15y = 10$
$3y + 2 - 3y = 2$
$2 = 2$
The system is dependent.
$\{(x, y) | x = 3y + 2\}$

47. $\begin{cases} 2x - y = -1 \\ y = -2x \end{cases}$

Substitute.
$2x - (-2x) = -1$
$4x = -1$
$x = -\frac{1}{4}$
$y = -2\left(-\frac{1}{4}\right) = \frac{1}{2}$
$\left\{\left(-\frac{1}{4}, \frac{1}{2}\right)\right\}$

49. $\begin{cases} 2x = 6 \\ y = 5 - x \end{cases}$

The first equation yields $x = 3$.
Substitute back.
$y = 5 - 3 = 2$
$\{(3, 2)\}$

51. $\begin{cases} \frac{x+5}{2} = \frac{6-4y}{3} \\ \frac{3x}{5} = \frac{21-7y}{10} \end{cases}$

$\rightarrow \begin{cases} 3x + 15 = 12 - 8y \\ 30x = 105 - 35y \end{cases}$

$\rightarrow \begin{cases} 3x + 8y = -3 \\ 6x = 21 - 7y \end{cases}$

$\rightarrow \begin{cases} 6x + 16y = -6 \\ 6x + 7y = 21 \end{cases}$

Subtract.
$9y = -27$
$y = -3$
$3x + 8(-3) = -3$
$3x - 24 = -3$
$3x = 21$
$x = 7$
$\{(7, -3)\}$

53. $\begin{cases} 4x - 7y = 7 \\ 12x - 21y = 24 \end{cases}$

$\rightarrow \begin{cases} 4x - 7y = 7 \\ 4x - 7y = 8 \end{cases}$

Inconsistent system.
{ }

55. $\begin{cases} \frac{2}{3}x - \frac{3}{4}y = -1 \\ -\frac{1}{6}x + \frac{3}{8}y = 1 \end{cases}$

$\rightarrow \begin{cases} 8x - 9y = -12 \\ -4x + 9y = 24 \end{cases}$

Add.
$4x = 12$
$x = 3$
Substitute back.
$-4(3) + 9y = 24$
$-12 + 9y = 24$
$9y = 36$
$y = 4$
$\{(3, 4)\}$

57. $\begin{cases} 0.7x - 0.2y = -1.6 \\ 0.2x - y = -1.4 \end{cases}$

$\rightarrow \begin{cases} 7x - 2y = -16 \\ x - 5y = -7 \end{cases}$

$\rightarrow \begin{cases} 7x - 2y = -16 \\ 7x - 35y = -49 \end{cases}$

Subtract.
$33y = 33$
$y = 1$
$x - 5(1) = -7$
$x - 5 = -7$
$x = -2$
$\{(-2, 1)\}$

59. $\begin{cases} 4x - 1.5y = 10.2 \\ 2x + 7.8y = -25.68 \end{cases}$

$\rightarrow \begin{cases} 4x - 1.5y = 10.2 \\ 4x + 15.6y = -51.36 \end{cases}$

Subtract.
$-17.1y = 61.56$
$y = -3.6$
$4x - 1.5(-3.6) = 10.2$
$4x + 5.4 = 10.2$
$4x = 4.8$
$x = 1.2$
$\{(1.2, -3.6)\}$

61. Point of intersection (5, 21)
5000 ties; $21

63. Supply is greater than demand.

65. $\begin{cases} y = 2.5x \\ y = 0.9x + 3000 \end{cases}$

Substitute.
$2.5x = 0.9x + 3000$
$1.6x = 3000$
$x = 1875$
$y = 2.5(1875) = 4687.5$
$\{(1875, 4687.5)\}$

67. Makes money because revenue is greater than cost at $x = 2000$.

69. For values of $x > 1875$ because the x-value at the intersection is 1875.

71. $\begin{cases} y = 67x + 971 \\ y = 44x + 1285 \end{cases}$

$67x + 971 = 44x + 1285$
$23x = 314$
$x = 13.65$
$1990 + 13 = 2003$

73. $3x - 4y + 2z = 5$
$3(1) - 4(2) + 2(5) = 3 - 8 + 10 = 5$
True

75. $-x - 5y + 3z = 15$
$-0 - 5(-1) + 3(5) = 0 + 5 + 15 = 20$
False

77.
$$\begin{array}{r} 3x + 2y - 5z = 10 \\ + -3x + 4y + z = 15 \\ \hline 6y - 4z = 25 \end{array}$$

79.
$$\begin{array}{r} 10x + 5y + 6z = 14 \\ + -9x + 5y - 6z = -12 \\ \hline x + 10y = 2 \end{array}$$

81. $\begin{cases} \frac{1}{x}+y=12 \\ \frac{3}{x}-y=4 \end{cases}$

Add.

$\frac{4}{x}=16$

$4=16x$

$x=\frac{4}{16}=\frac{1}{4}$ so $\frac{1}{x}=4$

Substitute back.

$4+y=12$

$y=8$

$\left\{\left(\frac{1}{4}, 8\right)\right\}$

83. $\begin{cases} \frac{1}{x}+\frac{1}{y}=5 \\ \frac{1}{x}-\frac{1}{y}=1 \end{cases}$

Add.

$\frac{2}{x}=6$

$2=6x$

$x=\frac{2}{6}=\frac{1}{3}$ so $\frac{1}{x}=3$

Substitute back.

$3-\frac{1}{y}=1$

$2=\frac{1}{y}$

$y=\frac{1}{2}$

$\left\{\left(\frac{1}{3}, \frac{1}{2}\right)\right\}$

85. $\begin{cases} \frac{2}{x}+\frac{3}{y}=-1 \\ \frac{3}{x}-\frac{2}{y}=18 \end{cases}$

$\rightarrow \begin{cases} \frac{6}{x}+\frac{9}{y}=-3 \\ \frac{6}{x}-\frac{4}{y}=36 \end{cases}$

Subtract.

$\frac{13}{y}=-39$

$13=-39y$

$y=-\frac{13}{39}=-\frac{1}{3}$ so $\frac{1}{y}=-3$

Substitute back.

$\frac{2}{x}+3(-3)=-1$

$\frac{2}{x}-9=-1$

$\frac{2}{x}=8$

$2=8x$

$x=\frac{2}{8}=\frac{1}{4}$

$\left\{\left(\frac{1}{4}, -\frac{1}{3}\right)\right\}$

87. $\begin{cases} \frac{2}{x}-\frac{4}{y}=5 \\ \frac{1}{x}-\frac{2}{y}=\frac{3}{2} \end{cases}$

$\rightarrow \begin{cases} \frac{2}{x}-\frac{4}{y}=5 \\ \frac{2}{x}-\frac{4}{y}=3 \end{cases}$

Inconsistent system.

{ }

Exercise Set 4.2

1. $\begin{cases} x+y=3 \\ 2y=10 \\ 3x+2y-3z=1 \end{cases}$

The second equation yields $y=5$.

Substitute.

$x+5=3$

$x=-2$

Substitute back.

$3(-2)+2(5)-3z=1$

$-6+10-3z=1$

$4-3z=1$

$3=3z$

$z=1$

$\{(-2, 5, 1)\}$

3. $\begin{cases} 2x+2y+z=1 \\ -x+y+2z=3 \\ x+2y+4z=0 \end{cases}$

Add equations 2 and 3.

$3y+6z=3$ or $y+2z=1$

Add twice equation 2 to equation 1.
$4y+5z=7$
$$\begin{cases} y+2z=1 \\ 4y+5z=7 \end{cases}$$
$$\rightarrow \begin{cases} 4y+8z=4 \\ 4y+5z=7 \end{cases}$$
Subtract.
$3z=-3$
$z=-1$
Substitute back.
$y+2(-1)=1$
$y-2=1$
$y=3$
Substitute back.
$x+2(3)+4(-1)=0$
$x+6-4=0$
$x+2=0$
$x=-2$
$\{(-2, 3, -1)\}$

5. $$\begin{cases} x-2y+z=-5 \\ -3x+6y-3z=15 \\ 2x-4y+2z=-10 \end{cases}$$
$$\rightarrow \begin{cases} x-2y+z=-5 \\ x-2y+z=-5 \\ x-2y+z=-5 \end{cases}$$
Dependent system.
$\{(x, y, z) | x-2y+z=-5\}$

7. $$\begin{cases} 4x-y+2z=5 \\ 2y+z=4 \\ 4x+y+3z=10 \end{cases}$$
Subtract equation 1 from equation 3.
$$\begin{cases} 2y+z=5 \\ 2y+z=4 \end{cases}$$
Inconsistent system.
{ }

9. Answers may vary.
One possibility is
$$\begin{cases} 3x=-3 \\ 2x+4y=6 \\ x-3y+z=-11 \end{cases}$$

11. $$\begin{cases} x+5z=0 \\ 5x+y=0 \\ y-3z=0 \end{cases}$$
Subtract equation 3 from equation 2.
$$\begin{cases} 5x+3z=0 \\ x+5z=0 \end{cases}$$
$$\rightarrow \begin{cases} 5x+3z=0 \\ 5x+25z=0 \end{cases}$$
Subtract.
$-22z=0$
$z=0$
Substitute back.
$5x+3(0)=0$
$5x=0$
$x=0$
Substitute back.
$5(0)+y=0$
$y=0$
$\{(0, 0, 0)\}$

13. $$\begin{cases} 6x-5z=17 \\ 5x-y+3z=-1 \\ 2x+y=-41 \end{cases}$$
Add equations 2 and 3.
$$\begin{cases} 7x+3z=-42 \\ 6x-5z=17 \end{cases}$$
$$\rightarrow \begin{cases} 35x+15z=-210 \\ 18x-15z=51 \end{cases}$$
Add.
$53x=-159$
$x=-3$
Substitute back.
$6(-3)-5z=17$
$18-5z=17$
$-5z=35$
$z=-7$
Substitute back.
$2(-3)+y=-41$
$-6+y=-41$
$y=-35$
$\{(-3, -35, -7)\}$

15. $\begin{cases} x+\ y+\ z=8 \\ 2x-\ y-\ z=10 \\ x-2y-3z=22 \end{cases}$

Add equations 1 and 2.
$3x = 18$ or $x = 6$
Add twice equation 1 to equation 3.
$3x - z = 38$
$3(6) - z = 38$
$18 - z = 38$
$-z = 20$
$z = -20$
Substitute back.
$6 + y + (-20) = 8$
$y - 14 = 8$
$y = 22$
$\{(6, 22, -20)\}$

17. $\begin{cases} x+2y-\ z=5 \\ 6x+\ y+\ z=7 \\ 2x+4y-2z=5 \end{cases}$

Add equations 1 and 2.
$7x + 3y = 12$
Add twice equation 2 to equation 3.
$\begin{cases} 14x+6y=19 \\ 7x+3y=12 \end{cases}$
$\rightarrow \begin{cases} 14x+6y=19 \\ 14x+6y=24 \end{cases}$
Inconsistent system.
$\{\ \}$

19. $\begin{cases} 2x-3y+\ z=2 \\ x-5y+5z=3 \\ 3x+\ y-3z=5 \end{cases}$

Add –2 times equation 2 to equation 1.
$7y - 9z = -4$
Add –3 times equation 2 to equation 3.
$16y - 18z = -4$
We now have the system:
$\begin{cases} 7y-\ 9z=-4 \\ 16y-18z=-4 \end{cases}$
$\begin{cases} -14y+18z=\ 8 \\ 16y-18z=-4 \end{cases}$
Add.
$2y = 4$
$y = 2$
Substitute back.
$7(2) - 9z = -4$
$-9z = -18$
$z = 2$
Substitute back.
$2x - 3(2) + 2 = 2$
$x = 3$
$\{(3, 2, 2)\}$

21. $\begin{cases} -2x-4y+\ 6z=-8 \\ x+2y-\ 3z=\ 4 \\ 4x+8y-12z=16 \end{cases}$

Add 2 times equation 2 to equation 1.
$0 = 0$
Add –4 times equation 2 to equation 3.
$0 = 0$
The system is dependent.
$\{(x, y, z) | x + 2y - 3z = 4\}$

23. $\begin{cases} 2x+2y-3z=\ 1 \\ y+2z=-14 \\ 3x-2y\quad\ =\ -1 \end{cases}$

Add equations 1 and 3.
$5x - 3z = 0$
Add twice equation 2 to equation 3.
$3x + 4z = -29$
$\begin{cases} 5x-3z=\ 0 \\ 3x+4z=-29 \end{cases}$
$\rightarrow \begin{cases} 20x-12z=\ 0 \\ 9x+12z=-87 \end{cases}$
Add.
$29x = -89$
$x = -3$
Substitute back.
$5(-3) - 3z = 0$
$3z = -15$
$z = -5$
Substitute back.
$y + 2(-5) = -14$
$y - 10 = -14$
$y = -4$
$\{(-3, -4, -5)\}$

25. $\begin{cases} \frac{3}{4}x - \frac{1}{3}y + \frac{1}{2}z = 9 \\ \frac{1}{6}x + \frac{1}{3}y - \frac{1}{2}z = 2 \\ \frac{1}{2}x - \quad y + \frac{1}{2}z = 2 \end{cases}$

$\rightarrow \begin{cases} 9x - 4y + 6z = 108 \\ x + 2y - 3z = 12 \\ x - 2y + z = 4 \end{cases}$

Add twice equation 2 to equation 1.
$11x = 132$
$x = 12$
Add equations 2 and 3.
$2x - 2z = 16$ or $x - z = 8$
$12 - z = 8$
$4 = z$
$z = 4$
Substitute back.
$12 - 2y + 4 = 4$
$12 - 2y = 0$
$12 = 2y$
$y = 6$
$\{(12, 6, 4)\}$

27. $\begin{cases} x + y + z = 1 \\ 2x - y + z = 0 \\ -x + 2y + 2z = -1 \end{cases}$

$$\begin{array}{r} 2x - y + z = 0 \\ -2x + 4y + 4z = -2 \\ \hline 3y + 5z = -2 \end{array}$$

$$\begin{array}{r} -2x - 2y - 2z = -2 \\ 2x - y + z = 0 \\ \hline -3y - z = -2 \end{array}$$

$$\begin{array}{r} 3y + 5z = -2 \\ -3y - z = -2 \\ \hline 4z = -4 \\ z = -1 \end{array}$$

$3y + 5(-1) = -2$
$3y = 3$
$y = 1$

$x + 1 + (-1) = 1$
$x = 1$
$\{(1, 1, -1)\}$
$\frac{1}{24} = \frac{x}{8} + \frac{y}{4} + \frac{z}{3}$
$\frac{1}{24} = \frac{1}{8} + \frac{1}{4} - \frac{1}{3}$
$\frac{1}{24} = \frac{3}{24} + \frac{6}{24} - \frac{8}{24}$
$\frac{1}{24} = \frac{1}{24}$ True

29. Let x = smaller number
Then $45 - x$ = larger number
$45 - x = 2x$
$45 = 3x$
$15 = x$
$45 - 15 = 30$
15 and 30

31. $2(x - 1) - 3x = x - 12$
$2x - 2 - 3x = x - 12$
$-2 - x = x - 12$
$10 = 2x$
$5 = x$
$\{5\}$

33. $-y - 5(y + 5) = 3y - 10$
$-y - 5y - 25 = 3y - 10$
$-6y - 25 = 3y - 10$
$-15 = 9y$
$-\frac{15}{9} = y$
$\left\{-\frac{15}{9}\right\}$

35. $\begin{cases} x + y \qquad - w = 0 \\ \quad y + 2z + w = 3 \\ x \qquad - z \qquad = 1 \\ 2x - y \qquad - w = -1 \end{cases}$

Add equations 2 and 4.
$\begin{cases} x \qquad - z = 1 \\ 2x \qquad + 2z = 2 \end{cases}$

$\rightarrow \begin{cases} 2x - 2z = 2 \\ 2x + 2z = 2 \end{cases}$
Add.
$4x = 4$
$x = 1$
Substitute back.
$1 - z = 1$
$z = 0$
Let $x = 1$ in equations 1 and 4.
$\begin{cases} y - w = -1 \\ -y - w = -3 \end{cases}$
Add.
$-2w = -4$
$w = 2$
Substitute back.
$y - 2 = -1$
$y = 1$
$\{(1, 1, 0, 2)\}$

37. $\begin{cases} x + y + z + w = 5 \\ 2x + y + z + w = 6 \\ x + y + z \quad = 2 \\ x + y \quad = 0 \end{cases}$
Subtract equation 1 from equation 2.
$x = 1$
Substitute in equation 4.
$1 + y = 0$
$y = -1$
Substitute both in equation 3.
$1 + (-1) + z = 2$
$z = 2$
Substitute back.
$1 + (-1) + 2 + w = 5$
$2 + w = 5$
$w = 3$
$\{(1, -1, 2, 3)\}$

Exercise Set 4.3

1. Let m = the first number and n = the second number. Then:
$\begin{cases} m = n + 2 \\ 2m = 3n - 4 \end{cases}$
Substitute.
$2(n + 2) = 3n - 4$
$2n + 4 = 3n - 4$
$8 = n$
$n = 8$
Substitute back.
$m = 8 + 2 = 10$
The numbers are 10 and 8.

3. Let p = the speed of the plane in still air and w = the speed of the wind. Then:
$\begin{cases} p + w = 560 \\ p - w = 480 \end{cases}$
Add. $2p = 1040$
$p = 520$
Substitute back.
$520 + w = 560$
$w = 40$
The speed of the plane in still air is 520 mph and the speed of the wind is 40 mph.

5. Let x = the number of quarts of 4% butterfat milk and y = the number of quarts of 10% butterfat milk. Then:
$\begin{cases} x + \quad y = 60 \\ .04x + .01y = .02(60) \end{cases}$
$\rightarrow \begin{cases} x + y = 60 \\ 4x + y = 120 \end{cases}$
Subtract.
$-3x = -60$
$x = 20$
Substitute back.
$20 + y = 60$
$y = 40$
20 quarts of 4% butterfat milk and 40 quarts of 1% butterfat milk should be used.

7. Let ℓ = the number of large frames purchased and s = the number of small frames purchased. Then:

$$\begin{cases} \ell + s = 22 \\ 15\ell + 8s = 239 \end{cases}$$

$$\rightarrow \begin{cases} 8\ell + 8s = 176 \\ 15\ell + 8s = 239 \end{cases}$$

Subtract.
$-7\ell = -63$
$\ell = 9$
Substitute back.
$9 + s = 22$
$s = 13$
She bought 9 large frames and 13 small frames.

9. Let m = the first number and n = the second number. Then:

$$\begin{cases} m = n - 2 \\ 2m = 3n + 4 \end{cases}$$

Substitute.
$2(n - 2) = 3n + 4$
$2n - 4 = 3n + 4$
$-8 = n$
$n = -8$
Substitute back.
$m = -8 - 2 = -10$
The numbers are –10 and –8.

11. Let x = the price of each tablet and y = the price of each pen. Then:

$$\begin{cases} 7x + 4y = 6.40 \\ 2x + 19y = 5.40 \end{cases}$$

$$\rightarrow \begin{cases} 14x + 8y = 12.80 \\ 14x + 133y = 37.80 \end{cases}$$

Subtract the first equation from the second.
$125y = 25$
$y = \frac{25}{125} = .20$
Substitute back.
$7x + 4(.20) = 6.40$
$7x + .80 = 6.40$
$7x = 5.60$
$x = .80$
Tablets cost \$.80 each and pens cost \$.20 each.

13. Let p = the speed of the plane in still air and w = the speed of the wind.
First note:
$\frac{2160 \text{ mi}}{3 \text{ hr}} = 720 \frac{\text{mi}}{\text{hr}}$ and
$\frac{2160 \text{ mi}}{4 \text{ hr}} = 540 \ \frac{\text{mi}}{\text{hr}}$
Now,

$$\begin{cases} p + w = 720 \\ p - w = 540 \end{cases}$$

Add.
$2p = 1260$
$p = 630$
Substitute back.
$630 + w = 720$
$w = 90$
The speed of the plane in still air is $630 \ \frac{\text{mi}}{\text{hr}}$ and the speed of the wind is $90 \ \frac{\text{mi}}{\text{hr}}$.

15. Let s = the length of the shortest side, ℓ = the length of the longest side, and w = the length of the two middle sides.
Now:

$$\begin{cases} s + 2m + \ell = 29 \\ \ell = 2s \\ m = s + 2 \end{cases}$$

Substituting, we get:
$s + 2(s + 2) + 2s = 29$
$s + 2s + 4 + 2s = 29$
$5s + 4 = 29$
$5s = 25$
$s = 5$
$\ell = 2(5) = 10$
and $m = 5 + 2 = 7$
The shortest side is 5 inches, the two middle sides are 7 inches and the longest side is 10 inches.

17. Let x = the first number, y = the second number, and z = the third number.
Then:

$$\begin{cases} x+y+z=40 \\ x \quad =y+5 \quad \text{or} \quad y=x-5 \\ x \quad =2z \quad \text{or} \quad z=\frac{1}{2}x \end{cases}$$

Substitute:

$x+x-5+\frac{1}{2}x=40$

$\frac{5}{2}x-5=40$

$\frac{5}{2}x=45$

$x=\frac{2}{5}(45)=18$

$y=18-5=13$

$z=\frac{1}{2}(18)=9$

The three numbers are 18, 13, and 9.

19. Let x = total weekly sales and y = total weekly salary. Then:

$$\begin{cases} y=200+0.05x \\ y=0.15x \end{cases}$$

Substitute:

$200+0.05x=0.15x$

$200=0.10x$

$x=\frac{200}{0.10}=2000$

Jack's salary would be the same, regardless of pay arrangement, at $2,000 worth of sales.

21. Let x = price for template
y = price for pencil
z = price for paper

$3x+y=6.45$

$2z+4y=7.50$

$z=3y$

Substitute last equation into second equation.

$2(3y)+4y=7.50$

$6y+4y=7.50$

$10y=7.50$

$y=0.75$

Substitute back.

$3(.75)=z$

$2.25=z$

$3x+0.75=6.45$

$3x=5.70$

$x=1.90$

$1.90 for template;
$0.75 for pencil;
$2.25 for paper

23. $C(x)=12x+15{,}000$

$R(x)=32x$

$12x+15{,}000=32x$

$15{,}000=20x$

$750=x$

750 units

25. $C(x)=0.8x+900$

$R(x)=2x$

$0.8x+900=2x$

$900=1.2x$

$750=x$

750 units

27. $C(x)=105x=70{,}000$

$R(x)=245x$

$105x+70{,}000=245x$

$70{,}000=140x$

$500=x$

500 units

29. **a.** $R(x)=31x$

b. $C(x)=15x+500$

c. $R(x)=C(x)$

$31x=15x+500$

$16x=500$

$x=31.25$

32 baskets

31. $x+2y=180$

$3x-10+y=180$ or $3x+y=190$

Solve first equation for x and substitute into second equation.

$x=180-2y$

$3(180-2y)+y=190$

$540-6y+y=190$

$-5y=-350$

$y=70$

Substitute back.

$x=180-2(70)$

$x=40$

$x=40°$ and $y=70°$

33.

Concentration	Amount	Solution
25	$2x$	$0.25(2x) = 0.5x$
40	x	$0.4x$
50	y	$0.5y$
32	200	$0.32(200) = 64$

$2x + x + y = 200$ or $3x + y = 200$
$0.5x + 0.4x + 0.5y = 64$ or $0.9x + 0.5y = 64$
Solve first equation for y and substitute into second equation.
$y = 200 - 3x$
$0.9x + 0.5(200 - 3x) = 64$
$0.9x + 100 - 1.5x = 64$
$-0.6x = -36$
$x = 60$ and $2x = 120$
Substitute back.
$y = 200 - 3(60) = 20$
120 liters of 25%
60 liters of 40%
20 liters of 50%

35. $y = ax^2 + bx + c$
$(1, 2)$: $2 = a + b + c$
$(2, 3)$: $3 = 4a + 2b + c$
$(-1, 6)$: $6 = a - b + c$

$$\begin{array}{r} -1 = 2a - \ \ c \\ -8 = 2a + 2c \\ \hline -9 = \quad -3c \\ 3 = \quad\quad c \end{array}$$

$c = 2a + 1$
$3 = 2a + 1$
$1 = a$
$2 = 1 + b + 3$
$-2 = b$
$a = 1,\ b = -2,\ c = 3$

37. $x = 180 - (z + 15)$
$x = 165 - z$

$y = 180 - (z - 13)$
$y = 193 - z$

$360 = x + y + z + 72$
$288 = x + y + z$

$$\begin{cases} z = 165 - x \\ z = 193 - y \\ x + y + z = 288 \end{cases}$$

$x + y + 165 - x = 288$
$y = 123$
$x + y + 193 - y = 288$
$x = 95$
$z = 165 - 95 = 70$
$(95°, 123°, 70°)$

39. $y = ax^2 + bx + c$
$(4, 2.47)$: $2.47 = 16a + 4b + c$
$(7, 0.6)$: $0.6 = 49a + 7b + c$
$(8, 1.1)$: $1.1 = 64a + 8b + c$

$$\begin{array}{r} 2.47 = \ \ 16a + 4b + c \\ -0.6 = -49a - 7b - c \\ \hline 1.87 = -33a - 3b \end{array}$$

$$\begin{array}{r} -0.6 = -49a - 7b - c \\ 1.1 = \ \ 64a + 8b + c \\ \hline 0.5 = \ \ 15a + \ \ b \\ b = 0.5 - 15a \end{array}$$

$1.87 = -33a - 3(0.5 - 15a)$
$1.87 = -33a - 1.5 + 45a$
$3.37 = 12a$
$0.28 = a$

$b = 0.5 - 15\,(0.28)$
$b = -3.70$

$c = 2.47 - 16a - 4b$
$c = 2.47 - 16(0.28) - 4(-3.70)$
$c = 12.78$
$(0.28, -3.70, 12.78)$
September, $x = 9$
$y = 0.28(9)^2 - 3.70(9) + 12.78$
$= 2.16$ inches

41. $\begin{cases} 2x + y + 3z = 7 \\ -4x + y + 2z = 4 \end{cases}$

$$\begin{array}{r} 4x + 2y + 6z = 14 \\ -4x + \ \ y + 2z = 4 \\ \hline 3y + 8z = 18 \end{array}$$

43. $\begin{cases} 2x-3y+2z=5 \\ x-9y+z=-1 \end{cases}$

$$\begin{array}{r} -6x+9y-6z=-15 \\ x-9y+z=-1 \\ \hline -5x \quad -5z=-16 \end{array}$$

45. $P(\text{green})=\frac{3}{8}$

47. $P(\text{red or blue})=\frac{5}{8}$

Exercise Set 4.4

1. $\begin{cases} x+y=1 \\ x-2y=4 \end{cases} \left(\begin{array}{cc|c} 1 & 1 & 1 \\ 1 & -2 & 4 \end{array}\right) \rightarrow$

$\left(\begin{array}{cc|c} 1 & 1 & 1 \\ 0 & -3 & 3 \end{array}\right) \rightarrow \left(\begin{array}{cc|c} 1 & 1 & 1 \\ 0 & 1 & -1 \end{array}\right) \rightarrow \left(\begin{array}{cc|c} 1 & 0 & 2 \\ 0 & 1 & -1 \end{array}\right)$

{(2, –1)}

3. $\begin{cases} 2y-4=0 \\ x+2y=0 \end{cases}$ or $\begin{cases} 2y=4 \\ x+2y=0 \end{cases}$

$\left(\begin{array}{cc|c} 0 & 2 & 4 \\ 1 & 2 & 0 \end{array}\right) \rightarrow \left(\begin{array}{cc|c} 1 & 2 & 0 \\ 0 & 2 & 4 \end{array}\right) \rightarrow \left(\begin{array}{cc|c} 1 & 2 & 0 \\ 0 & 1 & 2 \end{array}\right)$

$\rightarrow \left(\begin{array}{cc|c} 1 & 0 & -4 \\ 0 & 1 & 2 \end{array}\right)$

{(–4, 2)}

5. $\begin{cases} x+y=3 \\ 2y=10 \\ 3x+2y-4z=12 \end{cases} \left(\begin{array}{ccc|c} 1 & 1 & 0 & 3 \\ 0 & 2 & 0 & 10 \\ 3 & 2 & -4 & 12 \end{array}\right)$

$\rightarrow \left(\begin{array}{ccc|c} 1 & 1 & 0 & 3 \\ 0 & 2 & 0 & 10 \\ 0 & -1 & -4 & 3 \end{array}\right) \rightarrow \left(\begin{array}{ccc|c} 1 & 1 & 0 & 3 \\ 0 & 1 & 0 & 5 \\ 0 & -1 & -4 & 3 \end{array}\right)$

$\rightarrow \left(\begin{array}{ccc|c} 1 & 0 & 0 & -2 \\ 0 & 1 & 0 & 5 \\ 0 & -1 & -4 & 3 \end{array}\right) \rightarrow \left(\begin{array}{ccc|c} 1 & 0 & 0 & -2 \\ 0 & 1 & 0 & 5 \\ 0 & 0 & -4 & 8 \end{array}\right)$

$\rightarrow \left(\begin{array}{ccc|c} 1 & 0 & 0 & -2 \\ 0 & 1 & 0 & 5 \\ 0 & 0 & 1 & -2 \end{array}\right)$

{(–2, 5, –2)}

7. $\begin{cases} 2y-z=-7 \\ x+4y+z=-4 \\ 5x-y+2z=13 \end{cases}$

$\left(\begin{array}{ccc|c} 0 & 2 & -1 & -7 \\ 1 & 4 & 1 & -4 \\ 5 & -1 & 2 & 13 \end{array}\right) \rightarrow \left(\begin{array}{ccc|c} 1 & 4 & 1 & -4 \\ 0 & 2 & -1 & -7 \\ 5 & -1 & 2 & 13 \end{array}\right)$

$\rightarrow \left(\begin{array}{ccc|c} 1 & 4 & 1 & -4 \\ 0 & 2 & -1 & -7 \\ 0 & -21 & -3 & 33 \end{array}\right) \rightarrow \left(\begin{array}{ccc|c} 1 & 4 & 1 & -4 \\ 0 & 1 & -\frac{1}{2} & -\frac{7}{2} \\ 0 & -21 & -3 & 33 \end{array}\right)$

$\rightarrow \left(\begin{array}{ccc|c} 1 & 0 & 3 & 10 \\ 0 & 1 & -\frac{1}{2} & -\frac{7}{2} \\ 0 & -21 & -3 & 33 \end{array}\right) \rightarrow \left(\begin{array}{ccc|c} 1 & 0 & 3 & 10 \\ 0 & 1 & -\frac{1}{2} & -\frac{7}{2} \\ 0 & 0 & -\frac{27}{2} & -\frac{81}{2} \end{array}\right)$

$\rightarrow \left(\begin{array}{ccc|c} 1 & 0 & 3 & 10 \\ 0 & 1 & -\frac{1}{2} & -\frac{7}{2} \\ 0 & 0 & 1 & 3 \end{array}\right) \rightarrow \left(\begin{array}{ccc|c} 1 & 0 & 0 & 1 \\ 0 & 1 & -\frac{1}{2} & -\frac{7}{2} \\ 0 & 0 & 1 & 3 \end{array}\right)$

$\rightarrow \left(\begin{array}{ccc|c} 1 & 0 & 0 & 1 \\ 0 & 1 & 0 & -2 \\ 0 & 0 & 1 & 3 \end{array}\right)$

{(1, –2, 3)}

9. $\begin{cases} x-2y=4 \\ 2x-4y=4 \end{cases} \left(\begin{array}{cc|c} 1 & -2 & 4 \\ 2 & -4 & 4 \end{array}\right)$

$\rightarrow \left(\begin{array}{cc|c} 1 & -2 & 4 \\ 0 & 0 & -4 \end{array}\right)$

Inconsistent system.
{ }

11. $\begin{cases} 3x-3y=9 \\ 2x-2y=6 \end{cases} \left(\begin{array}{cc|c} 3 & -3 & 9 \\ 2 & -2 & 6 \end{array}\right)$

$\rightarrow \left(\begin{array}{cc|c} 1 & -1 & 3 \\ 2 & -2 & 6 \end{array}\right) \rightarrow \left(\begin{array}{cc|c} 1 & -1 & 3 \\ 0 & 0 & 0 \end{array}\right)$

Dependent system.

$\{(x, y)|x-y=3\}$

13. $\begin{cases} x-4=0 \\ x+y=1 \end{cases}$ or $\begin{cases} x \quad =4 \\ x+y=1 \end{cases}$

$\left(\begin{array}{cc|c} 1 & 0 & 4 \\ 1 & 1 & 1 \end{array}\right) \rightarrow \left(\begin{array}{cc|c} 1 & 0 & 4 \\ 0 & 1 & -3 \end{array}\right)$

$\{(4, -3)\}$

15. $\begin{cases} x+y+z=2 \\ 2x \quad +z=5 \\ \quad 3y+z=2 \end{cases} \left(\begin{array}{ccc|c} 1 & 1 & 1 & 2 \\ 2 & 0 & -1 & 5 \\ 0 & 3 & 1 & 2 \end{array}\right)$

$\rightarrow \left(\begin{array}{ccc|c} 1 & 1 & 1 & 2 \\ 0 & -2 & -3 & 1 \\ 0 & 3 & 1 & 2 \end{array}\right) \rightarrow \left(\begin{array}{ccc|c} 1 & 1 & 1 & 2 \\ 0 & 1 & \frac{3}{2} & -\frac{1}{2} \\ 0 & 3 & 1 & 2 \end{array}\right)$

$\rightarrow \left(\begin{array}{ccc|c} 1 & 0 & -\frac{1}{2} & \frac{5}{2} \\ 0 & 1 & \frac{3}{2} & -\frac{1}{2} \\ 0 & 3 & 1 & 2 \end{array}\right) \rightarrow \left(\begin{array}{ccc|c} 1 & 0 & -\frac{1}{2} & \frac{5}{2} \\ 0 & 1 & \frac{3}{2} & -\frac{1}{2} \\ 0 & 0 & -\frac{7}{2} & \frac{7}{2} \end{array}\right)$

$\rightarrow \left(\begin{array}{ccc|c} 1 & 0 & -\frac{1}{2} & \frac{5}{2} \\ 0 & 1 & \frac{3}{2} & -\frac{1}{2} \\ 0 & 0 & 1 & -1 \end{array}\right) \rightarrow \left(\begin{array}{ccc|c} 1 & 0 & 0 & 2 \\ 0 & 1 & \frac{3}{2} & -\frac{1}{2} \\ 0 & 0 & 1 & -1 \end{array}\right)$

$\rightarrow \left(\begin{array}{ccc|c} 1 & 0 & 0 & 2 \\ 0 & 1 & 0 & 1 \\ 0 & 0 & 1 & -1 \end{array}\right)$

$\{(2, 1, -1)\}$

17. $\begin{cases} 5x-2y=27 \\ -3x+5y=18 \end{cases} \left(\begin{array}{cc|c} 5 & -2 & 27 \\ -3 & 5 & 18 \end{array}\right)$

$\rightarrow \left(\begin{array}{cc|c} 1 & -\frac{2}{5} & \frac{27}{5} \\ -3 & 5 & 18 \end{array}\right) \rightarrow \left(\begin{array}{cc|c} 1 & -\frac{2}{5} & \frac{27}{5} \\ 0 & \frac{19}{5} & \frac{171}{5} \end{array}\right)$

$\rightarrow \left(\begin{array}{cc|c} 1 & -\frac{2}{5} & \frac{27}{5} \\ 0 & 1 & 9 \end{array}\right) \rightarrow \left(\begin{array}{cc|c} 1 & 0 & 9 \\ 0 & 1 & 9 \end{array}\right)$

$\{(9, 9)\}$

19. $\begin{cases} 4x-7y=7 \\ 12x-21y=24 \end{cases} \left(\begin{array}{cc|c} 4 & -7 & 7 \\ 12 & -21 & 24 \end{array}\right)$

$\rightarrow \left(\begin{array}{cc|c} 1 & -\frac{7}{4} & \frac{7}{4} \\ 12 & -21 & 24 \end{array}\right) \rightarrow \left(\begin{array}{cc|c} 1 & -\frac{7}{4} & \frac{7}{4} \\ 0 & 0 & 3 \end{array}\right)$

Inconsistent system.

{ }

21. $\begin{cases} 4x-y+2z=5 \\ \quad 2y \ +z=4 \\ 4x+y+3z=10 \end{cases} \left(\begin{array}{ccc|c} 4 & -1 & 2 & 5 \\ 0 & 2 & 1 & 4 \\ 4 & 1 & 3 & 10 \end{array}\right)$

$\rightarrow \left(\begin{array}{ccc|c} 1 & -\frac{1}{4} & \frac{1}{2} & \frac{5}{4} \\ 0 & 2 & 1 & 4 \\ 4 & 1 & 3 & 10 \end{array}\right) \rightarrow \left(\begin{array}{ccc|c} 1 & -\frac{1}{4} & \frac{1}{2} & \frac{5}{4} \\ 0 & 2 & 1 & 4 \\ 0 & 2 & 1 & 5 \end{array}\right)$

$\rightarrow \left(\begin{array}{ccc|c} 1 & -\frac{1}{4} & \frac{1}{2} & \frac{5}{4} \\ 0 & 1 & \frac{1}{2} & 2 \\ 0 & 2 & 1 & 5 \end{array}\right) \rightarrow \left(\begin{array}{ccc|c} 1 & 0 & \frac{5}{8} & \frac{7}{4} \\ 0 & 1 & \frac{1}{2} & 2 \\ 0 & 2 & 1 & 5 \end{array}\right)$

$\rightarrow \left(\begin{array}{ccc|c} 1 & 0 & \frac{5}{8} & \frac{7}{4} \\ 0 & 1 & \frac{1}{2} & 2 \\ 0 & 0 & 0 & 1 \end{array}\right)$

Inconsistent system.

{ }

23. $\begin{cases} 4x \ +y \ +z=3 \\ -x \ +y-2z=-11 \\ x+2y+2z=-1 \end{cases} \left(\begin{array}{ccc|c} 4 & 1 & 1 & 3 \\ -1 & 1 & -2 & -11 \\ 1 & 2 & 2 & -1 \end{array}\right)$

$\rightarrow \left(\begin{array}{ccc|c} 1 & \frac{1}{4} & \frac{1}{4} & \frac{3}{4} \\ -1 & 1 & -2 & -11 \\ 1 & 2 & 2 & -1 \end{array}\right) \rightarrow \left(\begin{array}{ccc|c} 1 & \frac{1}{4} & \frac{1}{4} & \frac{3}{4} \\ 0 & \frac{5}{4} & -\frac{7}{4} & -\frac{41}{4} \\ 1 & 2 & 2 & -1 \end{array}\right)$

$$\to\left(\begin{array}{ccc|c} 1 & \frac{1}{4} & \frac{1}{4} & \frac{3}{4} \\ 0 & \frac{5}{4} & -\frac{7}{4} & -\frac{41}{4} \\ 0 & \frac{7}{4} & \frac{7}{4} & -\frac{7}{4} \end{array}\right) \to \left(\begin{array}{ccc|c} 1 & \frac{1}{4} & \frac{1}{4} & \frac{3}{4} \\ 0 & 1 & -\frac{7}{5} & -\frac{41}{5} \\ 0 & \frac{7}{4} & \frac{7}{4} & -\frac{7}{4} \end{array}\right)$$

$$\to\left(\begin{array}{ccc|c} 1 & 0 & \frac{3}{5} & \frac{14}{5} \\ 0 & 1 & -\frac{7}{5} & -\frac{41}{5} \\ 0 & \frac{7}{4} & \frac{7}{4} & -\frac{7}{4} \end{array}\right) \to \left(\begin{array}{ccc|c} 1 & 0 & \frac{3}{5} & \frac{14}{5} \\ 0 & 1 & -\frac{7}{5} & -\frac{41}{5} \\ 0 & 0 & \frac{21}{5} & \frac{63}{5} \end{array}\right)$$

$$\to\left(\begin{array}{ccc|c} 1 & 0 & \frac{3}{5} & \frac{14}{5} \\ 0 & 1 & -\frac{7}{5} & -\frac{41}{5} \\ 0 & 0 & 1 & 3 \end{array}\right) \to \left(\begin{array}{ccc|c} 1 & 0 & 0 & 1 \\ 0 & 1 & -\frac{7}{5} & -\frac{41}{5} \\ 0 & 0 & 1 & 3 \end{array}\right)$$

$$\to\left(\begin{array}{ccc|c} 1 & 0 & 0 & 1 \\ 0 & 1 & 0 & -4 \\ 0 & 0 & 1 & 3 \end{array}\right)$$

$\{(1, -4, 3)\}$

25. Function

27. Not a function

29. $(-1)(-5) - (6)(3) = 5 - 18 = -13$

31. $(4)(-10) - (2)(-2) = -40 + 4 = -36$

33. $(-3)(-3) - (-1)(-9) = 9 - 9 = 0$

Exercise Set 4.5

1. $\begin{vmatrix} 3 & 5 \\ -1 & 7 \end{vmatrix} = 3(7) - 5(-1)$
$= 21 + 5 = 26$

3. $\begin{vmatrix} 9 & -2 \\ 4 & -3 \end{vmatrix} = 9(-3) - 4(-2)$
$= -27 + 8 = -19$

5. $\begin{vmatrix} -2 & 9 \\ 4 & -18 \end{vmatrix} = -2(-18) - 9(4)$
$= 36 - 36 = 0$

7. $\begin{cases} 2y - 4 = 0 \\ x + 2y = 5 \end{cases} \to \begin{cases} 2y = 4 \\ x + 2y = 5 \end{cases}$

$D = \begin{vmatrix} 0 & 2 \\ 1 & 2 \end{vmatrix} = 0(2) - 2(1) = 0 - 2 = -2$

$D_x = \begin{vmatrix} 4 & 2 \\ 5 & 2 \end{vmatrix} = 4(2) - 2(5) - 8 - 10 = -2$

$D_y = \begin{vmatrix} 0 & 4 \\ 1 & 5 \end{vmatrix} = 0(5) - 4(1) = 0 - 4 = -4$

$x = \frac{-2}{-2} = 1$ and $y = \frac{-4}{-2} = 2$

$\{(1, 2)\}$

9. $\begin{cases} 3x + y = 1 \\ 2y = 2 - 6x \end{cases} \to \begin{cases} 3x + y = 1 \\ 6x + 2y = 2 \end{cases}$

$D = \begin{vmatrix} 3 & 1 \\ 6 & 2 \end{vmatrix} = 3(2) - 6(1) = 6 - 6 = 0$

Thus, this system cannot be solved using Cramer's rule. Since equation 2 is 2 times equation 1, the system is dependent.

$\{(x, y) | 3x + y = 1\}$

11. $\begin{cases} 5x - 2y = 27 \\ -3x + 5y = 18 \end{cases}$

$D = \begin{vmatrix} 5 & -2 \\ -3 & 5 \end{vmatrix} = 5(5) - (-2)(-3) = 19$

$D_x = \begin{vmatrix} 27 & -2 \\ 18 & 5 \end{vmatrix} = 27(5) - (-2)18 = 171$

$D_y = \begin{vmatrix} 5 & 27 \\ -3 & 18 \end{vmatrix} = 5(18) - 27(-3) = 171$

$x = \frac{171}{19} = 9$ and $y = \frac{171}{19} = 9$

$\{(9, 9)\}$

13. $\begin{vmatrix} 2 & 1 & 0 \\ 0 & 5 & -3 \\ 4 & 0 & 2 \end{vmatrix} = 2\begin{vmatrix} 5 & -3 \\ 0 & 2 \end{vmatrix} - 1\begin{vmatrix} 0 & -3 \\ 4 & 2 \end{vmatrix} + 0\begin{vmatrix} 0 & 5 \\ 4 & 0 \end{vmatrix}$

$= 2[5(2) - (-3)(0)] - [0(2) - 4(-3)] + 0$
$= 2(10) - 12 = 8$

15. $\begin{vmatrix} 4 & -6 & 0 \\ -2 & 3 & 0 \\ 4 & -6 & 1 \end{vmatrix}$

$= 0\begin{vmatrix} -2 & 3 \\ 4 & -6 \end{vmatrix} - \begin{vmatrix} 4 & -6 \\ 4 & -6 \end{vmatrix} + 1\begin{vmatrix} 4 & -6 \\ -2 & 3 \end{vmatrix}$
$= 0 - 0 + [4(3) - (-6)(-2)] = 0$

17. $\begin{vmatrix} 3 & 6 & -3 \\ -1 & -2 & 3 \\ 4 & -1 & 6 \end{vmatrix}$

$= 3\begin{vmatrix} -2 & 3 \\ -1 & 6 \end{vmatrix} - 6\begin{vmatrix} -1 & 3 \\ 4 & 6 \end{vmatrix} + (-3)\begin{vmatrix} -1 & -2 \\ 4 & -1 \end{vmatrix}$
$= 3[-2(6) - 3(-1)] - 6[-1(6) - 3(4)]$
$\quad - 3[(-1)(-1) - (-2)4]$
$= 3(-9) - 6(-18) - 3(9)$
$= -27 + 108 - 27$
$= 54$

19. $\begin{cases} 3x \qquad + z = -1 \\ -x - 3y + z = 7 \\ \qquad 3y + z = 5 \end{cases}$

$D = \begin{vmatrix} 3 & 0 & 1 \\ -1 & -3 & 1 \\ 0 & 3 & 1 \end{vmatrix}$

$= 3\begin{vmatrix} -3 & 1 \\ 3 & 1 \end{vmatrix} - 0\begin{vmatrix} -1 & 1 \\ 0 & 1 \end{vmatrix} + 1\begin{vmatrix} -1 & -3 \\ 0 & 3 \end{vmatrix}$
$= 3[-3(1) - 1(3)] - 0 + [-1(3) - (-3)0]$
$= 3(-6) - 3 = -21$

$D_x = \begin{vmatrix} -1 & 0 & 1 \\ 7 & -3 & 1 \\ 5 & 3 & 1 \end{vmatrix}$

$= -1\begin{vmatrix} -3 & 1 \\ 3 & 1 \end{vmatrix} - 0 \cdot \begin{vmatrix} 7 & 1 \\ 5 & 1 \end{vmatrix} + 1\begin{vmatrix} 7 & -3 \\ 5 & 3 \end{vmatrix}$
$= -[-3(1) - 1(3)] - 0 + [7(3) - (-3)5]$
$= 6 + 36 = 42$

$D_y = \begin{vmatrix} 3 & -1 & 1 \\ -1 & 7 & 1 \\ 0 & 5 & 1 \end{vmatrix}$

$= 3\begin{vmatrix} 7 & 1 \\ 5 & 1 \end{vmatrix} - (-1)\begin{vmatrix} -1 & 1 \\ 0 & 1 \end{vmatrix} + 1 \cdot \begin{vmatrix} -1 & 7 \\ 0 & 5 \end{vmatrix}$
$= 3[7(1) - 1(5)] + 1[(-1)1 - 1(0)]$
$\quad + [-1(5) - 7(0)]$
$= 3(2) + (-1) + (-5) = 0$

$D_z = \begin{vmatrix} 3 & 0 & -1 \\ -1 & -3 & 7 \\ 0 & 3 & 5 \end{vmatrix}$

$= 3\begin{vmatrix} -3 & 7 \\ 3 & 5 \end{vmatrix} - 0\begin{vmatrix} -1 & 7 \\ 0 & 5 \end{vmatrix} + (-1)\begin{vmatrix} -1 & -3 \\ 0 & 3 \end{vmatrix}$
$= 3[-3(5) - 7(3)] - 0 - [-1(3) - (-3)0]$
$= 3(-36) - (-3) = -105$

$x = \frac{42}{-21} = -2, \ y = \frac{0}{-21} = 0,$

$z = \frac{-105}{-21} = 5$

{(–2, 0, 5)}

21. $\begin{cases} x + y + z = 8 \\ 2x - y - z = 10 \\ x - 2y + 3z = 22 \end{cases}$

$D = \begin{vmatrix} 1 & 1 & 1 \\ 2 & -1 & -1 \\ 1 & -2 & 3 \end{vmatrix}$

$= 1\begin{vmatrix} -1 & -1 \\ -2 & 3 \end{vmatrix} - 1\begin{vmatrix} 2 & -1 \\ 1 & 3 \end{vmatrix} + 1\begin{vmatrix} 2 & -1 \\ 1 & -2 \end{vmatrix}$
$= (-3 - 2) - (6 - (-1)) + (-4 - (-1))$
$= -5 - 7 - 3 = -15$

$D_x = \begin{vmatrix} 8 & 1 & 1 \\ 10 & -1 & -1 \\ 22 & -2 & 3 \end{vmatrix}$

$= 8\begin{vmatrix} -1 & -1 \\ -2 & 3 \end{vmatrix} - 1\begin{vmatrix} 10 & -1 \\ 22 & 3 \end{vmatrix} + 1\begin{vmatrix} 10 & -1 \\ 22 & -2 \end{vmatrix}$
$= 8(-3 - 2) - (30 - (-22)) + (-20 - (-22))$
$= 8(-5) - 52 + 2 = -90$

$$D_y = \begin{vmatrix} 1 & 8 & 1 \\ 2 & 10 & -1 \\ 1 & 22 & 3 \end{vmatrix}$$

$$= 1\begin{vmatrix} 10 & -1 \\ 22 & 3 \end{vmatrix} - 8\begin{vmatrix} 2 & -1 \\ 1 & 3 \end{vmatrix} + 1\begin{vmatrix} 2 & 10 \\ 1 & 22 \end{vmatrix}$$

$= [30 - (-22)] - 8[6 - (-1)] + [44 - 10]$
$= 52 - 8(7) + 34$
$= 52 - 56 + 34 = 30$

$$D_z = \begin{vmatrix} 1 & 1 & 8 \\ 2 & -1 & 10 \\ 1 & -2 & 22 \end{vmatrix}$$

$$= 1\begin{vmatrix} -1 & 10 \\ -2 & 22 \end{vmatrix} - 1\begin{vmatrix} 2 & 10 \\ 1 & 22 \end{vmatrix} + 8\begin{vmatrix} 2 & -1 \\ 1 & -2 \end{vmatrix}$$

$= [-22 - (-20)] - [44 - 10]$
$\quad + 8[-4 - (-1)]$
$= -2 - 34 + 8(-3)$
$= -36 - 24 = -60$

$x = \frac{-90}{-15} = 6, \; y = \frac{30}{-15} = -2,$

$z = \frac{-60}{-15} = 4$

$\{(6, -2, 4)\}$

23. $\begin{vmatrix} 10 & -1 \\ -4 & 2 \end{vmatrix} = 10(2) - (-1)(-4)$

$= 20 - 4 = 16$

25. $\begin{vmatrix} 1 & 0 & 4 \\ 1 & -1 & 2 \\ 3 & 2 & 1 \end{vmatrix} = 1\begin{vmatrix} -1 & 2 \\ 2 & 1 \end{vmatrix} - 0\begin{vmatrix} 1 & 2 \\ 3 & 1 \end{vmatrix} + 4\begin{vmatrix} 1 & -1 \\ 3 & 2 \end{vmatrix}$

$= 1[-1 - 4] - 0 + 4[2 - (-3)]$
$= -5 + 4(5) = -5 + 20 = 15$

27. $\begin{vmatrix} \frac{3}{4} & \frac{5}{2} \\ -\frac{1}{6} & \frac{7}{3} \end{vmatrix} = \frac{3}{4} \cdot \frac{7}{3} - \frac{5}{2}\left(-\frac{1}{6}\right)$

$= \frac{21}{12} + \frac{5}{12} = \frac{26}{12} = \frac{13}{6}$

29. $\begin{vmatrix} 4 & -2 & 2 \\ 6 & -1 & 3 \\ 2 & 1 & 1 \end{vmatrix}$

$$= 4\begin{vmatrix} -1 & 3 \\ 1 & 1 \end{vmatrix} - (-2) \cdot \begin{vmatrix} 6 & 3 \\ 2 & 1 \end{vmatrix} + 2\begin{vmatrix} 6 & -1 \\ 2 & 1 \end{vmatrix}$$

$= 4[-1 - 3] + 2[6 - 6] + 2[6 - (-2)]$
$= 4(-4) + 2 \cdot 0 + 2(8)$
$= -16 + 0 + 16 = 0$

31. $\begin{vmatrix} -2 & 5 & 4 \\ 5 & -1 & 3 \\ 4 & 1 & 2 \end{vmatrix}$

$$= -2\begin{vmatrix} -1 & 3 \\ 1 & 2 \end{vmatrix} - 5 \cdot \begin{vmatrix} 5 & 3 \\ 4 & 2 \end{vmatrix} + 4\begin{vmatrix} 5 & -1 \\ 4 & 1 \end{vmatrix}$$

$= -2[-2 - 3] - 5 \cdot [10 - 12] + 4[5 - (-4)]$
$= -2(-5) - 5(-2) + 4(9)$
$= 10 + 10 + 36 = 56$

33. If the elements of a single row (or column) of a determinant are all zero, the value of the determinant will be zero. To see this, consider expanding on that row (or column) containing all zeros.

35. $\begin{vmatrix} 1 & x \\ 2 & 7 \end{vmatrix} = -3$

$(1)(7) - 2x = -3$
$7 - 2x = -3$
$2x = 10$
$x = 5$

37. $\begin{cases} 2x - 5y = 4 \\ x + 2y = -7 \end{cases}$

$D = \begin{vmatrix} 2 & -5 \\ 1 & 2 \end{vmatrix} = 2(2) - (-5)(1)$

$= 4 + 5 = 9$

$D_x = \begin{vmatrix} 4 & -5 \\ -7 & 2 \end{vmatrix} = 4(2) - (-5)(-7)$

$= 8 - 35 = -27$

$D_y = \begin{vmatrix} 2 & 4 \\ 1 & -7 \end{vmatrix} = 2(-7) - 4(1)$

$= -14 - 4 = -18$

$x = \frac{-27}{9} = -3$, and $y = \frac{-18}{9} = -2$

$\{(-3, -2)\}$

39. $\begin{cases} 4x + 2y = 5 \\ 2x + y = -1 \end{cases}$

$D = \begin{vmatrix} 4 & 2 \\ 2 & 1 \end{vmatrix} = 4(1) - 2(2)$

$= 4 - 4 = 0$

Thus, Cramer's rule cannot be used to solve the system. Multiply equation 2 by 2 yielding the new system:

$\begin{cases} 4x + 2y = 5 \\ 4x + 2y = -2 \end{cases}$

Therefore, the system is inconsistent.

{ }

41. $\begin{cases} 2x + 2y + z = 1 \\ -x + y + 2z = 3 \\ x + 2y + 4z = 0 \end{cases}$

$D = \begin{vmatrix} 2 & 2 & 1 \\ -1 & 1 & 2 \\ 1 & 2 & 4 \end{vmatrix}$

$= 2\begin{vmatrix} 1 & 2 \\ 2 & 4 \end{vmatrix} - 2\begin{vmatrix} -1 & 2 \\ 1 & 4 \end{vmatrix} + 1\begin{vmatrix} -1 & 1 \\ 1 & 2 \end{vmatrix}$

$= 2(4 - 4) - 2(-4 - 2) + (-2 - 1)$

$= 2(0) - 2(-6) + (-3)$

$= 0 + 12 - 3 = 9$

$D_x = \begin{vmatrix} 1 & 2 & 1 \\ 3 & 1 & 2 \\ 0 & 2 & 4 \end{vmatrix}$

$= 1\begin{vmatrix} 1 & 2 \\ 2 & 4 \end{vmatrix} - 3\begin{vmatrix} 2 & 1 \\ 2 & 4 \end{vmatrix} + 0\begin{vmatrix} 2 & 1 \\ 1 & 2 \end{vmatrix}$

$= (4 - 4) - 3(8 - 2) + 0$

$= 0 - 3(6) = -18$

$D_y = \begin{vmatrix} 2 & 1 & 1 \\ -1 & 3 & 2 \\ 1 & 0 & 4 \end{vmatrix}$

$= 1\begin{vmatrix} 1 & 1 \\ 3 & 2 \end{vmatrix} - 0\begin{vmatrix} 2 & 1 \\ -1 & 2 \end{vmatrix} + 4\begin{vmatrix} 2 & 1 \\ -1 & 3 \end{vmatrix}$

$= (2 - 3) - 0 + 4[6 - (-1)] = -1 + 4(7)$

$= -1 + 28 = 27$

$D_z = \begin{vmatrix} 2 & 2 & 1 \\ -1 & 1 & 3 \\ 1 & 2 & 0 \end{vmatrix}$

$= 1\begin{vmatrix} 2 & 1 \\ 1 & 3 \end{vmatrix} - 2\begin{vmatrix} 2 & 1 \\ -1 & 3 \end{vmatrix} + 0\begin{vmatrix} 2 & 2 \\ -1 & 1 \end{vmatrix}$

$= (6 - 1) - 2[6 - (-1)] + 0$

$= 5 - 2(7) = 5 - 14 = -9$

$x = -\frac{18}{9} = -2$, $y = \frac{27}{9} = 3$,

$= \frac{-9}{9} = -1$

$\{(-2, 3, -1)\}$

43. $\begin{cases} \frac{2}{3}x - \frac{3}{4}y = -1 \\ -\frac{1}{6}x + \frac{3}{4}y = \frac{5}{2} \end{cases}$

$D = \begin{vmatrix} \frac{2}{3} & -\frac{3}{4} \\ -\frac{1}{6} & \frac{3}{4} \end{vmatrix} = \frac{2}{3} \cdot \frac{3}{4} - \left(-\frac{3}{4}\right)\left(-\frac{1}{6}\right)$

$= \frac{1}{2} - \frac{1}{8} = \frac{3}{8}$

$D_x = \begin{vmatrix} -1 & -\frac{3}{4} \\ \frac{5}{2} & \frac{3}{4} \end{vmatrix} = (-1)\frac{3}{4} - \left(-\frac{3}{4}\right)\frac{5}{2}$

$= -\frac{3}{4} + \frac{15}{8} = \frac{9}{8}$

$D_y = \begin{vmatrix} \frac{2}{3} & -1 \\ -\frac{1}{6} & \frac{5}{2} \end{vmatrix} = \frac{2}{3} \cdot \frac{5}{2} - (-1)\left(-\frac{1}{6}\right)$

$= \frac{5}{3} - \frac{1}{6} = \frac{3}{2}$

Thus, $x = \frac{\frac{9}{8}}{\frac{3}{8}} = 3$ and $y = \frac{\frac{3}{2}}{\frac{3}{8}} = 4$

$\{3, 4\}$

45. $\begin{cases} .7x - .2y = -1.6 \\ .2x - y = -1.4 \end{cases}$

$D = \begin{vmatrix} .7 & -.2 \\ .2 & -1 \end{vmatrix}$

$= .7(-1) - (-.2)(.2)$

$= -.7 + .04 = -.66$

$D_x = \begin{vmatrix} -1.6 & -.2 \\ -1.4 & -1 \end{vmatrix}$

$= (-1.6)(-1) - (-.2)(-1.4)$

$= 1.6 - .28 = 1.32$

$$D_y = \begin{vmatrix} .7 & -1.6 \\ .2 & -1.4 \end{vmatrix}$$
$= (.7)(-1.4) - (-1.6)(.2)$
$= -.98 + .32 = -.66$
$x = \dfrac{1.32}{-.66} = -2$ and $y = \dfrac{-.66}{-.66} = 1$
$\{(-2, 1)\}$

47. $\begin{cases} -2x + 4y - 2z = 6 \\ x - 2y + z = -3 \\ 3x - 6y + 3z = -9 \end{cases}$

$$D = \begin{vmatrix} -2 & 4 & -2 \\ 1 & -2 & 1 \\ 3 & -6 & 3 \end{vmatrix}$$
$= -2\begin{vmatrix} -2 & 1 \\ -6 & 3 \end{vmatrix} - 4\begin{vmatrix} 1 & 1 \\ 3 & 3 \end{vmatrix} + (-2)\begin{vmatrix} 1 & -2 \\ 3 & -6 \end{vmatrix}$
$= -2[-6 - (-6)] - 4[3 - 3] - 2[-6 - (-6)]$
$= -2(0) - 4(0) - 2(0) = 0$
Therefore, Cramer's rule will not provide the solution. Note that equation 1 is –2 times equation 2 and that equation 3 is 3 times equation 2. Thus, the system is dependent.
$\{(x, y, z) | x - 2y + z = -3\}$

49. $\begin{cases} x - 2y + z = -5 \\ 3y + 2z = 4 \\ 3x - y = -2 \end{cases}$

$$D = \begin{vmatrix} 1 & -2 & 1 \\ 0 & 3 & 2 \\ 3 & -1 & 0 \end{vmatrix}$$
$= 1\begin{vmatrix} 3 & 2 \\ -1 & 0 \end{vmatrix} - 0\begin{vmatrix} -2 & 1 \\ -1 & 0 \end{vmatrix} + 3\begin{vmatrix} -2 & 1 \\ 3 & 2 \end{vmatrix}$
$= [0 - (-2)] - 0 + 3[-4 - 3]$
$= 2 + 3(-7) = -19$

$$D_x = \begin{vmatrix} -5 & -2 & 1 \\ 4 & 3 & 2 \\ -2 & -1 & 0 \end{vmatrix}$$
$= 1\begin{vmatrix} 4 & 3 \\ -2 & -1 \end{vmatrix} - 2\begin{vmatrix} -5 & -2 \\ -2 & -1 \end{vmatrix} + 0\begin{vmatrix} 1 & -2 \\ 0 & 3 \end{vmatrix}$
$= [-4 - (-6)] - 2[5 - 4] + 0$
$= 2 - 2(1) = 0$

$$D_y = \begin{vmatrix} 1 & -5 & 1 \\ 0 & 4 & 2 \\ 3 & -2 & 0 \end{vmatrix}$$
$= 1\begin{vmatrix} 4 & 2 \\ -2 & 0 \end{vmatrix} - 0\begin{vmatrix} -5 & 1 \\ -2 & 0 \end{vmatrix} + 3\begin{vmatrix} -5 & 1 \\ 4 & 2 \end{vmatrix}$
$= [0 - (-4)] - 0 + 3[-10 - 4]$
$= 4 + 3(-14) = 4 - 42 = -38$

$$D_z = \begin{vmatrix} 1 & -2 & -5 \\ 0 & 3 & 4 \\ 3 & -1 & -2 \end{vmatrix}$$
$= 1 \cdot \begin{vmatrix} 3 & 4 \\ -1 & -2 \end{vmatrix} - 0\begin{vmatrix} -2 & -5 \\ -1 & -2 \end{vmatrix} + 3\begin{vmatrix} -2 & -5 \\ 3 & 4 \end{vmatrix}$
$= [-6 - (-4)] - 0 + 3[-8 - (-15)]$
$= -2 + 3(7) = 19$
$x = \dfrac{0}{-19} = 0,\ y = \dfrac{-38}{-19} = 2,$
$z = \dfrac{19}{-19} = -1$
$\{(0, 2, -1)\}$

51. $5x - 6 + x - 12 = 6x - 18$

53. $2(3x - 6) + 3(x - 1)$
$= 6x - 12 + 3x - 3 = 9x - 15$

55. $f(x) = 5x - 6$ or $y = 5x - 6$

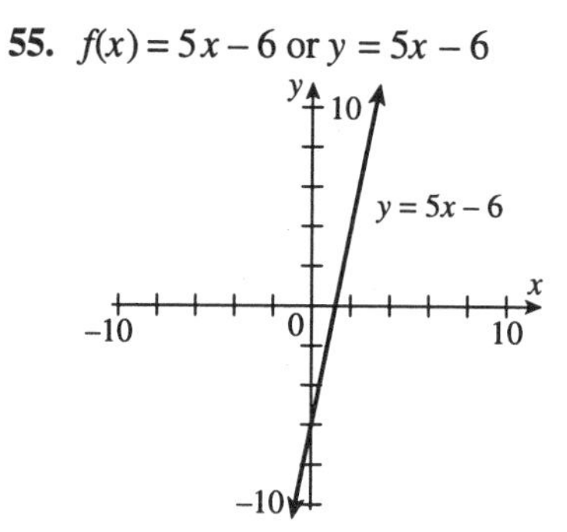

57. $h(x) = 3$ or $y = 3$

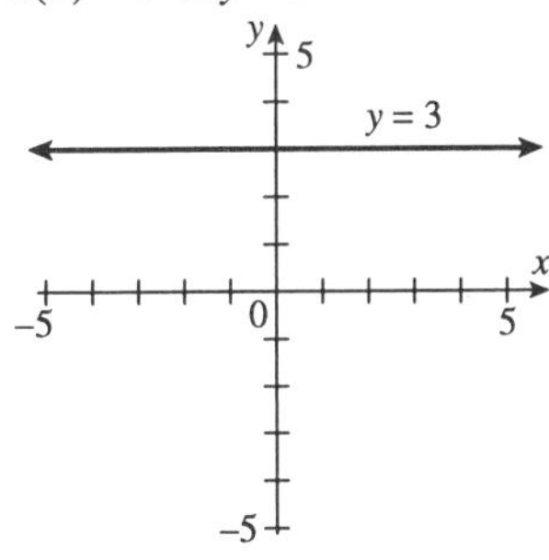

59. $\begin{vmatrix} 5 & 0 & 0 & 0 \\ 0 & 4 & 2 & -1 \\ 1 & 3 & -2 & 0 \\ 0 & -3 & 1 & 2 \end{vmatrix} = 5\begin{vmatrix} 4 & 2 & -1 \\ 3 & -2 & 0 \\ -3 & 1 & 2 \end{vmatrix}$

$-0\begin{vmatrix} 0 & 2 & -1 \\ 1 & -2 & 0 \\ 0 & 1 & 2 \end{vmatrix} + 0\begin{vmatrix} 0 & 4 & -1 \\ 1 & 3 & 0 \\ 0 & -3 & 2 \end{vmatrix}$

$-0\begin{vmatrix} 0 & 4 & 2 \\ 1 & 3 & -2 \\ 0 & -3 & 1 \end{vmatrix}$

$= 5\left[(-1)\begin{vmatrix} 3 & -2 \\ -3 & 1 \end{vmatrix} = 0 \cdot \begin{vmatrix} 4 & 2 \\ -3 & 1 \end{vmatrix} + 2\begin{vmatrix} 4 & 2 \\ 3 & -2 \end{vmatrix}\right]$

$= 5[-(3-6) - 0 + 2(-8-6)]$

$= 5[3 + 2(-14)]$

$= 5[3 - 28] = 5[-25] = -125$

61. $\begin{vmatrix} 4 & 0 & 2 & 5 \\ 0 & 3 & -1 & 1 \\ 0 & 0 & 2 & 0 \\ 0 & 0 & 0 & 1 \end{vmatrix} = 4\begin{vmatrix} 3 & -1 & 1 \\ 0 & 2 & 0 \\ 0 & 0 & 1 \end{vmatrix} - 0\begin{vmatrix} 0 & 2 & 5 \\ 0 & 2 & 0 \\ 0 & 0 & 1 \end{vmatrix}$

$+0\begin{vmatrix} 0 & 2 & 5 \\ 3 & -1 & 1 \\ 0 & 0 & 1 \end{vmatrix} - 0 \cdot \begin{vmatrix} 0 & 2 & 5 \\ 3 & -1 & 1 \\ 0 & 2 & 0 \end{vmatrix}$

$= 4\left[3 \cdot \begin{vmatrix} 2 & 0 \\ 0 & 1 \end{vmatrix} - 0\begin{vmatrix} -1 & 1 \\ 0 & 1 \end{vmatrix} + 0\begin{vmatrix} -1 & 1 \\ 2 & 0 \end{vmatrix}\right]$

$= 4[3(2-0) - 0 + 0] = 12(2) = 24$

Chapter 4 - Review

1. $\begin{cases} 3x + 10y = 1 \\ x + 2y = -1 \end{cases}$

a.

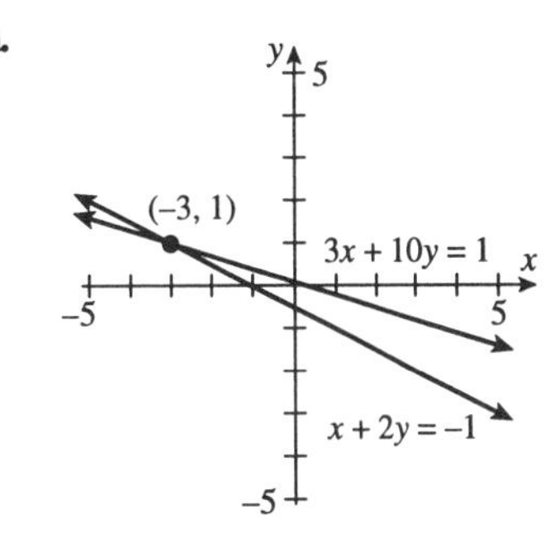

b. $x = -2y - 1$

$3(-2y - 1) + 10y = 1$

$-6y - 3 + 10y = 1$

$4y - 3 = 1$

$4y = 4$ so $y = 1$

Substitute back.

$x = -2(1) - 1 = -2 - 1 = -3$

$\{(-3, 1)\}$

c. $\rightarrow \begin{cases} 3x + 10y = 1 \\ 3x + 6y = -3 \end{cases}$

Subtract.

$4y = 4$ so $y = 1$

Substitute back.

$x = -2(1) - 1 = -2 - 1 = -3$

$\{(-3, 1)\}$

2. $\begin{cases} y = \frac{1}{2}x + \frac{2}{3} \\ 4x + 6y = 4 \end{cases}$

By substitution

$4x + 6\left(\frac{1}{2}x + \frac{2}{3}\right) = 4$

$4x + 3x + 4 = 4$

$7x = 0$

$x = 0$

$y = \frac{2}{3}$

By elimination

$$\begin{array}{r} -3x+6y=4 \\ \underline{4x+6y=4} \\ -7x \quad =0 \\ x \quad =0 \end{array}$$

$0+6y=4,\quad y=\frac{4}{6}=\frac{2}{3}$

By graphing,

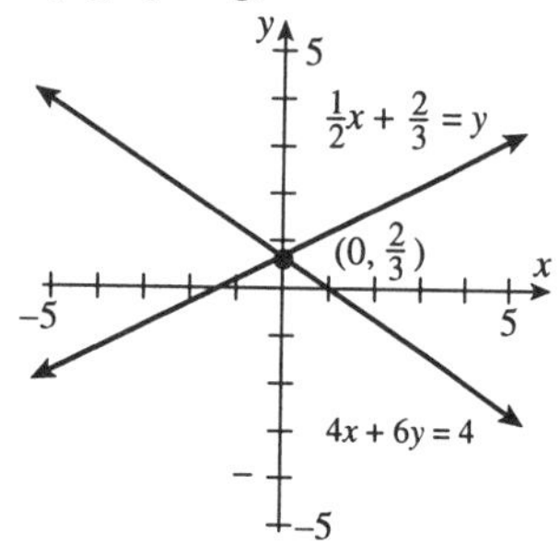

3. $\begin{cases} 2x-\ \ 4y=22 \\ 5x-10y=16 \end{cases}$

a.

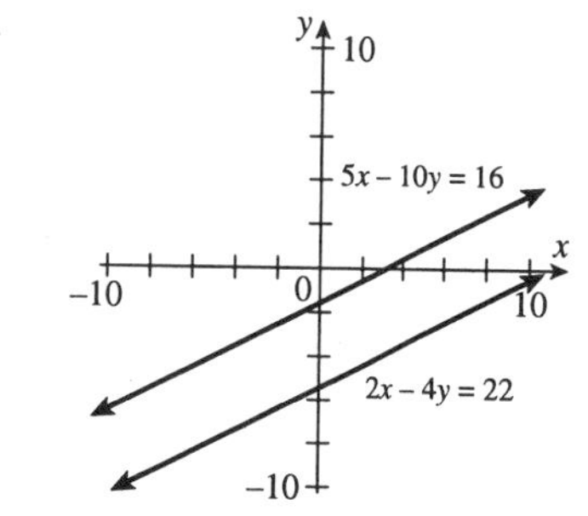

b. $x-2y=11$
$x=2y+11$
Substitute.
$5(2y+11)-10y=16$
$10y+55-10y=16$
$55=16$
Inconsistent system.
{ }

c. $\rightarrow\begin{cases} x-2y=11 \\ x-2y=\frac{16}{5} \end{cases}$

Thus, $11=\frac{16}{5}$

Inconsistent system.
{ }

4. $\begin{cases} 3x-6y=12 \\ \quad\ \ 2y=x-4 \end{cases}$

By substitution,
$x=2y+4$
$3(2y+4)-6y=12$
$6y+12-6y=12$
$0=0$
Solution is $\{(x, y)|3x-6y=12\}$
By elimination,

$$\begin{array}{r} 3x-6y=\ \ 12 \\ \underline{-x+2y=\ -4} \\ 3x-6y=\ \ 12 \\ \underline{3(-x+2y)=-4(3)} \\ 3x-6y=\ \ 12 \\ \underline{-3x+6y=\ \ 12} \\ 0=\ \ 0 \end{array}$$

Solution is $\{(x, y)|3x-6y=12\}$
By graphing,

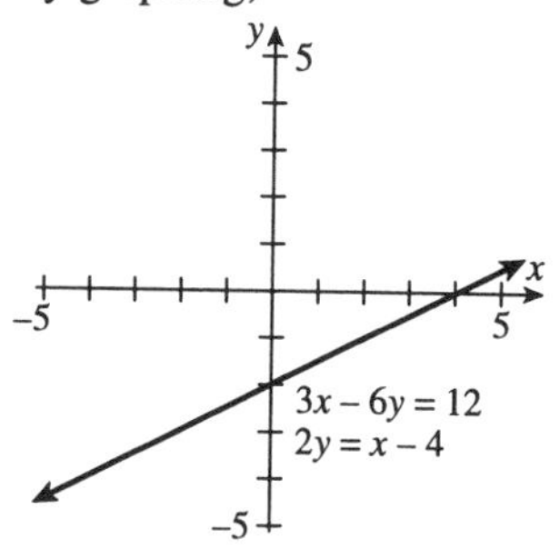

5. $\begin{cases} \frac{1}{2}x-\frac{3}{4}y=-\frac{1}{2} \\ \frac{1}{8}x+\frac{3}{4}y=\frac{19}{8} \end{cases}$

a.

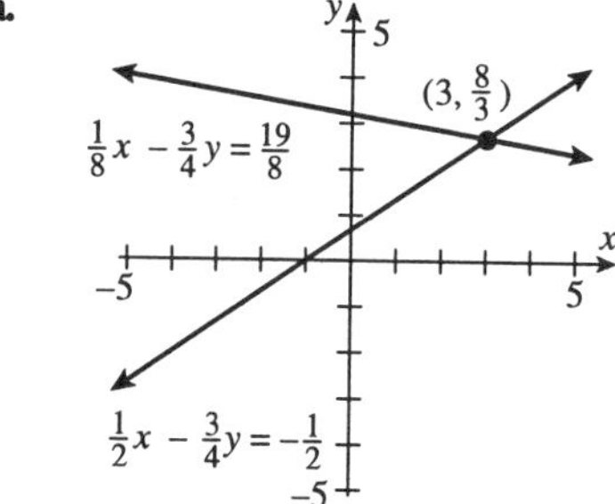

b. $\frac{3}{4}y = \frac{1}{2}x + \frac{1}{2}$

Substitute.

$\frac{1}{8}x + \frac{1}{2}x + \frac{1}{2} = \frac{19}{8}$

$x + 4x + 4 = 19$

$5x + 4 = 19$

$5x = 15$, so $x = 3$

Substitute back.

$\frac{3}{4}y = \frac{1}{2}(3) + \frac{1}{2}$

$\frac{3}{4}y = 2$

$y = \frac{8}{3}$

$\left\{\left(3, \frac{8}{3}\right)\right\}$

c. Add.

$\frac{5}{8}x = \frac{15}{8}$

$5x = 15$, so $x = 3$

Substitute back.

$\frac{1}{2}(3) - \frac{3}{4}y = -\frac{1}{2}$

$-\frac{3}{4}y = -2$

$-3y = -8$

$y = \frac{8}{3}$

$\left\{\left(3, \frac{8}{3}\right)\right\}$

6. $y = 32x$

$y = 15x + 25{,}500$

$32x = 15x + 25{,}500$

$17x = 25{,}500$

$x = 1500$

1500 backpacks

7. $\begin{cases} x + z = 4 \\ 2x - y = 4 \\ x + y - z = 0 \end{cases}$

Adding (2) and (3) gives $3x - z = 4$

Adding (1) and (4) gives $4x = 8$ or $x = 2$

Substitution in (1) gives $2 + z = 4$ or $z = 2$.

Substitution in (3) gives $2 + y - 2 = 0$ or $y = 0$.

8. $\begin{cases} 2x + 5y = 4 \\ x - 5y + z = -1 \\ 4x - z = 11 \end{cases}$

Add equations 2 and 3.

$5x - 5y = 10$

We now have the system

$\begin{cases} 2x + 5y = 4 \\ 5x - 5y = 10 \end{cases}$

Add.

$7x = 14$, so $x = 2$

Substitute back.

$2(2) + 5y = 4$

$4 + 5y = 4$

$5y = 0$, so $y = 0$

Substitute back.

$4(2) - z = 11$

$8 - z = 11$

$8 - 11 = z$

$z = -3$

$\{(2, 0, -3)\}$

9. $\begin{cases} 4y + 2z = 5 \\ 2x + 8y = 5 \\ 6x + 4z = 1 \end{cases}$

From (1) we have $4y = 5 - 2z$.

Substitution in (2) gives

$2x + 2(5 - 2z) = 5$ or

$2x - 4z = -5$ (4)

Adding (4) to (3) gives $8x = -4$ or

$x = -\frac{1}{2}$, and $2\left(-\frac{1}{2}\right) - 4z = -5$

$-4z = -4$ or $z = 1$.

Substitution in (1) gives

$4y + 2(1) = 5$ or $y = \frac{3}{4}$.

10. $\begin{cases} 5x + 7y = 9 \\ 14y - z = 28 \\ 4x + 2z = -4 \end{cases}$

$2x + z = -2$ or $z = -2x - 2$

Substitute back.

$14y - (-2x - 2) = 28$

$14y + 2x + 2 = 28$

$2x + 14y = 26$

$x + 7y = 13$

We now have the system:

We now have the system:

$$\begin{cases} x+7y=13 \\ 5x+7y=9 \end{cases}$$

Subtract.

$-4x = 4$, so $x = -1$

Substitute back.

$-1 + 7y = 13$

$7y = 14$, so $y = 2$

Substitute back.

$z = -2(-1) - 2 = 2 - 2 = 0$

$\{(-1, 2, 0)\}$

11. $$\begin{cases} 3x - 2y + 2z = 5 \\ -x + 6y + z = 4 \\ 3x + 14y + 7z = 20 \end{cases}$$

Multiplying (1) by 3 and adding to (2) gives

$8x + 7z = 19$ (4)

Multiplying (1) by 7 and adding to (3) gives

$24x + 21z = 55$ (5)

Multiplying (4) by -3 and adding to (5) gives

$0 = -2$

There is no solution.

12. $$\begin{cases} x+2y+3z=11 \\ y+2z=3 \\ 2x+2z=10 \end{cases}$$

$x + z = 5$, so $x = -z + 5$

Substitute.

$-z + 5 + 2y + 3z = 11$

$2y + 2z + 5 = 11$

$y + z = 3$ with equation 2 forms the system

$$\begin{cases} y+2z=3 \\ y+z=3 \end{cases}$$

$$\rightarrow \begin{cases} y+2z=3 \\ -y-z=-3 \end{cases}$$

Add.

$z = 0$

Substitute back.

$2y + 0 = 6$

$y = 3$

Substitute back.

$x + 2(3) + 3(0) = 11$

$x = 5$

$\{(5, 3, 0)\}$

13. $$\begin{cases} 7x-3y+2z=0 \\ 4x-4y-z=2 \\ 5x+2y+3z=1 \end{cases}$$

Multiplying (2) by 2 and adding to (1) gives

$15x - 11y = 4$ (4)

Multiplying (2) by 3 and adding to (3) gives

$17x - 10y = 7$ (5)

Multiplying (4) by 10 and (5) by -11 and adding gives $-37x = -37$ or $x = 1$.

Substitution in (4) gives

$15 - 11y = 4$ or $y = 1$.

Substitution in (1) gives

$7 - 3 + 2z = 0$ or $z = -2$.

14. $$\begin{cases} x-3y-5z=-5 \\ 4x-2y+3z=13 \\ 5x+3y+4z=22 \end{cases}$$

Multiply equation 1 by -4 and add to equation 2.

$10y + 23z = 33$

Multiply equation 1 by -5 and add to equation 3.

$18y + 29z = 47$

We now have the system

$$\begin{cases} 10y+23z=33 \\ 18y+29z=47 \end{cases} \rightarrow \begin{cases} 90y+207z=297 \\ 90y+145z=235 \end{cases}$$

Subtract.

$62z = 62$, so $z = 1$

Substitute back.

$10y + 23(1) = 33$

$10y + 23 = 33$

$10y = 10$, so $y = 1$

Substitute back.

$x - 3(1) - 5(1) = -5$

$x - 3 - 5 = -5$

$x - 3 = 0$, so $x = 3$

$\{(3, 1, 1)\}$

15. x is the first number
y is the second number
z is the third number

$$\begin{cases} x+y+z=98 \\ x+y=z+2 \\ y=4x \end{cases}$$

Substitution of (3) into (1) and (2) gives
$5x+z=98$ (4)
$5x-z=2$ (5)
Adding, we have
$10x=100$, or $x=10$.
From (3)
$y=4(10)=40$ and $10+40=z+2$, or
$z=48$.

16. Let p = the number of pennies,
n = the number of nickels, and
d = the number of dimes. Then

$$\begin{cases} p+n+d=95 \\ .01p+.05n+.10d=4.03 \\ p=2d \\ n=d-1 \end{cases}$$

Substitute into equation 1.
$2d+d-1+d=95$
$4d-1=95$
$4d=96$
$d=24$
Substitute back.
$p=2(24)=48$
$n=24-1=23$
Check value.
$.01(48)+.05(23)+.10(24)$
$=.48+1.15+2.40=4.03$ as desired.
Alice has 48 pennies, 23 nickels, and 24 dimes.

17. x is the first number
y is the second number

$$\begin{cases} x=3y \\ 2(x+y)=168 \end{cases}$$

Substitution of (1) into (2) gives
$2(3y+y)=168$
$8y=168$
$y=21$
$x=3(21)=63$

18. Let p = Pat's age, and s = Sue's age. Then

$$\begin{cases} s=p+16 \\ s+15=2(p+15) \end{cases}$$

Substitute.
$p+16+15=2(p+15)$
$p+31=2p+30$
$1=p$
Substitute.
$s=1+16=17$
Sue is 17 years old and Pat is 1 year old.

19. x is the speed of the first car
y is the speed of the second car

$$\begin{cases} 4x+4y=492 \\ y=x+7 \end{cases}$$

Substitution of (2) in (1) gives
$4x+4(x+7)=492$
$8x+28=492$
$8x=464$ or $x=58$ mph.
$y=58+7=65$ mph.

20. Let w = the width of the foundation and
ℓ = the length of the foundation. Then

$$\begin{cases} \ell=3w \\ 2w+2\ell=296 \end{cases}$$

Substitute.
$2w+2(3w)=296$
$2w+6w=296$
$8w=296$
$w=37$
Substitute back. $\ell=3(37)=111$.
The foundation is 37 feet wide and 111 feet long.

21. x is the # of liters of 10% solution
y is the # of liters of 60% solution

$$\begin{cases} x+y=50 \\ .10x+.60y=.40(50) \end{cases}$$

Substitution of (1) in (2) gives
$.10x+.60(50-x)=.40(50)$
$10x+3000-60x=2000$
$-50x=-1000$ or $x=20$
$y=50-20=30$

22. Let c = # of lbs of chocolates used.
n = # of lbs of nuts used, and r = # of lbs of raisins used. Then

$$\begin{cases} r = 2n \\ c+n+r=45 \\ 3.00c+2.70n+2.25r=2.80(45) \end{cases}$$

Substitute equation 1 into equation 2.
$c+n+2n=45$
$c+3n=45$
$c=-3n+45$
Now substitute for c and r in equation 3.
$3.00(-3n+45)+2.70n+2.25(2n)=126$
$-9n+135+2.7n+4.5n=126$
$-1.8n+135=126$
$-1.8n=-9$
$n=5$
Substitute back.
$r=2(5)=10$
$c=-3(5)+45=-15+45=30$
She should used 30 lbs of creme-filled chocolates, 5 lbs of chocolate-covered nuts, and 10 lbs of chocolate-covered raisins.

23. Let x = the number of pennies
Let y = the number of nickels
Let z = the number of dimes

$$\begin{cases} x+y+z=53 \\ .01x+.05y+.10z=2.77 \\ y=z+4 \end{cases}$$

Substituting (3) into (1) and (2) gives
$x+2z=49$ (4)
$x+15z=257$ (5)
Subtracting (4) from (5) gives
$13z=208$ or $z=16$
$y=16+4=20$
$x+2(16)=49$ or $x=17$

24. Let ℓ = the rate of interest on the larger investment and s = the rate of interest on the smaller investment, both expressed as decimals. Then

$$\begin{cases} 10{,}000\ell+4000s=1250 \\ \ell=s+.02 \end{cases}$$

Substitute.
$10{,}000(s+.02)+4000s=1250$
$10{,}000s+200+4000s=1250$
$14{,}000s+200=1250$
$14{,}000s=1050$
so $s=\dfrac{1050}{14{,}000}=.075$ and
$\ell=.075+.02=.095.$
The interest rate on the larger investment is 9.5% and the rate on the smaller investment is 7.5%.

25.

$$\begin{cases} x+y+z=73 \\ x=y \\ z=x+7 \end{cases}$$

Substitution of (2) & (3) in (1) gives
$x+x+x+7=73$
$3x=66$
$x=22$ cm
$y=22$ cm
$z=22+7=29$ cm

26. Let f = the first number,
s = the second number, and
t = the third number. Then

$$\begin{cases} f+s+t=295 \\ f=s+5 \text{ or } s=f-5 \\ f=2t \text{ or } t=\dfrac{f}{2} \end{cases}$$

Substitute.
$f+f-5+\dfrac{f}{2}=295$
$\dfrac{5}{2}f-5=295$
$\dfrac{5}{2}f=300$
$f=120$
Substitute back.
$s=120-5=115$
$t=\dfrac{120}{2}=60$
The first number is 120, the second is 115, and the third is 60.

27. $\begin{cases} 3x+10y=\ \ 1 \\ \ \ x+\ \ 2y=-1 \end{cases}$

The augmented matrix of the system is

$\left(\begin{array}{cc|c} 3 & 10 & 1 \\ 1 & 2 & -1 \end{array}\right)$. Interchanging row 1 and row 2 we have $\left(\begin{array}{cc|c} 1 & 2 & -1 \\ 3 & 10 & 1 \end{array}\right)$.

Multiplying row 1 by –3 and adding to row 2 gives $\left(\begin{array}{cc|c} 1 & 2 & -1 \\ 0 & 4 & 4 \end{array}\right)$. Dividing row 2 by 4 gives $\left(\begin{array}{cc|c} 1 & 2 & -1 \\ 0 & 1 & 1 \end{array}\right)$. This is the augmented matrix of the system

$x+2y=-1$
$y=1$

Substitution gives $x + 2(1) = -1$ or $x = -3$.

28. $\begin{cases} 3x-6y=12 \\ \ \ \ \ \ \ 2y=x-4 \end{cases} \rightarrow \begin{cases} 3x-6y=12 \\ -x+2y=-4 \end{cases}$

$\left(\begin{array}{cc|c} 3 & -6 & 12 \\ -1 & 2 & -4 \end{array}\right) \rightarrow \left(\begin{array}{cc|c} 1 & -2 & 4 \\ -1 & 2 & -4 \end{array}\right)$

$\rightarrow \left(\begin{array}{cc|c} 1 & -2 & 4 \\ 0 & 0 & 0 \end{array}\right)$

Dependent system.

$\{(x, y) | x - 2y = 4\}$

29. $\begin{cases} 3x-2y=-8 \\ 6x+5y=11 \end{cases}$

The augmented matrix of the system is

$\left(\begin{array}{cc|c} 3 & -2 & -8 \\ 6 & 5 & 11 \end{array}\right)$. Multiplying row 1 by –2 and adding to row 2 gives $\left(\begin{array}{cc|c} 3 & -2 & -8 \\ 6 & 5 & 11 \end{array}\right)$.

Dividing row 1 by 3 and row 2 by 9 gives

$\left(\begin{array}{cc|c} 1 & -\frac{2}{3} & -\frac{8}{3} \\ 0 & 1 & 3 \end{array}\right)$. This is the augmented matrix of the system $x-\frac{2}{3}y=-\frac{8}{3}$

Substitution gives

$x-\frac{2}{3}(3)=-\frac{8}{3}$ or $x=-\frac{2}{3}$

30. $\begin{cases} 6x-6y=-5 \\ 10x-2y=1 \end{cases} \left(\begin{array}{cc|c} 6 & -6 & -5 \\ 10 & -2 & 1 \end{array}\right)$

$\rightarrow \left(\begin{array}{cc|c} 1 & -1 & -\frac{5}{6} \\ 10 & -2 & 1 \end{array}\right) \rightarrow \left(\begin{array}{cc|c} 1 & -1 & -\frac{5}{6} \\ 0 & 8 & \frac{28}{3} \end{array}\right)$

$\rightarrow \left(\begin{array}{cc|c} 1 & -1 & -\frac{5}{6} \\ 0 & 1 & \frac{7}{6} \end{array}\right) \rightarrow \left(\begin{array}{cc|c} 1 & 0 & \frac{1}{3} \\ 0 & 1 & \frac{7}{6} \end{array}\right)$

$\left\{\left(\frac{1}{3}, \frac{7}{6}\right)\right\}$

31. $\begin{cases} 3x-6y=0 \\ 2x+4y=5 \end{cases}$

The augmented matrix of the system is

$\left(\begin{array}{cc|c} 3 & -6 & 0 \\ 2 & 4 & 5 \end{array}\right)$. Dividing row 1 by 3 gives

$\left(\begin{array}{cc|c} 1 & -2 & 4 \\ 0 & 2 & 5 \end{array}\right)$.

Multiplying row 1 by –2 and adding to row 2 gives $\left(\begin{array}{cc|c} 1 & -2 & 0 \\ 0 & 8 & 5 \end{array}\right)$. Dividing row 2 by 8 gives $\left(\begin{array}{cc|c} 1 & -2 & 0 \\ 0 & 1 & \frac{5}{8} \end{array}\right)$. This is the augmented matrix of the system

$x-2y=0$
$y=\frac{5}{8}$

Substitution gives

$x-2\left(\frac{5}{8}\right)=0$ or $x=\frac{5}{4}$.

32. $\begin{cases} 5x-3y=10 \\ -2x+y=-1 \end{cases} \left(\begin{array}{cc|c} 5 & -3 & 10 \\ -2 & 1 & -1 \end{array}\right)$

$\to \left(\begin{array}{cc|c} 1 & -\frac{3}{5} & 2 \\ -2 & 1 & -1 \end{array}\right) \to \left(\begin{array}{cc|c} 1 & -\frac{3}{5} & 2 \\ 0 & -\frac{1}{5} & 3 \end{array}\right)$

$\to \left(\begin{array}{cc|c} 1 & -\frac{3}{5} & 2 \\ 0 & 1 & -15 \end{array}\right) \to \left(\begin{array}{cc|c} 1 & 0 & -7 \\ 0 & 1 & -15 \end{array}\right)$

$\{(-7, -15)\}$

33. $\begin{cases} .2x-.3y=-0.7 \\ .5x+.3y=1.4 \end{cases}$

The augmented matrix of the system is

$\left(\begin{array}{cc|c} .2 & -.3 & -.7 \\ .5 & .3 & 1.4 \end{array}\right)$

Multiplying row 1 by 5 we have

$\left(\begin{array}{cc|c} .2 & -.3 & -.7 \\ .5 & .3 & 1.4 \end{array}\right)$. Multiplying row 1 by 5

we have $\left(\begin{array}{cc|c} 1 & -1.5 & -3.5 \\ .5 & .3 & 1.4 \end{array}\right)$. Multiplying

row 1 by –.5 and adding to row 2 gives

$\left(\begin{array}{cc|c} 1 & -1.5 & -3.5 \\ 0 & 1.05 & 3.15 \end{array}\right)$. Dividing row 2 by 1.05

gives $\left(\begin{array}{cc|c} 1 & -1.5 & -3.5 \\ 0 & 1 & 3 \end{array}\right)$. This is the

augmented matrix of the system

$x-1.5y=-3.5$

$y=3$

Substitution gives

$x-1.5(3)=-3.5$ or $x=1$.

34. $\begin{cases} 3x+2y=8 \\ 3x-y=5 \end{cases} \left(\begin{array}{cc|c} 3 & 2 & 8 \\ 3 & -1 & 5 \end{array}\right)$

$\to \left(\begin{array}{cc|c} 1 & \frac{2}{3} & \frac{8}{3} \\ 3 & -1 & 5 \end{array}\right) \to \left(\begin{array}{cc|c} 1 & \frac{2}{3} & \frac{8}{3} \\ 0 & -3 & -3 \end{array}\right)$

$\to \left(\begin{array}{cc|c} 1 & \frac{2}{3} & \frac{8}{3} \\ 0 & 1 & 1 \end{array}\right) \to \left(\begin{array}{cc|c} 1 & 0 & 2 \\ 0 & 1 & 1 \end{array}\right)$

$\{(2, 1)\}$

35. $\begin{cases} x+\quad z=4 \\ 2x-y\quad =0 \\ x+y-z=0 \end{cases}$

The augmented matrix of the system is

$\left(\begin{array}{ccc|c} 1 & 0 & 1 & 4 \\ 2 & -1 & 0 & 0 \\ 1 & 1 & -1 & 0 \end{array}\right)$. Multiplying row 1 by –2

and by –1 and adding to rows 2 and 3

respectively gives $\left(\begin{array}{ccc|c} 1 & 0 & 1 & 4 \\ 0 & -1 & -2 & -8 \\ 0 & 1 & -2 & -4 \end{array}\right)$.

Adding row 2 to row 3 and then multiplying row 2 by –1 gives

$\left(\begin{array}{ccc|c} 1 & 0 & 1 & 4 \\ 0 & 1 & 2 & 8 \\ 0 & 0 & -4 & -12 \end{array}\right)$. Dividing row 3 by –4

gives $\left(\begin{array}{ccc|c} 1 & 0 & 1 & 4 \\ 0 & 1 & 2 & 8 \\ 0 & 0 & 1 & 3 \end{array}\right)$. This is the augmented

matrix of the system

$x+z=4$

$y+2z=8$

$z=3$

Substitution gives

$y+2(3)=8$ or $y=2$ and $x+3=4$ or $x=1$.

36. $\begin{cases} 2x+5y\quad = 4 \\ x-5y+z=-1 \\ 4x\quad -z=11 \end{cases} \left(\begin{array}{ccc|c} 2 & 5 & 0 & 4 \\ 1 & -5 & 1 & -1 \\ 4 & 0 & -1 & 11 \end{array}\right)$

$\to \left(\begin{array}{ccc|c} 1 & \frac{5}{2} & 0 & 2 \\ 1 & -5 & 1 & -1 \\ 4 & 0 & -1 & 11 \end{array}\right) \to \left(\begin{array}{ccc|c} 1 & \frac{5}{2} & 0 & 2 \\ 0 & -\frac{15}{2} & 1 & -3 \\ 4 & 0 & -1 & 11 \end{array}\right)$

$\to \left(\begin{array}{ccc|c} 1 & \frac{5}{2} & 0 & 2 \\ 0 & -\frac{15}{2} & 1 & -3 \\ 0 & -10 & -1 & 3 \end{array}\right) \to \left(\begin{array}{ccc|c} 1 & \frac{5}{2} & 0 & 2 \\ 0 & 1 & -\frac{2}{15} & \frac{2}{5} \\ 0 & -10 & -1 & 3 \end{array}\right)$

$$\rightarrow\left(\begin{array}{ccc|c} 1 & 0 & \frac{1}{3} & 1 \\ 0 & 1 & -\frac{2}{15} & \frac{2}{5} \\ 0 & -10 & -1 & 3 \end{array}\right) \rightarrow \left(\begin{array}{ccc|c} 1 & 0 & \frac{1}{3} & 1 \\ 0 & 1 & -\frac{2}{15} & \frac{2}{5} \\ 0 & 0 & -\frac{7}{3} & 7 \end{array}\right)$$

$$\rightarrow\left(\begin{array}{ccc|c} 1 & 0 & \frac{1}{3} & 1 \\ 0 & 1 & -\frac{2}{15} & \frac{2}{5} \\ 0 & 0 & 1 & -3 \end{array}\right) \rightarrow \left(\begin{array}{ccc|c} 1 & 0 & 0 & 0 \\ 0 & 1 & -\frac{2}{15} & \frac{2}{5} \\ 0 & 0 & 1 & -3 \end{array}\right)$$

$$\rightarrow\left(\begin{array}{ccc|c} 1 & 0 & 0 & 2 \\ 0 & 1 & 0 & 0 \\ 0 & 0 & 1 & -3 \end{array}\right)$$

{(2, 0, –3)}

37. $\begin{cases} 3x - y \quad = 11 \\ x + \ 2z = 13 \\ \quad y - z = -7 \end{cases}$

The augmented matrix of the system

is $\left(\begin{array}{ccc|c} 3 & -1 & 0 & 11 \\ 1 & 0 & 2 & 13 \\ 0 & 1 & -1 & -7 \end{array}\right)$.

Interchanging row 1 with row 2, and then row 2 with row 3 gives

$\left(\begin{array}{ccc|c} 1 & 0 & 2 & 13 \\ 0 & 1 & -1 & -7 \\ 3 & -1 & 0 & 11 \end{array}\right)$.

Multiplying row 1 by –3 and adding to row 3 gives $\left(\begin{array}{ccc|c} 1 & 0 & 2 & 13 \\ 0 & 1 & -1 & -7 \\ 0 & -1 & -6 & -28 \end{array}\right)$. Adding row 2 to row 3 gives $\left(\begin{array}{ccc|c} 1 & 0 & 2 & 13 \\ 0 & 1 & -1 & -7 \\ 0 & 0 & -7 & -35 \end{array}\right)$.

Dividing row 3 by – 7 gives

$\left(\begin{array}{ccc|c} 1 & 0 & 2 & 13 \\ 0 & 1 & -1 & -7 \\ 0 & 0 & 1 & 5 \end{array}\right)$. This is the augmented matrix of the system

$x + 2z = 13$
$y - z = -7$
$z = 5$

Substitution gives
$y - 5 = -7$ or $y = -2$ and $x + 2(5) = 13$ or $x = 3$.

38. $\begin{cases} 5x + 7y + 3z = 9 \\ \quad 14y - z = 28 \\ 4x + \quad 2z = -4 \end{cases} \left(\begin{array}{ccc|c} 5 & 7 & 3 & 9 \\ 0 & 14 & -1 & 28 \\ 4 & 0 & 2 & -4 \end{array}\right)$

$$\rightarrow\left(\begin{array}{ccc|c} 1 & \frac{7}{5} & \frac{3}{5} & \frac{9}{5} \\ 0 & 14 & -1 & 28 \\ 4 & 0 & 2 & -4 \end{array}\right) \rightarrow \left(\begin{array}{ccc|c} 1 & \frac{7}{5} & \frac{3}{5} & \frac{9}{5} \\ 0 & 14 & -1 & 28 \\ 0 & -\frac{28}{5} & -\frac{2}{5} & -\frac{56}{5} \end{array}\right)$$

$$\rightarrow\left(\begin{array}{ccc|c} 1 & \frac{7}{5} & \frac{3}{5} & \frac{9}{5} \\ 0 & 1 & -\frac{1}{14} & 2 \\ 0 & -\frac{28}{5} & -\frac{2}{5} & -\frac{56}{5} \end{array}\right)$$

$$\rightarrow\left(\begin{array}{ccc|c} 1 & 0 & \frac{7}{10} & -1 \\ 0 & 1 & -\frac{1}{14} & 2 \\ 0 & -\frac{28}{5} & -\frac{2}{5} & -\frac{56}{5} \end{array}\right)$$

$$\rightarrow\left(\begin{array}{ccc|c} 1 & 0 & \frac{7}{10} & -1 \\ 0 & 1 & -\frac{1}{14} & 2 \\ 0 & 0 & -\frac{4}{5} & 0 \end{array}\right) \rightarrow \left(\begin{array}{ccc|c} 1 & 0 & \frac{7}{10} & -1 \\ 0 & 1 & -\frac{1}{14} & 2 \\ 0 & 0 & 1 & 0 \end{array}\right)$$

$$\rightarrow\left(\begin{array}{ccc|c} 1 & 0 & 0 & -1 \\ 0 & 1 & -\frac{1}{14} & 2 \\ 0 & 0 & 1 & 0 \end{array}\right) \rightarrow \left(\begin{array}{ccc|c} 1 & 0 & 0 & -1 \\ 0 & 1 & 0 & 2 \\ 0 & 0 & 1 & 0 \end{array}\right)$$

{(–1, 2, 0)}

39. $\begin{cases} 7x - 3y + 2z = 0 \\ 4x - 4y - \ z = 2 \\ 5x + 2y + 3z = 1 \end{cases}$

The augmented matrix of the system is

$\left(\begin{array}{ccc|c} 7 & -3 & 2 & 0 \\ 4 & -4 & -1 & 2 \\ 5 & 2 & 3 & 1 \end{array}\right)$.

Interchanging rows 1 & 2 and dividing row 1 by 4 gives $\left(\begin{array}{ccc|c} 1 & -1 & -\frac{1}{4} & \frac{1}{2} \\ 7 & -3 & 2 & 0 \\ 5 & 2 & 3 & 1 \end{array}\right)$.

Multiplying row 1 by –7 and by –5 and adding to rows 2 and 3 respectively gives $\left(\begin{array}{ccc|c} 1 & -1 & -\frac{1}{4} & \frac{1}{2} \\ 0 & 4 & \frac{15}{4} & -\frac{7}{2} \\ 0 & 7 & \frac{17}{4} & -\frac{3}{2} \end{array}\right)$. Dividing row 2 by 4 gives $\left(\begin{array}{ccc|c} 1 & -1 & -\frac{1}{4} & \frac{1}{2} \\ 0 & 1 & \frac{15}{16} & -\frac{7}{8} \\ 0 & 7 & \frac{17}{4} & -\frac{3}{2} \end{array}\right)$. Multiplying row 2 by –7 and adding to row 3 gives $\left(\begin{array}{ccc|c} 1 & -1 & -\frac{1}{4} & \frac{1}{2} \\ 0 & 1 & \frac{15}{16} & -\frac{7}{8} \\ 0 & 0 & -\frac{37}{16} & \frac{37}{8} \end{array}\right)$. Multiplying row 3 by $-\frac{16}{37}$ gives $\left(\begin{array}{ccc|c} 1 & -1 & -\frac{1}{4} & \frac{1}{2} \\ 0 & 1 & \frac{15}{16} & -\frac{7}{8} \\ 0 & 0 & 1 & -2 \end{array}\right)$.

This is the augmented matrix of the system

$$\begin{aligned} x - y - \tfrac{1}{4}z &= \tfrac{1}{2} \\ y + \tfrac{15}{16}z &= -\tfrac{7}{8} \\ z &= -2 \end{aligned}$$

Substitution gives

$y = \frac{15}{16}(-2) = -\frac{7}{8}$ or $y = 1$ and

$x = 1 - \frac{1}{4}(-2) = \frac{1}{2}$ or $x = 1$.

40. $\begin{cases} x + 2y + 3z = 14 \\ \quad y + 2z = 3 \\ 2x - \quad 2z = 10 \end{cases}$ $\left(\begin{array}{ccc|c} 1 & 2 & 3 & 14 \\ 0 & 1 & 2 & 3 \\ 2 & 0 & -2 & 10 \end{array}\right)$

$$\rightarrow \left(\begin{array}{ccc|c} 1 & 2 & 3 & 14 \\ 0 & 1 & 2 & 3 \\ 0 & -4 & -8 & -18 \end{array}\right) \rightarrow \left(\begin{array}{ccc|c} 1 & 0 & -1 & 8 \\ 0 & 1 & 2 & 3 \\ 0 & -4 & -8 & -18 \end{array}\right)$$

$$\rightarrow \left(\begin{array}{ccc|c} 1 & 0 & -1 & 8 \\ 0 & 1 & 2 & 3 \\ 0 & 0 & 0 & -6 \end{array}\right)$$

No solution exists.

{ }

41. $\begin{vmatrix} -1 & 3 \\ 5 & 2 \end{vmatrix} = -2 - 15 = -17$

42. $\begin{vmatrix} 3 & -1 \\ 2 & 5 \end{vmatrix} = 3(5) - (-1)2 = 15 + 2 = 17$

43. $\begin{vmatrix} 2 & -1 & -3 \\ 1 & 2 & 0 \\ 3 & -2 & 2 \end{vmatrix}$

$= 2\begin{vmatrix} 2 & 0 \\ -2 & 2 \end{vmatrix} + 1\begin{vmatrix} 1 & 0 \\ 3 & 2 \end{vmatrix} - 3\begin{vmatrix} 1 & 2 \\ 3 & -2 \end{vmatrix}$

$= 2(4) + 1(2) - 3(-8) = 34$

44. $\begin{vmatrix} -2 & 3 & 1 \\ 4 & 4 & 0 \\ 1 & -2 & 3 \end{vmatrix}$

$= 1 \cdot \begin{vmatrix} 4 & 4 \\ 1 & -2 \end{vmatrix} - 0 \cdot \begin{vmatrix} -2 & 3 \\ 1 & -2 \end{vmatrix} + 3\begin{vmatrix} -2 & 3 \\ 4 & 4 \end{vmatrix}$

$= (-8 - 4) - 0 + 3(-8 - 12)$

$= -12 + 3(-20) = -12 - 60 = -72$

45. $\begin{cases} 3x - 2y = -8 \\ 6x + 5y = 11 \end{cases}$

$D = \begin{vmatrix} 3 & -2 \\ 6 & 5 \end{vmatrix} = 27$

$D_x = \begin{vmatrix} -8 & -2 \\ 11 & 5 \end{vmatrix} = -18$

$D_y = \begin{vmatrix} 3 & -8 \\ 6 & 11 \end{vmatrix} = 81$

$x = \frac{D_x}{D} = \frac{-18}{27} = -\frac{2}{3}$

$y = \frac{D_y}{D} = \frac{81}{27} = 3$

46. $\begin{cases} 6x - 6y = -5 \\ 10x - 2y = 1 \end{cases}$

$D = \begin{vmatrix} 6 & -6 \\ 10 & -2 \end{vmatrix} = 6(-2) - (-6)(10)$

$= -12 + 60 = 48$

$D_x = \begin{vmatrix} -5 & -6 \\ 1 & -2 \end{vmatrix} = (-5)(-2) - (-6)(1)$
$= 10 + 6 = 16$
$D_y = \begin{vmatrix} 6 & -5 \\ 10 & 1 \end{vmatrix} = 6(1) - (-5)(10)$
$= 6 + 50 = 56$
$x = \frac{16}{48} = \frac{1}{3}$ and $y = \frac{56}{48} = \frac{7}{6}$
$\left\{\left(\frac{1}{3}, \frac{7}{6}\right)\right\}$

47. $\begin{cases} 3x + 10y = 1 \\ x + 2y = -1 \end{cases}$
$D = \begin{vmatrix} 3 & 10 \\ 1 & 2 \end{vmatrix} = -4$
$D_x = \begin{vmatrix} 1 & 10 \\ -1 & 2 \end{vmatrix} = 12$
$D_y = \begin{vmatrix} 3 & 1 \\ 1 & -1 \end{vmatrix} = -4$
$x = \frac{D_x}{D} = \frac{12}{-4} = -3$
$y = \frac{D_y}{D} = \frac{-4}{-4} = 1$

48. $\begin{cases} y = \frac{1}{2}x + \frac{2}{3} \\ 4x + 6y = 4 \end{cases} \rightarrow \begin{cases} -\frac{1}{2}x + y = \frac{2}{3} \\ 4x + 6y = 4 \end{cases}$
$D = \begin{vmatrix} -\frac{1}{2} & 1 \\ 4 & 6 \end{vmatrix} = \left(-\frac{1}{2}\right)6 - 1(4)$
$= -3 - 4 = -7$
$D_x = \begin{vmatrix} \frac{2}{3} & 1 \\ 4 & 6 \end{vmatrix} = \left(\frac{2}{3}\right)6 - 1(4)$
$= 4 - 4 = 0$
$D_y = \begin{vmatrix} -\frac{1}{2} & \frac{2}{3} \\ 4 & 4 \end{vmatrix} = \left(-\frac{1}{2}\right)4 - \left(\frac{2}{3}\right)4$
$= -2 - \frac{8}{3} = -\frac{14}{3}$
$x = \frac{0}{-7} = 0$ and $y = \frac{-\frac{14}{3}}{-7} = \frac{2}{3}$
$\left\{\left(0, \frac{2}{3}\right)\right\}$

49. $\begin{cases} 2x - 4y = 22 \\ 5x - 10y = 16 \end{cases}$
$D = \begin{vmatrix} 2 & -4 \\ 5 & -10 \end{vmatrix} = 0$
Cannot be solved by Cramer's rule. Multiplying (1) by –5 and (2) by 2 and adding gives 0 = –78. No solution.

50. $\begin{cases} 3x - 6y = 12 \\ 2y = x - 4 \end{cases} \rightarrow \begin{cases} 3x - 6y = 12 \\ -x + 2y = -4 \end{cases}$
$D = \begin{vmatrix} 3 & -6 \\ -1 & 2 \end{vmatrix} = 3(2) - (-6)(-1)$
$= 6 - 6 = 0$
Therefore, Cramer's Rule cannot be used to solve this system. Since equation 1 is –3 times equation 2, the system is dependent.
$\{(x, y) | -x + 2y = -4)\}$

51. $\begin{cases} x + z = 4 \\ 2x - y = 0 \\ x + y - z = 0 \end{cases}$
$D = \begin{vmatrix} 1 & 0 & 1 \\ 2 & -1 & 0 \\ 1 & 1 & -1 \end{vmatrix}$
$= 1 \cdot \begin{vmatrix} -1 & 0 \\ 1 & -1 \end{vmatrix} + 1 \cdot \begin{vmatrix} 2 & -1 \\ 1 & 1 \end{vmatrix}$
$= (1)(1) + (1)(3) = 4$
$D_x = \begin{vmatrix} 4 & 0 & 1 \\ 0 & -1 & 0 \\ 0 & 1 & -1 \end{vmatrix}$
$= 4 \cdot \begin{vmatrix} -1 & 0 \\ 1 & -1 \end{vmatrix} = 4(1) = 4$
$D_y = \begin{vmatrix} 1 & 4 & 1 \\ 2 & 0 & 0 \\ 1 & 0 & -1 \end{vmatrix}$
$= -4 \cdot \begin{vmatrix} 2 & 0 \\ 1 & -1 \end{vmatrix} = (-4)(-2) = 8$

$$D_z = \begin{vmatrix} 1 & 0 & 4 \\ 2 & -1 & 0 \\ 1 & 1 & 0 \end{vmatrix}$$
$$= 4 \cdot \begin{vmatrix} 2 & -1 \\ 1 & 1 \end{vmatrix} = 4(3) = 12$$
$$x = \frac{D_x}{D} = \frac{4}{4} = 1$$
$$y = \frac{D_y}{D} = \frac{8}{4} = 2$$
$$z = \frac{D_z}{D} = \frac{12}{4} = 3$$

52. $\begin{cases} 2x + 5y \quad = 4 \\ x - 5y + z = -1 \\ 4x - \quad z = 11 \end{cases}$

$$D = \begin{vmatrix} 2 & 5 & 0 \\ 1 & -5 & 1 \\ 4 & 0 & -1 \end{vmatrix}$$
$$= 0\begin{vmatrix} 1 & -5 \\ 4 & 0 \end{vmatrix} - 1 \cdot \begin{vmatrix} 2 & 5 \\ 4 & 0 \end{vmatrix} + (-1)\begin{vmatrix} 2 & 5 \\ 1 & -5 \end{vmatrix}$$
$$= 0 - (0 - 20) - (-10 - 5) = 20 + 15 = 35$$
$$D_x = \begin{vmatrix} 4 & 5 & 0 \\ -1 & -5 & 1 \\ 11 & 0 & -1 \end{vmatrix}$$
$$= 0 \cdot \begin{vmatrix} -1 & -5 \\ 11 & 0 \end{vmatrix} - 1 \cdot \begin{vmatrix} 4 & 5 \\ 11 & 0 \end{vmatrix} + (-1)\begin{vmatrix} 4 & 5 \\ -1 & -5 \end{vmatrix}$$
$$= 0 - (0 - 55) - [-20 - (-5)]$$
$$= 55 + 15 = 70$$
$$D_y = \begin{vmatrix} 2 & 4 & 0 \\ 1 & -1 & 1 \\ 4 & 11 & -1 \end{vmatrix}$$
$$= 0 \cdot \begin{vmatrix} 1 & -1 \\ 4 & 11 \end{vmatrix} - 1 \cdot \begin{vmatrix} 2 & 4 \\ 4 & 11 \end{vmatrix} + (-1)\begin{vmatrix} 2 & 4 \\ 1 & -1 \end{vmatrix}$$
$$= 0 - (22 - 16) - (-2 - 4) = -6 + 6 = 0$$
$$D_z = \begin{vmatrix} 2 & 5 & 4 \\ 1 & -5 & -1 \\ 4 & 0 & 11 \end{vmatrix}$$
$$= 4 \cdot \begin{vmatrix} 1 & -5 \\ 4 & 0 \end{vmatrix} - (-1) \cdot \begin{vmatrix} 2 & 5 \\ 4 & 0 \end{vmatrix} + 11 \cdot \begin{vmatrix} 2 & 5 \\ 1 & -5 \end{vmatrix}$$
$$= 4[0 - (-20)] + 1 \cdot (0 - 20) + 11 \cdot (-10 - 5)$$
$$= 4(20) - 20 + 11(-15) = 60 - 165 = -105$$
$$x = \frac{70}{35} = 2, \ y = \frac{0}{35} = 0, \ z = \frac{-105}{35} = -3$$
$\{(2, 0, -3)\}$

53. $\begin{cases} x + 3y - z = 5 \\ 2x - y - 2z = 3 \\ x + 2y + 3z = 4 \end{cases}$

$$D = \begin{vmatrix} 1 & 3 & -1 \\ 2 & -1 & -2 \\ 1 & 2 & 3 \end{vmatrix}$$
$$= 1 \cdot \begin{vmatrix} -1 & -2 \\ 2 & 3 \end{vmatrix} - 3 \cdot \begin{vmatrix} 2 & -2 \\ 1 & 3 \end{vmatrix} - 1 \cdot \begin{vmatrix} 2 & -1 \\ 1 & 2 \end{vmatrix}$$
$$= (1)(1) - 3(8) - 1(5) = -28$$
$$D_x = \begin{vmatrix} 5 & 3 & -1 \\ 3 & -1 & -2 \\ 4 & 2 & 3 \end{vmatrix}$$
$$= 5 \cdot \begin{vmatrix} -1 & -2 \\ 2 & 3 \end{vmatrix} - 3 \cdot \begin{vmatrix} 3 & -2 \\ 4 & 3 \end{vmatrix} - 1 \cdot \begin{vmatrix} 3 & -1 \\ 4 & 2 \end{vmatrix}$$
$$= 5(1) - 3(17) - 1(10) = -56$$
$$D_y = \begin{vmatrix} 1 & 5 & -1 \\ 2 & 3 & -2 \\ 1 & 4 & 3 \end{vmatrix}$$
$$= 1 \cdot \begin{vmatrix} 3 & -2 \\ 4 & 3 \end{vmatrix} - 5 \cdot \begin{vmatrix} 2 & -2 \\ 1 & 3 \end{vmatrix} - 1 \cdot \begin{vmatrix} 2 & 3 \\ 1 & 4 \end{vmatrix}$$
$$= 1(17) - 5(8) - 1(5) = -28$$
$$D_z = \begin{vmatrix} 1 & 3 & 5 \\ 2 & -1 & 3 \\ 1 & 2 & 4 \end{vmatrix}$$
$$= 1 \cdot \begin{vmatrix} -1 & 3 \\ 2 & 4 \end{vmatrix} - 3 \cdot \begin{vmatrix} 2 & 3 \\ 1 & 4 \end{vmatrix} + 5 \cdot \begin{vmatrix} 2 & -1 \\ 1 & 2 \end{vmatrix}$$
$$= 1(-10) - 3(5) + 5(5) = 0$$
$$x = \frac{D_x}{D} = \frac{-56}{-28} = 2$$
$$y = \frac{D_y}{D} = \frac{-28}{-28} = 1$$
$$z = \frac{D_z}{D} = 0$$

54. $\begin{cases} 2x - \quad z = 1 \\ 3x - y + 2z = 3 \\ x + y + 3z = -2 \end{cases}$

$$D = \begin{vmatrix} 2 & 0 & -1 \\ 3 & -1 & 2 \\ 1 & 1 & 3 \end{vmatrix}$$

$$= 2 \cdot \begin{vmatrix} -1 & 2 \\ 1 & 3 \end{vmatrix} - 0 \cdot \begin{vmatrix} 3 & 2 \\ 1 & 3 \end{vmatrix} + (-1) \cdot \begin{vmatrix} 3 & -1 \\ 1 & 1 \end{vmatrix}$$

$= 2(-3-2) - 0 - [3-(-1)]$
$= 2(-5) - 4 = -10 - 4 = -14$

$$D_x = \begin{vmatrix} 1 & 0 & -1 \\ 3 & -1 & 2 \\ -2 & 1 & 3 \end{vmatrix}$$

$$= 1 \cdot \begin{vmatrix} -1 & 2 \\ 1 & 3 \end{vmatrix} - 0 \cdot \begin{vmatrix} 3 & 2 \\ -2 & 3 \end{vmatrix} + (-1) \cdot \begin{vmatrix} 3 & -1 \\ -2 & 1 \end{vmatrix}$$

$= (-3-2) - 0 - (3-2) = -5 - 1 = -6$

$$D_y = \begin{vmatrix} 2 & 1 & -1 \\ 3 & 3 & 2 \\ 1 & -2 & 3 \end{vmatrix}$$

$$= 2 \cdot \begin{vmatrix} 3 & 2 \\ -2 & 3 \end{vmatrix} - 1 \begin{vmatrix} 3 & 2 \\ 1 & 3 \end{vmatrix} + (-1) \cdot \begin{vmatrix} 3 & 3 \\ 1 & -2 \end{vmatrix}$$

$= 2[9-(-4)] - (9-2) - (6-3)$
$= 2(13) - 7 + 9 = 26 + 2 = 28$

$$D_z = \begin{vmatrix} 2 & 0 & 1 \\ 3 & -1 & 3 \\ 1 & 1 & -2 \end{vmatrix}$$

$$= 2 \cdot \begin{vmatrix} -1 & 3 \\ 1 & -2 \end{vmatrix} - 0 \cdot \begin{vmatrix} 3 & 3 \\ 1 & -2 \end{vmatrix} + 1 \cdot \begin{vmatrix} 3 & -1 \\ 1 & 1 \end{vmatrix}$$

$= 2(2-3) - 0 + [3-(-1)]$
$= 2(-1) + 4 = -2 + 4 = 2$

$x = \frac{-6}{-14} = \frac{3}{7},\ y = \frac{28}{-14} = -2,$

$z = \frac{2}{-14} = \frac{-1}{7}$

$\left\{\left(\frac{3}{7}, -2, -\frac{1}{7}\right)\right\}$

55. $\begin{cases} x + 2y + 3z = 14 \\ \quad y + 2z = 3 \\ 2x - \quad 2z = 10 \end{cases}$

$$D = \begin{vmatrix} 1 & 2 & 3 \\ 0 & 1 & 2 \\ 2 & 0 & -2 \end{vmatrix}$$

$$= 1 \cdot \begin{vmatrix} 1 & 2 \\ 0 & -2 \end{vmatrix} + 2 \cdot \begin{vmatrix} 2 & 3 \\ 1 & 2 \end{vmatrix}$$

$= 1(-2) + 2(1) = 0$

Cannot be solved by Cramer's rule.
Substituting (3) and (2) into (1) gives
$(5 + z) + 2(3 - 2z) + 3z = 14$
$11 = 14$
No solution.

56. $\begin{cases} 5x + 7y \quad = 9 \\ \quad 14y - z = 28 \\ 4x + \quad 2z = -4 \end{cases}$

$$D = \begin{vmatrix} 5 & 7 & 0 \\ 0 & 14 & -1 \\ 4 & 0 & 2 \end{vmatrix}$$

$$= 5 \cdot \begin{vmatrix} 14 & -1 \\ 0 & 2 \end{vmatrix} - 7 \cdot \begin{vmatrix} 0 & -1 \\ 4 & 2 \end{vmatrix} + 0 \cdot \begin{vmatrix} 0 & 14 \\ 4 & 0 \end{vmatrix}$$

$= 5(28-0) - 7[0-(-4)] + 0$
$= 140 - 28 = 112$

$$D_x = \begin{vmatrix} 9 & 7 & 0 \\ 28 & 14 & -1 \\ -4 & 0 & 2 \end{vmatrix}$$

$$= 9 \cdot \begin{vmatrix} 14 & -1 \\ 0 & 2 \end{vmatrix} - 7 \cdot \begin{vmatrix} 28 & -1 \\ -4 & 2 \end{vmatrix} + 0 \cdot \begin{vmatrix} 28 & 14 \\ -4 & 0 \end{vmatrix}$$

$= 9(28-0) - 7(56-4) + 0$
$= 252 - 7(52) = 252 - 364 = -112$

$$D_y = \begin{vmatrix} 5 & 9 & 0 \\ 0 & 28 & -1 \\ 4 & -4 & 2 \end{vmatrix}$$

$$= 5 \cdot \begin{vmatrix} 28 & -1 \\ -4 & 2 \end{vmatrix} - 9 \cdot \begin{vmatrix} 0 & -1 \\ 4 & 2 \end{vmatrix} + 0 \cdot \begin{vmatrix} 0 & 28 \\ 4 & -4 \end{vmatrix}$$

$= 5(56-4) - 9[0-(-4)] + 0$
$= 5(52) - 9(4) = 260 - 36 = 224$

$$D_z = \begin{vmatrix} 5 & 7 & 9 \\ 0 & 14 & 28 \\ 4 & 0 & -4 \end{vmatrix}$$
$$= 5 \cdot \begin{vmatrix} 14 & 28 \\ 0 & -4 \end{vmatrix} - 0 \cdot \begin{vmatrix} 7 & 9 \\ 0 & -4 \end{vmatrix} + 4 \begin{vmatrix} 7 & 9 \\ 14 & 28 \end{vmatrix}$$
$= 5(-56 - 0) - 0 + 4(196 - 126)$
$= -280 + 4(70) = -280 + 280 = 0$
$x = \frac{-112}{112} = -1, \quad y = \frac{224}{112} = 2,$
$z = \frac{0}{112} = 0$
$\{(-1, 2, 0)\}$

Chapter 4 - Test

1. $\begin{vmatrix} 4 & -7 \\ 2 & 5 \end{vmatrix} = 4(5) - (-7)(2)$
$= 20 + 14 = 34$

2. $\begin{vmatrix} 4 & 0 & 2 \\ 1 & -3 & 5 \\ 0 & -1 & 2 \end{vmatrix}$
$$= 4 \cdot \begin{vmatrix} -3 & 5 \\ -1 & 2 \end{vmatrix} - 1 \cdot \begin{vmatrix} 0 & 2 \\ -1 & 2 \end{vmatrix} + 0 \cdot \begin{vmatrix} 0 & 2 \\ -3 & 5 \end{vmatrix}$$
$= 4[-6 - (-5)] - [0 - (-2)] + 0$
$= 4(-1) - 2 = -4 - 2 = -6$

3. $\begin{cases} 2x - y = -1 \\ 5x + 4y = 17 \end{cases}$

a.

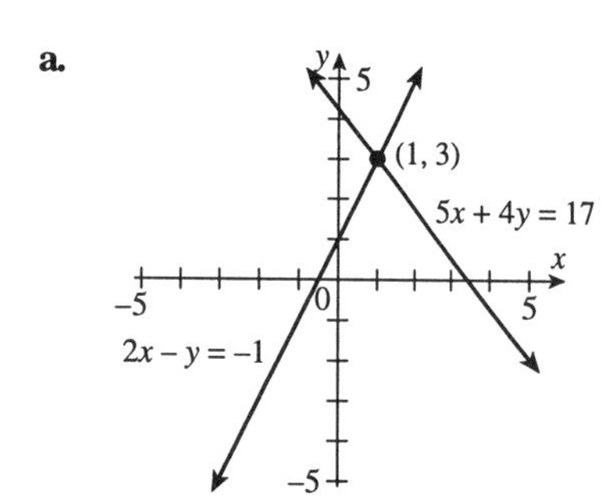

b. $\rightarrow \begin{cases} 8x - 4y = -4 \\ 5x + 4y = 17 \end{cases}$
Add.
$13x = 13$ so $x = 1$
Substitute back.
$2(1) - y = -1$
$2 - y = -1$
$3 = y$ or $y = 3$
$\{(1, 3)\}$

4. $\begin{cases} 7x - 14y = 5 \\ x = 2y \end{cases}$

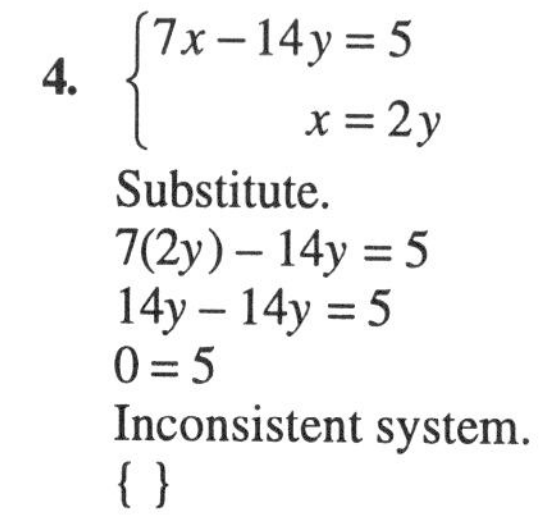

Substitute.
$7(2y) - 14y = 5$
$14y - 14y = 5$
$0 = 5$
Inconsistent system.
$\{ \}$

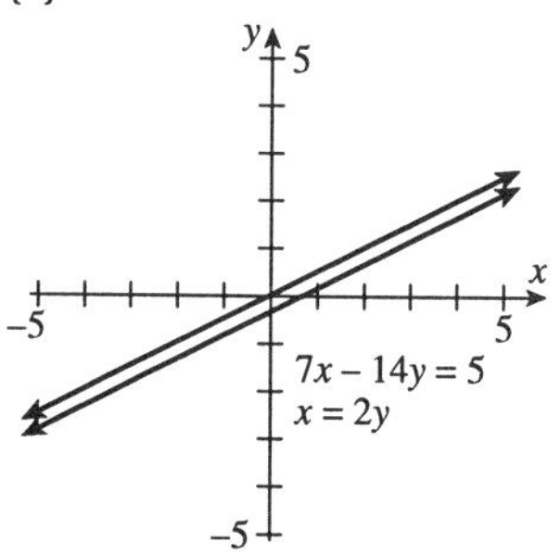

5. $\begin{cases} 4x - 7y = 29 \\ 2x + 5y = -11 \end{cases} \rightarrow \begin{cases} 4x - 7y = 29 \\ 4x + 10y = -22 \end{cases}$
Subtract.
$-17y = 51$ so $y = -3$
Substitute back.
$4x - 7(-3) = 29$
$4x + 21 = 29$
$4x = 8$
$x = 2$
$\{(2, -3)\}$

6. $\begin{cases} 15x + 6y = 15 \\ 10x + 4y = 10 \end{cases} \rightarrow \begin{cases} 5x + 2y = 5 \\ 5x + 2y = 5 \end{cases}$
The system is dependent.
$\{(x, y) | 5x + 2y = 5\}$

7. $\begin{cases} 2x - 3y = 4 \\ 3y + 2z = 2 \\ x - z = -5 \end{cases}$

Add equations 1 and 2.
$2x + 2z = 6$ or $x + z = 3$
We now have the system
$\begin{cases} x + z = 3 \\ x - z = -5 \end{cases}$
Add.
$2x = -2$ so $x = -1$
Substitute back.
$-1 + z = 3$ so $z = 4$
Substitute back.
$2(-1) - 3y = 4$
$-2 - 3y = 4$
$-3y = 6$ so $y = -2$
$\{(-1, -2, 4)\}$

8. $\begin{cases} 3x - 2y - z = -1 \\ 2x - 2y = 4 \\ 2x - 2z = -12 \end{cases}$

Subtract equation 2 from equation 1.
$x - z = -5$
Divide equation 3 by 2.
$x - z = -6$
Inconsistent system.
{ }

9. $\begin{cases} \frac{x}{2} + \frac{y}{4} = -\frac{3}{4} \\ x + \frac{3}{4}y = -4 \end{cases}$

Multiply first equation by –8, and second equation by 4.

$$\begin{array}{r} -4x - 2y = 6 \\ +\ \ 4x + 3y = -16 \\ \hline y = -10 \end{array}$$

Substitute back.
$x + \frac{3}{4}(-10) = -4$
$x = -4 + \frac{15}{2}$
$x = \frac{7}{2}$
$\left\{\left(\frac{7}{2}, -10\right)\right\}$

10. $\begin{cases} 3x - y = 7 \\ 2x + 5y = -1 \end{cases}$

$D = \begin{vmatrix} 3 & -1 \\ 2 & 5 \end{vmatrix} = 3(5) - (-1)(2)$
$= 15 + 2 = 17$
$D_x = \begin{vmatrix} 7 & -1 \\ -1 & 5 \end{vmatrix} = 7(5) - (-1)(-1)$
$= 35 - 1 = 34$
$D_y = \begin{vmatrix} 3 & 7 \\ 2 & -1 \end{vmatrix} = 3(-1) - 7(2)$
$= -3 - 14 = -17$
$x = \frac{34}{17} = 2$ and $y = \frac{-17}{17} = -1$
$\{(2, -1)\}$

11. $\begin{cases} 4x - 3y = -6 \\ -2x + y = 0 \end{cases}$

$D = \begin{vmatrix} 4 & -3 \\ -2 & 1 \end{vmatrix} = 4(1) - (-3)(-2)$
$= 4 - 6 = -2$
$D_x = \begin{vmatrix} -6 & -3 \\ 0 & 1 \end{vmatrix} = (-6)(1) - (-3)(0)$
$= -6 - 0 = -6$
$D_y = \begin{vmatrix} 4 & -6 \\ -2 & 0 \end{vmatrix} = 4(0) - (-6)(-2)$
$= 0 - 12 = -12$
$x = \frac{-6}{-2} = 3$ and $y = \frac{-12}{-2} = 6$
$\{(3, 6)\}$

12. $\begin{cases} x + y + z = 4 \\ 2x + 5y = 1 \\ x - y - 2z = 0 \end{cases}$

$D = \begin{vmatrix} 1 & 1 & 1 \\ 2 & 5 & 0 \\ 1 & -1 & -2 \end{vmatrix}$
$= 1 \cdot \begin{vmatrix} 2 & 5 \\ 1 & -1 \end{vmatrix} - 0 \cdot \begin{vmatrix} 1 & 1 \\ 1 & -1 \end{vmatrix} + (-2) \cdot \begin{vmatrix} 1 & 1 \\ 2 & 5 \end{vmatrix}$
$= (-2 - 5) - 0 - 2(5 - 2)$
$= -7 - 2(3) = -7 - 6 = -13$

$$D_x = \begin{vmatrix} 4 & 1 & 1 \\ 1 & 5 & 0 \\ 0 & -1 & -2 \end{vmatrix}$$

$$= 1 \cdot \begin{vmatrix} 1 & 5 \\ 0 & -1 \end{vmatrix} - 0 \cdot \begin{vmatrix} 4 & 1 \\ 0 & -1 \end{vmatrix} + (-2) \cdot \begin{vmatrix} 4 & 1 \\ 1 & 5 \end{vmatrix}$$

$= (-1 - 0) - 0 - 2(20 - 1)$
$= -1 - 2(19) = -1 - 38 = -39$

$$D_y = \begin{vmatrix} 1 & 4 & 1 \\ 2 & 1 & 0 \\ 1 & 0 & -2 \end{vmatrix}$$

$$= 1 \cdot \begin{vmatrix} 2 & 1 \\ 1 & 0 \end{vmatrix} - 0 \cdot \begin{vmatrix} 1 & 4 \\ 1 & 0 \end{vmatrix} + (-2) \cdot \begin{vmatrix} 1 & 4 \\ 2 & 1 \end{vmatrix}$$

$= (0 - 1) - 0 - 2(1 - 8) = -1 - 2(-7)$
$= -1 + 14 = 13$

$$D_z = \begin{vmatrix} 1 & 1 & 4 \\ 2 & 5 & 1 \\ 1 & -1 & 0 \end{vmatrix}$$

$$= 1 \cdot \begin{vmatrix} 1 & 4 \\ 5 & 1 \end{vmatrix} - (-1) \cdot \begin{vmatrix} 1 & 4 \\ 2 & 1 \end{vmatrix} + 0 \cdot \begin{vmatrix} 1 & 1 \\ 2 & 5 \end{vmatrix}$$

$= (1 - 20) + 1 \cdot (1 - 8) + 0$
$= -19 - 7 = -26$

$x = \frac{-39}{-13} = 3, \; y = \frac{13}{-13} = -1,$

$z = \frac{-26}{-13} = 2$

$\{(3, -1, 2)\}$

13. $\begin{cases} 3x + 2y + 3z = 3 \\ x - \quad\; z = 9 \\ 4y + z = -4 \end{cases}$

$$D = \begin{vmatrix} 3 & 2 & 3 \\ 1 & 0 & -1 \\ 0 & 4 & 1 \end{vmatrix}$$

$$= -1 \cdot \begin{vmatrix} 2 & 3 \\ 4 & 1 \end{vmatrix} + 0 \cdot \begin{vmatrix} 3 & 3 \\ 0 & 1 \end{vmatrix} - (-1) \cdot \begin{vmatrix} 3 & 2 \\ 0 & 4 \end{vmatrix}$$

$= -1(2 - 12) + 0 + 1 \cdot (12 - 0)$
$= -(-10) + 12 = 10 + 12 = 22$

$$D_x = \begin{vmatrix} 3 & 2 & 3 \\ 9 & 0 & -1 \\ -4 & 4 & 1 \end{vmatrix}$$

$$= -9 \cdot \begin{vmatrix} 2 & 3 \\ 4 & 1 \end{vmatrix} + 0 \cdot \begin{vmatrix} 3 & 3 \\ -4 & 1 \end{vmatrix} - (-1) \cdot \begin{vmatrix} 3 & 2 \\ -4 & 4 \end{vmatrix}$$

$= -9(2 - 12) + 0 + 1 \cdot [12 - (-8)]$
$= -9(-10) + 12 + 8 = 90 + 20 = 110$

$$D_y = \begin{vmatrix} 3 & 3 & 3 \\ 1 & 9 & -1 \\ 0 & -4 & 1 \end{vmatrix}$$

$$= 3 \cdot \begin{vmatrix} 9 & -1 \\ -4 & 1 \end{vmatrix} - 1 \cdot \begin{vmatrix} 3 & 3 \\ -4 & 1 \end{vmatrix} + 0 \cdot \begin{vmatrix} 3 & 3 \\ 9 & -1 \end{vmatrix}$$

$= 3(9 - 4) - [3 - (-12)] + 0$
$= 3(5) - (3 + 12) = 15 - 15 = 0$

$$D_z = \begin{vmatrix} 3 & 2 & 3 \\ 1 & 0 & 9 \\ 0 & 4 & -4 \end{vmatrix}$$

$$= 3 \cdot \begin{vmatrix} 0 & 9 \\ 4 & -4 \end{vmatrix} - 1 \cdot \begin{vmatrix} 2 & 3 \\ 4 & -4 \end{vmatrix} + 0 \cdot \begin{vmatrix} 2 & 3 \\ 0 & 9 \end{vmatrix}$$

$= 3(0 - 36) - (-8 - 12) + 0$
$= 3(-36) - (-20) = -108 + 20 = -88$

$x = \frac{110}{22} = 5, \; y = \frac{0}{22} = 0,$

$z = \frac{-88}{22} = -4$

$\{(5, 0, -4)\}$

14. $\begin{cases} x - y = -2 \\ 3x - 3y = -6 \end{cases} \left(\begin{array}{cc|c} 1 & -1 & -2 \\ 3 & -3 & -6 \end{array}\right)$

$\rightarrow \left(\begin{array}{cc|c} 1 & -1 & -2 \\ 0 & 0 & 0 \end{array}\right)$

Dependent system.
$\{(x, y) | x - y = -2\}$

15. $\begin{cases} x + 2y = -1 \\ 2x + 5y = -5 \end{cases} \left(\begin{array}{cc|c} 1 & 2 & -1 \\ 2 & 5 & -5 \end{array}\right)$

$\rightarrow \left(\begin{array}{cc|c} 1 & 2 & -1 \\ 0 & 1 & -3 \end{array}\right) \rightarrow \left(\begin{array}{cc|c} 1 & 0 & 5 \\ 0 & 1 & -3 \end{array}\right)$

$\{(5, -3)\}$

16. $\begin{cases} x - y - z = 0 \\ 3x - y - 5z = -2 \\ 2x + 3y = -5 \end{cases}$ $\left(\begin{array}{ccc|c} 1 & -1 & 1 & 0 \\ 3 & -1 & -5 & -2 \\ 2 & 3 & 0 & -5 \end{array}\right)$

$\rightarrow \left(\begin{array}{ccc|c} 1 & -1 & -1 & 0 \\ 0 & 2 & -2 & -2 \\ 2 & 3 & 0 & -5 \end{array}\right) \rightarrow \left(\begin{array}{ccc|c} 1 & -1 & -1 & 0 \\ 0 & 2 & -2 & -2 \\ 0 & 5 & 2 & -5 \end{array}\right)$

$\rightarrow \left(\begin{array}{ccc|c} 1 & -1 & -1 & 0 \\ 0 & 1 & -1 & -1 \\ 0 & 5 & 2 & -5 \end{array}\right) \rightarrow \left(\begin{array}{ccc|c} 1 & 0 & -2 & -1 \\ 0 & 1 & -1 & -1 \\ 0 & 5 & 2 & -5 \end{array}\right)$

$\rightarrow \left(\begin{array}{ccc|c} 1 & 0 & -2 & -1 \\ 0 & 1 & -1 & -1 \\ 0 & 0 & 7 & 0 \end{array}\right) \rightarrow \left(\begin{array}{ccc|c} 1 & 0 & -2 & -1 \\ 0 & 1 & -1 & -1 \\ 0 & 0 & 1 & 0 \end{array}\right)$

$\rightarrow \left(\begin{array}{ccc|c} 1 & 0 & 0 & -1 \\ 0 & 1 & -1 & -1 \\ 0 & 0 & 1 & 0 \end{array}\right) \rightarrow \left(\begin{array}{ccc|c} 1 & 0 & 0 & -1 \\ 0 & 1 & 0 & -1 \\ 0 & 0 & 1 & 0 \end{array}\right)$

$\{(-1, -1, 0)\}$

17. $\begin{cases} 2x - y + 3z = 4 \\ 3x - 3z = -2 \\ -5x + y = 0 \end{cases}$ $\left(\begin{array}{ccc|c} 2 & -1 & 3 & 4 \\ 3 & 0 & -3 & -2 \\ -5 & 1 & 0 & 0 \end{array}\right)$

$\rightarrow \left(\begin{array}{ccc|c} 1 & -\frac{1}{2} & \frac{3}{2} & 2 \\ 3 & 0 & -3 & -2 \\ -5 & 1 & 0 & 0 \end{array}\right)$

$\rightarrow \left(\begin{array}{ccc|c} 1 & -\frac{1}{2} & \frac{3}{2} & 2 \\ 0 & \frac{3}{2} & -\frac{15}{2} & -8 \\ -5 & 1 & 0 & 0 \end{array}\right)$

$\rightarrow \left(\begin{array}{ccc|c} 1 & -\frac{1}{2} & \frac{3}{2} & 2 \\ 0 & \frac{3}{2} & -\frac{15}{2} & -8 \\ 0 & -\frac{3}{2} & \frac{15}{2} & 10 \end{array}\right)$

$\rightarrow \left(\begin{array}{ccc|c} 1 & -\frac{1}{2} & \frac{3}{2} & 2 \\ 0 & 1 & -5 & -\frac{16}{3} \\ 0 & -\frac{3}{2} & \frac{15}{2} & 10 \end{array}\right)$

$\rightarrow \left(\begin{array}{ccc|c} 1 & 0 & -1 & -\frac{2}{3} \\ 0 & 1 & -5 & -\frac{16}{3} \\ 0 & -\frac{3}{2} & \frac{15}{2} & 10 \end{array}\right)$

$\rightarrow \left(\begin{array}{ccc|c} 1 & 0 & -1 & -\frac{2}{3} \\ 0 & 1 & -5 & -\frac{16}{3} \\ 0 & 0 & 0 & 2 \end{array}\right)$

Inconsistent system.

{ }

18. $R(x) = 38x$

$C(x) = 18x + 5500$

$38x = 18x + 5500$

$20x = 5500$

$x = 275$

275 frames

19. Let x = double occupancy rooms

Then $80 - x$ = single occupancy rooms

$90x + 80(80 - x) = 6930$

$90x + 6400 - 80x = 6930$

$10x = 530$

$x = 53$

$80 - 53 = 27$

53 double occupancy rooms

27 single occupancy rooms

20. Let x = amount of 10% solution

Then $20 - x$ = amount of 20% solution

$0.10x + 0.20(20 - x) = 20(0.175)$

$0.10x + 4 - 0.20x = 3.5$

$0.5 = 0.10x$

$5 = x$

$20 - 5 = 15$

5 gallons of 10% solution

15 gallons of 20% solution

Chapter 4 - Cumulative Review

1. a. $5 + y \geq 7$

b. $11 \neq z$

c. $20 < 5 - 2x$

2. a. $2 - 8 = -6$

b. $-8 - (-1) = -8 + 1 = -7$

c. $-11 - 5 = -16$

d. $10.7 - (-9.8) = 10.7 + 9.8 = 20.5$

e. $\frac{2}{3}-\frac{1}{2}=\frac{4}{6}-\frac{3}{6}=\frac{1}{6}$

f. $1-0.06=0.94$

g. $4-7=-3$

3. $\frac{y}{3}-\frac{y}{4}=\frac{1}{6}$
$4y-3y=2$
$y=2$
$\{2\}$

4. $25(0.16)=4$

5. $3x+4\geq 2x-6$
$3x-2x\geq -6-4$
$x\geq -10$
$[-10, \infty)$

−40 −30 −20 −10 0 10 20 30 40

6. $2<4-x<7$
$2-4<-x<7-4$
$-2<-x<3$
$2>x>-3$
$(-3, 2)$

7. $|5w+3|=7$

$5w+3=7$ or $5w+3=-7$
$5w=4$ or $5w=-10$
$w=\frac{4}{5}$ or $w=-2$

$\left\{-2, \frac{4}{5}\right\}$

8. $5x-2y=10$
Find ordered pair solutions.
Possible points:

x	y
0	−5
2	0
1	−2.5

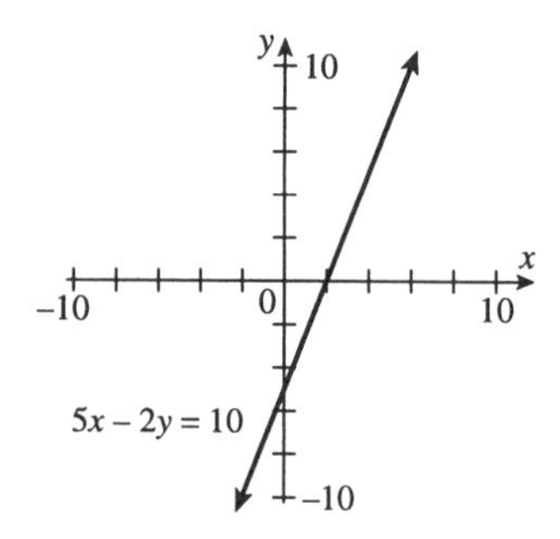

9. a. Domain: $[-4, 4]$
Range: $[-2, 2]$
Not a function

b. Domain: $(-\infty, \infty)$
Range: $[0, \infty)$
Function

c. Domain: $(-\infty, \infty)$
Range: $(-\infty, \infty)$
Function

10. $x=-2y$

Let $x=0$	Let $x=-2$	Let $x=2$
$0=-2y$	$-2=-2y$	$2=-2y$
$y=0$	$y=1$	$y=-1$

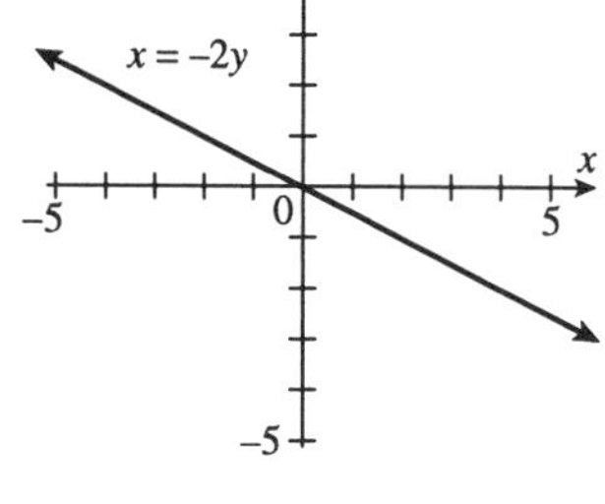

11. $m=\frac{5-3}{2-0}=\frac{2}{2}=1$

12. Slope of vertical line is undefined

13. $m=\frac{-5-0}{-4-4}=\frac{-5}{-8}=\frac{5}{8}$
$y-0=\frac{5}{8}(x-4)$
$y=\frac{5}{8}x-\frac{5}{2}$

14. $2x - y < 6$

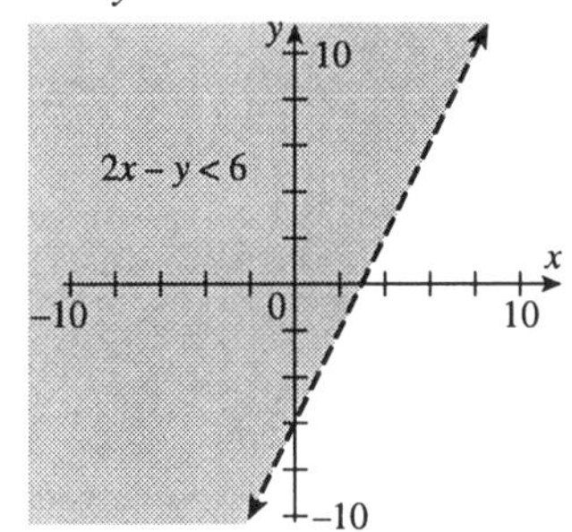

15. $\begin{cases} 2x + 4y = -6 \\ x - 2y = -5 \end{cases}$

$x = 2y - 5$
$2(2y - 5) + 4y = -6$
$4y - 10 + 4y = -6$
$8y = 4$
$y = \frac{1}{2}$
$x = 2\left(\frac{1}{2}\right) - 5 = 1 - 5 = -4$
$\left\{\left(-4, \frac{1}{2}\right)\right\}$

16. $\begin{cases} 3x + \frac{y}{2} = 2 \\ 6x + y = 5 \end{cases}$

Multiply first equation by −2. Then add.

$$\begin{array}{r} -6x - y = -4 \\ \underline{6x + y = 5} \\ 0 = 1 \end{array}$$

No solution
{ }

17. $\begin{cases} 3x - y + z = -15 \\ x + 2y - z = 1 \\ 2x + 3y - 2z = 0 \end{cases}$

Add first two equations.

$$\begin{array}{l} 3x - y + z = -15 \\ \underline{x + 2y - z = 1} \\ 4x + y = -14 \end{array}$$

Multiply second equation by −2.
Add to third equation.

$$\begin{array}{r} -2x - 4y + 2z = -2 \\ \underline{2x + 3y - 2z = 0} \\ -y = -2 \\ y = 2 \end{array}$$

Substitute back.
$4x + 2 = -14$
$4x = -16$
$x = -4$
$3(-4) - 2 + z = -15$
$-12 - 2 + z = -15$
$z = -1$
$\{(-4, 2, -1)\}$

18. Let x = speed of slower car
Then $x + 5$ = speed of faster car
$d = rt$
$297 = 3x + 3(x + 5)$
$297 = 3x + 3x + 15$
$282 = 6x$
$47 = x$
$47 + 5 = 52$
47 mph and 52 mph

19. $\begin{cases} x + 2y + z = 2 \\ -2x - y + 2z = 5 \\ x + 3y - 2z = -8 \end{cases}$

Multiply first equation by −1. Add first and third equations.

$$\begin{array}{r} -x - 2y - z = -2 \\ \underline{x + 3y - 2z = -8} \\ y - 3z = -10 \\ y = 3z - 10 \end{array}$$

Multiply third equation by 2.
Add second and third equations.

$$\begin{array}{r} -2x - y + 2z = 5 \\ \underline{2x + 6y - 4z = -16} \\ 5y - 2z = -11 \end{array}$$

Substitute.
$5(3z - 10) - 2z = -11$
$15z - 50 - 2z = -11$
$13z = 39$
$z = 3$
$y = 3(3) - 10 = -1$
$x + 2(-1) + (3) = 2$

$x-2+3=2$
$x=1$
$\{(1,-1,3)\}$

20. a. $\begin{vmatrix} 0 & 5 & 1 \\ 1 & 3 & -1 \\ -2 & 2 & 4 \end{vmatrix}$

$=0\cdot\begin{vmatrix} 3 & -1 \\ 2 & 4 \end{vmatrix}-1\cdot\begin{vmatrix} 5 & 1 \\ 2 & 4 \end{vmatrix}-2\cdot\begin{vmatrix} 5 & 1 \\ 3 & -1 \end{vmatrix}$
$=0-(20-2)-2(-5-3)$
$=0-18+16=-2$

b. $\begin{vmatrix} 0 & 5 & 1 \\ 1 & 3 & -1 \\ -2 & 2 & 4 \end{vmatrix}$

$=-1\cdot\begin{vmatrix} 5 & 1 \\ 2 & 4 \end{vmatrix}+3\cdot\begin{vmatrix} 0 & 1 \\ -2 & 4 \end{vmatrix}+1\cdot\begin{vmatrix} 0 & 5 \\ -2 & 2 \end{vmatrix}$
$=-(20-2)+3(0+2)+(0+10)$
$=-18+6+10=-2$

21. $\begin{cases} x-2y+z=4 \\ 3x+y-2z=3 \\ 5x+5y+3z=-8 \end{cases}$

$D=\begin{vmatrix} 1 & -2 & 1 \\ 3 & 1 & -2 \\ 5 & 5 & 3 \end{vmatrix}$

$=1\cdot\begin{vmatrix} 1 & -2 \\ 5 & 3 \end{vmatrix}+2\cdot\begin{vmatrix} 3 & -2 \\ 5 & 3 \end{vmatrix}+1\cdot\begin{vmatrix} 3 & 1 \\ 5 & 5 \end{vmatrix}$
$=(3+10)+2(9+10)+(15-5)$
$=13+38+10=61$

$D_x=\begin{vmatrix} 4 & -2 & 1 \\ 3 & 1 & -2 \\ -8 & 5 & 3 \end{vmatrix}$

$=4\cdot\begin{vmatrix} 1 & -2 \\ 5 & 3 \end{vmatrix}+2\cdot\begin{vmatrix} 3 & -2 \\ -8 & 3 \end{vmatrix}+1\cdot\begin{vmatrix} 3 & 1 \\ -8 & 5 \end{vmatrix}$
$=4(3+10)+2(9-16)+1(15+8)$
$=4(13)+2(-7)+23=61$

$D_y=\begin{vmatrix} 1 & 4 & 1 \\ 3 & 3 & -2 \\ 5 & -8 & 3 \end{vmatrix}$

$=1\cdot\begin{vmatrix} 3 & -2 \\ -8 & 3 \end{vmatrix}-4\cdot\begin{vmatrix} 3 & -2 \\ 5 & 3 \end{vmatrix}+1\cdot\begin{vmatrix} 3 & 3 \\ 5 & -8 \end{vmatrix}$
$=(9-16)-4(9+10)+(-24-15)$
$=-7-76-39=-122$

$D_z=\begin{vmatrix} 1 & -2 & 4 \\ 3 & 1 & 3 \\ 5 & 5 & -8 \end{vmatrix}$

$=1\cdot\begin{vmatrix} 1 & 3 \\ 5 & -8 \end{vmatrix}+2\cdot\begin{vmatrix} 3 & 3 \\ 5 & -8 \end{vmatrix}+4\cdot\begin{vmatrix} 3 & 1 \\ 5 & 5 \end{vmatrix}$
$=(-8-15)+2(-24-15)+4(15-5)$
$=-23-78+40=-61$

$x=\frac{61}{61}=1,\ y=\frac{-122}{61}=-2,$
$z=\frac{-61}{61}=-1$
$\{(1,-2,-1)\}$

Chapter 5

Section 5.1 Scientific Calculator Explorations

1. $(3\times10^{11})(2\times10^{32})=6\times10^{43}$

3. $(5.2\times10^{23})(7.3\times10^{4})=3.796\times10^{28}$

Mental Math

1. $5x^{-1}y^{-2}=\frac{5}{xy^2}$

3. $a^2b^{-1}c^{-5}=\frac{a^2}{bc^5}$

5. $\frac{y^{-2}}{x^{-4}}=\frac{x^4}{y^2}$

Exercise Set 5.1

1. $4^2\cdot4^3=4^{2+3}=4^5$

3. $x^5\cdot x^3=x^{5+3}=x^8$

5. $-7x^3\cdot20x^9=-140x^{3+9}=-140x^{12}$

7. $(4xy)(-5x)=-20x^2y$

9. $(-4x^3p^2)(4y^3x^3)$
$=-16x^{3+3}y^3p^2=-16x^6y^3p^2$

11. $-8^0=-1$

13. $(4x+5)^0=1$

15. $(5x)^0+5x^0=1+5(1)=1+5=6$

17. Answers may vary.

19. $\frac{a^5}{a^2}=a^{5-2}=a^3$

21. $\frac{x^9y^6}{x^8y^6}=x^{9-8}=x^1=x$

23. $-\frac{26z^{11}}{2z^7}=-13z^{11-7}=-13z^4$

25. $\frac{-36a^5b^7c^{10}}{6ab^3c^4}$
$=-6a^{5-1}b^{7-3}c^{10-4}=-6a^4b^4c^6$

27. $4^{-2}=\frac{1}{4^2}=\frac{1}{16}$

29. $\frac{x^7}{x^{15}}=\frac{1}{x^{15-7}}=\frac{1}{x^8}$

31. $5a^{-4}=5\frac{1}{a^4}=\frac{5}{a^4}$

33. $\frac{x^{-2}}{x^5}=\frac{1}{x^{5-(-2)}}=\frac{1}{x^7}$

35. $\frac{8r^4}{2r^{-4}}=4r^{4-(-4)}=4r^8$

37. $\frac{x^{-9}x^4}{x^{-5}}=\frac{x^{-9+4}}{x^{-5}}=\frac{x^{-5}}{x^{-5}}=1$

39. $x^5\cdot x^{7a}=x^{5+7a}$

41. $\frac{x^{3t-1}}{x^t}=x^{3t-1-t}=x^{2t-1}$

43. $x^{4a}\cdot x^7=x^{4a+7}$

45. $\frac{z^{6x}}{z^7}=z^{6x-7}$

47. $\frac{x^{3t}\cdot x^{4t-1}}{x^t}=\frac{x^{3t+4t-1}}{x^t}=x^{7t-1-t}$
$=x^{6t-1}$

49. $4^{-1}+3^{-2}=\frac{1}{4^1}+\frac{1}{3^2}$
$=\frac{1}{4}+\frac{1}{9}$
$=\frac{9+4}{36}=\frac{13}{36}$

51. $4x^0 + 5 = 4 \cdot 1 + 5 = 9$

53. $x^7 \cdot x^8 = x^{7+8} = x^{15}$

55. $2x^3 \cdot 5x^7 = 2 \cdot 5x^{3+7} = 10x^{10}$

57. $\frac{z^{12}}{z^{15}} = \frac{1}{z^{15-12}} = \frac{1}{z^3}$

59. $\frac{y^{-3}}{y^{-7}} = y^{-3-(-7)} = y^4$

61. $3x^{-1} = 3 \cdot \frac{1}{x} = \frac{3}{x}$

63. $3^0 - 3t^0 = 1 - 3 \cdot 1 = -2$

65. $\frac{r^4}{r^{-4}} = r^{4-(-4)} = r^8$

67. $\frac{x^{-7}y^{-2}}{x^2y^2} = \frac{1}{x^{2-(-7)}y^{2+2}} = \frac{1}{x^9y^4}$

69. $\frac{2a^{-6}b^2}{18ab^{-5}} = \frac{b^{2-(-5)}}{9a^{1-(-6)}} = \frac{b^7}{9a^7}$

71. $\frac{(24x^8)x}{20x^{-7}} = \frac{6x^{8+1}}{5x^{-7}} = \frac{6x^{9-(-7)}}{5} = \frac{6x^{16}}{5}$

73. $31,250,000 = 3.125 \times 10^7$

75. $0.016 = 1.6 \times 10^{-2}$

77. $67,413 = 6.7413 \times 10^4$

79. $0.0125 = 1.25 \times 10^{-2}$

81. $0.000053 = 5.3 \times 10^{-5}$

83. $3.6 \times 10^{-9} = 0.0000000036$

85. $9.3 \times 10^7 = 93,000,000$

87. $1.278 \times 10^6 = 1,278,000$

89. $7.35 \times 10^{12} = 7,350,000,000,000$

91. $4.03 \times 10^{-7} = 0.000000403$

93. Rewrite the number as the product of a number between 1 and 10 (including 1 but not 10) and 10 to an appropriate integer power.

95. $918,000,000$ km $= 9.18 \times 10^8$ km

97. $48,000,000,000 = 4.8 \times 10^{10}$

99. $0.001 = 1.0 \times 10^{-3}$

101. $(5 \cdot 2)^2 = (10)^2 = 100$

103. $\left(\frac{3}{4}\right)^3 = \frac{3^3}{4^3} = \frac{27}{64}$

105. $(2^3)^2 = 8^2 = 64$

107. $(2^{-1})^4 = \left(\frac{1}{2}\right)^4 = \frac{1^4}{2^4} = \frac{1}{16}$

Section 5.2

Mental Math

1. $(x^4)^5 = x^{4(5)} = x^{20}$

3. $x^4 \cdot x^5 = x^{4+5} = x^9$

5. $(y^6)^7 = y^{6(7)} = y^{42}$

7. $(z^4)^5 = z^{4(5)} = z^{20}$

9. $(z^{-6})^{-3} = z^{-6(-3)} = z^{18}$

Exercise Set 5.2

1. $(3^{-1})^2 = 3^{-1(2)} = 3^{-2} = \frac{1}{3^2} = \frac{1}{9}$

3. $(x^4)^{-9} = x^{4(-9)} = x^{-36} = \frac{1}{x^{36}}$

5. $(y)^{-5} = \frac{1}{y^5}$

7. $(3x^2y^3)^2 = 3^2(x^2)^2(y^3)^2$
$= 9x^{2(2)}y^{3(2)}$
$= 9x^4y^6$

9. $\left(\frac{2x^5}{y^{-3}}\right)^4 = \frac{2^4(x^5)^4}{(y^{-3})^4} = \frac{16x^{5(4)}}{y^{-3(4)}}$
$= \frac{16x^{20}}{y^{-12}} = 16x^{20}y^{12}$

11. $(a^2bc^{-3})^{-6} = (a^2)^{-6}b^{-6}(c^{-3})^{-6}$
$= a^{2(-6)}b^{-6}c^{-3(-6)}$
$= a^{-12}b^{-6}c^{18} = \frac{c^{18}}{a^{12}b^6}$

13. $\left(\frac{x^7y^{-3}}{z^{-4}}\right)^{-5} = \frac{(x^7)^{-5}(y^{-3})^{-5}}{(z^{-4})^{-5}}$
$= \frac{x^{-35}y^{15}}{z^{20}} = \frac{y^{15}}{x^{35}z^{20}}$

15. $\left(\frac{a^{-4}}{a^{-5}}\right)^{-2} = \frac{(a^{-4})^{-2}}{(a^{-5})^{-2}} = \frac{a^8}{a^{10}}$
$= \frac{1}{a^{10-8}} = \frac{1}{a^2}$

17. $\left(\frac{2a^{-2}b^5}{4a^2b^7}\right)^{-2}$
$= \left(\frac{1}{2a^4b^2}\right)^{-2}$
$= \frac{(1)^{-2}}{(2)^{-2}(a^4)^{-2}(b^2)^{-2}} = 4a^8b^4$

19. $\frac{4^{-1}x^2yz}{x^{-2}yz^3} = \frac{x^{2-(-2)}}{4z^{3-1}} = \frac{x^4}{4z^2}$

21. Yes; $a = \pm 1$

23. $(5^{-1})^3 = 5^{-3} = \frac{1}{5^3} = \frac{1}{125}$

25. $(x^7)^{-9} = x^{-63} = \frac{1}{x^{63}}$

27. $\left(\frac{7}{8}\right)^3 = \frac{7^3}{8^3} = \frac{343}{512}$

29. $(4x^2)^2 = 4^2(x^2)^2 = 16x^4$

31. $(-2^{-2}y)^3 = (-2^{-2})^3y^3$
$= -2^{-6}y^3 = \frac{-y^3}{2^6} = \frac{-y^3}{64}$

33. $\left(\frac{4^{-4}}{y^3x}\right)^{-2} = \frac{(4^{-4})^{-2}}{(y^3)^{-2}x^{-2}} = \frac{4^8}{y^{-6}x^{-2}}$
$= 4^8x^2y^6$

35. $\left(\frac{6p^6}{p^{12}}\right)^2 = \left(\frac{6}{p^{12-6}}\right)^2 = \frac{6^2}{(p^6)^2} = \frac{36}{p^{12}}$

37. $(-8y^3xa^{-2})^{-3}$
$= (-8)^{-3}(y^3)^{-3}x^{-3}(a^{-2})^{-3}$
$= -\frac{y^{-9}a^6}{8^3x^3} = -\frac{a^6}{512x^3y^9}$

39. $\left(\frac{x^{-2}y^{-2}}{a^{-3}}\right)^{-7} = \frac{(x^{-2})^{-7}(y^{-2})^{-7}}{(a^{-3})^{-7}}$
$= \frac{x^{14}y^{14}}{a^{21}}$

41. $\left(\frac{3x^5}{6x^4}\right)^4 = \left(\frac{x^{5-4}}{2}\right)^4 = \frac{(x^1)^4}{2^4}$
$= \frac{x^4}{16}$

43. $\left(\frac{1}{4}\right)^{-3} = (4^{-1})^{-3} = 4^3 = 64$

45. $\frac{(y^3)^{-4}}{y^3} = \frac{y^{-12}}{y^3} = \frac{1}{y^{3-(-12)}} = \frac{1}{y^{15}}$

47. $\frac{8p^7}{4p^9} = \frac{2}{p^{9-7}} = \frac{2}{p^2}$

49. $(4x^6y^5)^{-2}(6x^4y^3)$
$= 4^{-2}(x^6)^{-2}(y^5)^{-2}(6x^4y^3)$
$= \frac{1}{4^2}x^{-12}y^{-10} \cdot 6x^4y^3$
$= \frac{6}{16}x^{-12+4}y^{-10+3}$
$= \frac{3x^{-8}y^{-7}}{8} = \frac{3}{8x^8y^7}$

51. $x^6(x^6bc)^{-6} = x^6(x^6)^{-6}b^{-6}c^{-6}$
$= \frac{x^6 \cdot x^{-36}}{b^6c^6}$
$= \frac{x^{-30}}{b^6c^6}$
$= \frac{1}{x^{30}b^6c^6}$

53. $\frac{2^{-3}x^2y^{-5}}{5^{-2}x^7y^{-1}} = \frac{5^2}{2^3x^{7-2}y^{-1-(-5)}}$
$= \frac{25}{8x^5y^4}$

55. $\left(\frac{2x^2}{y^4}\right)^3\left(\frac{2x^5}{y}\right)^{-2} = \frac{2^3(x^2)^3 2^{-2}(x^5)^{-2}}{(y^4)^3y^{-2}}$
$= \frac{8x^6x^{-10}}{2^2y^{12}y^{-2}}$
$= \frac{8x^{6-10}}{4y^{12-2}}$
$= \frac{2x^{-4}}{y^{10}} = \frac{2}{x^4y^{10}}$

57. $(x^{3a+6})^3 = x^{(3a+6)3} = x^{9a+18}$

59. $\frac{x^{4a}(x^{4a})^3}{x^{4a-2}} = \frac{x^4a \cdot x^{12a}}{x^{4a-2}}$
$= \frac{x^{4a+12a}}{x^{4a-2}}$
$= x^{16a-(4a-2)}$
$= x^{12a+2}$

61. $(b^{5x-2})^{2x} = b^{(5x-2)2x} = b^{10x^2-4x}$

63. $\frac{(y^{2a})^8}{y^{a-3}} = \frac{y^{16a}}{y^{a-3}} = y^{16a-a+3} = y^{15a+3}$

65. $\left(\frac{2x^{3t}}{x^{2t-1}}\right)^4 = \frac{2^4(x^{3t})^4}{(x^{2t-1})^4}$
$= \frac{16x^{12t}}{x^{8t-4}}$
$= 16x^{12t-8t+4}$
$= 16x^{4t+4}$

67. $(5\times10^{11})(2.9\times10^{-3})$
$= 5\times2.9\times10^{11-3}$
$= 14.5\times10^8 = 1.45\times10^9$

69. $(2\times10^5)^3 = 2^3\times10^{5\cdot3} = 8.0\times10^{15}$

71. $\frac{3.6\times10^{-4}}{9\times10^2}$
$= 0.4\times10^{-4-2} = 0.4\times10^{-6} = 4\times10^{-7}$

73. $\frac{0.0069}{0.023} = \frac{6.9\times10^{-3}}{2.3\times10^{-2}}$
$= 3\times10^{-3-(-2)}$
$= 3\times10^{-1}$

75. $\frac{18,200\times100}{91,000} = \frac{1,820,000}{91,000}$
$= \frac{1.82\times10^6}{9.1\times10^4}$
$= 0.2\times10^{6-4}$
$= 2\times10^{2-1}$
$= 2\times10^1$

77. $\dfrac{6,000\times 0.006}{0.009\times 400}$
$=\dfrac{(6\times 10^3)(6\times 10^{-3})}{(9\times 10^{-3})(4\times 10^2)}$
$=\dfrac{36\times 10^{3-3}}{36\times 10^{-3+2}}$
$=\dfrac{10^0}{10^{-1}}$
$=10^{0-(-1)}$
$=10^1$
$=1.0\times 10^1$

79. $\dfrac{0.00064\times 2,000}{16,000}$
$=\dfrac{(6.4\times 10^{-4})(2\times 10^3)}{1.6\times 10^4}$
$=\dfrac{12.8\times 10^{-4+3}}{1.6\times 10^4}$
$=8\times 10^{-1-4}$
$=8\times 10^{-5}$

81. $\dfrac{66,000\times 0.001}{0.002\times 0.003}$
$=\dfrac{(6.6\times 10^4)(1\times 10^{-3})}{(2\times 10^{-3})(3\times 10^{-3})}$
$=\dfrac{6.6\times 10^{4-3}}{6\times 10^{-3-3}}$
$=\dfrac{1.1\times 10^1}{10^{-6}}$
$=1.1\times 10^{1-(-6)}$
$=1.1\times 10^7$

83. $\dfrac{1.25\times 10^{15}}{(2.2\times 10^{-2})(6.4\times 10^{-5})}$
$=\dfrac{1.25\times 10^{15}}{14.08\times 10^{-7}}$
$=0.08877840909\times 10^{22}$
$=8.877840909\times 10^{20}$

85. $200,000\times 10^{-8}=2\times 10^5\times 10^{-8}$
$=2\times 10^{-3}=0.002$ sec.

87. $D=\dfrac{M}{V}$ or $M=DV$
$M=(3.12\times 10^{-2})(4.269\times 10^{14})$
$=1.331928\times 10^{13}$ tons

89. Recall $V=s^3$
$=\left(\dfrac{2x^{-2}}{y}\right)^3$
$=\dfrac{2^3(x^{-2})^3}{y^3}$
$=\dfrac{8x^{-6}}{y^3}=\dfrac{8}{x^6y^3}\text{ m}^3$

91. $(4\times 10^{-2})(6.452\times 10^{-4})$
$=25.808\times 10^{-6}$
$=2.5808\times 10^{-5}$ square meters

93. Answers may vary.

95. $12m-14-15m-1=-3m-15$

97. $-9y-(5-6y)=-9y-5+6y=-3y-5$

99. $5(x-3)-4(2x-5)$
$=5x-15-8x+20=-3x+5$

Section 5.3 Graphing Calculator Explorations

1. $(2x^2+7x+6)+(x^3-6x^2-14)$
$=x^3-4x^2+7x-8$

3. $(1.8x^2-6.8x-1.7)-(3.9x^2-3.6x)$
$=-2.1x^2-3.2x-1.7$

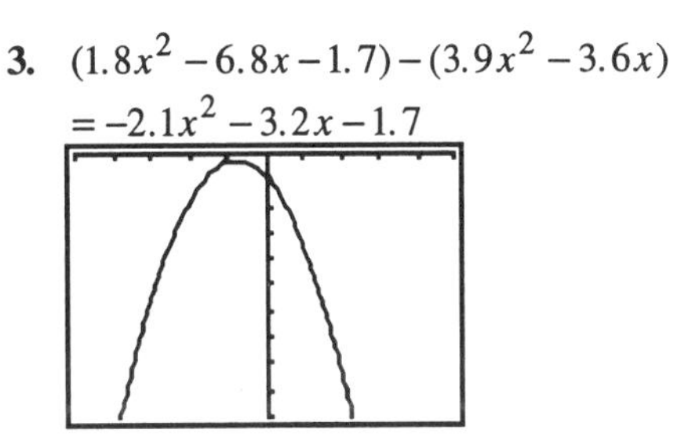

5. $(1.29x - 5.68)$
$+(7.69x^2 - 2.55x + 10.98)$
$= 7.69x^2 - 1.26x + 5.3$

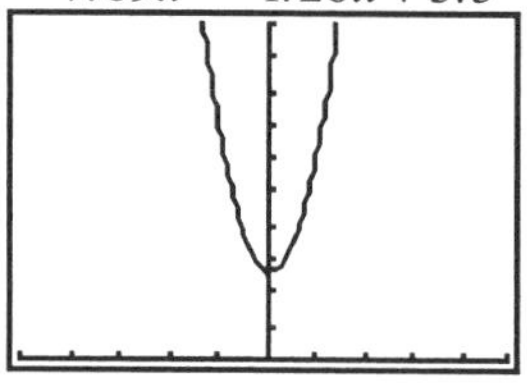

Exercise Set 5.3

1. 4 has degree 0.

3. $5x^2$ has degree 2.

5. $-3xy^2$ has degree $1 + 2 = 3$.

7. $6x + 3$ has degree 1 and is a binomial.

9. $3x^2 - 2x + 5$ has degree 2 and is a trinomial.

11. $-xyz$ has degree $1 + 1 + 1 = 3$ and is a monomial.

13. $x^2y - 4xy^2 + 5x + y$ has degree $2 + 1 = 3$ and is none of these.

17. $P(x) = x^2 + x + 1$
$P(7) = 7^2 + 7 + 1 = 49 + 7 + 1 = 57$

19. $Q(x) = 5x^2 - 1$
$Q(-10) = 5(-10)^2 - 1$
$= 5(100) - 1 = 500 - 1 = 499$

21. $P(x) = x^2 + x + 1$
$P(0) = 0^2 + 0 + 1 = 1$

23. $P(t) = -16t^2 + 1053$
$P(2) = -16(2)^2 + 1053$
$= -16(4) + 1053$
$= -64 + 1053 = 989$ feet

25. $P(t) = -16t^2 + 1053$
$P(6) = -16(6)^2 + 1053$
$= -16(36) + 1053 = -576 + 1053$
$= 477$ feet

27. $5y + y = 6y$

29. $4x + 7x - 3 = 11x - 3$

31. $4xy + 2x - 3xy - 1 = xy + 2x - 1$

33. $(9y^2 - 8) + (9y^2 - 9) = 18y^2 - 17$

35.
$$\begin{array}{r} x^2 + xy - y^2 \\ +\underline{2x^2 - 4xy + 7y^2} \\ 3x^2 - 3xy + 6y^2 \end{array}$$

37.
$$\begin{array}{r} x^2 - 6x + 3 \\ +\underline{\quad 2x + 5} \\ x^2 - 4x + 8 \end{array}$$

39.
$$\begin{array}{r} 9y^2 - 7y + 5 \\ -\underline{8y^2 - 7y + 2} \\ y^2 \qquad + 3 \end{array}$$

41.
$$\begin{array}{r} 4x^2 + 2x \\ -\underline{6x^2 - 3x} \\ -2x^2 + 5x \end{array}$$

43.
$$\begin{array}{r} 3x^2 - 4x + 8 \\ -\underline{5x^2 \qquad - 7} \\ -2x^2 - 4x + 15 \end{array}$$

45.
$$\begin{array}{r} 5x - 11 \\ +\underline{-x - 2} \\ 4x - 13 \end{array}$$

47.
$$\begin{array}{r} 7x^2 + x + 1 \\ -\underline{6x^2 + x - 1} \\ x^2 \qquad + 2 \end{array}$$

49. $(7x^3-4x+8)+(5x^3+4x+8x)$
$=7x^3-4x+8+5x^3+12x$
$=12x^3+8x+8$

51.
$$\begin{array}{r} 9x^3-2x^2+4x\ -7 \\ -2x^3-6x^2-4x\ +3 \\ \hline 7x^3+4x^2+8x-10 \end{array}$$

53.
$$\begin{array}{r} y^2+4yx\ +7 \\ +\ -19y^2+7yx\ +7 \\ \hline -18y^2+11yx+14 \end{array}$$

55.
$$\begin{array}{r} 3x^3-b+2a-6 \\ -4x^3+b+6a-6 \\ \hline -x^3\quad +8a-12 \end{array}$$

57.
$$\begin{array}{r} 4x^2-6x+2 \\ -\ -x^2+3x+5 \\ \hline 5x^2-9x-3 \end{array}$$

59.
$$\begin{array}{r} -3x+8 \\ +\ -3x^2+3x-5 \\ \hline -3x^2\qquad +3 \end{array}$$

61.
$$\begin{array}{r} 4x^2+7xy^2-3 \\ +2x^3\ -x^2+\ xy^2 \\ \hline 2x^3+3x^2+8xy^2-3 \end{array}$$

63.
$$\begin{array}{r} 6y^2-6y+4 \\ -\ -y^2-6y+7 \\ \hline 7y^2\qquad -3 \end{array}$$

65.
$$\begin{array}{r} 3x^2+15x+\ 8 \\ +(2x^2+\ 7x+\ 8) \\ \hline 5x^2+22x+16 \end{array}$$

67.
$$\begin{array}{r} 5q^4-2q^2-3q \\ +\ -6q^4+3q^2\qquad +5 \\ \hline -q^4+\ q^2-3q+5 \end{array}$$

69.
$$\begin{array}{r} 7x^2+\ 4x+9 \\ +\ 8x^2+\ 7x-8 \\ \hline 15x^2+11x+1 \\ -\qquad 3x+7 \\ \hline 15x^2+\ 8x-6 \end{array}$$

71.
$$\begin{array}{r} 4x^4-7x^2+3 \\ +\ -3x^4\qquad +2 \\ \hline x^4-7x^2+5 \end{array}$$

73.
$$\begin{array}{r} 8x^{2y}-7x^y+\ 3 \\ +\ -4x^{2y}+9x^y-14 \\ \hline 4x^{2y}+2x^y-11 \end{array}$$

75. $P(4000) = 45(4000) - 100{,}000$
$= 180{,}000 - 100{,}000$
$= \$80{,}000$

77. $f(x)=4.31x^2+29.17x+343.44$

a. $f(10)=4.31(10)^2+29.17(10)$
$+343.44=\$1066.14$

b. $f(20)=4.31(20)^2+29.17(20)$
$+343.44=\$2650.84$

c. $f(30)=4.31(30)^2+29.17(30)$
$+343.44=\$5097.54$

d. No; $f(x)$ is not linear

79. $C(x) = 0.8x + 10{,}000$
$C(20{,}000) = 0.8(20{,}000) + 10{,}000$
$= \$26{,}000$

81. $P(t)=-16t^2+300t$

a. $P(1)=-16(1)^2+300(1)=284$ feet

b. $P(2)=-16(2)^2+300(2)=536$ feet

c. $P(3)=-16(3)^2+300(3)=756$ feet

d. $P(4)=-16(4)^2+300(4)=944$ feet

83. $0 = -16t^2 + 300t$
$0 = -4t(4t - 75)$
$4t - 75 = 0$
$4t = 75$
$t = \frac{75}{4} = 18.75$
19 seconds

85.
$$\begin{array}{r} 5x^2 \qquad -7 \\ + \qquad 3x+3 \\ \hline 5x^2+3x-4 \end{array}$$

87.
$$\begin{array}{r} 3x+3 \\ - \quad 4x^2-6x+3 \\ \hline -4x^2+9x \end{array}$$

89. $-5(3x + 3) = -15x - 15$
$$\begin{array}{r} -15x-15 \\ - \quad 4x^2-6x+\ 3 \\ \hline -4x^2-9x-18 \end{array}$$

91. $2(4x^2 - 6x + 3) = 8x^2 - 12x + 6$
$7(5x^2 - 7) = 35x^2 - 49$
$$\begin{array}{r} 8x^2-12x \quad +6 \\ +35x^2 \qquad -49 \\ \hline 43x^2-12x-43 \end{array}$$

93. $f(x) = 3x^2 - 2$; A

95. $g(x) = -2x^3 - 3x^2 + 3x - 2$; D

97. $2(x + 5y) + 2(3x^2 - x + 2y)$
$= 2x + 10y + 6x^2 - 2x + 4y$
$= (6x^2 + 14y)$ units

99. $5(3x - 2) = 15x - 10$

101. $-2(x^2 - 5x + 6) = -2x^2 + 10x - 12$

103. $P(x) = 2x - 3$

a. $P(a) = 2a - 3$

b. $P(-x) = 2(-x) - 3 = -2x - 3$

c. $P(x + h) = 2(x + h) - 3 = 2x + 2h - 3$

105. $P(x) = 4x$

a. $P(a) = 4a$

b. $P(-x) = 4(-x) = -4x$

c. $P(x + h) = 4(x + h) = 4x + 4h$

107. $P(x) = 4x - 1$

a. $P(a) = 4a - 1$

b. $P(-x) = 4(-x) - 1 = -4x - 1$

c. $P(x + h) = 4(x + h) - 1$
$= 4x + 4h - 1$

Section 5.4 Graphing Calculator Explorations

1. $(x + 4)(x - 4) = x^2 - 16$

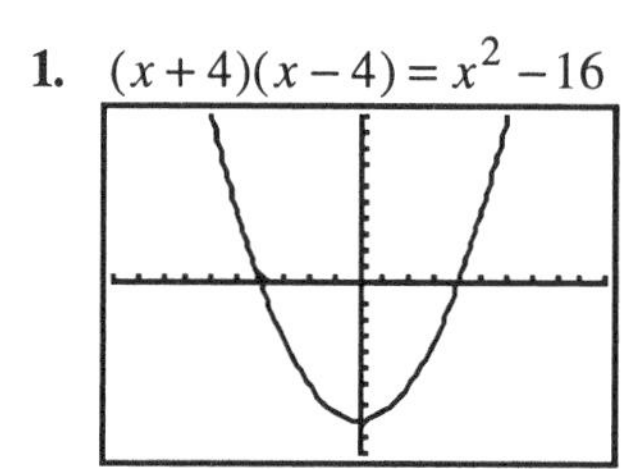

3. $(3x - 7)^2 = 9x^2 - 42x + 49$

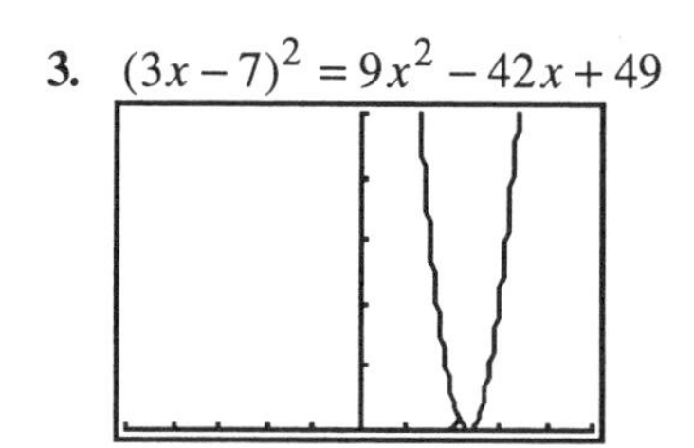

5. $(5x + 1)(x^2 - 3x - 2)$
$= 5x^3 - 14x^2 - 13x - 2$

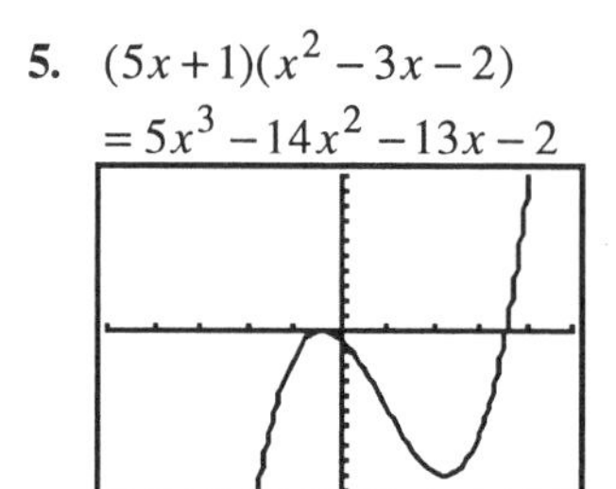

Exercise Set 5.4

1. $(-4x^3)(3x^2) = -12x^5$

3.
$$\begin{array}{l} 4x+7 \\ \underline{3x\qquad\quad} \\ 12x^2+21x \end{array}$$

5. $-6xy(4x+y)$
$= -6xy(4x) - 6xy - 6xy(y)$
$= -24x^2y - 6xy^2$

7. $-4ab(xa^2 + ya^2 - 3)$
$= -4ab(xa^2) - 4ab(ya^2) - 4ab(-3)$
$= -4xa^3b - 4ya^3b + 12ab$

9.
$$\begin{array}{r} 2x+4 \\ x-3 \\ \hline -6x-12 \\ 2x^2+4x \quad\;\; \\ \hline 2x^2-2x-12 \end{array}$$

11.
$$\begin{array}{r} x^3 \qquad\quad -x+2 \\ 2x+3 \\ \hline 3x^3 \qquad -3x+6 \\ 2x^4 \qquad -2x^2+4x \qquad\; \\ \hline 2x^4+3x^3-2x^2+\;\; x+6 \end{array}$$

13.
$$\begin{array}{r} 3x-2 \\ 5x+1 \\ \hline 3x-2 \\ 15x^2-10x \quad\;\; \\ \hline 15x^2-\;\;7x-2 \end{array}$$

15.
$$\begin{array}{r} 3m^2+2m-1 \\ 5m+2 \\ \hline 6m^2+4m-2 \\ 15m^3+10m^2-5m \quad\;\; \\ \hline 15m^3+16m^2-\;\;m-2 \end{array}$$

17. Answers may vary.

19.
$$\begin{array}{r} x-\;3 \\ x+\;4 \\ \hline 4x-12 \\ x^2-3x \quad\;\; \\ \hline x^2+\;\;x-12 \end{array}$$

21.
$$\begin{array}{r} 5x+8y \\ 2x-\;\;y \\ \hline -5xy-8y^2 \\ 10x^2+16xy \quad\;\;\; \\ \hline 10x^2+11xy-8y^2 \end{array}$$

23.
$$\begin{array}{r} 3x-1 \\ x+3 \\ \hline 9x-3 \\ 3x^2-\;\;x \quad\;\; \\ \hline 3x^2+8x-3 \end{array}$$

25.
$$\begin{array}{r} 3x+\frac{1}{2} \\ 3x-\frac{1}{2} \\ \hline -\frac{3}{2}x-\frac{1}{4} \\ 9x^2+\frac{3}{2}x \quad\;\;\; \\ \hline 9x^2 \qquad\;\; -\frac{1}{4} \end{array}$$

27. $(x+4)^2 = x^2 + 2(x)4 + 4^2$
$= x^2 + 8x + 16$

29. $(6y-1)(6y+1) = (6y)^2 - 1^2$
$= 36y^2 - 1$

31. $(3x-y)^2 = (3x)^2 - 2(3x)y + y^2$
$= 9x^2 - 6xy + y^2$

33. $(3b-6y)(3b+6y) = 9b^2 - 36y^2$

35. $(3+(4b+1))^2$
$= 3^2 + 2(3)(4b+1) + (4b+1)^2$
$= 9 + 6(4b+1) + (4b)^2 + 2(4b)1 + 1^2$
$= 9 + 24b + 6 + 16b^2 + 8b + 1$
$= 16b^2 + 32b + 16$

37. $((2s-3)-1)((2s-3)+1)$
$= (2s-3)^2 - 1^2$
$= (2s)^2 - 2(2s)3 + 3^2 - 1$
$= 4s^2 - 12s + 9 - 1$
$= 4s^2 - 12s + 8$

39. $((xy+4)-6)^2$
$= (xy+4)^2 - 2(xy+4)6 + 6^2$
$= (xy)^2 + 2(xy)4 + 4^2 - 12(xy+4) + 36$
$= x^2y^2 + 8xy + 16 - 12xy - 48 + 36$
$= x^2y^2 - 4xy + 4$

41. Answers may vary.

43.
$$\begin{array}{r} 3x+1 \\ 3x+5 \\ \hline 15x+5 \\ 9x^2+3x \\ \hline 9x^2+18x+5 \end{array}$$

45.
$$\begin{array}{rrrrrr} & & 2x^3 & & & +5 \\ & & & 5x^2 & +4x & +1 \\ \hline & & 2x^3 & & & +5 \\ & 8x^4 & & & +20x & \\ 10x^5 & & & +25x^2 & & \\ \hline 10x^5 & +8x^4 & +2x^3 & +25x^2 & +20x & +5 \end{array}$$

47. $(7x+3)(7x-3) = (7x)^2 - 3^2$
$= 49x^2 - 9$

49.
$$\begin{array}{r} 3x^2+4x-4 \\ 3x+6 \\ \hline 18x^2+24x-24 \\ 9x^3+12x^2-12x \\ \hline 9x^3+30x^2+12x-24 \end{array}$$

51.
$$\begin{array}{r} 4x+\frac{1}{3} \\ 4x-\frac{1}{2} \\ \hline -2x-\frac{1}{6} \\ 16x^2+\frac{4}{3}x \phantom{-\frac{1}{6}} \\ \hline 16x^2-\frac{2}{3}x-\frac{1}{6} \end{array}$$

53. $(6x+1)^2 = (6x)^2 + 2(6x)1 + 1^2$
$= 36x^2 + 12x + 1$

55. $(x^2+2y)(x^2-2y) = (x^2)^2 - (2y)^2$
$= x^4 - 4y^2$

57.
$$\begin{array}{r} 5a^2b^2-6a-6b \\ -6a^2b^2 \\ \hline -30a^4b^4+36a^3b^2+36a^2b^3 \end{array}$$

59.
$$\begin{array}{r} a-4 \\ 2a-4 \\ \hline -4a+16 \\ 2a^2-8a \\ \hline 2a^2-12a+16 \end{array}$$

61. $(7ab+3c)(7ab-3c) = (7ab)^2 - (3c)^2$
$= 49a^2b^2 - 9c^2$

63. $(m-4)^2 = m^2 - 2(m)4 + 4^2$
$= m^2 - 8m + 16$

65. $(3x+1)^2 = (3x)^2 + 2(3x)1 + 1^2$
$= 9x^2 + 6x + 1$

67.
$$\begin{array}{r} y-4 \\ y-3 \\ \hline -3y+12 \\ y^2-4y \quad \\ \hline y^2-7y+12 \end{array}$$

69.
$$\begin{array}{r} x+y \\ 2x-1 \\ \hline -x-y \\ 2x^2+2xy \qquad \\ \hline 2x^2+2xy-x-y \\ \text{then} \qquad x+1 \\ \hline 2x^2+2xy-x-y \\ 2x^3+2x^2y+x^2-xy \qquad \\ \hline 2x^3+2x^2y+x^2+xy-x-y \end{array}$$

71.
$$\begin{array}{r} 3x^2+2x-1 \\ 3x^2+2x-1 \\ \hline -3x^2-2x+1 \\ 6x^3+4x^2-2x \quad \\ 9x^4+6x^3-3x^2 \qquad \\ \hline 9x^4+12x^3-2x^2-4x+1 \end{array}$$

73.
$$\begin{array}{r} 4x^2-2x+5 \\ 3x+1 \\ \hline 4x^2-2x+5 \\ 12x^3-6x^2+15x \quad \\ \hline 12x^3-2x^2+13x+5 \end{array}$$

75.
$$\begin{array}{r} x+5 \\ 5x \\ \hline 5x^2+25x \end{array}$$

77. $(x^2-2)^2=(x^2-2)(x^2-2)$
$=(x^2)(x^2)-4(x^2)+(-2)(-2)$
$=x^4-4x^2+4$

79.
$$\begin{array}{r} x^2 \qquad -2 \\ x+5 \\ \hline 5x^2 \qquad -10 \\ x^3 \qquad -2x \quad \\ \hline x^3+5x^2-2x-10 \end{array}$$

81. a. $(3x+5)+(3x+7)=3x+3x+5+7$
$=6x+12$

b.
$$\begin{array}{r} 3x+5 \\ 3x+7 \\ \hline 21x+35 \\ 9x^2+15x \quad \\ \hline 9x^2+36x+35 \end{array}$$

83. $V=\pi r^2h$
$V=\pi(y-3)^2(7y)$
$V=\pi(y^2-6y+9)(7y)$
$V=\pi(7y^3-42y^2+63y)\text{ cm}^3$

85. $f(x)=x^2-3x$
$f(c)=c^2-3c$

87. $f(x)=x^2-3x$
$f(a+5)=(a+5)^2-3(a+5)$
$=a^2+10a+25-3a-15$
$=a^2+7a+10$

89. $f(x)=x^2-3x$
$f(a-b)=(a-b)^2-3(a-b)$
$=a^2-2ab+b^2-3a+3b$

91. $g(x)=x^2+2x+1$

a. $g(a+h)=(a+h)^2+2(a+h)+1$
$=a^2+2ah+h^2+2a+2h+1$

b. $g(a)=a^2+2a+1$

c. $g(a+h)-g(a)$
$= a^2+2ah+h^2+2a+2h+1-a^2$
$-2a-1$
$= 2ah+h^2+2h$

93. $-3yz^n(2y^3z^{2n}-1)$
$=(-3)(2)y^{1+3}z^{n+2n}+(-3)(-1)yz^n$
$=-6y^4z^{3n}+3yz^n$

95. $(x^a+y^{2b})(x^a-y^{2b})$
$= x^{a+a}-x^ay^{2b}+x^ay^{2b}-y^{2b+2b}$
$= x^{2a}-y^{4b}$

97. $y=\frac{3}{2}x-1$
$m=\frac{3}{2}$

99. $x+7y=2$
$7y=-x+2$
$y=-\frac{1}{7}x+\frac{2}{7}$
$m=-\frac{1}{7}$

101. Since any vertical line only crosses the graph once, it is a function.

Section 5.5

Mental Math

1. $6=2\cdot 3$
$12=2\cdot 2\cdot 3$
$\text{GCF}=2\cdot 3=6$

3. $15x=3\cdot 5\cdot x$
$10=2\cdot 5$
$\text{GCF}=5$

5. $13x=13\cdot x$
$2x=2\cdot x$
$\text{GCF}=x$

7. $7x=7\cdot x$
$14x=2\cdot 7\cdot x$
$\text{GCF}=7x$

Exercise Set 5.5

1. a^8, a^5, and a^3 have a GCF a^3.

3. $x^2y^3z^3$, y^2z^3, xy^2z^2 have GCF y^2z^2.

5. $6x^3y$, $9x^2y^2$, $12x^2y$ have GCF $3x^2y$.

7. $10x^3yz^3$, $20x^2z^5$, $45xz^3$ have GCF $5xz^3$.

9. $18x-12=6(3x-2)$

11. $4y^2-16xy^3=4y^2(1-4xy)$

13. $6x^5-8x^4+2x^3=2x^3(3x^2-4x+1)$

15. $8a^3b^3-4a^2b^2+4ab+16ab^2$
$=4ab(2a^2b^2-ab+1+4b)$

17. $6(x+3)+5a(x+3)$
$=(6+5a)(x+3)$

19. $2x(z+7)+(z+7)$
$=(2x+1)(z+7)$

21. $3x(x^2+5)-2(x^2+5)$
$=(3x-2)(x^2+5)$

23. Answers may vary.

25. $ab+3a+2b+6$
$=a(b+3)+2(b+3)$
$=(a+2)(b+3)$

27. $ac+4a-2c+8$
$=a(c+4)-2(c+4)$
$=(a-2)(c+4)$

29. $2xy-3x-4y+6$
$=x(2y-3)-2(2y-3)$
$=(x-2)(2y-3)$

31. $12xy-8x-3y+2$
$=4x(3y-2)-(3y-2)$
$=(4x-1)(3y-2)$

33. $2\pi r^2 + 2\pi rh = 2\pi r(r + h)$

35. $A = P + PRT$
$A = P(1 + RT)$

37. $h(t) = -16t^2 + 64t$

a. $h(t) = -16t(t - 4)$

b. $h(1) = -16(1)^2 + 64(1) = 48$
$h(1) = -16(1)(1 - 4) = 48$
48 feet

c. Answers may vary.

39. $6x^3 + 9 = 3(2x^3 + 3)$

41. $x^3 + 3x^2 = x^2(x + 3)$

43. $8a^3 - 4a = 4a(2a^2 - 1)$

45. $-20x^2y + 16x^3y = -4xy(5x - 4y^2)$

47. $10a^2b^3 + 5ab^2 - 15ab^3$
$= 5ab^2(2ab + 1 - 3b)$

49. $9abc^2 + 6a^2bc - 6ab + 3bc$
$= 3b(3ac^2 + 2a^2c - 2a + c)$

51. $4x(y - 2) - 3(y - 2) = (4x - 3)(y - 2)$

53. $6xy + 10x + 9y + 15$
$= 2x(3y + 5) + 3(3y + 5)$
$= (2x + 3)(3y + 5)$

55. $xy + 3y - 5x - 15$
$= y(x + 3) - 5(x + 3)$
$= (y - 5)(x + 3)$

57. $6ab - 2a - 9b + 3$
$= 2a(3b - 1) - 3(3b - 1)$
$= (2a - 3)(3b - 1)$

59. $12xy + 18x + 2y + 3$
$= 6x(2y + 3) + 1(2y + 3)$
$= (6x + 1)(2y + 3)$

61. $2m(n - 8) - (n - 8)$
$= (2m - 1)(n - 8)$

63. $15x^3y^2 - 18x^2y^2 = 3x^2y^2(5x - 6)$

65. $2x^2 + 3xy + 4x + 6y$
$= x(2x + 3y) + 2(2x + 3y)$
$= (x + 2)(2x + 3y)$

67. $5x^2 + 5xy - 3x - 3y$
$= 5x(x + y) - 3(x + y)$
$= (5x - 3)(x + y)$

69. $x^3 + 3x^2 + 4x + 12$
$= x^2(x + 3) + 4(x + 3)$
$= (x^2 + 4)(x + 3)$

71. $x^3 - x^2 - 2x + 2 = x^2(x - 1) - 2(x - 1)$
$= (x^2 - 2)(x - 1)$

73. None

a. $(2 - x)(3 - y) = 6 - 2y - 3x + xy$
$= xy - 3x - 2y + 6$

b. $(-2 + x)(-3 + y) = 6 - 2y - 3x + xy$
$= xy - 3x - 2y + 6$

c. $(x - 2)(y - 3) = xy - 3x - 2y + 6$

d. $(-x + 2)(-y + 3) = xy - 3x - 2y + 6$

75. $(5x^2)(11x^5)$
$= 5 \cdot 11 \cdot x^2 \cdot x^5 = 55x^7$

77. $(5x^2)^3$
$= 5x^2 \cdot 5x^2 \cdot 5x^2$
$= 5 \cdot 5 \cdot 5 \cdot x^2 \cdot x^2 \cdot x^2$
$= 125x^6$

79. $(x + 2)(x - 5)$
$= x^2 - 5x + 2x - 10$
$= x^2 - 3x - 10$

81. $(x+3)(x+2)$
$= x^2 + 2x + 3x + 6$
$= x^2 + 5x + 6$

83. $(y-3)(y-1)$
$= y^2 - y - 3y + 3$
$= y^2 - 4y + 3$

85. $x^{3n} - 2x^{2n} + 5x^n$
$= x^n(x^{2n} - 2x^n + 5)$

87. $6x^{8a} - 2x^{5a} - 4x^{3a}$
$= 2x^{3a}(3x^{5a} - x^{2a} - 2)$

Section 5.6

Mental Math

1. $10 = 2 \cdot 5$
$7 = 2 + 5$
2 and 5

3. $24 = 2 \cdot 2 \cdot 2 \cdot 3 = 8 \cdot 3$
$11 = 8 + 3$
8 and 3

Exercise Set 5.6

1. $x^2 + 9x + 18 = (x+6)(x+3)$

3. $x^2 - 12x + 32 = (x-4)(x-8)$

5. $x^2 + 10x - 24 = (x+12)(x-2)$

7. $x^2 - 2x - 24 = (x-6)(x+4)$

9. $3x^2 - 18x + 24 = 3(x^2 - 6x + 8)$
$= 3(x-2)(x-4)$

11. $4x^2z + 28xz + 40z$
$= 4z(x^2 + 7x + 10)$
$= 4z(x+2)(x+5)$

13. $2x^2 + 30x - 108$
$= 2(x^2 + 15x - 54)$
$= 2(x+18)(x-3)$

15. $x^2 + bx + 6$
$6 = 2 \cdot 3$ or $6 = (-2)(-3)$
$6 = 1 \cdot 6$ or $6 = (-1)(-6)$
$(x+2)(x+3) = x^2 + 5x + 6$
$(x-2)(x-3) = x^2 - 5x + 6$
$(x+1)(x+6) = x^2 + 7x + 6$
$(x-1)(x-6) = x^2 - 7x + 6$
$b = \pm 5$ and ± 7

17. $5x^2 + 16x + 3 = (5x+1)(x+3)$

19. $2x^2 - 11x + 12 = (2x-3)(x-4)$

21. $2x^2 + 25x - 20$ is prime.

23. $4x^2 - 12x + 9 = (2x-3)^2$

25. $12x^2 + 10x - 50$
$= 2(6x^2 + 5x - 25)$
$= 2(3x-5)(2x+5)$

27. $3y^4 - y^3 - 10y^2$
$= y^2(3y^2 - y - 10)$
$= y^2(3y+5)(y-2)$

29. $6x^3 + 8x^2 + 24x$
$= 2x(3x^2 + 4x + 12)$

31. $x^2 + 8xz + 7z^2 = (x+z)(x+7z)$

33. $2x^2 - 5xy - 3y^2 = (2x+y)(x-3y)$

35. $x^2 - x - 12$
$a = 1, b = -1, c = -12$
Now $ac = -12$. The two numbers are -4 and 3.
$x^2 - x - 12$
$= x^2 - 4x + 3x - 12$
$= x(x-4) + 3(x-4)$
$= (x+3)(x-4)$

37. $28y^2+22y+4$
$=2(14y^2+11y+2)$
$a=14$, $b=11$, and $c=2$.
Now $ac=28$. The numbers are 4 and 7.
$14y^2+11y+2$
$=14y^2+7y+4y+2$
$=7y(2y+1)+2(2y+1)$
$=(7y+2)(2y+1)$
So, $28y^2+22y+4=2(7y+2)(2y+1)$

39. $2x^2+15x-27$
$a=2$, $b=15$, $c=-27$
Now $ac=-54$.
The numbers are 18 and –3.
$2x^2+15x-27$
$=2x^2+18x-3x-27$
$=2x(x+9)-3(x+9)$
$=(2x-3)(x+9)$

41. $3x^2+bx+5$
$3=1\cdot 3$ or $3=(-1)(-3)$
$5=1\cdot 5$ or $5=(-1)(-5)$
$(3x+1)(x+5)=3x^2+16x+5$
$(3x-1)(x-5)=3x^2-16x+5$
$(-3x+1)(-x+5)=3x^2-16x+5$
$(-3x-1)(-x-5)=3x^2+16x+5$
$(3x+5)(x+1)=3x^2+8x+5$
$(3x-5)(x-1)=3x^2-8x+5$
$(-3x+5)(-x+1)=3x^2-8x+5$
$(-3x-5)(-x-1)=3x^2+8x+5$
$b=\pm 8$ and ± 16

43. $x^4+x^2-6=(x^2+3)(x^2-2)$

45. $(5x+1)^2+8(5x+1)+7$
$=[(5x+1)+1][(5x+1)+7]$
$=(5x+2)(5x+8)$

47. $x^6-7x^3+12=(x^3-4)(x^3-3)$

49. $(a+5)^2-5(a+5)-24$
$=[(a+5)-8][(a+5)+3]$
$=(a-3)(a+8)$

51. $3x^3-2x^2-8x$
$=x(3x^2-2x-8)$
$=x(3x+4)(x-2)$

53. $x^2-24x-81=(x-27)(x+3)$

55. $x^2-15x-54=(x-18)(x+3)$

57. $3x^2-6x+3$
$=3(x^2-2x+1)=3(x-1)^2$

59. $3x^2-5x-2=(3x+1)(x-2)$

61. $8x^2-26x+15=(4x-3)(2x-5)$

63. $18x^4+21x^3+6x^2$
$=3x^2(6x^2+7x+2)$
$=3x^2(3x+2)(2x+1)$

65. $3a^2+12ab+12b^2$
$=3(a^2+4ab+4b^2)$
$=3(a+2b)^2$

67. x^2+4x+5 is prime.

69. $2(x+4)^2+3(x+4)-5$
$=[2(x+4)+5][(x+4)-1]$
$=[2x+8+5][x+3]$
$=(2x+13)(x+3)$

71. $6x^2-49x+30=(3x-2)(2x-15)$

73. $x^4-5x^2-6=(x^2-6)(x^2+1)$

75. $6x^3-x^2-x$
$=x(6x^2-x-1)$
$=x(3x+1)(2x-1)$

77. $12a^2-29ab+15b^2=(4a-3b)(3a-5b)$

79. $9x^2+30x+25=(3x+5)^2$

81. $3x^2y - 11xy + 8y$
$= y(3x^2 - 11x + 8)$
$= y(3x - 8)(x - 1)$

83. $2x^2 + 2x - 12$
$= 2(x^2 + x - 6)$
$= 2(x + 3)(x - 2)$

85. $(x - 4)^2 + 3(x - 4) - 18$
$= [(x - 4) + 6][(x - 4) - 3]$
$= (x + 2)(x - 7)$

87. $2x^6 + 3x^3 - 9 = (2x^3 - 3)(x^3 + 3)$

89. $72xy^4 - 24xy^2z + 2xz^2$
$= 2x(36y^4 - 12y^2z + z^2)$
$= 2x(6y^2 - z)^2$

91. $x^4 + 6x^3 + 5x^2$
$= x^2(x^2 + 6x + 5)$
$= x^2(x + 5)(x + 1)$

93. $30x^3 + 9x^2 - 3x$
$= 3x(10x^2 + 3x - 1)$
$= 3x(5x - 1)(2x + 1)$

95.

$$\begin{array}{r} x^2 + 2x + 4 \\ x - 2 \\ \hline -2x^2 - 4x - 8 \\ x^3 + 2x^2 + 4x \quad \\ \hline x^3 \qquad\qquad\quad - 8 \end{array}$$

Answer is $x^3 - 8$

97. $P(0) = 3(0)^2 + 2(0) - 9 = -9$

99. $P(-1) = 3(-1)^2 + 2(-1) - 9$
$= 3 - 2 - 9 = -8$

101. $x^{2n} + 10x^n + 16 = (x^n + 2)(x^n + 8)$

103. $x^{2n} - 3x^n - 18 = (x^n - 6)(x^n + 3)$

105. $2x^{2n} + 11x^n + 5 = (2x^n + 1)(x^n + 5)$

107. $4x^{2n} - 12x^n + 9 = (2x^n - 3)^2$

Exercise Set 5.7

1. $x^2 + 6x + 9 = (x + 3)^2$

3. $4x^2 - 12x + 9 = (2x - 3)^2$

5. $3x^2 - 24x + 48$
$= 3(x^2 - 8x + 16)$
$= 3(x - 4)^2$

7. $9y^2x^2 + 12yx^2 + 4x^2$
$= x^2(9y^2 + 12y + 4)$
$= x^2(3y + 2)^2$

9. $x^2 - 25 = (x + 5)(x - 5)$

11. $9 - 4z^2 = (3 + 2z)(3 - 2z)$

13. $(y + 2)^2 - 49$
$= [(y + 2) - 7][(y + 2) + 7]$
$= (y - 5)(y + 9)$

15. $64x^2 - 100$
$= 4(16x^2 - 25)$
$= 4(4x + 5)(4x - 5)$

17. $x^3 + 27 = x^3 + 3^3$
$= (x + 3)(x^2 - 3x + 9)$

19. $z^3 - 1 = z^3 - 1^3$
$= (z - 1)(z^2 + z + 1)$

21. $m^3 + n^3 = (m + n)(m^2 - mn + n^2)$

23. $x^3y^2 - 27y^2$
$= (x^3 - 27)y^2$
$= (x^3 - 3^3)y^2$
$= (x - 3)(x^2 + 3x + 9)y^2$

25. $a^3b+8b^4=(a^3+8b^3)b$
$=(a^3+(2b)^3)b$
$=(a+2b)(a^2-2ab+4b^2)b$

27. $125y^3-8x^3$
$=(5y)^3-(2x)^3$
$=(5y-2x)(25y^2+10xy+4x^2)$

29. $(x^2+6x+9)-y^2$
$=(x+3)^2-y^2$
$=(x+3+y)(x+3-y)$

31. $(x^2-10x+25)-y^2$
$=(x-5)^2-y^2$
$=(x-5-y)(x-5+y)$

33. $(4x^2+4x+1)-z^2$
$=(2x+1)^2-z^2$
$=(2x+1-z)(2x+1+z)$

35. $9x^2-49=(3x+7)(3x-7)$

37. x^4-81
$=(x^2+9)(x^2-9)$
$=(x^2+9)(x+3)(x-3)$

39. $(x^2+8x+16)-4y^2$
$=(x+4)^2-(2y)^2$
$=(x+4-2y)(x+4+2y)$

41. $(x+2y)^2-9$
$=(x+2y-3)(x+2y+3)$

43. $x^3-1=x^3-1^3$
$=(x-1)(x^2+x+1)$

45. $x^3+125=x^3+5^3$
$=(x+5)(x^2-5x+25)$

47. $4x^2+25$ is prime.

49. $4a^2+12a+9$
$=(2a+3)(2a+3)$
$=(2a+3)^2$

51. $18x^2y-2y=2y(9x^2-1)$
$=2y(3x-1)(3x+1)$

53. x^6-y^3
$=(x^2)^3-y^3$
$=(x^2-y)(x^4+x^2y+y^2)$

55. $(x^2+16x+64)-x^4$
$=(x+8)^2-(x^2)^2$
$=(x+8-x^2)(x+8+x^2)$

57. $3x^6y^2+81y^2$
$=3y^2(x^6+27)$
$=3y^2((x^2)^3+3^3)$
$=3y^2(x^2+3)(x^4-3x^2+9)$

59. $(x+y)^3+125$
$=(x+y)^3+5^3$
$=[(x+y)+5][(x+y)^2-5(x+y)+25]$
$=(x+y+5)(x^2+2xy+y^2$
$-5x-5y+25)$

61. $(2x+3)^3-64=(2x+3)^3-4^3$
$=[(2x+3)-4]$
$[(2x+3)^2+4(2x+3)+16]$
$=(2x-1)(4x^2+12x+9+8x+12+16)$
$=(2x-1)(4x^2+20x+37)$

63. $A=\pi R^2-\pi r^2$
$=\pi(R^2-r^2)$
$=\pi(R-r)(R+r)$

65. The coating of the candy is the difference between the volume of the outer sphere $\frac{4}{3}\pi R^3$ and the inner sphere $\frac{4}{3}\pi(6)^3$.
Thus,
$V = \frac{4}{3}\pi R^3 - \frac{4}{3}\pi(6)^3$
$V = \frac{4}{3}\pi(R^3 - 6^3)$
$V = \frac{4}{3}\pi(R-6)(R^2+6R+36)$

67. $x^2 - 8x + 16 - y^2$
$= (x-4)^2 - y^2$
$= (x-4-y)(x-4+y)$

69. $x^4 - x = x(x^3 - 1)$
$= x(x-1)(x^2+x+1)$

71. $14x^2y - 2xy = 2xy(7x-1)$

73. $4x^2 - 16 = 4(x^2 - 4)$
$= 4(x+2)(x-2)$

75. $3x^2 - 8x - 11 = (3x-11)(x+1)$

77. $4x^2 + 8x - 12$
$= 4(x^2 + 2x - 3)$
$= 4(x+3)(x-1)$

79. $4x^2 + 36x + 81 = (2x+9)^2$

81. $8x^3 + 27y^3$
$= (2x)^3 + (3y)^3$
$= (2x+3y)(4x^2 - 6xy + 9y^2)$

83. $64x^2y^3 - 8x^2$
$= 8x^2(8y^3 - 1)$
$= 8x^2[(2y)^3 - 1]$
$= 8x^2(2y-1)(4y^2+2y+1)$

85. $(x+5)^3 + y^3$
$= [(x+5)+y]$
$\cdot\ [(x+5)^2 - (x+5)y + y^2]$
$= (x+5+y)$
$\cdot\ (x^2 + 10x + 25 - xy - 5y + y^2)$

87. $(5a-3)^2 - 6(5a-3) + 9$
$= [(5a-3)-3]^2 = (5a-6)^2$

89. $\left(\frac{6}{2}\right)^2 = 3^2 = 9$
$x^2 + 6x + 9 = (x+3)^2$

91. $\left(\frac{14}{2}\right)^2 = 7^2 = 49$
$m^2 - 14m + 49 = (m-7)^2$

93. $\pm 2\sqrt{16} = \pm 2 \cdot 4 = \pm 8$
$x^2 + 8x + 16 = (x+4)^2$
$x^2 - 8x + 16 = (x-4)^2$

95. a. $(x^3)^2 - 1^2$
$= (x^3+1)(x^3-1)$
$= (x+1)(x^2-x+1)$
$\cdot(x-1)(x^2+x+1)$

b. $(x^2)^3 - 1^3$
$= (x+1)(x-1)(x^4+x^2+1)$

c. Answers may vary.

97. $x + 7 = 0$
$x = -7$
$\{-7\}$

99. $5x - 15 = 0$
$5x = 15$
$x = 3$
$\{3\}$

101. $3x = 0$
$x = 0$
$\{0\}$

103. $-4x - 16 = 0$
$-4x = 16$
$x = -4$
$\{-4\}$

105. $x^{2n} - 36$
$= (x^n)^2 - 6^2$
$= (x^n - 6)(x^n + 6)$

107. $25x^{2n} - 81$
$= (5x^n)^2 - 9^2$
$= (5x^n - 9)(5x^n + 9)$

109. $x^{4n} - 625$
$= (x^{2n})^2 - (25)^2$
$= (x^{2n} - 25)(x^{2n} + 25)$
$= [(x^n)^2 - 5^2](x^{2n} + 25)$
$= (x^n - 5)(x^n + 5)(x^{2n} + 25)$

Section 5.8

Mental Math

1. $(x - 3)(x + 5) = 0$
$x - 3 = 0$ or $x + 5 = 0$
$x = 3$ or $x = -5$
$\{3, -5\}$

3. $(z - 3)(z + 7) = 0$
$z - 3 = 0$ or $z + 7 = 0$
$z = 3$ or $z = -7$
$\{3, -7\}$

5. $x(x - 9) = 0$
$x = 0$ or $x - 9 = 0$
$x = 9$
$\{0, 9\}$

Section 5.8 Graphing Calculator Explorations

1. $x^2 + 3x - 2 = 0$
$\{-3.562, 0.562\}$

3. $2.3x^2 - 4.4x - 5.6 = 0$
$\{-0.874, 2.787\}$

5. $0.09x^2 - 0.13x - 0.08 = 0$
$\{-0.465, 1.910\}$

Exercise Set 5.8

1. $(x + 3)(3x - 4) = 0$
$x + 3 = 0$ or $3x - 4 = 0$
$x = -3$ or $3x = 4$
$x = \frac{4}{3}$
$\left\{-3, \frac{4}{3}\right\}$

3. $3(2x - 5)(4x + 3) = 0$
$2x - 5 = 0$ or $4x + 3 = 0$
$2x = 5$ or $4x = -3$
$x = \frac{5}{2}$ or $x = -\frac{3}{4}$
$\left\{\frac{5}{2}, -\frac{3}{4}\right\}$

5. $x^2 + 11x + 24 = 0$
$(x + 3)(x + 8) = 0$
$x + 3 = 0$ or $x + 8 = 0$
$x = -3$ or $x = -8$
$\{-3, -8\}$

7. $12x^2 + 5x - 2 = 0$
$(4x - 1)(3x + 2) = 0$
$4x - 1 = 0$ or $3x + 2 = 0$
$4x = 1$ or $3x = -2$
$x = \frac{1}{4}$ or $x = -\frac{2}{3}$
$\left\{\frac{1}{4}, -\frac{2}{3}\right\}$

9. $z^2 + 9 = 10z$
$z^2 - 10z + 9 = 0$
$(z - 1)(z - 9) = 0$
$z - 1 = 0$ or $z - 9 = 0$
$z = 1$ or $z = 9$
$\{1, 9\}$

11. $x(5x + 2) = 3$
$5x^2 + 2x - 3 = 0$
$(5x - 3)(x + 1) = 0$
$5x - 3 = 0$ or $x + 1 = 0$
$5x = 3$ or $x = -1$
$x = \frac{3}{5}$
$\left\{\frac{3}{5}, -1\right\}$

13. $x^2 - 6x = x(8 + x)$
$x^2 - 6x = 8x + x^2$
$0 = 14x$
$x = 0$
$\{0\}$

15. $\frac{z^2}{6} - \frac{z}{2} - 3 = 0$
$z^2 - 3z - 18 = 0$
$(z - 6)(z + 3) = 0$
$z - 6 = 0$ or $z + 3 = 0$
$z = 6$ or $z = -3$
$\{6, -3\}$

17. $\frac{x^2}{2} + \frac{x}{20} = \frac{1}{10}$
$10x^2 + x = 2$
$10x^2 + x - 2 = 0$
$(5x - 2)(2x + 1) = 0$
$5x - 2 = 0$ or $2x + 1 = 0$
$5x = 2$ or $2x = -1$
$x = \frac{2}{5}$ or $x = -\frac{1}{2}$
$\left\{\frac{2}{5}, -\frac{1}{2}\right\}$

19. $\frac{4t^2}{5} = \frac{t}{5} + \frac{3}{10}$
$8t^2 = 2t + 3$
$8t^2 - 2t - 3 = 0$
$(4t - 3)(2t + 1) = 0$
$4t - 3 = 0$ or $2t + 1 = 0$
$4t = 3$ or $2t = -1$
$t = \frac{3}{4}$ or $t = -\frac{1}{2}$
$\left\{\frac{3}{4}, -\frac{1}{2}\right\}$

21. $(x + 2)(x - 7)(3x - 8) = 0$
$x + 2 = 0$ or $x - 7 = 0$ or $3x - 8 = 0$
$x = -2$ or $x = 7$ or $3x = 8$
$x = \frac{8}{3}$
$\left\{-2, 7, \frac{8}{3}\right\}$

23. $y^3 = 9y$
$y^3 - 9y = 0$
$y(y^2 - 9) = 0$
$y(y + 3)(y - 3) = 0$
$y = 0$ or $y + 3 = 0$ or $y - 3 = 0$
$y = -3$ or $y = 3$
$\{0, -3, 3\}$

25. $x^3 - x = 2x^2 - 2$
$x^3 - 2x^2 - x + 2 = 0$
$x^2(x - 2) - (x - 2) = 0$
$(x^2 - 1)(x - 2) = 0$
$(x + 1)(x - 1)(x - 2) = 0$
$x + 1 = 0$ or $x - 1 = 0$ or $x - 2 = 0$
$x = -1$ or $x = 1$ or $x = 2$
$\{-1, 1, 2\}$

27. Answers may vary.

29. $(2x + 7)(x - 10) = 0$
$2x + 7 = 0$ or $x - 10 = 0$
$2x = -7$ or $x = 10$
$x = -\frac{7}{2}$
$\left\{-\frac{7}{2}, 10\right\}$

31. $3x(x - 5) = 0$
$3x = 0$ or $x - 5 = 0$
$x = 0$ or $x = 5$
$\{0, 5\}$

33. $x^2 - 2x - 15 = 0$
$(x - 5)(x + 3) = 0$
$x - 5 = 0$ or $x + 3 = 0$
$x = 5$ or $x = -3$
$\{5, -3\}$

35. $12x^2 + 2x - 2 = 0$
$6x^2 + x - 1 = 0$
$(3x - 1)(2x + 1) = 0$
$3x - 1 = 0$ or $2x + 1 = 0$
$3x = 1$ or $2x = -1$
$x = \frac{1}{3}$ or $x = -\frac{1}{2}$
$\left\{\frac{1}{3}, -\frac{1}{2}\right\}$

37. $w^2 - 5w = 36$
$w^2 - 5w - 36 = 0$
$(w-9)(w+4) = 0$
$w - 9 = 0$ or $w + 4 = 0$
$w = 9$ or $w = -4$
$\{9, -4\}$

39. $25x^2 - 40x + 16 = 0$
$(5x-4)^2 = 0$
$5x - 4 = 0$
$5x = 4$
$x = \frac{4}{5}$
$\left\{\frac{4}{5}\right\}$

41. $2r^3 + 6r^2 = 20r$
$r^3 + 3r^2 = 10r$
$r^3 + 3r^2 - 10r = 0$
$r(r^2 + 3r - 10) = 0$
$r(r+5)(r-2) = 0$
$r = 0$ or $r + 5 = 0$ or $r - 2 = 0$
$r = -5$ or $r = 2$
$\{0, -5, 2\}$

43. $z(5z-4)(z+3) = 0$
$z = 0$ or $5z - 4 = 0$ or $z + 3 = 0$
$5z = 4$ or $z = -3$
$z = \frac{4}{5}$
$\left\{0, \frac{4}{5}, -3\right\}$

45. $2z(z+6) = 2z^2 + 12z - 8$
$2z^2 + 12z = 2z^2 + 12z - 8$
$0 = -8$
No solution exists.
$\{\ \}$

47. $(x-1)(x+4) = 24$
$x^2 + 3x - 4 = 24$
$x^2 + 3x - 28 = 0$
$(x+7)(x-4) = 0$
$x + 7 = 0$ or $x - 4 = 0$
$x = -7$ or $x = 4$
$\{-7, 4\}$

49. $\frac{x^2}{4} - \frac{5}{2}x + 6 = 0$
$x^2 - 10x + 24 = 0$
$(x-4)(x-6) = 0$
$x - 4 = 0$ or $x - 6 = 0$
$x = 4$ or $x = 6$
$\{4, 6\}$

51. $y^2 + \frac{1}{4} = -y$
$4y^2 + 1 = -4y$
$4y^2 + 4y + 1 = 0$
$(2y+1)^2 = 0$
$2y + 1 = 0$
$2y = -1$
$y = -\frac{1}{2}$
$\left\{-\frac{1}{2}\right\}$

53. $y^3 + 4y^2 = 9y + 36$
$y^3 + 4y^2 - 9y - 36 = 0$
$(y^2 - 9)(y+4) = 0$
$(y+3)(y-3)(y+4) = 0$
$y + 3 = 0$ or $y - 3 = 0$ or $y + 4 = 0$
$y = -3$ or $y = 3$ or $y = -4$
$\{-3, 3, -4\}$

55. $2x^3 = 50x$
$x^3 = 25x$
$x^3 - 25x = 0$
$x(x^2 - 25) = 0$
$x(x+5)(x-5) = 0$
$x = 0$ or $x + 5 = 0$ or $x - 5 = 0$
$x = -5$ or $x = 5$
$\{0, -5, 5\}$

57. $x^2 + (x+1)^2 = 61$
$x^2 + x^2 + 2x + 1 = 61$
$2x^2 + 2x - 60 = 0$
$x^2 + x - 30 = 0$
$(x+6)(x-5) = 0$
$x + 6 = 0$ or $x - 5 = 0$
$x = -6$ or $x = 5$
$\{-6, 5\}$

59. $m^2(3m-2)=m$
$3m^3-2m^2=m$
$3m^3-2m^2-m=0$
$m(3m^2-2m-1)=0$
$m(3m+1)(m-1)=0$
$m=0$ or $3m+1=0$ or $m-1=0$
$3m=-1$ or $m=1$
$m=-\frac{1}{3}$
$\left\{0, -\frac{1}{3}, 1\right\}$

61. $3x^2=-x$
$3x^2+x=0$
$x(3x+1)=0$
$x=0$ or $3x+1=0$
$x=0$ or $3x=-1$
$x=-\frac{1}{3}$
$\left\{0, -\frac{1}{3}\right\}$

63. $x(x-3)=x^2+5x+7$
$x^2-3x=x^2+5x+7$
$-7=8x$
$x=-\frac{7}{8}$
$\left\{-\frac{7}{8}\right\}$

65. $3(t-8)+2t=7+t$
$3t-24+2t=7+t$
$5t-24=7+t$
$4t=31$
$t=\frac{31}{4}$
$\left\{\frac{31}{4}\right\}$

67. $-3(x-4)+x=5(3-x)$
$-3x+12+x=15-5x$
$-2x+12=15-5x$
$3x=3$
$x=1$
$\{1\}$

69. Answers may vary.

71. Let n = the one number and $n+5$ = the other number.
Then: $n(n+5)=66$
$n^2+5n-66=0$
$(n+11)(n-6)=0$
$n+11=0$ or $n-6=0$
$n=-11$ or $n=6$
There are two solutions: -11 and -6 and 6 and 11.

73. Let d = the amount of cable needed. Then from the Pythagorean theorem,
$d^2=45^2+60^2=5625$
So, $d=\sqrt{5625}=75$ ft.

75. $C(x)=x^2-15x+50$
$9500=x^2-15x+50$
$0=x^2-15x-9450$
$0=(x-105)(x+90)$
$x-105=0$ or $x+90=0$
$x=105$ or $x=-90$
Disregard the negative.
105 units

77. Let x = one leg of a right triangle and $x-3$ = the other leg of the right triangle. Then from the Pythagorean theorem,
$15^2=x^2+(x-3)$
$225=x^2+x^2-6x+9$
$0=2x^2-6x-216$
$0=x^2-3x-108$
$0=(x-12)(x+9)$
$x-12=0$ or $x+9=0$
$x=12$ or $x=-9$
Discarding the extraneous solution of -9, we find that one leg of the right triangle is 12 cm and the other leg is 9 cm.

79. Let x be the width of the pine bark. Then the area of the garden with the pine bark is
$(20+2x)(30+2x)=4x^2+100x+600$.
The area of the garden itself is
$(20)(30)=600$. Thus,
$(4x^2+100x+600)-600=336$
$4x^2+100x-336=0$
$x^2+25x-84=0$

$(x-3)(x+28)=0$
$x-3=0$ or $x+28=0$
$x=3$ or $x=-28$
Disregard the negative. The answer is 3 ft.

81. $h(t)=-16t^2+80t$
$96=-16t^2+80t$
$0=-16t^2+80t-96$
$0=t^2-5t+6$
$(t-2)(t-3)=0$
$t=2, t=3$
ascending 2 sec, descending 3 sec

83. $g(x)=(x+1)(x-6)$; D

85. $F(x)=(x+1)(x-2)(x+5)$; A

87. $H(x)=2x^2-7x-4$; C

89. Answers may vary;
Ex. $p(x)=x^2-13x+42$

91. Answers may vary;
Ex. $f(x)=x^2-x-12$

93. (–4, 0), (0, 0), (3, 0); function

95. (–5, 0), (5, 0), (0, –4); function

97. Answers may vary.

Section 5.9

Mental Math

1. $f(x)=2x^2+7x+10$; upward

3. $f(x)=-x^2+5$; downward

Exercise Set 5.9

1. a. Domain, $(-\infty, \infty)$; Range, $(-\infty, 5]$

b. x-intercepts, (–2, 0), (6, 0);
y-intercept, (0, 5)

c. (0, 5)

d. There is no such point.

e. –2, 6

f. $-2<x<6$

g. {–2, 6}

3. a. Domain, $(-\infty, \infty)$; Range, $[-4, \infty)$

b. x-intercepts, (–3, 0), (1, 0);
y-intercept, (0, –3)

c. There is no such point.

d. (–1, –4)

e. –3, 1

f. $x<-3$ or $x>1$

g. {–3, 1}

5. a. Domain, $(-\infty, \infty)$; Range, $(-\infty, \infty)$

b. x-intercepts, (–2, 0), (0, 0), (2, 0);
y-intercept, (0, 0)

c. There is no such point.

d. There is no such point.

e. –2, 0, 2

f. between $x=-2$ and 0; $x>2$

g. {–2, 0, 2}

7. $f(x)=2x^2$

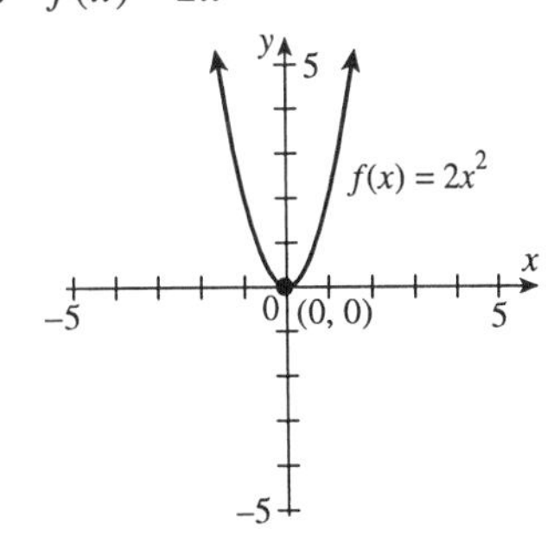

9. $f(x) = x^2 + 1$

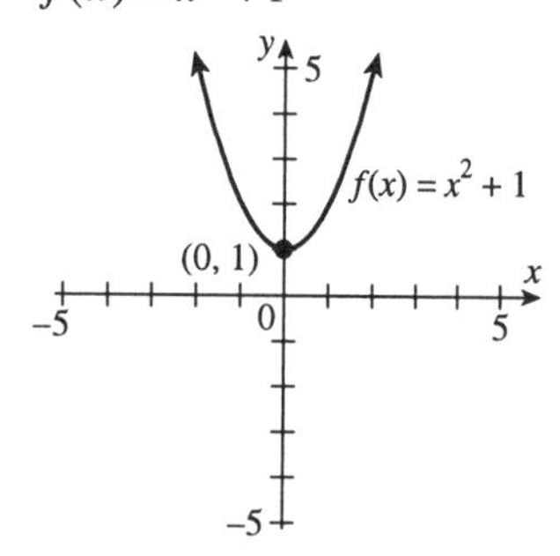

11. $f(x) = -x^2$

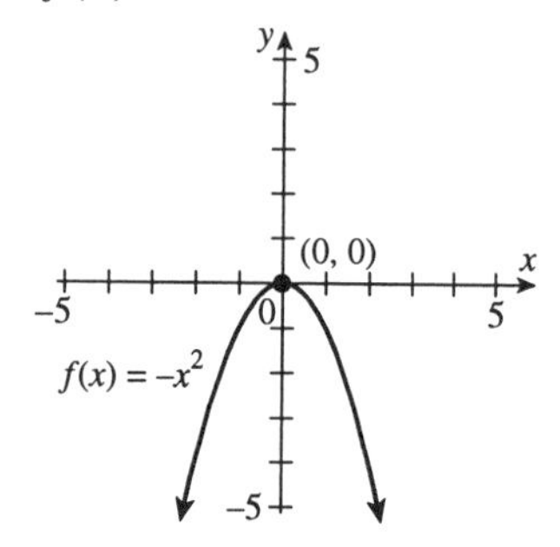

13. $f(x) = x^2 + 8x + 7,$
$a = 1$, and $b = 8$
So, $h = \dfrac{-8}{2(1)} = -4$ and

$k = f(-4) = (-4)^2 + 8(-4) + 7 = -9$
thus, $V(-4, -9)$

15. $f(x) = 3x^2 + 6x + 4$, $a = 3$, and $b = 6$.
So, $h = -\dfrac{6}{2(3)} = -1$ and

$k = f(-1) = 3(-1)^2 + 6(-1) + 4 = 1;$
thus $V(-1, 1)$.

17. $f(x) = -x^2 + 10x + 5$, $a = -1$ and
$b = 10$. So, $h = -\dfrac{10}{2(-1)} = 5$ and

$k = f(5) = -5^2 + 10(5) + 5 = 30;$
thus $V(5, 30)$.

19. Two

21. Zero

23. One x-intercept and one y-intercept

25. $f(x) = x^2 + 6x + 5$
$\dfrac{-b}{2a} = \dfrac{-6}{2(1)} = -3$
$f(-3) = 9 - 18 + 5 = -4$
The vertex is $(-3, -4)$.
If $x^2 + 6x + 5 = 0$
$(x + 1)(x + 5) = 0$
$x + 1 = 0$ or $x + 5 = 0$
$x = -1$ or $x = -5$
The x-intercepts are -1 and -5.
If $x = 0$, then $y = f(0) = 5$.
The y-intercept is 5.

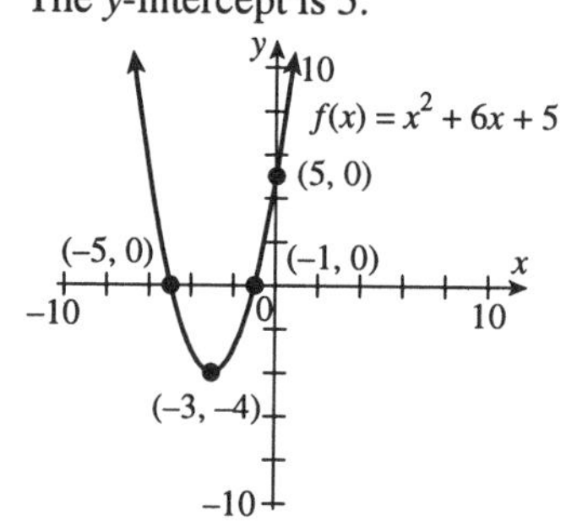

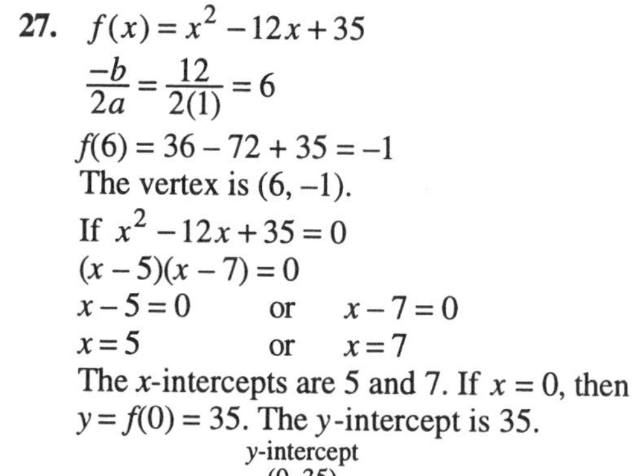

27. $f(x) = x^2 - 12x + 35$
$\dfrac{-b}{2a} = \dfrac{12}{2(1)} = 6$
$f(6) = 36 - 72 + 35 = -1$
The vertex is $(6, -1)$.
If $x^2 - 12x + 35 = 0$
$(x - 5)(x - 7) = 0$
$x - 5 = 0$ or $x - 7 = 0$
$x = 5$ or $x = 7$
The x-intercepts are 5 and 7. If $x = 0$, then $y = f(0) = 35$. The y-intercept is 35.

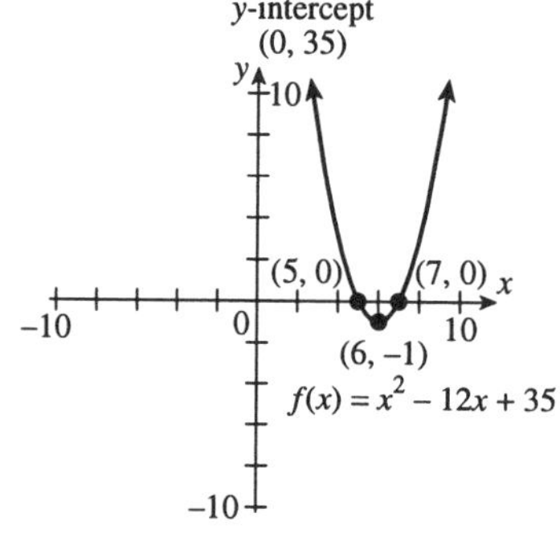

29. $f(x) = -3x^2 + 6x$
$\frac{-b}{2a} = \frac{-6}{2(-3)} = 1$
$f(1) = -3 + 6 = 3$
The vertex is (1, 3).
If $-3x^2 + 6x = 0$
$3x(-x + 2) = 0$
$3x = 0$ or $-x + 2 = 0$
$x = 0$ or $x = 2$
The x-intercepts are 0 and 2. If $x = 0$, then $y = f(0) = 0$. The y-intercept is 0.

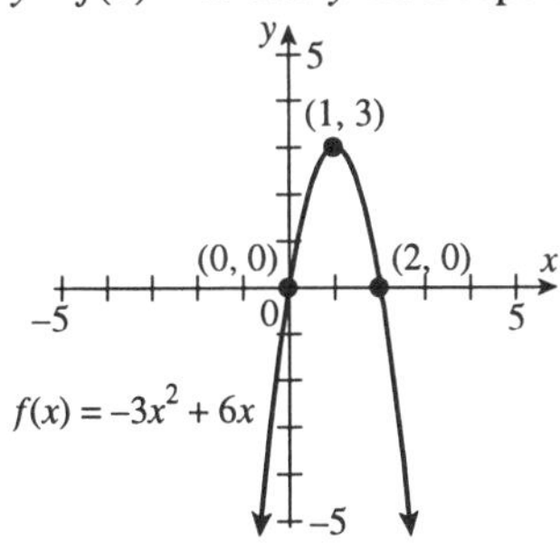

31. $f(x) = 2x^3 - 5x^2 - 3x$
If $2x^3 - 5x^2 - 3x = 0$
$x(2x^2 - 5x - 3) = 0$
$x(2x + 1)(x - 3) = 0$
$x = 0$ or $2x + 1 = 0$ or $x - 3 = 0$
$2x = -1$ or $x = 3$
$x = -\frac{1}{2}$
The x-intercepts are $-\frac{1}{2}$, 0, and 3.
The y-intercept is 0.

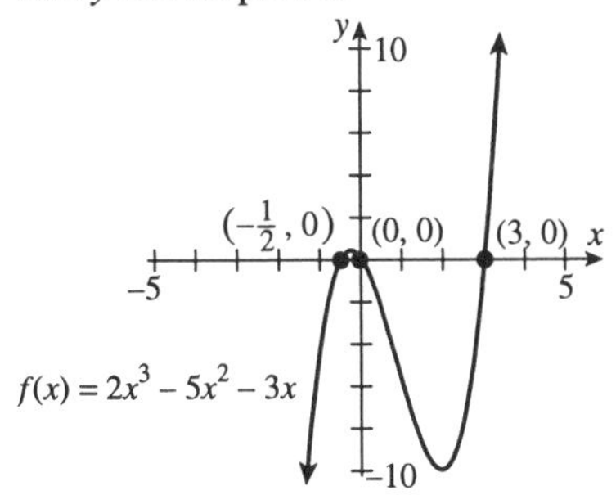

33. $f(x) = x^3 + x^2 - 4x - 4$
If $x^3 + x^2 - 4x - 4 = 0$
$x^2(x+1) - 4(x+1) = 0$
$(x+1)(x^2 - 4) = 0$
$(x + 1)(x + 2)(x - 2) = 0$
$x + 1 = 0$ or $x + 2 = 0$ or $x - 2 = 0$
$x = -1$ or $x = -2$ or $x = 2$
The x-intercepts are –2, –1, and 2. If $x = 0$, then $y = f(0) = -4$. The y-intercept is –4.

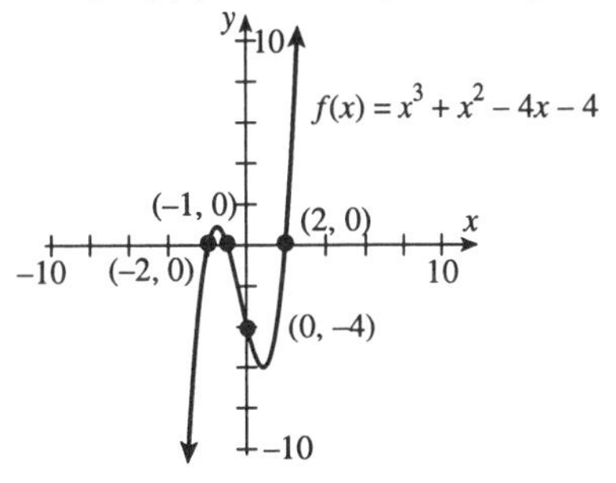

35. No

37. $f(x) = x^2 + 2x - 3$
$\frac{-b}{2a} = \frac{-2}{2(1)} = -1$
$f(-1) = 1 - 2 - 3 = -4$
The vertex is (–1, –4).
If $x^2 + 2x - 3 = 0$
$(x - 1)(x + 3) = 0$
$x - 1 = 0$ or $x + 3 = 0$
$x = 1$ or $x = -3$
The x-intercepts are –3 and 1. If $x = 0$, the $y = f(0) = -3$. The y-intercept is –3.

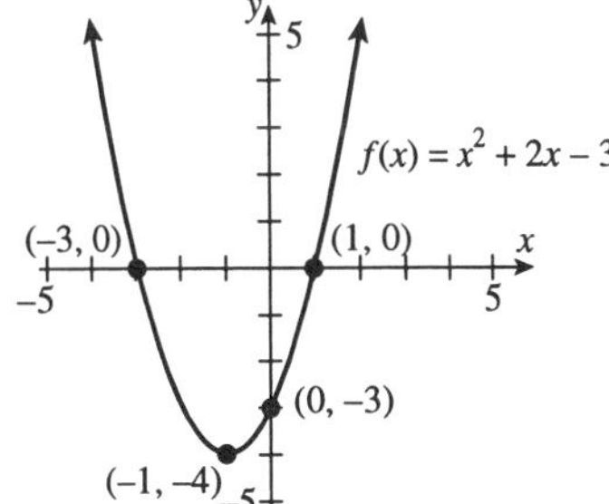

39. $f(x) = x^3 - 4x^2 + 3x$
If $x^3 - 4x^2 + 3x = 0$
$x(x^2 - 4x + 3) = 0$
$x(x-1)(x-3) = 0$
$x = 0$ or $x - 1 = 0$ or $x - 3 = 0$
$x = 1$ or $x = 3$
The x-intercepts are 0, 1, and 3.
The y-intercept is 0.

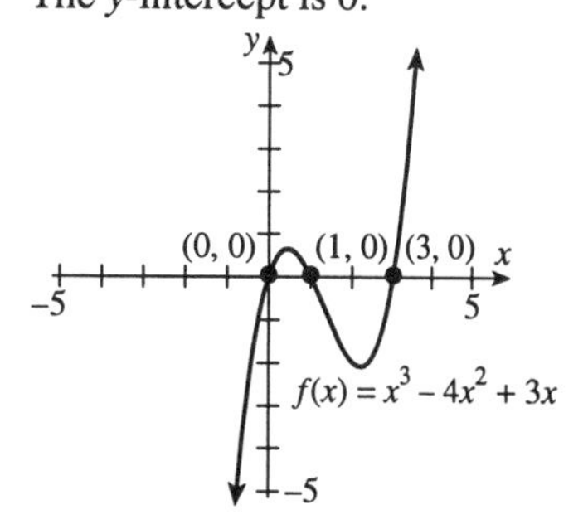

41. $f(x) = x^2 + 4$
$\frac{-b}{2a} = 0$
$f(0) = 4$
The vertex is (0, 4).
There are no x-intercepts.
The y-intercept is 0.

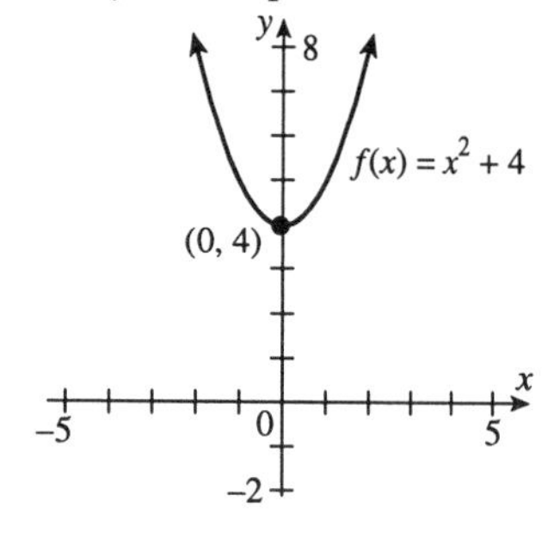

43. $f(x) = 3x^2 - 12x$
$\frac{-b}{2a} = \frac{12}{2(3)} = 2$
$f(2) = 12 - 24 = -12$
The vertex is (2, −12).
If $3x^2 - 12x = 0$
$3x(x-4) = 0$
$3x = 0$ or $x - 4 = 0$
$x = 0$ or $x = 4$
The x-intercepts are 0 and 4.
The y-intercept is 0.

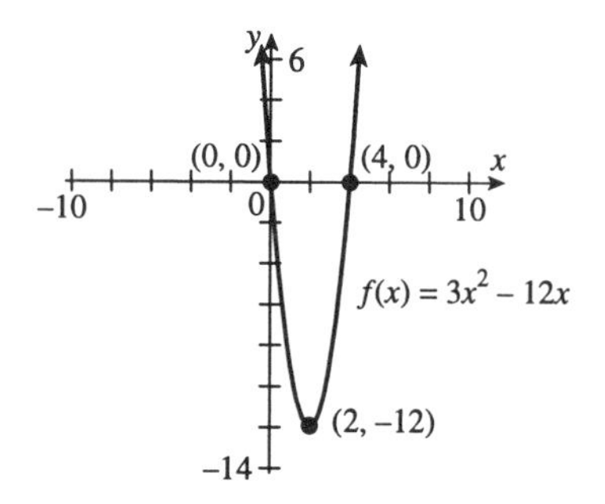

45. $f(x) = x^3 + x^2 - 12x$
If $x^3 + x^2 - 12x = 0$
$x(x^2 + x - 12) = 0$
$x(x-3)(x+4) = 0$
$x = 0$ or $x - 3 = 0$ or $x + 4 = 0$
$x = 3$ or $x = -4$
The x-intercepts are −4, 0, and 3.
The y-intercept is 0.

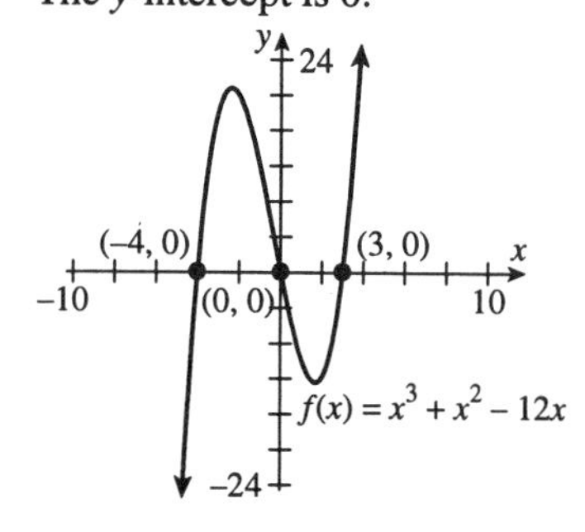

47. $f(x) = x^3 + x^2 - 9x - 9$
If $x^3 + x^2 - 9x - 9 = 0$
$x^2(x+1) - 9(x+1) = 0$
$(x+1)(x^2 - 9) = 0$
$(x+1)(x+3)(x-3) = 0$
$x + 1 = 0$ or $x + 3 = 0$ or $x - 3 = 0$
$x = -1$ or $x = -3$ or $x = 3$
The x-intercepts are −3, −1, and 3. If $x = 0$, then $y = f(0) = -9$. The y-intercept is −9.

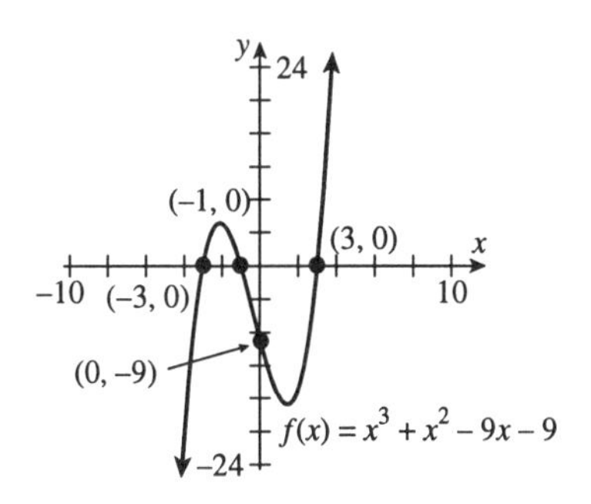

49. $f(x) = x^2 - 2x + 1$

$\frac{-b}{2a} = \frac{2}{2(1)} = 1$

$f(1) = 1 - 2 + 1 = 0$

The vertex is (1, 0).

If $x^2 - 2x + 1 = 0$

$(x - 1)(x - 1) = 0$

$x - 1 = 0$

$x = 1$

The x-intercept is 1. If $x = 0$, then $y = f(0) = 1$. The y-intercept is 1.

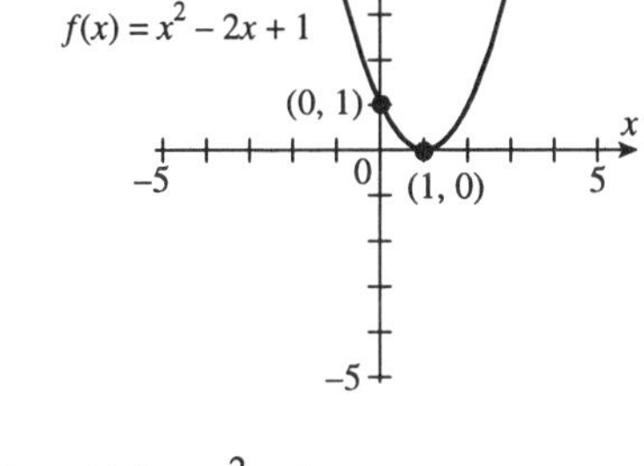

51. $f(x) = x^2 + 6x$

$\frac{-b}{2a} = \frac{-6}{2(1)} = -3$

$f(-3) = 9 - 18 = -9$

The vertex is (–3, –9).

If $x^2 + 6x = 0$

$x(x + 6) = 0$

$x = 0$ or $x + 6 = 0$

$x = -6$

The x-intercepts are –6 and 0.

The y-intercept is 0.

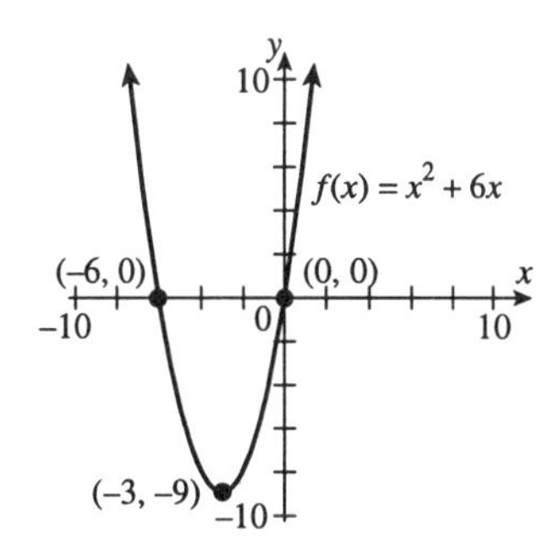

53. $f(x) = (x + 2)(x - 2)$

If $(x + 2)(x - 2) = 0$

$x + 2 = 0$ or $x - 2 = 0$

$x = -2$ or $x = 2$

The x-intercepts are –2 and 2.

If $x = 0$, then $y = f(0) = -4$.

The y-intercept is –4.

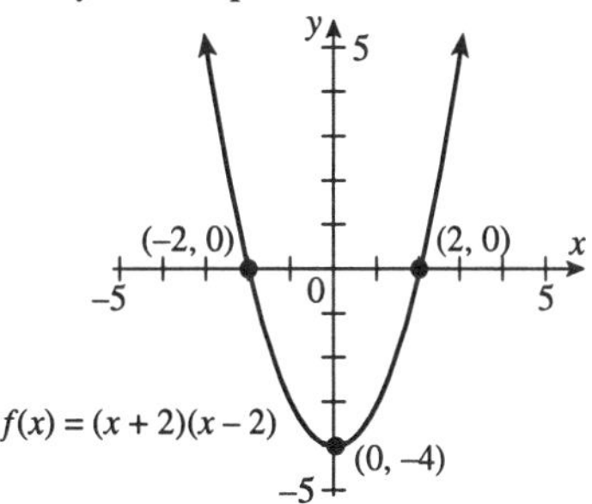

55. $f(x) = -x^3 + 25x$

If $-x^3 + 25x = 0$

$x(-x^2 + 25) = 0$

$x(5 - x)(5 + x) = 0$

$x = 0$ or $5 - x = 0$ or $5 + x = 0$

$x = 5$ or $x = -5$

The x-intercepts are –5, 0, and 5.

The y-intercept is 0.

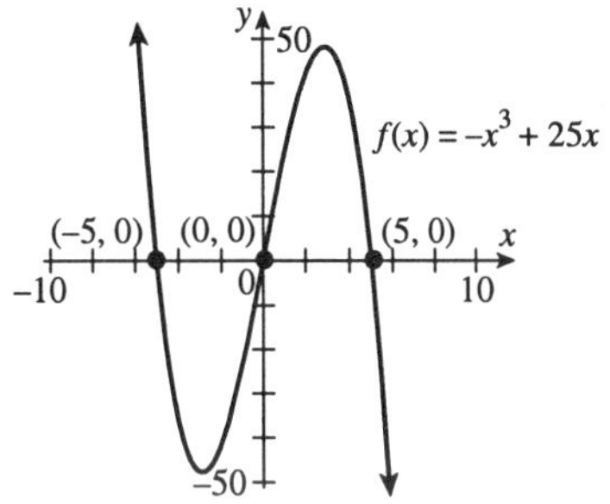

57. $f(x) = x^2 + 2x + 4$

$\frac{-b}{2a} = \frac{-2}{2(1)} = -1$

$f(-1) = 1 - 2 + 4 = 3$

The vertex is $(-1, 3)$.

There are no x-intercepts.

If $x = 0$, then $y = f(0) = 4$.

The y-intercept is 4.

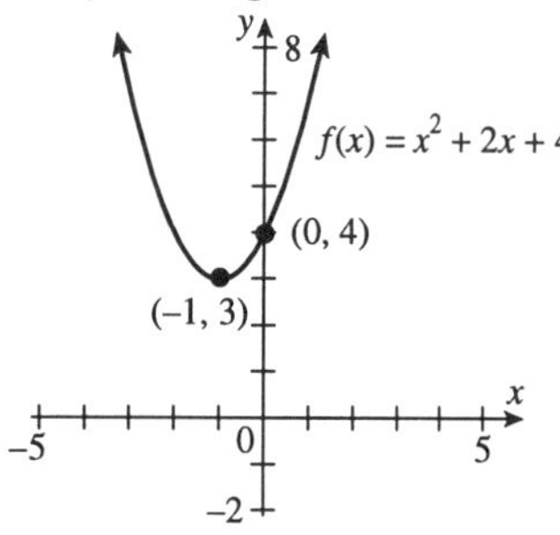

59. $f(x) = 3x(x-3)(x+5)$

If $3x(x-3)(x+5) = 0$

$3x = 0$ or $x - 3 = 0$ or $x + 5 = 0$

$x = 0$ or $x = 3$ or $x = -5$

The x-intercepts are -5, 0, and 3.

The y-intercept is 0.

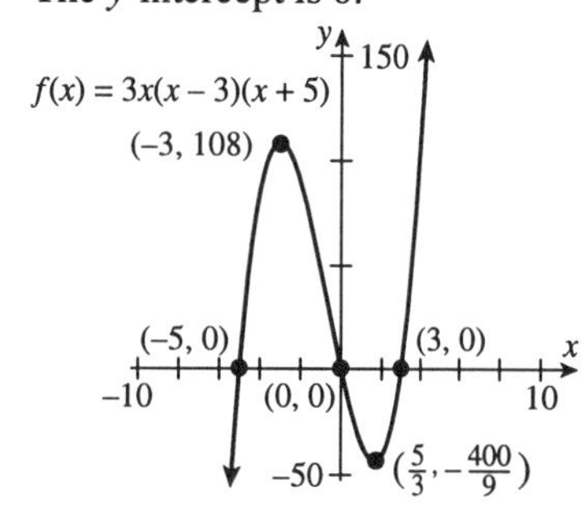

61. $h(x) = (x-4)(x-2)(2x+1)(x+3)$

$x - 4 = 0$ or $x - 2 = 0$

$x = 4$ or $x = 2$

or $2x + 1 = 0$ or $x + 3 = 0$

$x = -\frac{1}{2}$ or $x = -3$

The x-intercepts are 4, 2, $-\frac{1}{2}$, and -3.

If $x = 0$, then $y = h(0) = 24$.

The y-intercept is 24.

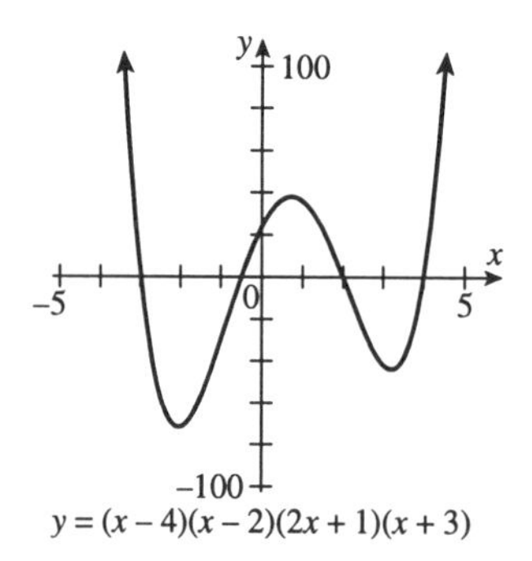

$y = (x-4)(x-2)(2x+1)(x+3)$

63. $f(x) = x^2 + 2x - 3$

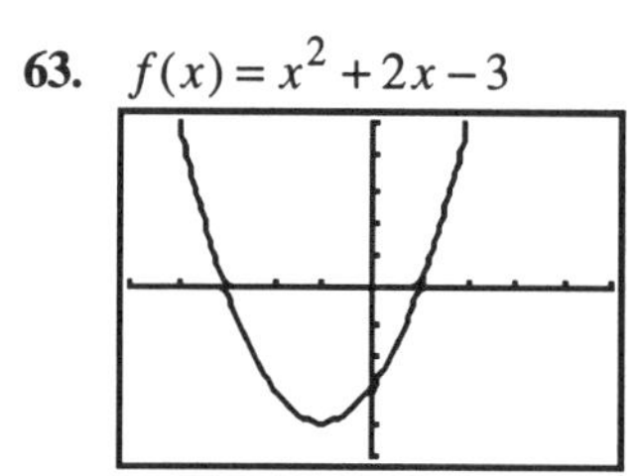

65. $f(x) = -x^3 + 25x$

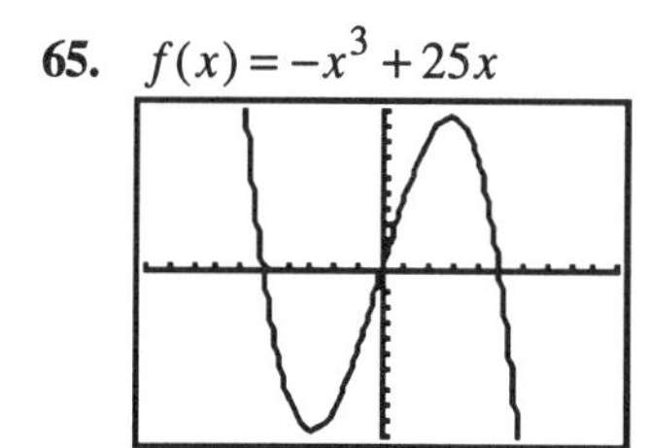

67. $G(x) = x^4 - 6.2x^2 - 6.2$

$x = -2.7,\ x = 2.7$

69. $-\frac{45}{100} = -\frac{9(5)}{20(5)} = -\frac{9}{20}$

71. $\frac{a^{14}b^2}{ab^4} = \frac{a^{14-1}}{b^{4-2}} = \frac{a^{13}}{b^2}$

73. $\frac{20x^{-3}y^5}{25y^{-2}x} = \frac{4(5)y^{5+2}}{5(5)x^{1+3}} = \frac{4y^7}{5x^4}$

Chapter 5 - Review

1. $(-2)^2 = (-2)(-2) = 4$

2. $(-3)^4 = (-3)\cdot(-3)\cdot(-3)\cdot(-3) = 81$

3. $-2^2 = -(2\cdot 2) = -4$

4. $-3^4 = -3 \cdot 3 \cdot 3 \cdot 3 = -81$

5. $8^0 = 1$

6. $-9^0 = -1$

7. $-4^{-2} = -\frac{1}{4^2} = -\frac{1}{16}$

8. $(-4)^{-2} = \frac{1}{(-4)^2} = \frac{1}{16}$

9. $-xy^2y^3xy^2z = -x^{1+1}y^{2+3+2}z = -x^2y^7z$

10. $(-4xy)(-3xy^2b) = 12x^2y^3b$

11. $a^{-14}a^5 = a^{-14+5} = a^{-9} = \frac{1}{a^9}$

12. $\frac{a^{16}}{a^{17}} = a^{16-17} = a^{-1} = \frac{1}{a}$

13. $\frac{x^{-7}}{x^4} = \frac{1}{x^{4-(-7)}} = \frac{1}{x^{11}}$

14. $\frac{9a(a^{-3})}{18a^{15}} = \frac{a^{1-3-15}}{2} = \frac{a^{-17}}{2} = \frac{1}{2a^{17}}$

15. $\frac{y^{6p-3}}{y^{6p+2}} = \frac{1}{y^{(6p+2)-(6p-3)}}$
$= \frac{1}{y^{6p+2-6p+3}} = \frac{1}{y^5}$

16. $36,890,000 = 3.689 \times 10^7$

17. $-0.000362 = -3.62 \times 10^{-4}$

18. $1.678 \times 10^{-6} = 0.000001678$

19. $4.1 \times 10^5 = 410,000$

20. $(8^5)^3 = 8^{5 \cdot 3} = 8^{15}$

21. $\left(\frac{a}{4}\right)^2 = \frac{a^2}{4^2} = \frac{a^2}{16}$

22. $(3x^3) = 3^3x^3 = 27x^3$

23. $(-4x)^{-2} = \frac{1}{(-4x)^2}$
$= \frac{1}{(-4)^2x^2} = \frac{1}{16x^2}$

24. $\left(\frac{6x}{5}\right)^2 = \frac{(6x)^2}{(5)^2} = \frac{36x^2}{25}$

25. $(8^6)^{-3} = 8^{-18} = \frac{1}{8^{18}}$

26. $\left(\frac{4}{3}\right)^{-2} = \frac{4^{-2}}{3^{-2}} = \frac{\frac{1}{4^2}}{\frac{1}{3^2}} = \frac{3^2}{4^2} = \frac{9}{16}$

27. $(-2x^3)^{-3} = \frac{1}{(-2x^3)^3}$
$= \frac{1}{(-2)^3(x^3)^3} = \frac{1}{-8x^9} = -\frac{1}{8x^9}$

28. $\left(\frac{8p^6}{4p^4}\right)^{-2} = (2p^2)^{-2} = 2^{-2}p^{-4} = \frac{1}{4p^4}$

29. $(-3x^{-2}y^2)^3 = (-3)^3(x^{-2})^3(y^2)^3$
$= -27x^{-6}y^6 = \frac{-27y^6}{x^6}$

30. $\left(\frac{x^{-5}y^{-3}}{z^3}\right)^{-5} = \frac{x^{25}y^{15}}{z^{-15}} = x^{25}y^{15}z^{15}$

31. $\frac{4^{-1}x^3yz}{x^{-2}yx^4} = \frac{x^{3-(-2)-4}z}{4}$
$= \frac{x^{5-4}z}{4} = \frac{xz}{4}$

32. $(5xyz)^{-4}(x^{-2})^{-3} = \frac{1}{(5xyz)^4}x^6$
$= \frac{x^6}{625x^4y^4z^4} = \frac{x^2}{625y^4z^4}$

33. $\dfrac{2(3yz)^{-3}}{y^{-3}} = \dfrac{2(3)^{-3}y^{-3}z^{-3}}{y^{-3}}$
$= \dfrac{2}{3^3 z^3} = \dfrac{2}{27z^3}$

34. $x^{4a}(3x^{5a})^3 = x^{4a}(3^3 x^{15a})$
$= 27x^{4a+15a} = 27x^{19a}$

35. $\dfrac{4y^{3x-3}}{2y^{2x+4}} = 2y^{(3x-3)-(2x+4)}$
$= 2y^{3x-3-2x-4} = 2y^{x-7}$

36. $\dfrac{(0.00012)(144,000)}{0.0003}$
$= \dfrac{(1.2\times 10^{-4})(1.44\times 10^{5})}{3\times 10^{-4}}$
$= 0.576\times 10^5 = 5.76\times 10^4$

37. $\dfrac{(-0.00017)(0.00039)}{3,000}$
$= \dfrac{(-1.7\times 10^{-4})(3.9\times 10^{-4})}{3\times 10^3}$
$= -2.21\times 10^{-4-4-3} = -2.21\times 10^{-11}$

38. $\dfrac{27x^5y^5}{18x^{-6}y^2}\cdot\dfrac{x^4y^{-2}}{x^{-2}y^3}$
$= \dfrac{3x^{-5+4}y^{5-2}}{2x^{-6-2}y^{2+3}} = \dfrac{3x^{-1}y^3}{2x^{-8}y^5}$
$= \dfrac{3}{2}x^{-1+8}y^{3-5} = \dfrac{3x^7}{2y^2}$

39. $\dfrac{3x^5}{y^{-4}}\cdot\dfrac{(3xy^{-3})^{-2}}{(z^{-3})^{-4}}$
$= \dfrac{3x^5\cdot 3^{-2}x^{-2}(y^{-3})^{-2}}{y^{-4}z^{12}}$
$= \dfrac{3^{1-2}x^{5-2}y^{6-(-4)}}{z^{12}}$
$= \dfrac{3^{-1}x^3y^{10}}{z^{12}} = \dfrac{x^3y^{10}}{3z^{12}}$

40. $\dfrac{(x^w)^2}{(x^{w-4})^{-2}} = x^{2w-(-2w+8)} = x^{4w-8}$

41. The degree of the polynomial $x^2y - 3xy^3z + 5x + 7y$ is the degree of the term $-3xy^3z$ which is 5.

42. $3x + 2$ has degree 1.

43. $4x + 8x - 6x^2 - 6x^2y$
$= (4+8)x - 6x^2 - 6x^2y$
$= 12x - 6x^2 - 6x^2y$

44. $-8xy^3 + 4xy^3 - 3x^3y$
$= (-8+4)xy^3 - 3x^3y$
$= -4xy^3 - 3x^3y$

45. $(3x+7y) + (4x^2 - 3x + 7) + (y-1)$
$= 4x^2 + (3-3)x + (7+1)y + (7-1)$
$= 4x^2 + 8y + 6$

46.
$$\begin{array}{rrr} 4x^2 & -6xy + & 9y^2 \\ -\;\; 8x^2 & -6xy - & y^2 \\ \hline -4x^2 & & +10y^2 \end{array}$$

47. $(3x^2 - 4b + 28) + (9x^2 - 30)$
$-(4x^2 - 6b + 20)$
$= 3x^2 - 4b + 28 + 9x^2 - 30$
$-4x^2 + 6b - 20$
$= (3+9-4)x^2 + (-4+6)b + (28-30-20)$
$= 8x^2 + 2b - 22$

48.
$$\begin{array}{rrrrr} & 4x^2 + & 9xy & & +18 \\ +\;-4x^3 & & +\;7xy & -9x & \\ \hline -4x^3 & +4x^2 & +16xy & -9x & +18 \end{array}$$

49. $(3x^2y-7xy-4)+(9x^2y+x)-(x-7)$
$=3x^2y-7xy-4+9x^2y+x-x+7$
$=(3+9)x^2y-7xy+(1-1)x+(-4+7)$
$=12x^2y-7xy+3$

50.
$$\begin{array}{r} x^2-5x+7 \\ -\quad x+4 \\ \hline x^2-6x+3 \end{array}$$

51.
$$\begin{array}{l} x^3 \quad +2xy^2-y \\ \quad x-4xy^2 \quad -7 \\ \hline x^3+x-2xy^2-y-7 \end{array}$$

52. $P(6)=9(6)^2-7(6)+8=290$

53. $P(x)=9x^2-7x+8$
$P(-2)=9(-2)^2-7(-2)+8$
$=36+14+8=58$

54. $P(-3)=9(-3)^2-7(-3)+8=110$

55. $(2x-1)+(x^2+2x-5)=x^2+4x-6$

56. $2(2x-1)-(x^2+2x-5)$
$=4x-2-x^2-2x+5$
$=-x^2+2x+3$

57. $2(x^2y+5)+2(2x^2y-6x+1)$
$=2x^2y+10+4x^2y-12x+2$
$=(6x^2y-12x+12)$ cm

58.
$$\begin{array}{r} 4x^2-6x+1 \\ -6x \\ \hline -24x^3+36x^2-6x \end{array}$$

59. $-4ab^2(3ab^3+7ab+1)$
$=(-4ab^2)(3ab^3)-4ab^2(7ab)-4ab^2(1)$
$=-12a^2b^5-28a^2b^3-4ab^2$

60. $(x-4)(2x+9)$
$=2x^2+9x-8x-36$
$=2x^2+x-36$

61. $(-3xa+4b)^2$
$=(-3xa)^2+2(-3xa)(4b)+(4b)^2$
$=9x^2a^2-24xab+16b^2$

62.
$$\begin{array}{r} 9x^2+4x+1 \\ 4x-3 \\ \hline -27x^2-12x-3 \\ 36x^3+16x^2+\quad 4x \\ \hline 36x^3-11x^2-8x-3 \end{array}$$

63. $(5x-9y)(3x+9y)$
$=15x^2+45xy-27xy-81y^2$
$=15x^2+18xy-81y^2$

64. $\left(x-\frac{1}{3}\right)\left(x+\frac{2}{3}\right)$
$=(x)(x)+(x)\left(\frac{2}{3}\right)-\left(\frac{1}{3}\right)(x)-\left(\frac{1}{3}\right)\left(\frac{2}{3}\right)$
$=x^2+\frac{2}{3}x-\frac{1}{3}x-\frac{2}{9}$
$=x^2+\frac{1}{3}x-\frac{2}{9}$

65. $(x^2+9x+1)^2$
$=(x^2+9x+1)(x^2+9x+1)$
$=x^2(x^2)+x^2(9x)+x^2(1)$
$+9x(x^2)+9x(9x)$
$+9x(1)+1(x^2)+1(9x)+(1)(1)$
$=x^4+9x^3+x^2+9x^3+81x^2$
$+9x+x^2+9x+1$
$=x^4+(9+9)x^3+(1+81+1)x^2$
$+(9+9)x+1$
$=x^4+18x^3+83x^2+18x+1$

66. $(3x-y)^2=(3x)^2-2(3x)y+y^2$
$=9x^2-6xy+y^2$

67. $(4x+9)^2$
$=(4x)^2+2(4x)(9)+(9)^2$
$=16x^2+72x+81$

68. $(x+3y)(x-3y)=x^2-(3y)^2$
$=x^2-9y^2$

69. $[4+(3a-b)][4-(3a-b)]$
$=4^2-(3a-b)^2$
$=16-(9a^2-6ab+b^2)$
$=-9a^2+6ab-b^2+16$

70.
$$\begin{array}{r} x^2+2x-5 \\ 2x-1 \\ \hline -\ x^2-2x+5 \\ 2x^3+4x^2-10x \\ \hline 2x^3+3x^2-12x+5 \end{array}$$

71. $(3y-7z)(3y+7z)$
$=(3y)^2-(7z)^2$
$=(9y^2-49z^2)$ square units

72. $4a^b(3a^{b+2}-7)$
$=4a^b(3a^{b+2})+4a^b(-7)$
$=12a^{b+b+2}-28a^b$
$=12a^{2b+2}-28a^b$

73. $(4xy^z-b)^2$
$=(4xy^z)^2-2(4xy^z)b+b^2$
$=4^2x^2(y^z)^2-8xy^zb+b^2$
$=16x^2y^{2z}-8xy^zb+b^2$

74. $(3x^a-4)(3x^a+4)$
$=(3x^a)(3x^a)+(3x^a)(4)-4(3x^a)-4(4)$
$=9x^{2a}+12x^a-12x^a-16$
$=9x^{2a}-16$

75. $16x^3-24x^2=8x^2(2x-3)$

76. $36y-24y^2=12y(3-2y)$

77. $6ab^2+8ab-4a^2b^2$
$=2ab(3b+4-2ab)$

78. $14a^2b^2-21ab^2+7ab$
$=7ab(2ab-3b+1)$

79. $6a(a+3b)-5(a+3b)$
$=(6a-5)(a+3b)$

80. $4x(x-2y)-5(x-2y)$
$=(x-2y)(4x-5)$

81. $xy-6y+3x-18$
$=(x-6)(y+3)$

82. $ab-8b+4a-32$
$=b(a-8)+4(a-8)$
$=(a-8)(b+4)$

83. $pq-3p-5q+15$
$=(p-5)(q-3)$

84. x^3-x^2-2x+2
$=(x^2-2)(x-1)$

85. $2xy-x^2=x(2y-x)$

86. $x^2-14x-72=(x-18)(x+4)$

87. $x^2+16x-80=(x-4)(x+20)$

88. $2x^2-18x+28$
$=2(x^2-9x+14)$
$=2(x-2)(x-7)$

89. $3x^2+33x+54$
$=3(x^2+11x+18)$
$=3(x+2)(x+9)$

90. $2x^3-7x^2-9=x(2x-9)(x+1)$

91. $3x^2+2x-16=(3x+8)(x-2)$

92. $6x^2+17x+10$
$=(6x+5)(x+2)$

93. $15x^2-91x+6$
$=15(x-1)(x-6)$

94. $4x^2+2x-12$
$=2(2x^2+x-6)$
$=2(2x-3)(x+2)$

95. $9x^2-12x-12$
$=3(3x^2-4x-4)$
$=3(x-2)(3x+2)$

96. $y^2(x+6)^2-2y(x+6)^2-3(x+6)^2$
$=(y^2-2y-3)(x+6)^2$
$=(y-3)(y+1)(x+6)^2$

97. Using the substitution $y=x+5$, we have
$(x+5)^2+6(x+5)+8$
$=y^2+6y+8=(y+2)(y+4)$
$=[(x+5)+2][(x+5)+4]$
$=(x+7)(x+9)$

98. $x^4-6x^2-16=(x^2-8)(x^2+2)$

99. Using the substitution $y=x^2$, we have
$x^4+8x^2-20=y^2+8y-20$
$=(y-2)(y+10)$
$=(x^2-2)(x^2+10)$

100. $x^2-100=(x+10)(x-10)$

101. $x^2-81=x^2-9^2=(x-9)(x+9)$

102. $2x^2-32=2(x^2-16)$
$=2(x+4)(x-4)$

103. $6x^2-54=6(x^2-9)=6(x^2-3^2)$
$=6(x-3)(x+3)$

104. $81-x^4=(9+x^2)(9-x^2)$
$=(9+x^2)(3+x)(3-x)$

105. $16-y^4=4^2-(y^2)^2$
$=(4+y^2)(4-y^2)$
$=(4+y^2)(2^2-y^2)$
$=(4+y^2)(2-y)(2+y)$

106. $(y+2)^2-25$
$=[(y+2)+5][(y+2)-5]$
$=(y+7)(y-3)$

107. $(x-3)^2-16$
$=(x-3)^2-4^2$
$=[(x-3)-4][(x-3)+4]$
$=(x-7)(x+1)$

108. x^3+216
$=x^3+6^3$
$=(x+6)(x^2-6x+36)$

109. $y^3+512=y^3+8^3$
$=(y+8)(y^2-8\cdot y+8^2)$
$=(y+8)(y^2-8y+64)$

110. $8-27y^3$
$=2^3-(3y)^3$
$=(2-3y)(4+6y+9y^2)$

111. $1-64y^3=1^3-(4y)^3$
$=(1-4y)(1+4y+(4y)^2)$
$=(1-4y)(1+4y+16y^2)$

112. $6x^4y+48xy$
$=6xy(x^3+8)$
$=6xy(x^3+2^3)$
$=6xy(x+2)(x^2-2x+4)$

113. $2x^5+16x^2y^3$
$=2x^2(x^3+8y^3)$
$=2x^2(x^3+(2y)^3)$
$=2x^2(x+2y)(x^2-x\cdot 2y+(2y)^2)$
$=2x^2(x+2y)(x^2-2xy+4y^2)$

114. $(x^2-2x+1)-y^2$
$=(x-1)^2-y^2$
$=(x-1-y)(x-1+y)$

115. $x^2-6x+9-4y^2$
$=(x-3)^2-4y^2$
$=(x-3)-(2y)^2$
$=(x-3-2y)(x-3+2y)$

116. $4x^2+12x+9$
$=(2x+3)(2x+3)=(2x+3)^2$

117. $16a^2-40ab+25b^2$
$=(4a-5b)(4a-5b)$
$=(4a-5b)^2$

118. $\pi R^2h-\pi r^2h$
$=\pi h(R^2-r^2)$
$=\pi h(R-r)(R+r)$ cubic units

119. $(3x-1)(x+7)=0$
$3x-1=0$ or $x+7=0$
$3x=1$ or $x=-7$
$x=\frac{1}{3}$
$\left\{\frac{1}{3},\ -7\right\}$

120. $3(x+5)(8x-3)=0$
$x+5=0$ or $8x-3=0$
$x=-5$ or $x=\frac{3}{8}$

121. $5x(x-4)(2x-9)=0$
$5x=0$ or $x-4=0$ or $2x-9=0$
$x=0$ or $x=4$ or $2x=9$
$x=\frac{9}{2}$
$\left\{0,\ 4,\ \frac{9}{2}\right\}$

122. $6(x+3)(x-4)(5x+1)=0$
$x+3=0$ or $x-4=0$ or $5x+1=0$
$x=-3$ or $x=4$ or $x=-\frac{1}{5}$

123. $2x^2=12x$
$2x^2-12x=0$
$2x(x-6)=0$
$2x=0$ or $x-6=0$
$x=0$ or $x=6$
$\{0, 6\}$

124. $4x^3-36x=0$
$4x(x^2-9)=0$
$4x(x-3)(x+3)=0$
$4x=0$ or $x-3=0$ or $x+3=0$
$x=0$ or $x=3$ or $x=-3$

125. $(1-x)(3x+2)=-4x$
$3x+2-3x^2-2x=-4x$
$-3x^2+x+2=-4x$
$0=3x^2-5x-2$
$0=(3x+1)(x-2)$
$3x+1=0$ or $x-2=0$
$3x=-1$ or $x=2$
$x=-\frac{1}{3}$
$\left\{-\frac{1}{3},\ 2\right\}$

126. $2x(x-12)=-40$
$2x^2-24x+40=0$
$2(x^2-12x+20)=0$
$2(x-2)(x-10)=0$
$x-2=0$ or $x-10=0$
$x=2$ or $x=10$

127. $3x^2 + 2x = 12 - 7x$
$3x^2 + 9x - 12 = 0$
$x^2 + 3x - 4 = 0$
$(x + 4)(x - 1) = 0$
$x + 4 = 0$ or $x - 1 = 0$
$x = -4$ or $x = 1$
$\{-4, 1\}$

128. $2x^2 + 3x = 35$
$2x^2 + 3x - 35 = 0$
$(2x - 7)(x + 5) = 0$
$2x - 7 = 0$ or $x + 5 = 0$
$x = \frac{7}{2}$ or $x = -5$

129. $x^3 - 18x = 3x^2$
$x^3 - 3x^2 - 18x = 0$
$x(x^2 - 3x - 18) = 0$
$x(x - 6)(x + 3) = 0$
$x = 0$ or $x - 6 = 0$ or $x + 3 = 0$
$x = 6$ or $x = -3$
$\{0, 6, -3\}$

130. $19x^2 - 42x = -x^3$
$x^3 + 19x^2 - 42x = 0$
$x(x^2 + 19x - 42) = 0$
$x(x - 2)(x + 21) = 0$
$x = 0$ or $x - 2 = 0$ or $x + 21 = 0$
$x = 0$ or $x = 2$ or $x = -21$

131. $12x = 6x^3 + 6x^2$
$6x^3 + 6x^2 - 12x = 0$
$6x(x^2 + x - 2) = 0$
$6x(x + 2)(x - 1) = 0$
$6x = 0$ or $x + 2 = 0$ or $x - 1 = 0$
$x = 0$ or $x = -2$ or $x = 1$
$\{0, -2, 1\}$

132. $8x^3 + 10x^2 = 3x$
$8x^3 + 10x^2 - 3x = 0$
$x(8x^2 + 10x - 3) = 0$
$x(4x - 1)(2x + 3) = 0$
$x = 0$ or $4x - 1 = 0$ or $2x + 3 = 0$
$x = 0$ or $x = \frac{1}{4}$ or $x = -\frac{3}{2}$

133. Let x be the number.
$x + 2x^2 = 105$
$2x^2 + x - 105 = 0$
$(2x + 15)(x - 7) = 0$
$2x + 15 = 0$ or $x - 7 = 0$
$x = -\frac{15}{2}$ or $x = 7$

134. Let x be the width. Then $5x - 2$ is the length.
$x(5x - 2) = 16$
$5x^2 - 2x - 16 = 0$
$(5x + 8)(x - 2) = 0$
$5x + 8 = 0$ or $x - 2 = 0$
$x = -\frac{8}{5}$ $x = 2$
Disregard the negative.
Width = 2 m
Length = 5(2) – 2 = 8 m

135. $h(t) = -16t^2 + 400$
$0 = -16t^2 + 400$
$0 = -16(t^2 - 25)$
$t^2 - 25 = 0$
$t^2 = 25$
$t = 5$ seconds

136. Domain; $(-\infty, \infty)$
Range; $(-\infty, 4]$

137. $(-4, 0), (2, 0), (0, 3)$

138. $(-1, 4)$

139. $-4 < x < 2$

140. $f(x) = x^2 + 6x + 9$
$f(x) = (x + 3)^2$
Solve $(x + 3)^2 = 0$
$x + 3 = 0$
$x = -3$
The x-intercept is –3. $h = \frac{-6}{2(1)} = -3$ and
$k = f(-3) = 0$ since –3 is an x-intercept.
Thus, $V(-3, 0)$.

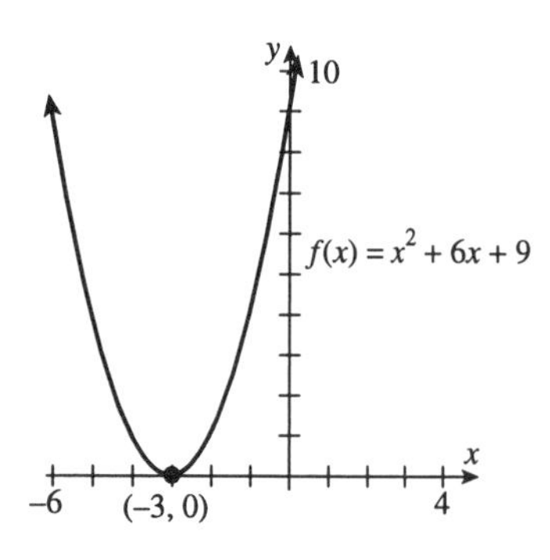

141. $f(x) = x^2 - 5x + 4$

$\frac{-b}{2a} = \frac{5}{2(1)} = \frac{5}{2}$

$f\left(\frac{5}{2}\right) = \left(\frac{5}{2}\right)^2 - 5\left(\frac{5}{2}\right) + 4 = -\frac{9}{4}$

The vertex is $\left(\frac{5}{2}, -\frac{9}{4}\right)$.

If $x^2 - 5x + 4 = 0$

$(x-1)(x-4) = 0$

$x - 1 = 0$ or $x - 4 = 0$

$x = 1$ or $x = 4$

The x-intercepts are 1 and 4. If $x = 0$, then $y = f(0) = 4$. The y-intercept is 4.

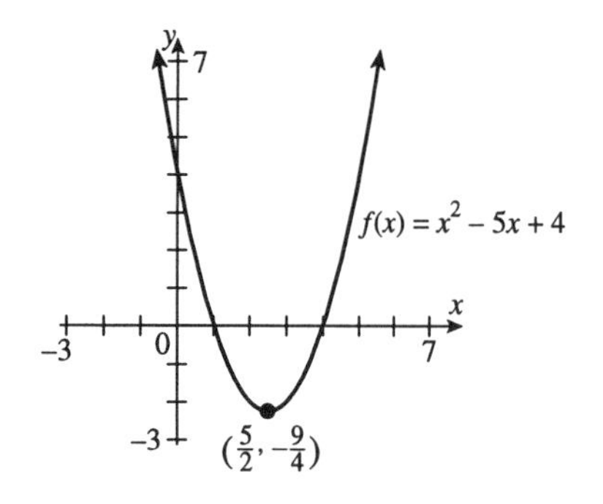

142. $f(x) = (x-1)(x^2 - 2x - 3)$

$= (x-1)(x-3)(x+1)$

Solve $(x-1)(x-3)(x+1) = 0$

$x - 1 = 0$ or $x - 3 = 0$ or $x + 1 = 0$

$x = 1$ or $x = 3$ or $x = -1$

The x-intercepts are 1, 3, and –1.

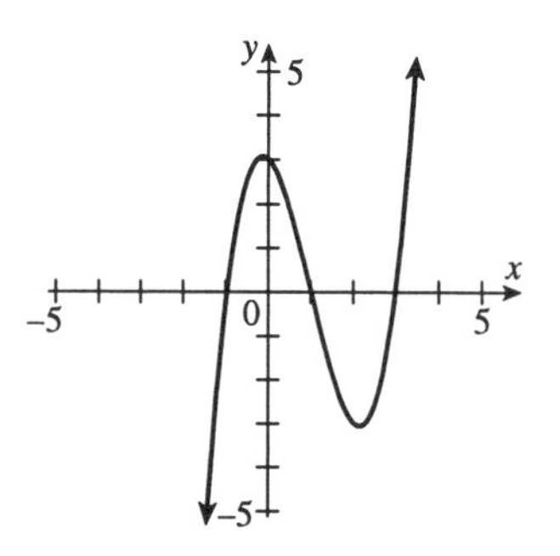

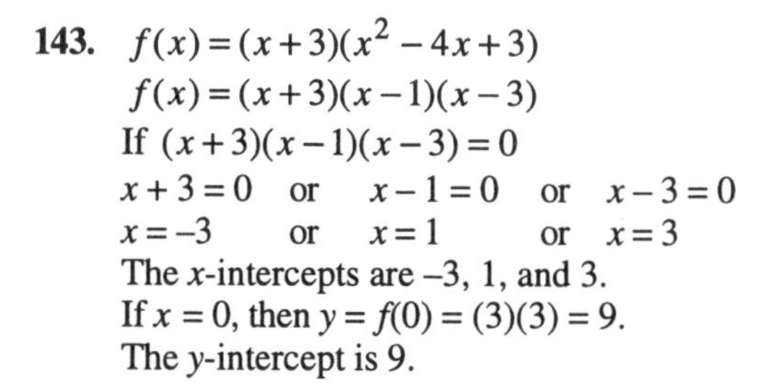

143. $f(x) = (x+3)(x^2 - 4x + 3)$

$f(x) = (x+3)(x-1)(x-3)$

If $(x+3)(x-1)(x-3) = 0$

$x + 3 = 0$ or $x - 1 = 0$ or $x - 3 = 0$

$x = -3$ or $x = 1$ or $x = 3$

The x-intercepts are –3, 1, and 3.

If $x = 0$, then $y = f(0) = (3)(3) = 9$.

The y-intercept is 9.

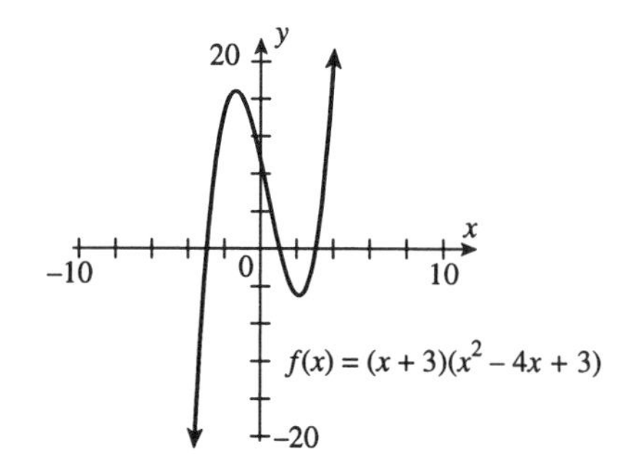

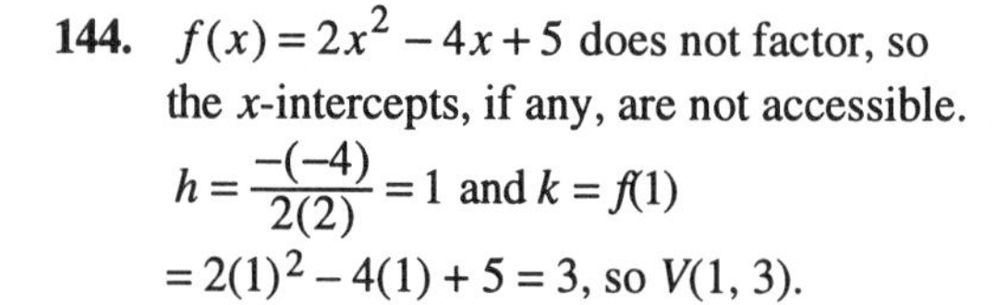

144. $f(x) = 2x^2 - 4x + 5$ does not factor, so the x-intercepts, if any, are not accessible.

$h = \frac{-(-4)}{2(2)} = 1$ and $k = f(1)$

$= 2(1)^2 - 4(1) + 5 = 3$, so $V(1, 3)$.

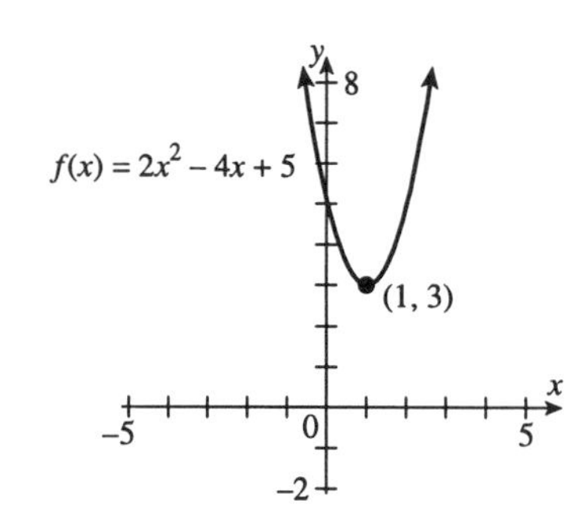

145. $f(x) = x^2 - 2x + 3$

$\dfrac{-b}{2a} = \dfrac{2}{2(1)} = 1$

$f(1) = 1 - 2 + 3 = 2$

The vertex is (1, 2). There are no x-intercepts. If $x = 0$, then $y = f(0) = 3$. The y-intercept is 3.

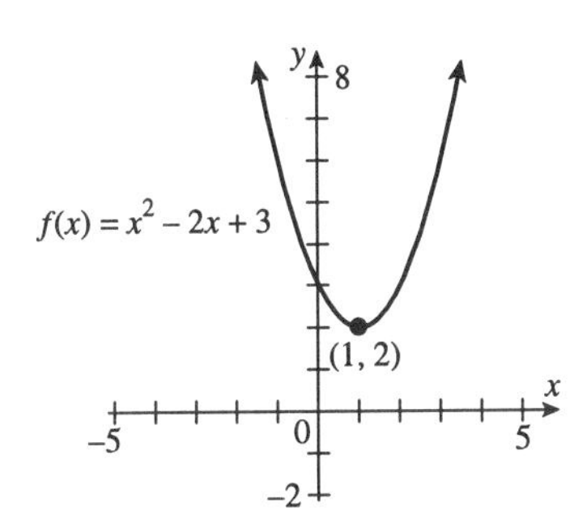

146. $f(x) = x^3 - 16x$

$= x(x^2 - 16)$

$= x(x + 4)(x - 4)$

Solve $x(x + 4)(x - 4) = 0$

$x = 0$ or $x + 4 = 0$ or $x - 4 = 0$

$x = -4$ or $x = 4$

The x-intercepts are 0, −4, and 4.

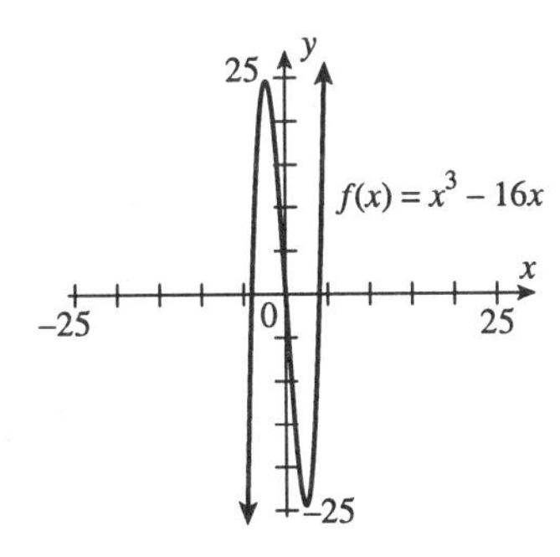

147. $f(x) = x^3 + 5x^2 + 6x$

$f(x) = x(x^2 + 5x + 6) = x(x + 3)(x + 2)$

If $x(x + 3)(x + 2) = 0$

$x = 0$ or $x + 3 = 0$ or $x + 2 = 0$

$x = 0$ or $x = -3$ or $x = -2$

The x-intercepts are −3, −2, and 0.
The y-intercept is 0.

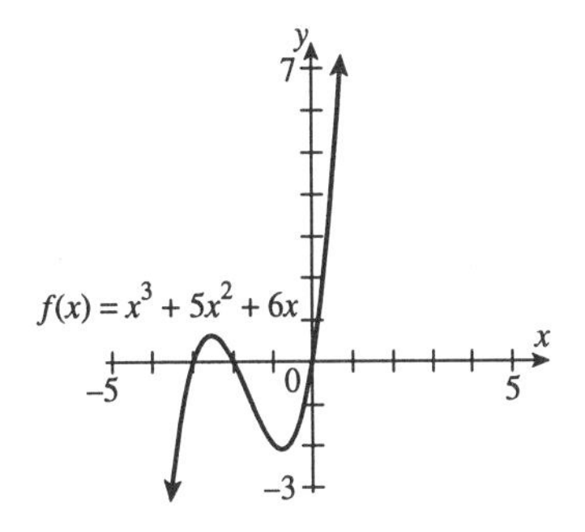

Chapter 5 - Test

1. $(-9x)^{-2} = \dfrac{1}{(9x)^2} = \dfrac{1}{81x^2}$

2. $-3xy^{-2}(4xy^2)z = (-3)(4)x^{1+1}y^{-2+2}z$
$= -12x^2z$

3. $\dfrac{6^{-1}a^2b^{-3}}{3^{-2}a^{-5}b^2} = \dfrac{3^2a^{2+5}}{6^1b^{2+3}} = \dfrac{9a^7}{6b^5} = \dfrac{3a^7}{2b^5}$

4. $\left(\dfrac{-xy^{-5}z}{xy^3}\right)^{-5} = \dfrac{-x^{-5}y^{25}z^{-5}}{x^{-5}y^{-15}}$

$= \dfrac{-x^{-5+5}y^{25+15}}{z^5} = -\dfrac{y^{40}}{z^5}$

5. $630,000,000 = 6.3 \times 10^8$

6. $0.01200 = 1.2 \times 10^{-2}$

7. $5.0 \times 10^{-6} = 0.000005$

8. $\dfrac{(2.4 \times 10^{-3})(1.2 \times 10^{-4})}{(3.2 \times 10^{-4})}$

$= \dfrac{(2.4)(1.2)}{(3.2)} \times 10^{-3-4+4}$

$= 0.9 \times 10^{-3} = 90 \times 10^{-4}$

$=0.0009$

9.
$$\begin{array}{r} 4x^3 - 3x - 4 \\ -\ \ 9x^3 + 8x + 5 \\ \hline -5x^3 - 11x - 9 \end{array}$$

10. $-3xy(4x+y)2$
$=-12x^2y-3xy^2$

11.
$$\begin{array}{r} 3x+4 \\ \times \quad 4x-7 \\ \hline -21x-28 \\ 12x^2+16x \quad \\ \hline 12x^2-5x-28 \end{array}$$

12. $(5a-2b)(5a+2b)$
$=(5a)^2-(2b)^2$
$=25a^2-4b^2$

13. $(6m+n)^2$
$=36m^2+12mn+n^2$

14.
$$\begin{array}{r} x^2-6x+4 \\ \times \quad 2x-1 \\ \hline -x^2+6x-4 \\ 2x^3-12x^2+8x \quad \\ \hline 2x^3-13x^2+14x-4 \end{array}$$

15. $16x^3y-12x^2y^4=4x^2y(4x-3y^3)$

16. $x^2-13x-30=(x-15)(x+2)$

17. $4y^2+20y+25=(2y+5)^2$

18. $6x^2-15x-9$
$=3(2x^2-5x-3)$
$=3(2x+1)(x-3)$

19. $4x^2-25=(2x+5)(2x-5)$

20. $x^3+64=x^3+4^3$
$=(x+4)(x^2-4x+16)$

21. $3x^2y-27y^3$
$=3y(x^2-9y^2)$
$=3y(x+3y)(x-3y)$

22. $6x^2+24=6(x^2+4)$

23. $x^2y-9y-3x^2+27$
$=(x^2-9)(y-3)$
$=(x+3)(x-3)(y-3)$

24. $3(n-4)(7n+8)=0$
$3(n-4)=0$ or $7n+8=0$
$n-4=0$ or $7n=-8$
$n=4$ or $n=-\frac{8}{7}$
$\left\{4, -\frac{8}{7}\right\}$

25. $(x+2)(x-2)=5(x+4)$
$x^2-4=5x+20$
$x^2-5x-24=0$
$(x-8)(x+3)=0$
$x-8=0$ or $x+3=0$
$x=8$ or $x=-3$
$\{-3, 8\}$

26. $2x^3+5x^2-8x-20=0$
$(2x+5)(x^2-4)=0$
$(2x+5)(x+2)(x-2)=0$
$2x+5=0$ or $x+2=0$ or $x-2=0$
$2x=-5$ or $x=-2$ or $x=2$
$x=-\frac{5}{2}$
$\left\{-\frac{5}{2}, -2, 2\right\}$

27. $x^2-(2y)^2$
$=(x-2y)(x+2y)$

28. $h(t)=-16t^2+96t+880$

a. $-16(1)^2+96(1)+880$
$=-16+96+880=960$
960 feet

b. $-16(5.1)^2+96(5.1)+880$
$=-416.16+489.6+880$
$=953.44$
953.44 feet

c. $0 = -16t^2 + 96t + 880$
$0 = (-16t - 80)(t - 11)$
$-16t - 80 = 0$ or $t - 11 = 0$
$t = -5$ or $t = 11$
Disregard the negative. 11 seconds

29. $f(x) = x^2 - 4x - 5$

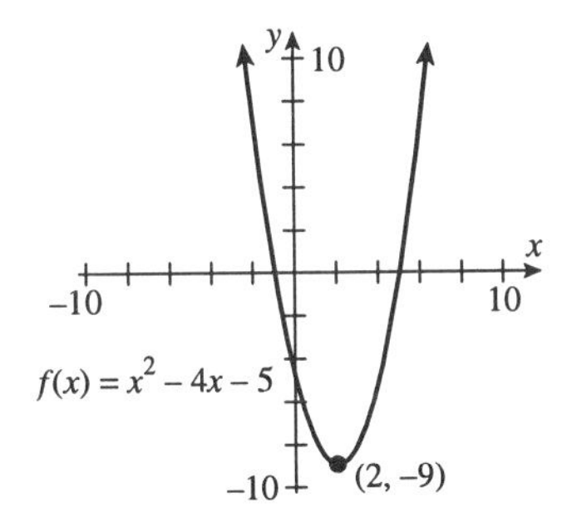

30. $f(x) = x^3 - 1$

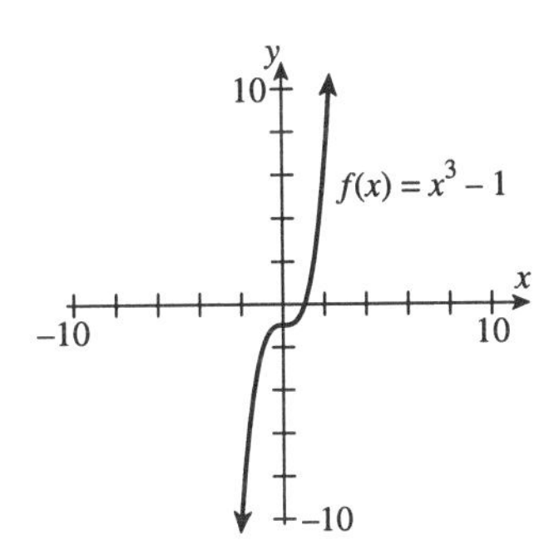

Chapter 5 - Cumulative Review

1. a. $8x$

b. $8x + 3$

c. $\frac{x}{-7}$

d. $2x - 1.6$

2. a. $|3| = 3$

b, $|-5| = 5$

c. $-|2| = -2$

d. $-|-8| = -8$

e. $|0| = 0$

3. a. $z - y = -3 - (-1) = -3 + 1 = -2$

b. $z^2 = (-3)^2 = 9$

c. $\frac{2x + y}{z} = \frac{2(2) + (-1)}{-3} = \frac{4 - 1}{-3} = \frac{3}{-3}$
$= -1$

4. $2(x - 3) = 5x - 9$
$2x - 6 = 5x - 9$
$3 = 3x$
$1 = x$
$\{1\}$

5. $-(x - 3) + 2 \le 3(2x - 5) + x$
$-x + 3 + 2 \le 6x - 15 + x$
$-x + 5 \le 7x - 15$
$20 \le 8x$
$\frac{5}{2} \le x$
$\left[\frac{5}{2}, \infty\right)$

6.

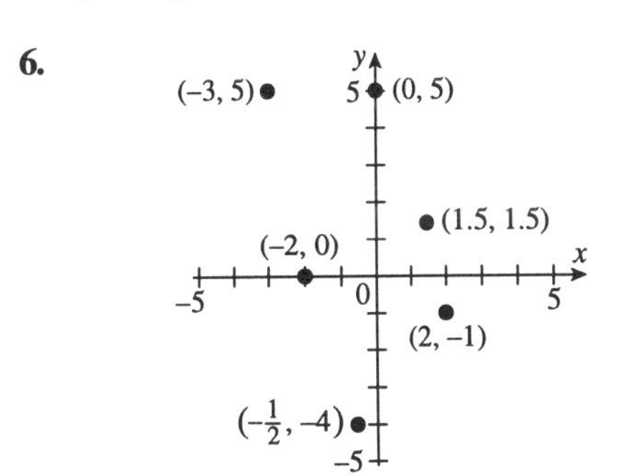

a. (2, −1), quadrant IV

b. (0, 5), not in any quadrant

c. (−3, 5), quadrant II

d. (−2, 0), not in any quadrant

e. $\left(-\frac{1}{2}, -4\right)$, quadrant III

f. (1.5, 1.5), quadrant I

7. a. Domain: $\{2, 0, 3\}$
Range: $\{3, 4, -1\}$

b. Domain: $\{-4, -3, -2, -1, 0, 1, 2, 3\}$
Range: $\{1\}$

c. Domain: {Eric, Miami, Escondido, Waco, Gary}
Range: $\{109, 359, 117, 104\}$

8. $x - 3y = 6$

If $x = 0$, then	If $y = 0$, then
$0 - 3y = 6$	$x - 3(0) = 6$
$-3y = 6$	$x - 0 = 6$
$y = -2$	$x = 6$
$(0, -2)$	$(6, 0)$

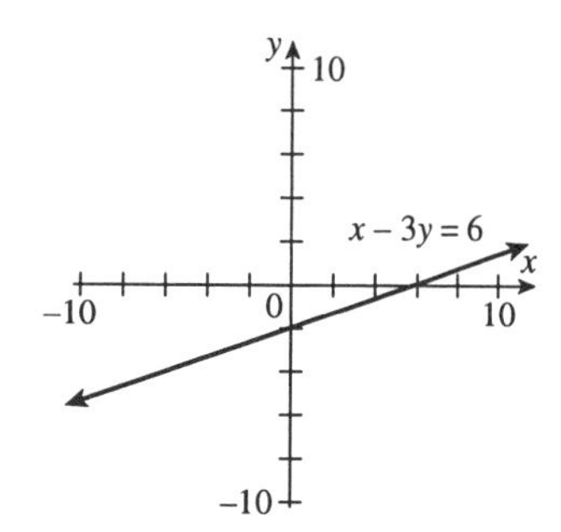

9. $f(x) = \frac{2}{3}x + 4$
$m = \frac{2}{3}$

10. $x \geq 1$ and $y \geq 2x - 1$

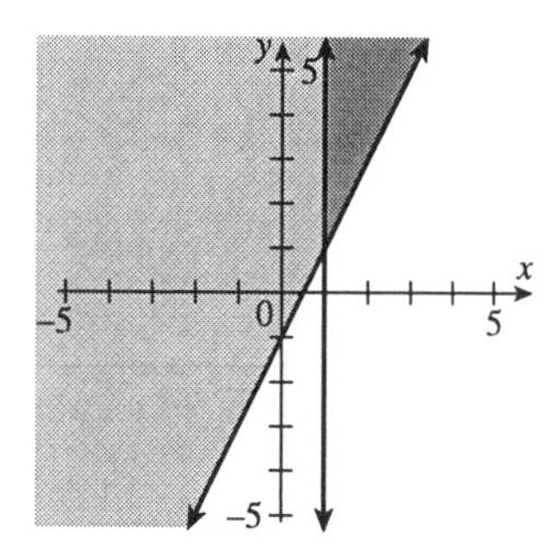

11. $\begin{cases} x - 5y = -12 \\ -x + y = 4 \end{cases}$
Add the equations.
$-4y = -8$
$y = 2$
Substitute back.
$x - 5(2) = -12$
$x - 10 = -12$
$x = -2$
$\{(-2, 2)\}$

12. $\left[\begin{array}{cc|c} 2 & -1 & 3 \\ 4 & -2 & 5 \end{array}\right]$
Multiply row 1 by -2 and add to row 2
$\left[\begin{array}{cc|c} 0 & 0 & -1 \\ 4 & -2 & 5 \end{array}\right]$
Inconsistent system; no solution

13. a. $\begin{vmatrix} -1 & 2 \\ 3 & -4 \end{vmatrix} = (-1)(-4) - (3)(2)$
$= 4 - 6 = -2$

b. $\begin{vmatrix} 2 & 0 \\ 7 & -5 \end{vmatrix} = 2(-5) - 7(0)$
$= -10 - 0 = -10$

14. a. $2^2 \cdot 2^5 = 2^{2+5} = 2^7$

b. $x^7 \cdot x^3 = x^{7+3} = x^{10}$

c. $y \cdot y^2 \cdot y^4 = y^{1+2+4} = y^7$

15. a. $(x^5)^7 = x^{5 \cdot 7} = x^{35}$

b. $(2^2)^3 = 2^{2 \cdot 3} = 2^6 = 64$

c. $(5^{-1})^2 = 5^{(-1)(2)} = 5^{-2} = \frac{1}{5^2} = \frac{1}{25}$

d. $(y^{-3})^{-4} = y^{(-3)(-4)} = y^{12}$

16. a. $7x^3 - 3x + 2$
Degree 3, trinomial

b. $-xyz$
Degree 3, monomial

c. $x^2 - 4$
Degree 2, binomial

d. $2xy + x^2y^2 - 5x^2 - 6$
Degree 4, polynomial

17. a. $2x(5x-4) = 10x^2 - 8x$

b. $-3x^2(4x^2 - 6x + 1)$
$= -12x^4 + 18x^3 - 3x^2$

c. $-xy(7x^2y + 3xy - 11)$
$= -7x^3y^2 - 3x^2y^2 + 11xy$

18. $2(x-5) + 3a(x-5) = (2+3a)(x-5)$

19. $12x^3y - 22x^2y + 8xy$
$= 2xy(6x^2 - 11x + 4)$
$= 2xy(2x-1)(3x-4)$

20. $x^3 + 8 = (x+2)(x^2 - 2x + 4)$

21. $2x^2 = \frac{17}{3}x + 1$
$6x^2 = 17x + 3$
$6x^2 - 17x - 3 = 0$
$(6x+1)(x-3) = 0$
$6x + 1 = 0$ or $x - 3 = 0$
$x = -\frac{1}{6}$ or $x = 3$
$\left\{-\frac{1}{6}, 3\right\}$

22. $h(t) = -16t^2 + 144t$
$0 = -16t^2 + 144t$
$0 = -16t(t-9)$
$t - 9 = 0$
$t = 9$ seconds

23. $f(x) = x^2 + 2x - 3$
$\frac{-b}{2a} = \frac{-2}{2(1)} = -1$
$(-1)^2 + 2(-1) - 3 = -4$
Vertex: $(-1, -4)$
$0 = x^2 + 2x - 3$
$0 = (x+3)(x-1)$
$x + 3 = 0$ or $x - 1 = 0$
$x = -3$ or $x = 1$
x-intercepts are -3 and 1

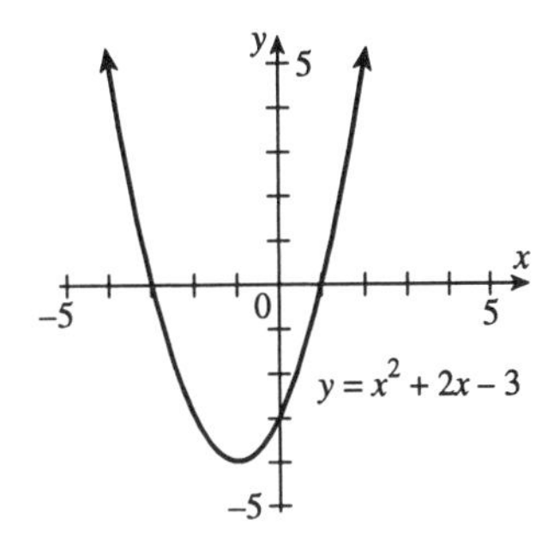

Chapter 6

Section 6.1

Graphing Calculator Explorations

1. $f(x)=\dfrac{x+1}{x^2-4}$

$f(x)=\dfrac{x+1}{(x+2)(x-2)}$

Domain: $\{x|x$ is a real number and $x\neq -2,\ x\neq 2\}$

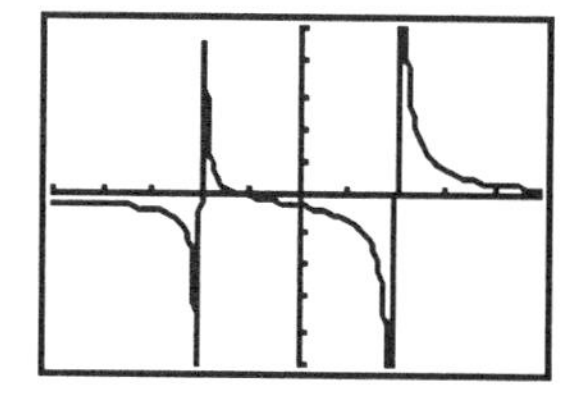

3. $h(x)=\dfrac{x^2}{2x^2+7x-4}$

$h(x)=\dfrac{x^2}{(2x-1)(x+4)}$

Domain: $\{x|x$ is a real number and $x\neq -4$ and $x\neq \frac{1}{2}\}$

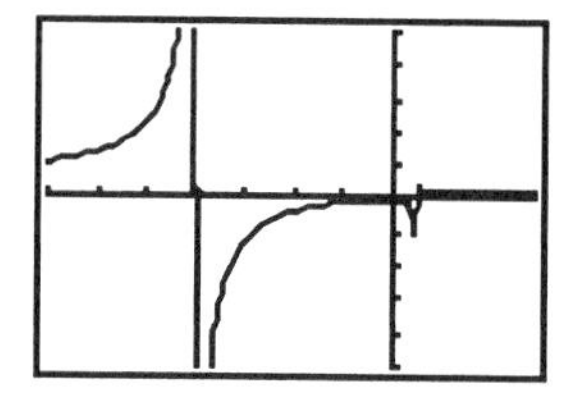

Exercise Set 6.1

1. $f(x)=\dfrac{x+8}{2x-1}$

$f(2)=\dfrac{2+8}{2(2)-1}=\dfrac{10}{4-1}=\dfrac{10}{3}$

$f(0)=\dfrac{0+8}{2(0)-1}=\dfrac{8}{-1}=-8$

$f(-1)=\dfrac{-1+8}{2(-1)-1}=\dfrac{7}{-3}=-\dfrac{7}{3}$

3. $g(x)=\dfrac{x^2+8}{x^3-25x}$

$g(3)=\dfrac{3^2+8}{3^3-25(3)}=\dfrac{9+8}{27-75}=\dfrac{17}{-48}$

$=-\dfrac{17}{48}$

$g(-2)=\dfrac{(-2)^2+8}{(-2)^3-25(-2)}=\dfrac{4+8}{-8+50}$

$=\dfrac{12}{42}=\dfrac{2}{7}$

$g(1)=\dfrac{1^2+8}{1^3-25(1)}=\dfrac{1+8}{1-25}=\dfrac{9}{-24}=-\dfrac{3}{8}$

5. $f(x)=\dfrac{5x-7}{4}$

Domain: $\{x|x$ is a real number$\}$

7. $s(t)=\dfrac{t^2+1}{2t}$

Undefined values when

$2t=0$

$t=\dfrac{0}{2}=0$

Domain: $\{t|t$ is a real number and $t\neq 0\}$

9. $f(x)=\dfrac{3x}{7-x}$

Undefined values when

$7-x=0$

$7=x$

Domain: $\{x|x$ is a real number and $x\neq 7\}$

11. $g(x)=\dfrac{2-3x^2}{3x^2+9x}$

Undefined values when

$3x^2+9x=0$

$3x(x+3)=0$

$3x=0$ or $x+3=0$

$x=0$ or $x=-3$

Domain: $\{x|x$ is a real number and $x\neq 0,\ x\neq -3\}$

13. $R(x)=\frac{3+2x}{x^3+x^2-2x}$
Undefined values when
$x^3+x^2-2x=0$
$x(x^2+x-2)=0$
$x(x+2)(x-1)=0$
$x=0$ or $x+2=0$ or $x-1=0$
$x=-2$ or $x=1$
Domain: $\{x|x$ is a real number and $x\neq 0,\ x\neq -2,\ x\neq 1\}$

15. $C(x)=\frac{x+3}{x^2-4}$
Undefined values when
$x^2-4=0$
$(x+2)(x-2)=0$
$x+2=0$ or $x-2=0$
$x=-2$ or $x=2$
Domain: $\{x|x$ is a real number $x\neq 2,\ x\neq -2\}$

17. Answers may vary.

19. $\frac{10x^3}{18x}=\frac{5x^{3-1}}{9}=\frac{5x^2}{9}$

21. $\frac{9x^6y^3}{18x^2y^5}=\frac{x^{6-2}}{2y^{5-3}}=\frac{x^4}{2y^2}$

23. $\frac{8q^2}{16q^3-16q^2}=\frac{8q^2}{16q^2(q-1)}=\frac{1}{2(q-1)}$

25. $\frac{x+5}{5+x}=1$

27. $\frac{x-1}{1-x^2}=\frac{-(1-x)}{(1+x)(1-x)}=\frac{-1}{1+x}$

29. $\frac{7-x}{x^2-14x+49}=\frac{-(x-7)}{(x-7)^2}=\frac{-1}{x-7}$

31. $\frac{4x-8}{3x-6}=\frac{4(x-2)}{3(x-2)}=\frac{4}{3}$

33. $\frac{2x-14}{7-x}=\frac{2(x-7)}{-(x-7)}=-2$

35. $\frac{x^2-2x-3}{x^2-6x+9}=\frac{(x-3)(x+1)}{(x-3)^2}=\frac{x+1}{x-3}$

37. $\frac{2x^2+12x+18}{x^2-9}$
$=\frac{2(x^2+6x+9)}{x^2-3^2}$
$=\frac{2(x+3)^2}{(x+3)(x-3)}$
$=\frac{2(x+3)}{x-3}$

39. $\frac{3x+6}{x^2+2x}=\frac{3(x+2)}{x(x+2)}=\frac{3}{x}$

41. $\frac{x+2}{x^2-4}=\frac{x+2}{(x+2)(x-2)}=\frac{1}{x-2}$

43. $\frac{2x^2-x-3}{2x^3-3x^2+2x-3}$
$=\frac{(2x-3)(x+1)}{(2x-3)(x^2+1)}$
$=\frac{x+1}{x^2+1}$

45. $\frac{x^4-16}{x^2+4}=\frac{(x^2+4)(x^2-4)}{x^2+4}=x^2-4$

47. $\frac{x^2+6x-40}{10+x}=\frac{(x+10)(x-4)}{x+10}=x-4$

49. $\frac{2x^2-7x-4}{x^2-5x+4}=\frac{(2x+1)(x-4)}{(x-1)(x-4)}=\frac{2x+1}{x-1}$

51. $\frac{x^3-125}{5-x}=\frac{x^3-5^3}{-(x-5)}$
$=\frac{(x-5)(x^2+5x+25)}{-(x-5)}$
$=-(x^2+5x+25)$
$=-x^2-5x-25$

53. $\frac{8x^3-27}{4x-6}=\frac{(2x)^3-3^3}{2(2x-3)}$
$=\frac{(2x-3)(4x^2+6x+9)}{2(2x-3)}$
$=\frac{4x^2+6x+9}{2}$

55. $\frac{x+5}{x^2+5}$ is in lowest terms.

57. First note $4y^3z=2y(2y^2z)$
$\frac{5}{2y}=\frac{5(2y^2z)}{2y(2y^2z)}=\frac{10y^2z}{4y^3z}$

59. First note $2x^2+9x-5=(2x-1)(x+5)$
$\frac{3x}{2x-1}=\frac{3x}{2x-1}\cdot\frac{x+5}{x+5}=\frac{3x^2+15x}{2x^2+9x-5}$

61. $\frac{x-1}{1}=\frac{(x-2)(x+2)}{1\cdot(x+2)}=\frac{x^2-4}{x+2}$

63. First note $6m^3=m(6m^2)$
$\frac{5}{m}=\frac{5(6m^2)}{m(6m^2)}=\frac{30m^2}{6m^3}$

65. First note $5m-10=5(m-2)$
$\frac{7}{m-2}=\frac{7(5)}{(m-2)5}=\frac{35}{5m-10}$

67. First note $y^2-16=(y-4)(y+4)$
$\frac{y+4}{y-4}=\frac{(y+4)(y+4)}{(y-4)(y+4)}=\frac{y^2+8y+16}{y^2-16}$

69. First note $x^2+4x+4=(x+2)(x+2)$
$\frac{12x}{x+2}=\frac{12x(x+2)}{(x+2)(x+2)}=\frac{12x^2+24x}{x^2+4x+4}$

71. First note $x^3+8=x^3+2^3$
$=(x+2)(x^2-2x+4)$
$\frac{1}{x+2}=\frac{1\cdot(x^2-2x+4)}{(x+2)(x^2-2x+4)}$
$=\frac{x^2-2x+4}{x^3+8}$

73. First note
$ab-3a+2b-6=(a+2)(b-3)$
$\frac{a}{a+2}=\frac{a(b-3)}{(a+2)(b-3)}=\frac{ab-3a}{ab-3a+2b-6}$

75. $f(x)=\frac{20x}{100-x}$

x	0	10	30	50	70	90	95	99
y	0	$\frac{20}{9}$	$\frac{60}{7}$	20	$\frac{140}{3}$	180	380	1980

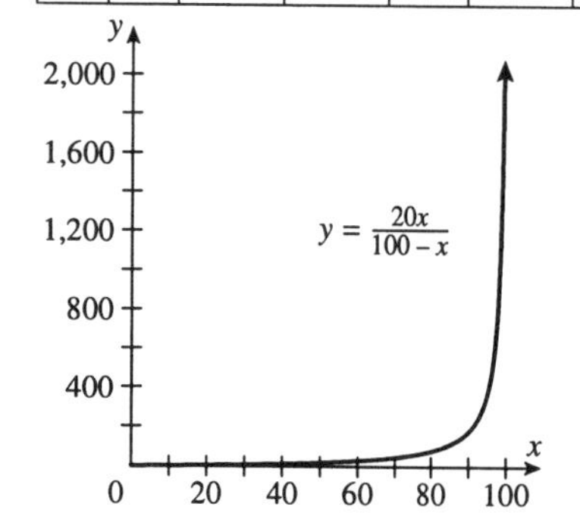

77. $f(x)=\frac{100,000x}{100-x}$

a. $\{x \mid 0\le x<100\}$

b. $f(30)=\frac{100,000(30)}{100-30}=\$42,857.14$

c. $f(60)=\frac{100,000(60)}{100-60}=\$150,000$

$f(80)=\frac{100,000(80)}{100-80}=\$400,000$

d. $f(90)=\frac{100,000(90)}{100-90}=\$900,000$

$f(95)=\frac{100,000(95)}{100-95}=\$1,900,000$

$f(99)=\frac{100,000(99)}{100-99}=\$9,900,000$

79. $\frac{6}{35}\cdot\frac{28}{9}=\frac{(3\cdot 2)(4\cdot 7)}{(5\cdot 7)(3\cdot 3)}=\frac{(21)(8)}{(21)(15)}=\frac{8}{15}$

81. $\frac{8}{35}\div\frac{4}{5}=\frac{8}{35}\cdot\frac{5}{4}=\frac{(4\cdot 2)(5)}{(5\cdot 7)(4)}=\frac{(20)(2)}{(20)(7)}$

$=\frac{2}{7}$

83. $\left(\frac{1}{2}\cdot\frac{1}{4}\right)\div\frac{3}{8}$

$=\frac{1}{8}\div\frac{3}{8}$

$=\frac{1}{8}\cdot\frac{8}{3}$

$=\frac{1}{3}$

85. $14+12.9+1.2=28.1\%$

87. $28.9+20.9=49.8\%$

89. $\frac{3+q^n}{q^n+3}=1$

91. $\frac{x^{2k}-9}{3+x^k}=\frac{(x^k-3)(x^k+3)}{3+x^k}=x^k-3$

93. $\frac{4x^k-12}{x^{2k}+4}$ is irreducible.

Exercise Set 6.2

1. $\frac{4}{x}\cdot\frac{x^2}{8}=\frac{1}{1}\cdot\frac{x}{2}=\frac{x}{2}$

3. $\frac{2a^2b}{6ac}\cdot\frac{3c^2}{4ab}=\frac{6a^2bc^2}{24a^2bc}=\frac{c}{4}$

5. $\frac{2x}{5}\cdot\frac{5x+10}{6(x+2)}$

$=\frac{(2x)\cdot 5(x+2)}{30(x+2)}=\frac{10x}{30}=\frac{x}{3}$

7. $\frac{2x-4}{15}\cdot\frac{6}{2-x}$

$=\frac{2(x-2)\cdot 6}{15[-(x-2)]}=\frac{12}{-15}=-\frac{4}{5}$

9. $\frac{18a-12a^2}{4a^2+4a+1}\cdot\frac{4a^2+8a+3}{4a^2-9}$

$=\frac{6a(3-2a)(2a+3)(2a+1)}{(2a+1)^2(2a+3)(2a-3)}$

$=\frac{(6a)[-(2a-3)]}{(2a+1)(2a-3)}=\frac{-6a}{2a+1}$

11. $\frac{x^2-6x-16}{2x^2-128}\cdot\frac{x^2+16x+64}{3x^2+30x+48}$

$=\frac{(x-8)(x+2)(x+8)^2}{2(x^2-64)[3(x^2+10x+16)]}$

$=\frac{(x-8)(x+2)(x+8)^2}{2(x+8)(x-8)3(x+8)(x+2)}$

$=\frac{1}{6}$

13. $\frac{4x+8}{x+1}\cdot\frac{2-x}{3x-15}\cdot\frac{2x^2-8x-10}{x^2-4}$

$=\frac{4(x+2)[-(x-2)]\cdot 2(x^2-4x-5)}{(x+1)[3(x-5)](x+2)(x-2)}$

$=\frac{-8(x-5)(x+1)}{3(x+1)(x-5)}$

$=-\frac{8}{3}$

15. Recall that $A = \ell \cdot w$. So,

$$A = \frac{x+2}{x} \cdot \frac{5x}{x^2-4} = \frac{(x+2)5}{(x+2)(x-2)} = \frac{5}{x-2} \text{ sq. m}$$

17. $\frac{4}{x} \div \frac{8}{x^2} = \frac{4}{x} \cdot \frac{x^2}{8} = \frac{1}{1} \cdot \frac{x}{2} = \frac{x}{2}$

19. $\frac{4ab}{3c^2} \div \frac{2a^2b}{6ac} = \frac{4ab}{3c^2} \cdot \frac{6ac}{2a^2b}$

$$= \frac{24a^2bc}{6a^2bc^2} = \frac{4}{c}$$

21. $\frac{2x}{5} \div \frac{6x+12}{5x+10} = \frac{2x}{5} \cdot \frac{5x+10}{6x+12}$

$$= \frac{2x[5(x+2)]}{5[6(x+2)]} = \frac{2x}{6} = \frac{x}{3}$$

23. $\frac{2(x+y)}{5} \div \frac{6(x+y)}{25}$

$$= \frac{2(x+y)}{5} \cdot \frac{25}{6(x+y)} = \frac{50(x+y)}{30(x+y)} = \frac{50}{30} = \frac{5}{3}$$

25. $\frac{x^2-6x+9}{x^2-x-6} \div \frac{x^2-9}{4}$

$$= \frac{(x-3)^2}{(x-3)(x+2)} \cdot \frac{4}{(x+3)(x-3)} = \frac{4}{(x+2)(x+3)}$$

27. $\frac{x^2-6x-16}{2x^2-128} \div \frac{x^2+10x+16}{x^2+16x+64}$

$$= \frac{(x-8)(x+2)}{2(x^2-64)} \cdot \frac{(x+8)^2}{(x+2)(x+8)} = \frac{(x-8)(x+8)}{2(x+8)(x-8)} = \frac{1}{2}$$

29. $\frac{14x^4}{y^5} \div \frac{2x^2}{y^7} \div \frac{2x^4}{7y^5}$

$$= \frac{14x^4}{y^5} \cdot \frac{y^7}{2x^2} \cdot \frac{7y^5}{2x^4} = \frac{98x^4y^{12}}{4x^6y^5} = \frac{49y^7}{2x^2}$$

31. $(f+g)(x) = x^2 + 5x + 1$

33. $(f-g)(x) = x^2 - 5x + 1$

35. $\left(\frac{g}{f}\right)(x) = \frac{5x}{x^2+1}$

37. $\frac{3xy^3}{4x^3y^2} \cdot \frac{-8x^3y^4}{9x^4y^7}$

$$= \frac{-24x^4y^7}{36x^7y^9} = \frac{-2}{3x^3y^2}$$

39. $\frac{8a}{3a^4b^2} \div \frac{4b^5}{6a^2b} = \frac{8a}{3a^4b^2} \cdot \frac{6a^2b}{4b^5}$

$$= \frac{48a^3b}{12a^4b^7} = \frac{4}{ab^6}$$

41. $\frac{a^2b}{a^2-b^2} \cdot \frac{a+b}{4a^3b}$

$$= \frac{a^2b(a+b)}{4a^3b(a+b)(a-b)} = \frac{1}{4a(a-b)}$$

43. $\frac{x^2-9}{4} \div \frac{x^2-6x+9}{x^2-x-6}$

$$= \frac{(x+3)(x-3)}{4} \cdot \frac{(x-3)(x+2)}{(x-3)^2} = \frac{(x+3)(x+2)}{4} = \frac{x^2+5x+6}{4}$$

45. $\frac{9x+9}{4x+8}\cdot\frac{2x+4}{3x^2-3}$
$=\frac{9(x+1)\cdot 2(x+2)}{4(x+2)\cdot 3(x^2-1)}$
$=\frac{18(x+1)}{12(x+1)(x-1)}$
$=\frac{3}{2(x-1)}$

47. $\frac{a+b}{ab}\div\frac{a^2-b^2}{4a^3b}$
$=\frac{a+b}{ab}\cdot\frac{4a^3b}{(a+b)(a-b)}$
$=\frac{4a^2}{a-b}$

49. $\frac{2x^2-4x-30}{5x^2-40x-75}\div\frac{x^2-8x+15}{x^2-6x+9}$
$=\frac{2(x^2-2x-15)}{5(x^2-8x-15)}\cdot\frac{(x-3)^2}{(x-5)(x-3)}$
$=\frac{2(x-5)(x+3)(x-3)}{5(x^2-8x-15)(x-5)}$
$=\frac{2(x+3)(x-3)}{5(x^2-8x-15)}=\frac{2x^2-18}{5(x^2-8x-15)}$

51. $\frac{2x^3-16}{6x^2+6x-36}\cdot\frac{9x+18}{3x^2+6x+12}$
$=\frac{2(x^3-8)\cdot 9(x+2)}{6(x^2+x-6)\cdot 3(x^2+2x+4)}$
$=\frac{18(x-2)(x^2+2x+4)(x+2)}{18(x+3)(x-2)(x^2+2x+4)}$
$=\frac{x+2}{x+3}$

53. $\frac{15b-3a}{b^2-a^2}\div\frac{a-5b}{ab+b^2}$
$=\frac{3(5b-a)}{(b+a)(b-a)}\cdot\frac{b(a+b)}{-(5b-a)}$
$=\frac{3}{b-a}\cdot\frac{b}{-1}$
$=\frac{3b}{a-b}$

55. $\frac{a^3+a^2b+a+b}{a^3+a}\cdot\frac{6a^2}{2a^2-2b^2}$
$=\frac{(a^2+1)(a+b)}{a(a^2+1)}\cdot\frac{6a^2}{2(a^2-b^2)}$
$=\frac{a+b}{1}\cdot\frac{3a}{(a+b)(a-b)}=\frac{3a}{a-b}$

57. $\frac{5a}{12}\cdot\frac{2}{25a^2}\cdot\frac{15a}{2}=\frac{1}{6}\cdot\frac{1}{5a}\cdot\frac{15a}{2}$
$=\frac{1}{6}\cdot\frac{1}{1}\cdot\frac{3}{2}=\frac{3}{12}=\frac{1}{4}$

59. $\frac{3x-x^2}{x^3-27}\div\frac{x}{x^2+3x+9}$
$=\frac{x(3-x)}{(x-3)(x^2+3x+9)}\cdot\frac{x^2+3x+9}{x}$
$=\frac{3-x}{x-3}=\frac{-(x-3)}{x-3}=-1$

61. $\frac{4a}{7}\div\left(\frac{a^2}{14}\cdot\frac{3}{a}\right)=\frac{4a}{7}\div\frac{3a}{14}$
$=\frac{4a}{7}\cdot\frac{14}{3a}=\frac{4}{1}\cdot\frac{2}{3}=\frac{8}{3}$

63. $\frac{8b+24}{3a+6}\div\frac{ab-2b+3a-6}{a^2-4a+4}$
$=\frac{8(b+3)}{3(a+2)}\cdot\frac{(a-2)^2}{(a-2)(b+3)}$
$=\frac{8(a-2)}{3(a+2)}$

65. $\frac{4}{x}\div\frac{3xy}{x^2}\cdot\frac{6x^2}{x^4}=\frac{4}{x}\cdot\frac{x^2}{3xy}\cdot\frac{6}{x^2}$
$=\frac{4}{3y}\cdot\frac{6}{x^2}=\frac{4}{y}\cdot\frac{2}{x^2}=\frac{8}{x^2y}$

67. $\frac{3x^2-5x-2}{y^2+y-2}\cdot\frac{y^2+4y-5}{12x^2+7x+1}\div\frac{5x^2-9x-2}{8x^2-2x-1}$
$=\frac{(3x+1)(x-2)(y+5)(y-1)(4x+1)(2x-1)}{(y+2)(y-1)(4x+1)(3x+1)(5x+1)(x-2)}$
$=\frac{(y+5)(2x-1)}{(y+2)(5x+1)}$

69. $\frac{5a^2-20}{3a^2-12a}\div\frac{a^3+2a^2}{2a^2-8a}\cdot\frac{9a^3+6a^2}{2a^2-4a}$
$=\frac{5(a^2-4)}{3a(a-4)}\cdot\frac{2a(a-4)}{a^2(a+2)}\cdot\frac{3a^2(3a+2)}{2a(a-2)}$
$=\frac{5(a+2)(a-2)(3a+2)}{a(a+2)(a-2)}$
$=\frac{5(3a+2)}{a}=\frac{15a+10}{a}$

71. $\frac{5x^4+3x^2-2}{x-1}\cdot\frac{x+1}{x^4-1}$
$=\frac{(5x^2-2)(x^2+1)(x+1)}{(x-1)(x^2+1)(x^2-1)}$
$=\frac{(5x^2-2)(x+1)}{(x-1)(x+1)(x-1)}$
$=\frac{5x^2-2}{(x-1)^2}$

73. $(f+g)(x)=x^2-2x+2$

75. $\left(\frac{f}{g}\right)(x)=-\frac{2x}{x^2+2}$

77. $(g-h)(x)=(x^2+2)-(4x+3)$
$=x^2-4x-1$

79. $(h+f)(x)=4x+3+(-2x)$
$=2x+3$

81. $f(a+b)=-2(a+b)=-2a-2b$

83. $\left(\frac{f}{h}\right)(x)=\frac{-2x}{4x+3}$

85. Answers may vary.

87. Since $A=b\cdot h$, $b=\frac{A}{h}$. Now,
$b=\frac{\frac{x^2+x-2}{x^3}}{\frac{x^2}{x-1}}$
$b=\frac{(x+2)(x-1)}{x^3}\cdot\frac{(x-1)}{x^2}$
$b=\frac{(x+2)(x-1)^2}{x^5}$ ft.

89. $P(x)=R(x)-C(x)$

91. $\frac{4}{5}+\frac{3}{5}$
$=\frac{7}{5}$

93. $\frac{5}{28}-\frac{2}{21}$
The LCD is 84.
$\frac{5}{28}\cdot\frac{3}{3}-\frac{2}{21}\cdot\frac{4}{4}$
$=\frac{15}{84}-\frac{8}{84}$
$=\frac{7}{84}=\frac{7\cdot1}{7\cdot12}=\frac{1}{12}$

95. $\frac{3}{8}+\frac{1}{2}-\frac{3}{16}$
The LCD is 16.
$\frac{3}{8}\cdot\frac{2}{2}+\frac{1}{2}\cdot\frac{8}{8}-\frac{3}{16}$
$=\frac{6}{16}+\frac{8}{16}-\frac{3}{16}$
$=\frac{14}{16}-\frac{3}{16}$
$=\frac{11}{16}$

97. $\frac{x^{2n}-4}{7x}\cdot\frac{14x^3}{x^n-2}$
$=\frac{(x^n+2)(x^n-2)2x^2}{x^n-2}$
$=2x^2(x^n+2)$

99. $\frac{y^{2n}+9}{10y}\cdot\frac{y^n-3}{y^{4n}-81}$
$=\frac{(y^{2n}+9)(y^n-3)}{10y(y^{2n}+9)(y^{2n}-9)}$
$=\frac{y^n-3}{10y(y^n+3)(y^n-3)}$
$=\frac{1}{10y(y^n+3)}$

101. $\frac{y^{2n}-y^n-2}{2y^n-4}\div\frac{y^{2n}-1}{1+y^n}$
$=\frac{(y^n-2)(y^n+1)}{2(y^n-2)}\cdot\frac{y^n+1}{(y^n+1)(y^n-1)}$
$=\frac{y^n+1}{2(y^n-1)}$

Exercise Set 6.3

1. $\frac{2}{x}-\frac{5}{x}=\frac{2-5}{x}=\frac{-3}{x}$

3. $\frac{2}{x-2}+\frac{x}{x-2}=\frac{2+x}{x-2}$

5. $\frac{x^2}{x+2}-\frac{4}{x+2}=\frac{x^2-4}{x+2}$
$=\frac{(x+2)(x-2)}{x+2}=x-2$

7. $\frac{2x-6}{x^2+x-6}+\frac{3-3x}{x^2+x-6}$
$=\frac{-x-3}{x^2+x-6}$
$=\frac{-(x+3)}{(x+3)(x-2)}$
$=\frac{-1}{x-2}$

9. Recall that $P=4\cdot s$.
$P=4\left(\frac{x}{x+5}\right)=\left(\frac{4x}{x+5}\right)$ feet
Recall that $A=s^2$.
$A=\left(\frac{x}{x+5}\right)^2=\frac{x^2}{(x+5)^2}$
$A=\left(\frac{x^2}{x^2+10x+25}\right)$ sq. ft.

11. The LCD of 7 and $5x$ is $7(5x)=35x$.

13. The LCD of x and $x+1$ is $x(x+1)$.

15. The LCD of $x+7$ and $x-7$ is $(x+7)(x-7)$.

17. The LCD of $3x+6=3(x+2)$ and $2x-4=2(x-2)$ is
$[3(x+2)][2(x-2)]$
$6(x+2)(x-2)$

19. The LCD of $a^2-b^2=(a+b)(a-b)$ and $a^2-2ab+b^2=(a-b)^2$ is:
$(a+b)(a-b)^2$.

21. Answers may vary.

23. $\frac{4}{3x}+\frac{3}{2x}=\frac{(4)2}{(3x)2}+\frac{(3)3}{(2x)3}$
$=\frac{8+9}{6x}=\frac{17}{6x}$

25. $\frac{5}{2y^2}-\frac{2}{7y}=\frac{(5)7}{(2y^2)7}-\frac{2(2y)}{(7y)2y}$
$=\frac{35-4y}{14y^2}$

27. $\frac{x-3}{x+4}-\frac{x+2}{x-4}$
$=\frac{(x-3)(x-4)}{(x+4)(x-4)}-\frac{(x+2)(x+4)}{(x-4)(x+4)}$
$=\frac{(x^2-7x+12)-(x^2+6x+8)}{(x+4)(x-4)}$
$=\frac{-13x+4}{(x+4)(x-4)}$

29. $\frac{1}{x-5}+\frac{x}{(x-5)(x+4)}$
$=\frac{1(x+4)}{(x-5)(x+4)}+\frac{x}{(x-5)(x+4)}$
$=\frac{x+4+x}{(x-5)(x+4)}=\frac{2x+4}{(x-5)(x+4)}$

31. $\frac{1}{a-b}+\frac{1}{b-a}=\frac{1}{a-b}+\frac{1}{-(a-b)}$
$=\frac{1}{a-b}-\frac{1}{a-b}=0$

33. $x+1+\frac{1}{x-1}$
$=\frac{(x+1)(x-1)}{x-1}+\frac{1}{x-1}$
$=\frac{x^2-1+1}{x-1}=\frac{x^2}{x-1}$

35. $\frac{5}{x-2}+\frac{x+4}{2-x}=\frac{5}{x-2}+\frac{x+4}{-(x-2)}$
$=\frac{5}{x-2}-\frac{x+4}{x-2}=\frac{5-(x+4)}{x-2}$
$=\frac{5-x-4}{x-2}=\frac{1-x}{x-2}$

37. $\frac{y+1}{y^2-6y+8}-\frac{3}{y^2-16}$
$=\frac{y+1}{(y-2)(y-4)}-\frac{3}{(y+4)(y-4)}$
$=\frac{(y+1)(y+4)}{(y-2)(y-4)(y+4)}$
$-\frac{3(y-2)}{(y-2)(y-4)(y+4)}$
$=\frac{y^2+5y+4-(3y-6)}{(y-2)(y-4)(y+4)}$
$=\frac{y^2+5y+4-3y+6}{(y-2)(y-4)(y+4)}$
$=\frac{y^2+2y+10}{(y-2)(y-4)(y+4)}$

39. $\frac{x+4}{3x^2+11x+6}+\frac{x}{2x^2+x-15}$
$=\frac{x+4}{(3x+2)(x+3)}+\frac{x}{(2x-5)(x+3)}$
$=\frac{(x+4)(2x-5)}{(3x+2)(x+3)(2x-5)}$
$+\frac{x(3x+2)}{(2x-5)(x+3)(3x+2)}$
$=\frac{2x^2+3x-20+3x^2+2x}{(3x+2)(x+3)(2x-5)}$
$=\frac{5x^2+5x-20}{(3x+2)(x+3)(2x-5)}$
$=\frac{5(x^2+x-4)}{(3x+2)(x+3)(2x-5)}$

41. $\frac{7}{x^2-x-2}+\frac{x}{x^2+4x+3}$
$=\frac{7}{(x-2)(x+1)}+\frac{x}{(x+1)(x+3)}$
$=\frac{7(x+3)}{(x-2)(x+1)(x+3)}$
$+\frac{x(x-2)}{(x+1)(x+3)(x-2)}$
$=\frac{7x+21+x^2-2x}{(x-2)(x+1)(x+3)}$
$=\frac{x^2+5x+21}{(x-2)(x+1)(x+3)}$

43. $\frac{2}{x+1}-\frac{3x}{3x+3}+\frac{1}{2x+2}$
$=\frac{2}{x+1}-\frac{3x}{3(x+1)}+\frac{1}{2(x+1)}$
$=\frac{2}{x+1}-\frac{x}{x+1}+\frac{1}{2(x+1)}$
$=\frac{2-x}{x+1}+\frac{1}{2(x+1)}$
$=\frac{2(2-x)}{2(x+1)}+\frac{1}{2(x+1)}$
$=\frac{4-2x+1}{2(x+1)}=\frac{5-2x}{2(x+1)}$

45. $\frac{3}{x+3}+\frac{5}{x^2+6x+9}-\frac{x}{x^2-9}$
$=\frac{3}{x+3}+\frac{5}{(x+3)^2}-\frac{x}{(x+3)(x-3)}$

$$= \frac{3(x+3)(x-3)}{(x+3)^2(x-3)} + \frac{5(x-3)}{(x+3)^2(x-3)} - \frac{x(x+3)}{(x+3)^2(x-3)}$$

$$= \frac{3x^2 - 27 + 5x - 15 - x^2 - 3x}{(x+3)^2(x-3)}$$

$$= \frac{2x^2 + 2x - 42}{(x+3)^2(x-3)}$$

$$= \frac{2(x^2 + x - 21)}{(x+3)^2(x-3)}$$

47. $\frac{4}{3x^2y^3} + \frac{5}{3x^2y^3}$

$$= \frac{4+5}{3x^2y^3} = \frac{9}{3x^2y^3} = \frac{3}{x^2y^3}$$

49. $\frac{x-5}{2x} - \frac{x+5}{2x} = \frac{x-5-x-5}{2x} = \frac{-10}{2x} = -\frac{5}{x}$

51. $\frac{3}{2x+10} + \frac{8}{3x+15}$

$$= \frac{3}{2(x+5)} + \frac{8}{3(x+5)}$$

$$= \frac{9}{6(x+5)} + \frac{16}{6(x+5)}$$

$$= \frac{9+16}{6(x+5)} = \frac{25}{6(x+5)}$$

53. $\frac{-2}{x^2-3x} - \frac{1}{x^3-3x^2}$

$$= \frac{-2}{x(x-3)} - \frac{1}{x^2(x-3)}$$

$$= \frac{-2x}{x^2(x-3)} - \frac{1}{x^2(x-3)}$$

$$= \frac{-2x-1}{x^2(x-3)}$$

55. $\frac{ab}{a^2-b^2} + \frac{b}{a+b}$

$$= \frac{ab}{(a+b)(a-b)} + \frac{b}{a+b}$$

$$= \frac{ab}{(a+b)(a-b)} + \frac{b(a-b)}{(a+b)(a-b)}$$

$$= \frac{ab + ab - b^2}{(a+b)(a-b)}$$

$$= \frac{2ab - b^2}{(a+b)(a-b)}$$

57. $\frac{5}{x^2-4} - \frac{3}{x^2+4x+4}$

$$= \frac{5}{(x+2)(x-2)} - \frac{3}{(x+2)^2}$$

$$= \frac{5(x+2)}{(x+2)^2(x-2)} - \frac{3(x-2)}{(x+2)^2(x-2)}$$

$$= \frac{5x+10-3x+6}{(x+2)^2(x-2)}$$

$$= \frac{2x+16}{(x+2)^2(x-2)}$$

59. $\frac{2}{a^2+2a+1} + \frac{3}{a^2-1}$

$$= \frac{2}{(a+1)^2} + \frac{3}{(a+1)(a-1)}$$

$$= \frac{2(a-1)}{(a+1)^2(a-1)} + \frac{3(a+1)}{(a+1)^2(a-1)}$$

$$= \frac{2a-2+3a+3}{(a+1)^2(a-1)}$$

$$= \frac{5a+1}{(a+1)^2(a-1)}$$

61. Answers may vary.

63. Answers may vary.

65. $\left(\frac{2}{3} - \frac{1}{x}\right) \cdot \left(\frac{3}{x} + \frac{1}{2}\right)$

$$= \left(\frac{2x}{3x} - \frac{3}{3x}\right) \cdot \left(\frac{6}{2x} + \frac{x}{2x}\right)$$

$$= \frac{2x-3}{3x} \cdot \frac{x+6}{2x}$$

$$= \frac{2x^2 + 9x - 18}{6x^2}$$

67. $\left(\frac{1}{x} + \frac{2}{3}\right) - \left(\frac{1}{x} - \frac{2}{3}\right)$

$$= \frac{1}{x} + \frac{2}{3} - \frac{1}{x} + \frac{2}{3} = \frac{4}{3}$$

69. $\left(\frac{2a}{3}\right)^2 \div \left(\frac{a^2}{a+1} - \frac{1}{a+1}\right)$

$= \frac{4a^2}{9} \cdot \frac{a+1}{(a+1)(a-1)}$

$= \frac{4a^2}{9} \div \frac{a^2-1}{a+1}$

$= \frac{4a^2}{9} \cdot \frac{1}{a-1}$

$= \frac{4a^2}{9(a-1)}$

71. $\left(\frac{2x}{3}\right)^2 \div \left(\frac{x}{3}\right)^2 = \left(\frac{\frac{2x}{3}}{\frac{x}{3}}\right)^2 = 2^2 = 4$

73. $\frac{x}{x^2-9} + \frac{3}{x^2-6x+9} - \frac{1}{x+3}$

$= \frac{x}{(x+3)(x-3)} + \frac{3}{(x-3)^2} - \frac{1}{(x+3)}$

$= \frac{x(x-3)+3(x+3)-(x-3)^2}{(x+3)(x-3)^2}$

$= \frac{x^2-3x+3x+9-x^2+6x-9}{(x-3)(x-3)^2}$

$= \frac{6x}{(x+3)(x-3)^2}$

75. $\left(\frac{x}{x+1} - \frac{x}{x-1}\right) \div \frac{x}{2x+2}$

$= \frac{x(x-1)-x(x+1)}{(x-1)(x+1)} \cdot \frac{2(x+1)}{x}$

$= \frac{x^2-x-x^2-x}{(x+1)(x-1)} \cdot \frac{2(x+1)}{x}$

$= \frac{-2x}{(x+1)(x-1)} \cdot \frac{2(x+1)}{x}$

$= \frac{-4}{x-1}$

77. $\frac{4}{x} \cdot \left(\frac{2}{x+2} - \frac{2}{x-2}\right)$

$= \frac{4 \cdot 2}{x}\left(\frac{1}{x+2} - \frac{1}{x-2}\right)$

$= \frac{8}{x}\left(\frac{x-2-(x+2)}{(x+2)(x-2)}\right)$

$= \frac{8}{x} \cdot \frac{x-2-x-2}{(x+2)(x-2)}$

$= \frac{8}{x} \cdot \frac{-4}{(x+2)(x-2)}$

$= \frac{-32}{x(x+2)(x-2)}$

83. $f(x) = \frac{3x}{x^2-4}$ $\quad g(x) = \frac{6}{x^2+2x}$

$(f+g)(x) = \frac{3x}{x^2-4} + \frac{6}{x^2+2x}$

$(f+g)(x) = \frac{3x}{(x+2)(x-2)} + \frac{6}{x(x+2)}$

The LCD is $x(x+2)(x-2)$.

$(f+g)(x)$

$= \frac{3x}{(x+2)(x-2)} \cdot \frac{x}{x} + \frac{6}{x(x+2)} \cdot \frac{(x-2)}{(x-2)}$

$(f+g)(x)$

$= \frac{3x^2}{x(x+2)(x-2)} + \frac{6(x-2)}{x(x+2)(x-2)}$

$(f+g)(x) = \frac{3x^2+6(x-2)}{x(x+2)(x-2)}$

$(f+g)(x) = \frac{3x^2+6x-12}{x(x+2)(x-2)}$

85. $\left(\frac{f}{g}\right)(x) = \frac{\frac{3}{x^2-4}}{\frac{6}{x^2+2x}}$

$\left(\frac{f}{g}\right)(x) = \frac{\frac{3}{(x+2)(x-2)}}{\frac{6}{x(x+2)}}$

$\left(\frac{f}{g}\right)(x) = \frac{3}{(x+2)(x-2)} \cdot \frac{x(x+2)}{6}$

$\left(\frac{f}{g}\right)(x) = \frac{x}{2(x-2)}$

87. $12\left(\frac{2}{3} + \frac{1}{6}\right) = \frac{24}{3} + \frac{12}{6} = 8 + 2 = 10$

89. $x^2\left(\frac{4}{x^2} + 1\right) = x^2 \cdot \frac{4}{x^2} + x^2 \cdot 1 = 4 + x^2$

91. $\sqrt{100} = 10$

93. $\sqrt[3]{8}$

$= 2$

95. $\sqrt[4]{81} = 3$

97. $a^2+b^2=c^2$
$3^2+4^2=c^2$
$9+16=c^2$
$25=c^2$
$5 \text{ meters} = c$

99. $x^{-1}+(2x)^{-1}=\frac{1}{x}+\frac{1}{2x}$
$=\frac{2}{2x}+\frac{1}{2x}=\frac{3}{2x}$

101. $4x^{-2}-3x^{-1}=\frac{4}{x^2}-\frac{3}{x}$
$=\frac{4}{x^2}-\frac{3x}{x^2}=\frac{4-3x}{x^2}$

103. $x^{-3}(2x+1)-5x^{-2}$
$=\frac{2x+1}{x^3}-\frac{5}{x^2}$
$=\frac{2x+1}{x^3}-\frac{5x}{x^3}$
$=\frac{2x+1-5x}{x^3}$
$=\frac{1-3x}{x^3}$

Exercise Set 6.4

1. $\frac{\frac{1}{3}}{\frac{2}{5}}=\frac{1}{3}\cdot\frac{5}{2}=\frac{5}{6}$

3. $\frac{\frac{4}{x}}{\frac{5}{2x}}=\frac{4}{x}\cdot\frac{2x}{5}=\frac{4}{1}\cdot\frac{2}{5}=\frac{8}{5}$

5. $\frac{\frac{10}{3x}}{\frac{5}{6x}}=\frac{10}{3x}\cdot\frac{6x}{5}=\frac{2}{1}\cdot\frac{2}{1}=4$

7. $\frac{1+\frac{2}{5}}{2+\frac{3}{5}}=\frac{\left(1+\frac{2}{5}\right)5}{\left(2+\frac{3}{5}\right)5}=\frac{5+2}{10+3}=\frac{7}{13}$

9. $\frac{\frac{4}{x-1}}{\frac{x}{x-1}}=\frac{4}{x-1}\cdot\frac{x-1}{x}=\frac{4}{x}$

11. $\frac{1-\frac{2}{x}}{x-\frac{4}{9x}}=\frac{\left(1-\frac{2}{x}\right)9x}{\left(x-\frac{4}{9x}\right)9x}=\frac{9x-18}{9x^2-4}$

13. $\frac{\frac{1}{x+1}-1}{\frac{1}{x-1}+1}=\frac{\left(\frac{1}{x+1}-1\right)(x+1)(x-1)}{\left(\frac{1}{x-1}+1\right)(x-1)(x+1)}$
$=\frac{[1-(x+1)](x-1)}{[1+(x-1)](x+1)}=\frac{(1-x-1)(x-1)}{(1+x-1)(x+1)}$
$=\frac{-x(x-1)}{x(x+1)}=\frac{-(x-1)}{x+1}=\frac{1-x}{x+1}$

15. $\frac{x^{-1}}{x^{-2}+y^{-2}}=\frac{x^{-1}(x^2y^2)}{(x^{-2}+y^{-2})(x^2y^2)}$
$=\frac{x^1y^2}{x^0y^2+x^2y^0}=\frac{xy^2}{y^2+x^2}$

17. $\frac{2a^{-1}+3b^{-2}}{a^{-1}-b^{-1}}=\frac{(2a^{-1}+3b^{-2})ab^2}{(a^{-1}-b^{-1})ab^2}$
$=\frac{2a^0b^2+3ab^0}{a^0b^2-ab^1}=\frac{2b^2+3a}{b^2-ab}$

19. $\frac{1}{x-x^{-1}}=\frac{1\cdot x}{(x-x^{-1})x}=\frac{x}{x^2-x^0}=\frac{x}{x^2-1}$

21. $\frac{\frac{x+1}{7}}{\frac{x+2}{7}}=\frac{x+1}{7}\cdot\frac{7}{x+2}=\frac{x+1}{x+2}$

23. $\frac{\frac{1}{2}-\frac{1}{3}}{\frac{3}{4}+\frac{2}{5}}=\frac{\left(\frac{1}{2}-\frac{1}{3}\right)60}{\left(\frac{3}{4}+\frac{2}{5}\right)60}=\frac{30-20}{45+24}=\frac{10}{69}$

25. $\frac{\frac{x+1}{3}}{\frac{2x-1}{6}}=\frac{x+1}{3}\cdot\frac{6}{2x-1}$
$\frac{x+1}{1}\cdot\frac{2}{2x-1}$
$=\frac{2(x+1)}{2x-1}$

27. $\frac{\frac{x}{3}}{\frac{2}{x+1}}=\frac{x}{3}\cdot\frac{x+1}{2}=\frac{x(x+1)}{6}$

29. $\dfrac{\frac{2}{x}+3}{\frac{4}{x^2}-9}=\dfrac{\left(\frac{2}{x}+3\right)x^2}{\left(\frac{4}{x^2}-9\right)x^2}=\dfrac{2x+3x^2}{4-9x^2}$

$\dfrac{x(2+3x)}{(2-3x)(2+3x)}=\dfrac{x}{2-3x}$

31. $\dfrac{1-\frac{x}{y}}{\frac{x^2}{y^2}-1}=\dfrac{\left(1-\frac{x}{y}\right)y^2}{\left(\frac{x^2}{y^2}-1\right)y^2}=\dfrac{y^2-xy}{x^2-y^2}$

$=\dfrac{y(y-x)}{(x+y)(x-y)}=\dfrac{y(-1)}{x+y}=\dfrac{-y}{x+y}$

33. $\dfrac{\frac{-2x}{x-y}}{\frac{y}{x^2}}=\dfrac{-2x}{x-y}\cdot\dfrac{x^2}{y}=\dfrac{-2x^3}{(x-y)y}$

35. $\dfrac{\frac{2}{x}+\frac{1}{x^2}}{\frac{y}{x^2}}=\dfrac{\left(\frac{2}{x}+\frac{1}{x^2}\right)x^2}{\left(\frac{y}{x^2}\right)x^2}=\dfrac{2x+1}{y}$

37. $\dfrac{\frac{x}{9}-\frac{1}{x}}{1+\frac{3}{x}}=\dfrac{\left(\frac{x}{9}-\frac{1}{x}\right)9x}{\left(1+\frac{3}{x}\right)9x}=\dfrac{x^2-9}{9x+27}$

$\dfrac{(x+3)(x-3)}{9(x+3)}=\dfrac{x-3}{9}$

39. $\dfrac{\frac{x-1}{x^2-4}}{1+\frac{1}{x-2}}$

$=\dfrac{\frac{x-1}{(x+2)(x-2)}(x+2)(x-2)}{\left(1+\frac{1}{x-2}\right)(x-2)(x+2)}$

$=\dfrac{x-1}{(x-2+1)(x+2)}$

$=\dfrac{x-1}{(x-1)(x+2)}=\dfrac{1}{x+2}$

41. $\dfrac{\frac{4}{5-x}+\frac{5}{x-5}}{\frac{2}{x}+\frac{3}{x-5}}=\dfrac{-\frac{4}{x-5}+\frac{5}{x-5}}{\frac{2(x-5)+3x}{x(x-5)}}$

$=\dfrac{\frac{1}{x-5}}{\frac{2x-10+3x}{x(x-5)}}=\dfrac{1}{x-5}\cdot\dfrac{x(x-5)}{5x-10}$

$=\dfrac{x}{5x-10}$

43. $\dfrac{\frac{x+2}{x}-\frac{2}{x-1}}{\frac{x+1}{x}+\frac{x+1}{x-1}}$

$=\dfrac{\frac{(x+2)(x-1)-2x}{x(x-1)}}{\frac{(x+1)(x-1)+(x+1)x}{x(x-1)}}$

$=\dfrac{x^2+x-2-2x}{x^2-1+x^2+x}$

$=\dfrac{x^2-x-2}{2x^2+x-1}$

$=\dfrac{(x-2)(x+1)}{(2x-1)(x+1)}=\dfrac{x-2}{2x-1}$

45. $\dfrac{\frac{x-2}{x+2}+\frac{x+2}{x-2}}{\frac{x-2}{x+2}-\frac{x+2}{x-2}}$

$=\dfrac{\frac{(x-2)(x-2)+(x+2)(x+2)}{(x+2)(x-2)}}{\frac{(x-2)(x-2)-(x+2)(x+2)}{(x+2)(x-2)}}$

$=\dfrac{x^2-4x+4+x^2+4x+4}{x^2-4x+4-x^2-4x-4}$

$=\dfrac{2x^2+8}{-8x}=-\dfrac{x^2+4}{4x}$

47. $\dfrac{\frac{2}{y^2}-\frac{5}{xy}-\frac{3}{x^2}}{\frac{2}{y^2}+\frac{7}{xy}+\frac{3}{x^2}}=\dfrac{\frac{2x^2-5xy-3y^2}{x^2y^2}}{\frac{2x^2+7xy+3y^2}{x^2y^2}}$

$=\dfrac{(2x+y)(x-3y)}{(2x+y)(x+3y)}=\dfrac{x-3y}{x+3y}$

49. $\dfrac{a^{-1}+1}{a^{-1}-1}=\dfrac{a(a^{-1}+1)}{a(a^{-1}-1)}=\dfrac{1+a}{1-a}$

51. $\dfrac{3x^{-1}+(2y)^{-1}}{x^{-2}}=\dfrac{2x^2y}{2x^2y}\cdot\dfrac{[3x^{-1}+(2y)^{-1}]}{x^{-2}}$

$=\dfrac{6xy+x^2}{2y}$

53. $\dfrac{2a^{-1}+(2a)^{-1}}{a^{-1}+2a^{-2}}=\dfrac{2a^2[2a^{-1}+(2a)^{-1}]}{2a^2(a^{-1}+2a^{-2})}$

$=\dfrac{4a+a}{2a+4}=\dfrac{5a}{2a+4}$

55. $\frac{5x^{-1}+2y^{-1}}{x^{-2}y^{-2}}=\frac{x^2y^2(5x^{-1}+2y^{-1})}{x^2y^2(x^{-2}y^{-2})}$
$=\frac{5xy^2+2x^2y}{1}=5xy^2+2x^2y$

57. $\frac{5x^{-1}-2y^{-1}}{25x^{-2}-4y^{-2}}$
$=\frac{5x^{-1}-2y^{-1}}{(5x^{-1}-2y^{-1})(5x^{-1}+2y^{-1})}$
$=\frac{1}{5x^{-1}+2y^{-1}}\cdot\frac{xy}{xy}=\frac{xy}{5y+2x}$

59. $(x^{-1}+y^{-1})^{-1}=\frac{1}{x^{-1}+y^{-1}}\cdot\frac{xy}{xy}=\frac{xy}{y+x}$

61. $\frac{x}{1-\frac{1}{1+\frac{1}{x}}}=\frac{x}{1-\frac{1}{\frac{x+1}{x}}}$...

$\frac{x}{1-\frac{1}{1+\frac{1}{x}}}=\frac{x}{1-\frac{x}{x+1}}=\frac{x}{\frac{x+1-x}{x+1}}$
$=x\cdot\frac{x+1}{1}=x^2+x$

63. $f(x)=\frac{1}{x}$

a. $f(a+h)=\frac{1}{a+h}$

b. $f(a)=\frac{1}{a}$

c. $\frac{f(a+h)-f(a)}{h}=\frac{\frac{1}{a+h}-\frac{1}{a}}{h}$

d. $\frac{f(a+h)-f(a)}{h}=\frac{\frac{1}{a+h}\cdot\frac{a}{a}-\frac{1}{a}\cdot\frac{(a+h)}{(a+h)}}{h}$
$\frac{f(a+h)-f(a)}{h}=\frac{\frac{a}{a(a+h)}-\frac{a+h}{a(a+h)}}{h}$
$\frac{f(a+h)-f(a)}{h}=\frac{\frac{a-(a+h)}{a(a+h)}}{h}$
$\frac{f(a+h)-f(a)}{h}=\frac{\frac{a-a-h}{a(a+h)}}{h}$
$\frac{f(a+h)-f(a)}{h}=\frac{\frac{-h}{a(a+h)}}{h}$
$\frac{f(a+h)-f(a)}{h}=\frac{-h}{a(a+h)}\cdot\frac{1}{h}$
$\frac{f(a+h)-f(a)}{h}=\frac{-1}{a(a+h)}$

65. $f(x)=\frac{3}{x+1}$

a. $f(a+h)=\frac{3}{a+h+1}$

b. $f(a)=\frac{3}{a+1}$

c. $\frac{f(a+h)-f(a)}{h}=\frac{\frac{3}{a+h+1}-\frac{3}{a+1}}{h}$

d. $\frac{f(a+h)-f(a)}{h}$
$=\frac{h(a+h+1)(a+1)\left[\frac{3}{(a+h+1)}-\frac{3}{(a+1)}\right]}{h(a+h+1)(a+1)\cdot[h]}$
$\frac{f(a+h)-f(a)}{h}$
$=\frac{3h(a+1)-3h(a+h+1)}{h^2(a+h+1)(a+1)}$
$\frac{f(a+h)-f(a)}{h}$
$=\frac{3ah+3h-3ah-3h^2-3h}{h^2(a+h+1)(a+1)}$
$\frac{f(a+h)-f(a)}{h}$
$=\frac{-3h^2}{h^2(a+h+1)(a+1)}$
$\frac{f(a+h)-f(a)}{h}=\frac{-3}{(a+h+1)(a+1)}$

67. $\frac{3x^3y^2}{12x}=\frac{x^{3-1}y^2}{4}=\frac{x^2y^2}{4}$

69. $\frac{144x^5y^5}{-16x^2y}=-9x^{5-2}y^{5-1}$
$=-9x^3y^4$

71. $|x-5|=9$

$x-5=9$ or $x-5=-9$

$x=14$ or $x=-4$

$\{-4, 14\}$

73. $|x-5|<9$

$x-5<9$ and $x-5>-9$

$x<14$ and $x>-4$

$(-4, 14)$

75. $\dfrac{1}{1-(1-x)^{-1}}=\dfrac{1-x}{(1-x)-1}$

$=\dfrac{1-x}{-x}=\dfrac{x-1}{x}$

77. $\dfrac{(x+2)^{-1}+(x-2)^{-1}}{(x^2-4)^{-1}}$

$=\dfrac{(x+2)^{-1}+(x-2)^{-1}}{[(x+2)(x-2)]^{-1}}$

$=\dfrac{(x+2)(x-2)[(x+2)^{-1}+(x-2)^{-1}]}{(x+2)(x-2)(x+2)^{-1}(x-2)^{-1}}$

$=\dfrac{x-2+x+2}{1}=2x$

79. $\dfrac{3(a+1)^{-1}+4a^{-2}}{(a^3+a^2)^{-1}}$

$=\dfrac{3(a+1)^{-1}+4a^{-2}}{[a^2(a+1)]^{-1}}$

$=\dfrac{(a+1)(a^2)[3(a+1)^{-1}+4a^{-2}]}{(a+1)(a^2)[a^{-2}(a+1)^{-1}]}$

$=\dfrac{3a^2+4(a+1)}{1}=3a^2+4a+4$

Exercise Set 6.5

1. $\dfrac{4a^2+8a}{2a}=\dfrac{4a^2}{2a}+\dfrac{8a}{2a}$

$=2a+4$

3. $\dfrac{12a^5b^2+16a^4b}{4a^4b}=\dfrac{12a^5b^2}{4a^4b}+\dfrac{16a^4b}{4a^4b}$

$=3ab+4$

5. $\dfrac{4x^2y^2+6xy^2-4y^2}{2y^2}$

$=\dfrac{4x^2y^2}{2y^2}+\dfrac{6xy^2}{2y^2}-\dfrac{4y^2}{2y^2}$

$=2x^2+3x-2$

7. $\dfrac{4x^2+8x+4}{4}=\dfrac{4x^2}{4}+\dfrac{8x}{4}+\dfrac{4}{4}$

$=x^2+2x+1$

9. $\dfrac{3x^4+6x-18}{3}$

$=\dfrac{3x^4}{3}+\dfrac{6x}{3}-\dfrac{18}{3}$

$=(x^4+2x-6)$ meters

11.
$$\begin{array}{r|l} & x+1 \\ x+2 & x^2+3x+3 \\ & \underline{x^2+2x} \\ & \quad x+3 \\ & \quad \underline{x+2} \\ & \quad\quad 1 \end{array}$$

Answer: $(x+1)+\dfrac{1}{x+2}$

13.
$$\begin{array}{r|l} & 2x-8 \\ x+1 & 2x^2-6x-8 \\ & \underline{2x^2+2x} \\ & \quad -8x-8 \\ & \quad \underline{-8x-8} \\ & \quad\quad 0 \end{array}$$

Answer: $2x-8$

15.
$$\begin{array}{r|l} & x-\frac{1}{2} \\ 2x+4 & 2x^2+3x-2 \\ & \underline{2x^2+4x} \\ & \quad -x-2 \\ & \quad \underline{-x-2} \\ & \quad\quad 0 \end{array}$$

Answer: $x-\dfrac{1}{2}$

17.
$$\begin{array}{r} 2x^2-\frac{1}{2}x+5 \\ 2x+4\overline{)4x^3+7x^2+8x+20} \\ \underline{4x^3+8x^2} \\ -x^2+8x \\ \underline{-x^2-2x} \\ 10x+20 \\ \underline{10x+20} \\ 0 \end{array}$$

Answer: $2x^2-\frac{1}{2}x+5$

19. Recall that $A=\ell\cdot w$ so
$w=\frac{A}{\ell}=\frac{15x^2-29x-14}{5x+2}$.

$$\begin{array}{r} 3x-7 \\ 5x+2\overline{)15x^2-29x-14} \\ \underline{15x^2+\ 6x} \\ -35x-14 \\ \underline{-35x-14} \\ 0 \end{array}$$

The width is $3x-7$ in.

21. $\frac{25a^2b^{12}}{10a^5b^7}=\frac{5b^{12-7}}{2a^{5-2}}=\frac{5b^5}{2a^3}$

23. $\frac{x^6y^6-x^3y^3}{x^3y^3}=\frac{x^6y^6}{x^3y^3}-\frac{x^3y^3}{x^3y^3}$
$=x^3y^3-1$

25.
$$\begin{array}{r} a+3 \\ a+1\overline{)a^2+4a+3} \\ \underline{a^2+\ a} \\ 3a+3 \\ \underline{3a+3} \\ 0 \end{array}$$

Answer: $a+3$

27.
$$\begin{array}{r} 2x+\ 5 \\ x-2\overline{)2x^2+\ x-10} \\ \underline{2x^2-4x} \\ 5x-10 \\ \underline{5x-10} \\ 0 \end{array}$$

Answer: $2x+5$

29. $\frac{-16y^3+24y^4}{-4y^2}=\frac{-16y^3}{-4y^2}+\frac{24y^4}{-4y^2}$
$=4y-6y^2$

31.
$$\begin{array}{r} 2x+23 \\ x-5\overline{)2x^2+13x+\ 15} \\ \underline{2x^2-10x} \\ 23x+\ 15 \\ \underline{23x-115} \\ 130 \end{array}$$

Answer: $(2x+23)+\frac{130}{x-5}$

33. $\frac{20x^2y^3+6xy^4-12x^3y^5}{2xy^3}$
$=\frac{20x^2y^3}{2xy^3}+\frac{6xy^4}{2xy^3}-\frac{12x^3y^5}{2xy^3}$
$=10x+3y-6x^2y^2$

35.
$$\begin{array}{r} 2x+4 \\ 3x+2\overline{)6x^2+16x+8} \\ \underline{6x^2+\ 4x} \\ 12x+8 \\ \underline{12x+8} \\ 0 \end{array}$$

Answer: $2x+4$

37.
$$\begin{array}{r} y+5 \\ 2y-3\overline{)\ 2y^2+7y-15} \\ \underline{2y^2-3y} \\ 10y-15 \\ \underline{10y-15} \\ 0 \end{array}$$

Answer: $y+5$

39.
$$\begin{array}{r} 2x+3 \\ 2x-3\overline{)\ 4x^2\ -9} \\ \underline{4x^2-6x} \\ 6x-9 \\ \underline{6x-9} \\ 0 \end{array}$$

Answer: $2x+3$

41.
$$\begin{array}{r}
2x^2-8x+38 \\
x+4\overline{)2x^3 \qquad +6x \quad -4} \\
\underline{2x^3+8x^2} \qquad\qquad \\
-8x^2+6x \qquad \\
\underline{-8x^2-32x} \qquad \\
38x-4 \\
\underline{38x+152} \\
-156
\end{array}$$

Answer: $(2x^2-8x+38)-\frac{156}{x+4}$

43.
$$\begin{array}{r}
3x+3 \\
x-1\overline{)3x^2 \qquad -4} \\
\underline{3x^2-3x} \\
3x-4 \\
\underline{3x-3} \\
-1
\end{array}$$

Answer: $(3x+3)-\frac{1}{x-1}$

45.
$$\begin{array}{r}
-2x^3+3x^2-x+4 \\
-x+5\overline{)2x^4-13x^3+16x^2-9x+20} \\
\underline{2x^4-10x^3} \qquad\qquad\qquad \\
-3x^3+16x^2 \qquad\quad \\
\underline{-3x^3+15x^2} \qquad\quad \\
x^2-9x \qquad \\
\underline{x^2-5x} \qquad \\
-4x+20 \\
\underline{-4x+20} \\
0
\end{array}$$

Answer: $-2x^3+3x^2-x+4$

47.
$$\begin{array}{r}
3x^3 \qquad +5x+4 \\
x^2-2\overline{)3x^5-x^3+4x^2-12x-8} \\
\underline{3x^5-6x^3} \qquad\qquad\qquad \\
5x^3+4x^2-12x-8 \\
\underline{5x^3 \qquad -10x} \qquad \\
4x^2-2x-8 \\
\underline{4x^2 \qquad -8} \\
-2x
\end{array}$$

Answer: $(3x^3+5x+4)-\frac{2x}{x^2-2}$

49. $\frac{3x^3-5}{3x^2}=\frac{3x^3}{3x^2}-\frac{5}{3x^2}=x-\frac{5}{3x^2}$

51. $P(x)=3x^3+2x^2-4x+3$
$P(1)=3(1)^3+2(1)^2-4(1)+3$
$=3+2-4+3=4$

$$\begin{array}{r}
3x^2+5x+1 \\
x-1\overline{)3x^3+2x^2-4x+3} \\
\underline{3x^3-3x^2} \qquad\qquad \\
5x^2-4x \qquad \\
\underline{5x^2-5x} \qquad \\
x+3 \\
\underline{x-1} \\
4
\end{array}$$

Remainder = 4

53. $P(x)=5x^4-2x^2+3x-6$
$P(-3)=5(-3)^4-2(-3)^2+3(-3)-6$
$=5(81)-2(9)-9-6$
$=405-18-9-6$
$=372$

$$\begin{array}{r}
5x^3-15x^2+43x-126 \\
x+3\overline{)5x^4 \qquad -2x^2 \qquad +3x-6} \\
\underline{5x^4+15x^3} \qquad\qquad\qquad\qquad \\
-15x^3-2x^2 \qquad\qquad \\
\underline{-15x^3-45x^2} \qquad\qquad \\
43x^2+3x \qquad \\
\underline{43x^2+129x} \qquad \\
-126x-6 \\
\underline{-126x-378} \\
372
\end{array}$$

Remainder = 372

55. Answers may vary.

57.
$$\begin{array}{r}
3x^2+10x+8 \\
x-2\overline{)4x^2-12x-12+3x^3} \\
\underline{-6x^2 \qquad\qquad -3x^3} \\
10x^2-12x \qquad\qquad \\
\underline{10x^2-20x} \qquad\qquad \\
8x-12 \qquad \\
\underline{8x-16} \qquad \\
4 \qquad
\end{array}$$

Definitely awkward.

$$\begin{array}{r} 3x^2+10x+\ 8 \\ x-2\overline{)\,3x^3+4x^2-12x-12} \\ \underline{3x^3-6x^2\qquad\qquad\quad} \\ 10x^2-12x\quad \\ \underline{10x^2-20x\quad} \\ 8x-12 \\ \underline{8x-16} \\ 4 \end{array}$$

Now, that's better.

Answer: $(3x^2+10x+8)+\frac{4}{x-2}$

59.

$$\begin{array}{r} 2x^2+\frac{1}{2}x-5 \\ x+2\overline{)\,2x^3-\frac{9}{2}x^2-4x-10} \\ \underline{2x^3\ +4x^2\qquad\qquad\quad} \\ \frac{1}{2}x^2-4x\quad \\ \underline{\frac{1}{2}x^2\ \ +x\quad} \\ -5x-10 \\ \underline{-5x-10} \\ 0 \end{array}$$

Answer: $2x^2+\frac{1}{2}x-5$

61.

$$\begin{array}{r} 2x^3-\frac{9}{2}x^2+10x+21 \\ x-2\overline{)\,2x^4+\frac{1}{2}x^3+\ \ x^2+\ \ \ x\qquad} \\ \underline{2x^4-4x^3\qquad\qquad\qquad\quad} \\ \frac{9}{2}x^3+\ x^2\qquad\qquad \\ \underline{\frac{9}{2}x^3-9x^2\qquad\qquad} \\ 10x^2+\ \ x\qquad \\ \underline{10x^2-20x\qquad} \\ 21x\qquad \\ \underline{21x-42} \\ 42 \end{array}$$

Answer: $\left(2x^3+\frac{9}{2}x^2+10x+21\right)+\frac{42}{x-2}$

63.

$$\begin{array}{r} 3x^4\qquad\quad -2x \\ 3x+2\overline{)\,9x^5+6x^4-6x^2-4x} \\ \underline{9x^5+6x^4\qquad\qquad\quad} \\ -6x^2-4x \\ \underline{-6x^2-4x} \\ 0 \end{array}$$

Answer: $3x^4-2x$

65. $(-5)^2=(-5)(-5)=25=(5)(5)=5^2$
Thus, the answer is $(-5)^2=5^2$

67. $3^4=3\cdot 3\cdot 3\cdot 3=81$
while $(-3)^4=(-3)(-3)(-3)(-3)=81$
so $3^4=(-3)^4$

69. $|x-1|\le 8$
$-8\le x-1\le 8$
$-7\le x\le 9$
$[-7, 9]$

71. $|4x+2|>10$
$4x+2<-10$ or $4x+2>10$
$4x<-12$ or $4x>8$
$x<-3$ or $x>2$
$(-\infty,-3)\cup(2,\infty)$

Exercise Set 6.6

1.

$$\begin{array}{r|rrr} 5 & 1 & 3 & -40 \\ & & 5 & 40 \\ \hline & 1 & 8 & 0 \end{array}$$

$x+8$

3.

$$\begin{array}{r|rrr} -6 & 1 & 5 & -6 \\ & & -6 & 6 \\ \hline & 1 & -1 & 0 \end{array}$$

$x-1$

5.

$$\begin{array}{r|rrrr} 2 & 1 & -7 & -13 & 5 \\ & & 2 & -10 & -46 \\ \hline & 1 & -5 & -23 & -41 \end{array}$$

$x^2-5x-23-\frac{41}{x-2}$

7.

$$\begin{array}{r|rrr} 2 & 4 & 0 & -9 \\ & & 8 & 16 \\ \hline & 4 & 8 & 7 \end{array}$$

$4x+8+\frac{7}{x-2}$

9. a. $P(2)=3(2)^2-4(2)-1$
$P(2)=12-8-1=3$

b.
$$\begin{array}{r|rrr} 2 & 3 & -4 & -1 \\ & & 6 & 4 \\ \hline & 3 & 2 & 3 \end{array}$$
So, $P(2)=3$

11. a. $P(-2)=4(-2)^4+7(-2)^2+9(-2)-1$
$P(-2)=64+28-18-1$
$P(-2)=73$

b.
$$\begin{array}{r|rrrrr} -2 & 4 & 0 & 7 & 9 & -1 \\ & & -8 & 16 & -46 & 74 \\ \hline & 4 & -8 & 23 & -37 & 73 \end{array}$$
So, $P(-2)=73$

13. a. $P(-1)=(-1)^5+3(-1)^4+3(-1)-7$
$P(-1)=-1+3-3-7$
$P(-1)=-8$

b.
$$\begin{array}{r|rrrrrr} -1 & 1 & 3 & 0 & 0 & 3 & -7 \\ & & -1 & -2 & 2 & -2 & -1 \\ \hline & 1 & 2 & -2 & 2 & 1 & -8 \end{array}$$
So, $P(-1)=-8$

15.
$$\begin{array}{r|rrrr} 3 & 1 & -3 & 0 & 2 \\ & & 3 & 0 & 0 \\ \hline & 1 & 0 & 0 & 2 \end{array}$$
$x^2+\dfrac{2}{x-3}$

17.
$$\begin{array}{r|rrr} -1 & 6 & 13 & 8 \\ & & -6 & -7 \\ \hline & 6 & 7 & 1 \end{array}$$
$6x+7+\dfrac{1}{x+1}$

19.
$$\begin{array}{r|rrrrr} 5 & 2 & -13 & 16 & -9 & 20 \\ & & 10 & -15 & 5 & -20 \\ \hline & 2 & -3 & 1 & -4 & 0 \end{array}$$
$2x^3-3x^2+x-4$

21.
$$\begin{array}{r|rrr} -3 & 3 & 0 & -15 \\ & & -9 & 27 \\ \hline & 3 & -9 & 12 \end{array}$$
$3x-9+\dfrac{12}{x+3}$

23.
$$\begin{array}{r|rrrr} \frac{1}{2} & 3 & -6 & 4 & 5 \\ & & \frac{3}{2} & -\frac{9}{4} & \frac{7}{8} \\ \hline & 3 & -\frac{9}{2} & \frac{7}{4} & \frac{47}{8} \end{array}$$
$$3x^2-\frac{9}{2}x+\frac{7}{4}+\frac{\frac{47}{8}}{x-\frac{1}{2}}$$
$$=3x^2-\frac{9}{2}x+\frac{7}{4}+\frac{47}{8x-4}$$

25.
$$\begin{array}{r|rrrr} \frac{1}{3} & 3 & 2 & -4 & 1 \\ & & 1 & 1 & -1 \\ \hline & 3 & 3 & -3 & 0 \end{array}$$
$3x^2+3x-3$

27.
$$\begin{array}{r|rrrr} -1 & 3 & 7 & -4 & 12 \\ & & -3 & -4 & 8 \\ \hline & 3 & 4 & -8 & 20 \end{array}$$
$3x^2+4x-8+\dfrac{20}{x+1}$

29.
$$\begin{array}{r|rrrr} 1 & 1 & 0 & 0 & -1 \\ & & 1 & 1 & 1 \\ \hline & 1 & 1 & 1 & 0 \end{array}$$
x^2+x+1

31. $\begin{array}{r|rrr} -6 & 1 & 0 & -36 \\ & & -6 & 36 \\ \hline & 1 & -6 & 0 \end{array}$

$x-6$

33. $\begin{array}{r|rrrr} 1 & 1 & 3 & -7 & 4 \\ & & 1 & 4 & -3 \\ \hline & 1 & 4 & -3 & 1 \end{array}$

Thus, $P(1)=1$

35. $\begin{array}{r|rrrr} -3 & 3 & -7 & -2 & 5 \\ & & -9 & 48 & -138 \\ \hline & 3 & -16 & 46 & -133 \end{array}$

Thus, $P(-3)=-133$

37. $\begin{array}{r|rrrrr} -1 & 4 & 0 & 1 & 0 & -2 \\ & & -4 & 4 & -5 & 5 \\ \hline & 4 & -4 & 5 & -5 & 3 \end{array}$

Thus, $P(-1)=3$

39. $\begin{array}{r|rrrrr} \frac{1}{3} & 2 & 0 & -3 & 0 & -2 \\ & & \frac{2}{3} & \frac{2}{9} & -\frac{25}{27} & -\frac{25}{81} \\ \hline & 2 & \frac{2}{3} & -\frac{25}{9} & -\frac{25}{27} & -\frac{187}{81} \end{array}$

Thus, $P\left(\frac{1}{3}\right)=\frac{-187}{81}$

41. $\begin{array}{r|rrrrrr} \frac{1}{2} & 1 & 1 & -1 & 0 & 0 & 3 \\ & & \frac{1}{2} & \frac{3}{4} & -\frac{1}{8} & -\frac{1}{16} & -\frac{1}{32} \\ \hline & 1 & \frac{3}{2} & -\frac{1}{4} & -\frac{1}{8} & -\frac{1}{16} & \frac{95}{32} \end{array}$

Thus, $P\left(\frac{1}{2}\right)=\frac{95}{32}$

43. Answers may vary.

45. $\begin{array}{r|rrrr} -3 & 1 & 3 & 4 & 12 \\ & & -3 & 0 & -12 \\ \hline & 1 & 0 & 4 & 0 \end{array}$

$(x+3)(x^2+4)=x^3+3x^2+4x+12$

47. $P(x)$ is equal to the remainder when $P(x)$ is divided by $x-c$. Therefore, $P(c)=0$.

49. Multiply (x^2-x+10) by $(x+3)$ and add the remainder, -2.

$(x^2-x+10)(x+3)$

$x^3-x^2+10x+3x^2-3x+30$

$x^3+2x^2+7x+30$

$\underline{\qquad\qquad\qquad -2}$

$x^3+2x^2+7x+28$

51. $V=lwh$

$w=\frac{V}{lh}$

$w=\frac{x^4+6x^3-7x^2}{x^2(x+7)}$

$w=\frac{x^4+6x^3-7x^2}{x^3+7x^2}$

$$\begin{array}{r} x-1 \\ x^3+7x^2 \overline{\big)\, x^4+6x^3-7x^2} \\ \underline{x^4+7x^3} \\ -1x^3-7x^2 \\ \underline{-1x^3-7x^2} \\ 0 \end{array}$$

The width is $(x-1)$ meters.

53. $4-2x=17-5x$

$5x-2x=17-4$

$3x=13$

$x=\frac{13}{3}$

$\left\{\frac{13}{3}\right\}$

55. $5x^2+10x=15$
$5x^2+10x-15=0$
$5(x^2+2x-3)=0$
$5(x+3)(x-1)=0$
$x+3=0$ or $x-1=0$
$x=-3$ or $x=1$
$\{-3, 1\}$

57. $\frac{2x}{9}+1=\frac{7}{9}$
$2x+9=7$
$2x=-2$
$x=-1$
$\{-1\}$

59. $8y^3+1=(2y+1)(4y^2-2y+1)$

61. $a^3-27=(a-3)(a^2+3a+9)$

63. $x^2-x+xy-y=x(x-1)+y(x-1)$
$=(x+y)(x-1)$

65. $2x^3-32x=2x(x^2-16)$
$=2x(x+4)(x-4)$

Exercise Set 6.7

1. $\frac{x}{2}-\frac{x}{3}=12$
$3x-2x=72$
$x=72$
$\{72\}$

3. $\frac{x}{3}=\frac{1}{6}+\frac{x}{4}$
$4x=2+3x$
$x=2$
$\{2\}$

5. $\frac{2}{x}+\frac{1}{2}=\frac{5}{x}$
$\frac{1}{2}=\frac{3}{x}$
$x=6$
$\{6\}$

7. $\frac{x+3}{x}=\frac{5}{x}$
$x+3=5$
$x=2$
$\{2\}$

9. $\frac{x+5}{x+3}=\frac{8}{x+3}$
$(x+3)\cdot\frac{x+5}{x+3}=(x+3)\cdot\frac{8}{x+3}$
$x+5=8$
$x=3$
$\{3\}$

11. $\frac{1}{x-1}+\frac{1}{x+1}=\frac{2}{x^2-1}$
$\frac{x+1+x-1}{(x-1)(x+1)}=\frac{2}{(x+1)(x-1)}$
$2x=2$
$x=1$
which we discard as extraneous.
No solution.
$\{\ \}$

13. $\frac{6}{x+3}=\frac{4}{x-3}$
$6(x-3)=4(x+3)$
$6x-18=4x+12$
$2x=30$
$x=15$
$\{15\}$

15. $\frac{3}{2x+3}-\frac{1}{2x-3}=\frac{4}{4x^2-9}$
$\frac{3(2x-3)-(2x+3)\cdot 1}{(2x+3)(2x-3)}=\frac{4}{(2x+3)(2x-3)}$
$6x-9-2x-3=4$
$4x-12=4$
$4x=16$
$x=4$
$\{4\}$

17. $\frac{2}{x^2-4}=\frac{1}{2x-4}$
$\frac{4}{2(x+2)(x-2)}=\frac{x+2}{2(x-2)(x+2)}$
$4=x+2$
$x=2$
which we discard as extraneous.
No solution.
$\{\ \}$

19. $\frac{12}{3x^2+12x}=1-\frac{1}{x+4}$

$\frac{12}{3x(x+4)}=\frac{3x(x+4)-3x}{3x(x+4)}$

$12=3x^2+12x-3x$

$12=3x^2+9x$

$4=x^2+3x$

$x^2+3x-4=0$

$(x+4)(x-1)=0$

$x+4=0$ or $x-1=0$

$x=-4$ or $x=1$

We discard –4 as extraneous.

$x=1$

$\{1\}$

21. $\frac{2}{x}=\frac{10}{5}$

$10=10x$

$x=1$

$\{1\}$

23. $7+\frac{6}{a}=5$

$\frac{6}{a}=-2$

$6=-2a$

$a=-3$

$\{-3\}$

25. $\frac{2}{x-5}+\frac{1}{2x}=\frac{5}{3x^2-15x}$

$\frac{2(2x)+x-5}{2x(x-5)}=\frac{5}{3x(x-5)}$

$\frac{3(4x+x-5)}{6x(x-5)}=\frac{2(5)}{6x(x-5)}$

$12x+3x-15=10$

$15x=25$

$x=\frac{25}{15}=\frac{5}{3}$

$\left\{\frac{5}{3}\right\}$

27. $\frac{x}{4}+\frac{5}{x}=3$

$\frac{x^2+20}{4x}=3$

$x^2+20=12x$

$x^2-12x+20=0$

$(x-10)(x-2)=0$

$x-10=0$ or $x-2=0$

$x=10$ or $x=2$

$\{2, 10\}$

29. $1-\frac{5}{y+7}=\frac{4}{y+7}$

$1=\frac{9}{y+7}$

$y+7=9$

$y=2$

$\{2\}$

31. $\frac{6x+7}{2x+9}=\frac{5}{3}$

$3(6x+7)=5(2x+9)$

$18x+21=10x+45$

$8x=24$

$x=3$

$\{3\}$

33. $\frac{2x+1}{4-x}=\frac{9}{4-x}$

$2x+1=9$

$2x=8$

$x=4$

which we discard as extraneous.

No solution.

$\{\ \}$

35. $\frac{12}{9-a^2}+\frac{3}{3+a}=\frac{2}{3-a}$

$\frac{12}{(3+a)(3-a)}+\frac{3(3-a)}{(3+a)(3-a)}=\frac{2(3+a)}{(3-a)(3+a)}$

$12+9-3a=6+2a$

$21-3a=6+2a$

$15=5a$

$3=a$

which we discard as extraneous.

No solution.

$\{\ \}$

37. $2+\frac{3}{x}=\frac{2x}{x+3}$

$\frac{2x+3}{x}=\frac{2x}{x+3}$

$(2x+3)(x+3)=2x\cdot x$

$2x^2+9x+9=2x^2$

$9x+9=0$

$9x=-9$

$x=-1$

$\{-1\}$

39. $\frac{36}{x^2-9}+1=\frac{2x}{x+3}$

$\frac{36+x^2-9}{(x+3)(x-3)}=\frac{2x(x-3)}{(x+3)(x-3)}$

$x^2+27=2x^2-6x$

$x^2-6x-27=0$

$(x-9)(x+3)=0$

$x-9=0$ or $x+3=0$

$x=9$ or $x=-3$

We discard –3 as extraneous.

$x=9$

$\{9\}$

41. $\frac{x^2-20}{x^2-7x+12}=\frac{3}{x-3}+\frac{5}{x-4}$

$\frac{x^2-20}{(x-3)(x-4)}=\frac{3(x-4)+5(x-3)}{(x-3)(x-4)}$

$x^2-20=3x-12+5x-15$

$x^2-20=8x-27$

$x^2-8x+7=0$

$(x-1)(x-7)=0$

$x-1=0$ or $x-7=0$

$x=1$ or $x=7$

$\{1, 7\}$

43. $\frac{3}{2x-5}+\frac{2}{2x+3}=0$

$\frac{3(2x+3)+2(2x-5)}{(2x-5)(2x+3)}=0$

$6x+9+4x-10=0$

$10x-1=0$

$10x=1$

$x=\frac{1}{10}$

$\left\{\frac{1}{10}\right\}$

45. $f(x)=20+\frac{4000}{x}$

$25=20+\frac{4000}{x}$

$25-20=\frac{4000}{x}$

$5=\frac{4000}{x}$

$5x=4000$

$x=\frac{4000}{5}$

$x=800$ sharpeners

47. $\frac{3}{x^2-25}=\frac{1}{x+5}+\frac{2}{x-5}$

$\frac{3}{(x+5)(x-5)}=\frac{x-5+2(x+5)}{(x+5)(x-5)}$

$3=x-5+2x+10$

$3=3x+5$

$3x=-2$

$x=-\frac{2}{3}$

$\left\{-\frac{2}{3}\right\}$

49. $\frac{5}{x^2-3x}\div\frac{4}{2x-6}$

$=\frac{5}{x(x-3)}\cdot\frac{2(x-3)}{4}$

$=\frac{5}{x}\cdot\frac{1}{2}=\frac{5}{2x}$

51. $\left(1-\frac{y}{x}\right)\div\left(1-\frac{x}{y}\right)$

$=\frac{x-y}{x}\div\frac{y-x}{y}$

$=\frac{x-y}{x}\cdot\frac{y}{-(x-y)}$

$=\frac{1}{x}\cdot\frac{y}{-1}=\frac{-y}{x}$

53. $\frac{2}{a-6}+\frac{3a}{a^2-5a-6}-\frac{a}{5a+5}$

$=\frac{2}{a-6}+\frac{3a}{(a-6)(a+1)}-\frac{a}{5(a+1)}$

$=\frac{2[5(a+1)]+3a(5)-a(a-6)}{5(a-6)(a+1)}$

$=\frac{10a+10+15a-a^2+6a}{5(a-6)(a+1)}$

$=\frac{-a^2+31a+10}{5(a-6)(a+1)}$

55. $\frac{5x-3}{2x}=\frac{10x+3}{4x+1}$

$(5x-3)(4x+1)=2x(10x+3)$

$20x^2-7x-3=20x^2+6x$

$-7x-3=6x$

$-3=13x$

$x=-\frac{3}{13}$

$\left\{-\frac{3}{13}\right\}$

57. $\frac{3}{4a-8}-\frac{a+2}{a^2-2a}$
$=\frac{3}{4(a-2)}-\frac{a+2}{a(a-2)}$
$=\frac{3a}{4a(a-2)}-\frac{4(a+2)}{4a(a-2)}$
$=\frac{3a-4a-8}{4a(a-2)}$
$=\frac{-a-8}{4a(a-2)}$

59. $\frac{x}{2x+6}+\frac{5}{x^2-9}$
$=\frac{x}{2(x+3)}+\frac{5}{(x+3)(x-3)}$
$=\frac{x(x-3)}{2(x+3)(x-3)}+\frac{2(5)}{2(x+3)(x-3)}$
$=\frac{x^2-3x+10}{2(x+3)(x-3)}$

61. $\frac{x-8}{x^2-x-2}+\frac{2}{x-2}=\frac{3}{x+1}$
$\frac{x-8}{(x-2)(x+1)}+\frac{2(x+1)}{(x-2)(x+1)}=\frac{3(x-2)}{(x+1)(x-2)}$
$x-8+2x+2=3x-6$
$3x-6=3x-6$
$-6=-6$
All real numbers except 2 and –1,
$\{x \mid x \text{ is a real number and } x \neq -1, x \neq 2\}$

63. $\frac{7}{3z-9}+\frac{5}{z}=\frac{7}{3(z-3)}+\frac{5}{z}$
$=\frac{7z+5[3(z-3)]}{3z(z-3)}$
$=\frac{7z+15z-45}{3z(z-3)}$
$=\frac{22z-45}{3z(z-3)}$

65. $x^{-2}-5x^{-1}-36=0$
$\frac{1}{x^2}-\frac{5}{x}-36=0$
$1-5x-36x^2=0$
$36x^2+5x-1=0$
$(9x-1)(4x+1)=0$
$9x-1=0$ or $4x+1=0$
$9x=1$ or $4x=-1$
$x=\frac{1}{9}$ or $x=-\frac{1}{4}$
$\left\{-\frac{1}{4}, \frac{1}{9}\right\}$

67. $6p^{-2}-5p^{-1}+1=0$
$\frac{6}{p^2}-\frac{5}{p}+1=0$
$6-5p+p^2=0$
$p^2-5p+6=0$
$(p-3)(p-2)=0$
$p-3=0$ or $p-2=0$
$p=3$ or $p=2$
$\{2, 3\}$

69. $\frac{-8.5}{x+1.9}=\frac{5.7}{x-3.6}$
$-8.5(x-3.6)=5.7(x+1.9)$
$-8.5x+30.6=5.7x+10.83$
$30.6-10.83=5.7x+8.5x$
$19.77=14.2x$
$1.39=x$
$\{1.39\}$

71. $\frac{12.2}{x}+17.3=\frac{9.6}{x}-14.7$
$\frac{12.2}{x}-\frac{9.6}{x}=-14.7-17.3$
$\frac{12.2-9.6}{x}=-32$
$2.6=-32x$
$-0.08=x$
$\{-0.08\}$

73. $\frac{2}{x}=\frac{10}{5}$
Verify $x=1$

75. $\frac{6x+7}{2x+9}=\frac{5}{3}$
Verify $x=3$

77. Let n = 1st integer
$n+1$ = 2nd integer.
$n+(n+1)=147$
$2n+1=147$

$2n = 146$
$n = 73$
The integers are 73 and 74.

79. Let x = the number.
$\frac{1}{x}$ = the reciprocal of the number.
$x + \frac{1}{x} = \frac{5}{2}$
$\frac{x^2 + 1}{x} = \frac{5}{2}$
$2x^2 + 2 = 5x$
$2x^2 - 5x + 2 = 0$
$(2x - 1)(x - 2) = 0$
$2x - 1 = 0$ or $x - 2 = 0$
$2x = 1$ or $x = 2$
$x = \frac{1}{2}$
The reciprocal pair is 2 and $\frac{1}{2}$.

81. $\sqrt{49} - (10 - 6)^2 = 7 - 4^2 = 7 - 16 = -9$

83. $(-4)^2 - 5^2 = 16 - 25 = -9$

85. 1%

87. 21% + 46% = 67%

89. $(x - 1)^2 + 3(x - 1) + 2 = 0$
Let $u = x - 1$
$u^2 + 3u + 2 = 0$
$(u + 1)(u + 2) = 0$
$u + 1 = 0$ or $u + 2 = 0$
$u = -1$ or $u = -2$
Substituting $x - 1$ for u,
$x - 1 = -1$ or $x - 1 = -2$
$x = 0$ or $x = -1$
$\{-1, 0\}$

91. $\left(\frac{3}{x-1}\right)^2 + 2\left(\frac{3}{x-1}\right) + 1 = 0$
Let $u = \frac{3}{x-1}$
$u^2 + 2u + 1 = 0$
$(u + 1)^2 = 0$
$u + 1 = 0$
$u = -1$
Substituting $\frac{3}{x-1}$ for u,
$\frac{3}{x-1} = -1$
$(x-1)\left(\frac{3}{x-1}\right) = -1(x-1)$
$3 = -x + 1$
$x = -2$
$\{-2\}$

Exercise Set 6.8

1. $F = \frac{9}{5}C + 32$
$F - 32 = \frac{9}{5}C$
$C = \frac{5}{9}(F - 32)$

3. $\frac{1}{R} = \frac{1}{R_1} + \frac{1}{R_2}$
$\frac{1}{R} = \frac{R_2 + R_1}{R_1 R_2}$
$R = \frac{R_1 R_2}{R_1 + R_2}$

5. $S = \frac{n(a + L)}{2}$
$2S = n(a + L)$
$n = \frac{2S}{a + L}$

7. $A = \frac{h(a + b)}{2}$
$2A = ah + bh$
$2A - ah = bh$
$\frac{2A - ah}{h} = b$

9. $\frac{P_1 V_1}{T_1} = \frac{P_2 V_2}{T_2}$
$(T_1 T_2)\left(\frac{P_1 V_1}{T_1}\right) = \left(\frac{P_2 V_2}{T_2}\right)(T_1 T_2)$
$P_1 V_1 T_2 = P_2 V_2 T_1$
$T_2 = \frac{P_2 V_2 T_1}{P_1 V_1}$

11. $f = \dfrac{f_1 f_2}{f_1 + f_2}$

$(f_1 + f_2)f = \left(\dfrac{f_1 f_2}{f_1 + f_2}\right)(f_1 + f_2)$

$f_1 f + f_2 f = f_1 f_2$

$f_1 f = f_1 f_2 - ff_2$

$f_1 f = f_2(f_1 - f)$

$\dfrac{f_1 f}{f_1 - f} = f_2$

13. $\lambda = \dfrac{2L}{n}$

$n\lambda = \left(\dfrac{2L}{n}\right)(n)$

$n\lambda = 2L$

$\dfrac{n\lambda}{2} = L$

15. $\dfrac{\theta}{w} = \dfrac{2L}{c}$

$cw\left(\dfrac{\theta}{w}\right) = \dfrac{2L}{c}(cw)$

$c\theta = 2Lw$

$c = \dfrac{2Lw}{\theta}$

17. Let n = the number.

$\dfrac{1}{n}$ = the reciprocal of the number.

$n + 5\left(\dfrac{1}{n}\right) = 6$

$n + \dfrac{5}{n} = 6$

$\dfrac{n^2 + 5}{n} = 6$

$n^2 + 5 = 6n$

$n^2 - 6n + 5 = 0$

$(n-1)(n-5) = 0$

$n - 1 = 0$ or $n - 5 = 0$

$n = 1$ or $n = 5$

The number is either 1 or 5.

19. Let x = the number.

$\dfrac{12 + x}{41 + 2x} = \dfrac{1}{3}$

$3(12 + x) = (41 + 2x) \cdot 1$

$36 + 3x = 41 + 2x$

$x = 5$

The number is 5.

21. $r = 2$ ohms and $r_1 = 3$ ohms.

$\dfrac{1}{r} = \dfrac{1}{r_1} + \dfrac{1}{r_2}$

$\dfrac{1}{2} = \dfrac{1}{3} + \dfrac{1}{r_2}$

$\dfrac{1}{2} = \dfrac{r_2 + 3}{3r_2}$

$3r_2 = 2(r_2 + 3)$

$3r_2 = 2r_2 + 6$

$r_2 = 6$

The other resistance is 6 ohms.

23. For three resistances r_1, r_2, and r_3 wired in a parallel circuit, the combined resistance r is given by

$\dfrac{1}{r} = \dfrac{1}{r_1} + \dfrac{1}{r_2} + \dfrac{1}{r_3}$

$\dfrac{1}{r} = \dfrac{1}{5} + \dfrac{1}{6} + \dfrac{1}{2}$

$\dfrac{1}{r} = \dfrac{6 + 5 + 15}{30}$

$\dfrac{1}{r} = \dfrac{26}{30} = \dfrac{13}{15}$

$r = \dfrac{15}{13}$

The combined resistance is $\dfrac{15}{13}$ ohms.

25. Let x = the number of hours needed to roof the house when two roofers are working together. Convert times to rates.

$\dfrac{1 \text{ roof}}{26 \text{ hrs}} = \dfrac{1}{26}\dfrac{\text{roof}}{\text{hr}}$

$\dfrac{1 \text{ roof}}{39 \text{ hrs}} = \dfrac{1}{39}\dfrac{\text{roof}}{\text{hr}}$

$\dfrac{1 \text{ roof}}{x \text{ hrs}} = \dfrac{1}{x}\dfrac{\text{roof}}{\text{hr}}$

Now, their combined rate can be expressed in two different but equivalent ways.

$\dfrac{1}{26} + \dfrac{1}{36} = \dfrac{1}{x}$

$\dfrac{3 + 2}{78} = \dfrac{1}{x}$

$\dfrac{5}{78} = \dfrac{1}{x}$ so $x = \dfrac{78}{5} = 15.6$

The roofers together can roof a house in 15.6 hrs.

27. Convert each time to a rate.

$\frac{1 \text{ task}}{20 \text{ min}} = \frac{1}{20} \frac{\text{task}}{\text{min}}$

$\frac{1 \text{ task}}{30 \text{ min}} = \frac{1}{30} \frac{\text{task}}{\text{min}}$

$\frac{1 \text{ task}}{60 \text{ min}} = \frac{1}{60} \frac{\text{task}}{\text{min}}$

Let x = amount of time required for all three computers to complete the task. So $\frac{1 \text{ task}}{x \text{ min}} = \frac{1}{x} \frac{\text{task}}{\text{min}}$. Summing we get,

$\frac{1}{20} + \frac{1}{30} + \frac{1}{60} = \frac{1}{x}$

$\frac{3+2+1}{60} = \frac{1}{x}$

$\frac{6}{60} = \frac{1}{x}$

$\frac{1}{10} = \frac{1}{x}$ so $x = 10$ min.

It takes 10 minutes for the three computers to complete the task.

29. Let r = the speed of the plane and $r - 150$ = the speed of the car. Recall that $d = rt$ or $t = \frac{d}{r}$. Then using that latter formula we get

$\frac{150}{r-150} = \frac{600}{r}$ or $\frac{1}{r-150} = \frac{4}{r}$

$r = 4(r - 150)$

$r = 4r - 600$

$600 = 3r$

$r = 200$

The speed of the plane is 200 mph.

31. Let r = the speed of the boat in still water. Recall that $d = rt$ or $t = \frac{d}{r}$. Using the latter equation we get $\frac{20}{r+5} = \frac{10}{r-5}$ where $r + 5$ is the rate of the boat traveling downstream and $r - 5$ is the rate of the boat traveling upstream. Now,

$\frac{2}{r+5} = \frac{1}{r-5}$

$2(r - 5) = r + 5$

$2r - 10 = r + 5$

$r = 15$

Thus, the speed of the boat in still water is 15 mph.

33. Let n = an integer and $n + 1$ = the next integer.

$\frac{1}{n} + \frac{1}{n+1} = -\frac{15}{56}$

$\frac{(n+1)+n}{n(n+1)} = -\frac{15}{56}$

$\frac{2n+1}{n^2+n} = -\frac{15}{56}$

$56(2n+1) = -15(n^2+n)$

$112n + 56 = -15n^2 - 15n$

$15n^2 + 127n + 56 = 0$

$(15n + 7)(n + 8) = 0$

$15n + 7 = 0$ or $n + 8 = 0$

$15n = -7$ or $n = -8$

$n = -\frac{7}{15}$

Discarding the noninteger solution, we find that –8 and –7 are the integers.

35. Convert times to rates.

$\frac{1 \text{ pond}}{45 \text{ min}} = \frac{1}{45} \frac{\text{pond}}{\text{min}}$

$\frac{1 \text{ pond}}{20 \text{ min}} = \frac{1}{20} \frac{\text{pond}}{\text{min}}$

Let x = the number of minutes required for the second hose to fill the pond. Then, $\frac{1 \text{ pond}}{x \text{ min}} = \frac{1}{x} \frac{\text{pond}}{\text{min}}$ Now,

$\frac{1}{20} = \frac{1}{45} + \frac{1}{x}$

$\frac{1}{20} = \frac{x+45}{45x}$

$45x = 20(x + 45)$

$45x = 20x + 900$

$25x = 900$

$x = \frac{900}{25} = 36$

Thus, the second hose alone can fill the pond in 36 minutes.

37. Let r = the rate of one train so $r + 15$ = the rate of the other train. Note that the trains are moving apart at $r + (r + 15) = 2r + 15$. Recall that $d = rt$ or $r = \frac{d}{t}$. Therefore,

$2r + 15 = \frac{630}{6}$

$2r + 15 = 105$

$2r = 90$

$r = 45$

Thus, one train travels at 45 mph and the other train at 60 mph.

39.

	distance	= rate	· time
First part	20	r	$\frac{20}{r}$
Cooldown	16	$r-2$	$\frac{16}{r-2}$

$\frac{20}{r}=\frac{16}{r-2}$
$20(r-2)=16r$
$20r-40=16r$
$20r-16r=40$
$4r=40$
$r=\frac{40}{4}=10$
The first part is at 10 mph, and the cooldown at 8 mph.

41.

	Hours to complete	Part of job completed in one hour
Machine #1	5	$\frac{1}{5}$
Machine #2	6	$\frac{1}{6}$
Machine #3	7.5	$\frac{1}{7.5}$
Together	t	$\frac{1}{t}$

$\frac{1}{5}+\frac{1}{6}+\frac{1}{7.5}=\frac{1}{t}$
$(225t)\left(\frac{1}{5}+\frac{1}{6}+\frac{1}{7.5}\right)=\frac{1}{t}(225t)$
$45t+37.5t+30t=225$
$112.5t=225$
$t=2$ hrs

43.

	distance	= rate	· time
with wind	465	$x+20$	$\frac{465}{x+20}$
against wind	345	$x-20$	$\frac{345}{x-20}$

$\frac{465}{x+20}=\frac{345}{x-20}$
$465(x-20)=345(x+20)$
$465x-9300=345x+6900$
$120x=16200$
$x=135$ mph

45.

	distance	= rate	· time
jogger 1	x	8	$\frac{x}{8}$
jogger 2	x	6	$\frac{x}{6}$

$\frac{x}{8}=\frac{x}{6}-\frac{1}{2}$
$24\left(\frac{x}{8}\right)=\left(\frac{x}{6}-\frac{1}{2}\right)(24)$
$3x=4x-12$
12 miles $=x$

47.

	Time	In one hour
Experienced	4	$\frac{1}{4}$
Apprentice	5	$\frac{1}{5}$
Together	x	$\frac{1}{x}$

$\frac{1}{4}+\frac{1}{5}=\frac{1}{x}$
$20x\left(\frac{1}{4}+\frac{1}{5}\right)=20x\left(\frac{1}{x}\right)$
$20x\left(\frac{1}{4}\right)+20x\left(\frac{1}{5}\right)=20$
$5x+4x=20$
$9x=20$
$x=\frac{20}{9}=2\frac{2}{9}$ hours

49.

	distance	= rate	· time
Car	240	$\frac{240}{x+1}$	$x+1$
Jet	1080	$\frac{1080}{x}$	x

$\frac{1080}{x}=6\left(\frac{240}{x+1}\right)$
$\frac{1080}{x}=\frac{1440}{x+1}$
$1080x+1080=1440x$
$1080=360x$
$\frac{360x}{360}=\frac{1080}{360}$
$x=3$
Jet = 3 hrs.
Car = 4 hrs.

51.

	Time	In one minute
Belt	2	$\frac{1}{2}$
Smaller	6	$\frac{1}{6}$
Together	x	$\frac{1}{x}$

$$\frac{1}{2}+\frac{1}{6}=\frac{1}{x}$$
$$6x\left(\frac{1}{2}+\frac{1}{6}\right)=6x\left(\frac{1}{x}\right)$$
$$6x\left(\frac{1}{2}\right)+6x\left(\frac{1}{6}\right)=6$$
$$3x+x=6$$
$$4x=6$$
$$x=\frac{6}{4}=\frac{3}{2}=1\frac{1}{2}\text{ minutes}$$

53. Let x = speed of car in still air

	distance = rate · time		
Into wind	10	$x-3$	$\frac{10}{x-3}$
With the wind	11	$x+3$	$\frac{11}{x+3}$

$$\frac{10}{x-3}=\frac{11}{x+3}$$
$$10\cdot(x+3)=11\cdot(x-3)$$
$$10x+30=11x-33$$
$$30+33=11x-10x$$
$$63\text{ mph}=x$$

55. $\frac{x}{5}=\frac{x+2}{3}$
$$3x=5(x+2)$$
$$3x=5x+10$$
$$-10=2x$$
$$x=-5$$
$$\{-5\}$$

57. $\frac{x-3}{2}=\frac{x-5}{6}$
$$6(x-3)=2(x-5)$$
$$6x-18=2x-10$$
$$4x=8$$
$$x=2$$
$$\{2\}$$

59. 25%

61. 25% + 59% = 84%

63. Answers may vary.

Exercise Set 6.9

1. $A=kB$

3. $X=\frac{k}{Z}$

5. $N=kP^2$

7. $T=\frac{k}{R}$

9. $P=kR$

11. $A=kB$
$$60=k(12)$$
$$k=5$$
$$A=5B$$
$$A=5(9)$$
$$A=45$$

13. $V=kT$
$$20=k(300)$$
$$k=\frac{20}{300}=\frac{1}{15}$$
$$V=\frac{1}{15}T$$
$$V=\frac{1}{15}(360)$$
$$V=24\text{ m}^3$$

15. $H=\frac{k}{J}$
$$4=\frac{k}{5}$$
$$k=20$$
$$H=\frac{20}{J}$$
$$H=\frac{20}{2}$$
$$H=10$$

17. $I=\frac{k}{R}$
$$40=\frac{k}{270}$$
$$k=10{,}800$$
$$I=\frac{10{,}800}{R}$$

$I = \frac{10,800}{150}$
$I = 72$ amps.

19. $x = kyz$

21. $r = kst^3$

23. $Q = kR$
$4 = k(20)$
$k = \frac{1}{5}$
$Q = \frac{1}{5}R = \frac{1}{5}(35) = 7$

25. $M = kP$
$8 = k(20)$
$k = \frac{8}{20} = \frac{2}{5}$
$M = \frac{2}{5}P = \frac{2}{5}(24) = \frac{48}{5}$

27. $B = \frac{k}{C}$
$12 = \frac{k}{3}$
$k = 36$
$B = \frac{36}{C} = \frac{36}{18} = 2$

29. $W = \frac{k}{X}$
$18 = \frac{k}{6}$
$k = 108$
$W = \frac{108}{X} = \frac{108}{40} = \frac{27}{10} = 2.7$

31. $W = kr^3$
$1.2 = k \cdot 2^3$
$k = \frac{1.2}{8} = 0.15$
$W = 0.15r^3 = 0.15(3)^3$
$W = 0.15(27) = 4.05$ lbs.

33. $C = \frac{k}{n}$
$1.20 = \frac{k}{4000}$
$k = 1.20(4000) = 4800$
$C = \frac{4800}{n} = \frac{4800}{6000} = \$0.80$

35. $C = kWt$
$60 = k(200)(2)$
$60 = 400k$
$k = \frac{60}{400} = \frac{3}{20} = 0.15$
$C = 0.15Wt = 0.15(240)(3) = 108$ cars

37. $F = kAv$
$20 = k(12)(10)$
$20 = 120k$
$k = \frac{1}{6}$
$F = \frac{1}{6}Av = \frac{1}{6}(8)12 = 16$ lbs.

39. $L = \frac{kd^4}{l^2}$
$16 = \frac{k(8^4)}{10^2}$
$16 = \frac{4096k}{100}$
$k = \frac{1600}{4096} = \frac{25}{64}$
$L = \frac{25d^4}{64l^2}$
$L = \frac{25(6^4)}{64(8^2)}$
$L = \frac{2025}{256}$ tons
$L = 7.91$ tons

41. $A_1 = kr^2$
Replace r by $3r$.
$A_2 = k(3r)^2$
$A_2 = k(9r^2)$
$A_2 = 9(kr^2)$
$A_2 = 9A_1$
Thus, the area is multiplied by 9.

43. $I_1 = \frac{k}{d^2}$
Replace d by $2d$.
$I_2 = \frac{k}{(2d)^2} = \frac{k}{4d^2} = \frac{1}{4}I_1$
Thus, the intensity is divided by 4.

45. $y_1 = kx$
$y_2 = k(2x)$
$y_2 = 2(kx)$
$y_2 = 2y_1$
It is multiplied by 2.

47.

x	$\frac{1}{4}$	$\frac{1}{2}$	1	2	4
$y = \frac{1}{x}$	4	2	1	$\frac{1}{2}$	$\frac{1}{4}$

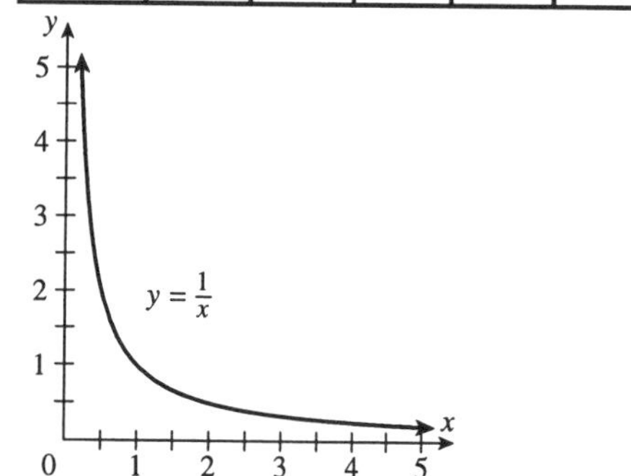

49.

x	$\frac{1}{4}$	$\frac{1}{2}$	1	2	4
$y = \frac{5}{x}$	20	10	5	$\frac{5}{2}$	$\frac{5}{4}$

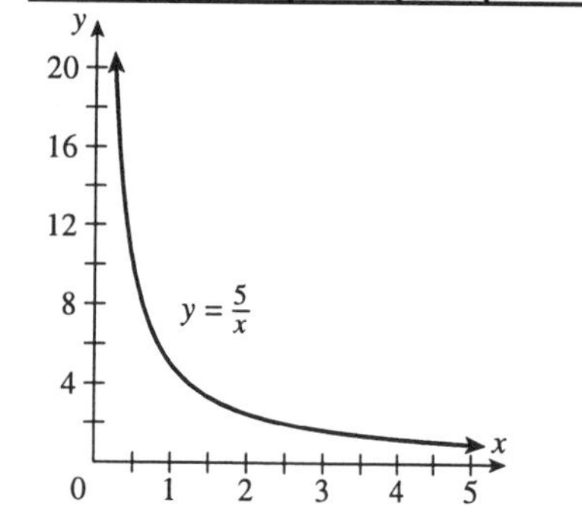

51. $r = 4$ in.
$C = 2\pi(4) = 8\pi$ in.
$A = \pi(4^2) = 16\pi \text{ in.}^2$

53. $r = 9$ cm.
$C = 2\pi(9) = 18\pi$ cm
$A = 2\pi(9^2) = 81\pi \text{ cm}^2$

55. $(-5, -2), (0, 7)$
$$m = \frac{7-(-2)}{0-(-5)} = \frac{7+2}{5} = \frac{9}{5}$$

57. $(2, 1), (2, -3)$
$$m = \frac{-3-1}{2-2} = \frac{-4}{0}$$
The slope is undefined.

59.

$f(x) = 2x - 3$

61.

$g(x) = |x|$

63.

$h(x) = x^2$

Chapter 6 - Review

1. $f(x)=\dfrac{3-5x}{7}$
Domain $\{x \mid x \text{ is a real number}\}$

2. $g(x)=\dfrac{2x+4}{11}$
Domain $\{x \mid x \text{ is a real number}\}$

3. $F(x)=\dfrac{-3x^2}{x-5}$
Undefined value when
$x-5=0$
$x=5$
Domain $\{x \mid x \text{ is a real number and } x \neq 5\}$

4. $h(x)=\dfrac{4x}{3x-12}$
Undefined values when
$3x-12=0$
$3x=12$
$x=4$
Domain $\{x \mid x \text{ is a real number and } x \neq 4\}$

5. $f(x)=\dfrac{x^3+2}{x^2+8x}$
Undefined values when
$x^2+8x=0$
$x(x+8)=0$
$x=0$ or $x+8=0$
$x=-8$
Domain $\{x \mid x \text{ is a real number and } x \neq 0,$
$x \neq -8\}$

6. $G(x)=\dfrac{20}{3x^2-48}$
Undefined value when
$3x^2-48=0$
$3(x^2-16)=0$
$3(x+4)(x-4)=0$
$x+4=0$ or $x-4=0$
$x=-4$ or $x=4$
Domain $\{x \mid x \text{ is a real number and } x \neq -4,$
$x \neq 4\}$

7. $\dfrac{15x^4}{45x^2}=\dfrac{15x^{4-2}}{15\cdot 3}=\dfrac{x^2}{3}$

8. $\dfrac{x+2}{2+x}=\dfrac{x+2}{x+2}=1$

9. $\dfrac{18m^6p^2}{10m^4p}=\dfrac{2\cdot 9m^{6-4}p^{2-1}}{2\cdot 5}=\dfrac{9m^2p}{5}$

10. $\dfrac{x-12}{12-x}=\dfrac{-1(12-x)}{12-x}=-1$

11. $\dfrac{5x-15}{25x-75}=\dfrac{5(x-3)}{5\cdot 5(x-3)}=\dfrac{1}{5}$

12. $\dfrac{22x+8}{11x+4}=\dfrac{2(11x+4)}{11x+4}=2$

13. $\dfrac{2x}{2x^2-2x}=\dfrac{2x}{2x(x-1)}=\dfrac{1}{x-1}$

14. $\dfrac{x+7}{x^2-49}=\dfrac{x+7}{(x+7)(x-7)}=\dfrac{1}{x-7}$

15. $\dfrac{2x^2+4x-30}{x^2+x-20}=\dfrac{2(x+5)(x-3)}{(x+5)(x-4)}$
$=\dfrac{2(x-3)}{x-4}$

16. $\dfrac{xy-3x+2y-6}{x^2+4x+4}=\dfrac{x(y-3)+2(y-3)}{(x+2)^2}$
$=\dfrac{(x+2)(y-3)}{(x+2)(x+2)}=\dfrac{y-3}{x+2}$

17. $C(x)=\dfrac{35x+4200}{x}$

(a) $C(50)=\dfrac{35(50)+4200}{50}$
$=\$119$

(b) $C(100)=\dfrac{35(100)+4200}{100}$
$=\$77$

(c) Decrease

18. $\dfrac{5}{x^3}\cdot\dfrac{x^2}{15}=\dfrac{1}{3x}$

19. $\frac{3x^4yz^3}{15x^2y^2}\cdot\frac{10xy}{z^6}=\frac{30x^{4+1}y^{1+1}z^3}{15x^2y^2z^6}$
$=\frac{2x^{5-2}}{z^{6-3}}=\frac{2x^3}{z^3}$

20. $\frac{4-x}{5}\cdot\frac{15}{2x-8}=\frac{4-x}{5}\cdot\frac{15}{2(x-4)}=\frac{-3}{2}$

21. $\frac{x^2-6x+9}{2x^2-18}\cdot\frac{4x+12}{5x-15}$
$=\frac{(x-3)^2}{2(x^2-9)}\cdot\frac{4(x+3)}{5(x-3)}$
$=\frac{x-3}{(x+3)(x-3)}\cdot\frac{2(x+3)}{5}$
$=\frac{1}{1}\cdot\frac{2}{5}=\frac{2}{5}$

22. $\frac{a-4b}{a^2+ab}\cdot\frac{b^2-a^2}{8b-2a}$
$=\frac{a-4b}{a(a+b)}\cdot\frac{(b-a)(b+a)}{2(4b-a)}$
$=\frac{-(b-a)}{2a}=\frac{a-b}{2a}$

23. $\frac{x^2-x-12}{2x^2-32}\cdot\frac{x^2+8x+16}{3x^2+21x+36}$
$=\frac{(x-4)(x+3)}{2(x^2-16)}\cdot\frac{(x+4)^2}{3(x^2+7x+12)}$
$=\frac{(x-4)(x+3)(x+4)^2}{6(x+4)(x-4)(x+3)(x+4)}$
$=\frac{1}{6}$

24. $\frac{2x^3+54}{5x^2+5x-30}\cdot\frac{6x+12}{3x^2-9x+27}$
$=\frac{2(x^3+27)}{5(x^2+x-6)}\cdot\frac{6(x+2)}{3(x^2-3x+9)}$
$=\frac{2(x+3)(x^2-3x+9)}{5(x-2)(x+3)}\cdot\frac{6(x+2)}{3(x^2-3x+9)}$
$=\frac{4(x+2)}{5(x-2)}$

25. $\frac{3}{4x}\div\frac{8}{2x^2}=\frac{3}{4x}\cdot\frac{2x^2}{8}=\frac{6x^2}{32x}=\frac{3x}{16}$

26. $\frac{4x+8y}{3}\div\frac{5x+10y}{9}$
$=\frac{4(x+2y)}{3}\cdot\frac{9}{5(x+2y)}=\frac{12}{5}$

27. $\frac{5ab}{14c^3}\div\frac{10a^4b^2}{6ac^5}$
$=\frac{5ab}{14c^3}\cdot\frac{6ac^5}{10a^4b^2}$
$=\frac{30a^2bc^5}{140a^4b^2c^3}=\frac{3c^{5-3}}{14a^{4-2}b^{2-1}}$
$=\frac{3c^2}{14a^2b}$

28. $\frac{2}{5x}\div\frac{4-18x}{6-27x}$
$=\frac{2}{5x}\cdot\frac{3(2-9x)}{2(2-9x)}=\frac{3}{5x}$

29. $\frac{x^2-25}{3}\div\frac{x^2-10x+25}{x^2-x-20}$
$=\frac{(x+5)(x-5)}{3}\cdot\frac{(x-5)(x+4)}{(x-5)^2}$
$=\frac{(x+5)(x+4)}{3}$

30. $\frac{a-4b}{a^2+ab}\div\frac{20b-5a}{b^2-a^2}$
$=\frac{a-4b}{a(a+b)}\cdot\frac{(b-a)(b+a)}{5(4b-a)}$
$=\frac{-(b-a)}{5a}=\frac{a-b}{5a}$

31. $\frac{7x+28}{2x+4}\div\frac{x^2+2x-8}{x^2-2x-8}$
$=\frac{7(x+4)}{2(x+2)}\cdot\frac{(x-4)(x+2)}{(x+4)(x-2)}$
$=\frac{7(x-4)}{2(x-2)}$

32. $\frac{3x+3}{x-1}\div\frac{x^2-6x-7}{x^2-1}$
$=\frac{3(x+1)}{x-1}\cdot\frac{(x-1)(x+1)}{(x+1)(x-7)}$
$=\frac{3(x+1)}{x-7}$

33. $\frac{2x-x^2}{x^3-8} \div \frac{x^2}{x^2+2x+4}$
$= \frac{x(2-x)}{(x-2)(x^2+2x+4)} \cdot \frac{x^2+2x+4}{x^2}$
$= \frac{-(x-2)}{x-2} \cdot \frac{1}{x} = -1 \cdot \frac{1}{x} = -\frac{1}{x}$

34. $\frac{5a^2-20}{a^3+2a^2+a+2} \div \frac{7a}{a^3+a}$
$= \frac{5(a+2)(a-2)}{(a^2+1)(a+2)} \cdot \frac{a(a^2+1)}{7a}$
$= \frac{5(a-2)}{7}$

35. $\frac{2a}{21} \div \frac{3a^2}{7} \cdot \frac{4}{a} = \frac{2a}{21} \cdot \frac{7}{3a^2} \cdot \frac{4}{a}$
$= \frac{56a}{63a^3} = \frac{8}{9a^2}$

36. $\frac{5x-15}{3-x} \cdot \frac{x+2}{10x+20} \cdot \frac{x^2-9}{x^2-x-6}$
$= \frac{5(x-3)(x+2)(x-3)(x+3)}{(3-x)10(x+2)(x+2)(x-3)}$
$= \frac{-(x+3)}{2(x+2)}$

37. $\frac{4a+8}{5a^2-20} \cdot \frac{3a^2-6a}{a+3} \div \frac{2a^2}{5a+15}$
$= \frac{4(a+2)}{5(a^2-4)} \cdot \frac{3a(a-2)}{a+3} \cdot \frac{5(a+3)}{2a^2}$
$= \frac{60a(a+2)(a-2)(a+3)}{10a^2(a+2)(a-2)(a+3)}$
$= \frac{6}{a}$

38. $(f+g)(x) = (x-5) + (2x+1) = 3x-4$

39. $(f-g)(x) = (x-5) - (2x+1)$
$= x-5-2x-1$
$= -x-6$

40. $(f \cdot g)(x) = (x-5)(2x+1)$
$= 2x^2-9x-5$

41. $\left(\frac{g}{f}\right)(x) = \frac{2x+1}{x-5}$

42. The LCD is (9)(2) = 18.

43. $\frac{5}{4x^2y^5}, \frac{3}{10x^2y^4}, \frac{x}{6y^4}$
The LCD is $60x^2y^5$.

44. The LCD is $2x(x-2)$.

45. $\frac{3}{5x}, \frac{2}{x-5}$
The LCD is $5x(x-5)$.

46. The first denominator is $5x^3$.
The second denominator is $(x-4)(x+7)$.
The third denominator is $10x(x-3)$.
The LCD is $10x^3(x-4)(x+7)(x-3)$.

47. $\frac{2}{15} + \frac{4}{15} = \frac{2+4}{15} = \frac{6}{15} = \frac{2}{5}$

48. $\frac{4}{x-4} + \frac{x}{x-4} = \frac{4+x}{x-4}$

49. $\frac{4}{3x^2} + \frac{2}{3x^2} = \frac{4+2}{3x^2} = \frac{6}{3x^2} = \frac{2}{x^2}$

50. $\frac{1}{x-2} - \frac{1}{2(2-x)}$
$= \frac{1(2)}{(x-2)(2)} + \frac{1}{2(x-2)}$
$= \frac{2+1}{2(x-2)} = \frac{3}{2(x-2)}$

51. $\frac{2x+1}{x^2+x-6} + \frac{2-x}{x^2+x-6}$
$= \frac{2x+1+2-x}{x^2+x-6}$
$= \frac{x+3}{(x+3)(x-2)}$
$= \frac{1}{x-2}$

52. $\frac{7}{2x} + \frac{5}{6x} = \frac{(7)\cdot(3)}{(2x)\cdot(3)} + \frac{5}{6x}$
$= \frac{21}{6x} + \frac{5}{6x} = \frac{26}{6x} = \frac{13}{3x}$

53. $\frac{1}{3x^2y^3}-\frac{1}{5x^4y}=\frac{5x^2}{15x^4y^3}-\frac{3y^2}{15x^4y^3}$
$=\frac{5x^2-3y^2}{15x^4y^3}$

54. $\frac{1}{10-x}+\frac{x-1}{x-10}$
$=\frac{-1}{x-10}+\frac{x-1}{x-10}=\frac{x-2}{x-10}$

55. $\frac{x-2}{x+1}-\frac{x-3}{x-1}$
$=\frac{(x-2)(x-1)-(x-3)(x+1)}{(x+1)(x-1)}$
$=\frac{x^2-3x+2-(x^2-2x-3)}{(x+1)(x-1)}$
$=\frac{x^2-3x+2-x^2+2x+3}{(x+1)(x-1)}$
$=\frac{-x+5}{x^2-1}$

56. $\frac{x}{9-x^2}-\frac{2}{5x-15}$
$=\frac{x}{(3-x)(3+x)}-\frac{2}{5(x-3)}$
$=\frac{-5x}{5(x-3)(x+3)}-\frac{2(x+3)}{5(x-3)(x+3)}$
$=\frac{-5x-2x-6}{5(x-3)(x+3)}$
$=\frac{-7x-6}{5(x-3)(x+3)}$

57. $2x+1-\frac{1}{x-3}$
$=\frac{(2x+1)(x-3)-1}{x-3}$
$=\frac{2x^2-5x-3-1}{x-3}$
$=\frac{2x^2-5x-4}{x-3}$

58. $\frac{2}{a^2-2a+1}+\frac{3}{a^2-1}$
$=\frac{2}{(a-1)^2}+\frac{3}{(a+1)(a-1)}$
$=\frac{2(a+1)}{(a-1)^2(a+1)}+\frac{3(a-1)}{(a-1)^2(a+1)}$
$=\frac{2a+2+3a-3}{(a-1)^2(a+1)}$
$=\frac{5a-1}{(a-1)^2(a+1)}$

59. $\frac{x}{9x^2+12x+16}-\frac{3x+4}{27x^3-64}$
$=\frac{x}{9x^2+12x+16}$
$-\frac{3x+4}{(3x-4)(9x^2+12x+16)}$
$=\frac{(3x-4)x-(3x+4)}{(3x-4)(9x^2+12x+16)}$
$=\frac{3x^2-4x-3x-4}{(3x-4)(9x^2+12x+16)}$
$=\frac{3x^2-7x-4}{27x^3-64}$

60. $\frac{2}{x-1}-\frac{3x}{3x-3}+\frac{1}{2x-2}$
$=\frac{2}{x-1}-\frac{3x}{3(x-1)}+\frac{1}{2(x-1)}$
$=\frac{4}{2(x-1)}-\frac{2x}{2(x-1)}+\frac{1}{2(x-1)}$
$=\frac{5-2x}{2(x-1)}$

61. $\frac{3}{2x}\left(\frac{2}{x+1}-\frac{2}{x-3}\right)$
$=\frac{3}{x}\left(\frac{1}{x+1}-\frac{1}{x-3}\right)$
$=\frac{3}{x}\left(\frac{x-3-(x+1)}{(x+1)(x-3)}\right)$
$=\frac{3}{x}\cdot\frac{x-3-x-1}{(x+1)(x-3)}$
$=\frac{3(-4)}{x(x+1)(x-3)}$
$=\frac{-12}{x(x+1)(x-3)}$

62. $\left(\frac{2}{x}-\frac{1}{5}\right)\cdot\left(\frac{2}{x}+\frac{1}{3}\right)=\left(\frac{10-x}{5x}\right)\cdot\left(\frac{6+x}{3x}\right)$
$=\frac{(10-x)(6+x)}{(5x)(3x)}=\frac{60+4x-x^2}{15x^2}$

63. $\frac{2}{x^2-16}-\frac{3x}{x^2+8x+16}+\frac{3}{x+4}$

$=\frac{2}{(x+4)(x-4)}-\frac{3x}{(x+4)^2}+\frac{3}{x+4}$

$=\frac{2(x+4)}{(x+4)^2(x-4)}-\frac{3x(x-4)}{(x+4)^2(x-4)}+\frac{3(x^2-16)}{(x+4)^2(x-4)}$

$=\frac{2x+8-3x^2+12x+3x^2-48}{(x+4)^2(x-4)}$

$=\frac{14x-40}{(x+4)^2(x-4)}$

64. $P=\frac{1}{x}+\frac{1}{x}+\frac{1}{x}+\frac{2}{x}+\frac{5}{2x}+\frac{2}{x}+\frac{3}{2x}$

$=\frac{1+1+1+2+2}{x}+\frac{5+3}{2x}$

$=\frac{7}{x}+\frac{8}{2x}$

$=\frac{14}{2x}+\frac{8}{2x}$

$=\frac{22}{2x}=\frac{11}{x}$

65. $\frac{\frac{2}{5}}{\frac{3}{5}}=\frac{2}{5}\cdot\frac{5}{3}=\frac{2}{1}\cdot\frac{1}{3}=\frac{2}{3}$

66. $\frac{\left(1-\frac{3}{4}\right)(4)}{\left(2+\frac{1}{4}\right)(4)}=\frac{4-3}{8+1}=\frac{1}{9}$

67. $\frac{\frac{1}{x}-\frac{2}{3x}}{\frac{5}{2x}-\frac{1}{3}}=\frac{\frac{3x-2x}{3x^2}}{\frac{15-2x}{6x}}$

$=\frac{x}{3x^2}\cdot\frac{6x}{15-2x}$

$=\frac{6x^2}{3x^2(15-2x)}$

$=\frac{2}{15-2x}$

68. $\frac{\frac{x^2}{15}}{\frac{x+1}{5x}}=\frac{x^2}{15}\div\frac{x+1}{5x}$

$=\frac{x^2}{15}\cdot\frac{5x}{x+1}=\frac{x^3}{3(x+1)}$

69. $\frac{\frac{3}{y^2}}{\frac{6}{y^3}}=\frac{3}{y^2}\cdot\frac{y^3}{6}=\frac{y}{2}$

70. $\frac{\frac{x+2}{3}}{\frac{5}{x-2}}=\frac{x+2}{3}\div\frac{5}{x+2}$

$=\frac{x+2}{3}\cdot\frac{x-2}{5}=\frac{(x+2)(x-2)}{15}$

71. $\frac{2-\frac{3}{2x}}{x-\frac{2}{5x}}$

$=\frac{\frac{4x-3}{2x}}{\frac{5x^2-2}{5x}}$

$=\frac{4x-3}{2x}\cdot\frac{5x}{5x^2-2}$

$=\frac{5(4x-3)}{2(5x^2-2)}=\frac{20x-15}{2(5x^2-2)}$

72. $\frac{\left(1+\frac{x}{y}\right)\cdot y^2}{\left(\frac{x^2}{y^2}-1\right)\cdot y^2}=\frac{y^2+xy}{x^2-y^2}$

$=\frac{y(y+x)}{(x+y)(x-y)}=\frac{y}{x-y}$

73. $\frac{\frac{5}{x}+\frac{1}{xy}}{\frac{3}{x^2}}=\frac{5y+1}{xy}\cdot\frac{x^2}{3}$

$=\frac{x(5y+1)}{3y}$

$=\frac{5xy+x}{3y}$

74. $\frac{\left(\frac{x}{3}-\frac{3}{x}\right)\cdot(3x)}{\left(1+\frac{3}{x}\right)\cdot(3x)}=\frac{x^2-9}{3x+9}$

$=\frac{(x+3)(x-3)}{3(x+3)}=\frac{x-3}{3}$

75. $\dfrac{\frac{1}{x-1}+1}{\frac{1}{x+1}-1}=\dfrac{\frac{1+x-1}{x-1}}{\frac{1-(x+1)}{x+1}}$
$=\dfrac{x}{x-1}\cdot\dfrac{x+1}{1-x-1}$
$=\dfrac{x(x+1)}{-x(x-1)}=\dfrac{x+1}{-(x-1)}$
$=\dfrac{1+x}{1-x}$

76. $\dfrac{(2)x}{\left(1-\frac{2}{x}\right)x}=\dfrac{2x}{x-2}$

77. $\dfrac{1}{1+\frac{2}{1-\frac{1}{x}}}=\dfrac{1}{1+\frac{2x}{x-1}}$
$=\dfrac{1}{\frac{x-1+2x}{x-1}}=\dfrac{x-1}{3x-1}$

78. $\dfrac{\frac{x^2+5x-6}{4x+3}}{\frac{(x+6)^2}{8x+6}}$
$=\dfrac{x^2+5x-6}{4x+3}\div\dfrac{(x+6)^2}{8x+6}$
$=\dfrac{(x-1)(x+6)}{4x+3}\cdot\dfrac{2(4x+3)}{(x+6)^2}$
$=\dfrac{2(x-1)}{x+6}$

79. $\dfrac{\frac{x-3}{x+3}+\frac{x+3}{x-3}}{\frac{x-3}{x+3}-\frac{x+3}{x-3}}$
$=\dfrac{\frac{(x-3)^2+(x+3)^2}{(x+3)(x-3)}}{\frac{(x-3)^2-(x+3)^2}{(x+3)(x-3)}}$
$=\dfrac{x^2-6x+9+x^2+6x+9}{x^2-6x+9-(x^2+6x+9)}$
$=\dfrac{2x^2+18}{x^2-6x+9-x^2-6x-9}$
$=\dfrac{2x^2+18}{-12x}=-\dfrac{x^2+9}{6x}$

80. $\dfrac{\frac{3}{x-1}-\frac{2}{1-x}}{\frac{2}{x-1}-\frac{2}{x}}$
$=\dfrac{\frac{3}{x-1}+\frac{2}{x-1}}{\frac{2x}{x(x-1)}-\frac{2(x-1)}{x(x-1)}}$
$=\dfrac{\frac{5}{x-1}}{\frac{2}{x(x-1)}}$
$=\dfrac{5}{x-1}\div\dfrac{2}{x(x-1)}$
$=\dfrac{5}{x-1}\cdot\dfrac{x(x-1)}{2}=\dfrac{5x}{2}$

81. $f(x)=\dfrac{3}{x}$

a. $f(a+h)=\dfrac{3}{a+h}$

b. $f(a)=\dfrac{3}{a}$

c. $\dfrac{f(a+h)-f(a)}{h}=\dfrac{\frac{3}{a+h}-\frac{3}{a}}{h}$

d. $\dfrac{\frac{3}{a+h}-\frac{3}{a}}{h}=\dfrac{\frac{3a}{a(a+h)}-\frac{3(a+h)}{a(a+h)}}{h}$
$=\dfrac{3a-3(a+h)}{a(a+h)}\cdot\dfrac{1}{h}$
$=\dfrac{3[a-(a+h)]}{ah(a+h)}$
$=\dfrac{3(-h)}{ah(a+h)}=\dfrac{-3}{a(a+h)}$

82. $\dfrac{3x^5yb^9}{9xy^7}=\dfrac{x^{5-1}b^9}{3y^{7-1}}=\dfrac{x^4b^9}{3y^6}$

83. $\dfrac{-9xb^4z^3}{-4axb^2}=\dfrac{(-1)(9)x^{1-1}b^{4-2}z^3}{(-1)(4)a}=\dfrac{9b^2z^3}{4a}$

84. $\dfrac{4xy+2x^2-9}{4xy}$

$=\dfrac{4xy}{4xy}+\dfrac{2x^2}{4xy}-\dfrac{9}{4xy}$

$=1+\dfrac{x^{2-1}}{2y}-\dfrac{9}{4xy}$

$=1+\dfrac{x}{2y}-\dfrac{9}{4xy}$

85. $\dfrac{12xb^2+16xb^4}{4xb^3}=\dfrac{12xb^2}{4xb^3}+\dfrac{16xb^4}{4xb^3}$

$=\dfrac{3\cdot 4x^{1-1}}{4b^{3-2}}+\dfrac{4\cdot 4x^{1-1}b^{4-3}}{4}=\dfrac{3}{b}+4b$

86.

$$\begin{array}{r}
3x^3+9x^2+2x+6 \\
x-3\overline{\smash{)}\,3x^4 \qquad -25x^2 \qquad -20} \\
\underline{3x^4-9x^3} \qquad\qquad\qquad \\
9x^3+25x^2 \qquad\qquad \\
\underline{9x^3-27x^2} \qquad\qquad \\
2x^2 \qquad -20 \\
\underline{2x^2-6x} \qquad\quad \\
6x-20 \\
\underline{6x-18} \\
-2
\end{array}$$

Answer: $(3x^3+9x^2+2x+6)-\dfrac{2}{x-3}$

87.

$$\begin{array}{r}
2x^3+6x^2+17x+56 \\
x-3\overline{\smash{)}\,2x^4+0x^3\quad -x^2\quad +5x-12} \\
\underline{2x^4-6x^3} \qquad\qquad\qquad\qquad \\
6x^3\quad -x^2 \qquad\qquad\quad \\
\underline{6x^3-18x^2} \qquad\qquad\quad \\
17x^2\quad +5x \qquad \\
\underline{17x^2-51x} \qquad \\
56x-12 \\
\underline{56x-168} \\
156
\end{array}$$

Answer: $2x^3+6x^2+17x+56+\dfrac{156}{x-3}$

88.

$$\begin{array}{r}
2x^3+2x-2 \\
x-\tfrac{1}{2}\overline{\smash{)}\,2x^4-x^3+2x^2-3x+1} \\
\underline{2x^4-x^3} \qquad\qquad\qquad\quad \\
2x^2-3x \qquad \\
\underline{2x^2-\ \ x} \qquad \\
-2x+1 \\
\underline{-2x+1} \\
0
\end{array}$$

Answer: $2x^3+2x-2$

89.

$$\begin{array}{r}
x^2+\tfrac{7}{2}x-\tfrac{1}{4} \\
x-\tfrac{1}{2}\overline{\smash{)}\,x^3+3x^2-2x+2} \\
\underline{x^3-\tfrac{1}{2}x^2} \qquad\qquad\quad \\
\tfrac{7}{2}x^2-2x \qquad \\
\underline{\tfrac{7}{2}x^2-\tfrac{7}{4}x} \qquad \\
-\tfrac{1}{4}x+2 \\
\underline{-\tfrac{1}{4}x+\tfrac{1}{8}} \\
\tfrac{15}{8}
\end{array}$$

Answer: $x^2+\dfrac{7}{2}x-\dfrac{1}{4}+\dfrac{15}{8\left(x-\frac{1}{2}\right)}$

90.

$$\begin{array}{r}
3x^2+2x-1 \\
x^2+x+2\overline{\smash{)}\,3x^4-5x^3+7x^2+3x-2} \\
\underline{3x^4+3x^3+6x^2} \qquad\qquad\quad \\
2x^3+\ \ x^2+3x \qquad \\
\underline{2x^3+2x^2+4x} \qquad \\
-x^2\quad -x-2 \\
\underline{-x^2\quad -x-2} \\
0
\end{array}$$

Answer: $3x^2+2x-1$

91.

$$\begin{array}{r}
3x^2+6 \\
3x^2-2x-5\overline{\smash{)}\,9x^4-6x^3+\ 3x^2-12x-30} \\
\underline{9x^4-6x^3-15x^2} \qquad\qquad\quad \\
18x^2-12x-30 \\
\underline{18x^2-12x-30} \\
0
\end{array}$$

Answer: $3x^2+6$

92.

$$\begin{array}{r|rrrr} 2 & 3 & 0 & 12 & -4 \\ & & 6 & 12 & 48 \\ \hline & 3 & 6 & 24 & 44 \end{array}$$

Answer: $(3x^2+6x+24)+\frac{44}{x-2}$

93.

$$\begin{array}{r} 3x^2-\frac{5}{2}x-\frac{1}{4} \\ x+\frac{3}{2}\overline{\big)\;3x^3+2x^2-\;\;4x-1} \\ -\left(3x^3+\frac{9}{2}x^2\right) \\ \hline -\frac{5}{2}x^2-\;\;4x \\ -\left(\frac{-5}{2}x^2-\frac{15}{4}x\right) \\ \hline -\frac{1}{4}x-1 \\ -\left(-\frac{1}{4}x-\frac{3}{8}\right) \\ \hline \frac{-5}{8} \end{array}$$

94.

$$\begin{array}{r|rrrrrr} -1 & 1 & 0 & 0 & 0 & 0 & -1 \\ & & -1 & 1 & -1 & 1 & -1 \\ \hline & 1 & -1 & 1 & -1 & 1 & -2 \end{array}$$

Answer: $(x^4-x^3+x^2-x+1)-\frac{2}{x+1}$

95.

$$\begin{array}{r} x^2+3x\;+9 \\ x-3\overline{\big)\;x^3\qquad\qquad\quad -81} \\ -(x^3-3x^2) \\ \hline 3x^2 \\ -(3x^2-9x) \\ \hline 9x-81 \\ -(9x-27) \\ \hline -54 \end{array}$$

96.

$$\begin{array}{r|rrrrr} 4 & 3 & 1 & -1 & 0 & -2 \\ & & 12 & 52 & 204 & 816 \\ \hline & 3 & 13 & 51 & 204 & 814 \end{array}$$

Answer: $(3x^3+13x^2+51x+204)+\frac{814}{x-4}$

97.

$$\begin{array}{r} 3x^3\;-6x^2+10x-20 \\ x+2\overline{\big)\;3x^4\qquad\quad -2x^2\qquad\quad +10} \\ -(3x^4+6x^3) \\ \hline -6x^3-\;\;2x^2 \\ -(-6x^3-12x^2) \\ \hline 10x^2 \\ -(10x^2+20x) \\ \hline -20x+10 \\ -(-20x-40) \\ \hline 50 \end{array}$$

98.

$$\begin{array}{r|rrrrrr} 4 & 3 & 0 & 0 & 0 & -9 & 7 \\ & & 12 & 48 & 192 & 768 & 3036 \\ \hline & 3 & 12 & 48 & 192 & 759 & 3043 \end{array}$$

Thus, $P(4)=3043$

99. $P(x)=3x^5-9x+7$

$P(-5)=3(-5)^5-9(-5)+7$

$P(-5)=3(-3125)+45+7=-9323$

100.

$$\begin{array}{r|rrrrrr} \frac{2}{3} & 3 & 0 & 0 & 0 & -9 & 7 \\ & & 2 & \frac{4}{3} & \frac{8}{9} & \frac{16}{27} & -\frac{454}{81} \\ \hline & 3 & 2 & \frac{4}{3} & \frac{8}{9} & -\frac{227}{27} & \frac{113}{81} \end{array}$$

101. $P(x)=3x^5-9x+7$

$P\left(-\frac{1}{2}\right)=3\left(-\frac{1}{2}\right)^5-9\left(-\frac{1}{2}\right)+7$

$P\left(-\frac{1}{2}\right)=-\frac{3}{32}+\frac{9}{2}+7$

$P\left(-\frac{1}{2}\right)=\frac{-3+144+244}{32}=\frac{365}{32}$

102.
$$\begin{array}{r} x^3+2x^2-6 \\ x-3\overline{)\,x^4-x^3-6x^2-6x+18} \\ \underline{x^4-3x^3} \\ 2x^3-6x^2 \\ \underline{2x^3-6x^2} \\ 0-6x+18 \\ \underline{-6x+18} \\ 0 \end{array}$$

(x^3+2x^2-6) miles

103. $\frac{2}{5}=\frac{x}{15}$
$30=5x$
$x=6$
$\{6\}$

104. $\frac{3}{x}+\frac{1}{3}=\frac{5}{x}$
$3x\left(\frac{3}{x}+\frac{1}{3}\right)=\left(\frac{5}{x}\right)(3x)$
$9+x=15$
$x=6$

105. $4+\frac{8}{x}=8$
$\frac{8}{x}=4$
$\frac{8}{4}=x$
$x=2$
$\{2\}$

106. $\frac{2x+3}{5x-9}=\frac{3}{2}$
$2(5x-9)\left(\frac{2x+3}{5x-9}\right)=\left(\frac{3}{2}\right)(2)(5x-9)$
$4x+6=15x-27$
$33=11x$
$3=x$

107. $\frac{1}{x-2}-\frac{3x}{x^2-4}=\frac{2}{x+2}$
$\frac{x+2-3x}{(x-2)(x+2)}=\frac{2(x-2)}{(x+2)(x-2)}$
$-2x+2=2x-4$
$6=4x$
$x=\frac{6}{4}=\frac{3}{2}$
$\left\{\frac{3}{2}\right\}$

108. $\frac{7}{x}-\frac{x}{7}=0$
$7x\left(\frac{7}{x}-\frac{x}{7}\right)=0(7x)$
$49-x^2=0$
$(7-x)(7+x)=0$
$7-x=0$ or $7+x=0$
$7=x$ or $x=-7$

109. $\frac{x-2}{x^2-7x+10}=\frac{1}{5x-10}-\frac{1}{x-5}$
$\frac{x-2}{(x-2)(x-5)}=\frac{1}{5(x-2)}-\frac{1}{x-5}$
$5(x-2)=x-5-5(x-2)$
$5x-10=x-5-5x+10$
$5x-10=-4x+5$
$9x=15$
$x=\frac{15}{9}=\frac{5}{3}$
$\left\{\frac{5}{3}\right\}$

110. $\frac{5}{x^2-7x}+\frac{4}{2x-14}$
$=\frac{5}{x(x-7)}+\frac{4}{2(x-7)}$
$=\frac{10}{2x(x-7)}+\frac{4x}{2x(x-7)}$
$=\frac{4x+10}{2x(x-7)}=\frac{2(2x+5)}{2x(x-7)}$
$=\frac{2x+5}{x(x-7)}$

111. $3-\frac{5}{x}-\frac{2}{x^2}=0$
$3x^2-5x-2=0$
$(3x+1)(x-2)=0$
$3x+1=0$ or $x-2=0$
$3x=-1$ or $x=2$
$x=-\frac{1}{3}$
$\left\{-\frac{1}{3},\ 2\right\}$

112. $\frac{4}{3-x}-\frac{7}{2x-6}+\frac{5}{x}$
$=\frac{-4}{x-3}-\frac{7}{2(x-3)}+\frac{5}{x}$
$=\frac{-8x}{2x(x-3)}-\frac{7x}{2x(x-3)}+\frac{10(x-3)}{2x(x-3)}$
$=\frac{-5x-30}{2x(x-3)}$

113. $A=\frac{h(a+b)}{2}$
$\frac{2A}{h}=a+b$
$a=\frac{2A}{h}-b$

114. $\frac{1}{R}=\frac{1}{R_1}+\frac{1}{R_2}$
$(RR_1R_2)\frac{1}{R}=\left(\frac{1}{R_1}+\frac{1}{R_2}\right)(RR_1R_2)$
$R_1R_2=RR_2+RR_1$
$R_1R_2-RR_2=RR_1$
$R_2=\frac{RR_1}{R_1-R}$

115. $I=\frac{E}{R+r}$
$R+r=\frac{E}{I}$
$R=\frac{E}{I}-r$

116. $A=P+Prt$
$A-P=Prt$
$\frac{A-P}{Pt}=r$

117. $H=\frac{kA(T_1-T_2)}{L}$
$LH=kA(T_1-T_2)$
$A=\frac{LH}{k(T_1-T_2)}$

118. $x+2\left(\frac{1}{x}\right)=3$
$x\left(x+\frac{2}{x}\right)=3(x)$
$x^2+2=3x$
$x^2-3x+2=0$
$(x-1)(x-2)=0$
$x-1=0$ or $x-2=0$
$x=1$ or $x=2$

119. Let x = the number added to the numerator, so $2x$ = the number added to the denominator. Then
$\frac{3+x}{7+2x}=\frac{10}{21}$
$21(3+x)=10(7+2x)$
$63+21x=70+20x$
$x=7$
7 is the number to be added to the numerator.

120. Let x = the numerator of the fraction, so $x+2$ = the denominator of the fraction. Then
$\frac{x-3}{(x+2)+5}=\frac{2}{3}$
$\frac{x-3}{x+7}=\frac{2}{3}$
$3(x-3)=2(x+7)$
$3x-9=2x+14$
$x=23$ and $x+2=25$
The original fraction was $\frac{23}{25}$.

121. Let n and $n+2$ represent two consecutive even integers. Then
$\frac{1}{n}+\frac{1}{n+2}=-\frac{9}{40}$
$\frac{n+2+n}{n(n+2)}=-\frac{9}{40}$
$\frac{2n+2}{n^2+2n}=-\frac{9}{40}$
$40(2n+2)=-9(n^2+2n)$
$80n+80=-9n^2-18n$
$9n^2+98+80=0$
$(9n+8)(n+10)=0$
$9n+8=0$ or $n+10=0$
$9n=-8$ or $n=-10$
$n=-\frac{8}{9}$
Discarding the extraneous solution $-\frac{8}{9}$, we find the two consecutive even integers to be −10 and -8.

122. $\frac{1}{4}+\frac{1}{5}+\frac{1}{6}=\frac{1}{t}$

$120t\left[\frac{1}{4}+\frac{1}{5}+\frac{1}{6}\right]=\left(\frac{1}{t}\right)(120t)$

$30t+24t+20t=120$

$74t=120$

$t=\frac{120}{74}=1\frac{23}{37}$ hrs.

123. Let n = the number of hours required for Tom to type the mailing labels. Convert times to rates:

$\frac{1 \text{ task}}{6 \text{ hrs}}=\frac{1}{6}\frac{\text{task}}{\text{hr}}$

$\frac{1 \text{ task}}{n \text{ hrs}}=\frac{1}{n}\frac{\text{task}}{\text{hr}}$

$\frac{1 \text{ task}}{4 \text{ hrs}}=\frac{1}{4}\frac{\text{task}}{\text{hr}}$

Expressing their combined rate in two ways, we have

$\frac{1}{6}+\frac{1}{n}=\frac{1}{4}$

$\frac{1}{n}=\frac{1}{4}-\frac{1}{6}=\frac{6-4}{24}$

$\frac{1}{n}=\frac{2}{24}=\frac{1}{12}$

$n=12$

Therefore, Tom can type the mailing labels in 12 hours.

124. $\frac{1}{2}-\frac{1}{2.5}=\frac{1}{t}$

$10t\left[\frac{1}{2}-\frac{1}{2.5}\right]=\frac{1}{t}(10t)$

$5t-4t=10$

$t=10$ hours

125. Let r = the speed of the car, so $r+430$ = the speed of the plane. Recall that $d=rt$ or $t=\frac{d}{r}$. Using this relationship, we get:

$\frac{210}{r}=\frac{1715}{r+430}$

$1715r=210(r+430)$

$1715r=210r+90{,}300$

$1505r=90{,}300$

$r=60$ so $r+430=490$

Thus, the speed of the plane is 490 mph.

126. $\frac{1}{R}=\frac{1}{r_1}+\frac{1}{r_2}$

$\frac{1}{\frac{30}{11}}=\frac{1}{5}+\frac{1}{r_2}$

$\frac{11}{30}=\frac{1}{5}+\frac{1}{r_2}$

$30r_2\left[\frac{11}{30}\right]=\left(\frac{1}{5}+\frac{1}{r_2}\right)(30r_2)$

$11r_2=6r_2+30$

$5r_2=30$

$r_2=6$ ohms

127. Let r = the speed of the current. Recall $d=rt$ or $t=\frac{d}{r}$. Using this relationship, we get:

$\frac{72}{32-r}=\frac{120}{32+r}$

$72(32+r)=120(32-r)$

$2304+72r=3840-120r$

$192r=1536$

$r=8$

Therefore, the speed of the current is 8 mph.

128.

	distance = rate · time		
with wind	445	$400+x$	$\frac{445}{400+x}$
against wind	355	$400-x$	$\frac{355}{400-x}$

$\frac{445}{400+x}=\frac{355}{400-x}$

$445(400-x)=355(400+x)$

$178000-445x=142000+355x$

$36000=800x$

$45 \text{ mph}=x$

129. Let r = the speed of the walker, so $r+3$ = the speed of the jogger. Recall that $d=rt$ or $t=\frac{d}{r}$. Using this relationship, we get:

$\frac{14}{r+3}=\frac{8}{r}$

$14r=8(r+3)$

$14r=8r+24$

$6r=24$

$r=4$

Thus, the speed of the walker is 4 mph.

130.

distance = rate · time		
382	$\frac{382}{6}$	6
382 − 112 = 270	$\frac{270}{6}$	6

The first train is traveling at the rate of $\frac{382}{6} = 63\frac{2}{3}$ mph.
The second train is traveling at the rate of $\frac{270}{6} = 45$ mph.

131. $\frac{3x-5}{14} = \frac{x}{4}$
$4(3x-5) = 14x$
$12x - 20 = 14x$
$-20 = 2x$
$x = -10$
$\{-10\}$

132. $\frac{2-5x}{12} = \frac{x}{8}$
$12x = 8(2-5x)$
$12x = 16 - 40x$
$52x = 16$
$x = \frac{16}{52} = \frac{4}{13}$

133. $A = kB$
$6 = k(14)$
$k = \frac{6}{14} = \frac{3}{7}$
so $A = \frac{3}{7}B = \frac{3}{7}(21) = 3(3) = 9$

134. $C = \frac{k}{D}$
$12 = \frac{k}{8}$
$96 = k$
$C = \frac{96}{24} = 4$

135. $P = \frac{k}{V}$
$1250 = \frac{k}{2}$
$k = 2500$ so $P = \frac{2500}{V}$
$800 = \frac{2500}{V}$
$V = \frac{2500}{800} = 3.125 \text{ ft}^3$

136. $A = kr^2$
$36 = k(3)^2$
$4 = k$
$A = 4(4)^2 = 64$ sq. in.

Chapter 6 - Test

1. $f(x) = \frac{5x^2}{1-x}$
Undefined value when
$1 - x = 0$
$1 = x$
Domain $\{x|x$ is a real number and $x \neq 1\}$

2. $g(x) = \frac{9x^2-9}{x^2+4x+3}$
Undefined values when
$x^2 + 4x + 3 = 0$
$(x+3)(x+1) = 0$
$x + 3 = 0$ or $x + 1 = 0$
$x = -3$ or $x = -1$
Domain
$\{x|x$ is a real number and $x \neq -3, x \neq -1\}$

3. $\frac{5x^7}{3x^4} = \frac{5x^{7-4}}{3} = \frac{5x^3}{3}$

4. $\frac{7x-21}{24-8x} = \frac{7(x-3)}{8(3-x)} = \frac{7(-1)}{8} = \frac{-7}{8}$

5. $\frac{x^2-4x}{x^2+5x-36} = \frac{x(x-4)}{(x+9)(x-4)} = \frac{x}{x+9}$

6. $\frac{x}{x-2}\cdot\frac{x^2-4}{5x}$
$=\frac{1}{x-2}\cdot\frac{(x-2)(x+2)}{5}$
$=\frac{1}{1}\cdot\frac{x+2}{5}=\frac{x+2}{5}$

7. $\frac{2x^3+16}{6x^2+12x}\cdot\frac{5}{x^2-2x+4}$
$=\frac{2(x^3+8)\cdot 5}{6x(x+2)(x^2-2x+4)}$
$=\frac{5(x+2)(x^2-2x+4)}{3x(x+2)(x^2-2x+4)}$
$=\frac{5}{3x}$

8. $\frac{26ab}{7c}\div\frac{13a^2c^5}{14a^4b^3}=\frac{26ab}{7c}\cdot\frac{14a^4b^3}{13a^2c^5}$
$=\frac{2ab}{c}\cdot\frac{2a^4b^3}{a^2c^5}=\frac{4a^5b^4}{a^2c^6}=\frac{4a^3b^4}{c^6}$

9. $\frac{3x^2-12}{x^2+2x-8}\div\frac{6x+18}{x+4}$
$=\frac{3(x^2-4)}{(x+4)(x-2)}\cdot\frac{x+4}{6(x+3)}$
$=\frac{(x+2)(x-2)}{x-2}\cdot\frac{1}{2(x+3)}$
$=\frac{x+2}{2(x+3)}$

10. $\frac{4x-12}{2x-9}\div\frac{3-x}{4x^2-81}\cdot\frac{x+3}{5x+15}$
$=\frac{4(x-3)}{2x-9}\cdot\frac{(2x+9)(2x-9)}{-(x-3)}\cdot\frac{x+3}{5(x+3)}$
$=\frac{4}{1}\cdot\frac{2x+9}{-1}\cdot\frac{1}{5}=\frac{8x+36}{-5}=\frac{-4(x+9)}{5}$

11. $\frac{5}{4x^3}+\frac{7}{4x^3}=\frac{5+7}{4x^3}=\frac{12}{4x^3}=\frac{3}{x^3}$

12. $\frac{3+2x}{10-x}+\frac{13+x}{x-10}$
$=\frac{3+2x-(13+x)}{10-x}$
$=\frac{3+2x-13-x}{10-x}$
$=\frac{-10+x}{-(-10+x)}$
$=\frac{1}{-1}=-1$

13. $\frac{3}{x^2-x-6}+\frac{2}{x^2-5x+6}$
$=\frac{3}{(x-3)(x+2)}+\frac{2}{(x-3)(x-2)}$
$=\frac{3(x-2)+2(x+2)}{(x-3)(x+2)(x-2)}$
$=\frac{3x-6+2x+4}{(x-3)(x+2)(x-2)}$
$=\frac{5x-2}{(x-3)(x+2)(x-2)}$

14. $\frac{5}{x-7}-\frac{2x}{3x-21}+\frac{x}{2x-14}$
$=\frac{5}{x-7}-\frac{2x}{3(x-7)}+\frac{x}{2(x-7)}$
$=\frac{6\cdot 5-2\cdot 2x+3\cdot x}{6(x-7)}$
$=\frac{30-4x+3x}{6(x-7)}=\frac{30-x}{6(x-7)}$

15. $\frac{3x}{5}\left(\frac{5}{x}-\frac{5}{2x}\right)=3x\left(\frac{1}{x}-\frac{1}{2x}\right)$
$=3x\cdot\frac{2-1}{2x}=3x\left(\frac{1}{2x}\right)=\frac{3}{2}$

16. $\frac{\frac{4x}{13}}{\frac{20x}{13}}=\frac{4x}{20x}=\frac{1}{5}$

17. $\frac{\frac{5}{x}-\frac{7}{3x}}{\frac{9}{8x}-\frac{1}{x}}=\frac{\frac{15-7}{3x}}{\frac{9-8}{8x}}=\frac{8}{3x}\cdot\frac{8x}{1}=\frac{64}{3}$

18. $\frac{\frac{x^2-5x+6}{x+3}}{\frac{x^2-4x+4}{x^2-9}}$
$=\frac{(x-3)(x-2)}{x+3}\cdot\frac{(x-3)(x+3)}{(x-2)(x-2)}$
$=\frac{(x-3)^2}{x-2}$

19. $\frac{4x^2y+9x+z}{3xz}$
$=\frac{4x^2y}{3xz}+\frac{9x}{3xz}+\frac{z}{3xz}$
$=\frac{4xy}{3z}+\frac{3}{z}+\frac{1}{3x}$

20.

$$\begin{array}{r|l} & x^5+5x^4+8x^3+16x^2+33x+63 \\ x-2 & x^6+3x^5-2x^4+x^2-3x+2 \\ & \underline{x^6-2x^5} \\ & 5x^5-2x^4 \\ & \underline{5x^5-10x^4} \\ & 8x^4 \\ & \underline{8x^4-16x^3} \\ & 16x^3+x^2 \\ & \underline{16x^3-32x^2} \\ & 33x^2-3x \\ & \underline{33x^2-66x} \\ & 63x+2 \\ & \underline{63x-126} \\ & 128 \end{array}$$

Answer: $(x^5+5x^4+8x^3+16x^2+33x+63)+\frac{128}{x-2}$

21.

–3	4	–3	2	–1	–1
		–12	45	–141	426
	4	–15	47	–142	425

Answer: $4x^3-15x^2+47x-142+\frac{425}{x+3}$

22.

–2	4	0	7	–2	–5
		–8	16	–46	96
	4	–8	23	–48	91

Thus, $P(-2)=91$

23. $(h-g)(x)=h(x)-g(x)$
$=(x^2-6x+5)-(x-7)$
$=x^2-6x+5-x+7$
$=x^2-7x+12$

24. $(h\cdot f)(x)=h(x)\cdot f(x)$
$=(x^2-6x+5)(x)$
$=x^3-6x^2+5x$

25. $\frac{5x+3}{3x-7}=\frac{19}{7}$
$19(3x-7)=7(5x+3)$
$57x-133=35x+21$
$22x=154$
$x=7$
$\{7\}$

26. $\frac{5}{x-5}+\frac{x}{x+5}=-\frac{29}{21}$
$\frac{5(x+5)+x(x-5)}{(x-5)(x+5)}=-\frac{29}{21}$
$\frac{5x+25+x^2-5x}{x^2-25}=-\frac{29}{21}$
$\frac{x^2+25}{x^2-25}=-\frac{29}{21}$

$21x^2 + 525 = -29x^2 + 725$
$50x^2 = 200$
$x^2 = 4$
$x = \pm 2$
$\{2, -2\}$

27. $\frac{x}{x-4} = 3 - \frac{4}{x-4}$
$\frac{x}{x-4} + \frac{4}{x-4} = 3$
$\frac{x+4}{x-4} = 3$
$x + 4 = 3(x-4)$
$x + 4 = 3x - 12$
$16 = 2x$
$x = 8$
$\{8\}$

28. $\frac{x+b}{a} = \frac{4x-7a}{b}$
$b(x+b) = a(4x-7a)$
$bx + b^2 = 4ax - 7a^2$
$7a^2 + b^2 = 4ax - bx$
$7a^2 + b^2 = (4a-b)x$
$x = \frac{7a^2+b^2}{4a-b}$

29. Let x = the number. Then
$(x+1)\frac{2}{x} = \frac{12}{5}$
$\frac{2(x+1)}{x} = \frac{12}{5}$
$12x = 10(x+1)$
$12x = 10x + 10$
$2x = 10$
$x = 5$
The number is 5.

30. Let n = the number of hours it takes Jan and her husband to weed the garden. Note that 1 hour and 30 minutes = $\frac{3}{2}$ hours.

Convert time to rates:
$\frac{1 \text{ task}}{2 \text{ hrs}} = \frac{1}{2} \frac{\text{task}}{\text{hr}}$
$\frac{1 \text{ task}}{\frac{3}{2} \text{ hrs}} = \frac{2}{3} \frac{\text{task}}{\text{hr}}$
$\frac{1 \text{ task}}{n \text{ hrs}} = \frac{1}{n} \frac{\text{task}}{\text{hr}}$

Expressing their combined rate in two ways, we get
$\frac{1}{2} + \frac{2}{3} = \frac{1}{n}$
$\frac{7}{6} = \frac{1}{n}$
$n = \frac{6}{7}$ hrs

It takes the two of them $\frac{6}{7}$ hours to weed the garden.

31. $W = \frac{k}{V}$
$20 = \frac{k}{12}$
$k = 240$
so $W = \frac{240}{V} = \frac{240}{15} = 16$

32. $Q = kRS^2$
$24 = k(3)4^2$
$24 = 48k$
$k = \frac{1}{2}$ so
$Q = \frac{1}{2}RS^2 = \frac{1}{2}(2)3^2 = 9$

33. $S = k\sqrt{d}$
$160 = k\sqrt{400}$
$k = \frac{160}{20} = 8$ so $S = 8\sqrt{d}$
$128 = 8\sqrt{d}$
$\sqrt{d} = 16$
$d = 256$
The height of the cliff is 256 feet.

Chapter 6 - Cumulative Review

1. a. $x + 5 = 20$

b. $2(3 + y) = 4$

c. $x - 8 = 2x$

d. $\frac{z}{9} = 3(z-5)$

2. **a.** $\sqrt[3]{27} = 3$

b. $\sqrt[5]{1} = 1$

c. $\sqrt[4]{16} = 2$

3. Let x = third side
Then $2x + 12$ = each equal side
$x + 2(2x + 12) = 149$
$x + 4x + 24 = 149$
$5x = 125$
$x = 25$
$2x + 12 = 2(25) + 12 = 62$
25 cm, 62 cm, 62 cm

4. $5x - 2y = 10$

x	y
0	–5
2	0
1	–2.5

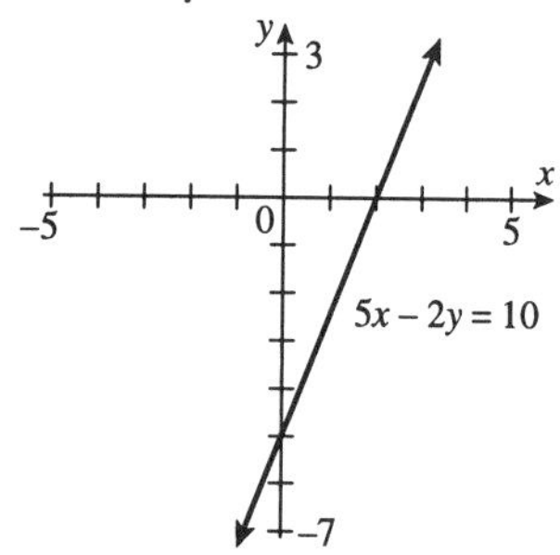

5. **a.** function

b. function

c. not a function

d. function

e. not a function

f. not a function

6. $y = -3$
Horizontal line with y-intercept at –3.

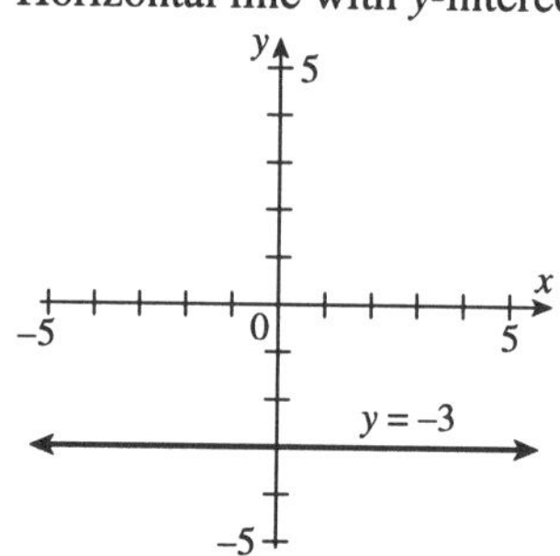

7. $2x + 3y = -6$
$3y = -2x - 6$
$y = -\frac{2}{3}x - 2$
$m = -\frac{2}{3}$
$y - 4 = -\frac{2}{3}(x - 4)$
$3y - 12 = -2x + 8$
$2x + 3y = 20$

8. $\left[\begin{array}{ccc|c} 2 & -4 & 8 & 2 \\ -1 & -3 & 1 & 11 \\ 1 & -2 & 4 & 0 \end{array}\right]$

Multiply third row by –2 and add to first row.

$\left[\begin{array}{ccc|c} 2 & -4 & 8 & 2 \\ -1 & -3 & 1 & 11 \\ 0 & 0 & 0 & 2 \end{array}\right]$

Inconsistent; No solution; { }

9. **a.** $(3x^6)(5x) = (3)(5)x^{6+1} = 15x^7$

b. $(-2x^3p^2)(4xp^{10})$
$= (-2)(4)x^{3+1}p^{2+10}$
$= -8x^4p^{12}$

10. a. $\left(\frac{3x^2y}{y^{-9}z}\right)^{-2} = \frac{3^{-2}x^{-4}y^{-2}}{y^{18}z^{-2}} = \frac{z^2}{9x^4y^{20}}$

b. $\left(\frac{3a^2}{2x^{-1}}\right)^3\left(\frac{x^{-3}}{4a^{-2}}\right)^{-1}$

$= \left(\frac{3^3a^6}{2^3x^{-3}}\right)\left(\frac{x^3}{4^{-1}a^2}\right)$

$= \frac{27(4)a^{6-2}x^{3+3}}{2(4)} = \frac{27a^4x^6}{2}$

11. $(2x-7)(3x-4)$

$= 6x^2 - 8x - 21x + 28$

$= 6x^2 - 29x + 28$

12. $2n^2 - 38n + 80 = 2(n^2 - 19n + 40)$

13. $x^3 = 4x$

$x^3 - 4x = 0$

$x(x^2 - 4) = 0$

$x(x+2)(x-2) = 0$

$x = 0$ or $x + 2 = 0$ or $x - 2 = 0$

$x = 0$ or $x = -2$ or $x = 2$

$\{0, -2, 2\}$

14. a. Domain; $(-\infty, \infty)$

Range; $(-\infty, 4]$

b. (–3, 0), (–1, 0), (1, 0), (5, 0), (0, –2)

c. (3, 4)

d. There is no such point.

e. –3, –1, 1, 5

f. $-3 < x < -1$ and $1 < x < 5$

15. $C(x) = \frac{2.6x + 10{,}000}{x}$

a. $C(100) = \frac{2.6(100) + 10{,}000}{100}$

$= \$102.60$

b. $C(1000) = \frac{2.6(1000) + 10{,}000}{1000}$

$= \$12.60$

16. a. $\frac{2x^2 + 3x - 2}{-4x - 8} \cdot \frac{16x^2}{4x^2 - 1}$

$= \frac{(2x-1)(x+2)}{-4(x+2)} \cdot \frac{4(4x^2)}{(2x-1)(2x+1)}$

$= -\frac{4x^2}{2x+1}$

b. $(ac - ad + bc - bd) \cdot \frac{a+b}{d-c}$

$= (a+b)(c-d) \cdot \frac{a+b}{-1(c-d)}$

$= -(a+b)^2$

17. $\frac{x^{-1} + 2xy^{-1}}{x^{-2} - x^{-2}y^{-1}}$

$= \frac{\frac{1}{x} + \frac{2x}{y}}{\frac{1}{x^2} - \frac{1}{x^2y}} = \frac{\frac{y+2x^2}{xy}}{\frac{y-1}{x^2y}}$

$= \frac{y+2x^2}{xy} \cdot \frac{x^2y}{y-1} = \frac{xy + 2x^3}{y-1}$

18. $\frac{7a^2b - 2ab^2}{2ab^2}$

$= \frac{7a^2b}{2ab^2} - \frac{2ab^2}{2ab^2} = \frac{7a}{2b} - 1$

19. $P(x) = 2x^3 - 4x^2 + 5$

a. $P(2) = 2(2)^3 - 4(2)^2 + 5$

$= 16 - 16 + 5 = 5$

b.

2	2	–4	0	5
		4	0	0
	2	0	0	5

The remainder is 5.

20. $\frac{x+6}{x-2} = \frac{2(x+2)}{x-2}$

$x + 6 = 2x + 4$

$2 = x$

No solution since $x = 2$ is undefined.

{ }

21. $I(x) = \frac{320}{x^2}$
$5 = \frac{320}{x^2}$
$5x^2 = 320$
$x^2 = 64$
$x = \pm 8$
Disregard the negative. The source of light is 8 feet away.

22. $y = kx$
$5 = k(30)$
$\frac{5}{30} = k$
$k = \frac{1}{6}$
When $x = 90$
$y = \frac{1}{6} \cdot 90$
$y = 15$

Chapter 7

Exercise Set 7.1

1. $\sqrt{100} = 10$

3. $\sqrt{\frac{1}{4}} = \frac{1}{2}$

5. $\sqrt{0.0001} = 0.01$

7. $-\sqrt{36} = -6$

9. $\sqrt{x^{10}} = x^5$

11. $\sqrt{16y^6} = \sqrt{16}\sqrt{y^6} = 4y^3$

13. $\sqrt[3]{\frac{1}{8}} = \frac{\sqrt[3]{1}}{\sqrt[3]{8}} = \frac{1}{2}$

15. $-\sqrt[4]{16} = -2$

17. $\sqrt[4]{-16}$ is not a real number

19. $\sqrt[3]{-125x^9} = \sqrt[3]{-125}\sqrt[3]{x^9} = -5x^3$

21. $\sqrt[5]{32x^5} = \sqrt[5]{32}\sqrt[5]{x^5} = 2x$

23. $\sqrt{z^2} = |z|$

25. $\sqrt[3]{x^3} = x$

27. $\sqrt{(x-5)^2} = |x-5|$

29. $\sqrt[4]{(2z)^4} = 2|z|$

31. $\sqrt{100(2x-y)^6}$
$= \sqrt{100}\sqrt{(2x-y)^6} = 10\left|(2x-y)^3\right|$

33. Answers may vary.

35. $-\sqrt{121} = -11$

37. $\sqrt[3]{8x^3} = 2x$

39. $\sqrt{y^{12}} = y^6$

41. $\sqrt{25a^2b^{20}} = \sqrt{25}\sqrt{a^2}\sqrt{b^{20}} = 5ab^{10}$

43. $\sqrt[3]{-27x^9} = \sqrt[3]{-27}\sqrt[3]{x^9} = -3x^3$

45. $\sqrt[4]{a^{16}b^4} = \sqrt[4]{a^{16}}\sqrt[4]{b^4} = a^4b$

47. $\sqrt[5]{-32x^{10}y^5} = \sqrt[5]{-32}\sqrt[5]{x^{10}}\sqrt[5]{y^5} = -2x^2y$

49. $\sqrt{\frac{25}{49}} = \frac{\sqrt{25}}{\sqrt{49}} = \frac{5}{7}$

51. $\sqrt{\frac{x^2}{4y^2}} = \frac{\sqrt{x^2}}{\sqrt{4}\sqrt{y^2}} = \frac{x}{2y}$

53. $-\sqrt[3]{\frac{z^{21}}{27x^3}} = -\frac{\sqrt[3]{z^{21}}}{\sqrt[3]{27}\sqrt[3]{x^3}} = -\frac{z^7}{3x}$

55. $\sqrt[4]{\frac{x^4}{16}} = \frac{\sqrt[4]{x^4}}{\sqrt[4]{16}} = \frac{x}{2}$

57. $\sqrt{9}$, rational, 3

59. $\sqrt{37}$, irrational, 6.083

61. $\sqrt{169}$, rational, 13

63. $\sqrt{4}$, rational, 2

65. $\sqrt[3]{x^{15}} = x^5$

67. $\sqrt{x^{12}} = x^6$

69. $\sqrt{81x^2} = 9|x|$

71. $-\sqrt{144y^{14}} = -12|y^7|$

73. $\sqrt{x^2 + 4x + 4} = |x+2|$

75. $f(x) = \sqrt{2x+3}$
$f(0) = \sqrt{2 \cdot 0 + 3} = \sqrt{3}$

77. $g(x) = \sqrt[3]{x-8}$
$g(7) = \sqrt[3]{7-8} = \sqrt[3]{-1} = -1$

79. $g(x) = \sqrt[3]{x-8}$
$g(-19) = \sqrt[3]{-19-8} = \sqrt[3]{-27} = -3$

81. $f(x) = \sqrt{2x+3}$
$f(2) = \sqrt{2 \cdot 2 + 3} = \sqrt{7}$

83. $f(x) = \sqrt{x} + 2$
Domain: $[0, \infty)$

85. $f(x) = \sqrt{x-3}$
Domain: $[3, \infty)$

x	$f(x)$
3	$\sqrt{3-3} = \sqrt{0} = 0$
4	$\sqrt{4-3} = \sqrt{1} = 1$
7	$\sqrt{7-3} = \sqrt{4} = 2$
12	$\sqrt{12-3} = \sqrt{9} = 3$

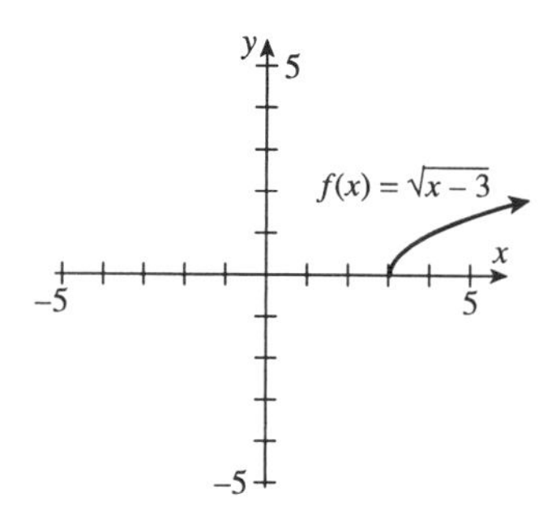

87. $f(x) = \sqrt[3]{x} + 1$
Domain: $(-\infty, \infty)$

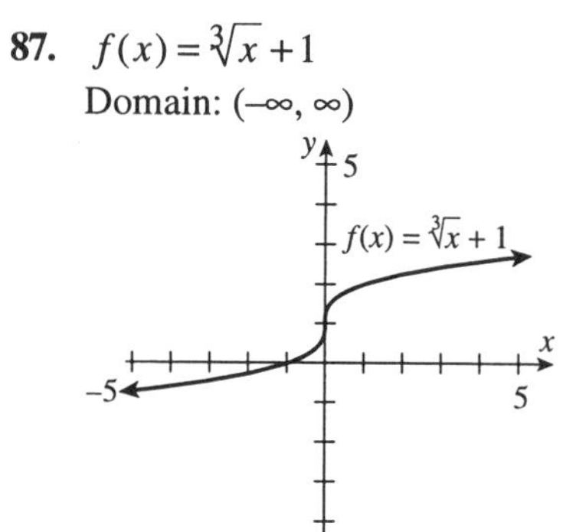

89. $g(x) = \sqrt[3]{x-1}$
Domain: $(-\infty, \infty)$

x	$f(x)$
1	$\sqrt[3]{1-1} = \sqrt[3]{0} = 0$
2	$\sqrt[3]{2-1} = \sqrt[3]{1} = 1$
0	$\sqrt[3]{0-1} = \sqrt[3]{-1} = -1$
9	$\sqrt[3]{9-1} = \sqrt[3]{8} = 2$
−7	$\sqrt[3]{-7-1} = \sqrt[3]{-8} = -2$

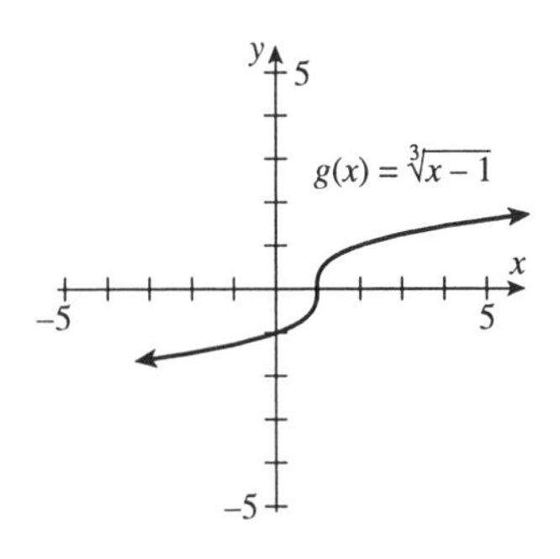

91. $f(x) = \sqrt{x} + 2$

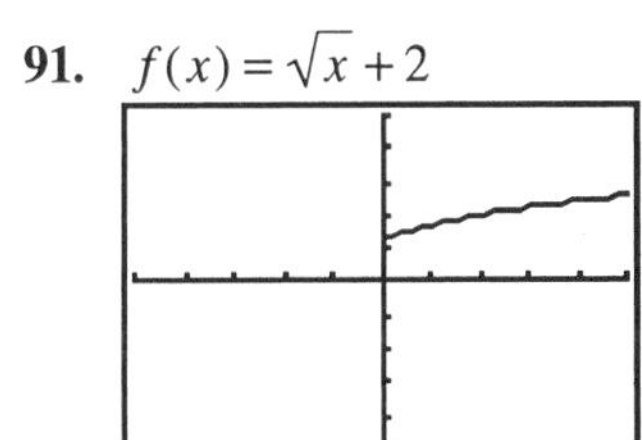

93. $f(x) = \sqrt{x - 3}$

95. $f(x) = \sqrt[3]{x} + 1$

97. $g(x) = \sqrt[3]{x - 1}$

99. $(-2x^3y^2)^5 = (-2)^5 x^{3 \cdot 5} y^{2 \cdot 5} = -32x^{15}y^{10}$

101. $(-3x^2y^3z^5)(20x^5y^7)$
$= (-3)(20)x^{2+5}y^{3+7}z^5$
$= -60x^7y^{10}z^5$

103. $\dfrac{7x^{-1}y}{14(x^5y^2)^{-2}} = \dfrac{7x^{-1}y}{14x^{-10}y^{-4}}$
$= \frac{1}{2}x^9y^5$

Exercise Set 7.2

1. $49^{1/2} = \sqrt{49} = 7$

3. $27^{1/3} = \sqrt[3]{27} = 3$

5. $\left(\frac{1}{16}\right)^{1/4} = \sqrt[4]{\frac{1}{16}} = \frac{1}{2}$

7. $169^{1/2} = \sqrt{169} = 13$

9. $2m^{1/3} = 2\sqrt[3]{m}$

11. $(9x^4)^{1/2} = \sqrt{9x^4} = 3x^2$

13. $(-27)^{1/3} = \sqrt[3]{-27} = -3$

15. $-16^{1/4} = -\sqrt[4]{16} = -2$

17. $16^{3/4} = (\sqrt[4]{16})^3 = 2^3 = 8$

19. $(-64)^{2/3} = \left[\sqrt[3]{-64}\right]^2 = (-4)^2 = 16$

21. $(-16)^{3/4} = \sqrt[4]{-16}$ is not a real number.

23. $(2x)^{3/5} = \sqrt[5]{(2x)^3}$ or $\left(\sqrt[5]{2x}\right)^3$

25. $(7x+2)^{2/3} = \sqrt[3]{(7x+2)^2}$ or $\left(\sqrt[3]{7x+2}\right)^2$

27. $\left(\frac{16}{9}\right)^{3/2} = \left(\sqrt{\frac{16}{9}}\right)^3 = \left(\frac{4}{3}\right)^3 = \frac{64}{27}$

29. $8^{-4/3} = \dfrac{1}{8^{4/3}} = \dfrac{1}{(8^{1/3})^4} = \dfrac{1}{2^4} = \dfrac{1}{16}$

31. $(-64)^{-2/3} = \frac{1}{(-64)^{2/3}}$
$= \frac{1}{\left[(-64)^{1/3}\right]^2} = \frac{1}{(-4)^2} = \frac{1}{16}$

33. $(-4)^{-3/2} = \frac{1}{(-4)^{3/2}} = \frac{1}{\left[(-4)^{1/2}\right]^3}$ is not a real number.

35. $x^{-1/4} = \frac{1}{x^{1/4}}$

37. $\frac{1}{a^{-2/3}} = a^{2/3}$

39. $\frac{5}{7x^{-3/4}} = \frac{5x^{3/4}}{7}$

41. Answers may vary.

43. $\sqrt{3} = 3^{1/2}$

45. $\sqrt[3]{y^5} = (y^5)^{1/3} = y^{5/3}$

47. $\sqrt[5]{4y^7} = (4y^7)^{1/5}$
$= 4^{1/5}(y^7)^{1/5} = 4^{1/5}y^{7/5}$

49. $\sqrt{(y+1)^3} = [(y+1)^3]^{1/2} = (y+1)^{3/2}$

51. $2\sqrt{x} - 3\sqrt{y} = 2x^{1/2} - 3y^{1/2}$

53. $\sqrt[6]{x^3} = x^{3/6} = x^{1/2} = \sqrt{x}$

55. $\sqrt[4]{16x^2} = 16^{1/4}x^{2/4} = 2x^{1/2} = 2\sqrt{x}$

57. $\sqrt[4]{(x+3)^2} = (x+3)^{2/4} = (x+3)^{1/2}$
$= \sqrt{x+3}$

59. $\sqrt[8]{x^4y^4} = x^{4/8}y^{4/8} = x^{1/2}y^{1/2} = \sqrt{xy}$

61. $a^{2/3}a^{5/3} = a^{2/3+5/3} = a^{7/3}$

63. $(4u^2v^{-6})^{3/2} = 4^{3/2}(u^2)^{3/2}(v^{-6})^{3/2}$
$(4^{1/2})^3u^3v^{-9} = \frac{2^3u^3}{v^9} = \frac{8u^3}{v^9}$

65. $\frac{b^{1/2}b^{3/4}}{-b^{1/4}} = -b^{1/2+3/4-1/4}$
$= -b^{1/2+1/2} = -b^1 = -b$

67. $y^{1/2}(y^{1/2} - y^{2/3}) = y^{1/2}y^{1/2} - y^{1/2}y^{2/3}$
$y^{1/2+1/2} - y^{1/2+2/3} = y^1 - y^{7/6}$
$= y - y^{7/6}$

69. $x^{2/3}(2x-2) = 2x^1x^{2/3} - 2x^{2/3}$
$= 2x^{5/3} - 2x^{2/3}$

71. $(2x^{1/3}+3)(2x^{1/3}-3) = (2x^{1/3})^2 - 3^2$
$= 2^2(x^{1/3})^2 - 9 = 4x^{2/3} - 9$

73. $x^{8/3} + x^{10/3} = x^{8/3}(1 + x^{2/3})$

75. $x^{2/5} - 3x^{1/5} = x^{1/5}(x^{1/5} - 3)$

77. $5x^{-1/3} + x^{2/3} = x^{-1/3}(5 + x^1)$
$= x^{-1/3}(5+x)$

79. $a^{1/3}$
$a^{1/3} \cdot a^{2/3} = a^{1/3+2/3} = a^{3/3} = a$

81. $x^{1/5}$
$\frac{x^{1/5}}{x^{-2/5}} = x^{1/5+2/5} = x^{3/5}$

83. $8^{1/4} = 1.6818$

85. $18^{3/5} = 5.6645$

87. $75 = 25 \cdot 3$

89. $48 = 16 \cdot 3$

91. $16 = 8 \cdot 2$

93. $54 = 27 \cdot 2$

Exercise Set 7.3

1. $\sqrt{7}\cdot\sqrt{2}=\sqrt{7\cdot 2}=\sqrt{14}$

3. $\sqrt[3]{4}\cdot\sqrt[3]{9}=\sqrt[3]{4\cdot 9}=\sqrt[3]{36}$

5. $\sqrt{2}\cdot\sqrt{3x}=\sqrt{2\cdot 3x}=\sqrt{6x}$

7. $\sqrt{\frac{7}{x}}\cdot\sqrt{\frac{2}{y}}=\sqrt{\frac{7\cdot 2}{x\cdot y}}=\sqrt{\frac{14}{xy}}$

9. $\sqrt[4]{4x^3}\cdot\sqrt[4]{5}=\sqrt[4]{4\cdot 5x^3}=\sqrt[4]{20x^3}$

11. $\frac{\sqrt{14}}{\sqrt{7}}=\sqrt{\frac{14}{7}}=\sqrt{2}$

13. $\frac{\sqrt[3]{24}}{\sqrt[3]{3}}=\sqrt[3]{\frac{24}{3}}=\sqrt[3]{8}=2$

15. $\frac{\sqrt{x^5y^3}}{\sqrt{xy}}=\sqrt{\frac{x^5y^3}{xy}}$
$=\sqrt{x^{5-1}y^{3-1}}=\sqrt{x^4y^2}=x^2y$

17. $\frac{8\sqrt[3]{54m^7}}{\sqrt[3]{2m}}=8\sqrt[3]{\frac{54m^7}{2m}}=8\sqrt[3]{27m^{7-1}}$
$=8\sqrt[3]{27m^6}=8\cdot 3m^2=24m^2$

19. $\sqrt{32}=\sqrt{16(2)}=\sqrt{16}\sqrt{2}=4\sqrt{2}$

21. $\sqrt[3]{192}=\sqrt[3]{64(3)}=\sqrt[3]{64}\sqrt[3]{3}=4\sqrt[3]{3}$

23. $5\sqrt{75}=5\sqrt{25(3)}=5\sqrt{25}\sqrt{3}$
$=5(5)\sqrt{3}=25\sqrt{3}$

25. $\sqrt{\frac{6}{49}}=\frac{\sqrt{6}}{\sqrt{49}}=\frac{\sqrt{6}}{7}$

27. $\sqrt{20}=\sqrt{4(5)}=\sqrt{4}\sqrt{5}=2\sqrt{5}$

29. $\sqrt[3]{\frac{4}{27}}=\frac{\sqrt[3]{4}}{\sqrt[3]{27}}=\frac{\sqrt[3]{4}}{3}$

31. $\sqrt{\frac{2}{49}}=\frac{\sqrt{2}}{\sqrt{49}}=\frac{\sqrt{2}}{7}$

33. $\sqrt{100x^5}=\sqrt{(100x^4)x}=\sqrt{100}\sqrt{x^4}\sqrt{x}$
$=10x^2\sqrt{x}$

35. $\sqrt[3]{16y^7}=\sqrt[3]{(8y^6)(2y)}$
$=\sqrt[3]{8}\sqrt[3]{y^6}\sqrt[3]{2y}$
$=2y^2\sqrt[3]{2y}$

37. $\sqrt[4]{a^8b^7}=\sqrt[4]{a^8b^4b^3}=\sqrt[4]{a^8}\sqrt[4]{b^4}\sqrt[4]{b^3}$
$=a^2b\sqrt[4]{b^3}$

39. $\sqrt{y^5}=\sqrt{y^4y}+\sqrt{y^4}\sqrt{y}=y^2\sqrt{y}$

41. $\sqrt{25a^2b^3}=\sqrt{25}\sqrt{a^2}\sqrt{b^3}$
$=5a\sqrt{b^2b}$
$=5a\sqrt{b^2}\sqrt{b}=5ab\sqrt{b}$

43. $\sqrt[5]{-32x^{10}y}=\sqrt[5]{-32}\sqrt[5]{x^{10}}\sqrt[5]{y}$
$=-2x^2\sqrt[5]{y}$

45. $\sqrt[3]{50x^{14}}=\sqrt[3]{x^{12}(50x^2)}$
$=\sqrt[3]{x^{12}}\sqrt[3]{50x^2}$
$=x^4\sqrt[3]{50x^2}$

47. $-\sqrt{32a^8b^7}=\sqrt{16a^8b^6(2b)}$
$=-\sqrt{16}\sqrt{a^8}\sqrt{b^6}\sqrt{2b}$
$=-4a^4b^3\sqrt{2b}$

49. $\sqrt{\frac{5x^2}{4y^2}}=\frac{\sqrt{5x^2}}{\sqrt{4y^2}}=\frac{\sqrt{5}\sqrt{x^2}}{\sqrt{4}\sqrt{y^2}}=\frac{\sqrt{5}x}{2y}$

51. $-\sqrt[3]{\dfrac{z^7}{27x^3}} = \dfrac{-\sqrt[3]{z^7}}{\sqrt[3]{27x^3}} = \dfrac{-\sqrt[3]{z^6 z}}{\sqrt[3]{27}\sqrt[3]{x^3}}$
$= \dfrac{-\sqrt[3]{z^6}\sqrt[3]{z}}{3x} = \dfrac{-z^2\sqrt[3]{z}}{3x}$

53. $\sqrt[4]{\dfrac{x^7}{16}} = \dfrac{\sqrt[4]{x^7}}{\sqrt[4]{16}} = \dfrac{\sqrt[4]{x^4 \cdot x^3}}{2} = \dfrac{\sqrt[4]{x^4}\sqrt[4]{x^3}}{2}$
$= \dfrac{x\sqrt[4]{x^3}}{2}$

55. $\sqrt{9x^7y^9} = \sqrt{9x^6y^8 \cdot xy}$
$= \sqrt{9x^6y^8}\sqrt{xy} = 3x^3y^4\sqrt{xy}$

57. $\sqrt[3]{125r^9s^{12}} = 5r^3s^4$

59. $\sqrt{\dfrac{x^2y}{100}} = \dfrac{\sqrt{x^2}\sqrt{y}}{\sqrt{100}} = \dfrac{x\sqrt{y}}{10}$

61. $\sqrt[4]{\dfrac{8}{x^8}} = \dfrac{\sqrt[4]{8}}{\sqrt[4]{x^8}} = \dfrac{\sqrt[4]{8}}{x^2}$

63. $\sqrt{2} \cdot \sqrt[3]{3} = 2^{1/2} \cdot 3^{1/3}$
$= 2^{3/6} \cdot 3^{2/6} = (2^3)^{1/6}(3^2)^{1/6}$
$= 8^{1/6} \cdot 9^{1/6} = 72^{1/6} = \sqrt[6]{72}$

65. $\sqrt[5]{7} \cdot \sqrt[3]{y} = 7^{1/5} \cdot y^{1/3}$
$= 7^{3/15} \cdot y^{5/15} = (7^3)^{1/15} \cdot (y^5)^{1/15}$
$= (343)^{1/5}(y^5)^{1/15} = (343y^5)^{1/15}$
$= \sqrt[15]{343y^5}$

67. $\sqrt[3]{x} \cdot \sqrt{x} = x^{1/3} \cdot x^{1/2}$
$= x^{2/6} \cdot x^{3/6} = x^{5/6} = \sqrt[6]{x^5}$

69. $\sqrt{5r}\sqrt[3]{s} = (5r)^{1/2} \cdot s^{1/3}$
$= (5r)^{3/6} \cdot s^{2/6} = (5^3r^3)^{1/6}(s^2)^{1/6}$
$= (125r^3s^2)^{1/6} = \sqrt[6]{125r^3s^2}$

71. $A = \pi r\sqrt{r^2 + h^2}$

a. $A = \pi \cdot 4\sqrt{4^2 + 3^2}$
$= 4\pi\sqrt{16+9} = 4\pi\sqrt{25}$
$= 4 \cdot 5\pi = 20\pi$ sq. cm

b. $A = \pi(6.8)\sqrt{(6.8)^2 + (7.2)^2}$
$= 6.8\pi\sqrt{46.24 + 51.84}$
$= 6.8\pi\sqrt{98.08}$
$= 211.57$ sq. feet

73. $A = \pi r\sqrt{r^2 + h^2}$
$A = \pi(25,200)\sqrt{(25,200)^2 + (4190)^2}$
$A \approx 2,022,426,050$ square feet

75. $F(x) = \sqrt{130 - x}$

a. $F(86) = \sqrt{130 - 86}$
$= \sqrt{44} = 6.63$ thousand tires

b. $F(79.99) = \sqrt{130 - 79.99}$
$= \sqrt{50.01} = 7.07$ thousand tires

c. Answers may vary.

77. $6x + 8x = 14x$

79. $(2x + 3)(x - 5)$
$= 2x^2 - 10x + 3x - 15$
$= 2x^2 - 7x - 15$

81. $9y^2 - 8y^2 = y^2$

83. $-3(x + 5) = -3x - 15$

Section 7.4

Mental Math

1. $2\sqrt{3} + 4\sqrt{3} = 6\sqrt{3}$

3. $8\sqrt{x} - 5\sqrt{x} = 3\sqrt{x}$

5. $7\sqrt[3]{x} + 5\sqrt[3]{x} = 12\sqrt[3]{x}$

Exercise Set 7.4

1. $\sqrt{8}-\sqrt{32}=\sqrt{4(2)}-16\sqrt{(2)}$
$=\sqrt{4}\sqrt{2}-\sqrt{16}\sqrt{2}$
$=2\sqrt{2}-4\sqrt{2}=-2\sqrt{2}$

3. $2\sqrt{2x^3}+4x\sqrt{8x}$
$=2\sqrt{x^2(2x)}+4x\sqrt{4(2x)}$
$=2\sqrt{x^2}\sqrt{2x}+4x\sqrt{4}\sqrt{2x}$
$=2x\sqrt{2x}+4x(2)\sqrt{2x}$
$=2x\sqrt{2x}+8x\sqrt{2x}$
$=10x\sqrt{2x}$

5. $2\sqrt{50}-3\sqrt{125}+\sqrt{98}$
$=2\sqrt{25(2)}-3\sqrt{25(5)}+\sqrt{49(2)}$
$=2\sqrt{25}\sqrt{2}-3\sqrt{25}\sqrt{5}+\sqrt{49}\sqrt{2}$
$=2(5)\sqrt{2}-3(5)\sqrt{5}+7\sqrt{2}$
$=10\sqrt{2}-15\sqrt{5}+7\sqrt{2}$
$=17\sqrt{2}-15\sqrt{5}$

7. $\sqrt[3]{16x}-\sqrt[3]{54x}$
$=\sqrt[3]{8(2x)}-\sqrt[3]{27(2x)}$
$=\sqrt[3]{8}\sqrt[3]{2x}-\sqrt[3]{27}\sqrt[3]{2x}$
$=2\sqrt[3]{2x}-3\sqrt[3]{2x}$
$=-\sqrt[3]{2x}$

9. $\sqrt{9b^3}-\sqrt{25b^3}+\sqrt{49b^3}$
$=\sqrt{9b^2(b)}-\sqrt{25b^2(b)}+\sqrt{49b^2(b)}$
$=\sqrt{9b^2}\sqrt{b}-\sqrt{25b^2}\sqrt{b}+\sqrt{49b^2}\sqrt{b}$
$=3b\sqrt{b}-5b\sqrt{b}+7b\sqrt{b}$
$=-2b\sqrt{b}+7b\sqrt{b}=5b\sqrt{b}$

11. $\frac{5\sqrt{2}}{3}+\frac{2\sqrt{2}}{5}$
$=\frac{5(5\sqrt{2})+3(2\sqrt{2})}{3(5)}$
$=\frac{25\sqrt{2}+6\sqrt{2}}{15}$
$=\frac{31\sqrt{2}}{15}$

13. $\sqrt[3]{\frac{11}{8}}-\frac{\sqrt[3]{11}}{6}$
$=\frac{\sqrt[3]{11}}{\sqrt[3]{8}}-\frac{\sqrt[3]{11}}{6}$
$=\frac{\sqrt[3]{11}}{2}-\frac{\sqrt[3]{11}}{6}$
$=\frac{3\sqrt[3]{11}-\sqrt[3]{11}}{6}$
$=\frac{2\sqrt[3]{11}}{6}$
$=\frac{\sqrt[3]{11}}{3}$

15. $\frac{\sqrt{20x}}{9}+\sqrt{\frac{5x}{9}}$
$=\frac{\sqrt{4(5x)}}{9}+\frac{\sqrt{5x}}{\sqrt{9}}$
$=\frac{\sqrt{4}\sqrt{5x}}{9}+\frac{\sqrt{5x}}{3}$
$=\frac{2\sqrt{5x}+3\sqrt{5x}}{9}$
$=\frac{5\sqrt{5x}}{9}$

17. $7\sqrt{9}-7+\sqrt{3}$
$=7(3)-7+\sqrt{3}$
$=21-7+\sqrt{3}$
$=14+\sqrt{3}$

19. $2+3\sqrt{y^2}-6\sqrt{y^2}+5$
$=7-3\sqrt{y^2}$
$=7-3y$

21. $3\sqrt{108}-2\sqrt{18}-3\sqrt{48}$
$=3\sqrt{36}\sqrt{3}-2\sqrt{9}\sqrt{2}-3\sqrt{16}\sqrt{3}$
$=3(6)\sqrt{3}-2(3)\sqrt{2}-3(4)\sqrt{3}$
$=18\sqrt{3}-6\sqrt{2}-12\sqrt{3}$
$=6\sqrt{3}-6\sqrt{2}$

23. $-5\sqrt[3]{625}+\sqrt[3]{40}$
$=-5\sqrt[3]{125}\sqrt[3]{5}+\sqrt[3]{8}\sqrt[3]{5}$
$=-5(5)\sqrt[3]{5}+2\sqrt[3]{5}$
$=-25\sqrt[3]{5}+2\sqrt[3]{5}$
$=-23\sqrt[3]{5}$

25. $\sqrt{9b^3}-\sqrt{25b^3}+\sqrt{16b^3}$
$=\sqrt{9b^2}\sqrt{b}-\sqrt{25b^2}\sqrt{b}+\sqrt{16b^2}\sqrt{b}$
$=3b\sqrt{b}-5b\sqrt{b}+4b\sqrt{b}$
$=(3-5+4)b\sqrt{b}$
$=2b\sqrt{b}$

27. $5y\sqrt{8y}+2\sqrt{50y^3}$
$=5y\sqrt{4}\sqrt{2y}+2\sqrt{25y^2}\sqrt{2y}$
$=5y(2)\sqrt{2y}+2(5y)\sqrt{2y}$
$=10y\sqrt{2y}+10y\sqrt{2y}$
$=20y\sqrt{2y}$

29. $\sqrt[3]{54xy^3}-5\sqrt[3]{2xy^3}+y\sqrt[3]{128x}$
$=\sqrt[3]{27y^3}\sqrt[3]{2x}-5\sqrt[3]{y^3}\sqrt[3]{2x}+y\sqrt[3]{64}\sqrt[3]{2x}$
$=3y\sqrt[3]{2x}-5y\sqrt[3]{2x}+y(4)\sqrt[3]{2x}$
$=-2y\sqrt[3]{2x}+4y\sqrt[3]{2x}$
$=2y\sqrt[3]{2x}$

31. $6\sqrt[3]{11}+8\sqrt{11}-12\sqrt{11}=6\sqrt[3]{11}-4\sqrt{11}$

33. $-2\sqrt[4]{x^7}+3\sqrt[4]{16x^7}$
$=-2\sqrt[4]{x^4}\sqrt[4]{x^3}+3\sqrt[4]{16x^4}\sqrt[4]{x^3}$
$=-2x\sqrt[4]{x^3}+3(2x)\sqrt[4]{x^3}$
$=-2x\sqrt[4]{x^3}+6x\sqrt[4]{x^3}$
$=4x\sqrt[4]{x^3}$

35. $\frac{4\sqrt{3}}{3}-\frac{\sqrt{12}}{3}$
$=\frac{4\sqrt{3}}{3}-\frac{\sqrt{4}\sqrt{3}}{\sqrt{3}}$
$=\frac{4\sqrt{3}-2\sqrt{3}}{3}$
$=\frac{2\sqrt{3}}{3}$

37. $\frac{\sqrt[3]{8x^4}}{7}+\frac{3x\sqrt[3]{x}}{7}$
$=\frac{\sqrt[3]{8x^3}\sqrt[3]{x}+3x\sqrt[3]{x}}{7}$
$=\frac{2x\sqrt[3]{x}+3x\sqrt[3]{x}}{7}$
$=\frac{5x\sqrt[3]{x}}{7}$

39. $\sqrt{\frac{28}{x^2}}+\sqrt{\frac{7}{4x^2}}$
$=\frac{\sqrt{28}}{\sqrt{x^2}}+\frac{\sqrt{7}}{\sqrt{4x^2}}$
$=\frac{\sqrt{4}\sqrt{7}}{x}+\frac{\sqrt{7}}{2x}$
$=\frac{2\sqrt{7}}{x}+\frac{\sqrt{7}}{2x}$
$=\frac{4\sqrt{7}+\sqrt{7}}{2x}$
$=\frac{5\sqrt{7}}{2x}$

41. $\sqrt[3]{\frac{16}{27}}-\frac{\sqrt[3]{54}}{6}$
$=\frac{\sqrt[3]{16}}{\sqrt[3]{27}}-\frac{\sqrt[3]{27}\sqrt[3]{2}}{6}$
$=\frac{\sqrt[3]{8}\sqrt[3]{2}}{3}-\frac{3\sqrt[3]{2}}{6}$
$=\frac{2(2)\sqrt[3]{2}}{6}-\frac{3\sqrt[3]{2}}{6}$
$=\frac{4\sqrt[3]{2}-3\sqrt[3]{2}}{6}$
$=\frac{\sqrt[3]{2}}{6}$

43. $-\frac{\sqrt[3]{2x^4}}{9}+\sqrt[3]{\frac{250x^4}{27}}$
$=\frac{-\sqrt[3]{x^3}\sqrt[3]{2x}}{9}+\frac{\sqrt[3]{250x^4}}{\sqrt[3]{27}}$
$=\frac{-x\sqrt[3]{2x}}{9}+\frac{\sqrt[3]{125x^3}\sqrt[3]{2x}}{3}$
$=\frac{-x\sqrt[3]{2x}}{9}+\frac{5x\sqrt[3]{2x}}{3}$
$=\frac{-x\sqrt[3]{2x}+15x\sqrt[3]{2x}}{9}$
$=\frac{14x\sqrt[3]{2x}}{9}$

45. $P=2\sqrt{12}+\sqrt{12}+2\sqrt{27}+3\sqrt{3}$
$=2\sqrt{4}\sqrt{3}+\sqrt{4}\sqrt{3}+2\sqrt{9}\sqrt{3}+3\sqrt{3}$
$=2\cdot 2\sqrt{3}+2\sqrt{3}+2\cdot 3\sqrt{3}+3\sqrt{3}$
$=(4+2+6+3)\sqrt{3}$
$=15\sqrt{3}$
The perimeter of the trapezoid is $15\sqrt{3}$ in.

47. $\sqrt{7}(\sqrt{5}+\sqrt{3})$
$=\sqrt{7}\sqrt{5}+\sqrt{7}\sqrt{3}$
$=\sqrt{35}+\sqrt{21}$

49. $(\sqrt{5}-\sqrt{2})^2$
$=\sqrt{5}^2-2\sqrt{5}\sqrt{2}+\sqrt{2}^2$
$=5-2\sqrt{10}+2$
$=7-2\sqrt{10}$

51. $\sqrt{3x}(\sqrt{3}-\sqrt{x})$
$=\sqrt{3x}\sqrt{3}-\sqrt{3x}\sqrt{x}$
$=\sqrt{3}\sqrt{x}\sqrt{3}-\sqrt{3}\sqrt{x}\sqrt{x}$
$=3\sqrt{x}-\sqrt{3}x$

53. $(2\sqrt{x}-5)(3\sqrt{x}+1)$
$=(2\sqrt{x})(3\sqrt{x})+(2\sqrt{x})1-5(3\sqrt{x})-5\cdot 1$
$=6x+2\sqrt{x}-15\sqrt{x}-5$
$=6x-13\sqrt{x}-5$

55. $(\sqrt[3]{a}-4)(\sqrt[3]{a}+5)$
$=(\sqrt[3]{a})^2+5\sqrt[3]{a}-4\sqrt[3]{a}-4\cdot 5$
$=\sqrt[3]{a^2}+\sqrt[3]{a}-20$

57. $6(\sqrt{2}-2)=6\sqrt{2}-6\cdot 2=6\sqrt{2}-12$

59. $\sqrt{2}(\sqrt{2}+x\sqrt{6})$
$=\sqrt{2}^2+\sqrt{2}(x\sqrt{6})$
$=2+\sqrt{2}(x\sqrt{2}\sqrt{3})$
$=2+2x\sqrt{3}$

61. $(2\sqrt{7}+3\sqrt{5})(\sqrt{7}-2\sqrt{5})$
$=2\sqrt{7}^2-(2\sqrt{7})(2\sqrt{5})+(3\sqrt{5})\sqrt{7}-3\cdot 2\sqrt{5}^2$
$=2\cdot 7-4\sqrt{35}+3\sqrt{35}-6\cdot 5$
$=14-\sqrt{35}-30$
$=-16-\sqrt{35}$

63. $(\sqrt{x}-y)(\sqrt{x}+y)$
$=\sqrt{x}^2-y^2$
$=x-y^2$

65. $(\sqrt{3}+x)^2$
$=\sqrt{3}^2+2\sqrt{3}x+x^2$
$=3+2\sqrt{3}x+x^2$

67. $(\sqrt{5x}-3\sqrt{2})(\sqrt{5x}-3\sqrt{3})$
$=(\sqrt{5x})^2-3\sqrt{3}\sqrt{5x}-3\sqrt{2}\sqrt{5x}+(3\sqrt{2})(3\sqrt{3})$
$=5x-3\sqrt{15x}-3\sqrt{10x}+9\sqrt{6}$

69. $(\sqrt[3]{4}+2)(\sqrt[3]{2}-1)$
$=\sqrt[3]{4}\sqrt[3]{2}-\sqrt[3]{4}\cdot 1+2\sqrt[3]{2}-2\cdot 1$
$=\sqrt[3]{8}-\sqrt[3]{4}+2\sqrt[3]{2}-2$
$=2-\sqrt[3]{4}+2\sqrt[3]{2}-2$
$=-\sqrt[3]{4}+2\sqrt[3]{2}$

71. $(\sqrt[3]{x}+1)(\sqrt[3]{x}-4\sqrt{x}+7)$
$=\sqrt[3]{x}(\sqrt[3]{x}-4\sqrt{x}+7)+1(\sqrt[3]{x}-4\sqrt{x}+7)$
$=\sqrt[3]{x}^2-\sqrt[3]{x}(4\sqrt{x})+\sqrt[3]{x}(7)+\sqrt[3]{x}-4\sqrt{x}+7$
$=\sqrt[3]{x}^2-4\sqrt[3]{\sqrt{x^2}}\sqrt{\sqrt[3]{x^3}}+8\sqrt[3]{x}-4\sqrt{x}+7$
$=\sqrt[3]{x^2}-4\sqrt[6]{x^2}\sqrt[6]{x^3}+8\sqrt[3]{x}-4\sqrt{x}+7$
$=\sqrt[3]{x^2}-4\sqrt[6]{x^5}+8\sqrt[3]{x}-4\sqrt{x}+7$

73. $P=2(3\sqrt{20})+2\sqrt{125}$
$=6\sqrt{4}\sqrt{5}+2\sqrt{25}\sqrt{5}$
$=6(2)\sqrt{5}+2(5)\sqrt{5}$
$=12\sqrt{5}+10\sqrt{5}$
$=22\sqrt{5}$ ft

75. $\frac{2x-14}{2}=\frac{2(x-7)}{2}=x-7$

77. $\frac{7x-7y}{x^2-y^2}=\frac{7(x-y)}{(x-y)(x+y)}=\frac{7}{x+y}$

79. $\frac{6a^2b-9ab}{3ab}=\frac{3ab(2a-3)}{3ab}=2a-3$

81. $\frac{-4+2\sqrt{3}}{6}=\frac{2(-2+\sqrt{3})}{2\cdot 3}=\frac{-2+\sqrt{3}}{3}$

Exercise Set 7.5

1. $\frac{1}{\sqrt{3}}=\frac{1}{\sqrt{3}}\cdot\frac{\sqrt{3}}{\sqrt{3}}=\frac{\sqrt{3}}{3}$

3. $\sqrt{\frac{1}{5}}=\sqrt{\frac{1}{5}\cdot\frac{5}{5}}=\sqrt{\frac{5}{25}}=\frac{\sqrt{5}}{\sqrt{25}}=\frac{\sqrt{5}}{5}$

5. $\frac{4}{\sqrt[3]{3}}\cdot\frac{\sqrt[3]{9}}{\sqrt[3]{9}}=\frac{4\sqrt[3]{9}}{\sqrt[3]{27}}=\frac{4\sqrt[3]{9}}{3}$

7. $\frac{3}{\sqrt{8x}}=\frac{3}{\sqrt{8x}}\cdot\frac{\sqrt{2x}}{\sqrt{2x}}=\frac{3\sqrt{2x}}{\sqrt{16x^2}}=\frac{3\sqrt{2x}}{4x}$

9. $\frac{3}{\sqrt[3]{4x^2}}=\frac{3}{\sqrt[3]{4x^2}}\cdot\frac{\sqrt[3]{2x}}{\sqrt[3]{2x}}=\frac{3\sqrt[3]{2x}}{\sqrt[3]{8x^3}}=\frac{3\sqrt[3]{2x}}{2x}$

11. $\frac{6}{2-\sqrt{7}}$
$=\frac{6}{2-\sqrt{7}}\cdot\frac{2+\sqrt{7}}{2+\sqrt{7}}$
$=\frac{6(2+\sqrt{7})}{2^2-\sqrt{7^2}}$
$=\frac{6(2+\sqrt{7})}{4-7}$
$=-\frac{6(2+\sqrt{7})}{3}$
$=-2(2+\sqrt{7})$

13. $\frac{-7}{\sqrt{x}-3}$
$=\frac{-7}{\sqrt{x}-3}\cdot\frac{\sqrt{x}+3}{\sqrt{x}+3}$
$=\frac{-7(\sqrt{x}+3)}{\sqrt{x^2}-3^2}$
$=\frac{-7(\sqrt{x}+3)}{x-9}$

15. $\frac{\sqrt{2}-\sqrt{3}}{\sqrt{2}+\sqrt{3}}$
$=\frac{\sqrt{2}-\sqrt{3}}{\sqrt{2}+\sqrt{3}}\cdot\frac{\sqrt{2}-\sqrt{3}}{\sqrt{2}-\sqrt{3}}$
$=\frac{\sqrt{2}^2-2\sqrt{2}\sqrt{3}+\sqrt{3}^2}{\sqrt{2}^2-\sqrt{3}^2}$
$=\frac{2-2\sqrt{6}+3}{2-3}$
$=\frac{5-2\sqrt{6}}{-1}$
$=-5+2\sqrt{6}$

17. $\dfrac{\sqrt{a}+1}{2\sqrt{a}-\sqrt{b}}$

$=\dfrac{\sqrt{a}+1}{2\sqrt{a}-\sqrt{b}}\cdot\dfrac{2\sqrt{a}+\sqrt{b}}{2\sqrt{a}+\sqrt{b}}$

$=\dfrac{\sqrt{a}(2\sqrt{a})+\sqrt{a}\sqrt{b}+1(2\sqrt{a})+1\sqrt{b}}{(2\sqrt{a})^2-\sqrt{b}^2}$

$=\dfrac{2a+\sqrt{ab}+2\sqrt{a}+\sqrt{b}}{4a-b}$

19. $\dfrac{9}{\sqrt{3a}}=\dfrac{9}{\sqrt{3a}}\cdot\dfrac{\sqrt{3a}}{\sqrt{3a}}=\dfrac{9\sqrt{3a}}{3a}=\dfrac{3\sqrt{3a}}{a}$

21. $\dfrac{3}{\sqrt[3]{2}}=\dfrac{3}{\sqrt[3]{2}}\cdot\dfrac{\sqrt[3]{4}}{\sqrt[3]{4}}=\dfrac{3\sqrt[3]{4}}{\sqrt[3]{8}}=\dfrac{3\sqrt[3]{4}}{2}$

23. $\dfrac{2\sqrt{3}}{\sqrt{7}}=\dfrac{2\sqrt{3}}{\sqrt{7}}\cdot\dfrac{\sqrt{7}}{\sqrt{7}}=\dfrac{2\sqrt{21}}{7}$

25. $\dfrac{8}{1+\sqrt{10}}$

$=\dfrac{8}{1+\sqrt{10}}\cdot\dfrac{1-\sqrt{10}}{1-\sqrt{10}}$

$\dfrac{8(1-\sqrt{10})}{1^2-\sqrt{10}^2}$

$=\dfrac{8(1-\sqrt{10})}{1-10}$

$=-\dfrac{8(1-\sqrt{10})}{9}$

27. $\dfrac{\sqrt{x}}{\sqrt{x}+\sqrt{y}}$

$=\dfrac{\sqrt{x}}{\sqrt{x}+\sqrt{y}}\cdot\dfrac{\sqrt{x}-\sqrt{y}}{\sqrt{x}-\sqrt{y}}$

$=\dfrac{\sqrt{x}\sqrt{x}-\sqrt{x}\sqrt{y}}{\sqrt{x}^2-\sqrt{y}^2}$

$\dfrac{x-\sqrt{xy}}{x-y}$

29. $\sqrt{\dfrac{5}{3}}=\dfrac{\sqrt{5}}{\sqrt{3}}\cdot\dfrac{\sqrt{5}}{\sqrt{5}}=\dfrac{\sqrt{25}}{\sqrt{15}}=\dfrac{5}{\sqrt{15}}$

31. $\sqrt{\dfrac{18}{5}}=\dfrac{\sqrt{18}}{\sqrt{5}}=\dfrac{\sqrt{9}\sqrt{2}}{\sqrt{5}}=\dfrac{3\sqrt{2}}{\sqrt{5}}$

$=\dfrac{3\sqrt{2}}{\sqrt{5}}\cdot\dfrac{\sqrt{2}}{\sqrt{2}}=\dfrac{3\sqrt{4}}{\sqrt{10}}=\dfrac{3(2)}{\sqrt{10}}=\dfrac{6}{\sqrt{10}}$

33. $\dfrac{\sqrt{4x}}{7}\cdot\dfrac{\sqrt{4x}}{\sqrt{4x}}=\dfrac{\sqrt{16x^2}}{7\sqrt{4x}}=\dfrac{4x}{7\sqrt{4x}}$

35. $\dfrac{\sqrt[3]{5y^2}}{\sqrt[3]{4x}}\cdot\dfrac{\sqrt[3]{5^2y}}{\sqrt[3]{5^2y}}=\dfrac{\sqrt[3]{5^3y^3}}{\sqrt[3]{4(5^2)xy}}=\dfrac{5y}{\sqrt[3]{100xy}}$

37. $\dfrac{(2-\sqrt{7})}{-5}\cdot\dfrac{(2+\sqrt{7})}{(2+\sqrt{7})}$

$=\dfrac{(2-\sqrt{7})(2+\sqrt{7})}{-5(2+\sqrt{7})}$

$=\dfrac{4-\sqrt{49}}{-10-5\sqrt{7}}$

$=\dfrac{4-7}{-10-5\sqrt{7}}$

$\dfrac{-3}{-10-5\sqrt{7}}$

$\dfrac{-1(3)}{-1\left(10+5\sqrt{7}\right)}$

$=\dfrac{3}{10+5\sqrt{7}}$

39. $\dfrac{(\sqrt{x}+3)}{\sqrt{x}}\cdot\dfrac{(\sqrt{x}-3)}{(\sqrt{x}-3)}$

$=\dfrac{(\sqrt{x}+3)(\sqrt{x}-3)}{\sqrt{x}(\sqrt{x}-3)}$

$=\dfrac{\sqrt{x^2}-9}{\sqrt{x^2}-3\sqrt{x}}$

$=\dfrac{x-9}{x-3\sqrt{x}}$

41. $\sqrt{\dfrac{2}{5}}=\dfrac{\sqrt{2}}{\sqrt{5}}\cdot\dfrac{\sqrt{2}}{\sqrt{2}}=\dfrac{\sqrt{4}}{\sqrt{10}}=\dfrac{2}{\sqrt{10}}$

43. $\dfrac{\sqrt{2x}}{11}\cdot\dfrac{\sqrt{2x}}{\sqrt{2x}}=\dfrac{\sqrt{4x^2}}{11\sqrt{2x}}=\dfrac{2x}{11\sqrt{2x}}$

45. $\sqrt[3]{\frac{7}{8}} = \frac{\sqrt[3]{7}}{\sqrt[3]{8}} = \frac{\sqrt[3]{7}}{2} \cdot \frac{\sqrt[3]{7^2}}{\sqrt[3]{7^2}} = \frac{\sqrt[3]{7^3}}{2\sqrt[3]{7^2}} = \frac{7}{2\sqrt[3]{7^2}}$

47. $\frac{(2-\sqrt{11})}{6} \cdot \frac{(2+\sqrt{11})}{(2+\sqrt{11})}$
$= \frac{(2-\sqrt{11})(2+\sqrt{11})}{6(2+\sqrt{11})}$
$= \frac{4-\sqrt{121}}{12+6\sqrt{11}}$
$= \frac{4-11}{12+6\sqrt{11}}$
$= \frac{-7}{12+6\sqrt{11}}$

49. $\frac{\sqrt[3]{3x^5}}{10} \cdot \frac{\sqrt[3]{3^2 x}}{\sqrt[3]{3^2 x}} = \frac{\sqrt[3]{3^3 x^6}}{10\sqrt[3]{3^2 x}} = \frac{3x^2}{10\sqrt[3]{9x}}$

51. $\sqrt{\frac{18x^4y^6}{3z}} = \sqrt{\frac{6x^4y^6}{z}} = \frac{x^2y^3\sqrt{6}}{\sqrt{z}}$
$= \frac{x^2y^3\sqrt{6}}{\sqrt{z}} \cdot \frac{\sqrt{6}}{\sqrt{6}} = \frac{x^2y^3\sqrt{36}}{\sqrt{6z}} = \frac{6x^2y^3}{\sqrt{6z}}$

53. $\frac{(\sqrt{2}-1)}{(\sqrt{2}+1)} \cdot \frac{(\sqrt{2}+1)}{(\sqrt{2}+1)}$
$= \frac{(\sqrt{2}-1)(\sqrt{2}+1)}{(\sqrt{2}+1)(\sqrt{2}+1)}$
$= \frac{\sqrt{4}-1}{\sqrt{4}-2\sqrt{2}+1}$
$= \frac{2-1}{2-2\sqrt{2}+1}$
$= \frac{1}{3-2\sqrt{2}}$

55. $\frac{(\sqrt{x}+1)}{(\sqrt{x}-1)} \cdot \frac{(\sqrt{x}-1)}{(\sqrt{x}-1)}$
$= \frac{(\sqrt{x}+1)(\sqrt{x}-1)}{(\sqrt{x}-1)(\sqrt{x}-1)}$
$= \frac{\sqrt{x^2}-1}{\sqrt{x^2}-2\sqrt{x}+1}$
$= \frac{x-1}{x-2\sqrt{x}+1}$

57. Answers may vary.

59. $r = \sqrt{\frac{A}{4\pi}}$
$= \frac{\sqrt{A}}{\sqrt{4\pi}}$
$= \frac{\sqrt{A}}{2\sqrt{\pi}}$
$= \frac{\sqrt{A}\sqrt{\pi}}{2\sqrt{\pi}\sqrt{\pi}}$
$= \frac{\sqrt{A\pi}}{2\pi}$

61. Answers may vary.

63. $2x - 7 = 3(x-4)$
$2x - 7 = 3x - 12$
$12 - 7 = 3x - 2x$
$5 = x$
$\{5\}$

65. $(x-6)(2x+1) = 0$
$x - 6 = 0$ or $2x + 1 = 0$
$x = 6$ or $x = -\frac{1}{2}$
$\left\{-\frac{1}{2}, 6\right\}$

67. $x^2 - 8x = -12$
$x^2 - 8x + 12 = 0$
$(x-6)(x-2) = 0$
$x - 6 = 0$ or $x - 2 = 0$
$x = 0$ or $x = 2$
$\{2, 6\}$

Section 7.6 Graphing Calculator Explorations

1. $\sqrt{x+7} = x$
$\{3.19\}$

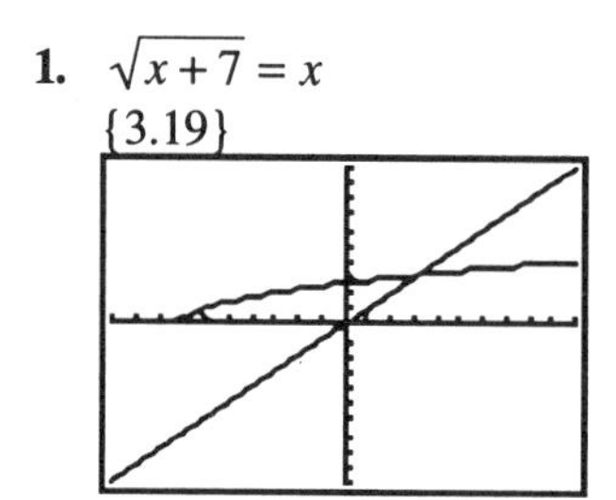

3. $\sqrt{2x+1}=\sqrt{2x+2}$

{ }

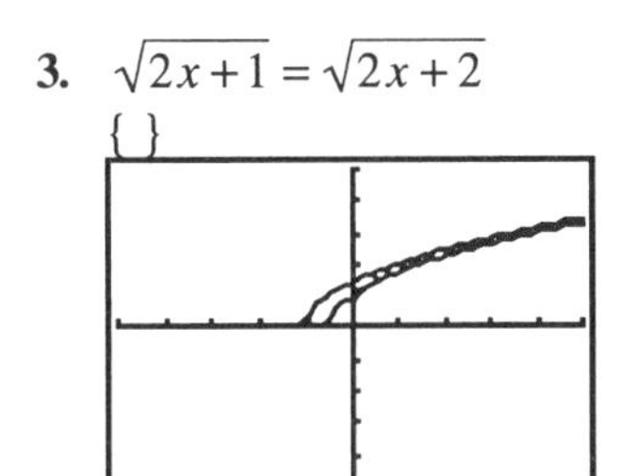

5. $1.2x=\sqrt{3.1x+5}$

{3.23}

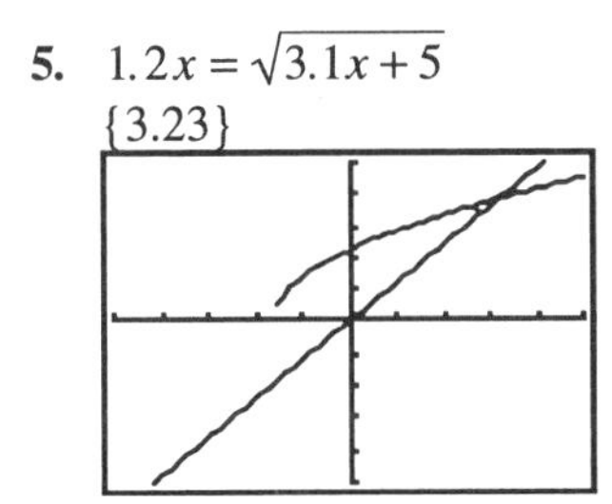

Exercise Set 7.6

1. $\sqrt{2x}=4$
$2x=4^2$
$2x=16$
$x=8$
$\{8\}$

3. $\sqrt{x-3}=2$
$x-3=2^2$
$x-3=4$
$x=7$
$\{7\}$

5. $\sqrt{2x}=-4$
No solution since a principle square root does not yield a high negative number.

7. $\sqrt{4x-3}-5=0$
$\sqrt{4x-3}=5$
$4x-3=5^2$
$4x-3=25$
$4x=28$
$x=7$
$\{7\}$

9. $\sqrt{2x-3}-2=1$
$\sqrt{2x-3}=3$
$2x-3=3^2$
$2x-3=9$
$2x=12$
$x=6$
$\{6\}$

11. $\sqrt[3]{6x}=-3$
$6x=(-3)^3$
$6x=-27$
$x=\frac{-27}{6}=-\frac{9}{2}$
$\left\{-\frac{9}{2}\right\}$

13. $\sqrt[3]{x-2}-3=0$
$\sqrt[3]{x-2}=3$
$x-2=3^3$
$x-2=27$
$x=29$
$\{29\}$

15. $\sqrt{13-x}=x-1$
$13-x=(x-1)^2$
$13-x=x^2-2x+1$
$0=x^2-x-12$
$0=(x-4)(x+3)$
$x-4=0$ or $x+3=0$
$x=4$ or $x=-3$
We discard the –3 as extraneous, leaving $x=4$ as the only solution. $\{4\}$

17. $x-\sqrt{4-3x}=-8$
$x+8=\sqrt{4-3x}$
$(x+8)^2=4-3x$
$x^2+16x+64=4-3x$
$x^2+19x+60=0$
$(x+4)(x+15)=0$
$x+4=0$ or $x+15=0$
$x=-4$ or $x=-15$
We discard the –15 as extraneous, leaving $x=-4$ as the only solution. $\{-4\}$

19. $\sqrt{y+5}=2-\sqrt{y-4}$
$y+5=(2-\sqrt{y-4})^2$
$y+5=4-4\sqrt{y-4}+y-4$
$y+5=y-4\sqrt{y-4}$
$5=-4\sqrt{y-4}$
$25=16(y-4)$
$\frac{25}{16}=y-4$
$y=\frac{89}{16}=5\frac{9}{16}$, which we discard as extraneous.
{ }

21. $\sqrt{x-3}+\sqrt{x+2}=5$
$\sqrt{x-3}=5-\sqrt{x+2}$
$x-3=25-10\sqrt{x+2}+x+2$
$-3=27-10\sqrt{x+2}$
$-30=-10\sqrt{x+2}$
$3=\sqrt{x+2}$
$9=x+2$
$7=x$
$\{7\}$

23. Let c = the length of the hypotenuse of the right triangle. By the Pythagorean theorem,
$c^2=6^2+3^2=36+9$
$c^2=45$
or $c=\sqrt{45}=\sqrt{9}\sqrt{5}=3\sqrt{5}$ ft.

25. Let b = the length of the unknown leg of the right triangle. By the Pythagorean theorem,
$7^2=3^2+b^2$
$49=9+b^2$
$b^2=40$
$b=\sqrt{40}=\sqrt{4}\sqrt{10}=2\sqrt{10}$ m

27. $\sqrt{3x-2}=5$
$3x-2=5^2$
$3x-2=25$
$3x=27$
$x=9$
$\{9\}$

29. $-\sqrt{2x}+4=-6$
$10=\sqrt{2x}$
$10^2=2x$
$100=2x$
$x=50$
$\{50\}$

31. $\sqrt{3x+1}+2=0$
$\sqrt{3x+1}=-2$
No solution exists since a principle square root does not yield a negative number.
{ }

33. $\sqrt[4]{4x+1}-2=0$
$\sqrt[4]{4x+1}=2$
$4x+1=2^4$
$4x+1=16$
$4x=15$
$x=\frac{15}{4}$
$\left\{\frac{15}{4}\right\}$

35. $\sqrt{4x-3}=5$
$4x-3=5^2$
$4x-3=25$
$4x=28$
$x=7$
$\{7\}$

37. $\sqrt[3]{6x-3}-3=0$
$\sqrt[3]{6x-3}=3$
$6x-3=3^3$
$6x-3=27$
$6x=30$
$x=5$
$\{5\}$

39. $\sqrt[3]{2x-3}-2=-5$
$\sqrt[3]{2x-3}=-3$
$2x-3=(-3)^3$
$2x-3=-27$
$2x=-24$
$x=-12$
$\{-12\}$

41. $\sqrt{x+4}=\sqrt{2x-5}$
$x+4=2x-5$
$9=x$
$x=9$
$\{9\}$

43. $x-\sqrt{1-x}=-5$
$x+5=\sqrt{1-x}$
$(x+5)^2=1-x$
$x^2+10x+25=1-x$
$x^2+11x+24=0$
$(x+3)(x+8)=0$
$x+3=0$ or $x+8=0$
$x=-3$ or $x=-8$
We discard –8 as extraneous.
$\{-3\}$

45. $\sqrt[3]{-6x-1}=\sqrt[3]{-2x-5}$
$-6x-1=-2x-5$
$4=4x$
$x=1$
$\{1\}$

47. $\sqrt{5x-1}-\sqrt{x}+2=3$
$\sqrt{5x-1}=\sqrt{x}+1$
$5x-1=x+2\sqrt{x}+1$
$4x-2=2\sqrt{x}$
$2x-1=\sqrt{x}$
$4x^2-4x+1=x$
$4x^2-5x+1=0$
$(4x-1)(x-1)=0$
$4x-1=0$ or $x-1=0$
$4x-1=0$ or $x=1$
$x=\frac{1}{4}$
Discard $\frac{1}{4}$ as extraneous.
$\{1\}$

49. $\sqrt{2x-1}=\sqrt{1-2x}$
$\sqrt{2x-1}=\sqrt{-(2x-1)}$
It follows that $2x-1=0$ (Otherwise one of the radicands would be negative).
So $2x=1$
$x=\frac{1}{2}$
$\left\{\frac{1}{2}\right\}$

51. Answers may vary.

53. Let b = the length of the unknown leg of the right triangle. By the Pythagorean theorem,
$(11\sqrt{5})^2=9^2+b^2$
$605=81+b^2$
$b^2=524$
$b=\sqrt{524}=\sqrt{4}\sqrt{131}=2\sqrt{131}\approx 22.9$ m

55. Let c = the length of the hypotenuse of the right triangle. By the Pythagorean theorem,
$c^2=7^2+(7.2)^2=100.84$
so $c=\sqrt{100.84}\approx 10.0$ mm

57. Let c = the length of the hypotenuse of the right triangle as shown in the figure. By the Pythagorean theorem,
$c^2=8^2+15^2=64+225=289$
so $c=\sqrt{289}=17$ ft.
Thus, 17 feet of cable is needed.

59. $x^2=(5)^2+(12)^2$
$x^2=25+144$
$x^2=169$
$x=\pm\sqrt{169}$
$x=\pm 13$
The answer is 13 feet.

61. $c^2 = a^2 + b^2$

$a = \sqrt{c^2 - b^2}$

$a = \sqrt{(2382)^2 - (2063)^2}$

$a = 1191$ feet

63. We need to solve the equation:

$80\sqrt[3]{n} + 500 = 1620$

$80\sqrt[3]{n} = 1120$

$\sqrt[3]{n} = 14$

$n = 14^3$

$n = 2744$

Thus, the company needs to make fewer than 2744 deliveries per day, or 2743 deliveries.

65. We are given $R = \sqrt{A^2 + B^2}$ with $A = 600$ and $R = 850$.

Substituting, we get

$850 = \sqrt{600^2 + B^2}$

$850^2 = 600^2 + B^2$

$B^2 = 850^2 - 600^2 = 722{,}500 - 360{,}000$
$= 362{,}500$

so $B = \sqrt{362{,}500} = 50\sqrt{145} \approx 602$ lbs.

Thus, tractor B is exerting approximately 602 lbs. of force.

67. $D(h) = 111.7\sqrt{h}$

$80 = 111.7\sqrt{h}$

$\frac{80}{111.7} = \sqrt{h}$

$0.7162 = \sqrt{h}$

$(0.7162)^2 = h$

$0.513K = h$

69. It is not a function.

71. It is not a function.

73. It is not a function.

75. $\dfrac{\frac{1}{y} + \frac{4}{5}}{-\frac{3}{20}}$

$= \dfrac{20\left(\frac{1}{y} + \frac{4}{5}\right)}{20y\left(-\frac{3}{20}\right)}$

$= \dfrac{20y\left(\frac{1}{y}\right) + 20y\left(\frac{4}{5}\right)}{20y\left(-\frac{3}{20}\right)}$

$= \dfrac{20 + 16y}{-3y} = -\dfrac{20 + 16y}{3y}$

77. $\dfrac{\frac{1}{y} + \frac{1}{x}}{\frac{1}{y} - \frac{1}{x}}$

$= \dfrac{xy\left(\frac{1}{y} + \frac{1}{x}\right)}{xy\left(\frac{1}{y} - \frac{1}{x}\right)}$

$= \dfrac{xy\left(\frac{1}{y}\right) + xy\left(\frac{1}{x}\right)}{xy\left(\frac{1}{y}\right) - xy\left(\frac{1}{x}\right)}$

$= \dfrac{x + y}{x - y}$

79. $\sqrt{(x^2 - x) + 7} = 2(x^2 - x) - 1$

Let $u = x^2 - x$.

$\sqrt{u + 7} = 2u - 1$

$(\sqrt{u + 7})^2 = (2u - 1)^2$

$u + 7 = 4u^2 - 4u + 1$

$0 = 4u^2 - 5u - 6$

$0 = (4u + 3)(u - 2)$

$4u + 3 = 0$ or $u - 2 = 0$

$4u = -3$ or $u = 2$

$u = -\frac{3}{4}$ or $u = 2$

Substituting $x^2 - x = u$

$x^2 - x = -\frac{3}{4}$ or $x^2 - x = 2$

$4(x^2 - x) = 4\left(-\frac{3}{4}\right)$ or $x^2 - x - 2 = 0$

$4x^2 - 4x = -3$ or $(x - 2)(x + 1) = 0$

$4x^2 - 4x + 3 = 0$ or $x - 2 = 0$ or $x + 1 = 0$

Can't factor or $x = 2$ or $x = -1$

$\{-1, 2\}$

81. $x^2+6x=4\sqrt{x^2+6x}$

Let $u=x^2+6x$.

$u=4\sqrt{u}$

$(u)^2=(4\sqrt{u})^2$

$u^2=16u$

$u^2-16u=0$

$u(u-16)=0$

$u=0$ or $u-16=0$

Substituting $x^2+6x=u$

$x^2+6x=0$ or $x^2+6x-16=0$

$x(x+6)=0$ or $(x+8)(x-2)=0$

$x=0$ or $x+6=0$ or $x+8=0$ or $x-2=0$

$x=0$ or $x=-6$ or $x=-8$ or $x=2$

All solutions check.

$\{-8, -6, 0, 2\}$

Section 7.7

Mental Math

1. $\sqrt{-81}=9i$

3. $\sqrt{-7}=i\sqrt{7}$

5. $-\sqrt{16}=-4$

7. $\sqrt{-64}=8i$

Exercise Set 7.7

1. $\sqrt{-24}=\sqrt{4}\sqrt{6}\sqrt{-1}=2i\sqrt{6}$

3. $-\sqrt{-36}=-\sqrt{36}\sqrt{-1}=-6i$

5. $8\sqrt{-63}=8\sqrt{9}\sqrt{7}\sqrt{-1}$
$=8\cdot 3\sqrt{7}i=24i\sqrt{7}$

7. $-\sqrt{54}=-\sqrt{9}\sqrt{6}=-3\sqrt{6}$

9. $(4-7i)+(2+3i)$
$=(4+2)+(-7+3)i$
$=6-4i$

11. $(6+5i)-(8-i)$
$=(6-8)+[5-(-1)]i$
$=-2+6i$

13. $6-(8+4i)$
$=6-8-4i$
$=-2-4i$

15. $6i(2-3i)$
$=6i(2)-6i(3i)$
$=12i-18i^2$
$=12i-18(-1)$
$=18+12i$

17. $(\sqrt{3}+2i)(\sqrt{3}-2i)$
$(\sqrt{3})^2-(2i)^2$
$\sqrt{3^2}+2^2=3+4=7$

19. $(4-2i)^2$
$=16-16i+4i^2$
$=16-16i+4(-1)$
$=16-4-16i$
$=12-16i$

21. $\frac{4}{i}\cdot\frac{-i}{-i}=\frac{-4i}{-i^2}=\frac{-4i}{-(-1)}=\frac{-4i}{1}=-4i$

23. $\frac{7}{4+3i}=\frac{7}{4+3i}\cdot\frac{4-3i}{4-3i}=\frac{28-21i}{4^2+3^2}$
$=\frac{28-21i}{16+9}=\frac{28-21i}{25}=\frac{28}{25}-\frac{21}{25}i$

25. $\frac{3+5i}{1+i}=\frac{3+5i}{1+i}\cdot\frac{1-i}{1-i}=\frac{3+5i-3i-5i^2}{1^2+1^2}$
$=\frac{3+2i-5(-1)}{1+1}=\frac{3+5+2i}{2}$
$=\frac{8+2i}{2}=4+i$

27. Answers may vary.

29. $\sqrt{-2}\sqrt{-7}=(\sqrt{2}i)(\sqrt{7}i)=\sqrt{14}i^2$
$\sqrt{14}(-1)=-\sqrt{14}$

31. $\sqrt{-5}\sqrt{-10}=(\sqrt{5}i)(\sqrt{10}i)=\sqrt{5}\sqrt{5}\sqrt{2}i^2$
$5\sqrt{2}(-1)=-5\sqrt{2}$

33. $\sqrt{16}\sqrt{-1} = 4i$

35. $\frac{\sqrt{-9}}{\sqrt{3}} = \frac{3i}{\sqrt{3}} = i\sqrt{3}$

37. $\frac{\sqrt{-80}}{\sqrt{-10}} = \frac{\sqrt{80}i}{\sqrt{10}i} = \sqrt{\frac{80}{10}}$
$= \sqrt{8} = \sqrt{4}\sqrt{2} = 2\sqrt{2}$

39. $i^8 = (i^4)^2 = 1^2 = 1$

41. $i^{21} = i^{20}i = (i^4)^5 i = 1^5 i = 1i = i$

43. $i^{11} = i^{10}i = (i^2)^5 i = (-1)^5 i = (-1)i = -i$

45. $i^{-6} = (i^2)^{-3} = (-1)^{-3} = \frac{1}{(-1)^3} = \frac{1}{-1} = -1$

47. $(7i)(-9i) = -63i^2 = -63(-1) = 63$

49. $(6-3i)-(4-2i) = 6-3i-4+2i = 2-i$

51. $(6-2i)(3+i) = 2(3-i)(3+i)$
$= 2(3^2+1^2) = 2(9+1) = 2(10) = 20$

53. $(8-\sqrt{-3})-(2+\sqrt{-12})$
$= 8 - i\sqrt{3} - 2 - i\sqrt{4}\sqrt{3}$
$= 6 - i\sqrt{3} - 2i\sqrt{3} = 6 - 3i\sqrt{3}$

55. $(1-i)(1+i) = 1^2 - i^2 = 1^2 + 1^2 = 1+1 = 2$

57. $\frac{16+15i}{-3i} = \frac{(16+15i)i}{-3i^2} = \frac{16i+15i^2}{-3(-1)}$
$= \frac{16i+15(-1)}{3} = \frac{-15}{3} + \frac{16}{3}i = -5 + \frac{16}{3}i$

59. $(9+8i)^2 = 81 + 144i + 64i^2$
$= 81 + 144i + 64(-1)$
$= 81 - 64 + 144i$
$= 17 + 144i$

61. $\frac{2}{3+i} = \frac{2}{3+i} \cdot \frac{3-i}{3-i} = \frac{2(3-i)}{3^2+1^2}$
$= \frac{6-2i}{9+1} = \frac{6}{10} - \frac{2}{10}i = \frac{3}{5} - \frac{1}{5}i$

63. $(5-6i)-4i = 5-6i-4i = 5-10i$

65. $\frac{2-3i}{2+i} = \frac{(2-3i)(2-i)}{(2+i)(2-i)} = \frac{4-6i-2i+3i^2}{2^2+1^2}$
$= \frac{4-8i+3(-1)}{4+1} = \frac{4-3-8i}{5} = \frac{1}{5} - \frac{8}{5}i$

67. $(2+4i)+(6-5i)$
$= (2+6)+(4-5)i$
$= 8 - i$

69. Since the unknown angle is the complement of the 50° angle, it must have measure 90° − 50° = 40°.

71.

1	1	−6	3	−4
		1	−5	−2
	1	−5	−2	−6

Answer: $x^2 - 5x - 2 - \frac{6}{x-1}$

73. 5

75. $5 + 9 = 14$

77. $\frac{5}{30} = 0.167$
16.7%

Chapter 7 - Review

1. $\sqrt{81} = 9$ because $9^2 = 81$.

2. $\sqrt[4]{81} = 3$

3. $\sqrt[3]{-8} = -2$ because $(-2)^3 = -8$.

4. $\sqrt[4]{-16}$ is not a real number since the 4th power of no real number is negative.

5. $-\sqrt{\frac{1}{49}} = -\frac{1}{7}$ because $\left(\frac{1}{7}\right)^2 = \frac{1}{49}$

6. $\sqrt{x^{64}} = (x^{64})^{1/2} = x^{32}$

7. $-\sqrt{36} = -6$ because $(6)^2 = 36$

8. $\sqrt[3]{64} = 4$

9. $\sqrt[3]{-a^6b^9} = \sqrt[3]{-1}\sqrt[3]{a^6}\sqrt[3]{b^9}$
$= (-1)a^2b^3 = -a^2b^3$

10. $\sqrt{16a^4b^{12}} = 4a^2b^6$

11. $\sqrt[5]{32a^5b^{10}} = \sqrt[5]{32}\sqrt[5]{a^5}\sqrt[5]{b^{10}} = 2ab^2$

12. $\sqrt[5]{-32x^{15}y^{20}} = -2x^3y^4$

13. $\sqrt{\dfrac{x^{12}}{36y^2}} = \dfrac{\sqrt{x^{12}}}{\sqrt{36y^2}} = \dfrac{x^6}{6y}$

14. $\sqrt[3]{\dfrac{27y^3}{z^{12}}} = \dfrac{\sqrt[3]{27y^3}}{\sqrt[3]{z^{12}}} = \dfrac{3y}{z^4}$

15. $\sqrt{(-x)^2} = |-x|$

16. $\sqrt[4]{(x^2-4)^4} = |x^2-4|$

17. $\sqrt[3]{(-27)^3} = -27$

18. $\sqrt[5]{(-5)^5} = -5$

19. $-\sqrt[5]{x^5} = -x$

20. $\sqrt[4]{16(2y+z)^{12}} = 2|2y+z|^3$

21. $\sqrt{25(x-y)^{10}} = \sqrt{25}\sqrt{(x-y)^{10}}$
$= 5|(x-y)^5|$

22. $\sqrt[5]{-y^5} = \sqrt[5]{-1}\sqrt[5]{y^5}$
$= (-1)y = -y$

23. $\sqrt[9]{-x^9} = -x$

24. $f(x) = \sqrt{x} + 3$
Domain: $[0, \infty)$

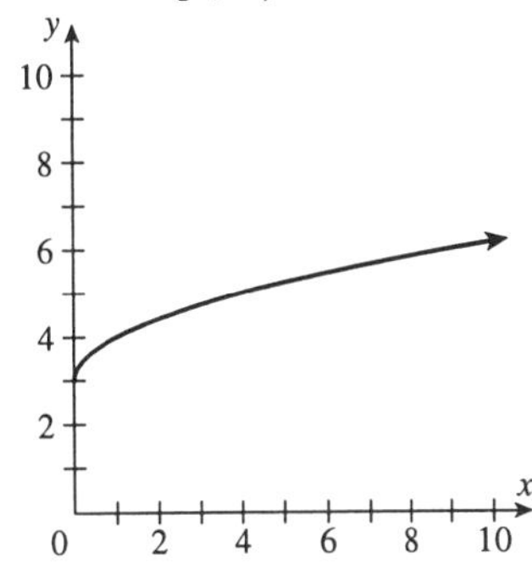

25. $g(x) = \sqrt[3]{x-3}$
Domain: $(-\infty, \infty)$

x	−5	2	3	4	11
$g(x)$	−2	−1	0	1	2

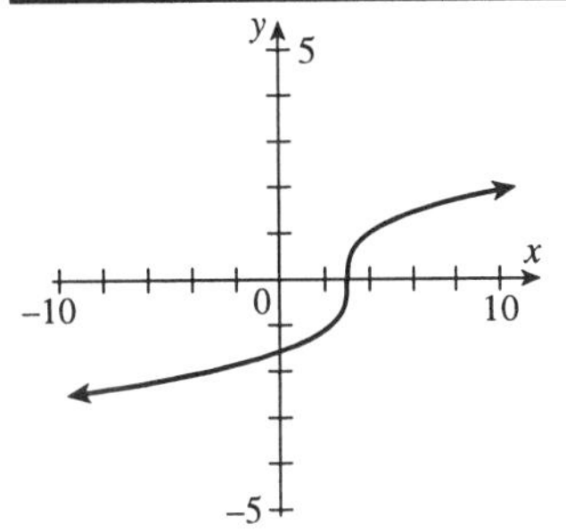

26. $\left(\frac{1}{81}\right)^{1/4} = \frac{1}{3}$

27. $\left(-\frac{1}{27}\right)^{1/3} = \dfrac{1}{(-27)^{1/3}} = \frac{1}{-3} = -\frac{1}{3}$

28. $(-27)^{-1/3} = \dfrac{1}{(-27)^{1/3}} = \frac{1}{-3} = -\frac{1}{3}$

29. $(-64)^{-1/3} = \dfrac{1}{(-64)^{1/3}} = \frac{1}{-4} = -\frac{1}{4}$

30. $-9^{3/2} = -(9^{1/2})^3 = -3^3 = -27$

31. $64^{-1/3} = \dfrac{1}{(64)^{1/3}} = \frac{1}{4}$

32. $(-25)^{5/2} = [(-25)^{1/2}]^5$ is not a real number, since the square of no real number is negative.

33. $\left(\frac{25}{49}\right)^{-3/2} = \frac{1}{\left(\frac{25}{49}\right)^{3/2}} = \frac{1}{\left(\left(\frac{25}{49}\right)^{1/2}\right)^3}$
$= \frac{1}{\left(\frac{5}{7}\right)^3} = \frac{1}{\frac{125}{343}} = \frac{343}{125}$

34. $\left(\frac{8}{27}\right)^{-2/3} = \frac{1}{\left(\frac{8}{27}\right)^{2/3}} = \frac{1}{\left[\left(\frac{8}{27}\right)^{1/3}\right]^2}$
$= \frac{1}{\left(\frac{2}{3}\right)^2} = \frac{1}{\frac{4}{9}} = \frac{9}{4}$

35. $\left(-\frac{1}{36}\right)^{-1/4} = \frac{1}{\left(-\frac{1}{36}\right)^{1/4}}$ is not a real number.

36. $\sqrt[3]{x^2} = x^{2/3}$

37. $\sqrt[5]{5x^2y^3} = (5x^2y^3)^{1/5}$
$= 5^{1/5}(x^2)^{1/5}(y^3)^{1/5}$
$= 5^{1/5}x^{2/5}y^{3/5}$

38. $y^{4/5} = \sqrt[5]{y^4}$

39. $5(xy^2z^5)^{1/3} = 5\sqrt[3]{xy^2z^5}$

40. $(x+2y)^{-1/2} = \frac{1}{(x+2y)^{1/2}} = \frac{1}{\sqrt{x+2y}}$

41. $a^{1/3}a^{4/3}a^{1/2} = a^{(1/3+4/3+1/2)}$
$= a^{(5/3+1/2)} = a^{(10/6+3/6)} = a^{13/6}$

42. $\frac{b^{1/3}}{b^{4/3}} = b^{1/3-4/3} = b^{-1} = \frac{1}{b}$

43. $(a^{-1/2}a^{-2})^3 = (a^{1/2-2})^3 = (a^{-3/2})^3$
$= a^{-9/2} = \frac{1}{a^{9/2}}$

44. $(x^{-3}y^6)^{1/3} = (x^{-3})^{1/3}(y^6)^{1/3}$
$= x^{-1}y^2 = \frac{y^2}{x}$

45. $\left(\frac{b^{3/4}}{a^{-1/2}}\right)^8 = \frac{b^6}{a^{-4}} = a^4b^6$

46. $\frac{x^{1/4}x^{-1/2}}{x^{2/3}} = \frac{x^{1/4-1/2}}{x^{2/3}} = \frac{x^{-1/4}}{x^{2/3}}$
$= \frac{1}{x^{2/3-(-1/4)}} + \frac{1}{x^{8+3/12}} = \frac{1}{x^{11/12}}$

47. $\left(\frac{49c^{5/3}}{a^{-1/4}b^{5/6}}\right)^{-1} = \frac{49^{-1}c^{-5/3}}{a^{1/4}b^{-5/6}}$
$= \frac{b^{5/6}}{49a^{1/4}c^{5/3}}$

48. $a^{-1/4}(a^{5/4} - a^{9/4})$
$= a^{-1/4+5/4} - a^{-1/4+9/4}$
$= a - a^2$

49. $\sqrt{20} = 4.472$

50. $\sqrt[3]{-39} = -3.391$

51. $\sqrt[4]{726} = 5.191$

52. $56^{1/3} = 3.826$

53. $-78^{3/4} = -26.246$

54. $105^{-2/3} = 0.045$

55. $\sqrt{3} \cdot \sqrt{8} = \sqrt{24} = \sqrt{4 \cdot 6} = 2\sqrt{6}$

56. $\sqrt[3]{7y} \cdot \sqrt[3]{x^2z} = \sqrt[3]{7x^2yz}$

57. $\frac{\sqrt{44x^3}}{\sqrt{11x}} = \sqrt{\frac{44x^3}{11x}} = \sqrt{4x^2} = 2x$

58. $\dfrac{\sqrt[4]{a^6b^{13}}}{\sqrt[4]{a^2b}} = \sqrt[4]{\dfrac{a^6b^{13}}{a^2b}}$
$= \sqrt[4]{a^{6-2}b^{13-1}} = \sqrt[4]{a^4b^{12}} = ab^3$

59. $\sqrt{60} = \sqrt{4 \cdot 15} = 2\sqrt{15}$

60. $-\sqrt{75} = -\sqrt{25 \cdot 3} = -5\sqrt{3}$

61. $\sqrt[3]{162} = \sqrt[3]{27 \cdot 6} = 3\sqrt[3]{6}$

62. $\sqrt[3]{-32} = \sqrt[3]{(-8)(4)} = -2\sqrt[3]{4}$

63. $\sqrt{36x^7} = \sqrt{36x^6 \cdot x} = 6x^3\sqrt{x}$

64. $\sqrt[3]{24a^5b^7} = \sqrt[3]{8a^3b^6 \cdot 3a^2b}$
$= 2ab^2\sqrt[3]{3a^2b}$

65. $\sqrt{\dfrac{p^{17}}{121}} = \dfrac{\sqrt{p^{16} \cdot p}}{\sqrt{121}} = \dfrac{p^8\sqrt{p}}{11}$

66. $\sqrt[3]{\dfrac{y^5}{27x^6}} = \dfrac{\sqrt[3]{y^3 \cdot y^2}}{\sqrt[3]{27x^6}} = \dfrac{y\sqrt[3]{y^2}}{3x^2}$

67. $\sqrt[4]{\dfrac{xy^6}{81}} = \dfrac{\sqrt[4]{y^4 \cdot xy^2}}{\sqrt[4]{81}} = \dfrac{y\sqrt[4]{xy^2}}{3}$

68. $\sqrt{\dfrac{2x^3}{49y^4}} = \dfrac{\sqrt{x^2 \cdot 2x}}{\sqrt{49y^4}} = \dfrac{x\sqrt{2x}}{7y^2}$

69. $\sqrt[3]{2} \cdot \sqrt{7} = 2^{1/3} \cdot 7^{1/2} = 2^{2/6} \cdot 7^{3/6}$
$= (2^2)^{1/6} \cdot (7^3)^{1/6} = 4^{1/6} \cdot 343^{1/6}$
$= (4 \cdot 343)^{1/6} = \sqrt[6]{1372}$

70. $\sqrt[3]{3} \cdot \sqrt[4]{x} = 3^{1/3} \cdot x^{1/4}$
$= 3^{4/12} \cdot x^{3/12} = (3^4)^{1/12}(x^3)^{1/12}$
$= (81)^{1/12}(x^3)^{1/12} = (81x^3)^{1/12}$
$= \sqrt[12]{81x^3}$

71. $r = \sqrt{\dfrac{A}{\pi}}$

a. $r = \sqrt{\dfrac{25}{\pi}} = \dfrac{\sqrt{25}}{\sqrt{\pi}} = \dfrac{5}{\sqrt{\pi}}$

$\dfrac{5}{\sqrt{\pi}}$ or $\dfrac{5\sqrt{\pi}}{\pi}$ meters

b. $r = \sqrt{\dfrac{104}{\pi}} = 5.75$ inches

72. $x\sqrt{75xy} - \sqrt{27x^3y}$
$= x\sqrt{25}\sqrt{3xy} - \sqrt{9x^2}\sqrt{3xy}$
$= 5x\sqrt{3xy} - 3x\sqrt{3xy}$
$= 2x\sqrt{3xy}$

73. $2\sqrt{32x^2y^3} - xy\sqrt{98y}$
$= 2\sqrt{16x^2y^2 \cdot 2y} - xy\sqrt{49 \cdot 2y}$
$= 2\sqrt{16x^2y^2}\sqrt{2y} - xy\sqrt{49}\sqrt{2y}$
$= 8xy\sqrt{2y} - 7xy\sqrt{2y} = xy\sqrt{2y}$

74. $\sqrt[3]{128} + \sqrt[3]{250}$
$= \sqrt[3]{64 \cdot 2} + \sqrt[3]{125 \cdot 2}$
$= \sqrt[3]{64}\sqrt[3]{2} + \sqrt[3]{125}\sqrt[3]{2}$
$= 4\sqrt[3]{2} + 5\sqrt[3]{2} = 9\sqrt[3]{2}$

75. $3\sqrt[4]{32a^5} - a\sqrt[4]{162a}$
$= 3\sqrt[4]{16a^4 \cdot 2a} - a\sqrt[4]{81 \cdot 2a}$
$= 3\sqrt[4]{16a^4}\sqrt[4]{2a} - a\sqrt[4]{81}\sqrt[4]{2a}$
$= 6a\sqrt[4]{2a} - 3a\sqrt[4]{2a} = 3a\sqrt[4]{2a}$

76. $\dfrac{5}{\sqrt{4}} + \dfrac{\sqrt{3}}{3} = \dfrac{5}{2} + \dfrac{\sqrt{3}}{3}$
$= \dfrac{5(3) + 2\sqrt{3}}{2 \cdot 3} = \dfrac{15 + 2\sqrt{3}}{6}$

77. $\sqrt{\frac{8}{x^2}}-\sqrt{\frac{50}{16x^2}}$
$=\frac{2\sqrt{2}}{x}-\frac{5\sqrt{2}}{4x}$
$=\frac{8\sqrt{2}-5\sqrt{2}}{4x}=\frac{3\sqrt{2}}{4x}$

78. $2\sqrt{50}-3\sqrt{125}+\sqrt{98}$
$=2\sqrt{25}\sqrt{2}-3\sqrt{25}\sqrt{5}+\sqrt{49}\sqrt{2}$
$=2(5)\sqrt{2}-3(5)\sqrt{5}+7\sqrt{2}$
$=10\sqrt{2}-15\sqrt{5}+7\sqrt{2}=17\sqrt{2}-15\sqrt{5}$

79. $2a\sqrt[4]{32b^5}-3b\sqrt[4]{162a^4b}+\sqrt[4]{2a^4b^5}$
$=2a\sqrt[4]{16b^4}\sqrt[4]{2b}-3b\sqrt[4]{81a^4}\sqrt[4]{2b}$
$+\sqrt[4]{a^4b^4}\sqrt[4]{2b}$
$=[(2a)(2b)-(3b)(3a)+ab]\sqrt[4]{2b}$
$=(4ab-9ab+ab)\sqrt[4]{2b}$
$=-4ab\sqrt[4]{2b}$

80. $\sqrt{3}(\sqrt{27}-\sqrt{3})=\sqrt{3}(\sqrt{9}\sqrt{3}-\sqrt{3})$
$=\sqrt{3}(3\sqrt{3}-\sqrt{3})$
$=\sqrt{3}(2\sqrt{3})=2(3)=6$

81. $(\sqrt{x}-3)^2=\left(\sqrt{x}\right)^2-2(3)\sqrt{x}+3^2$
$=x-6\sqrt{x}+9$

82. $(\sqrt{5}-5)(2\sqrt{5}+2)$
$=2\sqrt{25}+2\sqrt{5}-10\sqrt{5}-10$
$=10-8\sqrt{5}-10=-8\sqrt{5}$

83. $(2\sqrt{x}-3\sqrt{y})(2\sqrt{x}+3\sqrt{y})$
$=(2\sqrt{x})^2-(3\sqrt{y})^2$
$=2^2\sqrt{x^2}-3^2\sqrt{y^2}=4x-9y$

84. $(\sqrt{a}+3)(\sqrt{a}-3)$
$=\left(\sqrt{a}\right)^2+3\sqrt{a}-3\sqrt{a}-3\cdot 3$
$=a-9$

85. $\left(\sqrt[3]{a}+2\right)^2=\sqrt[3]{a}^2+2(2)\sqrt[3]{a}+2^2$
$=\sqrt[3]{a^2}+4\sqrt[3]{a}+4$

86. $(\sqrt[3]{5x}+9)(\sqrt[3]{5x}-9)$
$=(\sqrt[3]{5x})^2-9^2=\sqrt[3]{25x^2}-81$

87. $(\sqrt[3]{a}+4)(\sqrt[3]{a^2}-4\sqrt[3]{a}+16)$
$=(\sqrt[3]{a})^3+4^3$
$=a+64$

88. $\frac{3}{\sqrt{7}}=\frac{3\sqrt{7}}{\sqrt{7}\cdot\sqrt{7}}=\frac{3\sqrt{7}}{7}$

89. $\sqrt{\frac{x}{12}}=\frac{\sqrt{x}}{\sqrt{12}}=\frac{\sqrt{x}\cdot\sqrt{12}}{\sqrt{12}\cdot\sqrt{12}}$
$=\frac{\sqrt{12x}}{12}=\frac{2\sqrt{3x}}{12}=\frac{\sqrt{3x}}{6}$

90. $\frac{5}{\sqrt[3]{4}}=\frac{5\sqrt[3]{16}}{\sqrt[3]{4}\sqrt[3]{16}}=\frac{5\sqrt[3]{8\cdot 2}}{\sqrt[3]{64}}$
$=\frac{10\sqrt[3]{2}}{4}=\frac{5\sqrt[3]{2}}{2}$

91. $\sqrt{\frac{24x^5}{3y^2}}=\frac{\sqrt{4x^4\cdot 6x}}{\sqrt{y^2\cdot 3}}=\frac{2x^2\sqrt{6x}}{y\sqrt{3}}$
$=\frac{2x^2\sqrt{6x}\cdot\sqrt{3}}{y\sqrt{3}\cdot\sqrt{3}}=\frac{2x^2\sqrt{18x}}{3y}$
$=\frac{6x^2\sqrt{2x}}{3y}=\frac{2x^2\sqrt{2x}}{y}$

92. $\sqrt[3]{\frac{15x^6y^7}{z^2}}=\frac{\sqrt[3]{15x^6y^7}}{\sqrt[3]{z^2}}$
$=\frac{\sqrt[3]{x^6y^6\cdot 15y}\sqrt[3]{z}}{\sqrt[3]{z^2}\cdot\sqrt[3]{z}}=\frac{x^2y^2\sqrt[3]{15yz}}{z}$

93. $\frac{5}{2-\sqrt{7}} = \frac{5}{2-\sqrt{7}} \cdot \frac{2+\sqrt{7}}{2+\sqrt{7}}$

$= \frac{5(2+\sqrt{7})}{2^2 - \sqrt{7}^2} = \frac{10+5\sqrt{7}}{4-7}$

$= \frac{10+5\sqrt{7}}{-3} = -\frac{10}{3} - \frac{5}{3}\sqrt{7}$

94. $\frac{3}{\sqrt{y}-2} \cdot \frac{\sqrt{y}+2}{\sqrt{y}+2}$

$= \frac{3\sqrt{y}+6}{y+2\sqrt{y}-2\sqrt{y}-4} = \frac{3\sqrt{y}+6}{y-4}$

95. $\frac{\sqrt{2}-\sqrt{3}}{\sqrt{2}+\sqrt{3}} = \frac{\sqrt{2}-\sqrt{3}}{\sqrt{2}+\sqrt{3}} \cdot \frac{\sqrt{2}-\sqrt{3}}{\sqrt{2}-\sqrt{3}}$

$= \frac{\sqrt{2}^2 - 2\sqrt{2}\sqrt{3} + \sqrt{3}^2}{\sqrt{2}^2 - \sqrt{3}^2}$

$= \frac{2-2\sqrt{6}+3}{2-3}$

$= \frac{5-2\sqrt{6}}{-1} = -5+2\sqrt{6}$

96. $\frac{\sqrt{11}}{3} = \frac{\sqrt{11}\cdot\sqrt{11}}{3\cdot\sqrt{11}} = \frac{11}{3\sqrt{11}}$

97. $\sqrt{\frac{18}{y}} = \frac{\sqrt{18}}{\sqrt{y}} = \frac{3\sqrt{2}}{\sqrt{y}}$

$= \frac{3\sqrt{2}\cdot\sqrt{2}}{\sqrt{y}\cdot\sqrt{2}} = \frac{3\cdot 2}{\sqrt{2y}} = \frac{6}{\sqrt{2y}}$

98. $\frac{\sqrt[3]{9}}{7} = \frac{\sqrt[3]{9}\cdot\sqrt[3]{3}}{7\cdot\sqrt[3]{3}} = \frac{3}{7\sqrt[3]{3}}$

99. $\sqrt{\frac{24x^5}{3y^2}} = \frac{\sqrt{4x^4\cdot 6x}}{\sqrt{y^2\cdot 3}}$

$= \frac{2x^2\sqrt{6x}}{y\sqrt{3}} = \frac{2x^2\sqrt{6x}\cdot\sqrt{6x}}{y\sqrt{3}\cdot\sqrt{6x}}$

$= \frac{2x^2(6x)}{y\sqrt{18x}} = \frac{12x^3}{3y\sqrt{2x}} = \frac{4x^3}{y\sqrt{2x}}$

100. $\sqrt[3]{\frac{xy^2}{10z}} = \frac{\sqrt[3]{xy^2}\cdot\sqrt[3]{x^2y}}{\sqrt[3]{10z}\cdot\sqrt[3]{x^2y}} = \frac{xy}{\sqrt[3]{10x^2yz}}$

101. $\frac{\sqrt{x}+5}{-3} = \frac{(\sqrt{x}+5)(\sqrt{x}-5)}{-3(\sqrt{x}-5)} = \frac{x-25}{-3\sqrt{x}+15}$

102. $\sqrt{y-7} = 5$
$y-7 = 25$
$y = 32$
$\{32\}$

103. $\sqrt{2x}+1 = -4$
$\sqrt{2x} = -5$
No solution exists since the principle square root of a number is not negative.
{ }

104. $(\sqrt[3]{2x-6})^3 = (4)^3$
$2x-6 = 64$
$2x = 70$
$x = 35$
It checks.

105. $\sqrt{x+6} = \sqrt{x+2}$
$x+6 = x+2$
$6 = 2$
Since the last equation is never satisfied, there is no solution.
{ }

106. $2x - 5\sqrt{x} = 3$
$2x-3 = 5\sqrt{x}$
$(2x-3)^2 = (5\sqrt{x})^2$
$4x^2 - 12x + 9 = 25x$
$4x^2 - 37x + 9 = 0$
$(4x-1)(x-9) = 0$
$4x-1 = 0$ or $x-9 = 0$
$4x = 1$ or $x = 9$
$x = \frac{1}{4}$
It doesn't check. It checks.

107. $\sqrt{x+9} = 2 + \sqrt{x-7}$
$x+9 = 4 + 4\sqrt{x-7} + x - 7$
$12 = 4\sqrt{x-7}$
$3 = \sqrt{x-7}$
$9 = x - 7$
$16 = x$
$\{16\}$

108. Let c = the length of the hypotenuse of the right triangle. By the Pythagorean theorem,
$c^2 = 3^2 + 3^2 = 9 + 9 = 9(2)$
so $c = \sqrt{9(2)} = \sqrt{9}\sqrt{2} = 3\sqrt{2}$

109. $x^2 = 7^2 + (8\sqrt{3})^2$
$x^2 = 49 + 192 = 241$
$x = \pm\sqrt{241}$
Answer is $\sqrt{241}$

110. $c^2 = a^2 + b^2$
$(65)^2 = (40)^2 + b^2$
$4225 - 1600 = b^2$
$2625 = b^2$
$\pm\sqrt{2625} = b$
$\pm 51.235 = b$
51.2 feet

111. $c^2 = 3^2 + 3^2$
$c^2 = 9 + 9$
$c^2 = 18$
$c = \pm\sqrt{18}$
$c = \pm 4.243$
4.24 feet

112. $-\sqrt{-8} = -\sqrt{4}\sqrt{2}\sqrt{-1} = -2\sqrt{2}i$

113. $-\sqrt{-6} = -\sqrt{6}\sqrt{-1} = -i\sqrt{6}$

114. $\sqrt{-4} + \sqrt{-16} = 2i + 4i = 6i$

115. $\sqrt{-2}\sqrt{-5} = i\sqrt{2} \cdot i\sqrt{5} = i^2\sqrt{2 \cdot 5} = -\sqrt{10}$

116. $(12 - 6i) + (3 + 2i)$
$= (12 + 3) + (-6 + 2)i$
$= 15 - 4i$

117. $(-8 - 7i) - (5 - 4i)$
$= (-8 - 5) + (-7 + 4)i$
$= -13 - 3i$

118. $(\sqrt{3} + \sqrt{2}) + (3\sqrt{2} - \sqrt{-8})$
$= \sqrt{3} + \sqrt{2} + 3\sqrt{2} = \sqrt{4}\sqrt{2}\sqrt{-1}$
$= (\sqrt{3} + 4\sqrt{2}) - 2\sqrt{2}i$

119. $2i(2 - 5i) = 4i - 10i^2$
$= 4i - 10(-1)$
$= 10 + 4i$

120. $-3i(6 - 4i) = -18i + 12i^2$
$= -18i + 12(-1)$
$= -12 - 18i$

121. $(3 + 2i)(1 + i)$
$= 3 + 2i + 3i + 2i^2$
$= 3 + 5i + 2(-1)$
$= (3 - 2) + 5i$
$= 1 + 5i$

122. $(2 - 3i)^2 = (2 - 3i)(2 - 3i)$
$= (2)^2 + 2(2)(-3i) + (-3i)^2$
$= 4 - 12i + 9i^2$
$= 4 - 12i - 9$
$= -5 - 12i$

123. $(\sqrt{6} - 9i)(\sqrt{6} + 9i)$
$= (\sqrt{6})^2 - (9i)^2$
$= 6 - 81i^2 = 6 + 81 = 87$

124. $\frac{2+3i}{2i} = \frac{(2+3i)i}{2i^2} = \frac{2i+3i^2}{2(-1)}$
$= \frac{2i+3(-1)}{-2} = \frac{-3+2i}{-2} = \frac{3}{2} - i$

125. $\frac{1+i}{-3i} \cdot \frac{(3i)}{(3i)} = \frac{3i+3i^2}{-9i^2} = \frac{3i+3(-1)}{-9(-1)}$
$= \frac{3(-1+i)}{9} = \frac{-1+i}{3} = -\frac{1}{3} + \frac{1}{3}i$

Chapter 7 - Test

1. $\sqrt{216} = \sqrt{36 \cdot 6} = 6\sqrt{6}$

2. $-\sqrt[4]{x^{64}} = -(x^{64})^{1/4} = -x^{16}$

3. $\left(\frac{1}{125}\right)^{1/3} = \frac{1}{5}$

4. $\left[\frac{1}{125}\right]^{-1/3} = \left[\left(\frac{1}{125}\right)^{-1}\right]^{1/3} = 125^{1/3} = 5$

5. $\left[\frac{8x^3}{27}\right]^{2/3} = \frac{8^{2/3}(x^3)^{2/3}}{27^{2/3}} = \frac{(8^{1/3})^2 x^2}{(27^{1/3})^2}$
$= \frac{2^2 x^2}{3^2} = \frac{4x^2}{9}$

6. $\sqrt[3]{-a^{18}b^9} = \sqrt[3]{-1}\sqrt[3]{a^{18}}\sqrt[3]{b^9}$
$= (-1)a^6b^3 = -a^6b^3$

7. $\left[\frac{64c^{4/3}}{a^{-2/3}b^{5/6}}\right]^{1/2} = \left[\frac{64a^{2/3}c^{4/3}}{b^{5/6}}\right]^{1/2}$
$= \frac{64^{1/2}\left(a^{2/3}\right)^{1/2}\left(c^{4/3}\right)^{1/2}}{\left(b^{5/6}\right)^{1/2}}$
$= \frac{8a^{1/3}c^{2/3}}{b^{5/12}}$

8. $a^{-2/3}(a^{5/4} - a^3) = a^{-2/3+5/4} - a^{-2/3+3}$
$= a^{-8+15/12} - a^{-2+9/3}$
$= a^{7/12} - a^{7/3}$

9. $\sqrt[4]{(4xy)^2} = |4xy|$

10. $\sqrt[3]{(-27)^3} = -27$

11. $\sqrt{\frac{9}{y}} = \frac{\sqrt{9}}{\sqrt{y}} = \frac{3}{\sqrt{y}} \cdot \frac{\sqrt{y}}{\sqrt{y}} = \frac{3\sqrt{y}}{y}$

12. $\frac{4-\sqrt{x}}{4+2\sqrt{x}} = \frac{(4-\sqrt{x})(2-\sqrt{x})}{2(2+\sqrt{x})(2-\sqrt{x})}$
$= \frac{8 - 2\sqrt{x} - 4\sqrt{x} + x}{2(2^2 - \sqrt{x}^2)}$
$= \frac{8 - 6\sqrt{x} + x}{2(4-x)}$

13. $\frac{\sqrt[3]{ab}}{\sqrt[3]{ab^2}} = \sqrt[3]{\frac{ab}{ab^2}} = \sqrt[3]{\frac{1}{b^1}}$
$= \frac{1}{\sqrt[3]{b}} \cdot \frac{\sqrt[3]{b^2}}{\sqrt[3]{b^2}}$
$= \frac{\sqrt[3]{b^2}}{\sqrt[3]{b^3}} = \frac{\sqrt[3]{b^2}}{b}$

14. $\frac{\sqrt{6}+x}{8} = \frac{(\sqrt{6}+x)(\sqrt{6}-x)}{8(\sqrt{6}-x)}$
$= \frac{(\sqrt{6})^2 - x^2}{8(\sqrt{6}-x)} = \frac{6-x^2}{8(\sqrt{6}-x)}$

15. $\sqrt{125x^3} - 3\sqrt{20x^3}$
$= \sqrt{25x^2}\sqrt{5x} - 3\sqrt{4x^2}\sqrt{5x}$
$= (5x - 3(2x))\sqrt{5x}$
$= (5x - 6x)\sqrt{5x} = -x\sqrt{5x}$

16. $\sqrt{3}(\sqrt{16} - \sqrt{2}) = \sqrt{3}(4 - \sqrt{2})$
$= 4\sqrt{3} - \sqrt{3}\sqrt{2}$
$= 4\sqrt{3} - \sqrt{6}$

17. $(\sqrt{x}+1)^2 = \sqrt{x}^2 + 2\sqrt{x} + 1$
$= x + 2\sqrt{x} + 1$

18. $(\sqrt{2} - 4)(\sqrt{3} + 1)$
$= \sqrt{2}\sqrt{3} - 4\sqrt{3} + \sqrt{2} - 4$
$= \sqrt{6} - 4\sqrt{3} + \sqrt{2} - 4$

19. $(\sqrt{5}+5)(\sqrt{5}-5) = \sqrt{5}^2 - 5^2$
$= 5 - 25 = -20$

20. $\sqrt{561} = 23.685$

21. $386^{-2/3} = 0.019$

22. $x = \sqrt{x-2} + 2$
$x - 2 = \sqrt{x-2}$
$(x-2)^2 = x - 2$
$x^2 - 4x + 4 = x - 2$
$x^2 - 5x + 6 = 0$
$(x-2)(x-3) = 0$
$x - 2 = 0$ or $x - 3 = 0$
$x = 2$ or $x = 3$
$\{2, 3\}$

23. $\sqrt{x^2 - 7} + 3 = 0$
$\sqrt{x^2 - 7} = -3$
No solution exists since a principle square root is not negatively valued.
{ }

24. $\sqrt{x+5} = \sqrt{2x-1}$
$x + 5 = 2x - 1$
$6 = x$
$x = 6$
$\{6\}$

25. $\sqrt{-2} = \sqrt{2}i$

26. $-\sqrt{-8} = -\sqrt{4}\sqrt{2}\sqrt{-1} = -2\sqrt{2}i$

27. $(12 - 6i) - (12 - 3i)$
$= 12 - 6i - 12 + 3i = -3i$

28. $(6 - 2i)(6 + 2i) = 6^2 + 2^2$
$= 36 + 4 = 40$

29. $(4 + 3i)^2 = 16 + 24i + 9i^2$
$= 16 + 24i + 9(-1)$
$= (16 - 9) + 24i$
$= 7 + 24i$

30. $\frac{1+4i}{1-i} = \frac{(1+4i)(1+i)}{(1-i)(1+i)}$
$= \frac{1 + 4i + i + 4i^2}{1^2 + 1^2}$
$= \frac{1 + 5i + 4(-1)}{1+1}$
$= \frac{(1-4) + 5i}{2}$
$= -\frac{3}{2} + \frac{5}{2}i$

31. Let x = the length of a leg of the isosceles right triangle. By the Pythagorean theorem,
$x^2 + x^2 = 5^2$
$2x^2 = 25$
$x^2 = \frac{25}{2}$
$x = \sqrt{\frac{25}{2}} = \frac{\sqrt{25}}{\sqrt{2}} = \frac{5}{\sqrt{2}} \cdot \frac{\sqrt{2}}{\sqrt{2}} = \frac{5\sqrt{2}}{2}$

32. $g(x) = \sqrt{x+2}$
Domain: $[-2, \infty)$

x	–2	–1	2	7
$g(x)$	0	1	2	3

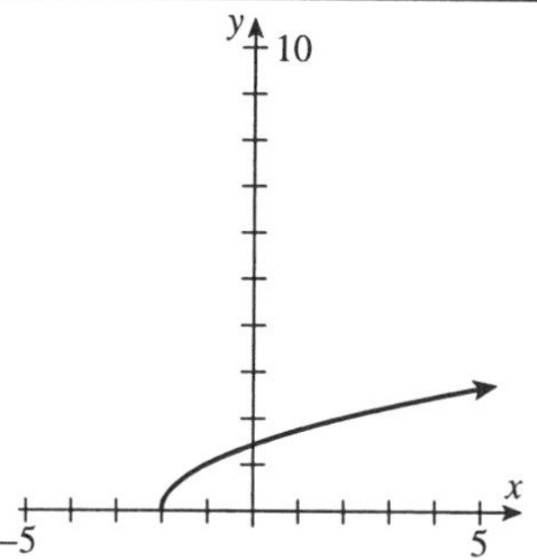

33. $V(300) = \sqrt{2.5(300)} \approx 27$

34. $V(r) = \sqrt{2.5r}$
$30 = \sqrt{2.5r}$
$900 = 2.5r$
$360 = r$

Chapter 7 - Cumulative Review

1. $4(9y) = (4 \cdot 9)y = 36y$

2. a. $x + (x + 1) = 2x + 1$

b. $x + 5x + 6x - 3 = 12x - 3$

3. $\left|\frac{x}{3}-1\right|-7 \geq -5$

$\left|\frac{x}{3}-1\right| \geq 2$

$\frac{x}{3}-1 \geq 2$ or $\frac{x}{3}-1 \leq -2$

$\frac{x}{3} \geq 3$ or $\frac{x}{3} \leq -1$

$x \geq 9$ or $x \leq -3$

$(-\infty, -3] \cup [9, \infty)$

4. $g(x) = 2x + 1$

The graph of $g(x) = 2x + 1$ is the graph of $f(x) = 2x$ shifted up one unit.

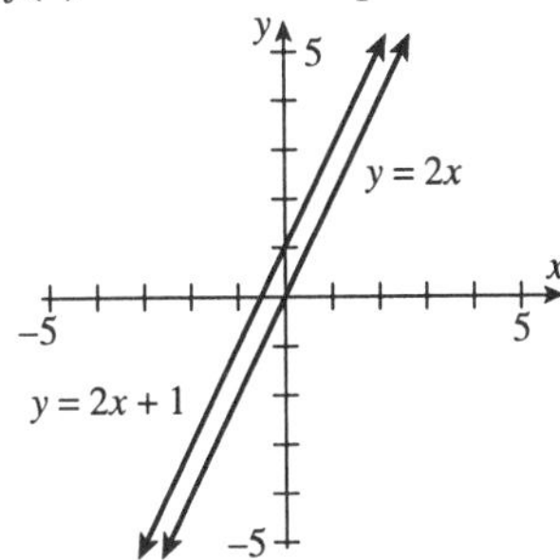

5. $\begin{cases} 3x - 2y = 10 \\ 4x - 3y = 15 \end{cases}$

Multiply row 1 by –4 and row 2 by 3 and add.

$-12x + 8y = -40$

$12x - 9y = 45$

$-y = 5$

$y = -5$

Substitute back.

$3x - 2(-5) = 10$

$3x + 10 = 10$

$3x = 0$

$x = 0$

$\{(0, -5)\}$

6.

Solution	Liters	Alcohol content
30%	x	$0.3(x)$
80%	$70 - x$	$0.8(70 - x)$
50%	70	$0.5(70)$

$0.3x + 0.8(70 - x) = 0.5(70)$

$0.3x + 56 - 0.8x = 35$

$-0.5x = -21$

$x = 42$

$70 - x = 28$

42 liters of 30% solution and 28 liters of 80% solution

7. $P(x) = 3x^2 - 2x - 5$

a. $P(1) = 3(1)^2 - 2(1) - 5$
$= 3 - 2 - 5 = -4$

b. $P(-2) = 3(-2)^2 - 2(-2) - 5$
$= 3(4) + 4 - 5 = 12 + 4 - 5 = 11$

8. $ab - 6a + 2b - 12$
$= a(b - 6) + 2(b - 6)$
$= (a + 2)(b - 6)$

9. $2x^2 + 11x + 15 = (2x + 5)(x + 3)$

10. a. $5p^2 + 5 + qp^2 + q$
$= 5(p^2 + 1) + q(p^2 + 1)$
$= (p^2 + 1)(5 + q)$

b. $9x^2 + 24x + 16$
$= (3x + 4)(3x + 4) = (3x + 4)^2$

c. $y^2 + 5$ is prime

11. a. $\frac{24x^6y^5}{8x^7y} = \frac{8 \cdot 3y^{5-1}}{8x^{7-6}} = \frac{3y^4}{x}$

b. $\frac{2x^2}{10x^3 - 2x^2} = \frac{2x^2}{2x^2(5x - 1)} = \frac{1}{5x - 1}$

12. a. $\frac{3x}{5y} \div \frac{9y}{x^5} = \frac{3x}{5y} \cdot \frac{x^5}{9y}$
$= \frac{3x^6}{45y^2} = \frac{x^6}{15y^2}$

b. $\dfrac{8m^2}{3m^2-12} \div \dfrac{40}{2-m}$
$= \dfrac{8m^2}{3m^2-12} \cdot \dfrac{2-m}{40}$
$= \dfrac{8m^2(-1)(m-2)}{3(m+2)(m-2)\cdot 8 \cdot 5}$
$= -\dfrac{m^2}{15(m+2)}$

13. a. $\frac{5}{7}+\frac{x}{7}=\frac{5+x}{7}$

b. $\frac{x}{4}+\frac{5x}{4}=\frac{x+5x}{4}=\frac{6x}{4}=\frac{3x}{2}$

c. $\dfrac{x^2}{x+7}-\dfrac{49}{x+7}=\dfrac{x^2-49}{x+7}$
$=\dfrac{(x+7)(x-7)}{x+7}=x-7$

d. $\dfrac{x}{3y^2}-\dfrac{x+1}{3y^2}=\dfrac{x-(x+1)}{3y^2}$
$=\dfrac{x-x-1}{3y^2}=-\dfrac{1}{3y^2}$

14. a. $\dfrac{\frac{5x}{x+2}}{\frac{10}{x-2}}=\dfrac{5x}{x+2}\cdot\dfrac{x-2}{10}=\dfrac{x(x-2)}{2(x+2)}$

b. $\dfrac{x+\frac{1}{y}}{y+\frac{1}{x}}=\dfrac{\frac{xy+1}{y}}{\frac{xy+1}{x}}=\dfrac{xy+1}{y}\cdot\dfrac{x}{xy+1}=\dfrac{x}{y}$

15.
$$\begin{array}{r|l} & 2x-3 \\ 3x-5 & 6x^2-19x+12 \\ & \underline{6x^2-10x} \\ & -9x+12 \\ & \underline{-9x+15} \\ & -3 \end{array}$$

Answer: $2x-3-\dfrac{3}{3x-5}$

16. $\frac{8x}{5}+\frac{3}{2}=\frac{3x}{5}$
$16x+15=6x$
$10x=-15$
$x=-\frac{3}{2}$
$\left\{-\frac{3}{2}\right\}$

17.

	Hours	Part complete in 1 hr
Melissa	4 hrs	1/4
Zack	5 hrs	1/5
Together	t	$1/t$

$\frac{1}{4}+\frac{1}{5}=\frac{1}{t}$
$5t+4t=20$
$9t=20$
$t=\frac{20}{9}$ or $2\frac{2}{9}$ hours; No

18. a. $\sqrt{36}=6$

b. $\sqrt{0}=0$

c. $\sqrt{\frac{4}{49}}=\frac{\sqrt{4}}{\sqrt{49}}=\frac{2}{7}$

d. $\sqrt{0.25}=0.5$

e. $\sqrt{x^6}=\sqrt{(x^3)^2}=x^3$

f. $\sqrt{9x^{10}}=\sqrt{(3x^5)^2}=3x^5$

g. $-\sqrt{81}=-9$

19. a. $4^{1/2}=\sqrt{4}=2$

b. $64^{1/3}=\sqrt[3]{64}=4$

c. $x^{1/4}=\sqrt[4]{x}$

d. $0^{1/6}=\sqrt[6]{0}=0$

e. $-9^{1/2}=-\sqrt{9}=-3$

f. $(81x^8)^{1/4} = \sqrt[4]{81x^8} = 3x^2$

g. $(5y)^{1/3} = \sqrt[3]{5y}$

20. a. $\sqrt{50} = \sqrt{25 \cdot 2} = \sqrt{25}\sqrt{2} = 5\sqrt{2}$

b. $\sqrt[3]{24} = \sqrt[3]{8 \cdot 3} = \sqrt[3]{8}\sqrt[3]{3} = 2\sqrt[3]{3}$

c. $\sqrt{26}$ cannot be simplified

d. $\sqrt[4]{32} = \sqrt[4]{16 \cdot 2} = \sqrt[4]{16}\sqrt[4]{2} = 2\sqrt[4]{2}$

21. a. $\sqrt{3}(5+\sqrt{30}) = 5\sqrt{30} + \sqrt{3 \cdot 30}$
$= 5\sqrt{30} + \sqrt{90} = 5\sqrt{30} + \sqrt{9 \cdot 10}$
$= 5\sqrt{30} + 3\sqrt{10}$

b. $(\sqrt{5}-\sqrt{6})(\sqrt{7}+1)$
$= \sqrt{5 \cdot 7} + \sqrt{5} - \sqrt{6 \cdot 7} - \sqrt{6}$
$= \sqrt{35} + \sqrt{5} - \sqrt{42} - \sqrt{6}$

c. $(7\sqrt{x}+5)(3\sqrt{x}-\sqrt{5})$
$= 7 \cdot 3\sqrt{x \cdot x} - 7\sqrt{5 \cdot x} + 15\sqrt{x} - 5\sqrt{5}$
$= 21\sqrt{x^2} - 7\sqrt{5x} + 15\sqrt{x} - 5\sqrt{5}$
$= 21x - 7\sqrt{5x} + 15\sqrt{x} - 5\sqrt{5}$

d. $(4\sqrt{3}-1)^2$
$= (4\sqrt{3})^2 + 2(4\sqrt{3})(-1) + (-1)^2$
$= 16 \cdot 3 - 8\sqrt{3} + 1 = 49 - 8\sqrt{3}$

e. $(\sqrt{2x}-5)(\sqrt{2x}+5)$
$= (\sqrt{2x})^2 - 5^2$
$= 2x - 25$

22. a. $\dfrac{\sqrt{27}}{\sqrt{5}} = \dfrac{\sqrt{27} \cdot \sqrt{5}}{\sqrt{5} \cdot \sqrt{5}} = \dfrac{\sqrt{135}}{\sqrt{25}}$
$= \dfrac{\sqrt{9 \cdot 15}}{5} = \dfrac{3\sqrt{15}}{5}$

b. $\dfrac{2\sqrt{16}}{\sqrt{9x}} = \dfrac{2 \cdot 4}{3\sqrt{x}} = \dfrac{8 \cdot \sqrt{x}}{3\sqrt{x} \cdot \sqrt{x}} = \dfrac{8\sqrt{x}}{3x}$

c. $\sqrt[3]{\dfrac{1}{2}} = \dfrac{\sqrt[3]{1}}{\sqrt[3]{2}} = \dfrac{1}{\sqrt[3]{2}} \cdot \dfrac{\sqrt[3]{4}}{\sqrt[3]{4}} = \dfrac{\sqrt[3]{4}}{\sqrt[3]{8}} = \dfrac{\sqrt[3]{4}}{2}$

23. $\sqrt{4-x} = x-2$
$(\sqrt{4-x})^2 = (x-2)^2$
$4 - x = x^2 - 4x + 4$
$0 = x^2 - 3x$
$0 = x(x-3)$
$x = 0$ or $x - 3 = 0$
$x = 3$

Check: $\sqrt{4-0} = 0-2$
$\sqrt{4} = -2$
False

Check: $\sqrt{4-3} = 3-2$
$\sqrt{1} = 1$
True

$\{3\}$

Chapter 8

Graphing Calculator Explorations 8.1

1. $x(x-5)=8$
$Y_1=x(x-5)$
$Y_2=8$

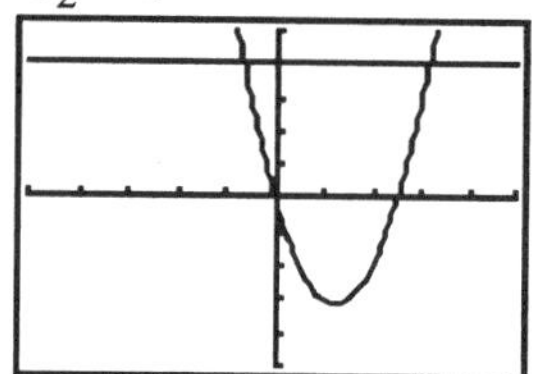

Solutions are −1.27 and 6.27.

3. $x^2+0.5x=0.3x+1$
$Y_1=x^2+0.5x$
$Y_2=0.3x+1$

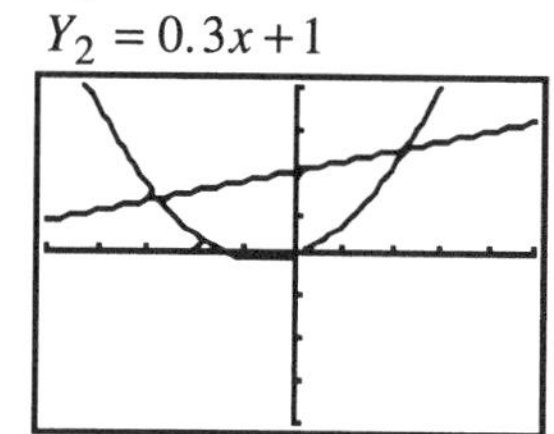

Solutions are −1.10 and 0.90.

5. $(2x-5)^2=-16$
$Y_1=(2x-5)^2$
$Y_2=-16$

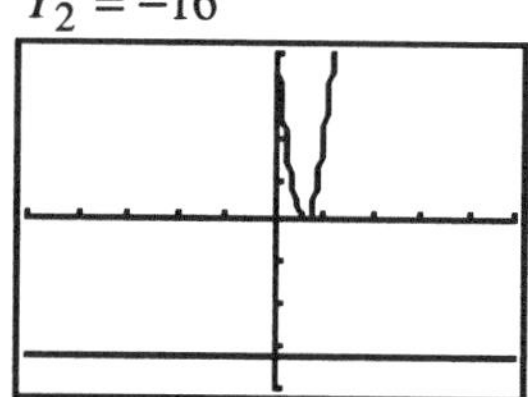

There are no real solutions since the graphs do not intersect.

Exercise Set 8.1

1. $x^2=16$
$x=\pm\sqrt{16}$
$x=\pm 4$
$\{-4, 4\}$

3. $x^2-7=0$
$x^2=7$
$x=\pm\sqrt{7}$
$\{-\sqrt{7}, \sqrt{7}\}$

5. $x^2=18$
$x=\pm\sqrt{18}$
$x=\pm\sqrt{9}\sqrt{2}$
$x=\pm 3\sqrt{2}$
$\{-3\sqrt{2}, 3\sqrt{2}\}$

7. $3z^2-30=0$
$3z^2=30$
$z^2=10$
$z=\pm\sqrt{10}$
$\{-\sqrt{10}, \sqrt{10}\}$

9. $(x+5)^2=9$
$x+5=\pm\sqrt{9}$
$x+5=\pm 3$
$x=-5\pm 3$
$x=-8, -2$
$\{-8, -2\}$

11. $(z-6)^2=18$
$z-6=\pm\sqrt{18}$
$z-6=\pm\sqrt{9}\sqrt{2}$
$z-6=\pm 3\sqrt{2}$
$\{6-3\sqrt{2}, 6+3\sqrt{2}\}$

13. $(2x-3)^2=8$
$2x-3=\pm\sqrt{8}$
$2x-3=\pm\sqrt{4}\sqrt{2}$
$2x-3=\pm 2\sqrt{2}$
$2x=3\pm 2\sqrt{2}$
$x=\frac{3}{2}\pm\sqrt{2}$
$\left\{\frac{3}{2}-\sqrt{2}, \frac{3}{2}+\sqrt{2}\right\}$

15. $x^2+9=0$
$x^2=-9$
$x=\pm\sqrt{-9}$
$x=\pm 3i$
$\{-3i, 3i\}$

17. $x^2-6=0$
$x^2=6$
$x=\pm\sqrt{6}$
$\{-\sqrt{6}, \sqrt{6}\}$

19. $2z^2+16=0$
$2z^2=-16$
$z^2=-8$
$z=\pm\sqrt{-8}$
$z=\pm\sqrt{4}\sqrt{2}\sqrt{-1}$
$z=\pm 2i\sqrt{2}$
$\{-2i\sqrt{2}, 2i\sqrt{2}\}$

21. $(x-1)^2=-16$
$x-1=\pm\sqrt{-16}$
$x-1=\pm 4i$
$x=1\pm 4i$
$\{1-4i, 1+4i\}$

23. $(z+7)^2=5$
$z+7=\pm\sqrt{5}$
$z=-7\pm\sqrt{5}$
$\{-7-\sqrt{5}, -7+\sqrt{5}\}$

25. $(x+3)^2=-8$
$x+3=\pm\sqrt{-8}$
$x+3=\pm\sqrt{4}\sqrt{2}\sqrt{-1}$
$x+3=\pm 2i\sqrt{2}$
$x=-3\pm 2i\sqrt{2}$
$\left\{-3-2i\sqrt{2}, -3+2i\sqrt{2}\right\}$

27. $x^2+16x+\left(\frac{16}{2}\right)^2$
$=x^2+16x+64=(x+8)^2$

29. $z^2-12z+\left(\frac{12}{2}\right)^2$
$=z^2-12z+36=(z-6)^2$

31. $p^2+9p+\left(\frac{9}{2}\right)^2$
$=p^2+9p+\frac{81}{4}=\left(p+\frac{9}{2}\right)^2$

33. $x^2+x+\left(\frac{1}{2}\right)^2$
$=x^2+x+\frac{1}{4}=\left(x+\frac{1}{2}\right)^2$

35. x^2+_+16
$\pm 2\sqrt{16}=\pm 2(4)=\pm 8$
$x^2-8x+16=(x-4)^2$
$x^2+8x+16=(x+4)^2$
Answer: $\pm 8x$

37. $z^2+\_+\frac{25}{4}$
$\pm 2\sqrt{\frac{25}{4}}=\pm 2\left(\frac{5}{2}\right)=\pm 5$
$z^2-5z+\frac{25}{4}=\left(z-\frac{5}{2}\right)^2$
$z^2+5z+\frac{25}{4}=\left(z+\frac{5}{2}\right)^2$
Answer: $\pm 5z$

39. $x^2+8x=-15$
$x^2+8x+\left(\frac{8}{2}\right)^2=-15+16$
$(x+4)^2=1$
$x+4=\pm\sqrt{1}$
$x=-4\pm 1$
$x=-5, -3$
$\{-5, -3\}$

41. $x^2+6x+2=0$
$x^2+6x+\left(\frac{6}{2}\right)^2=-2+9$
$(x+3)^2=7$
$x+3=\pm\sqrt{7}$

$x = -3 \pm \sqrt{7}$

$\{-3 - \sqrt{7}, -3 + \sqrt{7}\}$

43. $x^2 + x - 1 = 0$

$x^2 + x + \left(\frac{1}{2}\right)^2 = 1 + \frac{1}{4}$

$\left(x + \frac{1}{2}\right)^2 = \frac{5}{4}$

$x + \frac{1}{2} = \pm\sqrt{\frac{5}{4}}$

$x + \frac{1}{2} = \pm\frac{\sqrt{5}}{2}$

$x = -\frac{1}{2} \pm \frac{\sqrt{5}}{2}$

$\left\{-\frac{1}{2} - \frac{\sqrt{5}}{2}, -\frac{1}{2} + \frac{\sqrt{5}}{2}\right\}$

45. $x^2 + x - 1 = 0$

$x^2 + x + \left(\frac{1}{2}\right)^2 = 1 + \left(\frac{1}{2}\right)^2$

$x^2 + x + \frac{1}{4} = 1 + \frac{1}{4}$

$\left(x + \frac{1}{2}\right)^2 = \frac{5}{4}$

$x + \frac{1}{2} = \pm\sqrt{\frac{5}{4}}$

$x = -\frac{1}{2} \pm \frac{\sqrt{5}}{2}$

$x = \frac{-1 \pm \sqrt{5}}{2}$

$\left\{\frac{-1 - \sqrt{5}}{2}, \frac{-1 + \sqrt{5}}{2}\right\}$

47. $3p^2 - 12p + 2 = 0$

$3p^2 - 12p = -2$

$p^2 - 4p = -\frac{2}{3}$

$p^2 - 4p + \left(\frac{4}{2}\right)^2 = -\frac{2}{3} + 4$

$(p-2)^2 = \frac{10}{3}$

$p - 2 = \pm\sqrt{\frac{10}{3}}$

$p - 2 = \pm\sqrt{\frac{10 \cdot 3}{3^2}}$

$p - 2 = \pm\frac{\sqrt{30}}{3}$

$p = 2 \pm \frac{\sqrt{30}}{3}$

$\left\{2 - \frac{\sqrt{30}}{3}, 2 + \frac{\sqrt{30}}{3}\right\}$

49. $4y^2 - 12y - 2 = 0$

$4y^2 - 12y = 2$

$y^2 - 3y = \frac{1}{2}$

$y^2 - 3y + \left(\frac{3}{2}\right)^2 = \frac{1}{2} + \frac{9}{4}$

$\left(y - \frac{3}{2}\right)^2 = \frac{11}{4}$

$y - \frac{3}{2} = \pm\sqrt{\frac{11}{4}}$

$y - \frac{3}{2} = \pm\frac{\sqrt{11}}{2}$

$y = \frac{3}{2} \pm \frac{\sqrt{11}}{2}$

$\left\{\frac{3}{2} - \frac{\sqrt{11}}{2}, \frac{3}{2} + \frac{\sqrt{11}}{2}\right\}$

51. $2x^2 + 7x = 4$

$x^2 + \frac{7}{2}x = 2$

$x^2 + \frac{7}{2}x + \left(\frac{7/2}{2}\right)^2 = 2 + \frac{49}{16}$

$\left(x + \frac{7}{4}\right)^2 = \frac{81}{16}$

$x + \frac{7}{4} = \pm\sqrt{\frac{81}{16}}$

$x = -\frac{7}{4} \pm \frac{9}{4}$

$x = -4, \frac{1}{2}$

$\left\{-4, \frac{1}{2}\right\}$

53. $x^2 - 4x - 5 = 0$
$x^2 - 4x = 5$
$x^2 - 4x + \left(\frac{4}{2}\right)^2 = 5 + 4$
$(x-2)^2 = 9$
$x - 2 = \pm\sqrt{9}$
$x - 2 = \pm 3$
$x = 2 \pm 3$
$x = -1, 5$
$\{-1, 5\}$

55. $x^2 + 8x + 1 = 0$
$x^2 + 8x = -1$
$x^2 + 8x + \left(\frac{8}{2}\right)^2 = -1 + 16$
$(x+4)^2 = 15$
$x + 4 = \pm\sqrt{15}$
$x = -4 \pm \sqrt{15}$
$\left\{-4 - \sqrt{15},\ -4 + \sqrt{15}\right\}$

57. $3y^2 + 6y - 4 = 0$
$3y^2 + 6y = 4$
$y^2 + 2y = \frac{4}{3}$
$y^2 + 2y + \left(\frac{2}{2}\right)^2 = \frac{4}{3} + 1$
$(y+1)^2 = \frac{7}{3}$
$y + 1 = \pm\sqrt{\frac{7}{3}}$
$y + 1 = \pm\frac{\sqrt{7}}{\sqrt{3}} \cdot \frac{\sqrt{3}}{\sqrt{3}}$
$y + 1 = \pm\frac{\sqrt{21}}{3}$
$y = -1 \pm \frac{\sqrt{21}}{3}$
$\left\{-1 - \frac{\sqrt{21}}{3},\ -1 + \frac{\sqrt{21}}{3}\right\}$

59. $2x^2 - 3x - 5 = 0$
$2x^2 - 3x = 5$
$x^2 - \frac{3}{2}x = \frac{5}{2}$
$x^2 - \frac{3}{2}x + \left(\frac{3/2}{2}\right)^2 = \frac{5}{2} + \frac{9}{16}$
$\left(x - \frac{3}{4}\right)^2 = \frac{49}{16}$
$x - \frac{3}{4} = \pm\frac{7}{4}$
$x = \frac{3}{4} \pm \frac{7}{4}$
$x = -1,\ \frac{5}{2}$
$\left\{-1,\ \frac{5}{2}\right\}$

61. $y^2 + 2y + 2 = 0$
$y^2 + 2y + \left(\frac{2}{2}\right)^2 = -2 + 1$
$(y+1)^2 = -1$
$y + 1 = \pm\sqrt{-1}$
$y + 1 = \pm i$
$y = -1 \pm i$
$\{-1 + i, -1 - i\}$

63. $x^2 - 6x + 3 = 0$
$x^2 - 6x + \left(\frac{6}{2}\right)^2 = -3 + 9$
$(x-3)^2 = 6$
$x - 3 = \pm\sqrt{6}$
$x = 3 \pm \sqrt{6}$
$\left\{3 - \sqrt{6},\ 3 + \sqrt{6}\right\}$

65. $2a^2 + 8a = -12$
$a^2 + 4a = -6$
$a^2 + 4a + \left(\frac{4}{2}\right)^2 = -6 + 4$
$(a+2)^2 = -2$
$a + 2 = \pm\sqrt{-2}$
$a + 2 = \pm\sqrt{2}i$
$a = -2 \pm \sqrt{2}i$
$\left\{-2 + \sqrt{2}i,\ -2 - \sqrt{2}i\right\}$

67. $5x^2+15x-1=0$

$5x^2+15x=1$

$x^2+3x=\frac{1}{5}$

$x^2+3x+\left(\frac{3}{2}\right)^2=\frac{1}{5}+\frac{9}{4}$

$\left(x+\frac{3}{2}\right)^2=\frac{49}{20}$

$x+\frac{3}{2}=\pm\sqrt{\frac{49}{20}}$

$=\frac{\pm\sqrt{49}}{\sqrt{4}\sqrt{5}}=\pm\frac{7}{2\sqrt{5}}$

$=\pm\frac{7\sqrt{5}}{2\cdot\sqrt{5^2}}=\pm\frac{7\sqrt{5}}{2\cdot 5}$

$x+\frac{3}{2}=\pm\frac{7\sqrt{5}}{10}$ so $x=-\frac{3}{2}\pm\frac{7\sqrt{5}}{10}$

$\left\{-\frac{3}{2}+\frac{7\sqrt{5}}{10},\ -\frac{3}{2}-\frac{7\sqrt{5}}{10}\right\}$

69. $2x^2-x+6=0$

$2x^2-x=-6$

$x^2-\frac{1}{2}x=-3$

$x^2-\frac{1}{2}x+\left(\frac{1/2}{2}\right)^2=-3+\frac{1}{16}$

$\left(x-\frac{1}{4}\right)^2=-\frac{47}{16}$

$x-\frac{1}{4}=\pm\sqrt{-\frac{47}{16}}=\pm\frac{\sqrt{47}\sqrt{-1}}{\sqrt{6}}$

$=\pm\frac{\sqrt{47}i}{4}$

$x=\frac{1}{4}\pm\frac{\sqrt{47}}{4}i$

$\left\{\frac{1}{4}+\frac{\sqrt{47}}{4}i,\ \frac{1}{4}-\frac{\sqrt{47}}{4}i\right\}$

71. $x^2+10x+28=0$

$x^2+10x=-28$

$x^2+10x+\left(\frac{10}{2}\right)^2=-28+25$

$(x+5)^2=-3$

$x+5=\pm\sqrt{-3}$

$x+5=\pm i\sqrt{3}$

$x=-5\pm i\sqrt{3}$

$\{-5-i\sqrt{3},\ -5+i\sqrt{3}\}$

73. $z^2+3z-4=0$

$z^2+3z=4$

$z^2+3z+\left(\frac{3}{2}\right)^2=4+\left(\frac{3}{2}\right)^2$

$\left(z+\frac{3}{2}\right)^2=\frac{25}{4}$

$z+\frac{3}{2}=\pm\sqrt{\frac{25}{4}}$

$z+\frac{3}{2}=\pm\frac{5}{2}$

$z=-\frac{3}{2}\pm\frac{5}{2}$

$z=-4,\ 1$

$\{-4, 1\}$

75. $2x^2-4x+3=0$

$2x^2-4x=-3$

$x^2-2x=-\frac{3}{2}$

$x^2-2x+\left(\frac{2}{2}\right)^2=-\frac{3}{2}+1$

$(x-1)^2=-\frac{1}{2}$

$x-1=\pm\sqrt{-\frac{1}{2}}$

$x-1=\pm\frac{\sqrt{-1}}{\sqrt{2}}$

$x-1=\pm\frac{i}{\sqrt{2}}\cdot\frac{\sqrt{2}}{\sqrt{2}}$

$x-1=\pm\frac{\sqrt{2}}{2}i$

$x=1\pm\frac{\sqrt{2}}{2}i$

$\left\{1-\frac{\sqrt{2}}{2}i,\ 1+\frac{\sqrt{2}}{2}i\right\}$

77. $3x^2+3x=5$

$x^2+x=\frac{5}{3}$

$x^2+x+\left(\frac{1}{2}\right)^2=\frac{5}{3}+\frac{1}{4}$

$\left(x+\frac{1}{2}\right)^2 = \frac{23}{12}$

$x+\frac{1}{2} = \pm\sqrt{\frac{23}{12}} = \pm\frac{\sqrt{23}}{\sqrt{4}\sqrt{3}}$

$x+\frac{1}{2} = \pm\frac{\sqrt{23}}{2\sqrt{3}} = \pm\frac{\sqrt{23}\sqrt{3}}{2\sqrt{3^2}} = \pm\frac{\sqrt{69}}{2\cdot 3}$

$x+\frac{1}{2} = \pm\frac{\sqrt{69}}{6}$

$x = -\frac{1}{2} \pm \frac{\sqrt{69}}{6}$

$\left\{-\frac{1}{2}-\frac{\sqrt{69}}{6}, -\frac{1}{2}+\frac{\sqrt{69}}{6}\right\}$

79. $A = P(1+r)^t$

$4320 = 3000(1+r)^2$

$\frac{4320}{3000} = (1+r)^2$

$1.44 = (1+r)^2$

$\pm\sqrt{1.44} = 1+r$

$\pm 1.2 = 1 + r$

$-1 \pm 1.2 = r$

$r = -1 + 1.2$ or $r = -1 - 1.2$

$r = 0.2$ or $r = -2.2$

Rate cannot be negative, so the return is $r = 0.2 = 20\%$.

81. $A = P(1+r)^t$

$1000 = 810(1+r)^2$

$\frac{1000}{810} = (1+r)^2$

$1.2346 = (1+r)^2$

$\pm\sqrt{1.2346} = 1+r$

$\pm 1.11 = 1 + r$

$-1 \pm 1.11 = r$

$r = -1 + 1.11$ or $r = -1 - 1.11$

$r = 0.11$ or $r = -2.11$

Rate cannot be negative, so $r = 0.11 = 11\%$.

83. Answers may vary.

85. Simple

87. $s(t) = 16t^2$

$1017 = 16t^2$

$63.5625 = t^2$

$\pm\sqrt{63.5625} = t$

$\pm 7.97 \approx t$

Time cannot be negative, so $t \approx 7.97$ seconds.

89. $s(t) = 16t^2$

$568 = 16t^2$

$35.5 = t^2$

$\pm\sqrt{35.5} = t$

$\pm 5.96 \approx t$

Time cannot be negative, so $t \approx 5.96$ seconds.

91. $A = \pi r^2$

$\frac{36\pi}{\pi} = \frac{\pi r^2}{\pi}$

$\sqrt{36} = \sqrt{r^2}$

$\pm 6 = r$

disregard –6

$r = 6$ inches

93. $x^2 + x^2 = 27^2$

$\frac{2x^2}{2} = \frac{729}{2}$

$\sqrt{x^2} = \sqrt{\frac{729}{2}}$

$x = \pm\frac{27}{\sqrt{2}}\cdot\frac{\sqrt{2}}{\sqrt{2}}$

$x = \pm\frac{27\sqrt{2}}{\sqrt{4}}$

$x = \pm\frac{27\sqrt{2}}{2}$

disregard the negative

$x = \frac{27\sqrt{2}}{2}$ inches

95. $p = -x^2 + 15$

$7 = -x^2 + 15$

$x^2 = 8$

$x = \pm\sqrt{8}$

$x \approx 2.828$

Demand cannot be negative, therefore the demand is approximately 2.828 units.

97. $\frac{3}{5}+\sqrt{\frac{16}{25}}=\frac{3}{5}+\frac{4}{5}=\frac{7}{5}$

99. $\frac{9}{10}-\sqrt{\frac{49}{100}}=\frac{9}{10}-\frac{7}{10}=\frac{2}{10}=\frac{1}{5}$

101. $\frac{10-20\sqrt{3}}{2}=\frac{10}{2}-\frac{20\sqrt{3}}{2}=5-10\sqrt{3}$

103. $\frac{12-8\sqrt{7}}{16}=\frac{12}{16}-\frac{8\sqrt{7}}{16}=\frac{3}{4}-\frac{\sqrt{7}}{2}$
$=\frac{3}{4}-\frac{2\sqrt{7}}{4}=\frac{3-2\sqrt{7}}{4}$

105. $\sqrt{b^2-4ac}$
$a=1,\ b=6,\ c=2$
$\sqrt{(6)^2-4(1)(2)}=\sqrt{36-8}=\sqrt{28}$
$=\sqrt{4}\sqrt{7}=2\sqrt{7}$

107. $\sqrt{b^2-4ac}$
$a=1,\ b=-3,\ c=-1$
$\sqrt{(-3)^2-4(1)(-1)}=\sqrt{9+4}=\sqrt{13}$

Exercise Set 8.2

1. $m^2+5m-6=0$
$a=1,\ b=5,\ c=-6$
$m=\frac{-5\pm\sqrt{5^2-4(1)(-6)}}{2(1)}$
$m=\frac{-5\pm\sqrt{25+24}}{2}=\frac{-5\pm\sqrt{49}}{2}$
$m=\frac{-5\pm 7}{2}=-6,\ 1$
$\{-6, 1\}$

3. $2y=5y^2-3$
$5y^2-2y-3=0$
$a=5,\ b=-2,\ c=-3$
$y=\frac{2\pm\sqrt{(-2)^2-4(5)(-3)}}{2(5)}$
$y=\frac{2\pm\sqrt{4+60}}{10}=\frac{2\pm\sqrt{64}}{10}$
$y=\frac{2\pm 8}{10}=-\frac{3}{5},\ 1$
$\left\{-\frac{3}{5},\ 1\right\}$

5. $x^2-6x+9=0$
$a=1,\ b=-6,\ c=9$
$x=\frac{6\pm\sqrt{(-6)^2-4(1)(9)}}{2(1)}$
$x=\frac{6\pm\sqrt{36-36}}{2}=\frac{6\pm\sqrt{0}}{2}=\frac{6}{2}=3$
$\{3\}$

7. $x^2+7x+4=0$
$a=1,\ b=7,\ c=4$
$x=\frac{-7\pm\sqrt{7^2-4(1)(4)}}{2(1)}$
$x=\frac{-7\pm\sqrt{49-16}}{2}=\frac{-7\pm\sqrt{33}}{2}$
$x=-\frac{7}{2}\pm\frac{\sqrt{33}}{2}$
$\left\{-\frac{7}{2}-\frac{\sqrt{33}}{2},\ -\frac{7}{2}+\frac{\sqrt{33}}{2}\right\}$

9. $8m^2-2m=7$
$8m^2-2m-7=0$
$a=8,\ b=-2,\ c=-7$
$m=\frac{2\pm\sqrt{(-2)^2-4(8)(-7)}}{2(8)}$
$m=\frac{2\pm\sqrt{4+224}}{16}=\frac{2\pm\sqrt{228}}{16}$
$m=\frac{2\pm\sqrt{4}\sqrt{57}}{16}=\frac{2\pm 2\sqrt{57}}{16}$
$m=\frac{1\pm\sqrt{57}}{8}=\frac{1}{8}\pm\frac{\sqrt{57}}{8}$
$\left\{\frac{1}{8}-\frac{\sqrt{57}}{8},\ \frac{1}{8}+\frac{\sqrt{57}}{8}\right\}$

11. $3m^2-7m=3$
$3m^2-7m-3=0$
$a=3,\ b=-7,\ c=-3$
$m=\frac{7\pm\sqrt{(-7)^2-4(3)(-3)}}{2(3)}$

$m = \frac{7 \pm \sqrt{49+36}}{6} = \frac{7 \pm \sqrt{85}}{6}$

$m = \frac{7}{6} \pm \frac{\sqrt{85}}{6}$

$\left\{\frac{7}{6} - \frac{\sqrt{85}}{6}, \frac{7}{6} + \frac{\sqrt{85}}{6}\right\}$

13. $\frac{1}{2}x^2 - x - 1 = 0$

$x^2 - 2x - 2 = 0$

$a = 1, b = -2, c = -2$

$x = \frac{2 \pm \sqrt{(-2)^2 - 4(1)(-2)}}{2(1)}$

$x = \frac{2 \pm \sqrt{4+8}}{2} = \frac{2 \pm \sqrt{12}}{2}$

$x = \frac{2 \pm \sqrt{4}\sqrt{3}}{2} = \frac{2 \pm 2\sqrt{3}}{2}$

$x = 1 \pm \sqrt{3}$

$\left\{1 - \sqrt{3}, 1 + \sqrt{3}\right\}$

15. $\frac{2}{5}y^2 + \frac{1}{5}y = \frac{3}{5}$

$2y^2 + y = 3$

$2y^2 + y - 3 = 0$

$a = 2, b = 1, c = -3$

$y = \frac{-1 \pm \sqrt{1^2 - 4(2)(-3)}}{2(2)}$

$y = \frac{-1 \pm \sqrt{1+24}}{4} = \frac{-1 \pm \sqrt{25}}{4}$

$y = \frac{-1 \pm 5}{4} = -\frac{3}{2}, 1$

$\left\{-\frac{3}{2}, 1\right\}$

17. $\frac{1}{3}y^2 - y - \frac{1}{6} = 0$

$2y^2 - 6y - 1 = 0$

$a = 2, b = -6, c = -1$

$y = \frac{6 \pm \sqrt{(-6)^2 - 4(2)(-1)}}{2(2)}$

$y = \frac{6 \pm \sqrt{36+8}}{4} = \frac{6 \pm \sqrt{44}}{4}$

$y = \frac{6 \pm \sqrt{4}\sqrt{11}}{4} = \frac{6 \pm 2\sqrt{11}}{4}$

$y = \frac{3 \pm \sqrt{11}}{2} = \frac{3}{2} \pm \frac{\sqrt{11}}{2}$

$\left\{\frac{3}{2} - \frac{\sqrt{11}}{2}, \frac{3}{2} + \frac{\sqrt{11}}{2}\right\}$

19. $m^2 + 5m - 6 = 0$

$(m+6)(m-1) = 0$

$m + 6 = 0$ or $m - 1 = 0$

$m = -6$ or $m = 1$

$\{-6, 1\}$

The results are the same.

21. $6 = -4x^2 + 3x$

$4x^2 - 3x + 6 = 0$

$a = 4, b = -3, c = 6$

$x = \frac{3 \pm \sqrt{3^2 - 4(4)(6)}}{2(4)}$

$x = \frac{3 \pm \sqrt{9-96}}{8} = \frac{3 \pm \sqrt{-87}}{8}$

$x = \frac{3 \pm \sqrt{87}i}{8} = \frac{3 \pm i\sqrt{87}}{8}$

$\left\{\frac{3 - i\sqrt{87}}{8}, \frac{3 + i\sqrt{87}}{8}\right\}$

23. $(x+5)(x-1) = 2$

$x^2 + 4x - 5 = 2$

$x^2 + 4x - 7 = 0$

$a = 1, b = 4, c = -7$

$x = \frac{-4 \pm \sqrt{4^2 - 4(1)(-7)}}{2(1)}$

$x = \frac{-4 \pm \sqrt{16+28}}{2} = \frac{-4 + \sqrt{44}}{2}$

$x = \frac{-4 \pm \sqrt{4}\sqrt{11}}{2} = \frac{-4 \pm 2\sqrt{11}}{2}$

$x = -2 \pm \sqrt{11}$

$\left\{-2 - \sqrt{11}, -2 + \sqrt{11}\right\}$

25. $10y^2 + 10y + 3 = 0$

$a = 10, b = 10, c = 3$

$y = \frac{-10 \pm \sqrt{10^2 - 4(10)(3)}}{2(10)}$

$y = \frac{-10 \pm \sqrt{100-120}}{20}$

$y = \frac{-10 \pm \sqrt{-20}}{20} = \frac{-10 \pm \sqrt{4}\sqrt{5}\sqrt{-1}}{20}$

$y = \frac{-10 \pm 2i\sqrt{5}}{20} = \frac{-5 \pm i\sqrt{5}}{10}$

$\left\{\frac{-5 - i\sqrt{5}}{10}, \frac{-5 + i\sqrt{5}}{10}\right\}$

27. $\frac{-b + \sqrt{b^2 - 4ac}}{2a} + \frac{-b - \sqrt{b^2 - 4ac}}{2a}$

$= \frac{-b + \sqrt{b^2 - 4ac} - b - \sqrt{b^2 - 4ac}}{2a}$

$= \frac{-2b}{2a} = -\frac{b}{a}$

29. $9x - 2x^2 + 5 = 0$

$-2x^2 + 9x + 5 = 0$

$a = -2, b = 9, c = 5$

$b^2 - 4ac = 9^2 - 4(-2)(5)$

$b^2 - 4ac = 81 + 40$

$b^2 - 4ac = 121 > 0$

Therefore, there are two real solutions.

31. $4x^2 + 12x = -9$

$4x^2 + 12x + 9 = 0$

$a = 4, b = 12, c = 9$

$b^2 - 4ac = 12^2 - 4(4)(9)$

$b^2 - 4ac = 144 - 144 = 0$

Therefore, there is 1 real solution.

33. $3x = -2x^2 + 7$

$2x^2 + 3x - 7 = 0$

$a = 2, b = 3, c = -7$

$b^2 - 4ac = 3^2 - 4(2)(-7)$

$b^2 - 4ac = 9 + 56$

$b^2 - 4ac = 65 > 0$

Therefore, there are 2 distinct real solutions.

35. $6 = 4x - 5x^2$

$5x^2 - 4x + 6 = 0$

$a = 5, b = -4, c = 6$

$b^2 - 4ac = (-4)^2 - 4(5)(6)$

$b^2 - 4ac = 16 - 120$

$b^2 - 4ac = -104 < 0$

Therefore, there are 2 distinct complex solutions.

37. $x^2 + 5x = -2$

$x^2 + 5x + 2 = 0$

$a = 1, b = 5, c = 2$

$x = \frac{-5 \pm \sqrt{5^2 - 4(1)(2)}}{2(1)}$

$x = \frac{-5 \pm \sqrt{25 - 8}}{2} = \frac{-5 \pm \sqrt{17}}{2}$

$x = -\frac{5}{2} \pm \frac{\sqrt{17}}{2}$

$\left\{-\frac{5}{2} - \frac{\sqrt{17}}{2}, -\frac{5}{2} + \frac{\sqrt{17}}{2}\right\}$

39. $(m + 2)(2m - 6) = 5(m - 1) - 12$

$2m^2 - 2m - 12 = 5m - 5 - 12$

$2m^2 - 7m + 5 = 0$

$a = 2, b = -7, c = 5$

$m = \frac{-(-7) \pm \sqrt{(-7)^2 - 4(2)(5)}}{2(2)}$

$m = \frac{7 \pm \sqrt{49 - 40}}{4}$

$m = \frac{7 \pm \sqrt{9}}{4}$

$m = \frac{7 \pm 3}{4}$

$m = \frac{10}{4}, \frac{4}{4}$

$\left\{\frac{5}{2}, 1\right\}$

41. $\frac{x^2}{3} - x = \frac{5}{3}$

$x^2 - 3x = 5$

$x^2 - 3x - 5 = 0$

$a = 1, b = -3, c = -5$

$x = \frac{3 \pm \sqrt{(-3)^2 - 4(1)(-5)}}{2(1)}$

$x = \frac{3 \pm \sqrt{9 + 20}}{2} = \frac{3 \pm \sqrt{29}}{2}$

$x = \frac{3}{2} \pm \frac{\sqrt{29}}{2}$

$\left\{\frac{3}{2} - \frac{\sqrt{29}}{2}, \frac{3}{2} + \frac{\sqrt{29}}{2}\right\}$

43. $x(6x + 2) - 3 = 0$

$6x^2 + 2x - 3 = 0$

$a = 6, b = 2, c = -3$

$x = \frac{-2 \pm \sqrt{2^2 - 4(6)(-3)}}{2(6)}$

$x = \frac{-2 \pm \sqrt{4 + 72}}{12} = \frac{-2 \pm \sqrt{76}}{12}$

$x = \frac{-2 \pm \sqrt{4}\sqrt{19}}{12} = \frac{-2 \pm 2\sqrt{19}}{12}$

$x = \frac{-1 \pm \sqrt{19}}{6} = -\frac{1}{6} \pm \frac{\sqrt{19}}{6}$

$\left\{-\frac{1}{6} - \frac{\sqrt{19}}{6}, -\frac{1}{6} + \frac{\sqrt{19}}{6}\right\}$

45. $x^2 + 6x + 13 = 0$

$a = 1, b = 6, c = 13$

$x = \frac{-6 \pm \sqrt{6^2 - 4(1)(13)}}{2(1)}$

$x = \frac{-6 \pm \sqrt{36 - 52}}{2} = \frac{-6 \pm \sqrt{-16}}{2}$

$x = \frac{-6 \pm 4i}{2} = -3 \pm 2i$

$\{-3 - 2i, -3 + 2i\}$

47. $\frac{2}{5}y^2 + \frac{1}{5}y + \frac{3}{5} = 0$

$2y^2 + y + 3 = 0$

$a = 2, b = 1, c = 3$

$y = \frac{-1 \pm \sqrt{1^2 - 4(2)(3)}}{2(2)}$

$y = \frac{-1 \pm \sqrt{1 - 24}}{4} = \frac{-1 \pm \sqrt{-23}}{4}$

$y = \frac{-1 \pm i\sqrt{23}}{4}$

$\left\{\frac{-1 - i\sqrt{23}}{4}, \frac{-1 + i\sqrt{23}}{4}\right\}$

49. $\frac{1}{2}y^2 = y - \frac{1}{2}$

$y^2 = 2y - 1$

$y^2 - 2y + 1 = 0$

$a = 1, b = -2, c = 1$

$y = \frac{2 \pm \sqrt{(-2)^2 - 4(1)(1)}}{2(1)}$

$y = \frac{2 \pm \sqrt{4 - 4}}{2} = \frac{2 \pm \sqrt{0}}{2} = 1$

$\{1\}$

51. $(n - 2)^2 = 15n$

$n^2 - 4n + 4 = 15n$

$n^2 - 19n + 4 = 0$

$a = 1, b = -19, c = 4$

$n = \frac{19 \pm \sqrt{(-19)^2 - 4(1)(4)}}{2(1)}$

$n = \frac{19 \pm \sqrt{361 - 16}}{2} = \frac{19 \pm \sqrt{345}}{2}$

$n = \frac{19}{2} \pm \frac{\sqrt{345}}{2}$

$\left\{\frac{19}{2} - \frac{\sqrt{345}}{2}, \frac{19}{2} + \frac{\sqrt{345}}{2}\right\}$

53. We are given $A = 400$. Let w = the width of the pen. So $\frac{5}{2}w$ = the length of the pen. Then since $A = l \cdot w$ we have

$400 = \left(\frac{5}{2}w\right)w$

$400 = \frac{5}{2}w^2$

$800 = 5w^2$

$w^2 = 160$

$w = \sqrt{160} = \sqrt{16}\sqrt{10} = 4\sqrt{10}$, so

$\frac{5}{2}w = \frac{5}{2}(4\sqrt{10}) = 5(2\sqrt{10}) = 10\sqrt{10}$.

The pen is $4\sqrt{10}$ ft by $10\sqrt{10}$ ft.

55. We are given that $A = 42$. Let h = the height of the triangle, so $2h$ = the base of the triangle. Then since $A = \frac{1}{2}bh$ we get

$42 = \frac{1}{2}(2h)(h)$

$42 = h^2$

$h = \pm\sqrt{42}$
but we discard $-\sqrt{42}$ as meaningless.
Then $2h = 2\sqrt{42}$. The base of the triangle is $2\sqrt{42}$ cm and its height is $\sqrt{42}$ cm.

57. We are given that $A = 1200$. Let w = the width of the poster, so $\frac{3}{2}w$ = the length of the poster. Then since $A = l \cdot w$, we have

$1200 = \left(\frac{3}{2}w\right)w$
$1200 = \frac{3}{2}w^2$
$2400 = 3w^2$
$w^2 = 800$
$w = \sqrt{800} = \sqrt{400}\sqrt{2} = 20\sqrt{2}$
So, $\frac{3}{2}w = \frac{3}{2}(20\sqrt{2}) = 3(10\sqrt{2}) = 30\sqrt{2}$
The width must be $20\sqrt{2}$ in. and the length must be $30\sqrt{2}$ in.

59. Let w = the width of the rectangle so $3w$ = the length of the rectangle. By the Pythagorean theorem,

a. $50^2 = w^2 + (3w)^2$
$2500 = w^2 + 9w^2$
$2500 = 10w^2$
$w^2 = 250$
$w = \sqrt{250} = \sqrt{25}\sqrt{10} = 5\sqrt{10}$
so $3w = 3(5\sqrt{10}) = 15\sqrt{10}$
The dimensions of the rectangle are $5\sqrt{10}$ cm by $15\sqrt{10}$ cm.

b. Recall that
$P = 2l + 2w$
$P = 2(15\sqrt{10}) + 2(5\sqrt{10})$
$P = 30\sqrt{10} + 10\sqrt{10}$
$P = 40\sqrt{10}$
The perimeter of the rectangle is $40\sqrt{10}$ cm.

61. $2x^2 - 6x + 3 = 0$
$a = 2,\ b = -6,\ c = 3$
$x = \frac{-(-6) \pm \sqrt{(-6)^2 - 4(2)(3)}}{2(2)}$
$= \frac{6 \pm \sqrt{36 - 24}}{4}$
$= \frac{6 \pm \sqrt{12}}{4} = \frac{6 \pm 2\sqrt{3}}{4} = \frac{3 \pm \sqrt{3}}{2}$
$\{0.6, 2.4\}$

63. $1.3x^2 - 2.5x - 7.9 = 0$
$a = 1.3,\ b = -2.5,\ c = -7.9$
$x = \frac{-(-2.5) \pm \sqrt{(-2.5)^2 - 4(1.3)(-7.9)}}{2(1.3)}$
$= \frac{2.5 \pm \sqrt{6.25 + 41.08}}{2.6}$
$= \frac{2.5 \pm \sqrt{47.33}}{2.6}$
$\{-1.7, 3.6\}$

65. $h(t) = -16t^2 - 20t + 180$
$0 = -16t^2 - 20t + 180$
$0 = -4(4t^2 + 5t - 45)$
$0 = 4t^2 + 5t - 45$
$t = \frac{-(5) \pm \sqrt{5^2 - 4(4)(-45)}}{2(4)}$
$t = \frac{-5 \pm \sqrt{25 + 720}}{8} = \frac{-5 \pm \sqrt{745}}{8}$
$t \approx \frac{-5 \pm 27.3}{8}$
$t \approx \frac{-5 + 27.3}{8}$ or $t \approx \frac{-5 - 27.3}{8}$
$t \approx 2.8$ or $t \approx -4.0$
Time cannot be negative, therefore the time until it strikes the ground is 2.8 seconds.

67. From Sunday to Monday

69. Wednesday

71. $f(x) = 3x^2 - 18x + 57$
$x = 4$ days from Sunday
$f(4) = 3(4)^2 - 18(4) + 57$
$f(4) = 3(16) - 18(4) + 57$

$f(4) = 48 - 72 + 57$
$f(4) = 33°\text{F}$
It does agree with the graph.

73. The results are the same.

75. $\sqrt{5x-2} = 3$
$(\sqrt{5x}-2)^2 = 3^2$
$5x - 2 = 9$
$5x = 11$
$x = \frac{11}{5}$
Checking
$\sqrt{5\left(\frac{11}{5}\right)-2} \stackrel{?}{=} 3$
$\sqrt{11-2} \stackrel{?}{=} 3$
$\sqrt{9} \stackrel{?}{=} 3$
$3 = 3$
The solution is $\frac{11}{5}$.
$\left\{\frac{11}{5}\right\}$

77. $\frac{1}{x} + \frac{2}{5} = \frac{7}{x}$
$5x \cdot \frac{1}{x} + 5x \cdot \frac{2}{5} = 5x \cdot \frac{7}{x}$
$5 + 2x = 35$
$2x = 30$
$x = 15$
$\{15\}$

79. $x^4 + x^2 - 20$
$= (x^2 + 5)(x^2 - 4)$
$= (x^2 + 5)(x + 2)(x - 2)$

81. $z^4 - 13z^2 + 36$
$= (z^2 - 9)(z^2 - 4)$
$= (z + 3)(z - 3)(z + 2)(z - 2)$

83. $3x^2 - \sqrt{12x} + 1 = 0$
$a = 3,\ b = -\sqrt{12},\ c = 1$
$$x = \frac{\sqrt{12} \pm \sqrt{(-\sqrt{12})^2 - 4(3)(1)}}{2(3)}$$
$$x = \frac{\sqrt{12} \pm \sqrt{12-12}}{6} = \frac{\sqrt{4}\sqrt{3} \pm \sqrt{0}}{6}$$
$$x = \frac{2\sqrt{3}}{6} = \frac{\sqrt{3}}{3}$$
$\left\{\frac{\sqrt{3}}{3}\right\}$

85. $x^2 + \sqrt{2}x + 1 = 0$
$a = 1,\ b = \sqrt{2},\ c = 1$
$$x = \frac{-\sqrt{2} \pm \sqrt{(\sqrt{2})^2 - 4(1)(1)}}{2(1)}$$
$$x = \frac{-\sqrt{2} \pm \sqrt{2-4}}{2} = \frac{-\sqrt{2} \pm \sqrt{-2}}{2}$$
$$x = \frac{-\sqrt{2} \pm i\sqrt{2}}{2}$$
$\left\{\frac{-\sqrt{2} - i\sqrt{2}}{2},\ \frac{-\sqrt{2} + i\sqrt{2}}{2}\right\}$

87. $2x^2 - \sqrt{3}x - 1 = 0$
$a = 2,\ b = -\sqrt{3},\ c = -1$
$$x = \frac{\sqrt{3} \pm \sqrt{(-\sqrt{3})^2 - 4(2)(-1)}}{2(2)}$$
$$x = \frac{\sqrt{3} + \sqrt{3+8}}{4} = \frac{\sqrt{3} \pm \sqrt{11}}{4}$$
$\left\{\frac{\sqrt{3} - \sqrt{11}}{4},\ \frac{\sqrt{3} + \sqrt{11}}{4}\right\}$

Exercise Set 8.3

1. $\frac{2}{x} + \frac{3}{x-1} = 1$
$\frac{2(x-1) + 3x}{x(x-1)} = 1$
$\frac{2x - 2 + 3x}{x^2 - x} = 1$
$5x - 2 = x^2 - x$
$x^2 - 6x + 2 = 0$
$a = 1$, $b = -6$, and $c = 2$
Thus,
$$x = \frac{6 \pm \sqrt{(-6)^2 - 4(1)(2)}}{2(1)}$$
$$= \frac{6 \pm \sqrt{36-8}}{2} = \frac{6 \pm \sqrt{28}}{2}$$
$$= \frac{6 \pm \sqrt{4}\sqrt{7}}{2} = \frac{6 \pm 2\sqrt{7}}{2}$$

$3 \pm \sqrt{7}$

$\{3+\sqrt{7},\ 3-\sqrt{7}\}$

3. $\frac{3}{x}+\frac{4}{x+2}=2$

$\frac{3(x+2)+4x}{x(x+2)}=2$

$\frac{3x+6+4x}{x^2+2x}=2$

$7x+6=2(x^2+2x)$

$7x+6=2x^2+4x$

$2x^2-3x-6=0$

$a=2$, $b=-3$, and $c=-6$

Thus,

$x=\frac{3\pm\sqrt{(-3)^2-4(2)(-6)}}{2(2)}$

$x=\frac{3\pm\sqrt{9+48}}{4}=\frac{3\pm\sqrt{57}}{4}$

$=\frac{3}{4}\pm\frac{\sqrt{57}}{4}$

$\left\{\frac{3}{4}+\frac{\sqrt{57}}{4},\ \frac{3}{4}-\frac{\sqrt{57}}{4}\right\}$

5. $\frac{7}{x^2-5x+6}=\frac{2x}{x-3}-\frac{x}{x-2}$

$\frac{7}{(x-3)(x-2)}=\frac{2x(x-2)-x(x-3)}{(x-3)(x-2)}$

$7=2x^2-4x-x^2+3x$

$7=x^2-x$

$0=x^2-x-7$

$a=1$, $b=-1$, and $c=-7$

Thus,

$x=\frac{1\pm\sqrt{(-1)^2-4(1)(-7)}}{2(1)}$

$=\frac{1\pm\sqrt{1+28}}{2}=\frac{1\pm\sqrt{29}}{2}$

$=\frac{1}{2}\pm\frac{\sqrt{29}}{2}$

$\left\{\frac{1}{2}+\frac{\sqrt{29}}{2},\ \frac{1}{2}-\frac{\sqrt{29}}{2}\right\}$

7. $y^3-1=0$

$(y-1)(y^2+y+1)=0$

$y-1=0$ or $y^2+y+1=0$

$y=1$

$a=1$, $b=1$, and $c=1$

Thus,

$y=\frac{-1\pm\sqrt{1^2-4(1)(1)}}{2(1)}$

$=\frac{-1\pm\sqrt{1-4}}{2}=\frac{-1\pm\sqrt{-3}}{2}$

$=\frac{-1\pm\sqrt{3}i}{2}=-\frac{1}{2}\pm\frac{\sqrt{3}}{2}i$

$\left\{1,\ -\frac{1}{2}+\frac{\sqrt{3}}{2}i,\ -\frac{1}{2}-\frac{\sqrt{3}}{2}i\right\}$

9. $x^4+27x=0$

$x(x^3+27)=0$

$x(x+3)(x^2-3x+9)=0$

$x=0$ or $x+3=0$ or $x^2-3x+9=0$

$x=-3$

$a=1$, $b=-3$, $c=9$

Thus,

$x=\frac{3\pm\sqrt{(-3)^2-4(1)(9)}}{2(1)}$

$=\frac{3\pm\sqrt{9-36}}{2}=\frac{3\pm\sqrt{-27}}{2}$

$=\frac{3\pm\sqrt{9}\sqrt{3}\sqrt{-1}}{2}=\frac{3\pm3\sqrt{3}i}{2}$

$=\frac{3}{2}\pm\frac{3\sqrt{3}}{2}i$

$\left\{0,\ -3,\ \frac{3}{2}+\frac{3\sqrt{3}}{2}i,\ \frac{3}{2}-\frac{3\sqrt{3}}{2}i\right\}$

11. $z^3=64$

$z^3-64=0$

$(z-4)(z^2+4z+16)=0$

$z-4=0$ or $z^2+4z+16=0$

$z=4$

$a=1$, $b=4$, and $c=16$

$$z = \frac{-4 \pm \sqrt{4^2 - 4(1)(16)}}{2(1)}$$
$$= \frac{-4 \pm \sqrt{16-64}}{2} = \frac{-4 \pm \sqrt{-48}}{2}$$
$$= \frac{-4 \pm \sqrt{16}\sqrt{3}\sqrt{-1}}{2} = \frac{-4 \pm 4\sqrt{3}i}{2}$$
$$= -2 \pm 2\sqrt{3}i$$
$$\left\{4,\ -2+2\sqrt{3}i,\ -2-2\sqrt{3}i\right\}$$

13. $p^4 - 16 = 0$
$(p^2+4)(p^2-4) = 0$
$(p+2i)(p-2i)(p+2)(p-2)$
$p+2i = 0$ or $p-2i = 0$
or $p+2 = 0$ or $p-2 = 0$
$p = -2i$ or $p = 2i$
$p = -2$ or $p = 2$
$\{-2i, 2i, -2, 2\}$

15. $4x^4 + 11x^2 - 3 = 0$
$(4x^2-1)(x^2+3) = 0$
$(2x+1)(2x-1)(x+\sqrt{3}i)(x-\sqrt{3}i) = 0$
$2x+1 = 0$ or $2x-1 = 0$ or $x+\sqrt{3}i = 0$ or
$x-\sqrt{3}i = 0$
$2x = -1$ or $2x = 1$ or $x = -\sqrt{3}i$ or $x = \sqrt{3}i$
$x = -\frac{1}{2}$ or $x = \frac{1}{2}$
$\left\{-\frac{1}{2},\ \frac{1}{2},\ -\sqrt{3}i,\ \sqrt{3}i\right\}$

17. $z^4 - 13z^2 + 36 = 0$
$(z^2-9)(z^2-4) = 0$
$(z+3)(z-3)(z+2)(z-2) = 0$
$z+3 = 0$ or $z-3 = 0$ or $z+2 = 0$
or $z-2 = 0$
$z = -3$ or $z = 3$ or $z = -2$ or $z = 2$
$\{-3, 3, -2, 2\}$

19. $x^{2/3} - 3x^{1/3} - 10 = 0$
$(x^{1/3}-5)(x^{1/3}+2) = 0$
$x^{1/3} - 5 = 0$ or $x^{1/3} + 2 = 0$
$x^{1/3} = 5$ or $x^{1/3} = -2$
$x = 5^3$ or $x = (-2)^3$
$x = 125$ or $x = -8$
$\{125, -8\}$

21. $(5n+1)^2 + 2(5n+1) - 3 = 0$
$[(5n+1)+3][(5n+1)-1] = 0$
$(5n+4)(5n) = 0$
$5n+4 = 0$ or $5n = 0$
$n = -\frac{4}{5}$ or $n = 0$
$\left\{-\frac{4}{5},\ 0\right\}$

23. $2x^{2/3} - 5x^{1/3} = 3$
$2x^{2/3} - 5x^{1/3} - 3 = 0$
$(2x^{1/3}+1)(x^{1/3}-3) = 0$
$2x^{1/3}+1 = 0$ or $x^{1/3}-3 = 0$
$2x^{1/3} = -1$ or $x^{1/3} = 3$
$x^{1/3} = -\frac{1}{2}$ or $x = 3^3$
$x = \left(-\frac{1}{2}\right)^3$ or $x = 27$
$x = -\frac{1}{8}$
$\left\{-\frac{1}{8},\ 27\right\}$

25. $1 + \frac{2}{3t-2} = \frac{8}{(3t-2)^2}$
$(3t-2)^2 + 2(3t-2) - 8 = 0$
$[(3t-2)+4][(3t-2)-2] = 0$
$(3t+2)(3t-4) = 0$
$3t+2 = 0$ or $3t-4 = 0$
$t = -\frac{2}{3}$ or $t = \frac{4}{3}$
$\left\{-\frac{2}{3},\ \frac{4}{3}\right\}$

27. $20x^{2/3} - 6x^{1/3} - 2 = 0$
$(5x^{1/3}+1)(4x^{1/3}-2) = 0$
$5x^{1/3}+1 = 0$ or $4x^{1/3}-2 = 0$
$5x^{1/3} = -1$ or $4x^{1/3} = 2$
$x^{1/3} = -\frac{1}{5}$ or $x^{1/3} = \frac{1}{2}$
$x = \left(-\frac{1}{5}\right)^3$ or $x = \left(\frac{1}{2}\right)^3$

$x = -\frac{1}{125}$ or $x = \frac{1}{8}$

$\left\{-\frac{1}{125}, \frac{1}{8}\right\}$

29. Answers may vary.

31. $a^4 - 5a^2 + 6 = 0$

$(a^2 - 3)(a^2 - 2) = 0$

$a^2 - 3 = 0$ or $a^2 - 2 = 0$

$a^2 = 3$ or $a^2 = 2$

$a = \pm\sqrt{3}$ or $a = \pm\sqrt{2}$

$\{\sqrt{3}, -\sqrt{3}, \sqrt{2}, -\sqrt{2}\}$

33. $\frac{2x}{x-2} + \frac{x}{x+3} = \frac{-5}{x+3}$

$\frac{2x}{x-2} = \frac{-x}{x+3} - \frac{5}{x+3}$

$\frac{2x}{x-2} = \frac{-x-5}{x+3}$

$2x(x+3) = (x-2)(-x-5)$

$2x^2 + 6x = -x^2 + 2x - 5x + 10$

$2x^2 + 6x = -x^2 - 3x + 10$

$3x^2 + 9x - 10 = 0$

$a = 3$, $b = 9$, and $c = -10$

Thus,

$$x = \frac{-9 \pm \sqrt{9^2 - 4(3)(-10)}}{2(3)}$$

$$= \frac{-9 \pm \sqrt{81+120}}{6} = \frac{-9 \pm \sqrt{201}}{6}$$

$$= -\frac{9}{6} \pm \frac{\sqrt{201}}{6} = -\frac{3}{2} \pm \frac{\sqrt{201}}{6}$$

$\left\{-\frac{3}{2} + \frac{\sqrt{201}}{6}, -\frac{3}{2} - \frac{\sqrt{201}}{6}\right\}$

35. $(p+2)^2 = 9(p+2) - 20$

$(p+2)^2 - 9(p+2) + 20 = 0$

$[(p+2) - 5][(p+2) - 4] = 0$

$(p-3)(p-2) = 0$

$p - 3 = 0$ or $p - 2 = 0$

$p = 3$ or $p = 2$

$\{2, 3\}$

37. $x^3 + 64 = 0$

$(x+4)(x^2 - 4x + 16) = 0$

$x + 4 = 0$ or $x^2 - 4x + 16 = 0$

$x = -4$

$a = 1$, $b = -4$, and $c = 16$

$$x = \frac{4 \pm \sqrt{(-4)^2 - 4(1)(16)}}{2(1)}$$

$$= \frac{4 \pm \sqrt{16-64}}{2} = \frac{4 \pm \sqrt{-48}}{2}$$

$$= \frac{4 \pm \sqrt{16}\sqrt{3}\sqrt{-1}}{2} = \frac{4 \pm 4\sqrt{3}i}{2}$$

$$= 2 \pm 2\sqrt{3}i$$

$\{-4, 2 + 2\sqrt{3}i, 2 - 2\sqrt{3}i\}$

39. $x^{2/3} - 8x^{1/3} + 15 = 0$

$(x^{1/3} - 5)(x^{1/3} - 3) = 0$

$x^{1/3} - 5 = 0$ or $x^{1/3} - 3 = 0$

$x^{1/3} = 5$ or $x^{1/3} = 3$

$x = 5^3$ or $x = 3^3$

$x = 125$ or $x = 27$

$\{125, 27\}$

41. $y^3 + 9y - y^2 - 9 = 0$

$(y^2 + 9)(y - 1) = 0$

$(y + 3i)(y - 3i)(y - 1) = 0$

$y + 3i = 0$ or $y - 3i = 0$ or $y - 1 = 0$

$y = -3i$ or $y = 3i$ or $y = 1$

$\{-3i, 3i, 1\}$

43. $2x^{2/3} + 3x^{1/3} - 2 = 0$

$(2x^{1/3} - 1)(x^{1/3} + 2) = 0$

$2x^{1/3} - 1 = 0$ or $x^{1/3} + 2 = 0$

$2x^{1/3} = 1$ or $x^{1/3} = -2$

$x^{1/3} = \frac{1}{2}$ or $x = (-2)^3$

$x = \left(\frac{1}{2}\right)^3 = \frac{1}{8}$ or $x = -8$

$\left\{\frac{1}{8}, -8\right\}$

45. $x^{-2} - x^{-1} - 6 = 0$

$\frac{1}{x^2} - \frac{1}{x} - 6 = 0$

$1 - x - 6x^2 = 0$

$-6x^2 - x + 1 = 0$

$(-3x + 1)(2x + 1) = 0$

$-3x+1=0$ or $2x+1=0$

$x=\frac{1}{3}$ or $x=-\frac{1}{2}$

$\left\{-\frac{1}{2}, \frac{1}{3}\right\}$

47. $2x^3-250=0$

$x^3-125=0$

$(x-5)(x^2+5x+25)=0$

$x-5=0$ or $x^2+5x+25=0$

$x=5$

$a=1$, $b=5$, and $c=25$

$$x=\frac{-5\pm\sqrt{5^2-4(1)(25)}}{2(1)}$$

$$=\frac{-5\pm\sqrt{25-100}}{2}$$

$$=\frac{-5\pm\sqrt{-75}}{2}=\frac{-5\pm\sqrt{25}\sqrt{3}\sqrt{-1}}{2}$$

$$=\frac{-5\pm5\sqrt{3}i}{2}=-\frac{5}{2}\pm\frac{5\sqrt{3}i}{2}$$

$$\left\{5, -\frac{5}{2}+\frac{5\sqrt{3}}{2}i, -\frac{5}{2}-\frac{5\sqrt{3}}{2}i\right\}$$

49. $\frac{x}{x-1}+\frac{1}{x+1}=\frac{2}{x^2-1}$

$$\frac{x(x+1)+x-1}{(x-1)(x+1)}=\frac{2}{(x+1)(x-1)}$$

$x^2+x+x-1=2$

$x^2+2x-1=0$

$x^2+2x-3=0$

$(x+3)(x-1)=0$

$x+3=0$ or $x-1=0$

$x=-3$ or $x=1$

We discard the 1 as extraneous.

$\{-3\}$

51. $p^4-p^2-20=0$

$(p^2-5)(p^2+4)=0$

$p^2-5=0$ or $p^2+4=0$

$p^2=5$ or $(p+2i)(p-2i)=0$

$p=\pm\sqrt{5}$ or $p+2i=0$ or $p-2i=0$

$p=-2i$ or $p=2i$

$\left\{\sqrt{5}, -\sqrt{5}, -2i, 2i\right\}$

53. $2x^3=-54$

$x^3=-27$

$x^3+27=0$

$(x+3)(x^2-3x+9)=0$

$x+3=0$ or $x^2-3x+9=0$

$x=-3$

$a=1$, $b=-3$, or $c=9$

$$x=\frac{3\pm\sqrt{(-3)^2-4(1)(9)}}{2(1)}$$

$$=\frac{3\pm\sqrt{9-36}}{2}=\frac{3\pm\sqrt{-27}}{2}$$

$$=\frac{3\pm\sqrt{9}\sqrt{3}\sqrt{-1}}{2}=\frac{3\pm3\sqrt{3}i}{2}$$

$$=\frac{3}{2}\pm\frac{3\sqrt{3}}{2}i$$

$$\left\{-3, \frac{3}{2}+\frac{3\sqrt{3}}{2}i, \frac{3}{2}-\frac{3\sqrt{3}}{2}i\right\}$$

55. $1=\frac{4}{x-7}+\frac{5}{(x-7)^2}$

$(x-7)^2=4(x-7)+5$

$(x-7)^2-4(x-7)-5=0$

$[(x-7)-5][(x-7)+1]=0$

$(x-12)(x-6)=0$

$x-12=0$ or $x-6=0$

$x=12$ or $x=6$

$\{6, 12\}$

57. $27y^4+15y^2=2$

$27y^4+15y^2-2=0$

$(9y^2-1)(3y^2+2)=0$

$(3y+1)(3y-1)(3y^2+2)=0$

$3y+1=0$ or $3y-1=0$ or $3y^2+2=0$

$3y=-1$ or $3y=1$ or $3y^2=-2$

$y=-\frac{1}{3}$ or $y=\frac{1}{3}$ or $y^2=-\frac{2}{3}$

$$y=\pm\sqrt{-\frac{2}{3}}=\pm\frac{\sqrt{-2}}{\sqrt{3}}=\pm\frac{\sqrt{2}i(\sqrt{3})}{\sqrt{3}^2}=\pm\frac{\sqrt{6}i}{3}$$

$$\left\{-\frac{1}{3}, \frac{1}{3}, \frac{\sqrt{6}}{3}i, \frac{-\sqrt{6}}{3}i\right\}$$

59.

	Hours	Part complete in one hour
Bill	$x-1$	$\frac{1}{x-1}$
Billy	x	$\frac{1}{x}$
Together	4	$\frac{1}{4}$

$\frac{1}{x-1}+\frac{1}{x}=\frac{1}{4}$

$\frac{x+(x-1)}{x(x-1)}=\frac{1}{4}$

$4(2x-1)=x(x-1)$

$8x-4=x^2-x$

$0=x^2-9x+4$

$a=1,\ b=-9,\ c=4$

$x=\frac{-(-9)\pm\sqrt{(-9)^2-4(1)(4)}}{2(1)}$

$x=\frac{9\pm\sqrt{65}}{2}$

$x=0.47$ or 8.53

8.53 hours

61. Let x = number

$x(x-4)=96$

$x^2-4x=96$

$x^2-4x-96=0$

$(x-12)(x+8)=0$

$x-12=0$ or $x+8=0$

$x=12$ or $x=-8$

12 and -8

63.

	Hours	Part complete in one hour
IBM	$x-1$	$\frac{1}{x-1}$
Toshiba	x	$\frac{1}{x}$
Together	8	$\frac{1}{8}$

$\frac{1}{x-1}+\frac{1}{x}=\frac{1}{8}$

$\frac{x+(x-1)}{x(x-1)}=\frac{1}{8}$

$8(2x-1)=x(x-1)$

$16x-8=x^2-x$

$0=x^2-17x+8$

$a=1,\ b=-17,\ c=8$

$x=\frac{-(-17)\pm\sqrt{(-17)^2-4(1)(8)}}{2(1)}$

$x=\frac{17\pm\sqrt{257}}{2}$

$x=0.5$ or $x=16.5$

$x=16.5$

$x-1=15.5$

Toshiba: 16.5 hr

IBM: 15.5 hr

65. Answers may vary.

67. $\frac{5x}{3}+2\le 7$

$\frac{5x}{3}\le 5$

$\frac{3}{5}\left(\frac{5x}{3}\right)\le\frac{3}{5}(5)$

$x\le 3$

$(-\infty, 3]$

69. $\frac{y-1}{15}>-\frac{2}{5}$

$15\left(\frac{y-1}{15}\right)>15\left(-\frac{2}{5}\right)$

$y-1>-6$

$y>-5$

$(-5, \infty)$

71. Domain $\{x \mid x$ is a real number$\}$

Range $\{y \mid y$ is a real number$\}$

It is a function.

73. Domain $\{x \mid x$ is a real number$\}$

Range$\{y \mid y\ge -1\}$

It is a function.

Exercise Set 8.4

1. $(x+1)(x+5)>0$

$(x+1)(x+5)=0$

$x+1=0$ or $x+5=0$

$x=-1$ or $x=-5$

Region	Interval	Test Point
A	$(-\infty, -5)$	-6
B	$(-5, -1)$	-2
C	$(-1, \infty)$	0
$x=-6$	$(-6+1)(-6+5)>0$	True
$x=-2$	$(-2+1)(-2+5)>0$	False
$x=0$	$(0+1)(0+5)>0$	True

The solution is $(-\infty, -5) \cup (-1, \infty)$.

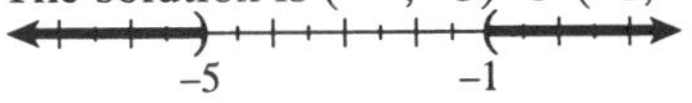

3. $(x-3)(x+4) \le 0$
$(x-3)(x+4)=0$
$x-3=0$ or $x+4=0$
$x=3$ or $x=-4$

Region	Interval	Test Point
A	$(-\infty, -4)$	-5
B	$(-4, 3)$	0
C	$(3, \infty)$	4
$x=-5$	$(-5-3)(-5+4) \le 0$	False
$x=0$	$(0-3)(0+4) \le 0$	True
$x=4$	$(4-3)(4+4) \le 0$	False

The solution is $[-4, 3]$.

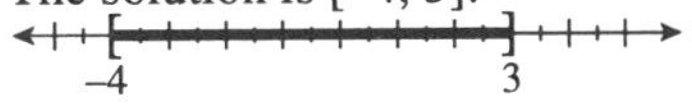

5. $x^2-7x+10 \le 0$
$(x-5)(x-2) \le 0$
$(x-5)(x-2)=0$
$x-5=0$ or $x-2=0$
$x=5$ or $x=2$

Region	Interval	Test Point
A	$(-\infty, 2)$	0
B	$(2, 5)$	3
C	$(5, \infty)$	6
$x=0$	$(0-5)(0-2) \le 0$	False
$x=3$	$(3-5)(3-2) \le 0$	True
$x=6$	$(6-5)(6-2) \le 0$	False

The solution is $[2, 5]$.

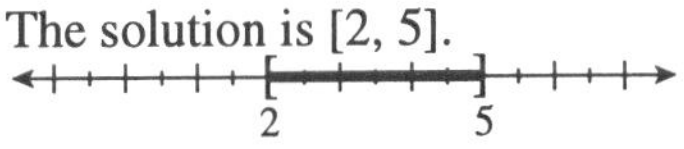

7. $3x^2+16x<-5$
$3x^2+16x+5<0$
$(3x+1)(x+5)<0$
$(3x+1)(x+5)=0$
$3x+1=0$ or $x+5=0$
$3x=-1$ or $x=-5$
$x=-\frac{1}{3}$

Region	Interval	Test Point
A	$(-\infty, -5)$	-6
B	$\left(-5, -\frac{1}{3}\right)$	-1
C	$\left(-\frac{1}{3}, \infty\right)$	0
$x=-6$	$[3(-6)+1](-6+5)<0$	False
$x=-1$	$[3(-1)+1](-1+5)<0$	True
$x=0$	$[3(0)+1](0+5)<0$	False

The solution is $\left(-5, -\frac{1}{3}\right)$.

9. $(x-6)(x-4)(x-2)>0$
$(x-6)(x-4)(x-2)=0$
$x-6=0$ or $x-4=0$ or $x-2=0$
$x=6$ or $x=4$ or $x=2$

Region	Interval	Test Point
A	$(-\infty, 2)$	1
B	$(2, 4)$	3
C	$(4, 6)$	5
D	$(6, \infty)$	7
$x=1$	$(1-6)(1-4)(1-2)>0$	False
$x=3$	$(3-6)(3-4)(3-2)>0$	True
$x=5$	$(5-6)(5-4)(5-2)>0$	False
$x=7$	$(7-6)(7-4)(7-2)>0$	True

The solution is $(2, 4) \cup (6, \infty)$.

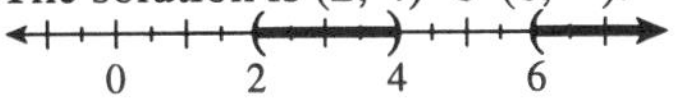

11. $x(x-1)(x+4) \le 0$
$x(x-1)(x+4) = 0$
$x = 0$ or $x - 1 = 0$ or $x + 4 = 0$
$x = 0$ or $x = 1$ or $x = -4$

Region	Interval	Test Point
A	$(-\infty, -4)$	-5
B	$(-4, 0)$	-1
C	$(0, 1)$	$\frac{1}{2}$
D	$(1, \infty)$	2
$x=-5$	$-5(-5-1)(-5+4) \le 0$	True
$x=-1$	$-1(-1-1)(-1+4) \le 0$	False
$x=\frac{1}{2}$	$\frac{1}{2}\left(\frac{1}{2}-1\right)\left(\frac{1}{2}+4\right) \le 0$	True
$x=2$	$2(2-1)(2+4) \le 0$	False

The solution is $(-\infty, -4] \cup [0, 1]$.

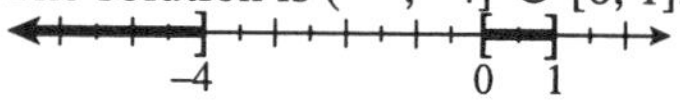

13. $(x^2-9)(x^2-4) > 0$
$(x+3)(x-3)(x+2)(x-2) > 0$
$(x+3)(x-3)(x+2)(x-2) = 0$
$x + 3 = 0$ or $x - 3 = 0$ or $x + 2 = 0$ or $x - 2 = 0$
$x = -3$ or $x = 3$ or $x = -2$ or $x = 2$

Region	Interval	Test Point
A	$(-\infty, -3)$	-4
B	$(-3, -2)$	$-\frac{5}{2}$
C	$(-2, 2)$	0
D	$(2, 3)$	$\frac{5}{2}$
E	$(3, \infty)$	4
$x=-4$	$(-4+3)(-4-3)(-4+2)(-4-2) > 0$	True
$x=-\frac{5}{2}$	$\left(-\frac{5}{2}+3\right)\left(-\frac{5}{2}-3\right)\left(-\frac{5}{2}+2\right)\left(-\frac{5}{2}-2\right) > 0$	False
$x=0$	$(0+3)(0-3)(0+2)(0-2) > 0$	True
$x=\frac{5}{2}$	$\left(\frac{5}{2}+3\right)\left(\frac{5}{2}-3\right)\left(\frac{5}{2}+2\right)\left(\frac{5}{2}-2\right) > 0$	False
$x=4$	$(4+3)(4-3)(4+2)(4-2) > 0$	True

The solution is $(-\infty, -3) \cup (-2, 2) \cup (3, \infty)$.

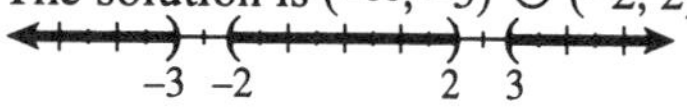

15. $\frac{x+7}{x-2}<0$

$x+7=0$ or $x-2=0$
$x=-7$ or $x=2$

Region	Interval	Test Point
A	$(-\infty, -7)$	-8
B	$(-7, 2)$	0
C	$(2, \infty)$	3
$x=-8$	$\frac{-8+7}{-8-2}<0$	False
$x=0$	$\frac{0+7}{0-2}<0$	True
$x=3$	$\frac{3+7}{3-2}<0$	False

The solution is $(-7, 2)$.

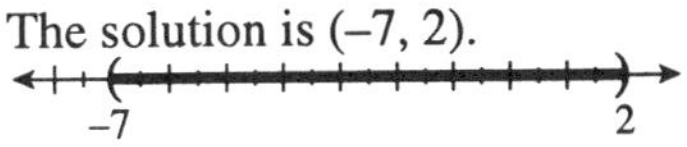

17. $\frac{5}{x+1}>0$

$x+1=0$
$x=-1$

Region	Interval	Test Point
A	$(-\infty, -1)$	-2
B	$(-1, \infty)$	0
$x=-2$	$\frac{5}{-2+1}>0$	False
$x=0$	$\frac{5}{0+1}>0$	True

The solution is $(-1, \infty)$.

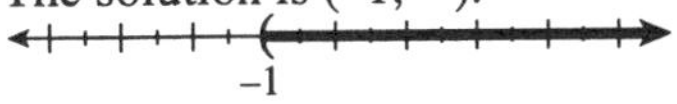

19. $\frac{x+1}{x-4}\geq 0$

$x+1=0$ or $x-4=0$
$x=-1$ or $x=4$

Region	Interval	Test Point
A	$(-\infty, -1)$	-2
B	$(-1, 4)$	0
C	$(4, \infty)$	5
$x=-2$	$\frac{-2+1}{-2-4}\geq 0$	True
$x=0$	$\frac{0+1}{0-4}\geq 0$	False
$x=5$	$\frac{5+1}{5-4}\geq 0$	True

The solution is $(-\infty, -1] \cup (4, \infty)$.

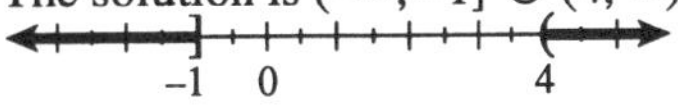

21. Answers may vary.

23. $\frac{3}{x-2}<4$

$$\frac{3}{x-2}-4<0$$
$$\frac{3-4(x-2)}{x-2}<0$$
$$\frac{3-4x+8}{x-2}<0$$
$$\frac{11-4x}{x-2}<0$$

$11-4x=0$ and $x-2=0$
$11=4x$ and $x=2$
$x=\frac{11}{4}$

Region	Interval	Test Point
A	$(-\infty, 2)$	1
B	$\left(2, \frac{11}{4}\right)$	$\frac{5}{2}$
C	$\left(\frac{11}{4}, \infty\right)$	3
$x=1$	$\frac{11-4(1)}{1-2}<0$	True
$x=\frac{5}{2}$	$\frac{11-4\left(\frac{5}{2}\right)}{\frac{5}{2}-2}<0$	False
$x=3$	$\frac{11-4(3)}{4-2}<0$	True

The solution is $(-\infty, 2)\cup\left(\frac{11}{4}, \infty\right)$.

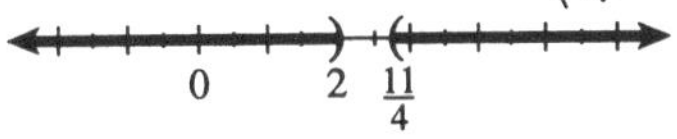

25. $\frac{x^2+6}{5x} \ge 1$

$\frac{x^2+6}{5x} - 1 \ge 0$

$\frac{x^2+6-5x}{5x} \ge 0$

$\frac{(x-2)(x-3)}{5x} \ge 0$

$x-2=0$ and $x-3=0$ and $5x=0$
$x=2$ and $x=3$ and $x=0$

Region	Interval	Test Point
A	$(-\infty, 0)$	-1
B	$(0, 2)$	1
C	$(2, 3)$	$\frac{5}{2}$
D	$(3, \infty)$	4
$x=-1$	$\frac{(-1-2)(-1-3)}{5(-1)} \ge 0$	False
$x=1$	$\frac{(1-2)(1-3)}{5(1)} \ge 0$	True
$x=\frac{5}{2}$	$\frac{\left(\frac{5}{2}-2\right)\left(\frac{5}{2}-3\right)}{5\left(\frac{5}{2}\right)} \ge 0$	False
$x=4$	$\frac{(4-2)(4-3)}{5(4)} \ge 0$	True

The solution is $(0, 2] \cup [3, \infty)$.

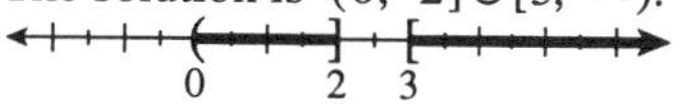

27. $(x-8)(x+7) > 0$
$(x-8)(x+7) = 0$
$x-8=0$ or $x+7=0$
$x=8$ or $x=-7$

Region	Interval	Test Point
A	$(-\infty, -7)$	-8
B	$(-7, 8)$	0
C	$(8, \infty)$	9
$x=-8$	$(-8-8)(-8+7) > 0$	True
$x=0$	$(0-8)(0+7) > 0$	False
$x=9$	$(9-8)(9+7) > 0$	True

The solution is $(-\infty, -7) \cup (8, \infty)$.

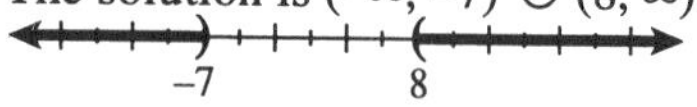

29. $(2x-3)(4x+5) \le 0$
$(2x-3)(4x+5) = 0$
$2x-3=0$ or $4x+5=0$
$2x=3$ or $4x=-5$
$x=\frac{3}{2}$ or $x=-\frac{5}{4}$

Region	Interval	Test Point
A	$\left(-\infty, -\frac{5}{4}\right)$	-2
B	$\left(-\frac{5}{4}, \frac{3}{2}\right)$	0
C	$\left(\frac{3}{2}, \infty\right)$	2
$x=-2$	$[2(-2)-3][4(-2)+5] \le 0$	False
$x=0$	$[2(0)-3][4(0)+5] \le 0$	True
$x=2$	$[2(2)-3][4(2)+5] \le 0$	False

The solution is $\left[-\frac{5}{4}, \frac{3}{2}\right]$.

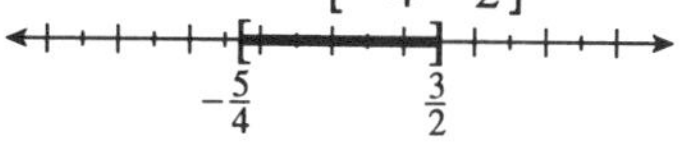

31. $x^2 > x$
$x^2 - x > 0$
$x(x-1) > 0$
$x(x-1) = 0$
$x=0$ or $x-1=0$
$x=1$

Region	Interval	Test Point
A	$(-\infty, 0)$	-1
B	$(0, 1)$	$\frac{1}{2}$
C	$(1, \infty)$	2
$x=-1$	$-1(-1-1) > 0$	True
$x=\frac{1}{2}$	$\frac{1}{2}\left(\frac{1}{2}-1\right) > 0$	False
$x=2$	$2(2-1) > 0$	True

The solution is $(-\infty, 0) \cup (1, \infty)$.

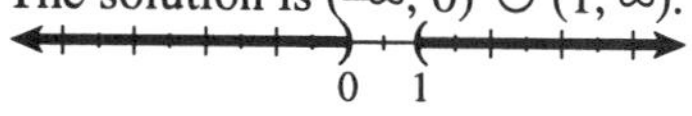

27. $(x-8)(x+7)>0$
$(x-8)(x+7)=0$
$x-8=0$ or $x+7=0$
$x=8$ or $x=-7$

Region	Interval	Test Point
A	$(-\infty, -7)$	-8
B	$(-7, 8)$	0
C	$(8, \infty)$	9
$x=-8$	$(-8-8)(-8+7)>0$	True
$x=0$	$(0-8)(0+7)>0$	False
$x=9$	$(9-8)(9+7)>0$	True

The solution is $(-\infty, -7) \cup (8, \infty)$.

-7 8

29. $(2x-3)(4x+5) \le 0$
$(2x-3)(4x+5)=0$
$2x-3=0$ or $4x+5=0$
$2x=3$ or $4x=-5$
$x=\frac{3}{2}$ or $x=-\frac{5}{4}$

Region	Interval	Test Point
A	$\left(-\infty, -\frac{5}{4}\right)$	-2
B	$\left(-\frac{5}{4}, \frac{3}{2}\right)$	0
C	$\left(\frac{3}{2}, \infty\right)$	2
$x=-2$	$[2(-2)-3][4(-2)+5] \le 0$	False
$x=0$	$[2(0)-3][4(0)+5] \le 0$	True
$x=2$	$[2(2)-3][4(2)+5] \le 0$	False

The solution is $\left[-\frac{5}{4}, \frac{3}{2}\right]$.

$-\frac{5}{4}$ $\frac{3}{2}$

31. $x^2 > x$
$x^2 - x > 0$
$x(x-1) > 0$
$x(x-1) = 0$
$x=0$ or $x-1=0$
$x=1$

Region	Interval	Test Point
A	$(-\infty, 0)$	-1
B	$(0, 1)$	$\frac{1}{2}$
C	$(1, \infty)$	2
$x=-1$	$-1(-1-1)>0$	True
$x=\frac{1}{2}$	$\frac{1}{2}\left(\frac{1}{2}-1\right)>0$	False
$x=2$	$2(2-1)>0$	True

The solution is $(-\infty, 0) \cup (1, \infty)$.

0 1

33. $(2x-8)(x+4)(x-6) \le 0$
$(2x-8)(x+4)(x-6) = 0$
$2x-8=0$ or $x+4=0$ or $x-6=0$
$2x=8$ or $x=-4$ or $x=6$
$x=4$

Region	Interval	Test Point
A	$(-\infty, -4)$	-5
B	$(-4, 4)$	0
C	$(4, 6)$	5
D	$(6, \infty)$	7
$x=-5$	$[2(-5)-8](-5+4)(-5-6) \le 0$	True
$x=0$	$[2(0)-8](0+4)(0-6) \le 0$	False
$x=5$	$[2(5)-8](5+4)(5-6) \le 0$	True
$x=7$	$[2(7)-8)](7+4)(7-6) \le 0$	False

The solution is $(-\infty, -4] \cup [4, 6]$.

-4 4 6

35. $6x^2 - 5x \ge 6$
$6x^2 - 5x - 6 \ge 0$
$(3x+2)(2x-3) \ge 0$
$(3x+2)(2x-3) = 0$
$3x+2=0$ or $2x-3=0$
$3x=-2$ or $2x=3$
$x=-\frac{2}{3}$ or $x=\frac{3}{2}$

Region	Interval	Test Point
A	$\left(-\infty, -\frac{2}{3}\right)$	-1
B	$\left(-\frac{2}{3}, \frac{3}{2}\right)$	0
C	$\left(\frac{3}{2}, \infty\right)$	2

$x=-1$	$[3(-1)+2][2(-1)-3] \geq 0$	True
$x=0$	$[3(0)+2][2(0)-3] \geq 0$	False
$x=2$	$[3(2)+2][2(2)-3] \geq 0$	True

The solution is $\left(-\infty, -\frac{2}{3}\right] \cup \left[\frac{3}{2}, \infty\right)$.

$-\frac{2}{3}$ $\frac{3}{2}$

37. $4x^3+16x^2-9x-36>0$

$4x^2(x+4)-9(x+4)>0$

$(4x^2-9)(x+4)>0$

$(2x-3)(2x+3)(x+4)>0$

$2x-3=0$ or $2x+3=0$ or $x+4=0$

$x=\frac{3}{2}$ or $x=-\frac{3}{2}$ or $x=-4$

Region	Interval	Test Point
A	$(-\infty, -4)$	-5
B	$\left(-4, -\frac{3}{2}\right)$	-3
C	$\left(-\frac{3}{2}, \frac{3}{2}\right)$	0
D	$\left(\frac{3}{2}, \infty\right)$	2

$x=-5$	$4(-5)^3+16(-5)^2-9(-5)-36=-91$ $-91>0$	False
$x=-3$	$4(-3)^3+16(-3)^2-9(-3)-36=27$ $27>0$	True
$x=0$	$4(0)^3+16(0)^2-9(0)-36=-36$ $-36>0$	False
$x=2$	$4(2)^3+16(2)^2-9(2)-36=42$ $42>0$	True

The solution is $\left(-4, -\frac{3}{2}\right) \cup \left(\frac{3}{2}, \infty\right)$.

$-4 \quad -\frac{3}{2} \quad 0 \quad \frac{3}{2}$

39. $x^4 - 26x^2 + 25 \geq 0$
$(x^2 - 25)(x^2 - 1) \geq 0$
$(x-5)(x+5)(x-1)(x+1) \geq 0$
$(x-5)(x+5)(x-1)(x+1) = 0$
$x - 5 = 0$ or $x + 5 = 0$ or $x - 1 = 0$ or $x + 1 = 0$
$x = 5$ or $x = -5$ or $x = 1$ or $x = -1$

Region	Interval	Test Point
A	$(-\infty, -5)$	-6
B	$(-5, -1)$	-3
C	$(-1, 1)$	0
D	$(1, 5)$	3
E	$(5, \infty)$	6

$x = -6$	$(-6-5)(-6+5)(-6-1)(-6+1) \geq 0$	True
$x = -3$	$(-3-5)(-3+5)(-3-1)(-3+1) \geq 0$	False
$x = 0$	$(0-5)(0+5)(0-1)(0+1) \geq 0$	True
$x = 3$	$(3-5)(3+5)(3-1)(3+1) \geq 0$	False
$x = 6$	$(6-5)(6+5)(6-1)(6+1) \geq 0$	True

The solution is $(-\infty, -5] \cup [-1, 1] \cup [5, \infty)$.

$-5 \ -4 \ -3 \ -2 \ -1 \ 0 \ 1 \ 2 \ 3 \ 4 \ 5$

41. $(2x-7)(3x+5) > 0$
$(2x-7)(3x+5) = 0$
$2x - 7 = 0$ or $3x + 5 = 0$
$2x = 7$ or $3x = -5$
$x = \frac{7}{2}$ or $x = -\frac{5}{3}$

Region	Interval	Test Point
A	$\left(-\infty, -\frac{5}{3}\right)$	-2
B	$\left(-\frac{5}{3}, \frac{7}{2}\right)$	0
C	$\left(\frac{7}{2}, \infty\right)$	4

$x = -2$	$[2(-2)-7][3(-2)+5] > 0$	True
$x = 0$	$[2(0)-7][3(0)+5] > 0$	False
$x = 4$	$[2(4)-7][2(4)+5] > 0$	True

The solution is $\left(-\infty, -\frac{5}{3}\right) \cup \left(\frac{7}{2}, \infty\right)$.

43. $\frac{x}{x-10} < 0$

$x = 0$ or $x - 10 = 0$

$x = 10$

Region	Interval	Test Point
A	$(-\infty, 0)$	-1
B	$(0, 10)$	1
C	$(10, \infty)$	11
$x=-1$	$\frac{-1}{-1-10} < 0$	False
$x=1$	$\frac{1}{1-10} < 0$	True
$x=11$	$\frac{11}{11-10} < 0$	False

The solution is $(0, 10)$.

45. $\frac{x-5}{x+4} \geq 0$

$x - 5 = 0$ or $x + 4 = 0$

$x = 5$ or $x = -4$

Region	Interval	Test Point
A	$(-\infty, -4)$	-5
B	$(-4, 5)$	0
C	$(5, \infty)$	6
$x=-5$	$\frac{-5-5}{-5+4} \geq 0$	True
$x=0$	$\frac{0-5}{0+4} \geq 0$	False
$x=6$	$\frac{6-5}{6+4} \geq 0$	True

The solution is $(-\infty, -4) \cup [5, \infty)$.

47. $\frac{x(x+6)}{(x-7)(x+1)} \geq 0$

$x = 0$ and $x + 6 = 0$ and $x - 7 = 0$ and $x + 1 = 0$

$x = 0$ and $x = -6$ and $x = 7$ and $x = -1$

Region	Interval	Test Point
A	$(-\infty, -6)$	-7
B	$(-6, -1)$	-2
C	$(-6, -1)$	$-\frac{1}{2}$
D	$(0, 7)$	1
E	$(7, \infty)$	8
$x=-7$	$\frac{-7(-7+6)}{(-7-7)(-7+1)} \geq 0$	True
$x=-2$	$\frac{-2(-2+6)}{(-2-7)(-2+1)} \geq 0$	False
$x=-\frac{1}{2}$	$\frac{-\frac{1}{2}\left(-\frac{1}{2}+6\right)}{\left(-\frac{1}{2}-7\right)\left(-\frac{1}{2}+1\right)} \geq 0$	True
$x=1$	$\frac{1(1+6)}{(1-7)(1+1)} \geq 0$	False
$x=8$	$\frac{8(8+6)}{(8-7)(8+1)} \geq 0$	True

The solution is $(-\infty, -6] \cup (-1, 0] \cup (7, \infty)$.

49. $\frac{-1}{x-1} > -1$

$\frac{1}{x-1} < 1$

$\frac{1}{x-1} - 1 < 0$

$\frac{1-(x-1)}{x-1} < 0$

$\frac{1-x+1}{x-1} < 0$

$\frac{2-x}{x-1} < 0$

$2 - x = 0$ or $x - 1 = 0$

$2 = x$ or $x = 1$

Region	Interval	Test Point
A	$(-\infty, 1)$	0
B	$(1, 2)$	$\frac{3}{2}$
C	$(2, \infty)$	3
$x=0$	$\frac{2-0}{0-1}<0$	True
$x=\frac{3}{2}$	$\frac{2-\frac{3}{2}}{\frac{3}{2}-1}<0$	False
$x=3$	$\frac{2-3}{3-1}<0$	True

The solution is $(-\infty, 1) \cup (2, \infty)$.

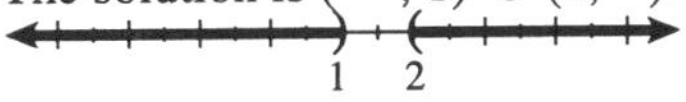

51. $\frac{x}{x+4} \le 2$

$\frac{x}{x+4} - 2 \le 0$

$\frac{x-2x-8}{x+4} \le 0$

$\frac{-x-8}{x+4} \le 0$

$-x-8=0$ and $x+4=0$

$-x=8$ and $x=-4$

$x=-8$

Region	Interval	Test Point
A	$(-\infty, -8)$	-9
B	$(-8, -4)$	-5
C	$(-4, \infty)$	0
$x=-9$	$\frac{-(-9)-8}{-9+4} \le 0$	True
$x=-5$	$\frac{-(-5)-8}{-5+4} \le 0$	False
$x=0$	$\frac{-0-8}{0+4} \le 0$	True

The solution is $(-\infty, -8] \cup (-4, \infty)$.

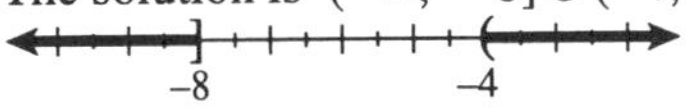

53. $\frac{z}{z-5} \ge 2z$

$\frac{z}{z-5} - 2z \ge 0$

$\frac{z-2z(z-5)}{z-5} \ge 0$

$\frac{z-2z^2+10z}{z-5} \ge 0$

$\frac{11z-2z^2}{z-5} \ge 0$

$\frac{z(11-2z)}{z-5} \ge 0$

$z=0$ and $11-2z=0$ and $z-5=0$

$11=2z$ and $z=5$

$\frac{11}{2}=z$

Region	Interval	Test Point
A	$(-\infty, 0)$	-1
B	$(0, 5)$	1
C	$\left(5, \frac{11}{2}\right)$	$\frac{21}{4}$
D	$\left(\frac{11}{2}, \infty\right)$	6
$z=-1$	$\frac{-1[11-2(-1)]}{-1-5} \ge 0$	True
$z=1$	$\frac{1[11-2(1)]}{1-5} \ge 0$	False
$z=\frac{21}{4}$	$\frac{\frac{21}{4}\left[11-2\left(\frac{21}{4}\right)\right]}{\frac{21}{4}-5} \ge 0$	True
$z=6$	$\frac{6[11-2(6)]}{6-5} \ge 0$	False

The solution is $(-\infty, 0] \cup \left(5, \frac{11}{2}\right)$.

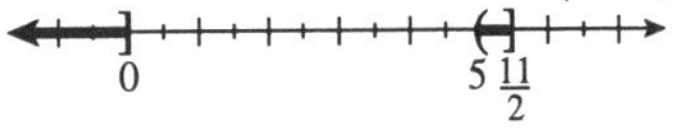

55. $\frac{(x+1)^2}{5x} > 0$

$(x+1)^2=0$ and $5x=0$

$x+1=0$ and $x=0$

$x=-1$

Region	Interval	Test Point
A	$(-\infty, -1)$	-2
B	$(-1, 0)$	$-\frac{1}{2}$
C	$(-1, \infty)$	1
$x=-2$	$\frac{(-2+1)^2}{5(-2)} > 0$	False
$x=-\frac{1}{2}$	$\frac{\left(-\frac{1}{2}+1\right)^2}{5\left(-\frac{1}{2}\right)} > 0$	False
$x=1$	$\frac{(1+1)^2}{5(1)} > 0$	True

The solution is $(0, \infty)$.

0

57. Let x = the number.

$\frac{1}{x}$ = the reciprocal of the number.

$x - \frac{1}{x} < 0$

$\frac{x^2 - 1}{x} < 0$

$\frac{(x+1)(x-1)}{x} < 0$

$x + 1 = 0$ and $x - 1 = 0$ and $x = 0$

$x = -1$ and $x = 1$ and $x = 0$

Region	Interval	Test Point
A	$(-\infty, -1)$	-2
B	$(-1, 0)$	$-\frac{1}{2}$
C	$(0, 1)$	$\frac{1}{2}$
D	$(1, \infty)$	2
$x = -2$	$\frac{(-2+1)(-2-1)}{-2} < 0$	True
$x = -\frac{1}{2}$	$\frac{\left(-\frac{1}{2}+1\right)\left(-\frac{1}{2}-1\right)}{-\frac{1}{2}} < 0$	False
$x = \frac{1}{2}$	$\frac{\left(\frac{1}{2}+1\right)\left(\frac{1}{2}-1\right)}{\frac{1}{2}} < 0$	True
$x = 2$	$\frac{(2+1)(2-1)}{2} < 0$	False

The solution is $(-\infty, -1) \cup (0, 1)$.

−1 0 1

59. $P(x) = -2x^2 + 26x - 44$

$-2x^2 + 26x - 44 > 0$

$-2(x^2 - 13x + 22) > 0$

$-2(x - 11)(x - 2) > 0$

$x - 11 = 0$ or $x - 2 = 0$

$x = 11$ or $x = 2$

Region	Interval	Test Point
A	$(-\infty, 2)$	0
B	$(2, 11)$	3
C	$(11, \infty)$	12
$x = 0$	$-2(0 - 11)(0 - 2) > 0$	False
$x = 3$	$-2(3 - 11)(3 - 2) > 0$	True
$x = 12$	$-2(12 - 11)(12 - 2) > 0$	False

The solution is $(2, 11)$.

2 11

61. The results are the same.

63. The results are the same.

65. $g(x) = |x| + 2$

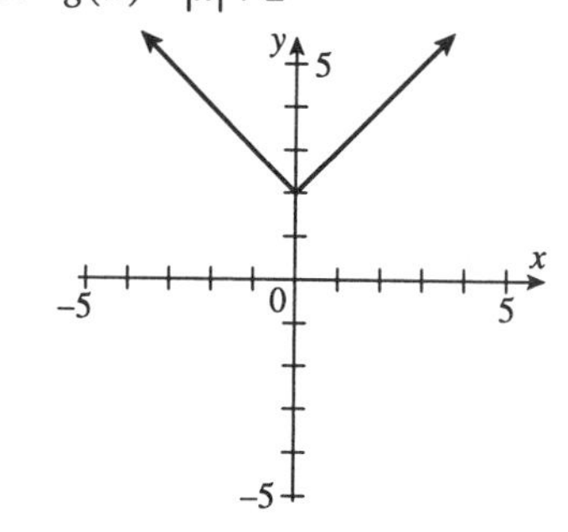

67. $F(x) = |x| - 1$

y 5

x

−5 5

−5

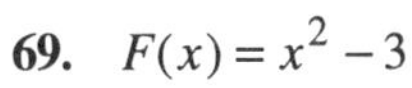

69. $F(x) = x^2 - 3$

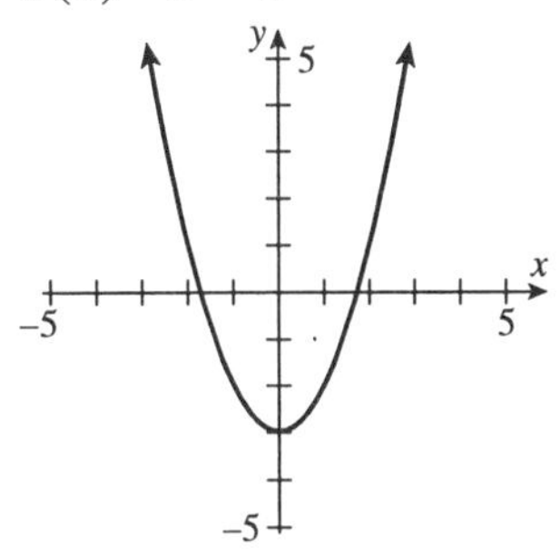

71. $H(x) = x^2 + 1$

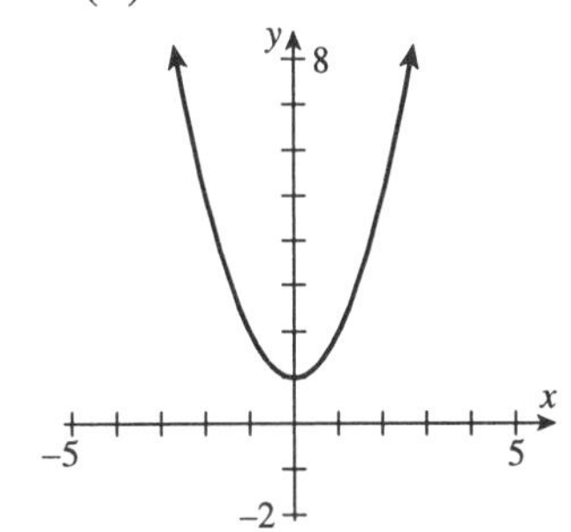

Graphing Calculator Explorations 8.5

1. $F(x) = \sqrt{x};\ G(x) = \sqrt{x} + 1$

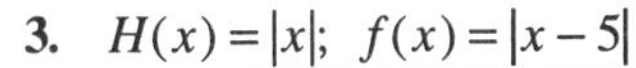

3. $H(x) = |x|;\ f(x) = |x - 5|$

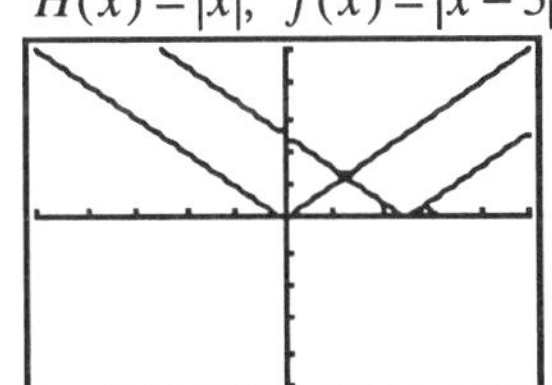

5. $f(x) = |x + 4|;\ F(x) = |x + 4| + 3$

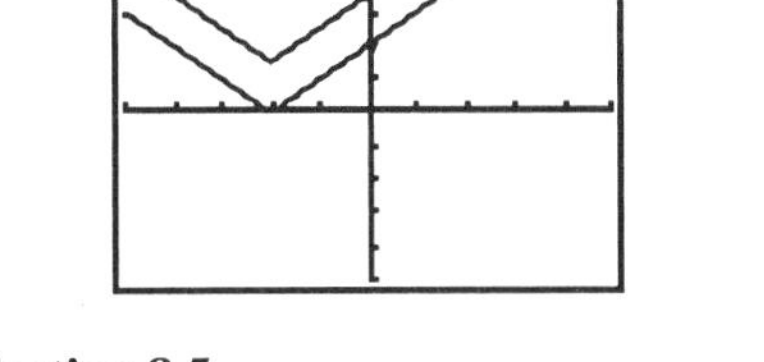

Section 8.5

Mental Math

1. $f(x) = x^2$; vertex: (0, 0)

3. $g(x) = (x - 2)^2$; vertex: (2, 0)

5. $f(x) = 2x^2 + 3$; vertex: (0, 3)

7. $g(x) = (x + 1)^2 + 5$; vertex: (−1, 5)

Exercise Set 8.5

1. $f(x) = x^2 - 1$

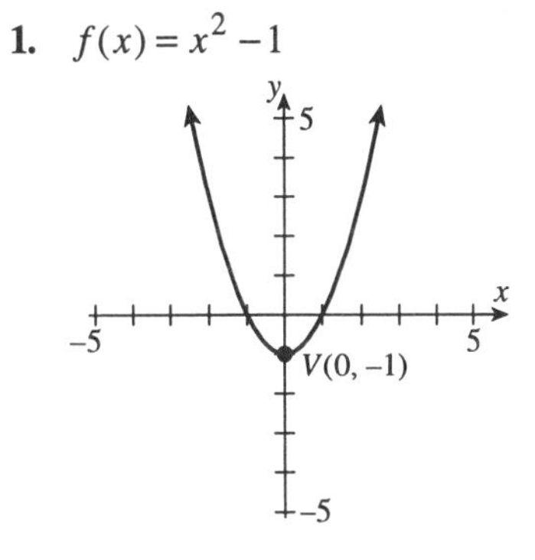

3. $h(x) = x^2 + 5$

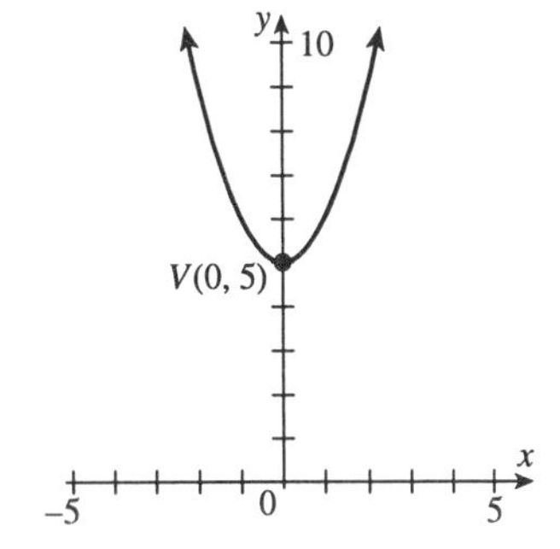

5. $g(x) = x^2 + 7$

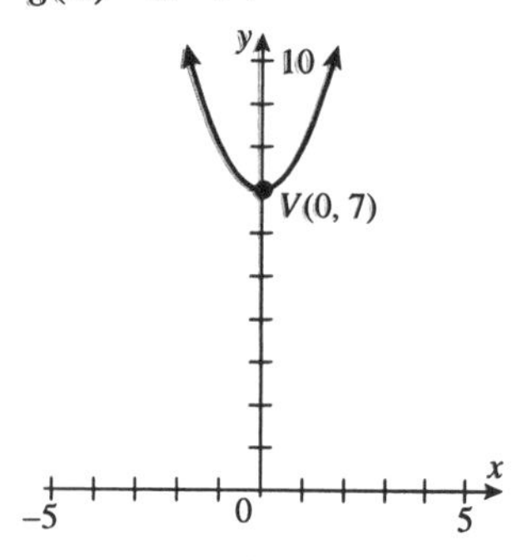

7. $f(x) = (x-5)^2$

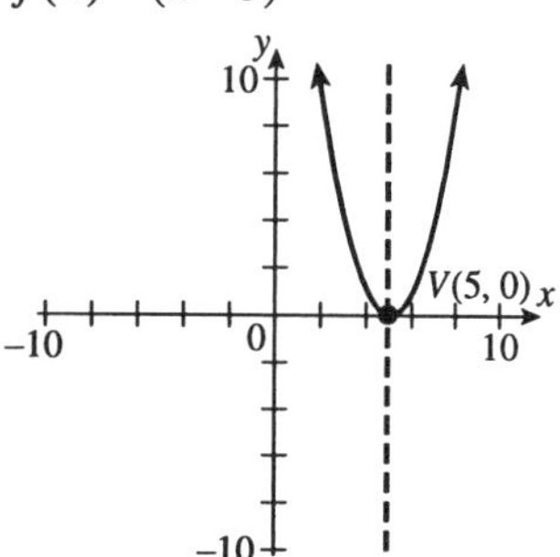

9. $h(x) = (x+2)^2$

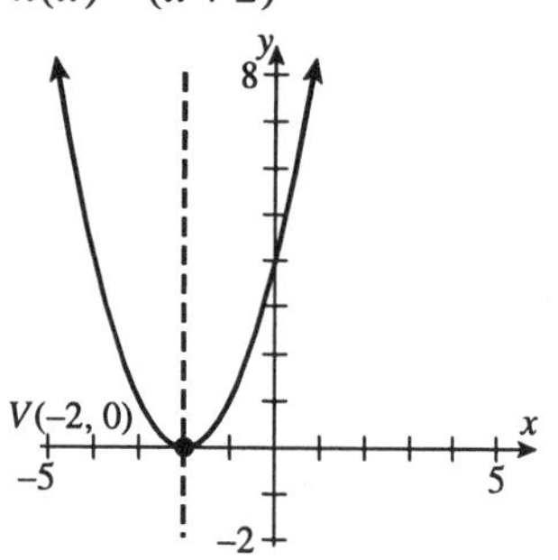

11. $G(x) = (x+3)^2$

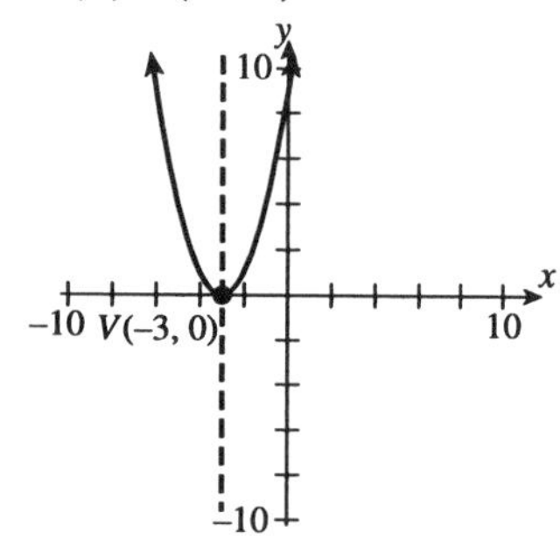

13. $f(x) = (x-2)^2 + 5$

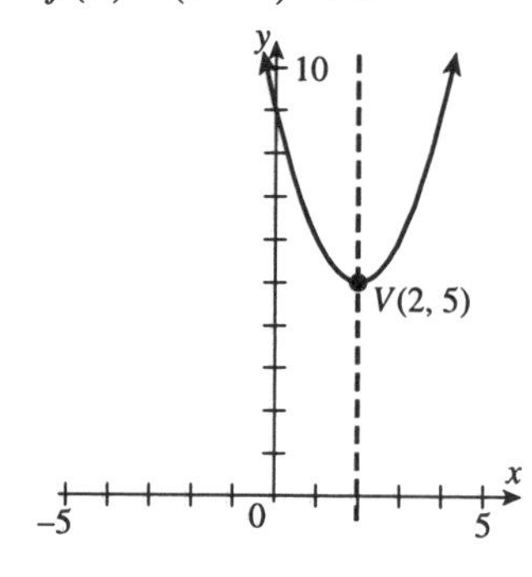

15. $h(x) = (x+1)^2 + 4$

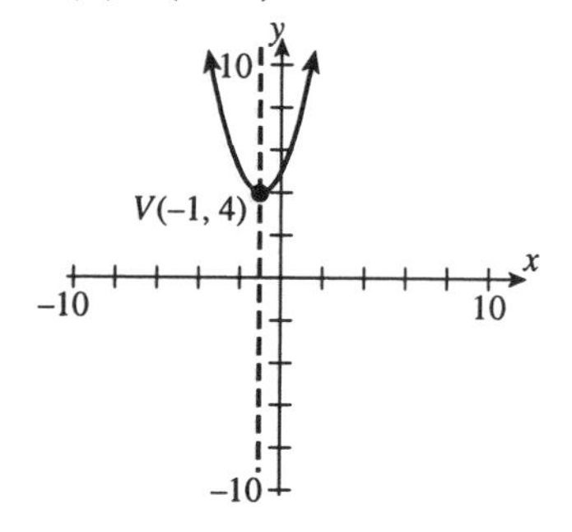

17. $g(x) = (x+2)^2 - 5$

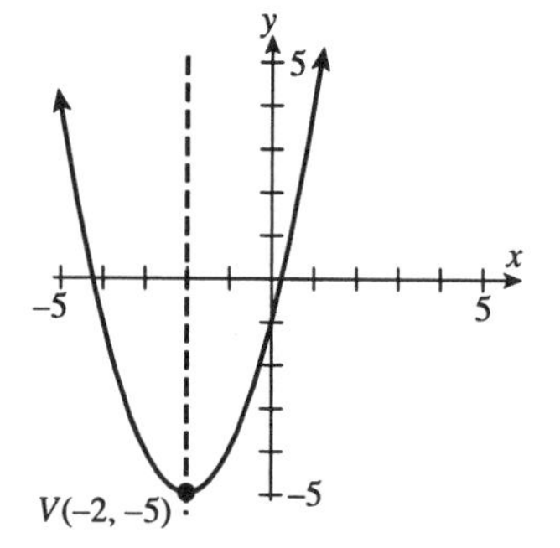

19. $g(x) = -x^2$

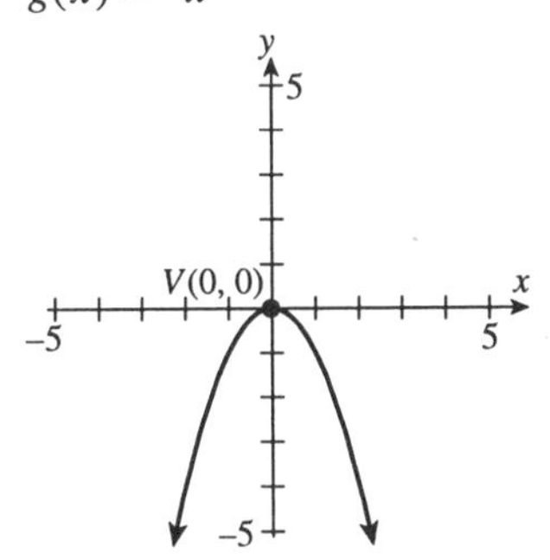

21. $h(x) = \frac{1}{3}x^2$

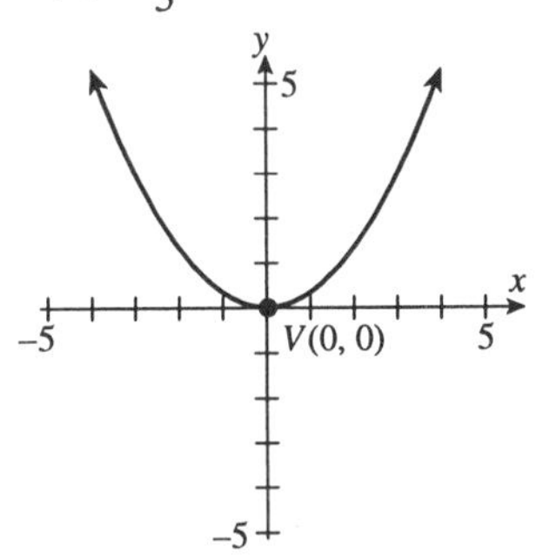

23. $H(x) = 2x^2$

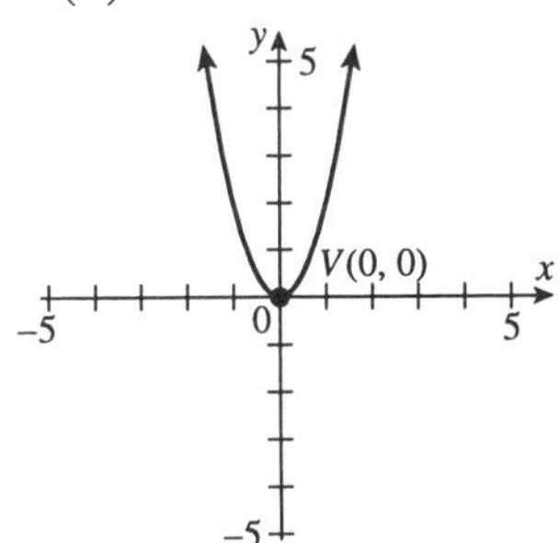

25. $f(x) = 2(x-1)^2 + 3$

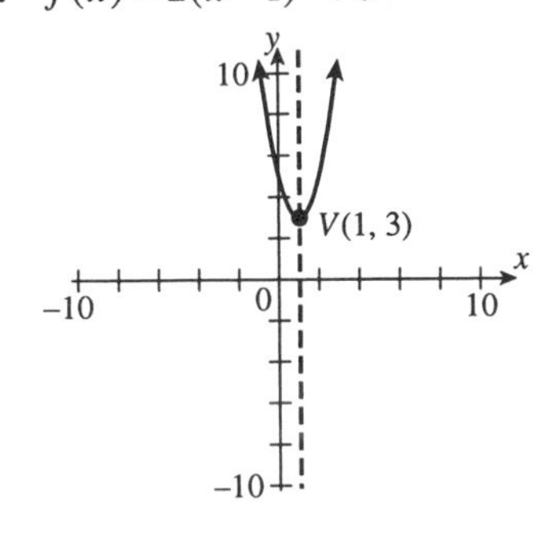

27. $h(x) = -3(x+3)^2 + 1$

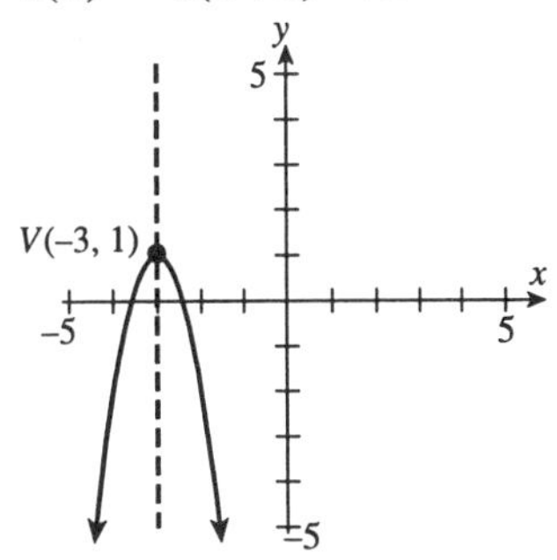

29. $H(x) = \frac{1}{2}(x-6)^2 - 3$

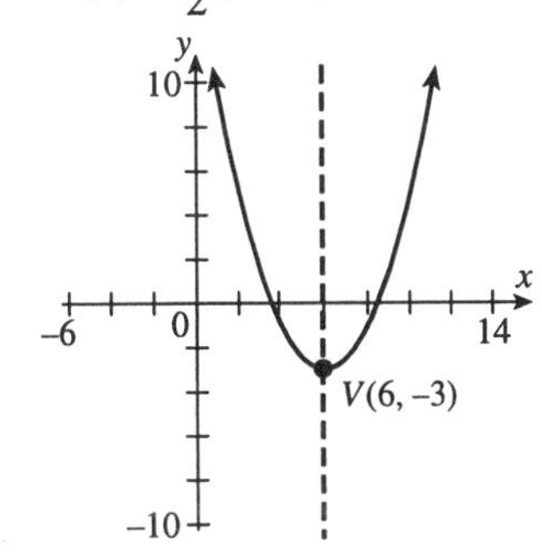

31. $f(x) = -(x-2)^2$

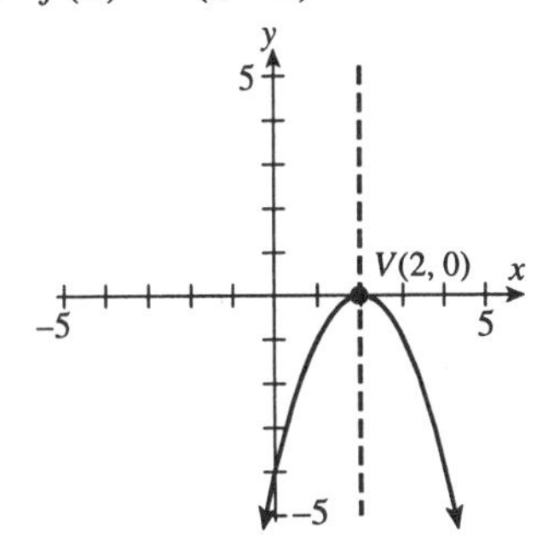

33. $F(x) = -x^2 + 4$

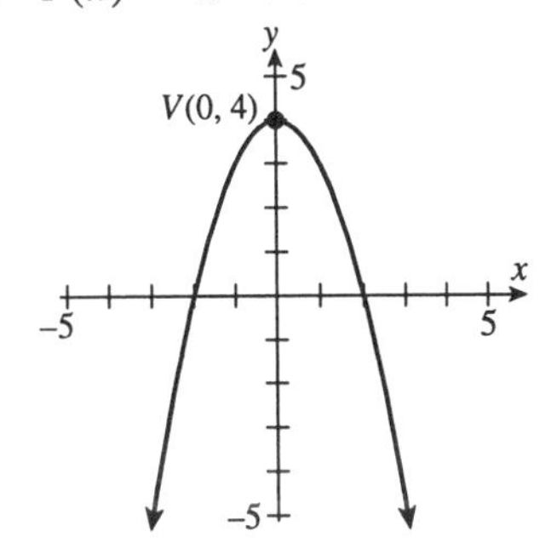

39. $F(x) = \left(x + \frac{1}{2}\right)^2 - 2$

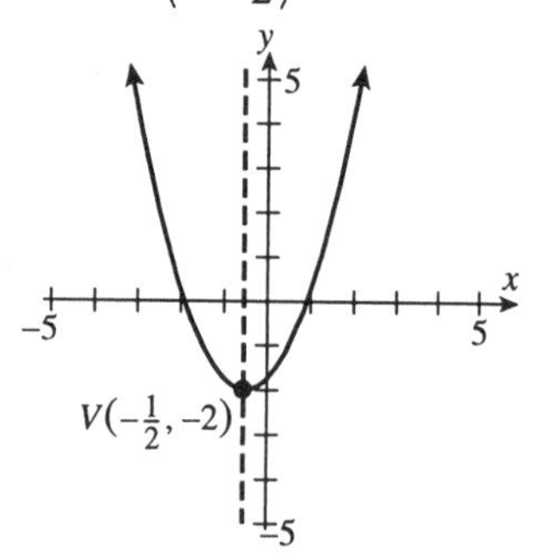

41. $F(x) = \frac{3}{2}(x+7)^2 + 1$

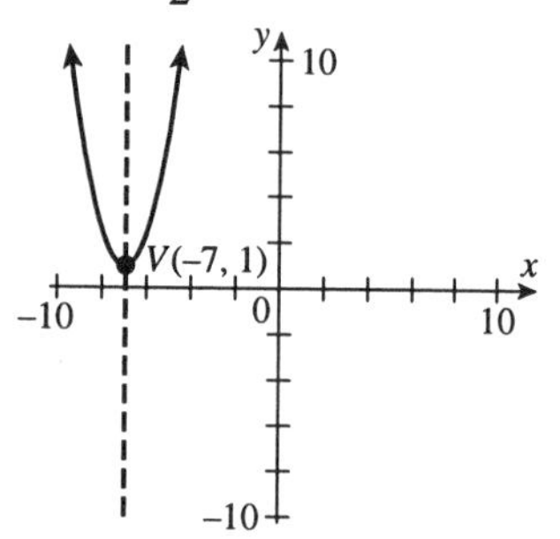

43. $f(x) = \frac{1}{4}x^2 - 9$

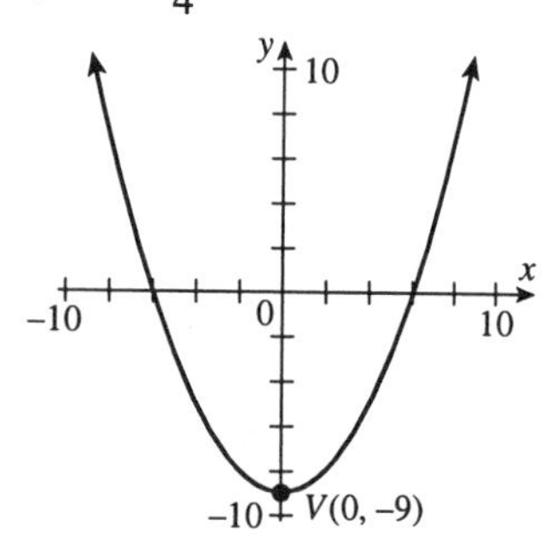

45. $G(x) = 5\left(x + \frac{1}{2}\right)^2$

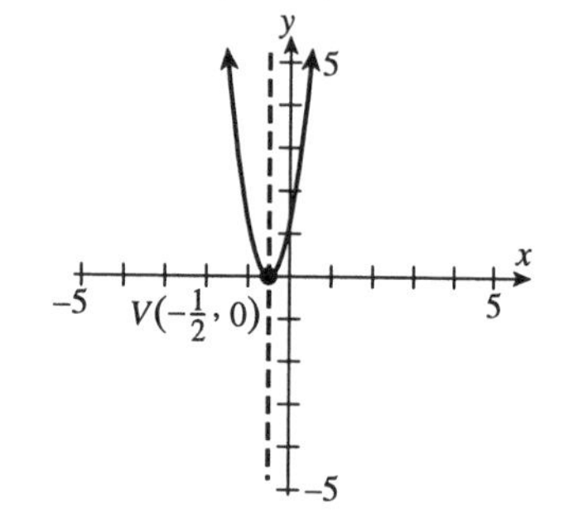

47. $f(x) = -(x-1)^2 - 1$

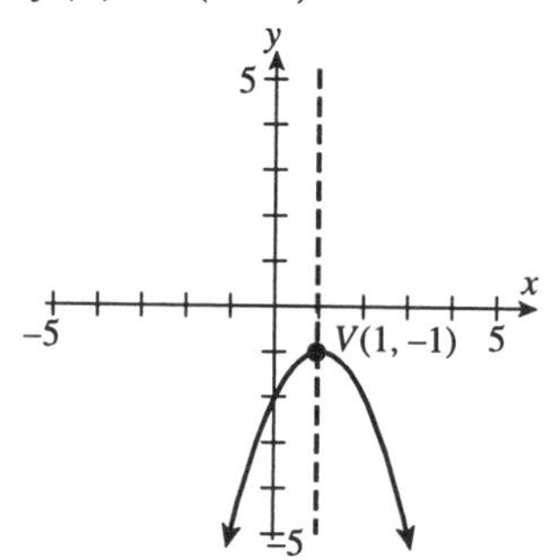

49. $g(x) = \sqrt{3}(x+5)^2 + \frac{3}{4}$

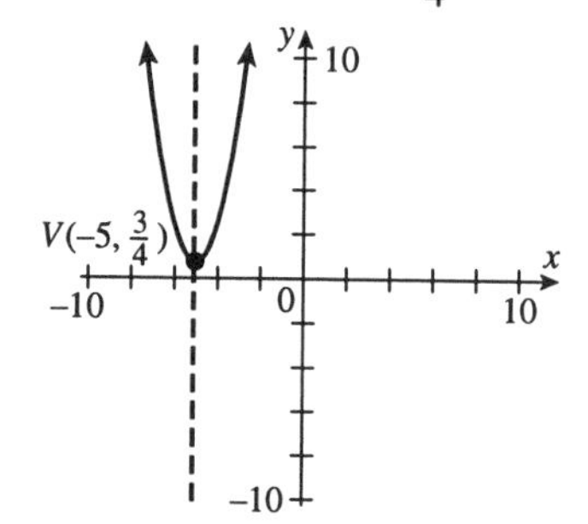

51. $h(x) = 10(x+4)^2 - 6$

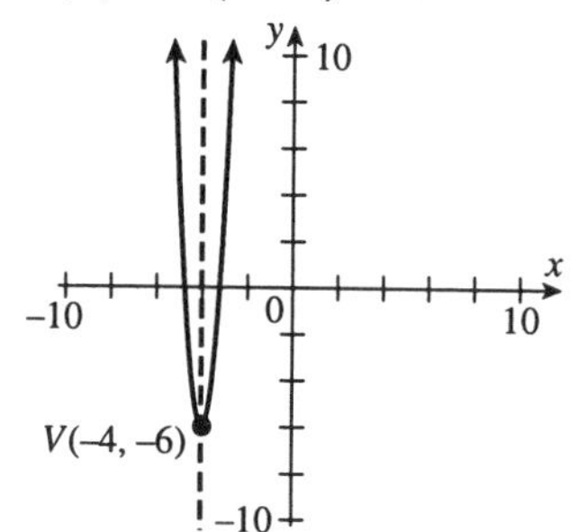

53. $f(x) = -2(x-4)^2 + 5$

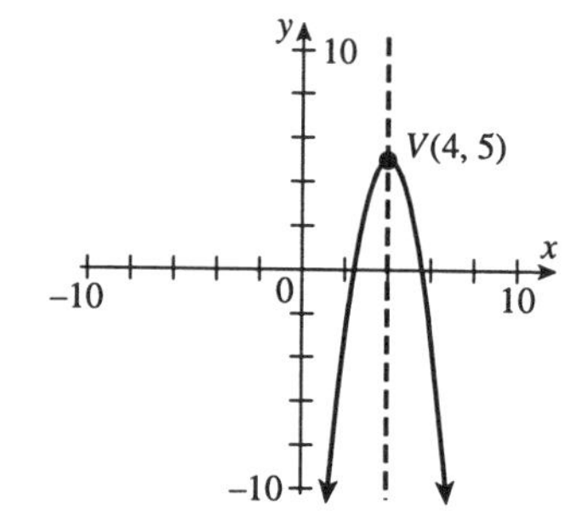

55. $f(x) = 5(x-2)^2 + 5$

57. $f(x) = 5[(x-(-3))]^2 + 6$

$f(x) = 5(x+3)^2 + 6$

59. $y = f(x) + 1$

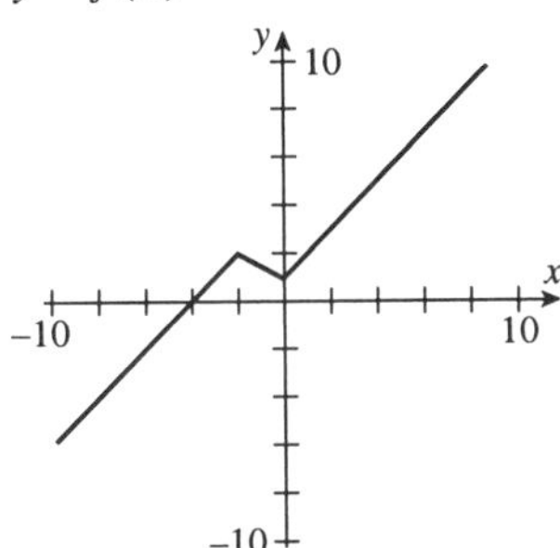

61. $y = f(x-3)$

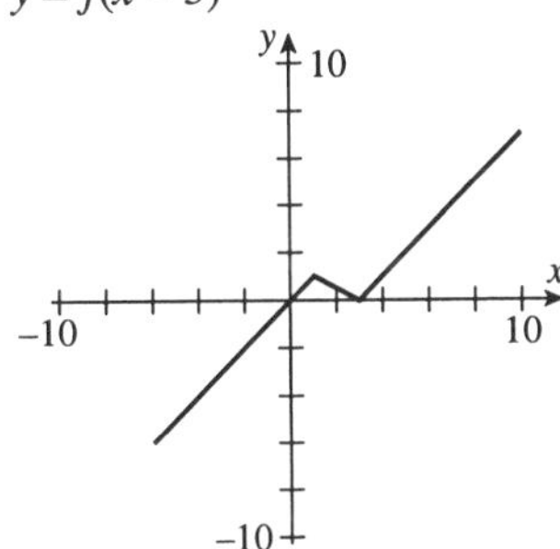

63. $y = f(x+2) + 2$

65. $x^2 + 8x$

$\left[\frac{1}{2}(8)\right]^2 = [4]^2 = 16$

$x^2 + 8x + 16$

67. $z^2 - 16z$

$\left[\frac{1}{2}(-16)\right]^2 = [-8]^2 = 64$

$z^2 - 16z + 64$

69. $y^2 + y$

$\left[\frac{1}{2}(1)\right]^2 = \left[\frac{1}{2}\right]^2 = \frac{1}{4}$

$y^2 + y + \frac{1}{4}$

71. $x^2 + 4x = 12$

$x^2 + 4x + 4 = 12 + 4$

$(x+2)^2 = 16$

$x + 2 = \pm\sqrt{16}$

$x + 2 = \pm 4$

$x = -2 \pm 4$

$x = -2 + 4$ or $x = -2 - 4$

$x = 2$ or $x = -6$

$\{-6, 2\}$

73. $z^2 + 10z - 1 = 0$

$z^2 + 10z = 1$

$z^2 + 10z + 25 = 1 + 25$

$(z+5)^2 = 26$

$z + 5 = \pm\sqrt{26}$

$z = -5 \pm \sqrt{26}$

$z = -5 + \sqrt{26}$ or $z = -5 - \sqrt{26}$

$\left\{-5 - \sqrt{26},\ -5 + \sqrt{26}\right\}$

75. $z^2 - 8z = 2$

$z^2 - 8z + 16 = 2 + 16$

$(z-4)^2 = 18$

$z - 4 = \pm\sqrt{18}$

$z - 4 = \pm 3\sqrt{2}$

$z = 4 \pm 3\sqrt{2}$

$z = 4 + 3\sqrt{2}$ or $z = 4 - 3\sqrt{2}$

$\left\{4 - 3\sqrt{2},\ 4 + 3\sqrt{2}\right\}$

Exercise Set 8.6

1. $f(x) = x^2 + 8x + 7$

$-\frac{b}{2a} = \frac{-8}{2(1)} = -4$ and

$f(-4) = (-4)^2 + 8(-4) + 7$

$f(-4) = 16 - 32 + 7 = -9$

Thus, $V(-4, -9)$.

3. $f(x) = -x^2 + 10x + 5$

$\frac{-b}{2a} = \frac{-10}{2(-1)} = 5$ and

$f(5) = -5^2 + 10(5) + 5$

$f(5) = -25 + 50 + 5 = 30$

Thus, $V(5, 30)$.

5. $f(x) = 5x^2 - 10x + 3$

$\frac{-b}{2a} = \frac{-(-10)}{2(5)} = 1$ and

$f(1) = 5(1)^2 - 10(1) + 3$

$f(1) = 5 - 10 + 3 = -2$

Thus, $V(1, -2)$.

7. $f(x) = -x^2 + x + 1$

$\frac{-b}{2a} = \frac{-1}{2(-1)} = \frac{1}{2}$ and

$f\left(\frac{1}{2}\right) = -\left(\frac{1}{2}\right)^2 + \frac{1}{2} + 1$

$f\left(\frac{1}{2}\right) = -\frac{1}{4} + \frac{1}{2} + 1 = \frac{5}{4}$

Thus, $V\left(\frac{1}{2}, \frac{5}{4}\right)$.

9. $f(x) = x^2 + 4x - 5$

$\frac{-b}{2a} = \frac{-4}{2(1)} = -2$ and

$f(-2) = (-2)^2 + 4(-2) - 5$

$f(-2) = 4 - 8 - 5 = -9$

Thus, $V(-2, -9)$.

The graph opens upward since $a = 1 > 0$.

$x^2 + 4x - 5 = 0$

$(x + 5)(x - 1) = 0$

$x + 5 = 0$ or $x - 1 = 0$

$x = -5$ or $x = 1$

The x-intercepts are -5 and 1.

$f(0) = -5$ is the y-intercept.

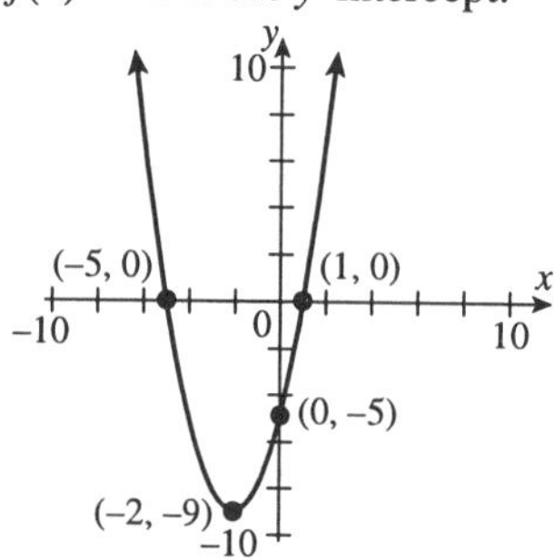

11. $f(x) = -x^2 + 2x - 1$

$\frac{-b}{2a} = \frac{-2}{2(-1)} = 1$ and

$f(1) = -1^2 + 2(1) - 1$

$f(1) = -1 + 2 - 1 = 0$

Thus, $V(1, 0)$.

The graph opens down since $a = -1 < 0$.

$-x^2 + 2x - 1 = 0$

$-(x^2 - 2x + 1) = 0$

$-(x - 1)^2 = 0$

$x = 1$

The x-intercept is 1.

$f(0) = -1$ is the y-intercept.

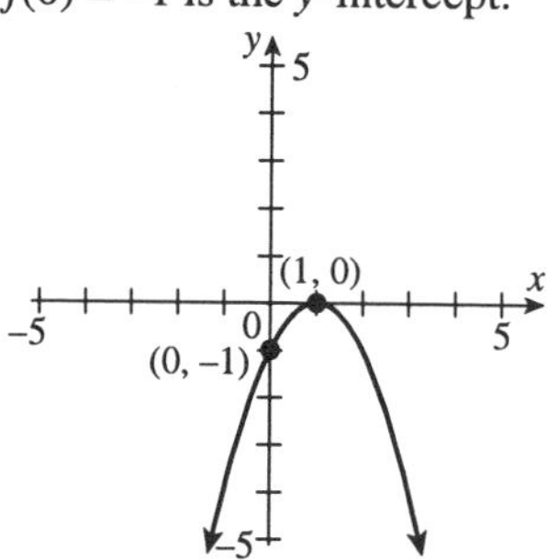

13. $f(x) = x^2 - 4$

$\frac{-b}{2a} = \frac{-0}{2(1)} = 0$ and $f(0) = -4$

Thus, $V(0, -4)$.

The graph opens upward since $a = 1 > 0$.

$x^2 - 4 = 0$

$(x + 2)(x - 2) = 0$

$x + 2 = 0$ or $x - 2 = 0$

$x = -2$ or $x = 2$

The x-intercepts are -2 and 2.

$f(0) = -4$ is the y-intercept.

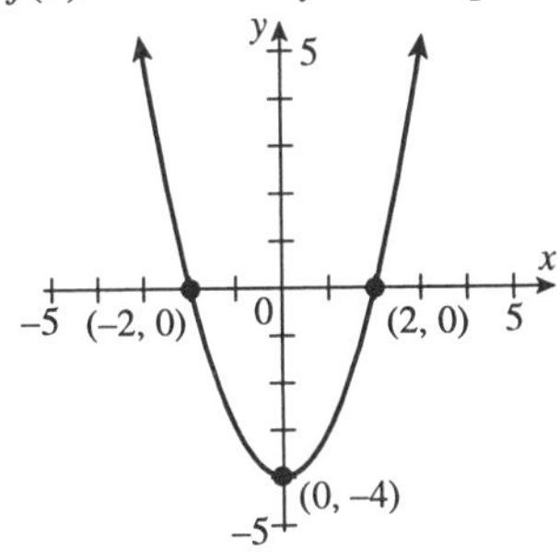

15. $f(x) = 4x^2 + 4x - 3$

$\frac{-b}{2a} = \frac{-4}{2(4)} = -\frac{1}{2}$ and

$f\left(-\frac{1}{2}\right) = 4\left(-\frac{1}{2}\right)^2 + 4\left(-\frac{1}{2}\right) - 3$

$f\left(-\frac{1}{2}\right) = 1 - 2 - 3 = -4$

Thus, $V\left(-\frac{1}{2}, -4\right)$.

The graph opens upward since $a = 4 > 0$.

$4x^2 + 4x - 3 = 0$

$(2x - 1)(2x + 3) = 0$

$2x - 1 = 0$ or $2x + 3 = 0$

$2x = 1$ or $2x = -3$

$x = \frac{1}{2}$ or $x = -\frac{3}{2}$

The x-intercepts are $\frac{1}{2}$ and $-\frac{3}{2}$.

$f(0) = -3$ is the y-intercept.

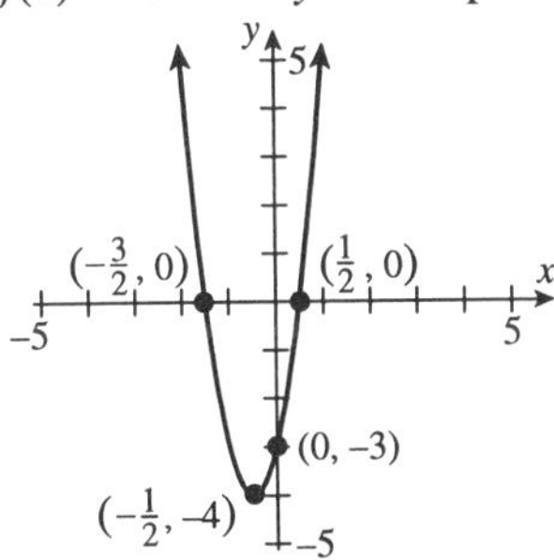

17. $f(x) = x^2 + 8x + 15$

$f(x) = x^2 + 8x + \left(\frac{8}{2}\right)^2 + 15 - 16$

$f(x) = (x + 4)^2 - 1$

Thus, $V(-4, -1)$.

The graph opens upward since $a = 1 > 0$.

$x^2 + 8x + 15 = 0$

$(x + 3)(x + 5) = 0$

$x + 3 = 0$ or $x + 5 = 0$

$x = -3$ or $x = -5$

The x-intercepts are -3 and -5.

$f(0) = 15$ is the y-intercept.

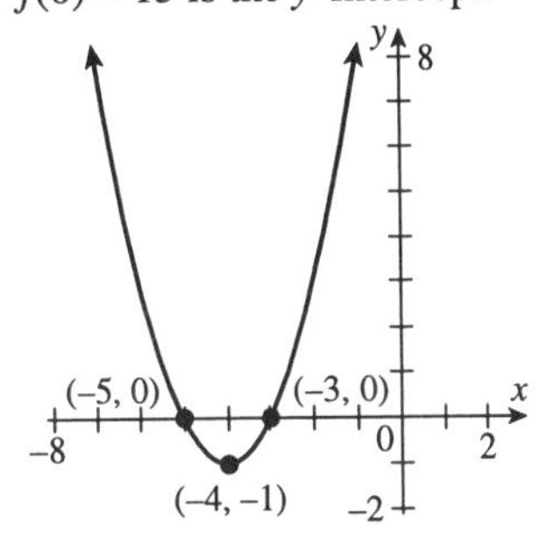

19. $f(x) = x^2 - 6x + 5$

$f(x) = x^2 - 6x + \left(\frac{6}{2}\right)^2 + 5 - 9$

$f(x) = (x - 3)^2 - 4$

Thus, $V(3, -4)$.

The graph opens upward since $a = 1 > 0$.

$x^2 - 6x + 5 = 0$

$(x - 1)(x - 5) = 0$

$x - 1 = 0$ or $x - 5 = 0$

$x = 1$ or $x = 5$

The x-intercepts are 1 and 5.

$f(0) = 5$ is the y-intercept.

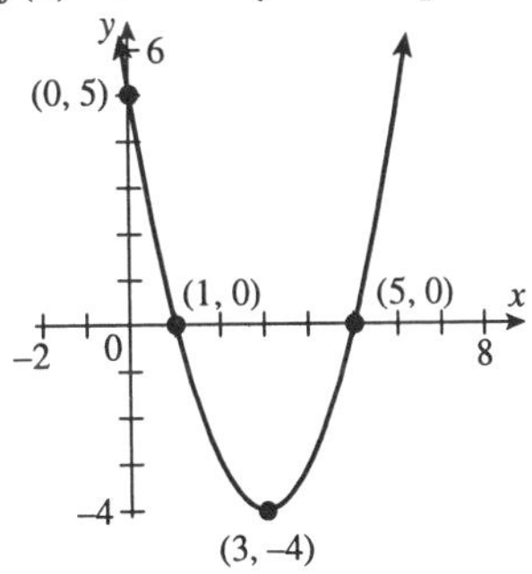

21. $f(x) = x^2 - 4x + 5$

$f(x) = x^2 - 4x + \left(\frac{4}{2}\right)^2 + 5 - 4$

$f(x) = (x - 2)^2 + 1$

Thus, $V(2, 1)$.

The graph opens upward since $a = 1 > 0$.

$(x-2)^2+1=0$
$(x-2)^2=-1$
Hence, there are no x-intercepts.
$f(0) = 5$ is the y-intercept.

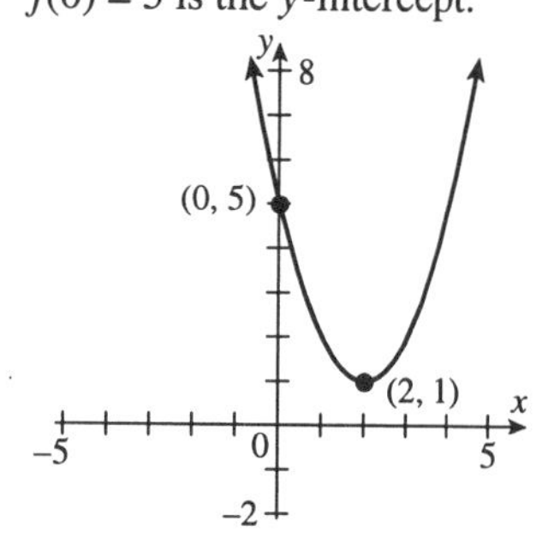

23. $f(x)=2x^2+4x+5$
$f(x)=2(x^2+2x)+5$
$f(x)=2\left[x^2+2x+\left(\frac{2}{2}\right)^2\right]+5-2$
$f(x)=2(x+1)^2+3$
Thus, $V(-1, 3)$.
The graph opens upward since $a=2>0$.
$f(0)=2(0+1)^2+3=5$ is the y-intercept.

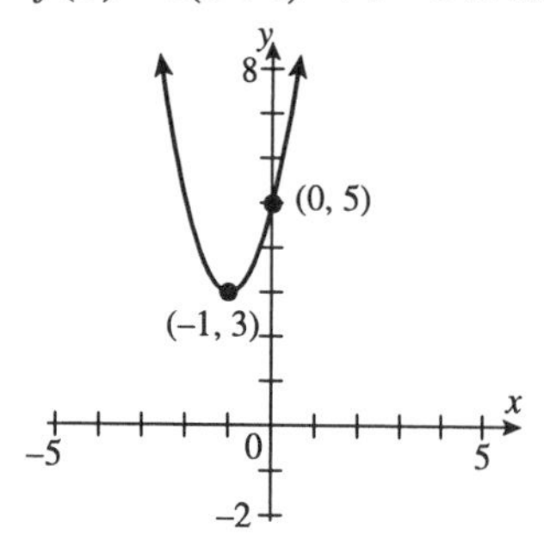

25. $f(x)=-2x^2+12x$
$f(x)=-2(x^2-6x)$
$f(x)=-2\left[x^2-6x+\left(\frac{6}{2}\right)^2\right]+18$
$f(x)=-2(x-3)^2+18$
Thus, $V(3, 18)$.
The graph opens downward since $a=3<0$.
$0=-2x^2+12x$
$0=-2x(x-6)$
$-2x=0$ or $x-6=0$
$x=0$ or $x=6$
The x-intercepts are 0 and 6.
$f(0) = 0$ is the y-intercept.

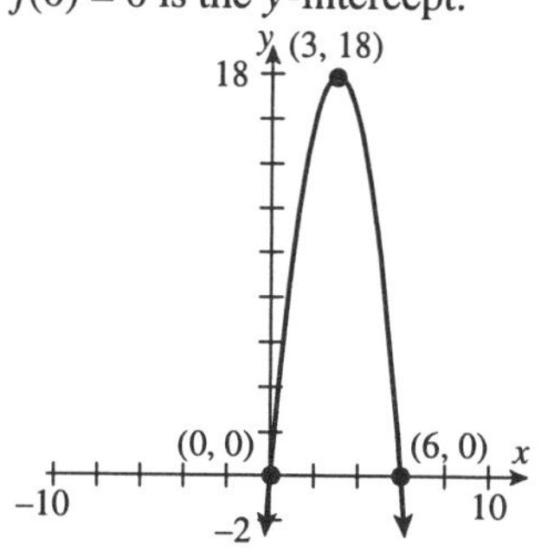

27. $f(x)=x^2+1$
$\frac{-b}{2a}=\frac{-0}{2(1)}=0$ and $f(0) = 1$.
Thus, $V(0, 1)$.
The graph opens upward since $a=1>0$.
$x^2+1=0$
$x^2=-1$, so there are no x-intercepts.
$f(0) = 1$ is the y-intercept.

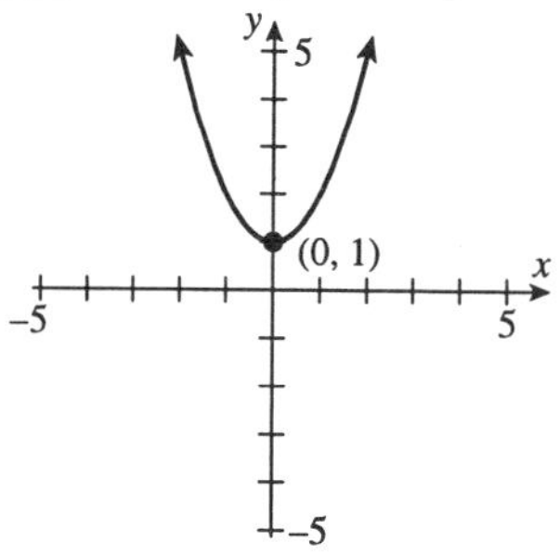

29. $f(x)=x^2-2x-15$
$\frac{-b}{2a}=\frac{-(-2)}{2(1)}=1$ and
$f(1)=1^2-2(1)-15$
$f(1) = 1 - 2 - 15 = -16$
Thus, $V(1, -16)$.
The graph opens upward since $a=1>0$.
$x^2-2x-15=0$
$(x-5)(x+3)=0$
$x-5=0$ or $x+3=0$
$x=5$ or $x=-3$
The x-intercepts are 5 and −3.

$f(0) = -15$ is the y-intercept.

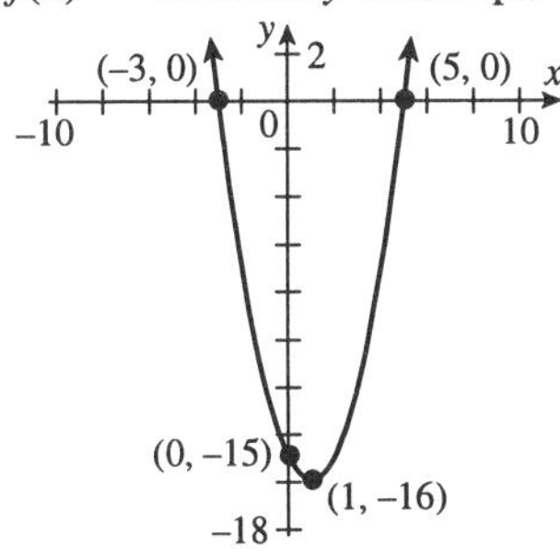

31. $f(x) = -5x^2 + 5x$

$\frac{-b}{2a} = \frac{-5}{2(-5)} = \frac{1}{2}$ and

$f\left(\frac{1}{2}\right) = -5\left(\frac{1}{2}\right)^2 + 5\left(\frac{1}{2}\right)$

$f\left(\frac{1}{2}\right) = -\frac{5}{4} + \frac{5}{2} = \frac{5}{4}$

Thus, $V\left(\frac{1}{2}, \frac{5}{4}\right)$.

The graph opens downward since $a = -5 < 0$.

$-5x^2 + 5x = 0$

$-5x(x-1) = 0$

$-5x = 0$ or $x - 1 = 0$

$x = 0$ or $x = 1$

The x-intercepts are 0 and 1.

$f(0) = 0$ is the y-intercept.

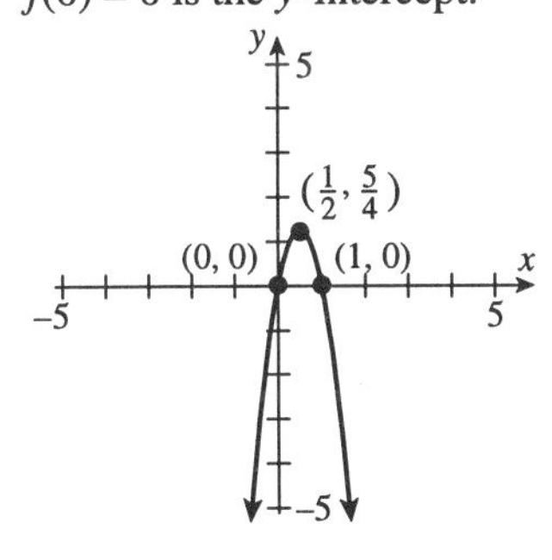

33. $f(x) = -x^2 + 2x - 12$

$\frac{-b}{2a} = \frac{-2}{2(-1)} = 1$ and

$f(1) = -1^2 + 2(1) - 12$

$f(1) = -1 + 2 - 12 = -11$

Thus, $V(1, -11)$.

The graph opens downward since $a = -1 < 0$.

$-x^2 + 2x - 12 = 0$

$x^2 - 2x = -12$

$x^2 - 2x + \left(\frac{2}{2}\right)^2 = -12 + 1$

$(x-1)^2 = -11$, so there are no x-intercepts. $f(0) = -12$ is the y-intercept.

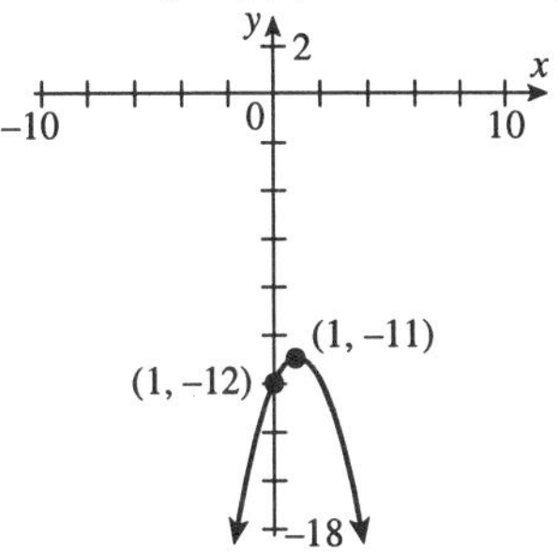

35. $f(x) = 3x^2 - 12x + 15$

$\frac{-b}{2a} = \frac{-(-12)}{2(3)} = 2$ and

$f(2) = 3(2)^2 - 12(2) + 15$

$f(2) = 12 - 24 + 15 = 3$

Thus, $V(2, 3)$.

The graph opens upward since $a = 3 > 0$.

$3x^2 - 12x + 15 = 0$

$x^2 - 4x + 5 = 0$

$x^2 - 4x + \left(\frac{4}{2}\right)^2 = -5 + 4$

$(x-2)^2 = -1$, so there are no x-intercepts.

$f(0) = 15$ is the y-intercept.

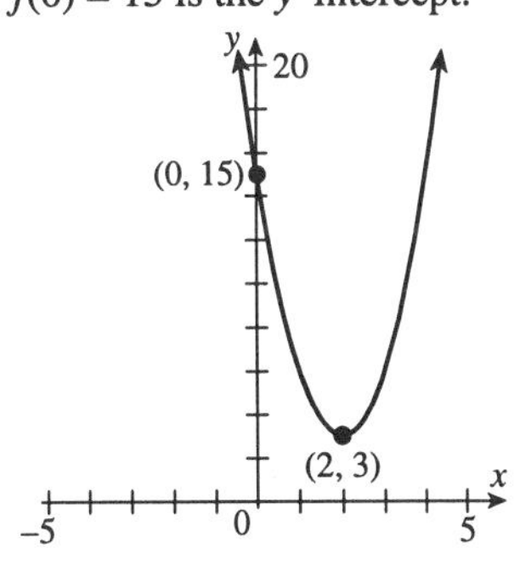

37. $f(x) = x^2 + x - 6$

$\frac{-b}{2a} = \frac{-1}{2(1)} = -\frac{1}{2}$ and

$f\left(-\frac{1}{2}\right)=\left(-\frac{1}{2}\right)^2+\left(-\frac{1}{2}\right)-6$

$f\left(-\frac{1}{2}\right)=\frac{1}{4}-\frac{1}{2}-6=-6\frac{1}{4}$ thus,

$V\left(-\frac{1}{2},\ -6\frac{1}{4}\right)$. The graph opens upward since $a=1>0$.

$x^2+x-6=0$
$(x+3)(x-2)=0$
$x+3=0$ or $x-2=0$
$x=-3$ or $x=2$

The x-intercepts are –3 and 2.
$f(0)=-6$ is the y-intercept.

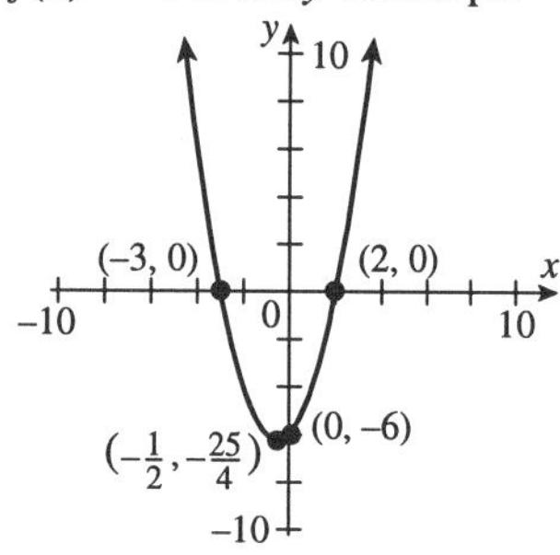

39. $f(x)=-2x^2-3x+35$

$\frac{-b}{2a}=\frac{-(-3)}{2(-2)}=-\frac{3}{4}$ and

$f\left(-\frac{3}{4}\right)=-2\left(-\frac{3}{4}\right)^2-3\left(-\frac{3}{4}\right)+35$

$f\left(-\frac{3}{4}\right)=-\frac{9}{8}+\frac{9}{4}+35=36\frac{1}{8}$

Thus, $V\left(-\frac{3}{4},\ 36\frac{1}{8}\right)$.

The graph opens downward since $a=-2<0$.

$-2x^2-3x+35=0$
$2x^2+3x-35=0$
$(2x-7)(x+5)=0$
$2x-7=0$ or $x+5=0$
$2x=7$ or $x=-5$
$x=\frac{7}{2}$

The x-intercepts are $\frac{7}{2}$ and –5.

$f(0)=35$ is the y-intercept.

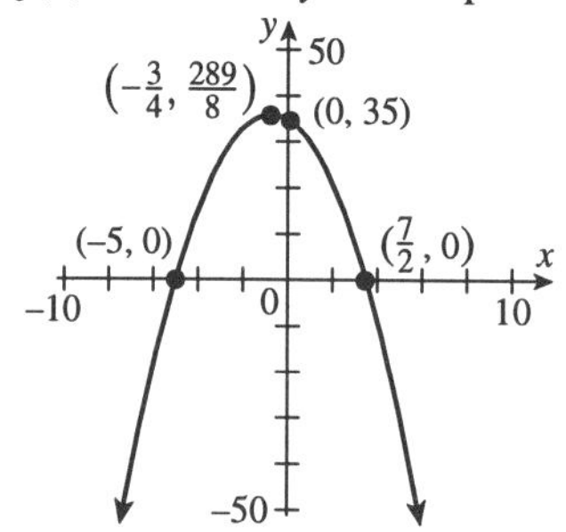

41. $h(t)=-16t^2+96t$

We seek the second coordinate of the vertex.

$\frac{-b}{2a}=\frac{-96}{2(-16)}=3$ and

$h(3)=-16(3)^2+96(3)$
$h(3)=-144+288=144$

The maximum height reached by the projectile is 144 ft.

43. $h(t)=-16t^2+32t$

$\frac{-b}{2a}=\frac{-32}{2(-16)}=1$

$h(1)=-16(1)^2+32(1)$
$h(1)=16$ feet

45. Let x = one number.
$60-x$ = other number
$f(x)=x(60-x)$
$f(x)=60x-x^2$
$f(x)=-x^2+60x$
$f(x)=-1(x^2-60x)$
$f(x)=-1(x^2-60x+900)+900$
$f(x)=-(x-30)^2+900$

The maximum will occur at the vertex which is (30, 900). The numbers are 30 and 30.

47. $f(x)=x^2+10x+15$

$\frac{-b}{2a}=\frac{-10}{2(1)}=-5$

$f(-5)=(-5)^2+10(-5)+15=-10$

Thus, $V(-5,-10)$.
The graph opens upward since $a=1>0$.

$x^2 + 10x + 15 = 0$

$x = \frac{-10 \pm \sqrt{(10)^2 - 4(1)(15)}}{2(1)}$

The x-intercepts are approximately –8.2 and –1.8.

$f(0) = 15$ is the y-intercept.

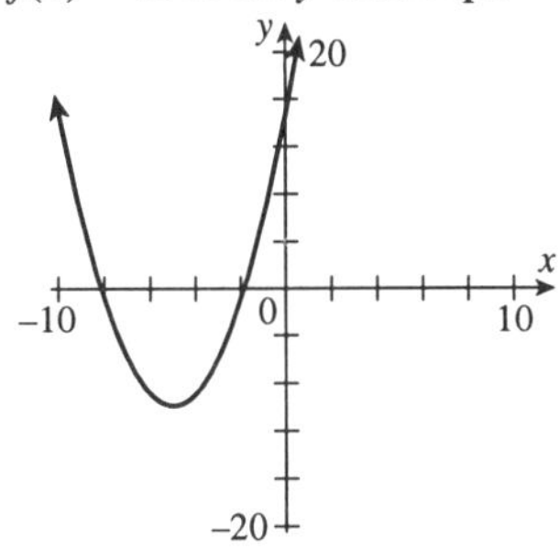

49. $f(x) = 3x^2 - 6x + 7$

$\frac{-b}{2a} = \frac{-(-6)}{2(3)} = 1$

$f(1) = 3(1)^2 - 6(1) + 7 = 4$

Thus, $V(1, 4)$.

The graph opens upward since $a = 3 > 0$.

$3x^2 - 6x + 7 = 0$

$x = \frac{-(-6) \pm \sqrt{(-6)^2 - 4(3)(7)}}{2(3)}$

There are no x-intercepts.

$f(0) = 7$ is the y-intercept.

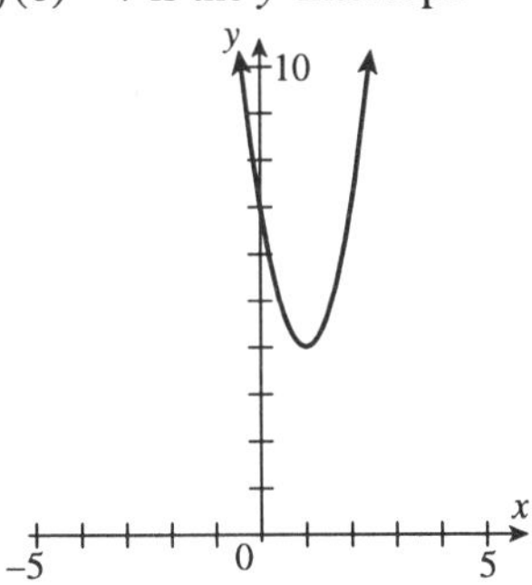

51. The graphs are the same.

53. The graphs are the same.

55. $f(x) = 2.3x^2 - 6.1x + 3.2$
minimum ≈ -0.84

57. $f(x) = -1.9x^2 + 5.6x - 2.7$
maximum ≈ 1.43

59.

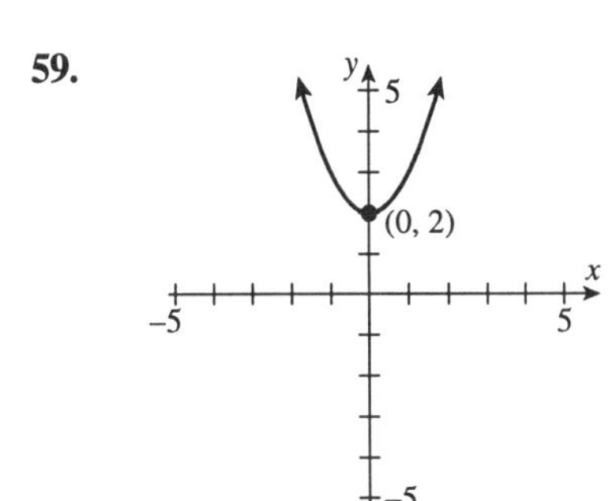

61.

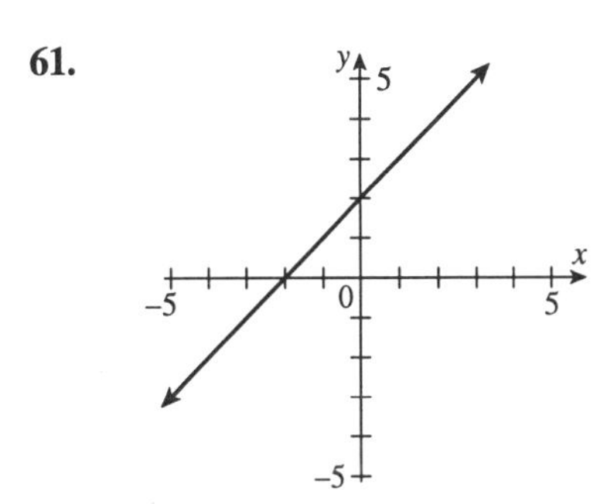

63.

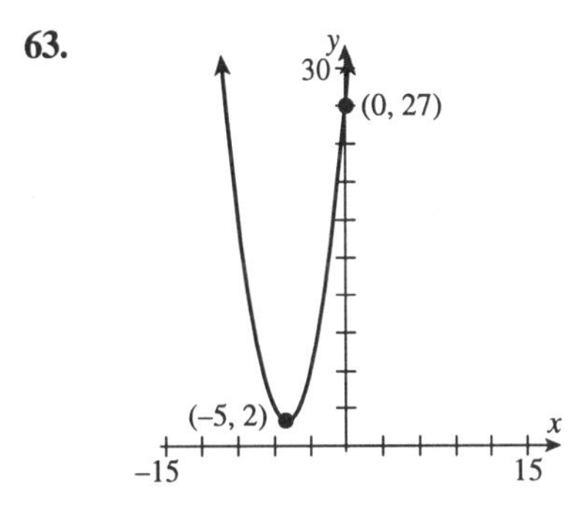

65.

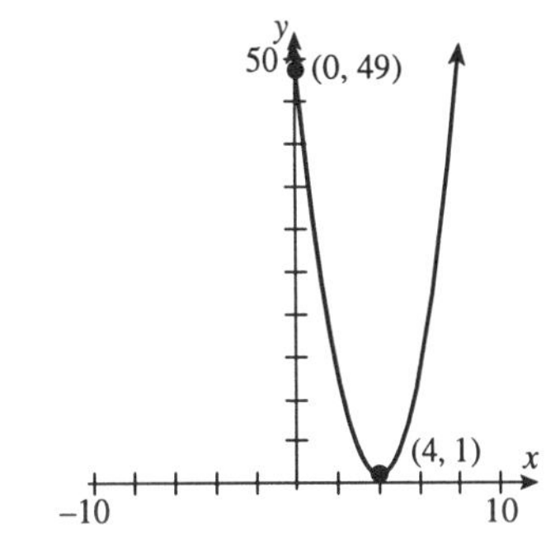

67.

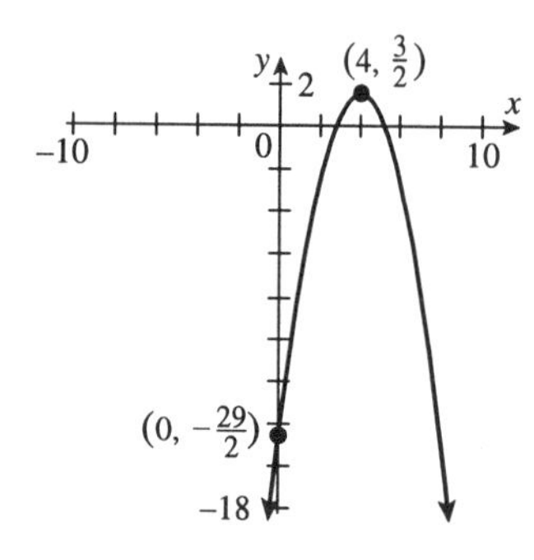

Chapter 8 - Review

1. $x^2 - 15x + 14 = 0$
$(x - 14)(x - 1) = 0$
$x - 14 = 0$ or $x - 1 = 0$
$x = 14$ or $x = 1$
$\{14, 1\}$

2. $x^2 - x - 30 = 0$
$(x + 5)(x - 6) = 0$
$x + 5 = 0$ or $x - 6 = 0$
$x = -5$ or $x = 6$

3. $10x^2 = 3x + 4$
$10x^2 - 3x - 4 = 0$
$(5x - 4)(2x + 1) = 0$
$5x - 4 = 0$ or $2x + 1 = 0$
$5x = 4$ or $2x = -1$
$x = \frac{4}{5}$ or $x = -\frac{1}{2}$
$\left\{\frac{4}{5}, -\frac{1}{2}\right\}$

4. $7a^2 = 29a + 30$
$7a^2 - 29a - 30 = 0$
$(7a + 6)(a - 5) = 0$
$7a + 6 = 0$ or $a - 5 = 0$
$a = -\frac{6}{7}$ or $a = 5$

5. $4m^2 = 196$
$m^2 = 49$
$m = \pm\sqrt{49}$
$m = \pm 7$
$\{7, -7\}$

6. $9y^2 = 36$
$y^2 = 4$
$y = \pm 2$

7. $(9n + 1)^2 = 9$
$9n + 1 = \pm\sqrt{9}$
$9n + 1 = \pm 3$
$9n = -1 \pm 3$
$n = \frac{-1 \pm 3}{9} = \frac{2}{9}, -\frac{4}{9}$
$\left\{\frac{2}{9}, -\frac{4}{9}\right\}$

8. $(5x - 2)^2 = 2$
$5x - 2 = \pm\sqrt{2}$
$5x = 2 \pm \sqrt{2}$
$x = \frac{2 \pm \sqrt{2}}{5}$

9. $z^2 + 3z + 1 = 0$
$z^2 + 3z = -1$
$z^2 + 3z + \left(\frac{3}{2}\right)^2 = -1 + \frac{9}{4}$
$\left(z + \frac{3}{2}\right)^2 = \frac{5}{4}$
$z + \frac{3}{2} = \pm\sqrt{\frac{5}{4}}$
$z + \frac{3}{2} = \frac{\pm\sqrt{5}}{\sqrt{4}}$
$z = -\frac{3}{2} \pm \frac{\sqrt{5}}{2}$
$\left\{-\frac{3}{2} + \frac{\sqrt{5}}{2}, -\frac{3}{2} - \frac{\sqrt{5}}{2}\right\}$

10. $x^2 + x + 7 = 0$
$x^2 + x + \frac{1}{4} = -7 + \frac{1}{4}$
$\left(x + \frac{1}{2}\right)^2 = -\frac{27}{4}$
$x + \frac{1}{2} = \pm\sqrt{\frac{-27}{4}} = \pm\frac{3i\sqrt{3}}{2}$
$x = \frac{-1 \pm 3i\sqrt{3}}{2}$

11. $(2x+1)^2 = x$
$4x^2+4x+1=x$
$4x^2+3x=-1$
$x^2+\frac{3}{4}x=-\frac{1}{4}$
$x^2+\frac{3}{4}x+\left(\frac{\frac{3}{4}}{2}\right)^2=-\frac{1}{4}+\frac{9}{64}$
$\left(x+\frac{3}{8}\right)^2=-\frac{7}{64}$
$x+\frac{3}{8}=\pm\sqrt{\frac{-7}{64}}$
$x+\frac{3}{8}=\pm\frac{\sqrt{7}i}{8}$
$x=-\frac{3}{8}\pm\frac{\sqrt{7}}{8}i$
$\left\{-\frac{3}{8}+\frac{\sqrt{7}}{8}i,\ -\frac{3}{8}-\frac{\sqrt{7}}{8}i\right\}$

12. $(3x-4)^2=10x$
$9x^2-24x+16-10x=0$
$9x^2-34x+16=0$
$x^2-\frac{34}{9}x+\frac{289}{81}=-\frac{16}{9}+\frac{289}{81}$
$\left(x-\frac{17}{9}\right)^2=\frac{145}{81}$
$x-\frac{17}{9}=\pm\sqrt{\frac{145}{81}}$
$x=\frac{17\pm\sqrt{145}}{9}$

13. $c^2=a^2+b^2$ where $a=b$
$c^2=a^2+a^2$
$(150)^2=2a^2$
$11250=a^2$
$\pm 75\sqrt{2}=a$
Disregard the negative
$75\sqrt{2}$ or 106.1 miles

14. Two complex solutions exist.

15. $b^2-4ac=48>0$
Two real solutions

16. Since 100 is a perfect square, two distinct real solutions exist.

17. $b^2-4ac=0$
One real solution

18. $x^2-16x+64=0$
$a=1$, $b=-16$, and $c=64$
$x=\frac{16\pm\sqrt{(-16)^2-4(1)(64)}}{2(1)}$
$x=\frac{16\pm\sqrt{256-256}}{2}=\frac{16\pm\sqrt{0}}{2}=8$
$\{8\}$

19. $x^2+5x=0$
$x=\frac{-5\pm\sqrt{(5)^2-4(1)(0)}}{2(1)}$
$x=\frac{-5\pm\sqrt{25}}{2}=\frac{-5\pm 5}{2}$
$x=0$ or $x=-5$

20. $x^2+11=0$
$a=1$, $b=0$, and $c=11$
$x=\frac{-0\pm\sqrt{0^2-4(1)(11)}}{2(1)}$
$=\frac{\pm\sqrt{-44}}{2}=\frac{\pm 2\sqrt{11}i}{2}=\pm\sqrt{11}i$
$\left\{\sqrt{11}i,\ -\sqrt{11}i\right\}$

21. $2x^2+3x=5$
$2x^2+3x-5=0$
$x=\frac{-3\pm\sqrt{9-4(2)(-5)}}{2(2)}$
$x=\frac{-3\pm\sqrt{49}}{4}=\frac{-3\pm 7}{4}$
$x=1$ or $x=-\frac{5}{2}$

22. $6x^2+7=5x$
$6x^2-5x+7=0$
$a=6$, $b=-5$, and $c=7$
$x=\frac{5\pm\sqrt{(-5)^2-4(6)(7)}}{2(6)}$

$$=\frac{5\pm\sqrt{25-168}}{12}=\frac{5\pm\sqrt{-143}}{12}$$
$$=\frac{5\pm\sqrt{143}i}{12}=\frac{5}{12}\pm\frac{\sqrt{143}}{12}i$$
$$\left\{\frac{5}{12}+\frac{\sqrt{143}}{12}i,\ \frac{5}{12}-\frac{\sqrt{143}}{12}i\right\}$$

23. $9a^2+4=2a$

$9a^2-2a+4=0$

$$a=\frac{2\pm\sqrt{4-4(9)(4)}}{2(9)}$$
$$=\frac{2\pm\sqrt{-140}}{18}$$
$$=\frac{2\pm 2i\sqrt{35}}{18}=\frac{1\pm i\sqrt{35}}{9}$$

24. $(5a-2)^2-a=0$

$25a^2-20a+4-a=0$

$25a^2-21a+4=0$

$a=25$, $b=-21$ and $c=4$

$$a=\frac{21\pm\sqrt{(-21)^2-4(25)(4)}}{2(25)}$$
$$=\frac{21\pm\sqrt{441-400}}{50}=\frac{21\pm\sqrt{41}}{50}$$
$$=\frac{21}{50}\pm\frac{\sqrt{41}}{50}$$
$$\left\{\frac{21}{50}+\frac{\sqrt{41}}{50},\ \frac{21}{50}-\frac{\sqrt{41}}{50}\right\}$$

25. $(2x-3)^2=x$

$4x^2-12x+9-x=0$

$4x^2-13x+9=0$

$$x=\frac{13\pm\sqrt{169-4(4)(9)}}{2(4)}$$
$$x=\frac{13\pm\sqrt{25}}{8}=\frac{13\pm 5}{8}$$
$$x=\frac{9}{4} \quad \text{or} \quad x=1$$

26. $d(t)=-16t^2+30t+6$

a. $d(1)=-16(1)^2+30(1)+6$
$=-16+30+6=20$ feet

b. $-16t^2+30t+6=0$

$8t^2-15t-3=0$

$a=8$, $b=-15$, and $c=-3$

$$x=\frac{15\pm\sqrt{(-15)^2-4(8)(-3)}}{2(8)}$$
$$x=\frac{15\pm\sqrt{225+96}}{16}$$
$$=\frac{15\pm\sqrt{321}}{16}$$
$$=\frac{15}{16}\pm\frac{\sqrt{321}}{16}$$

Discarding the negative value as extraneous, we find that

$t=\frac{15}{16}+\frac{\sqrt{321}}{16}\approx 2.1$ seconds.

27. Let x = length of leg

$x+6$ = length of hypotenuse

$x^2+x^2=(x+6)^2$

$2x^2=x^2+12x+36$

$x^2-12x-36=0$

$$x=\frac{12\pm\sqrt{144-4(1)(-36)}}{2}$$
$$x=\frac{12\pm\sqrt{288}}{2}$$
$$x=\frac{12\pm 12\sqrt{2}}{2}=6\pm 6\sqrt{2}$$

The length of each leg is $(6+6\sqrt{2})$ cm.

28. $x^3=27$

$x^3-27=0$

$(x-3)(x^2+3x+9)=0$

$x-3=0$ or $x^2+3x+9=0$

$x=3$

$a=1$, $b=3$, and $c=9$

$$x=\frac{-3\pm\sqrt{3^2-4(1)(9)}}{2(1)}$$
$$=\frac{-3\pm\sqrt{9-36}}{2}=\frac{-3\pm\sqrt{-27}}{2}$$
$$=\frac{-3\pm 3\sqrt{3}i}{2}=-\frac{3}{2}\pm\frac{3\sqrt{3}i}{2}$$
$$\left\{3,\ -\frac{3}{2}+\frac{3\sqrt{3}}{2}i,\ -\frac{3}{2}-\frac{3\sqrt{3}}{2}i\right\}$$

29. $y^3 = -64$

$y^3 + 64 = 0$

$(y+4)(y^2 - 4y + 16) = 0$

$y + 4 = 0$ or $y^2 - 4y + 16 = 0$

$y = -4$ $\quad y = \frac{4 \pm \sqrt{16 - 4(1)(16)}}{2}$

$y = \frac{4 \pm \sqrt{-48}}{2}$

$y = \frac{4 \pm 4i\sqrt{3}}{2}$

$y = 2 \pm 2i\sqrt{3}$

30. $\frac{5}{x} + \frac{6}{x-2} = 3$

$\frac{5(x-2) + 6x}{x(x-2)} = 3$

$5x - 10 + 6x = 3x(x-2)$

$11x - 10 = 3x^2 - 6x$

$3x^2 - 17x + 10 = 0$

$(3x - 2)(x - 5) = 0$

$3x - 2 = 0$ or $x - 5 = 0$

$3x = 2$ or $x = 5$

$x = \frac{2}{3}$

$\left\{\frac{2}{3}, 5\right\}$

31. $\frac{7}{8} = \frac{8}{x^2}$

$7x^2 = 64$

$x^2 = \frac{64}{7}$

$x = \pm\sqrt{\frac{64}{7}} = \frac{\pm 8\sqrt{7}}{7}$

32. $x^4 - 21x^2 - 100 = 0$

$(x^2 - 25)(x^2 + 4) = 0$

$x^2 - 25 = 0$ or $x^2 + 4 = 0$

$x^2 = 25$ or $x^2 = -4$

$x = \pm\sqrt{25}$ or $x = \pm\sqrt{-4}$

$x = \pm 5$ or $x = \pm 2i$

$\{5, -5, 2i, -2i\}$

33. $5(x+3)^2 - 19(x+3) = 4$

Let $u = x + 3$. Then

$5u^2 - 19u - 4 = 0$

$(5u + 1)(u - 4) = 0$

$5u + 1 = 0$ or $u - 4 = 0$

$u = -\frac{1}{5}$ or $u = 4$

$x + 3 = -\frac{1}{5}$ or $x + 3 = 4$

$x = -\frac{16}{5}$ or $x = 1$

34. $x^{2/3} - 6x^{1/3} + 5 = 0$

$(x^{1/3} - 1)(x^{1/3} - 5) = 0$

$x^{1/3} - 1 = 0$ or $x^{1/3} - 5 = 0$

$x^{1/3} = 1$ or $x^{1/3} = 5$

$x = 1^3 = 1$ or $x = 5^3 = 125$

$\{1, 125\}$

35. $x^{2/3} - 6x^{1/3} + 8 = 0$

Let $m = x^{1/3}$. Then

$m^2 - 6m + 8 = 0$

$(m - 2)(m - 4) = 0$

$m - 2 = 0$ or $m - 4 = 0$

$m = 2$ or $m = 4$

$x^{1/3} = 2$ or $x^{1/3} = 4$

$x = 2^3 = 8$ or $x = 4^3 = 64$

36. $a^6 - a^2 = a^4 - 1$

$a^2(a^4 - 1) - (a^4 - 1) = 0$

$(a^2 - 1)(a^4 - 1) = 0$

$(a^2 - 1)(a^2 - 1)(a^2 + 1) = 0$

$(a^2 - 1)^2(a^2 + 1) = 0$

$[(a+1)(a-1)]^2(a^2 + 1) = 0$

$(a+1)^2(a-1)^2(a^2 + 1) = 0$

$(a+1)^2 = 0$ or $(a-1)^2 = 0$ or $a^2 + 1 = 0$

$a + 1 = 0$ or $a - 1 = 0$ or $a^2 = -1$

$a = -1$ or $a = 1$ or $a = \pm\sqrt{-1} = \pm i$

$\{-1, 1, i, -i\}$

37. $y^{-2} + y^{-1} = 20$

$\frac{1}{y^2} + \frac{1}{y} = 20$

$1+y=20y^2$
$0=20y^2-y-1$
$0=(5y+1)(4y-1)$
$5y+1=0$ or $4y-1=0$
$y=-\frac{1}{5}$ or $y=\frac{1}{4}$
$\left\{-\frac{1}{5}, \frac{1}{4}\right\}$

38.

	Hours	Job complete in one hour
Al	x	$\frac{1}{x}$
Tim	$x-1$	$\frac{1}{x-1}$
Together	5	$\frac{1}{5}$

$\frac{1}{x}+\frac{1}{x-1}=\frac{1}{5}$
$\frac{(x-1)+x}{x(x-1)}=\frac{1}{5}$
$\frac{2x-1}{x^2-x}=\frac{1}{5}$
$10x-5=x^2-x$
$0=x^2-11x+5$
$a=1$, $b=-11$, and $c=5$
$x=\frac{-(-11)\pm\sqrt{(-11)^2-4(1)(5)}}{2(1)}$
$x=\frac{11\pm\sqrt{101}}{2}$
$x\approx 0.475$ or 10.525
$x=10.5$, $x-1=9.5$
Al; 10.5 hours
Tim; 9.5 hours

39. Let x = the number so $\frac{1}{x}$ = the reciprocal of the number. Then
$x-\frac{1}{x}=-\frac{24}{5}$
$\frac{x^2-1}{x}=-\frac{24}{5}$
$5(x^2-1)=-24x$
$5x^2-5=-24x$
$5x^2+24x-5=0$
$(5x-1)(x+5)=0$
$5x-1=0$ or $x+5=0$
$5x=1$ or $x=-5$
$x=\frac{1}{5}$
Discarding the positive value as extraneous, we find the number to be –5.

40. $A=P(1+r)^2$
$2717=2500(1+r)^2$
$1.0868=(1+r)^2$
$\pm\sqrt{1.0868}=1+r$
Disregard the negative.
$1.042497=1+r$
$0.042497=r$
$r=4.25\%$

41. $2x^2-50\le 0$
$2x^2\le 50$
$x^2\le 25$
$-5\le x\le 5$
$[-5, 5]$

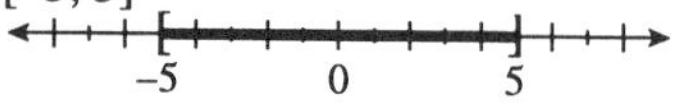

42. $\frac{1}{4}x^2<\frac{1}{16}$
$x^2<\frac{1}{4}$
$x^2-\frac{1}{4}<0$
$\left(x+\frac{1}{2}\right)\left(x-\frac{1}{2}\right)<0$

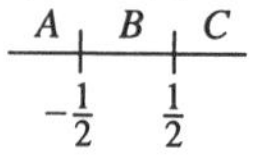

Region	Test Point	$\left(x-\frac{1}{2}\right)\left(x+\frac{1}{2}\right)<0$
A	–1	$\left(-\frac{3}{2}\right)\left(-\frac{1}{2}\right)>0$
B	0	$\left(-\frac{1}{2}\right)\left(\frac{1}{2}\right)<0$
C	1	$\left(\frac{1}{2}\right)\left(\frac{3}{2}\right)>0$

$\left(-\frac{1}{2}, \frac{1}{2}\right)$ satisfies the inequality.

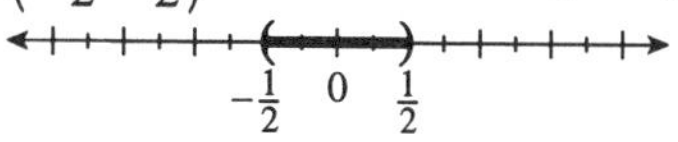

43. $(2x-3)(4x+5) \ge 0$
First solve
$(2x-3)(4x+5)=0$
$2x-3=0$ or $4x+5=0$
$2x=3$ or $4x=-5$
$x=\frac{3}{2}$ or $x=-\frac{5}{4}$
Now select three appropriate test points.
$x=-2$: $[2(-2)-3][4(-2)+5]=(-)(-)=+$
$x=2$: $[2\cdot 2-3][4\cdot 2+5]=(+)(+)=+$
$x=0$: $[2(0)-3][4(0)+5]=(-)(+)=-$
Thus, the solutions are $x \ge \frac{3}{2}$ and $x \le -\frac{5}{4}$
$\left(-\infty, -\frac{5}{4}\right] \cup \left[\frac{3}{2}, \infty\right)$

$-\frac{5}{4}$ 0 $\frac{3}{2}$

44. $(x^2-16)(x^2-1)>0$
$(x+4)(x-4)(x+1)(x-1)>0$

A B C D E
–4 –1 1 4

Region	Test Point	$(x^2-16)(x^2-1)>0$
A	–5	$(9)(24)>0$
B	–2	$(-12)(3)<0$
C	0	$(-16)(-1)>0$
D	2	$(-12)(3)<0$
E	5	$(9)(24)>0$

$(-\infty, -4) \cup (-1, 1) \cup (4, \infty)$ satisfies the inequality.

–4 –1 1 4

45. $\frac{x-5}{x-6}<0$
First solve both
$x-5=0$ and $x-6=0$
$x=5$ and $x=6$
Now select three appropriate test points.
$x=4$: $\frac{4-5}{4-6}=\frac{-}{-}=+$
$x=\frac{11}{2}$: $\frac{\frac{11}{2}-5}{\frac{11}{2}-6}=\frac{+}{-}=-$
$x=7$: $\frac{7-5}{7-6}=\frac{+}{+}=+$
Thus, the solutions are $5<x<6$.
$(5, 6)$

5 6

46. $\frac{x(x+5)}{4x-3} \ge 0$

A B C D
–5 0 $\frac{3}{4}$

Region	Test Point	$\frac{x(x+5)}{4x-3} \ge 0$
A	–6	$\frac{(-6)(-1)}{-27} \le 0$
B	–1	$\frac{(-1)(4)}{-7} \ge 0$
C	$\frac{1}{2}$	$\frac{\frac{1}{2}\left(5\frac{1}{2}\right)}{(-1)} \le 0$
D	1	$\frac{1(6)}{1} \ge 0$

$[-5, 0] \cup \left(\frac{3}{4}, \infty\right)$ satisfies the inequality.

–5 0 $\frac{3}{4}$

47. $\frac{(4x+3)(x-5)}{x(x+6)}>0$
First solve both:
$(4x+3)(x-5)=0$ and $x(x+6)=0$
$4x+3=0$ or $x-5=0$ and
$x=0$ or $x+6=0$
$4x=-3$ or $x=5$ and $x=0$ or $x=-6$
$x=-\frac{3}{4}$
Now select five appropriate test points.
$x=-7$: $\frac{[4(-7)+3][-7-5]}{-7(-7+6)}$
$=\frac{(-)(-)}{(-)(-)}=\frac{+}{+}=+$
$x=-1$: $\frac{[4(-1)+3][-1-5]}{-1(-1+6)}$
$=\frac{(-)(-)}{(-)(+)}=\frac{+}{-}=-$

$x=-1$: $\dfrac{[4(-1)+3][-1-5]}{-1(-1+6)}$
$=\dfrac{(-)(-)}{(-)(+)}=\dfrac{+}{-}=-$

$x=-\frac{1}{2}$: $\dfrac{\left[4\left(-\frac{1}{2}\right)+3\right]\left[-\frac{1}{2}-5\right]}{-\frac{1}{2}\left(-\frac{1}{2}+6\right)}$
$=\dfrac{(+)(-)}{(-)(+)}=\dfrac{-}{-}=+$

$x=1$: $\dfrac{[4(1)+3][1-5]}{1(1+6)}$
$=\dfrac{(+)(-)}{(+)(+)}=\dfrac{-}{+}=-$

$x=6$: $\dfrac{[4(6)+3][6-5]}{6(6+6)}$
$=\dfrac{(+)(+)}{(+)(+)}=\dfrac{+}{+}=+$

Thus, the solutions are $x<-6$.
$-\frac{3}{4}<x<0$, and $x>5$

$(-\infty,\ -6)\cup\left(-\frac{3}{4},\ 0\right)\cup(5,\ \infty)$

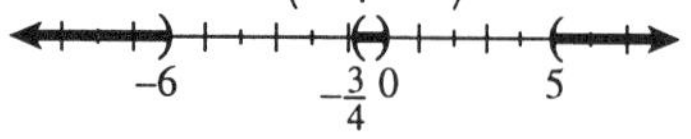

48. $(x+5)(x-6)(x+2)\le 0$

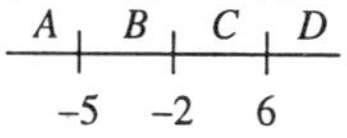

Region	Test Point	$(x+5)(x-6)(x+2)\le 0$
A	-6	$(-1)(-12)(-4)\le 0$
B	-3	$(2)(-9)(-1)\ge 0$
C	0	$(5)(-6)(2)\le 0$
D	7	$(12)(1)(9)\ge 0$

$(-\infty,\ -5]\cup[-2,\ 6]$ satisfies the inequality.

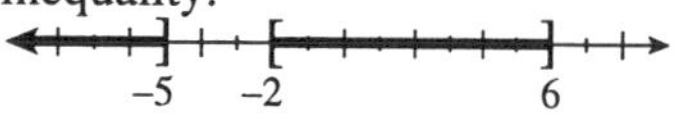

49. $x^3+3x^2-25x-75>0$
$x^2(x+3)-25(x+3)>0$
$(x+3)(x^2-25)>0$
$(x+3)(x+5)(x-5)>0$

Region	A	B	C	D
x-values	$(-\infty, -5)$	$(-5, -3)$	$(-3, 5)$	$(5, \infty)$

Region	Test Point	$(x+3)(x+5)(x-5)>0$
A	-6	$(-3)(-1)(-11)<0$
B	-4	$(-1)(1)(-9)>0$
C	0	$(3)(5)(-5)<0$
D	6	$(9)(11)(1)>0$

$(-5,\ -3)\cup(5,\ \infty)$ satisfies the inequality

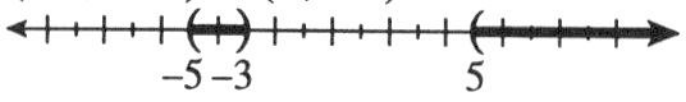

50. $\dfrac{x^2+4}{3x}\le 1$

First solve both

$3x=0$ and $\dfrac{x^2+4}{3x}=1$
$x=0$ $\quad x^2+4=3x$
$x^2-3x+4=0$
has no real solutions

Now select two appropriate test points.

$x=-5$: $\dfrac{(-5)^2+4}{3(-5)}=\dfrac{+}{-}=-$

$x=1$: $\dfrac{1^2+4}{3(1)}=\dfrac{+}{+}=+$

Thus, the solutions are
$x<0$
$(-\infty, 0)$

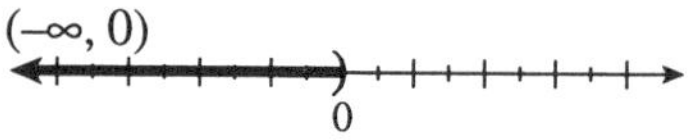

51. $\dfrac{(5x+6)(x-3)}{x(6x-5)}<0$

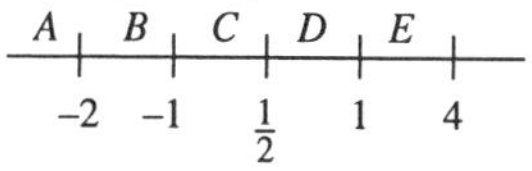

Region	Test Point	$\frac{(5x+6)(x-3)}{x(6x-5)} < 0$
A	-2	$\frac{(-1)(-5)}{(-2)(-17)} > 0$
B	-1	$\frac{(1)(-4)}{(-1)(-11)} < 0$
C	$\frac{1}{2}$	$\frac{\left(8\frac{1}{2}\right)\left(-2\frac{1}{2}\right)}{\frac{1}{2}(-2)} > 0$
D	1	$\frac{(11)(-2)}{1(1)} < 0$
E	4	$\frac{(26)(1)}{4(19)} > 0$

$\left(-\frac{6}{5}, 0\right) \cup \left(\frac{5}{6}, 3\right)$ satisfies the inequality.

52. $\frac{3}{x-2} > 2$
$3 > 2x - 4$
$0 > 2x - 7$
$x = 2$ is undefined
$x = \frac{7}{2}$ is a critical point

Region	A	B	C
x-values	$(-\infty, 2)$	$\left(2, \frac{7}{2}\right)$	$\left(\frac{7}{2}, \infty\right)$

Region	Test Point	$\frac{3}{x-2} > 2$
A	0	$-\frac{3}{2} < 2$
B	3	$3 > 2$
C	4	$\frac{3}{2} < 2$

$\left(2, \frac{7}{2}\right)$ satisfies the inequality

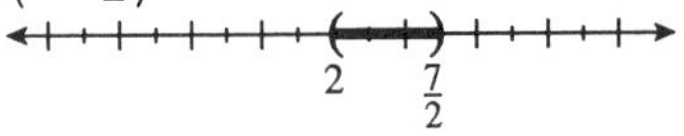

53. $f(x) = x^2 - 4$
$\frac{-b}{2a} = \frac{0}{2(1)} = 0$
$f(0) = 0^2 - 4 = -4$
Vertex: $(0, -4)$
Axis of symmetry: $x = 0$

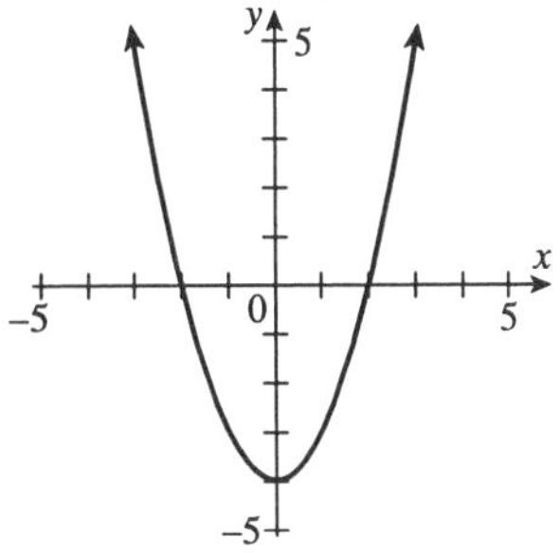

54. $g(x) = x^2 + 7$
$\frac{-b}{2a} = \frac{0}{2(1)} = 0$
$g(0) = 0^2 + 7$
Vertex: $(0, 7)$
Axis of symmetry: $x = 0$

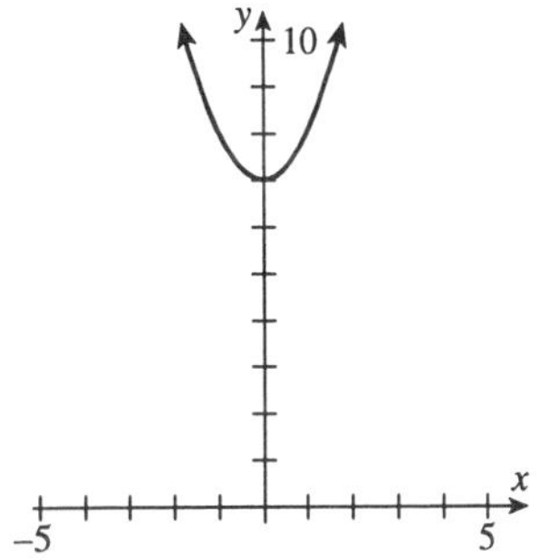

55. $H(x) = 2x^2$
Vertex: $(0, 0)$
Axis of symmetry: $x = 0$

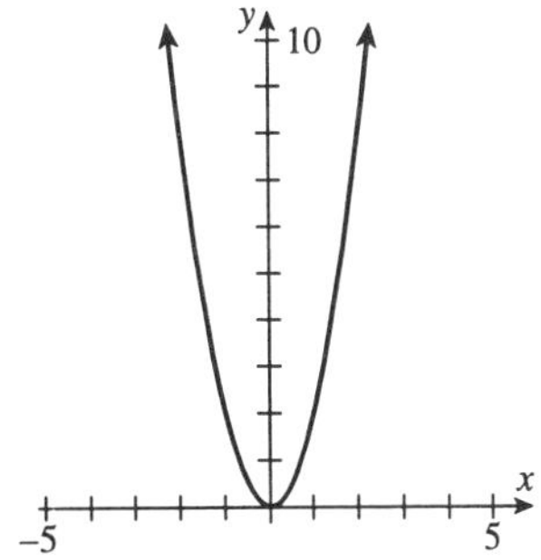

56. $h(x) = -\frac{1}{3}x^2$
Vertex: $(0, 0)$

Axis of symmetry: $x = 0$

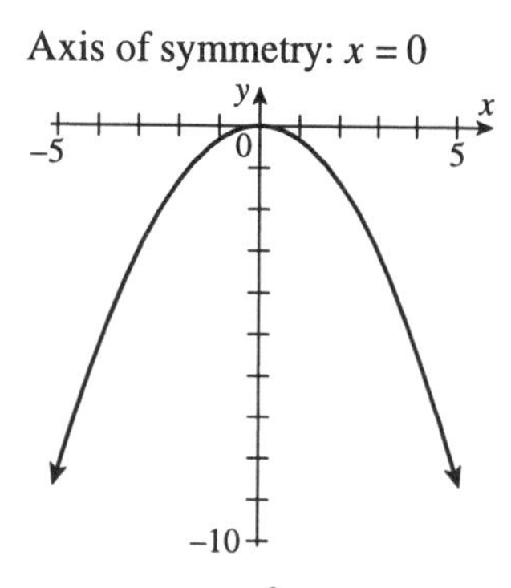

57. $F(x) = (x-1)^2$

$F(x) = x^2 - 2x + 1$

$\frac{-b}{2a} = \frac{-(-2)}{2(1)} = 1$

$F(1) = 0$

Vertex: (1, 0)

Axis of symmetry: $x = 1$

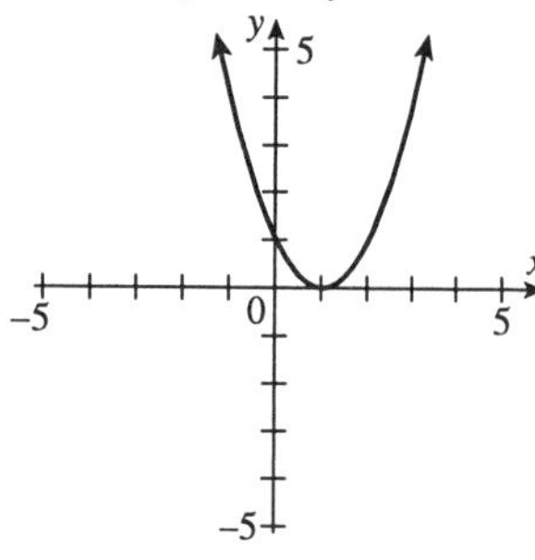

58. $G(x) = (x+5)^2$

$G(x) = x^2 + 10x + 25$

$\frac{-b}{2a} = \frac{-10}{2(1)} = -5$

$G(-5) = 0$

Vertex: (−5, 0)

Axis of symmetry: $x = -5$

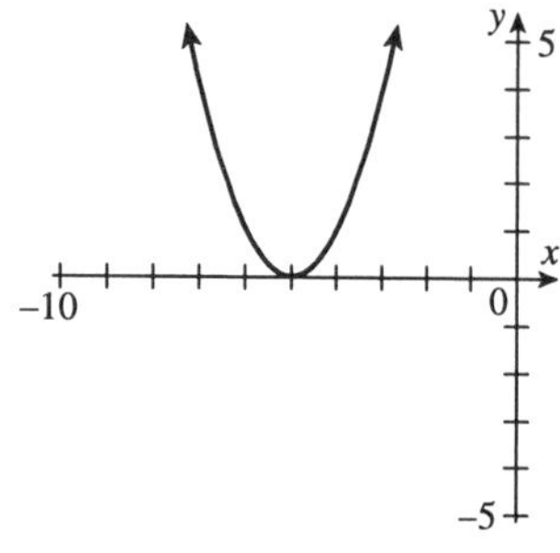

59. $f(x) = (x-4)^2 - 2$

Vertex: (4, −2)

Axis of symmetry: $x = 4$

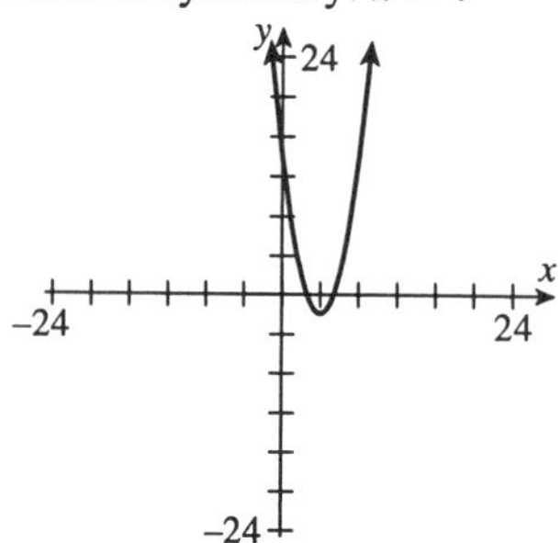

60. $y = -3(x-1)^2 + 1$

Vertex: (1, 1)

y-intercept: −2

x-intercepts: $\frac{3 \pm \sqrt{3}}{3}$

Axis of symmetry: $x = 1$

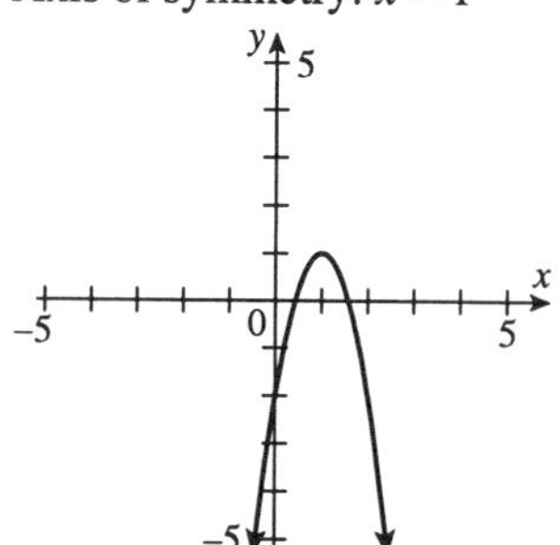

61. $f(x) = x^2 + 10x + 25$ or

$f(x) = (x+5)^2 + 0$

Thus, $V(-5, 0)$.

$(x+5)^2 = 0 \Rightarrow x + 5 = 0$

$x = -5$ is the x-intercept.

$f(0) = 25$ is the y-intercept.

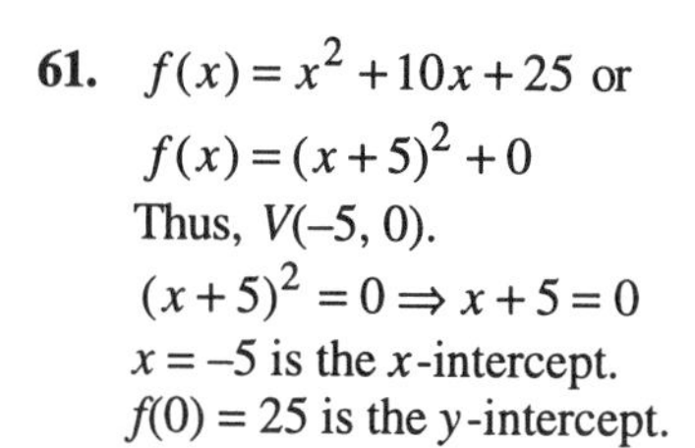

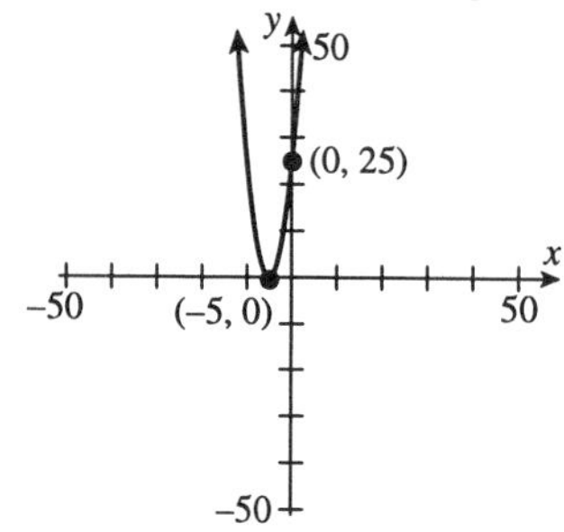

62. $y = -x^2 + 6x - 9$

$\frac{-b}{2a} = \frac{-6}{2(-1)} = 3$

$y = -(3)^2 + 6(3) - 9 = 0$

Vertex: (3, 0)

$y(0) = -9$

y-intercept is -9

x-intercept is 3

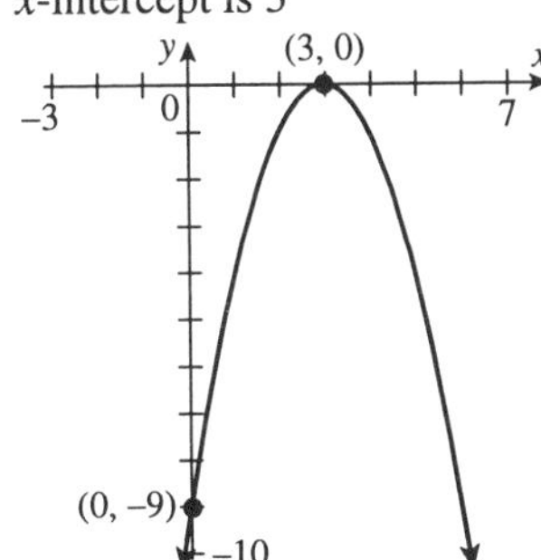

63. $f(x) = 4x^2 - 1$ or

$f(x) = 4(x - 0)^2 - 1$

Thus, $V(0, -1)$.

$4x^2 - 1 = 0$

$4x^2 = 1$

$x^2 = \frac{1}{4}$

$x = \pm\frac{1}{2}$

are the x-intercepts.

$f(0) = -1$ is the y-intercept.

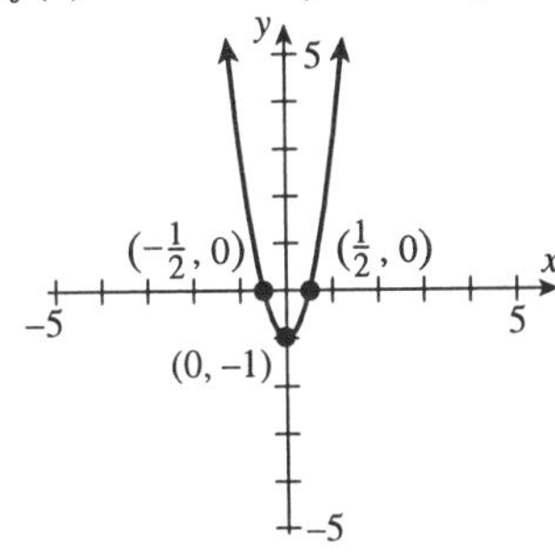

64. $y = -5x^2 + 5$

$\frac{-b}{2a} = 0$

$y = 0 + 5$

Vertex: (0, 5)

y-intercept is 5

x-intercepts are -1 and 1

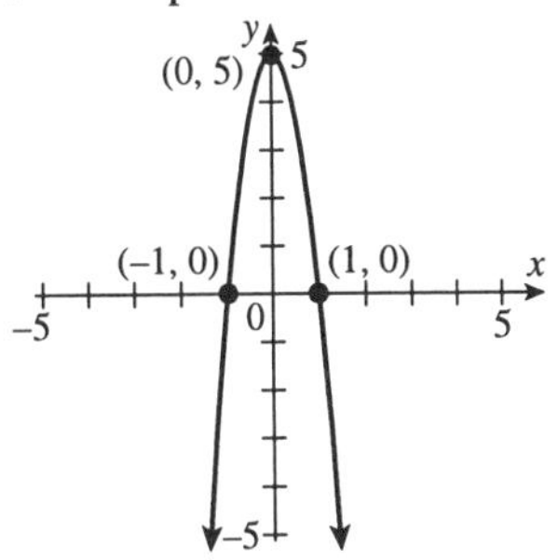

65. $f(x) = -3x^2 - 5x + 4$

$\frac{-b}{2a} = \frac{-(-5)}{2(-3)} = -\frac{5}{6}$

$f\left(\frac{5}{6}\right) = -3\left(-\frac{5}{6}\right)^2 - 5\left(-\frac{5}{6}\right) + 4 = \frac{73}{12}$

Vertex: $\left(-\frac{5}{6}, \frac{73}{12}\right)$

The graph opens down because $a = -3 < 0$.

$f(0) = 4$ is the y-intercept.

$$x = \frac{-(-5) \pm \sqrt{(-5)^2 - 4(-3)(4)}}{2(-3)}$$

$$x = \frac{5 \pm \sqrt{73}}{-6}$$

The x-intercepts are approximately -2.26 and 0.59.

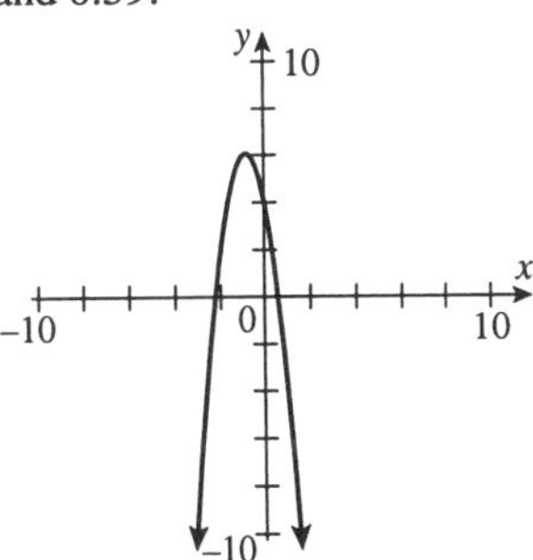

66. $h(t) = -16t^2 + 120t + 300$

a. $350 = -16t^2 + 120t + 300$

$0 = -16t^2 + 120t - 50$

$0 = -8t^2 + 60t - 25$

$a = -8,\ b = 60,\ c = -25$

$t = \frac{-60 \pm \sqrt{(60)^2 - 4(-8)(-25)}}{2(-8)}$

$t = \frac{-60 \pm \sqrt{2800}}{-16}$

$t \approx 0.4$ seconds and 7.1 seconds

b. The object will be at 350 feet on the way up and on the way down.

67. Let x = one number so $420 - x$ = the other number. Let $f(x)$ represent their product. Thus,

$f(x) = x(420 - x)$

$f(x) = -x^2 + 420x$

$\frac{-b}{2a} = \frac{-420}{2(-1)} = 210$

Therefore, the numbers are both 210.

68. $x = a(y-k)^2 + h$

$x = a(y-7)^2 - 3$

$0 = a(49) - 3$

$\frac{3}{49} = a$

$x = \frac{3}{49}(y-7)^2 - 3$

$y = a(x-h)^2 + k$

$y = a(x+3)^2 + 7$

$0 = a(0+3)^2 + 7$

$y - 7 = 9a$

$-\frac{7}{9} = a$

$y = -\frac{7}{9}(x+3)^2 + 7$

Chapter 8 - Test

1. $5x^2 - 2x = 7$

$5x^2 - 2x - 7 = 0$

$(5x - 7)(x + 1) = 0$

$5x - 7 = 0$ or $x + 1 = 0$

$5x = 7$ or $x = -1$

$x = \frac{7}{5}$

$\left\{\frac{7}{5}, -1\right\}$

2. $(x+1)^2 = 10$

$x + 1 = \pm\sqrt{10}$

$x = -1 \pm \sqrt{10}$

$\left\{-1-\sqrt{10}, -1+\sqrt{10}\right\}$

3. $m^2 - m + 8 = 0$

$a = 1$, $b = -1$, and $c = 8$

$m = \frac{1 \pm \sqrt{(-1)^2 - 4(1)(8)}}{2(1)}$

$= \frac{1 \pm \sqrt{1-32}}{2} = \frac{1 \pm \sqrt{-31}}{2}$

$= \frac{1 \pm \sqrt{31}i}{2} = \frac{1}{2} \pm \frac{\sqrt{31}}{2}i$

$\left\{\frac{1}{2} + \frac{\sqrt{31}}{2}i, \frac{1}{2} - \frac{\sqrt{31}}{2}i\right\}$

4. $u^2 - 6u + 2 = 0$

$a = 1$, $b = -6$, $c = 2$

$u = \frac{6 \pm \sqrt{(-6)^2 - 4(1)(2)}}{2(1)}$

$= \frac{6 \pm \sqrt{36-8}}{2} = \frac{6 \pm \sqrt{28}}{2}$

$= \frac{6 \pm 2\sqrt{7}}{2} = 3 \pm \sqrt{7}$

$\left\{3+\sqrt{7}, 3-\sqrt{7}\right\}$

5. $7x^2 + 8x + 1 = 0$

$(7x + 1)(x + 1) = 0$

$7x + 1 = 0$ or $x + 1 = 0$

$7x = -1$ or $x = -1$

$x = -\frac{1}{7}$

$\left\{-\frac{1}{7}, -1\right\}$

6. $a^2 - 3a = 5$

$a^2 - 3a - 5 = 0$

$a = 1$, $b = -3$, and $c = -5$

$a = \frac{3 \pm \sqrt{(-3)^2 - 4(1)(-5)}}{2(1)}$

$= \frac{3 \pm \sqrt{9+20}}{2} = \frac{3 \pm \sqrt{29}}{2}$

$= \frac{3}{2} \pm \frac{\sqrt{29}}{2}$

$\left\{\frac{3}{2} + \frac{\sqrt{29}}{2},\ \frac{3}{2} - \frac{\sqrt{29}}{2}\right\}$

7. $\frac{4}{x+2} + \frac{2x}{x-2} = \frac{6}{x^2-4}$

$\frac{4(x-2)+2x(x+2)}{(x+2)(x-2)} = \frac{6}{(x+2)(x-2)}$

$4x-8+2x^2+4x=6$

$2x^2+8x-8=6$

$2x^2+8x-14=0$

$x^2+4x-7=0$

$a=1$, $b=4$, and $c=-7$

$x = \frac{-4 \pm \sqrt{4^2-4(1)(-7)}}{2(1)}$

$= \frac{-4 \pm \sqrt{16+28}}{2} = \frac{-4 \pm \sqrt{44}}{2}$

$= \frac{-4 \pm 2\sqrt{11}}{2} = -2 \pm \sqrt{11}$

$\left\{-2+\sqrt{11},\ -2-\sqrt{11}\right\}$

8. $x^4-8x^2-9=0$

$(x^2-9)(x^2+1)=0$

$x^2-9=0$ or $x^2+1=0$

$x^2=9$ or $x^2=-1$

$x=\pm\sqrt{9}$ or $x=\pm\sqrt{-1}$

$x=\pm 3$ or $x=\pm i$

$\{3, -3, i, -i\}$

9. $x^6+1=x^4+x^2$

$x^6-x^4-x^2+1=0$

$x^4(x^2-1)-(x^2-1)=0$

$(x^4-1)(x^2-1)=0$

$(x^2+1)(x^2-1)(x^2-1)=0$

$(x^2+1)(x^2-1)^2=0$

$(x^2+1)[(x+1)(x-1)]^2=0$

$(x^2+1)(x+1)^2(x-1)^2=0$

$x^2+1=0$ or $(x+1)^2=0$ or $(x-1)^2=0$

$x^2=-1$ or $x+1=0$ or $x-1=0$

$x=\pm\sqrt{-1}$ or $x=-1$ or $x=1=\pm i$

$\{1, -1, i, -i\}$

10. $(x+1)^2-15(x+1)+56=0$

Let $u=x+1$

$u^2-15u+56=0$

$(u-7)(u-8)=0$

$u-7=0$ or $u-8=0$

$u=7$ or $u=8$

Now substitute back

$x+1=7$ or $x+1=8$

$x=6$ or $x=7$

$\{6, 7\}$

11. $x^2-6x=-2$

$x^2-6x+\left(\frac{6}{2}\right)^2=-2+9$

$(x-3)^2=7$

$x-3=\pm\sqrt{7}$

$x=3\pm\sqrt{7}$

$\left\{3+\sqrt{7},\ 3-\sqrt{7}\right\}$

12. $2a^2+5=4a$

$2a^2-4a=-5$

$a^2-2a=-\frac{5}{2}$

$a^2-2a+\left(\frac{2}{2}\right)^2=-\frac{5}{2}+1$

$(a-1)^2=-\frac{3}{2}$

$a-1=\pm\sqrt{\frac{-3}{2}}=\pm\frac{\sqrt{3}i}{\sqrt{2}}$

$a-1=\frac{\pm\sqrt{6}i}{2}$

$a=1\pm\frac{\sqrt{6}}{2}i$

$\left\{1+\frac{\sqrt{6}}{2}i,\ 1-\frac{\sqrt{6}}{2}i\right\}$

13. $2x^2-7x>15$

$2x^2-7x-15>0$

$(2x+3)(x-5)>0$

$2x+3=0$ or $x-5=0$

$x=-\frac{3}{2}$ or $x=5$

Now select three appropriate test points.

$x = -2$: $2(-2)^2 - 7(-2) - 15 = 8 > 0$

$x = 0$: $2(0)^2 - 7(0) - 15 = -15 < 0$

$x = 6$: $2(6)^2 - 7(6) - 15 = 15 > 0$

Thus, the solutions are

$\left(-\infty, -\frac{3}{2}\right) \cup (5, \infty)$

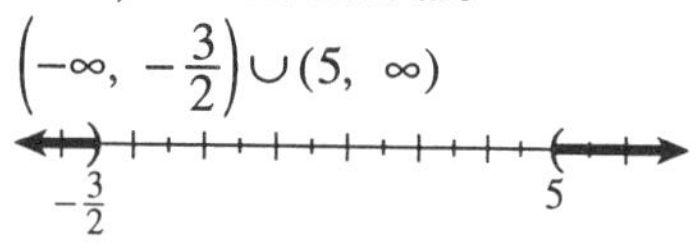

14. $(x^2 - 16)(x^2 - 25) > 0$

Let $u = x^2$.

$(u - 16)(u - 25) > 0$

First solve $(u - 16)(u - 25) = 0$

$u - 16 = 0$ or $u - 25 = 0$

$u = 16$ or $u = 25$

Now select three appropriate test points.

$u = 15$: $(15 - 16)(15 - 25) = (-)(-) = +$

$u = 20$: $(20 - 16)(20 - 25) = (+)(-) = -$

$u = 30$: $(30 - 16)(30 - 25) = (+)(+) = +$

Thus, the solutions are $u < 16$ or $u > 25$.

Now substitute back

$x^2 < 16$ or $x^2 > 25x$

$-4 < x < 4$ or $x < -5$ or $x > 5$

$(-\infty, -5) \cup (-4, 4) \cup (5, \infty)$

15. $\frac{5}{x+3} < 1$

One critical point is $x = -3$

$5 < x + 3$

$0 < x - 2$

Another critical point is $x = 2$.

Now select three appropriate test points.

$x = -4$: $\frac{5}{-4+3} < 1$

$x = 0$: $\frac{5}{0+3} > 1$

$x = 3$: $\frac{5}{3+3} < 1$

$(-\infty, -3) \cup (2, \infty)$

16. $\frac{7x - 14}{x^2 - 9} \leq 0$

First solve both

$7x - 14 = 0$ and $x^2 - 9 = 0$.

$7x = 14$ and $x^2 = 9$

$x = 2$ and $x = \pm\sqrt{9} = \pm 3$

Now select four appropriate test points.

$x = -4$: $\frac{7(-4) - 14}{(-4)^2 - 9} = \frac{-}{+} = -$

$x = 0$: $\frac{7(0) - 14}{0^2 - 9} = \frac{-}{-} = +$

$x = \frac{5}{2}$: $\frac{7\left(\frac{5}{2}\right) - 14}{\left(\frac{5}{2}\right)^2 - 9} = \frac{+}{-} = -$

$x = 4$: $\frac{7(4) - 14}{4^2 - 9} = \frac{+}{+} = +$

Thus, the solutions are $x < -3$ and $2 \leq x < 3$.

$(-\infty, -3) \cup [2, 3)$

17. $f(x) = 3x^2$

vertex: (0, 0)

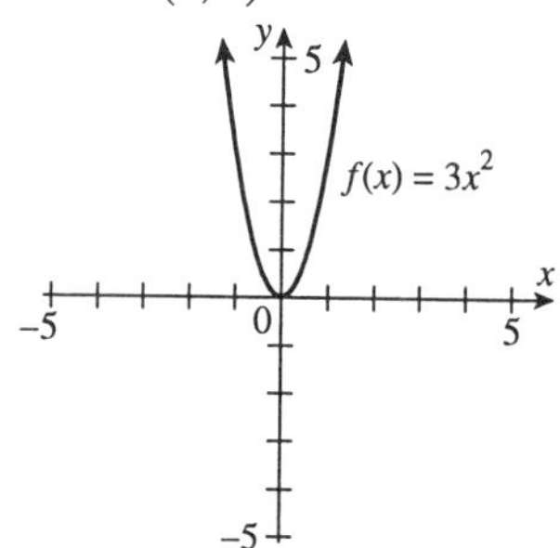

18. $G(x) = -2(x - 1)^2 + 5$

vertex: (1, 5)

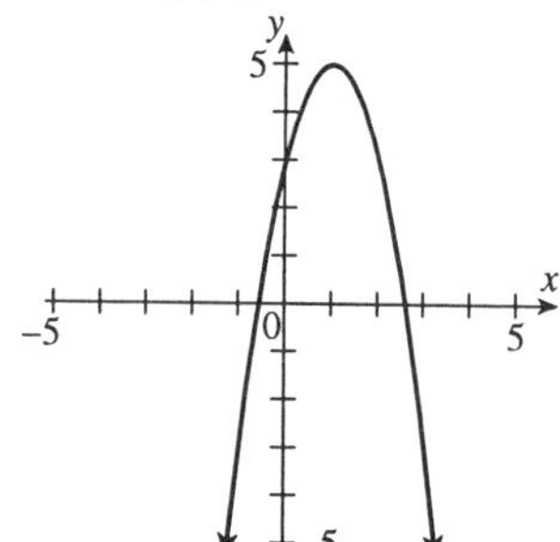

19. $h(x) = x^2 - 4x + 4$

$\frac{-b}{2a} = \frac{-(-4)}{2(1)} = 2$

$h(2) = 2^2 - 4(2) + 4 = 0$
vertex: (2, 0)
$h(0) = 4$ is the y-intercept.
$0 = x^2 - 4x + 4 = (x-2)^2$
$x = 2$ is the x-intercept

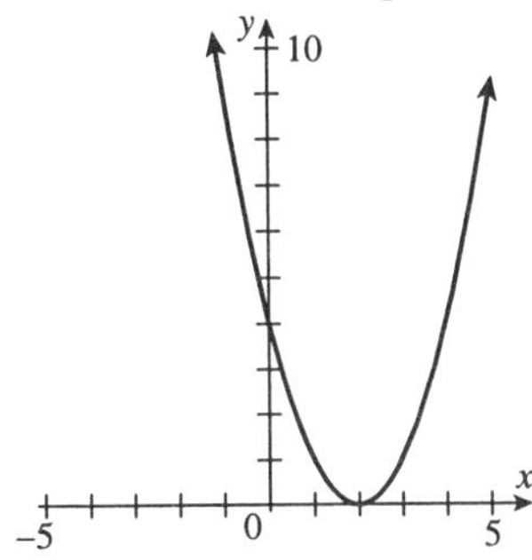

20. $F(x) = 2x^2 - 8x + 9$

$\frac{-b}{2a} = \frac{-(-8)}{2(2)} = 2$

$F(2) = 2(2)^2 - 8(2) + 9 = 1$
Vertex: (2, 1)
$F(0) = 9$ is the y-intercept
There are no x-intercepts

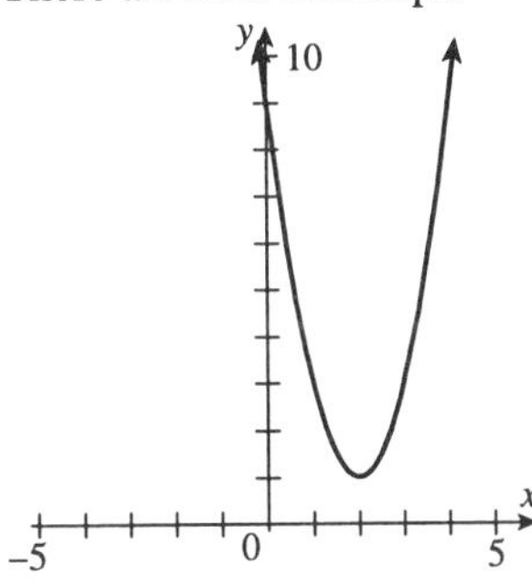

21. $c^2 = a^2 + b^2$

$(10)^2 = x^2 + (x-4)^2$

$100 = x^2 + x^2 - 8x + 16$

$0 = 2x^2 - 8x - 84$

$0 = x^2 - 4x - 42$

$a = 1,\ b = -4,\ c = -42$

$x = \frac{-(-4) \pm \sqrt{(-4)^2 - 4(1)(-42)}}{2(1)}$

$x = \frac{4 \pm \sqrt{16 + 168}}{2}$

$x = \frac{4 \pm \sqrt{184}}{2}$

$x = \frac{4 \pm 2\sqrt{46}}{2}$

$x = 2 \pm \sqrt{46}$

Disregard the negative result.

$2 + \sqrt{46}$ or 8.8 feet

22.

	Hours	Job complete in 1 hr
Dave	$x-2$	$\frac{1}{x-2}$
Sandy	x	$\frac{1}{x}$
Together	4	$\frac{1}{4}$

$\frac{1}{x-2} + \frac{1}{x} = \frac{1}{4}$

$\frac{x + (x-2)}{x(x-2)} = \frac{1}{4}$

$\frac{2x-2}{x^2 - 2x} = \frac{1}{4}$

$8x - 8 = x^2 - 2x$

$0 = x^2 - 10x + 8$

$a = 1,\ b = -10,\ c = 8$

$x = \frac{-(-10) \pm \sqrt{(-10)^2 - 4(1)(8)}}{2(1)}$

$x = \frac{10 \pm \sqrt{68}}{2}$

$x = \frac{10 \pm 2\sqrt{17}}{2}$

$x = 5 \pm \sqrt{17}$

$x + (x-2)$ must add to 4.

Therefore $x = 5 + \sqrt{17}$.

$5 + \sqrt{17}$ or 9.12 hours

23. $s(t) = -16t^2 + 32t + 256$

a. $\frac{-b}{2a} = \frac{-32}{2(-16)} = 1$

$s(1) = -16(1)^2 + 32(1) + 256 = 272$
Vertex: (1, 272)
Maximum height = 272 feet

b. $0 = -16t^2 + 32t + 256$
$0 = -t^2 + 2t + 16$
$a = -1,\ b = 2,\ c = 16$
$x = \dfrac{-2 \pm \sqrt{2^2 - 4(-1)(16)}}{2(-1)}$
$x = \dfrac{-2 \pm \sqrt{68}}{-2}$
$x = \dfrac{-2 \pm 2\sqrt{17}}{-2}$
$x = 1 \pm \sqrt{17}$
$x \approx -3.12$ and 5.12
Disregard the negative.
5.12 seconds

Chapter 8 - Cumulative Review

1. a. $(-8)(-1) = 8$

b. $(-2)\frac{1}{6} = -\frac{1}{3}$

c. $3(-3) = -9$

d. $(0)(11) = 0$

e. $\left(\frac{1}{5}\right)\left(-\frac{10}{11}\right) = -\frac{2}{11}$

f. $(7)(1)(-2)(-3) = 42$

g. $8(-2)(0) = 0$

2. $|2x| + 5 = 7$
$|2x| = 2$
$2x = 2$ or $2x = -2$
$x = 1$ or $x = -1$
$\{-1, 1\}$

3. a. $f(1) = 7(1)^2 - 3(1) + 1$
$= 7 - 3 + 1 = 5$

b. $g(1) = 3(1) - 2 = 3 - 2 = 1$

c. $f(-2) = 7(-2)^2 - 3(-2) + 1$
$= 7(4) + 6 + 1$
$= 28 + 6 + 1 = 35$

d. $g(0) = 3(0) - 2 = 0 - 2 = -2$

4. a. $\dfrac{x^{-9}}{x^2} = \dfrac{1}{x^{2+9}} = \dfrac{1}{x^{11}}$

b. $\dfrac{p^4}{p^{-3}} = p^{4+3} = p^7$

c. $\dfrac{2^{-3}}{2^{-1}} = \dfrac{1}{2^{3-1}} = \dfrac{1}{2^2} = \dfrac{1}{4}$

d. $\dfrac{2x^{-7}y^2}{10xy^{-5}} = \dfrac{y^{2+5}}{5x^{1+7}} = \dfrac{y^7}{5x^8}$

e. $\dfrac{(3x^{-3})(x^2)}{x^6} = \dfrac{3}{x^{6+3-2}} = \dfrac{3}{x^7}$

5. a. $-12x^2 + 7x^2 - 6x = -5x^2 - 6x$

b. $3xy - 2x + 5xy - x = 8xy - 3x$

6. $16x^2 - 24xy + 9y^2$
$= (4x + 3y)(4x + 3y) = (4x + 3y)^2$

7. $f(x) = 3x^2 - 12x + 13$
$\dfrac{-b}{2a} = \dfrac{-(-12)}{2(3)} = \dfrac{12}{6} = 2$
$f(2) = 3(2)^2 - 12(2) + 13 = 1$
Vertex: (2, 1)
The graph opens upward since $a = 3 > 0$.
The graph has no x-intercepts.
$f(0) = 13$ is the y-intercept.

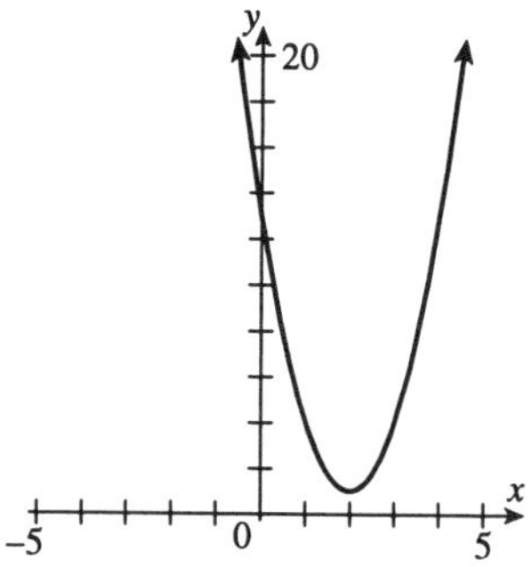

8. a. $\dfrac{3x}{2y} \cdot \dfrac{5xy^2}{5xy^2} = \dfrac{15x^2y^2}{10xy^3}$

b. $\frac{3x+1}{x-5}\cdot\frac{2x-1}{2x-1}=\frac{(3x+1)(2x-1)}{(x-5)(2x-1)}$
$=\frac{6x^2-x-1}{2x^2-11x+5}$

9. a. $(f+g)(x)=x-1+2x-3=3x-4$

b. $(f-g)(x)=(x-1)-(2x-3)$
$=x-1-2x+3=-x+2$

c. $(f\cdot g)(x)=(x-1)(2x-3)$
$=2x^2-5x+3$

d. $\left(\frac{f}{g}\right)(x)=\frac{x-1}{2x-3},\ x\neq\frac{3}{2}$

10. a. $\frac{x}{x-3}-5=\frac{x}{x-3}-\frac{5(x-3)}{x-3}$
$=\frac{x-(5x-15)}{x-3}=\frac{x-5x+15}{x-3}$
$=\frac{-4x+15}{x-3}$

b. $\frac{7}{x-y}+\frac{3}{y-x}=\frac{7}{x-y}-\frac{3}{x-y}$
$=\frac{7-3}{x-y}=\frac{4}{x-y}$

11.

$$\begin{array}{r|l} & 2x-5 \\ \hline x+2 & 2x^2-\ \ x-10 \\ & \underline{2x^2+4x} \\ & \quad -5x-10 \\ & \quad \underline{-5x-10} \\ & \qquad\quad 0 \end{array}$$

Answer: $2x-5$

12.

$$\begin{array}{c|cccc} 3 & 2 & -1 & -13 & 1 \\ & & 6 & 15 & 6 \\ \hline & 2 & 5 & 2 & 7 \end{array}$$

Answer: $2x^2+5x+2+\frac{7}{x-3}$

13. a. $\sqrt[4]{81}=3$

b. $\sqrt[5]{-32}=-2$

c. $-\sqrt{25}=-5$

d. $\sqrt[4]{-81}$ is not a real number

e. $\sqrt[3]{64x^3}=4x$

14. a. $\sqrt[5]{x}=x^{1/5}$

b. $\sqrt[3]{17x^2y^5}=17^{1/3}x^{2/3}y^{5/3}$

c. $\sqrt{x-5a}=(x-5a)^{1/2}$

d. $3\sqrt{2p}-5\sqrt[3]{p^2}$
$=3(2p)^{1/2}-5p^{2/3}$

15. a. $\frac{\sqrt{45}}{4}-\frac{\sqrt{5}}{3}$
$=\frac{3\sqrt{45}}{12}-\frac{4\sqrt{5}}{12}$
$=\frac{3\sqrt{9\cdot 5}-4\sqrt{5}}{12}$
$=\frac{9\sqrt{5}-4\sqrt{5}}{12}$
$=\frac{5\sqrt{5}}{12}$

b. $\sqrt[3]{\frac{7x}{8}}+2\sqrt[3]{7x}$
$=\frac{\sqrt[3]{7x}}{\sqrt[3]{8}}+2\sqrt[3]{7x}$
$=\frac{\sqrt[3]{7x}}{2}+2\sqrt[3]{7x}$
$=\frac{\sqrt[3]{7x}+4\sqrt[3]{7x}}{2}$
$=\frac{5\sqrt[3]{7x}}{2}$

16. $\frac{\sqrt{x}+2}{5}\cdot\frac{(\sqrt{x}-2)}{(\sqrt{x}-2)}$
$=\frac{x-4}{5(\sqrt{x}-2)}$

17. $\sqrt{2x+5}+\sqrt{2x}=3$
$\sqrt{2x+5}=3-\sqrt{2x}$
$2x+5=9-6\sqrt{2x}+2x$

$6\sqrt{2x} = 4$
$3\sqrt{2x} = 2$
$9(2x) = 4$
$18x = 4$
$x = \frac{2}{9}$
$\left\{\frac{2}{9}\right\}$

18. a. $c^2 = a^2 + b^2$
$(50)^2 = (20)^2 + b^2$
$2500 - 400 = b^2$
$2100 = b^2$
$\pm\sqrt{2100} = b$
Disregard the negative
$\sqrt{2100} \approx 45.8$ feet

b. $75\left(\frac{3}{5}\right) = 45$ feet
Yes

19. $(x+1)^2 = 12$
$x + 1 = \pm\sqrt{12}$
$x = -1 \pm \sqrt{4 \cdot 3}$
$x = -1 \pm 2\sqrt{3}$
$\left\{-1 + 2\sqrt{3},\ -1 - 2\sqrt{3}\right\}$

20. $2x^2 - 4x = 3$
$2x^2 - 4x - 3 = 0$
$a = 2,\ b = -4,\ c = -3$
$x = \frac{-(-4) \pm \sqrt{(-4)^2 - 4(2)(-3)}}{2(2)}$
$x = \frac{4 \pm \sqrt{16 + 24}}{4}$
$x = \frac{4 \pm \sqrt{40}}{4}$
$x = \frac{4 \pm 2\sqrt{10}}{4}$
$x = \frac{2 \pm \sqrt{10}}{2}$
$\left\{\frac{2 + \sqrt{10}}{2},\ \frac{2 - \sqrt{10}}{2}\right\}$

21. $p^4 - 3p^2 - 4 = 0$
$(p^2 - 4)(p^2 + 1) = 0$
$p^2 - 4 = 0$ or $p^2 + 1 = 0$
$p^2 = 4$ or $p^2 = -1$
$p = \pm\sqrt{4}$ or $p = \pm\sqrt{-1}$
$p = \pm 2$ or $p = \pm i$
$\{-2, 2, -i, i\}$

22. $\frac{x+2}{x-3} \le 0$
$x = -2$ and $x = 3$ are critical points

Region	Test Point	$\frac{x+2}{x-3} \le 0$
A	-3	$\frac{-1}{-6} > 0$
B	0	$\frac{2}{-3} < 0$
C	4	$\frac{6}{1} > 0$

$[-2, 3)$ satisfies the inequality

23. $F(x) = (x - 3)^2 + 1$
Vertex: $(3, 1)$
Axis of symmetry: $x = 3$

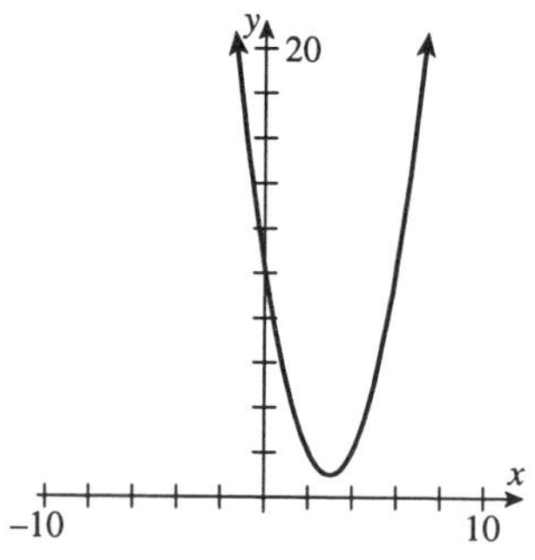

24. $f(t) = -16t^2 + 20t$
$\frac{-b}{2a} = \frac{-20}{2(-16)} = \frac{5}{8}$
$f\left(\frac{5}{8}\right) = -16\left(\frac{5}{8}\right)^2 + 20\left(\frac{5}{8}\right) = 6\frac{1}{4}$
Vertex: $\left(\frac{5}{8},\ 6\frac{1}{4}\right)$
Maximum height is $6\frac{1}{4}$ feet in $\frac{5}{8}$ second

Chapter 9

Section 9.1 Graphing Calculator Box

1. $x^2 + y^2 = 55$

$y^2 = 55 - x^2$

$y = \pm\sqrt{55 - x^2}$

3. $7x^2 + 7y^2 - 89 = 0$

$7y^2 = 89 - 7x^2$

$y^2 = \frac{89 - 7x^2}{7}$

$y = \pm\sqrt{\frac{89 - 7x^2}{7}}$

$Y_1 = \sqrt{\frac{89 - 7x^2}{7}}$

$Y_2 = -\sqrt{\frac{89 - 7x^2}{7}}$

Mental Math

1. $y = x^2 - 7x + 5$; upward

3. $x = -y^2 - y + 2$; to the left

5. $y = -x^2 + 2x + 1$; downward

Exercise Set 9.1

1. $x = 3y^2$

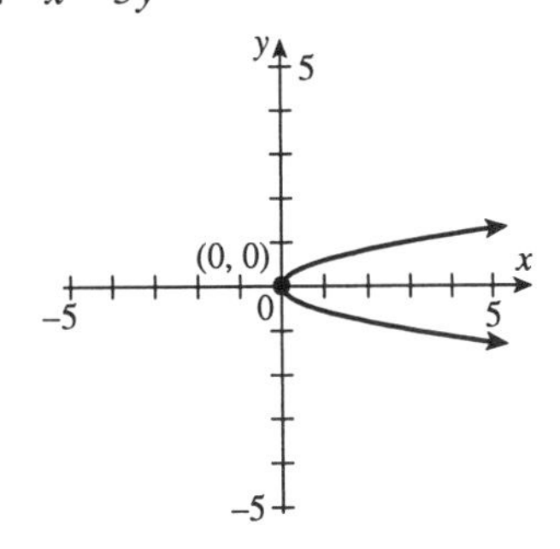

3. $x = (y - 2)^2 + 3$

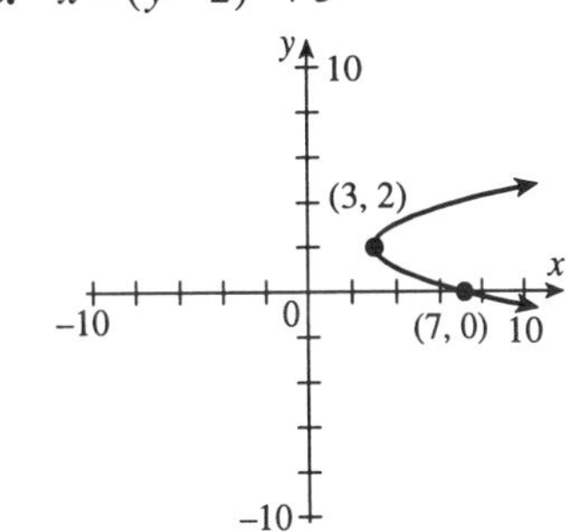

5. $y = 3(x - 1)^2 + 5$

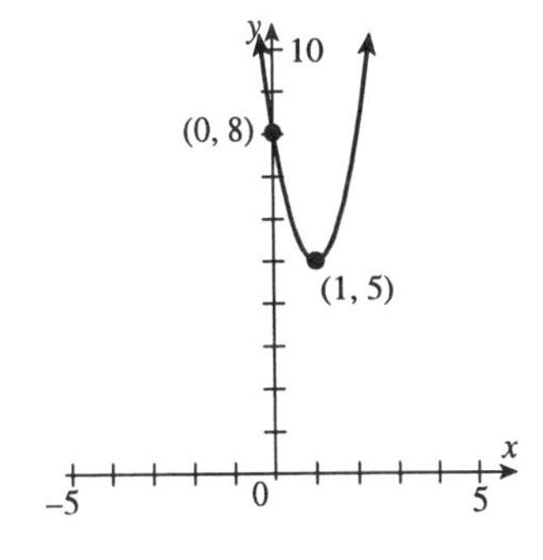

7. $x = y^2 + 6y + 8$

$x = y^2 + 6y + \left(\frac{6}{2}\right)^2 + 8 - 9$

$x = (y + 3)^2 - 1$

Thus, $V(-1, -3)$.

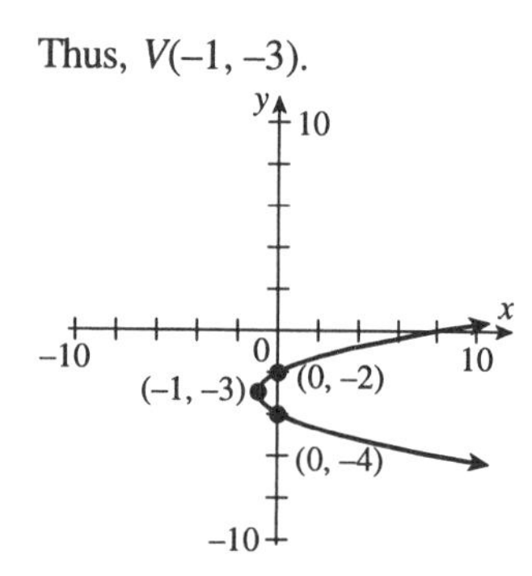

9. $y = x^2 + 10x + 20$

$y = x^2 + 10x + \left(\frac{10}{2}\right)^2 + 20 - 25$

$y = (x+5)^2 - 5$

Thus, $V(-5, -5)$.

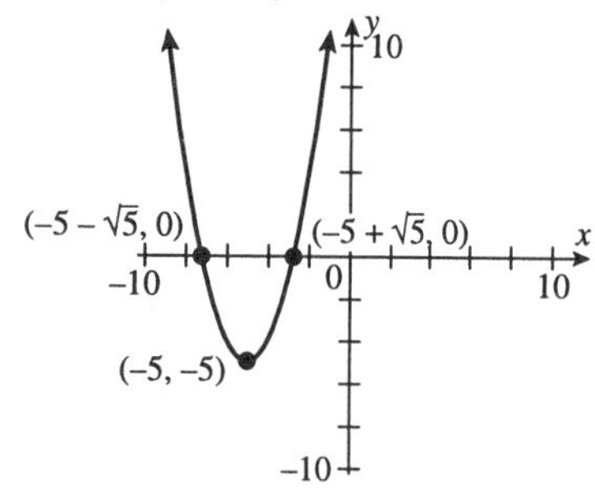

11. $x = -2y^2 + 4y + 6$

$x = -2(y^2 - 2y) + 6$

$x = -2\left[y^2 - 2y + \left(\frac{2}{2}\right)^2\right] + 6 + 2$

$x = -2(y-1)^2 + 8$

Thus, $V(8, 1)$.

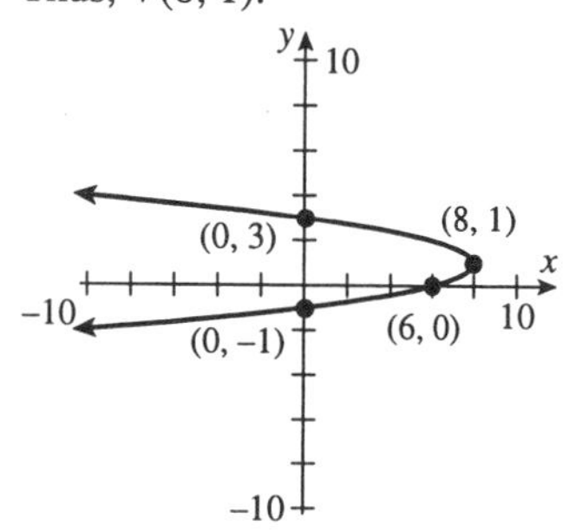

13. $(5, 1), (8, 5)$

$d = \sqrt{(8-5)^2 + (5-1)^2}$

$d = \sqrt{9+16}$

$d = \sqrt{25}$

$d = 5$

15. $(-3, 2), (1, -3)$

$d = \sqrt{[1-(-3)]^2 + (-3-2)^2}$

$d = \sqrt{16+25}$

$d = \sqrt{41}$

17. $(-9, 4), (-8, 1)$

$d = \sqrt{[-8-(-9)]^2 + (1-4)^2}$

$d = \sqrt{(-8+9)^2 + (-3)^2}$

$d = \sqrt{1^2 + 9}$

$d = \sqrt{10}$

19. $(0, -\sqrt{2}), (\sqrt{3}, 0)$

$d = \sqrt{(\sqrt{3}-0)^2 + [0-(-\sqrt{2})]^2}$

$d = \sqrt{(\sqrt{3})^2 + (\sqrt{2})^2}$

$d = \sqrt{3+2}$

$d = \sqrt{5}$

21. $(1.7, -3.6), (-8.6, 5.7)$

$d = \sqrt{(-8.6-1.7)^2 + [5.7-(-3.6)]^2}$

$d = \sqrt{(-10.3)^2 + (9.3)^2}$

$d = \sqrt{192.58}$

$d = 13.88$

23. $(2\sqrt{3}, \sqrt{6}), (-\sqrt{3}, 4\sqrt{6})$

$d = \sqrt{(-\sqrt{3}-2\sqrt{3})^2 + (4\sqrt{6}-\sqrt{6})^2}$

$d = \sqrt{(-3\sqrt{3})^2 + (3\sqrt{6})^2}$

$d = \sqrt{27+54}$

$d = \sqrt{81}$

$d = 9$

25. $(6, -8), (2, 4)$

$\left(\frac{6+2}{2}, \frac{-8+4}{2}\right)$

$(4, -2)$

27. $(-2, -1), (-8, 6)$

$\left(\frac{-2+(-8)}{2}, \frac{-1+6}{2}\right)$

$\left(-5, \frac{5}{2}\right)$

29. $(7, 3), (-1, -3)$

$\left(\frac{7-1}{2}, \frac{3-3}{2}\right)$

$(3, 0)$

31. $\left(\frac{1}{2}, \frac{3}{8}\right), \left(-\frac{3}{2}, \frac{5}{8}\right)$

$\left(\frac{\frac{1}{2}-\frac{3}{2}}{2}, \frac{\frac{3}{8}+\frac{5}{8}}{2}\right)$

$\left(-\frac{1}{2}, \frac{1}{2}\right)$

33. $(\sqrt{2}, 3\sqrt{5}), (\sqrt{2}, -2\sqrt{5})$

$\left(\frac{\sqrt{2}+\sqrt{2}}{2}, \frac{3\sqrt{5}-2\sqrt{5}}{2}\right)$

$\left(\sqrt{2}, \frac{\sqrt{5}}{2}\right)$

35. $(4.6, -3.5), (7.8, -9.8)$

$\left(\frac{4.6+7.8}{2}, \frac{-3.5-9.8}{2}\right)$

$(6.2, -6.65)$

37. $x^2 + y^2 = 9$

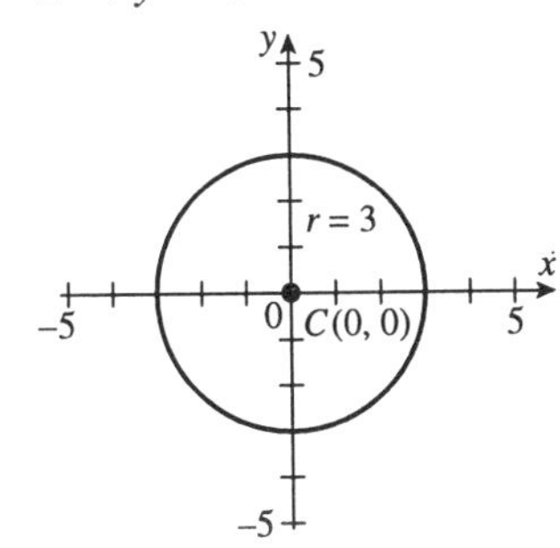

39. $x^2 + (y-2)^2 = 1$

y
5
r = 1
C(0, 2)
x
–5
0
5
–5

41. 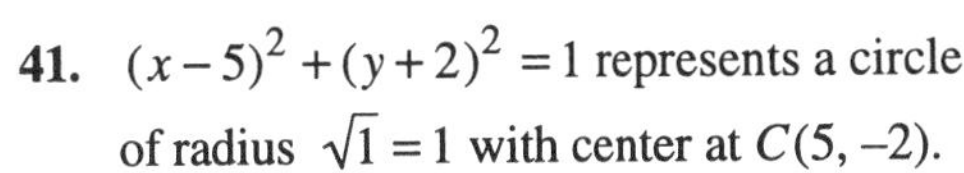
$(x-5)^2 + (y+2)^2 = 1$ represents a circle of radius $\sqrt{1} = 1$ with center at $C(5, -2)$.

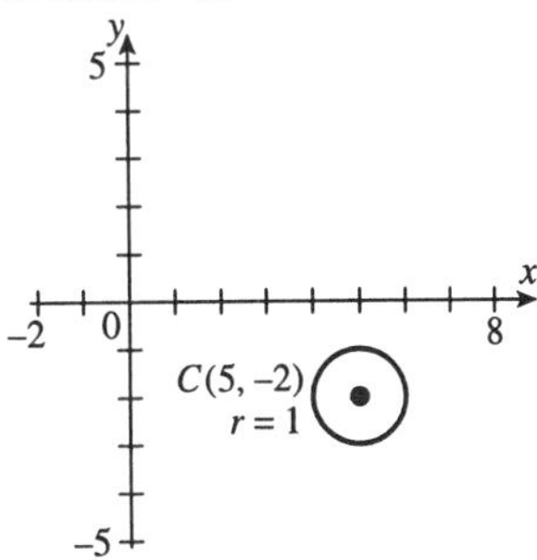

43. $x^2 + y^2 + 6y = 0$

$x^2 + y^2 + 6y + \left(\frac{6}{2}\right)^2 = 9$

$(x-0)^2 + (y+3)^2 = 3^2$

Thus, $C(0, -3)$ and $r = 3$.

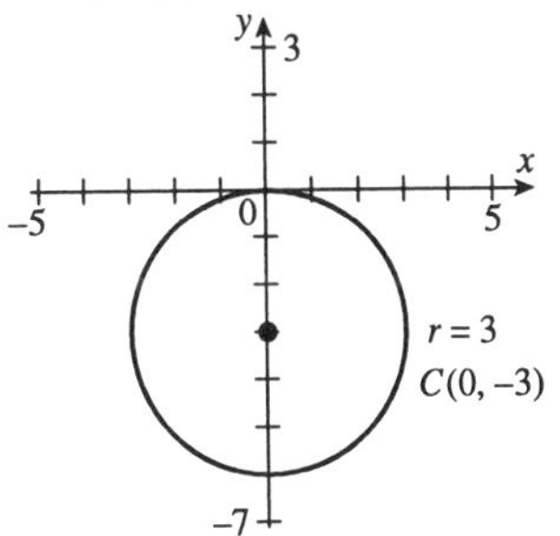

45. $x^2 + y^2 + 2x - 4y = 4$

$x^2 + 2x + \left(\frac{2}{2}\right)^2 + y^2 - 4y + \left(\frac{4}{2}\right)^2$

$= 4 + 1 + 4$

$(x+1)^2 + (y-2)^2 = 9$

$(x+1)^2 + (y-2)^2 = 3^2$
Thus, $C(-1, 2)$ and $r = 3$.

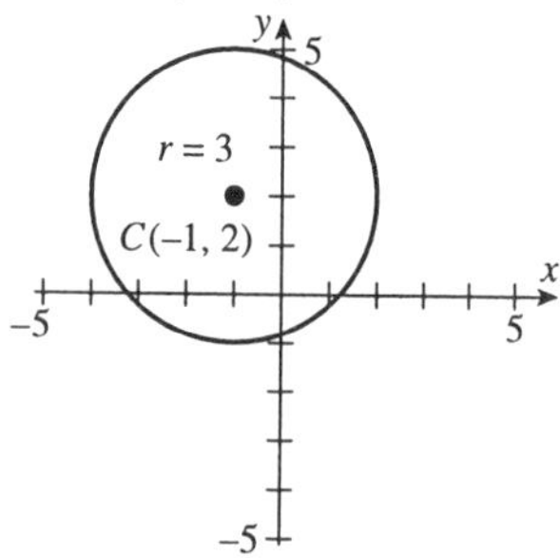

47. $C(2, 3)$; $r = 6$
$(x-2)^2 + (y-3)^2 = 6^2$
$(x-2)^2 + (y-3)^2 = 36$

49. $C(0, 0)$; $r = \sqrt{3}$
$(x-0)^2 + (y-0)^2 = (\sqrt{3})^2$
$x^2 + y^2 = 3$

51. $C(-5, 4)$; $r = 3\sqrt{5}$
$[x-(-5)]^2 + (y-4)^2 = (3\sqrt{5})^2$
$(x+5)^2 + (y-4)^2 = 45$

53. Answers may vary.

55. $x = y^2 - 3$
$V(-3, 0)$

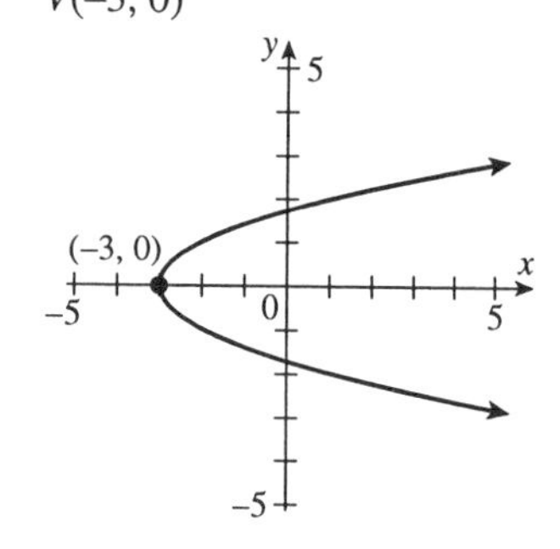

57. $y = (x-2)^2 - 2$
$V(2, -2)$

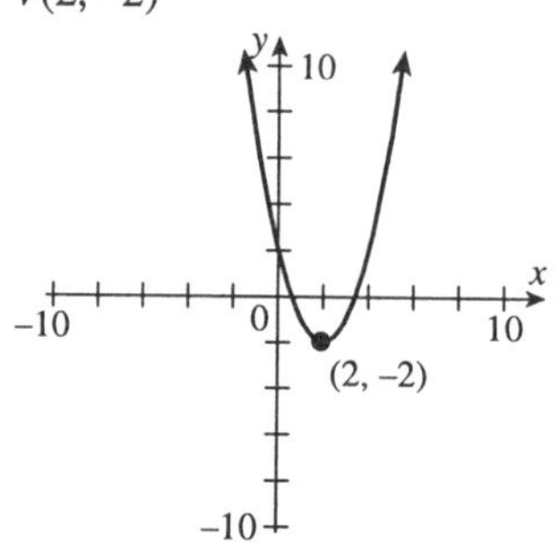

59. $x^2 + y^2 = 1$
$C(0, 0)$; $r = 1$

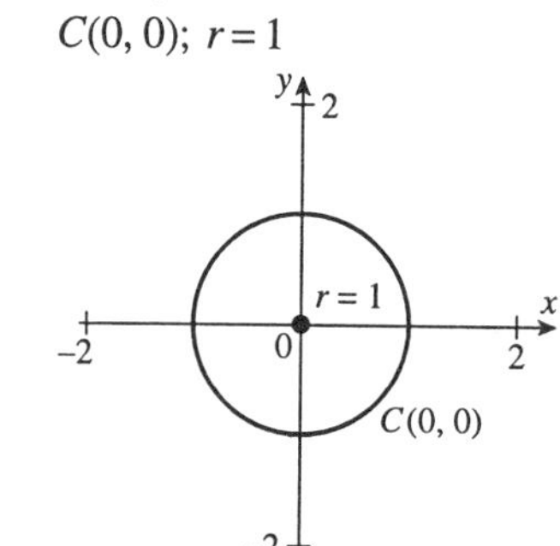

61. $x = (y+3)^2 - 1$
$V(-1, -3)$

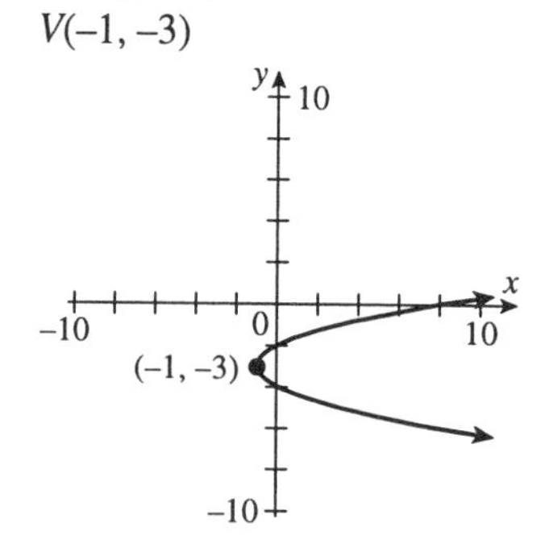

63. $(x-2)^2+(y-2)^2=16$

$C(2, 2)$; $r=4$

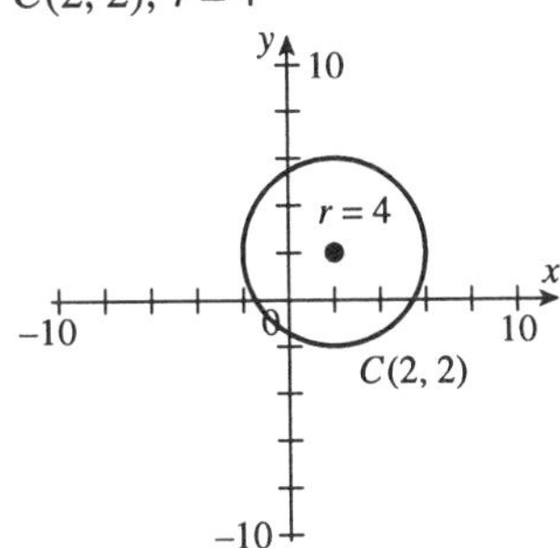

65. $x=-(y-1)^2$

$V(0, 1)$

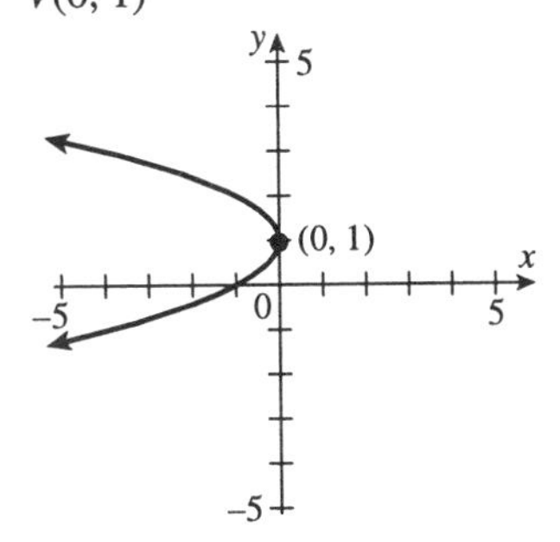

67. $(x-4)^2+y^2=7$

$C(4, 0)$; $r=\sqrt{7}$

y
10
C(4, 0)
x
−10
0
10
r = √7
−10

69. $y=5(x+5)^2+3$

$V(-5, 3)$

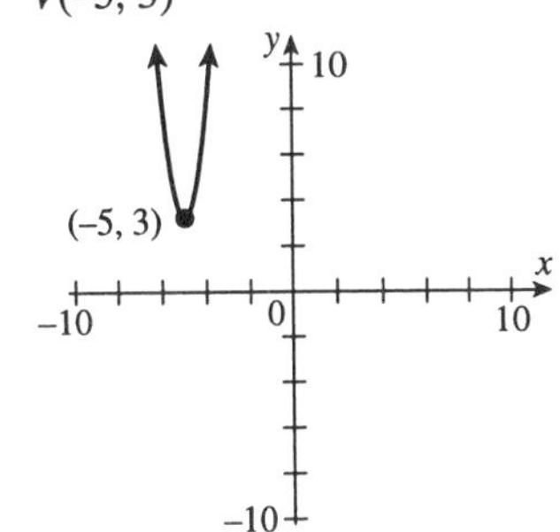

71. $\frac{x^2}{8}+\frac{y^2}{8}=2$

$C(0, 0)$; $r=4$

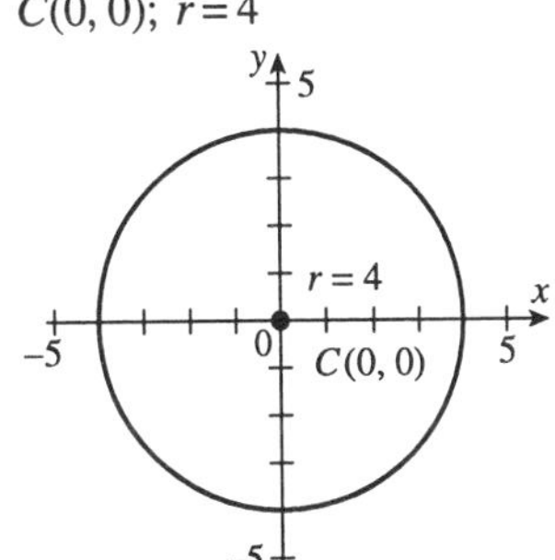

73. $y=x^2+7x+6$

$y=x^2+7x+\frac{49}{4}+6-\frac{49}{4}$

$y=\left(x+\frac{7}{2}\right)^2-\frac{25}{4}$

$V\left(-\frac{7}{2}, -\frac{25}{4}\right)$

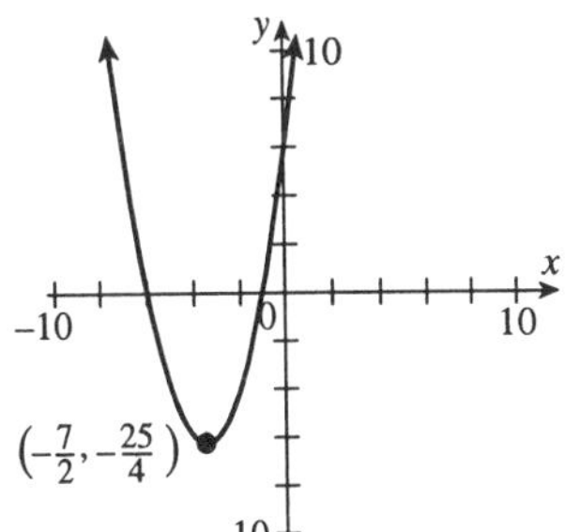

75. $x^2+y^2+2x+12y-12=0$

$x^2+2x+1+y^2+12y+36=12+1+36$

$(x+1)^2+(y+6)^2=49$

$C(-1,-6)$; $r=7$

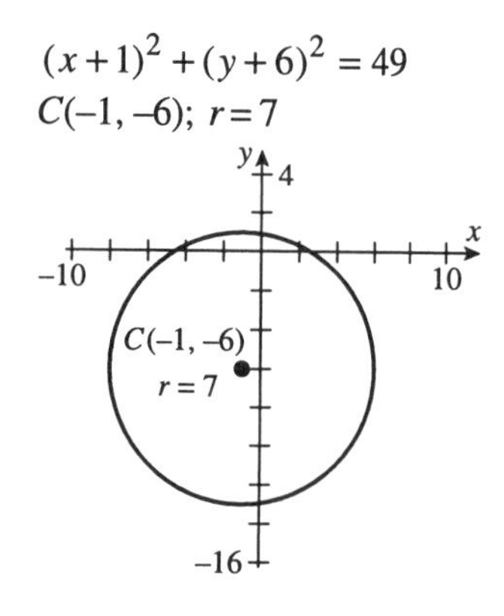

77. $x=y^2+8y-4$

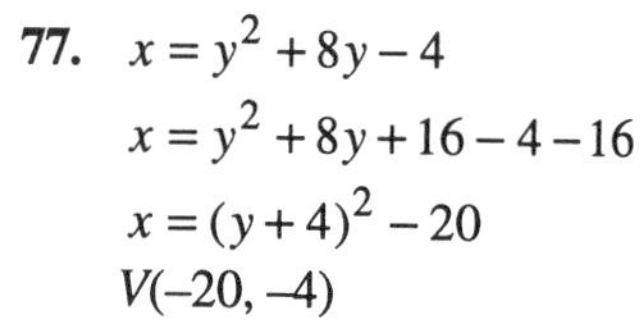

$x=y^2+8y+16-4-16$

$x=(y+4)^2-20$

$V(-20,-4)$

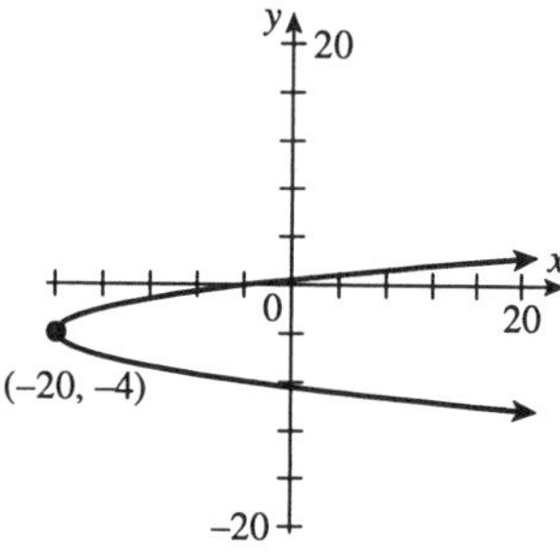

79. $x^2-10y+y^2+4=0$

$x^2+y^2-10y+25=-4+25$

$x^2+(y-5)^2=21$

$C(0, 5)$; $r=\sqrt{21}$

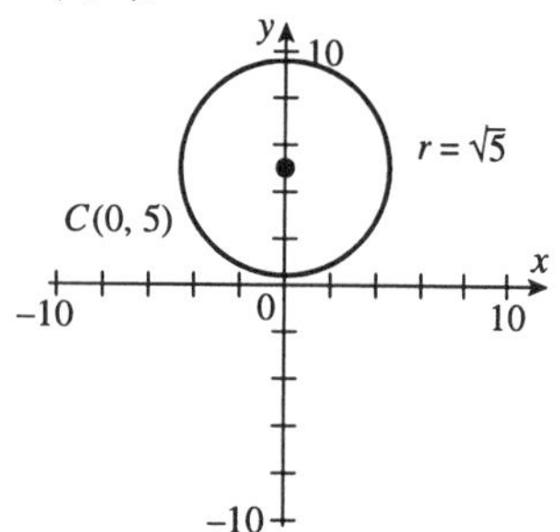

81. $x=-3y^2+30y$

$x=-3(y^2-10y+25)+75$

$x=-3(y-5)^2+75$

$V(75, 5)$

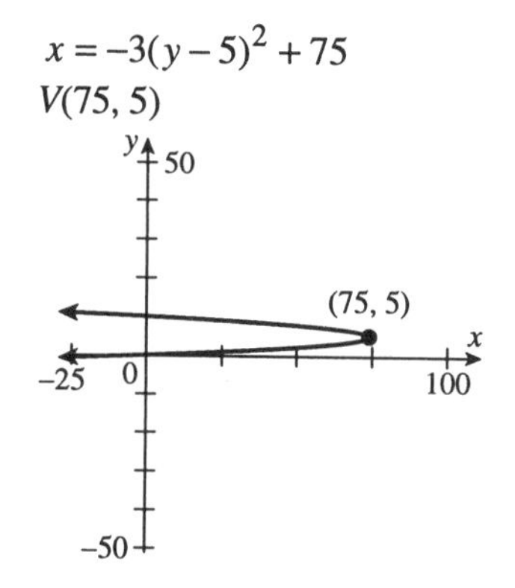

83. $5x^2+5y^2=25$

$x^2+y^2=5$

$C(0, 0)$; $r=\sqrt{5}$

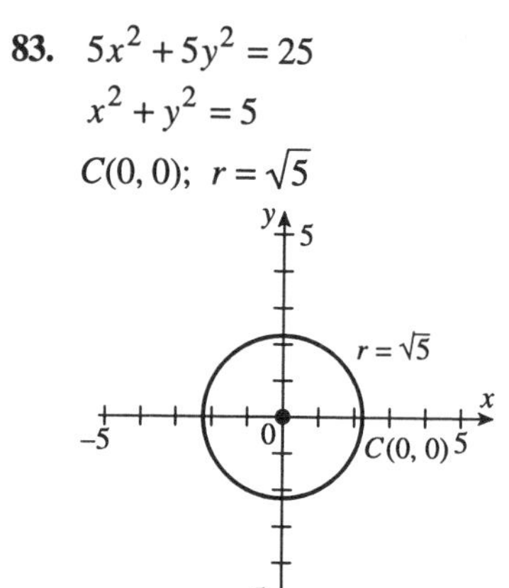

85. $y=5x^2-20x+16$

$y=5(x^2-4x+4)+16-20$

$y=5(x-2)^2-4$

$V(2,-4)$

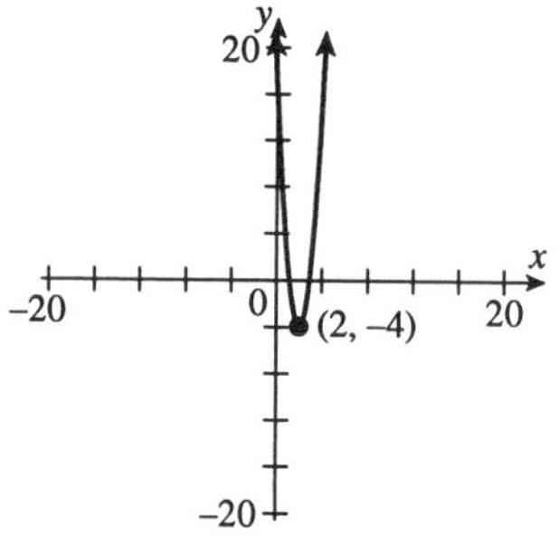

87. The distance between the points (5, 1) and (2, 6) is

$d=\sqrt{(5-2)^2+(1-6)^2}$

$d = \sqrt{3^2 + 5^2} = \sqrt{9+25}$
$d = \sqrt{34}$ units.
The distance between the points (5, 1) and (0, –2) is
$d = \sqrt{(5-0)^2 + [1-(-2)]^2}$
$d = \sqrt{5^2 + 3^2} = \sqrt{25+9}$
$d = \sqrt{34}$ units
Therefore, the triangle with vertices (2, 6), (0, –2), and (5, 1) is an isosceles triangle.

89. Setting up a coordinate system with the axis of symmetry as the y-axis and the base as the x-axis. The parabola would pass through the points (–50, 0), (0, 40), and (50, 0).
Using the equation for a parabola:
$y = ax^2 + bx + c$
Substituting the x and y coordinates for the known points yields a system of equations.

$$\begin{cases} 0 = a(-50)^2 + b(-50) + c \\ 40 = a(0)^2 + b(0) + c \\ 0 = a(50)^2 + b(50) + c \end{cases}$$

$$\begin{cases} 0 = 2500a - 50b + c \\ 40 = c \\ 0 = 2500a + 50b + c \end{cases}$$

Solving $c = 40$:

$$\begin{cases} 0 = 2500a - 50b + 40 \\ 0 = 2500a + 50b + 40 \end{cases}$$

Adding the two equations:
$0 = 5000a + 80$
$-80 = 5000a$
$\frac{-80}{5000} = a$
$-\frac{2}{125} = a$
$0 = 2500\left(-\frac{2}{125}\right) - 50b + 40$
$0 = -40 - 50b + 40$
$0 = -50b$
$0 = b$
Thus the equation of the parabola is
$y = -\frac{2}{125}x^2 + 40.$

91. Answers are the same.

93. Answers are the same.

95. $y = -3x + 3$

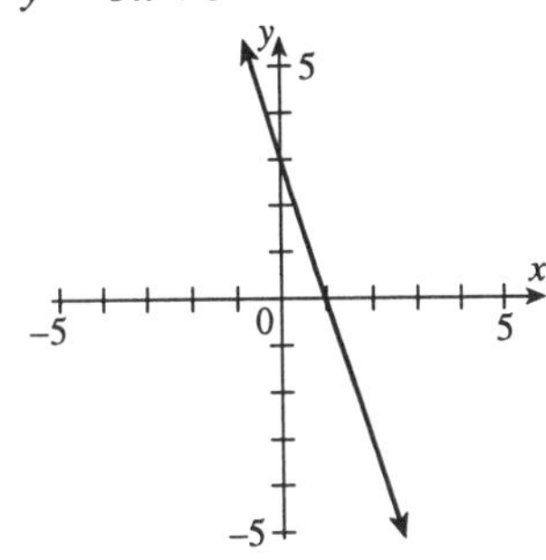

97. $x = -2$

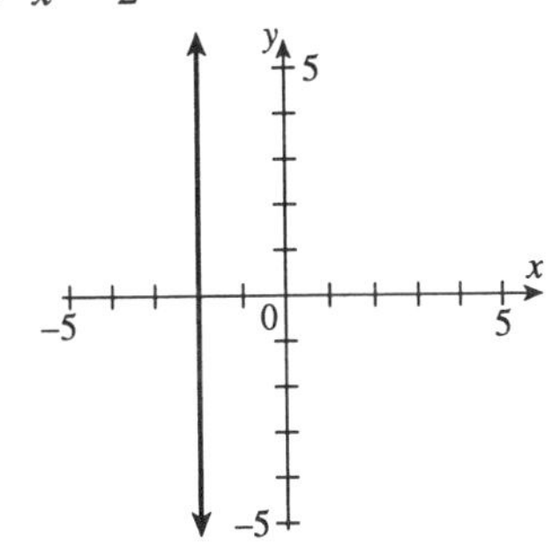

99. $\frac{\sqrt{5}}{\sqrt{8}} = \frac{\sqrt{5}}{2\sqrt{2}} = \frac{\sqrt{5}}{2\sqrt{2}} \cdot \frac{\sqrt{2}}{\sqrt{2}}$
$= \frac{\sqrt{10}}{2\sqrt{4}} = \frac{\sqrt{10}}{2(2)} = \frac{\sqrt{10}}{4}$

101. $\frac{10}{\sqrt{5}} = \frac{10}{\sqrt{5}} \cdot \frac{\sqrt{5}}{\sqrt{5}} = \frac{10\sqrt{5}}{\sqrt{25}}$
$= \frac{10\sqrt{5}}{5} = 2\sqrt{5}$

Graphing Calculator Box 9.2

1. $10x^2 + y^2 = 32$
$y^2 = 32 - 10x^2$
$y = \pm\sqrt{32 - 10x^2}$
$Y_1 = \sqrt{32 - 10x^2}$

$Y_2 = -\sqrt{32 - 10x^2}$

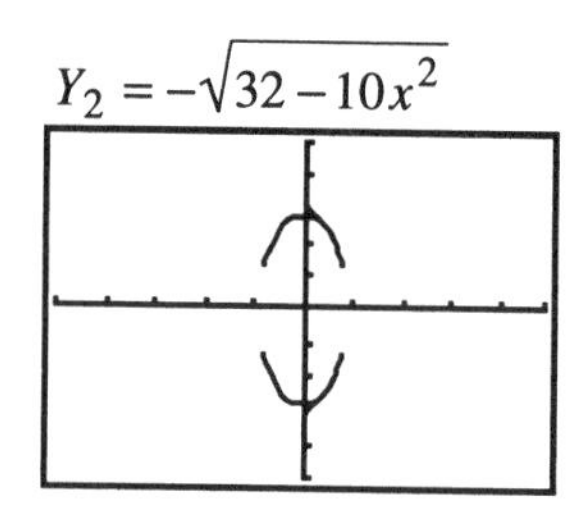

3. $7.3x^2 + 15.5y^2 = 95.2$

$15.5y^2 = 95.2 - 7.3x^2$

$y^2 = \dfrac{95.2 - 7.3x^2}{15.5}$

$y = \pm\sqrt{\dfrac{95.2 - 7.3x^2}{15.5}}$

$Y_1 = \sqrt{\dfrac{95.2 - 7.3x^2}{15.5}}$

$Y_2 = -\sqrt{\dfrac{95.2 - 7.3x^2}{15.5}}$

Exercise Set 9.2

1. $\dfrac{x^2}{4} + \dfrac{y^2}{25} = 1$

$\dfrac{x^2}{2^2} + \dfrac{y^2}{5^2} = 1$

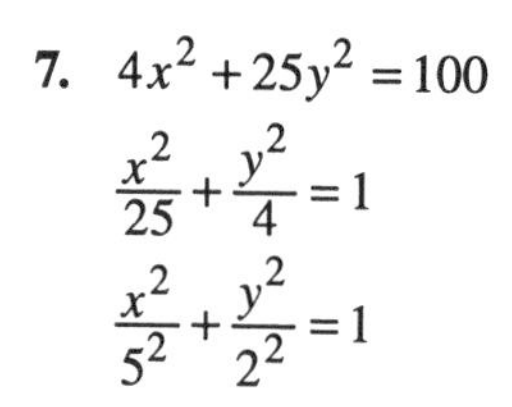

3. $\dfrac{x^2}{16} + \dfrac{y^2}{9} = 1$

$\dfrac{x^2}{4^2} + \dfrac{y^2}{3^2} = 1$

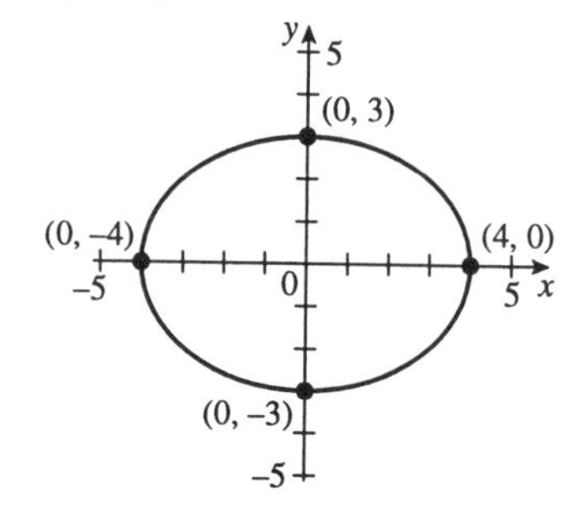

5. $9x^2 + 4y^2 = 36$

$\dfrac{x^2}{4} + \dfrac{y^2}{9} = 1$

$\dfrac{x^2}{2^2} + \dfrac{y^2}{3^2} = 1$

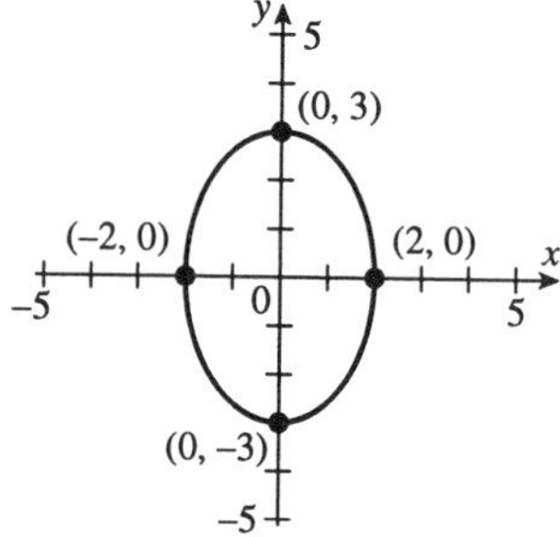

7. $4x^2 + 25y^2 = 100$

$\dfrac{x^2}{25} + \dfrac{y^2}{4} = 1$

$\dfrac{x^2}{5^2} + \dfrac{y^2}{2^2} = 1$

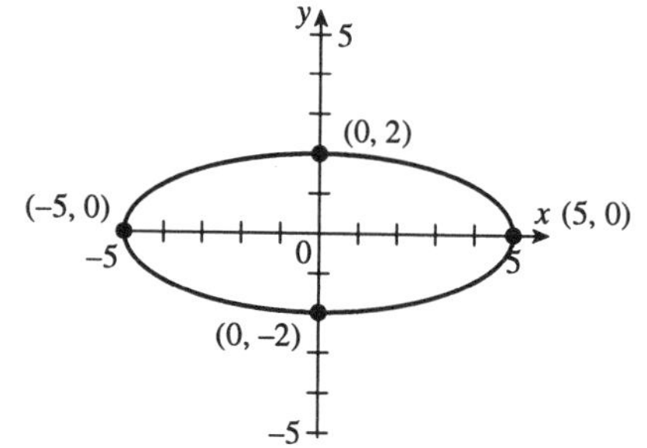

9. $\dfrac{(x+1)^2}{36}+\dfrac{(y-2)^2}{49}=1$

$\dfrac{(x+1)^2}{6^2}+\dfrac{(y-2)^2}{7^2}=1$

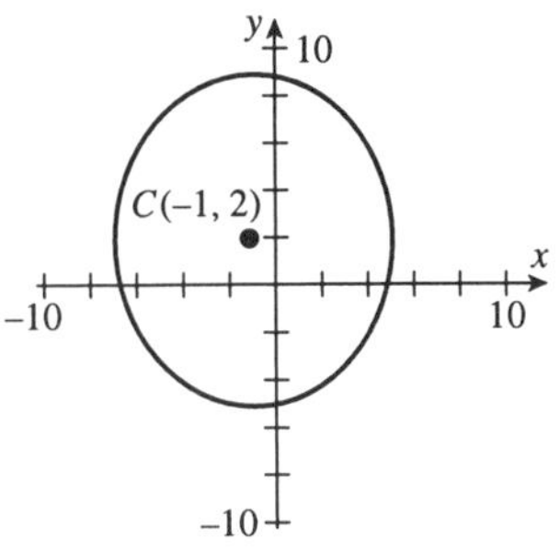

11. $\dfrac{(x-1)^2}{4}+\dfrac{(y-1)^2}{25}=1$

$\dfrac{(x-1)^2}{2^2}+\dfrac{(y-1)^2}{5^2}=1$

13. $\dfrac{x^2}{4}-\dfrac{y^2}{9}=1$

$\dfrac{x^2}{2^2}-\dfrac{y^2}{3^2}=1$

15. $\dfrac{y^2}{25}-\dfrac{x^2}{16}=1$

$\dfrac{y^2}{5^2}-\dfrac{x^2}{4^2}=1$

17. $x^2-4y^2=16$

$\dfrac{x^2}{16}-\dfrac{y^2}{4}=1$

$\dfrac{x^2}{4^2}-\dfrac{y^2}{2^2}=1$

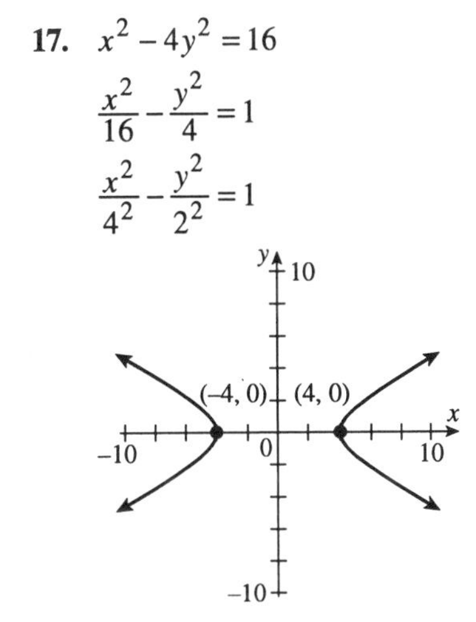

19. $16y^2-x^2=16$

$\dfrac{y^2}{1}-\dfrac{x^2}{16}=1$

$\dfrac{y^2}{1^2}-\dfrac{x^2}{4^2}=1$

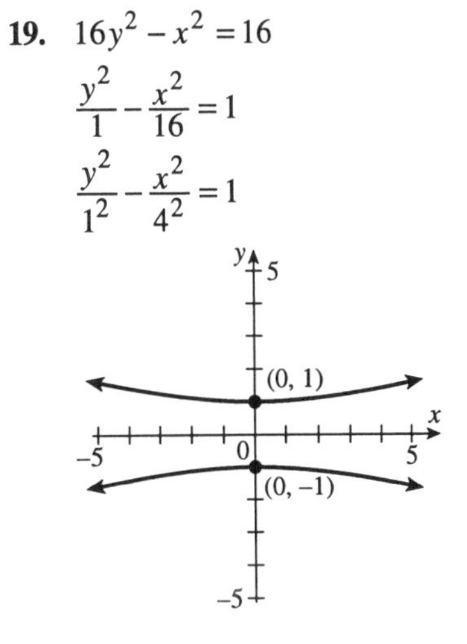

21. Answers may vary.

23. $y = x^2 + 4$
parabola: $V(0, 4)$
opens upward

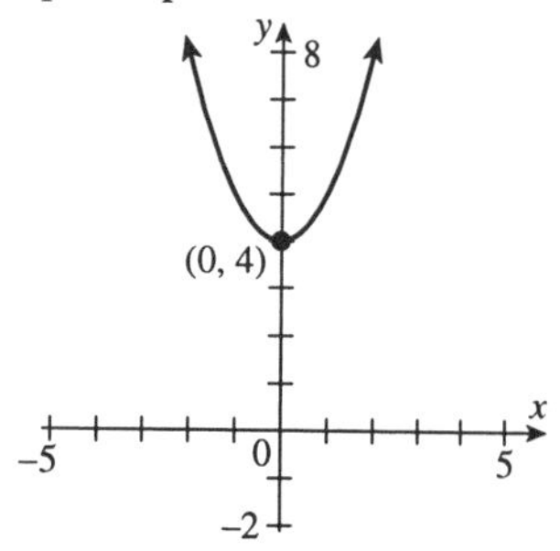

25. $\frac{x^2}{4} + \frac{y^2}{9} = 1$
ellipse: center (0, 0)
$a = 3$, $b = 2$
y-intercepts at ±3
x-intercepts at ±2

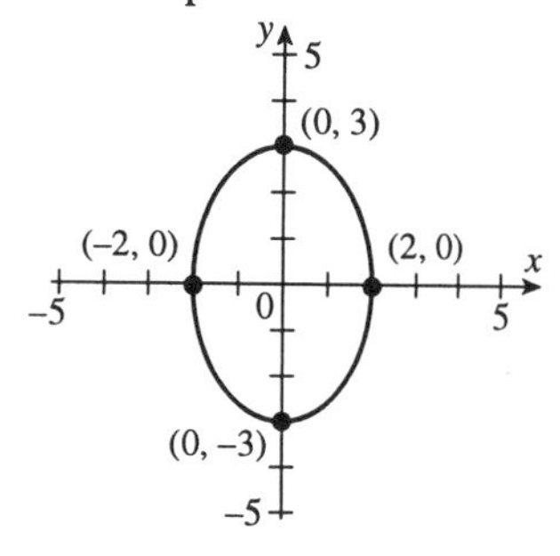

27. $\frac{x^2}{16} - \frac{y^2}{4} = 1$
hyperbola: center (0, 0)
$a = 4$, $b = 2$
x-intercepts at ±4

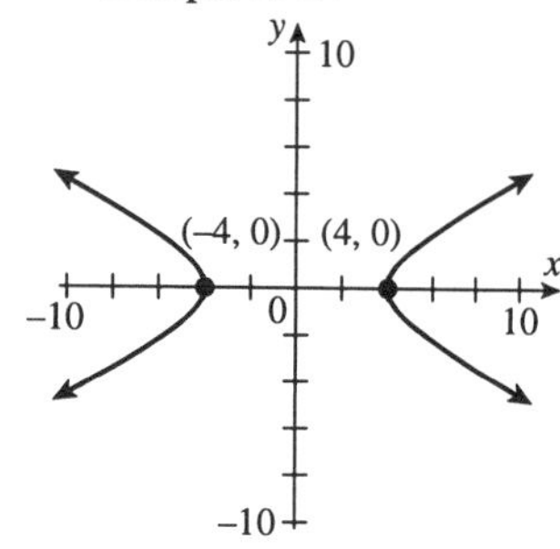

29. $x^2 + y^2 = 16$
circle: center (0, 0)
radius = 4

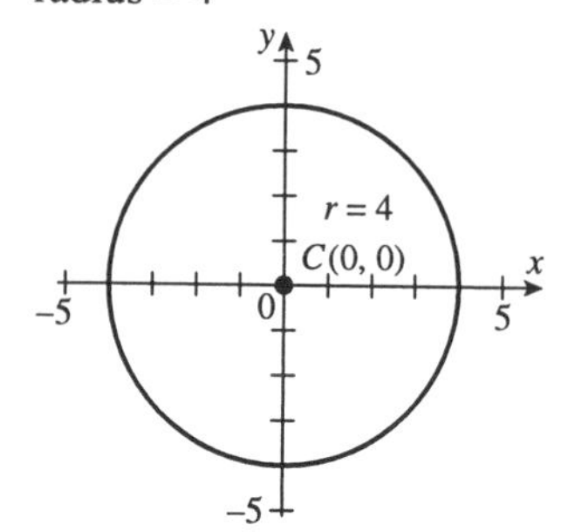

31. $x = -y^2 + 6y$
parabola: $\frac{-b}{2a} = \frac{-6}{2(-1)} = 3$
$x = -9 + 18 = 9$
$V(9, 3)$
opens to the left

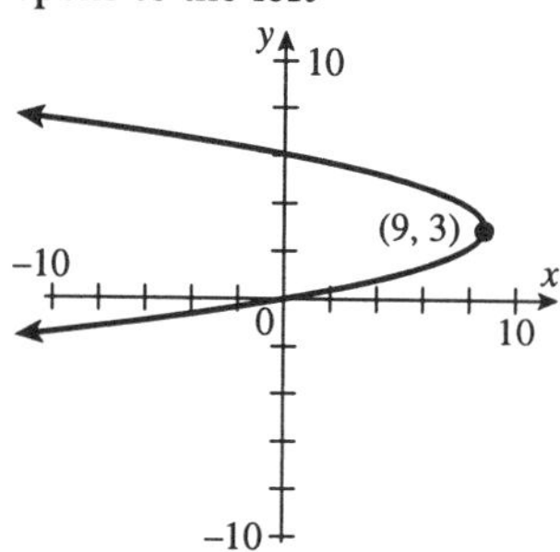

33. $9x^2 + 4y^2 = 36$
$\frac{x^2}{4} + \frac{y^2}{9} = 1$
ellipse: center (0, 0)
$a = 3$, $b = 2$
y-intercepts at ±3
x-intercepts at ±2

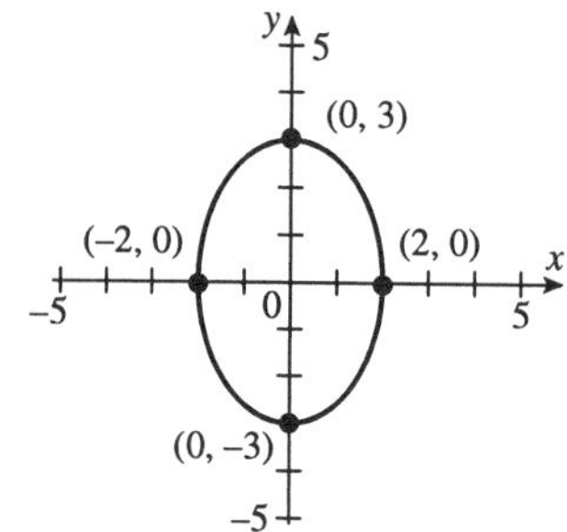

35. $y^2 = x^2 + 16$

$\frac{y^2}{16} - \frac{x^2}{16} = 1$

hyperbola: center (0, 0)

$a = 4,\ b = 4$

y-intercepts at ±4

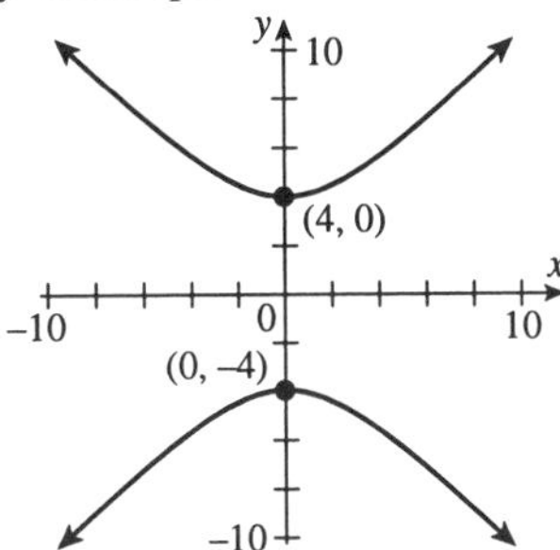

37. $y = -2x^2 + 4x - 3$

parabola: $\frac{-b}{2a} = \frac{-4}{2a} = 1$

$y = -2 + 4 - 3 = -1$

$V(1, -1)$

opens downward

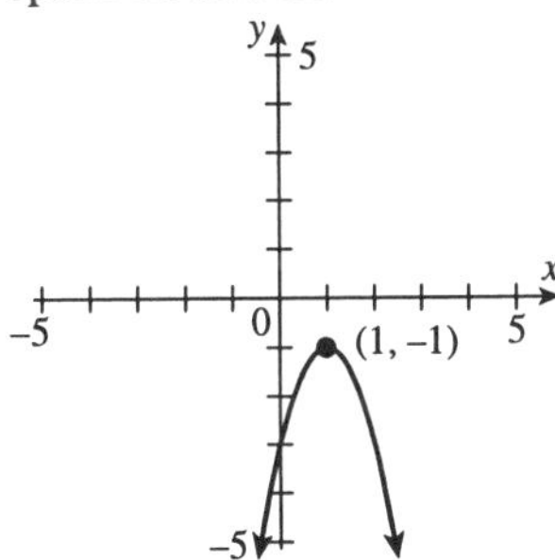

39. $\frac{(x - 1,782,000,000)^2}{(3.42)(10^{23})} + \frac{(y - 356,400,000)^2}{(1.368)(10^{22})} = 1$

center (1,782,000,000, 356,400,000)

41. Answers are the same.

43. $x < 5$ or $x < 1$

$(-\infty, 5)$

45. $2x - 1 \ge 7$ and $-3x \le -6$

$x \ge 4$ and $x \ge 2$

$[4, \infty)$

47. $2x^3 - 4x^3 = -2x^3$

49. $(-5x^2)(x^2) = -5x^4$

51. $\frac{(x+2)^2}{9} - \frac{(y-1)^2}{4} = 1$

center (−2, 1)

$a = 3,\ b = 2$

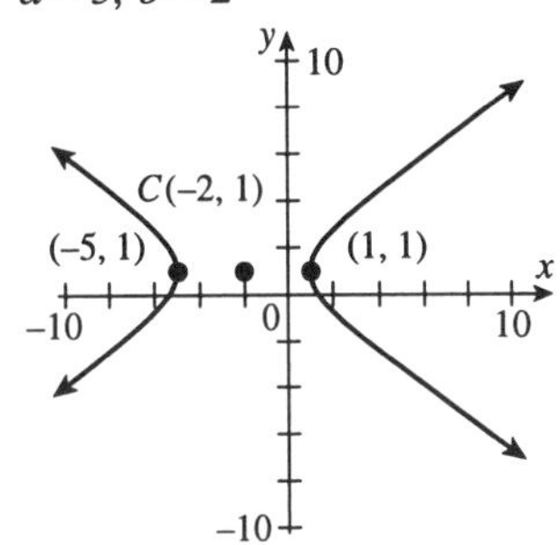

53. $\frac{(y+4)^2}{4} - \frac{x^2}{25} = 1$

center (0, −4)

$a = 2,\ b = 5$

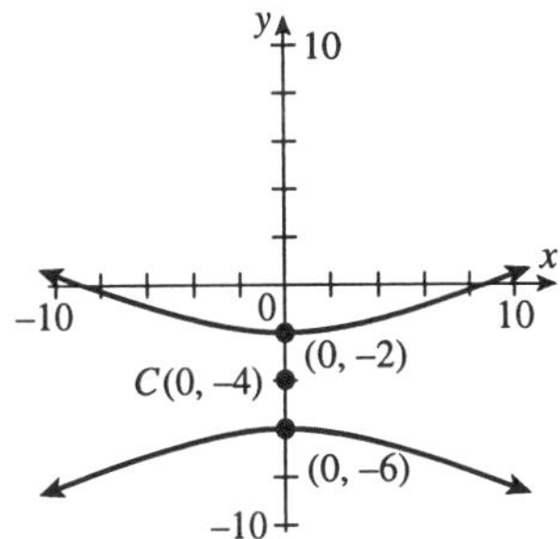

55. $\frac{(x-3)^2}{9} - \frac{(y-2)^2}{4} = 1$

center (3, 2)

$a = 3,\ b = 2$

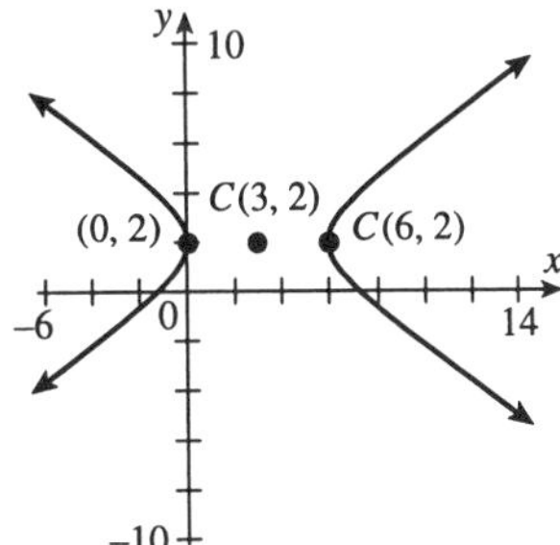

Exercise Set 9.3

1. $\begin{cases} x^2 + y^2 = 25 \\ 4x + 3y = 0 \end{cases}$

Solve equation 2 for *y*.

$3y = -4x$

$y = \frac{-4x}{3}$

Substitute.

$x^2 + \left(-\frac{4x}{3}\right)^2 = 25$

$x^2 + \frac{16x^2}{9} = 25$

$\frac{25}{9}x^2 = 25$

$\frac{x^2}{9} = 1$

$x^2 = 9$

$x = \pm\sqrt{9} = \pm 3$

$x = 3$: $\quad y = -\frac{4}{3}(3) = -4$

$x = -3$ $\quad y = -\frac{4}{3}(-3) = 4$

$\{(3, -4), (-3, 4)\}$

3. $\begin{cases} x^2 + 4y^2 = 10 \\ y = x \end{cases}$

Substitute.

$x^2 + 4x^2 = 10$

$5x^2 = 10$

$x^2 = 2$

$x = \pm\sqrt{2}$

$x = \sqrt{2};\ y = \sqrt{2}$

$x = -\sqrt{2};\ y = -\sqrt{2}$

$\left\{(\sqrt{2}, \sqrt{2}), (-\sqrt{2}, -\sqrt{2})\right\}$

5. $\begin{cases} y^2 = 4 - x \\ x - 2y = 4 \end{cases}$

$-2y = 4 - x$

Substitute.

$y^2 = -2y$

$y^2 + 2y = 0$

$y(y + 2) = 0$

$y = 0$ or $y + 2 = 0$

$y = -2$

$y = 0$: $\quad x - 2(0) = 4$

$x = 4$

$y = -2$: $\quad x - 2(-2) = 4$

$x + 4 = 4$

$x = 0$

$\{(4, 0), (0, -2)\}$

7. $\begin{cases} x^2 + y^2 = 9 \\ 16x^2 - 4y^2 = 64 \end{cases}$

$\begin{cases} x^2 + y^2 = 9 \\ 4x^2 - y^2 = 16 \end{cases}$

Add.

$5x^2 = 25$

$x^2 = 5$

$x = \pm\sqrt{5}$

Substitute back.

$5 + y^2 = 9$

$y^2 = 4$

$y = \pm 2$

$\left\{\left(\sqrt{5}, 2\right), \left(\sqrt{5}, -2\right), \left(-\sqrt{5}, 2\right), \left(-\sqrt{5}, -2\right)\right\}$

9. $\begin{cases} x^2 + 2y^2 = 2 \\ x - y = 2 \end{cases}$

$x = y + 2$

Substitute.

$(y + 2)^2 + 2y^2 = 2$

$y^2 + 4y + 4 + 2y^2 = 2$

$3y^2 + 4y + 4 = 2$

$3y^2 + 4y + 2 = 0$

$b^2 - 4ac = 4^2 - 4(3)(2)$

$b^2 - 4ac = 16 - 24 = -8 < 0$

Therefore, no real solutions exist.

{ }

11. $\begin{cases} y = x^2 - 3 \\ 4x - y = 6 \end{cases}$

Substitute.

$4x - (x^2 - 3) = 6$

$4x - x^2 + 3 = 6$
$0 = x^2 - 4x + 3$
$0 = (x-3)(x-1)$
$x - 3 = 0$ or $x - 1 = 0$
$x = 3$ or $x = 1$
$x = 3$ $y = 3^2 - 3 = 9 - 3 = 6$
$x = 1$ $y = 1^2 - 3 = 1 - 3 = -2$
$\{(3, 6), (1, -2)\}$

13. $\begin{cases} y = x^2 \\ 3x + y = 10 \end{cases}$
Substitute.
$3x + x^2 = 10$
$x^2 + 3x - 10 = 0$
$(x+5)(x-2) = 0$
$x + 5 = 0$ or $x - 2 = 0$
$x = -5$ or $x = 2$
$x = -5$ $y = (-5)^2 = 25$
$x = 2$ $y = 2^2 = 4$
$\{(-5, 25), (2, 4)\}$

15. $\begin{cases} y = 2x^2 + 1 \\ x + y = -1 \end{cases}$
Substitute.
$x + 2x^2 + 1 = -1$
$2x^2 + x + 1 = -1$
$2x^2 + x + 2 = 0$
$b^2 - 4ac = 1^2 - 4(2)(2)$
$b^2 - 4ac = 1 - 16 = -15 < 0$
Therefore, no real solutions exist.
$\{\ \}$

17. $\begin{cases} y = x^2 - 4 \\ y = x^2 - 4x \end{cases}$
Substitute.
$x^2 - 4 = x^2 - 4x$
$-4 = -4x$
$x = 1$
$y = 1^2 - 4 = -3$
$\{(1, -3)\}$

19. $\begin{cases} 2x^2 + 3y^2 = 14 \\ -x^2 + y^2 = 3 \end{cases}$
$y^2 = x^2 + 3$
Substitute.
$2x^2 + 3(x^2 + 3) = 14$
$2x^2 + 3x^2 + 9 = 14$
$5x^2 + 9 = 14$
$5x^2 = 5$
$x^2 = 1$
$x = \pm 1$
Substitute back.
$y^2 = 1 + 3$
$y^2 = 4$
$y = \pm 2$
$\{(1, -2), (1, 2), (-1, -2), (-1, 2)\}$

21. $\begin{cases} x^2 + y^2 = 1 \\ x^2 + (y+3)^2 = 4 \end{cases}$
Subtract equation 1 from equation 2.
$(y+3)^2 - y^2 = 3$
$y^2 + 6y + 9 - y^2 = 3$
$6y + 9 = 3$
$6y = -6$
$y = -1$
Substitute back.
$x^2 + (-1)^2 = 1$
$x^2 + 1 = 1$
$x^2 = 0$
$x = 0$
$\{(0, -1)\}$

23. $\begin{cases} y = x^2 + 2 \\ y = -x^2 + 4 \end{cases}$
Substitute.
$x^2 + 2 = -x^2 + 4$
$2x^2 = 2$
$x^2 = 1$
$x = \pm 1$
Substitute back.
$y = 1 + 2 = 3$
$\{(1, 3), (-1, 3)\}$

25. $\begin{cases} 3x^2 + y^2 = 9 \\ 3x^2 - y^2 = 9 \end{cases}$

Subtract.

$2y^2 = 0$

$y^2 = 0$

$y = 0$

Substitute back.

$3x^2 + 0 = 9$

$3x^2 = 9$

$x^2 = 3$

$x = \pm\sqrt{3}$

$\{(\sqrt{3},\ 0),\ (-\sqrt{3},\ 0)\}$

27. $\begin{cases} x^2 + 3y^2 = 6 \\ x^2 - 3y^2 = 10 \end{cases}$

Add.

$2x^2 = 16$

$x^2 = 8$

$x = \pm\sqrt{8}$

$x = \pm 2\sqrt{2}$

Substitute back.

$8 + 3y^2 = 6$

$3y^2 = -2$

$y^2 = -\frac{2}{3}$

Therefore, no real solutions exist.

{ }

29. $\begin{cases} x^2 + y^2 = 36 \\ y = \frac{1}{6}x^2 - 6 \end{cases}$

$y + 6 = \frac{1}{6}x^2$

$x^2 = 6(y + 6)$

Substitute.

$6(y + 6) + y^2 = 36$

$6y + 36 + y^2 = 36$

$6y + y^2 = 0$

$y(6 + y) = 0$

$y = 0$ or $6 + y = 0$

$y = -6$

$y = 0$: $x^2 + 0^2 = 36$

$x^2 = 36$

$x = \pm 6$

$y = -6$ $x^2 + (-6)^2 = 36$

$x^2 + 36 = 36$

$x^2 = 0$

$x = 0$

$\{(6, 0), (-6, 0), (0, -6)\}$

31. 0, 1, 2, 3, or 4

33. $\begin{cases} x^2 + y^2 = 130 \\ x^2 - y^2 = 32 \end{cases}$

Add.

$2x^2 = 162$

$x^2 = 81$

$x = \pm 9$

Substitute back.

$9^2 + y^2 = 130$ $(-9)^2 + y^2 = 130$

$y^2 = 49$ $y^2 = 49$

$y = \pm 7$ $y = \pm 7$

Answer: 9 and 7

9 and –7

–9 and 7

–9 and –7

35. $\begin{cases} x + y = 68 \\ xy = 285 \end{cases}$

Solve for x.

$x = 68 - y$

Substitute in second equation.

$(68 - y)y = 285$

$68y - y^2 = 285$

$0 = y^2 - 68y + 285$

$0 = (y - 19)(y - 15)$

$y = 19$ or $y = 15$

Substitute back.

$x + 19 = 68$

$x = 15$

15 cm by 19 cm

37. $p = -0.01x^2 - 0.2x + 9$

$p = 0.01x^2 - 0.1x + 3$

$-0.01x^2 - 0.2x + 9 = 0.01x^2 - 0.1x + 3$

$0 = 0.02x^2 + 0.1x - 6$
$0 = x^2 + 5x - 300$
$0 = (x + 20)(x - 15)$
$x + 20 = 0$ or $x - 15 = 0$
$x = -20$ or $x = 15$
Disregard the negative
$p = -0.01(15)^2 - 0.2(15) + 9$
$p = 3.75$
15 thousand compact discs
price, $3.75

39. The results are the same.

41. The results are the same.

43. $x > -3$

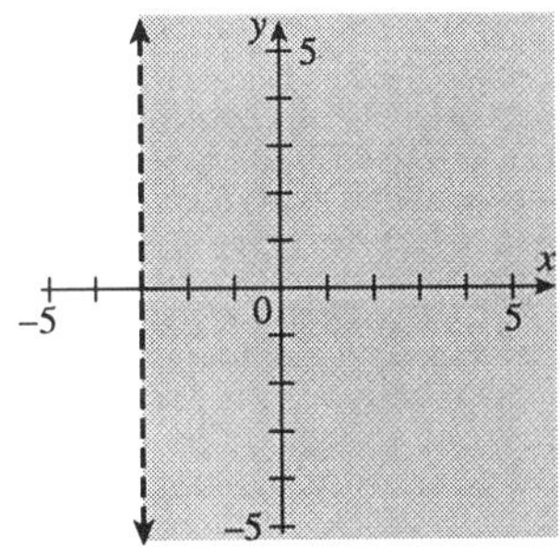

45. $y < 2x - 1$

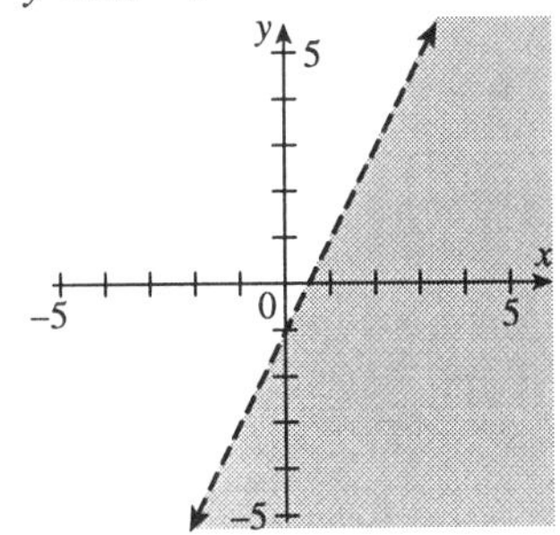

47. $P = x + (2x - 5) + (5x - 20)$
$P = (8x - 25)$ inches

49. $P = 2(x^2 + 3x + 1) + 2x^2$
$P = 2x^2 + 6x + 2 + 2x^2$
$P = (4x^2 + 6x + 2)$ meters

Exercise Set 9.4

1. $y < x^2$

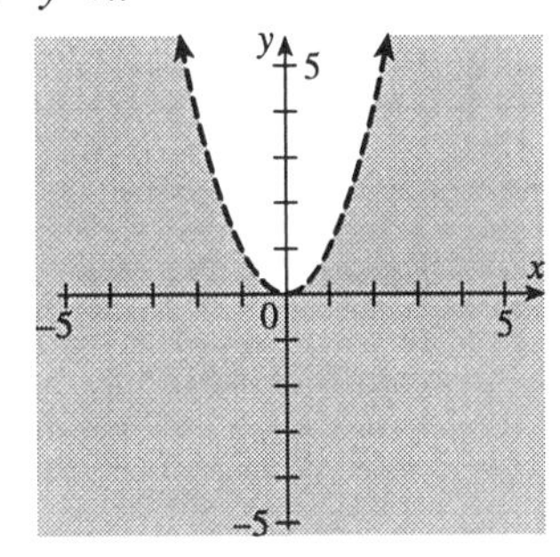

3. $x^2 + y^2 \geq 16$

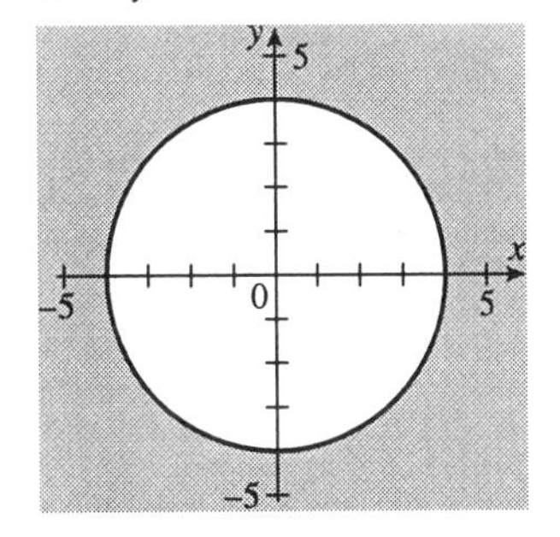

5. $\frac{x^2}{4} - y^2 < 1$

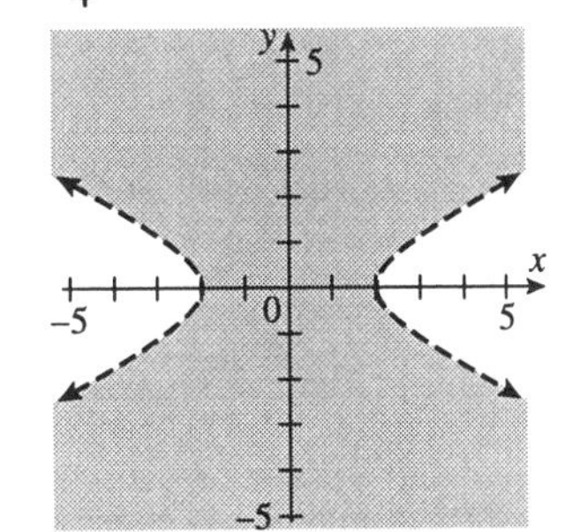

7. $y > (x-1)^2 - 3$

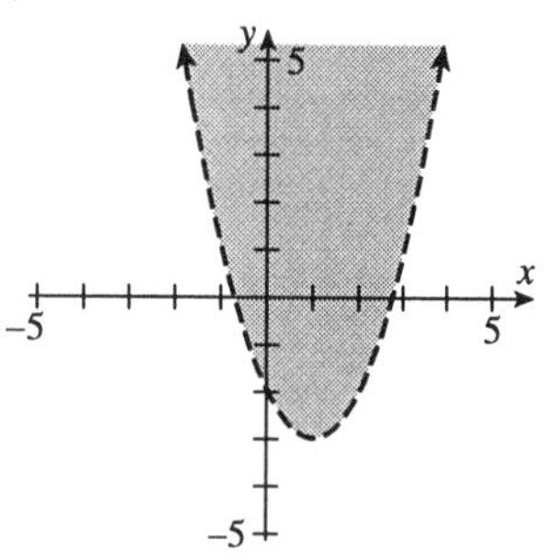

9. $x^2 + y^2 \le 9$

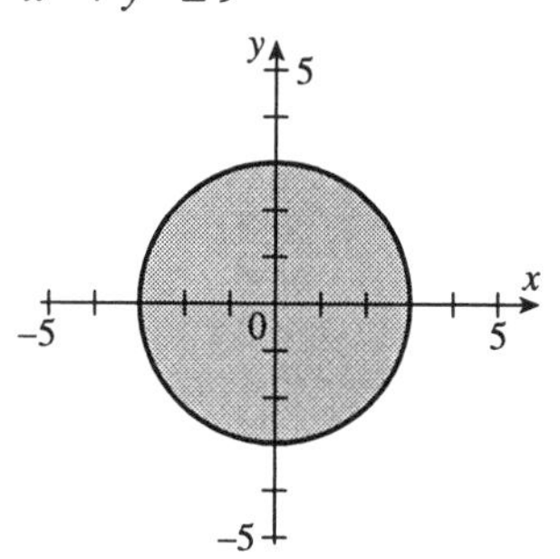

11. $y > -x^2 + 5$

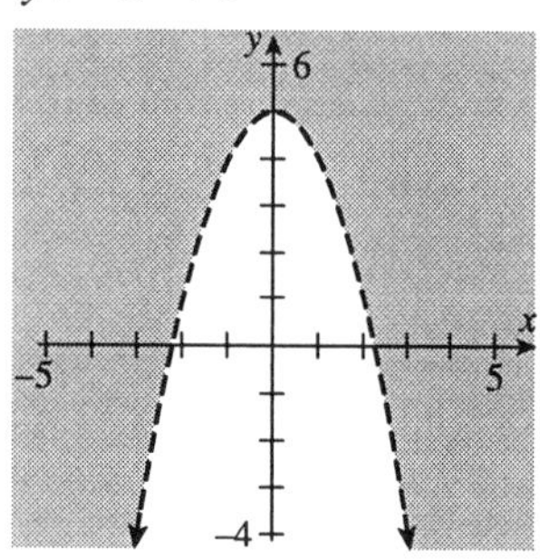

13. $\frac{x^2}{4} + \frac{y^2}{9} \le 1$

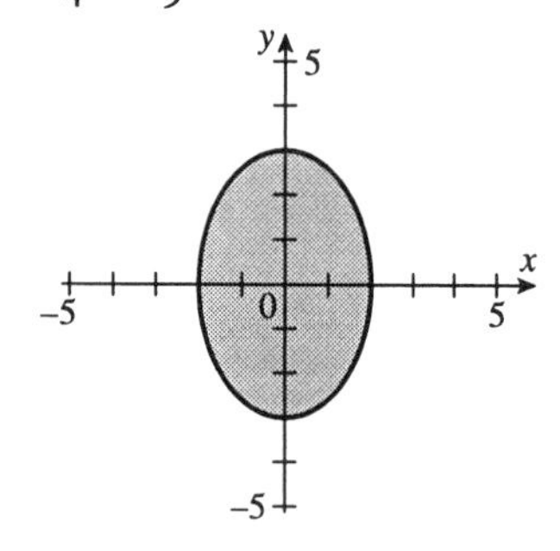

15. $\frac{y^2}{4} - x^2 \le 1$

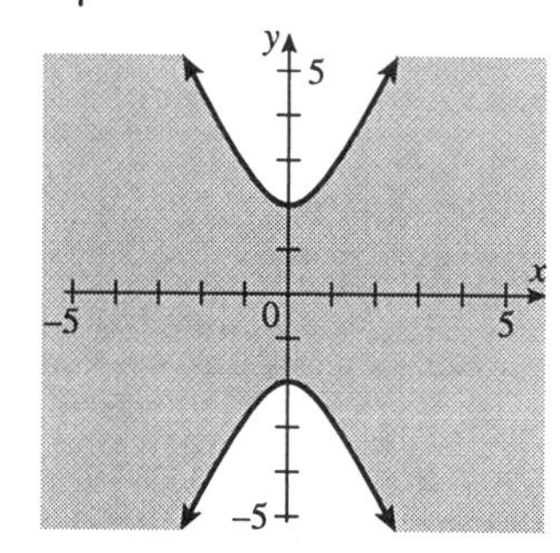

17. $y < (x-2)^2 + 1$

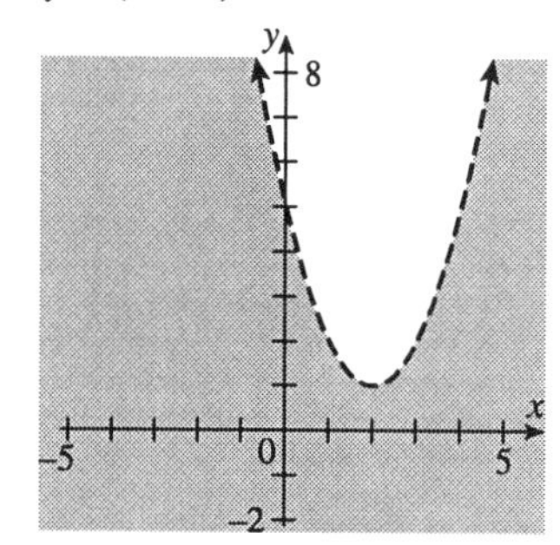

19. $y \le x^2 + x - 2$

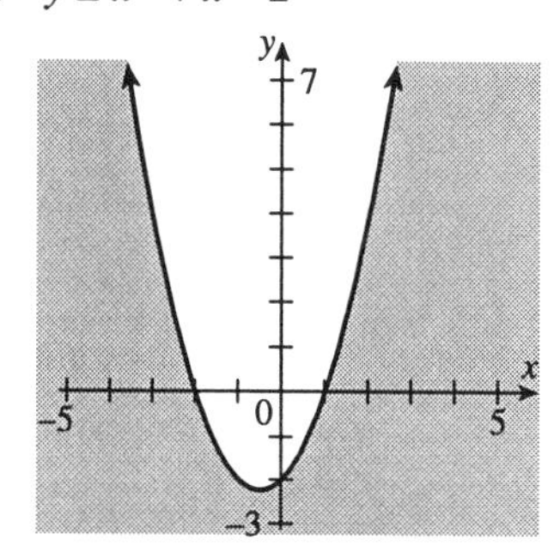

21. Answers may vary.

23. $\begin{cases} 2x - y < 2 \\ y \le -x \end{cases}$

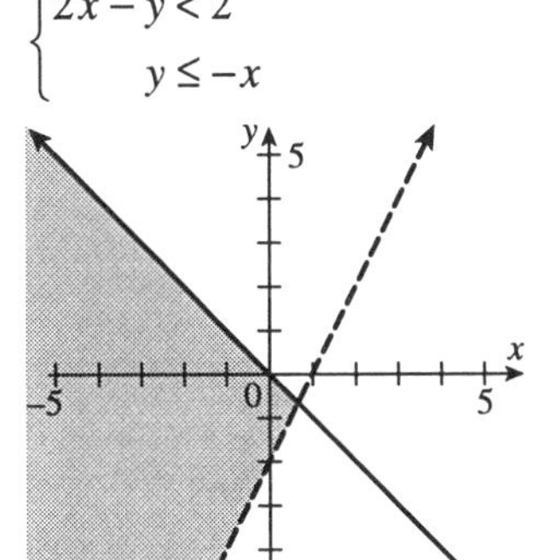

25. $\begin{cases} 4x + 3y \ge 12 \\ x^2 + y^2 < 16 \end{cases}$

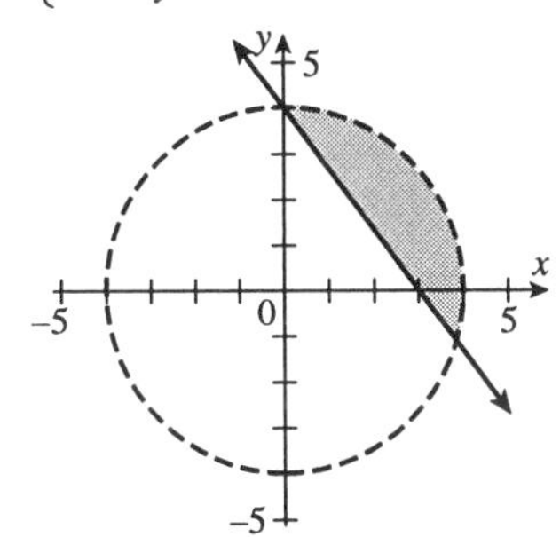

27. $\begin{cases} x^2 + y^2 \le 9 \\ x^2 + y^2 \ge 1 \end{cases}$

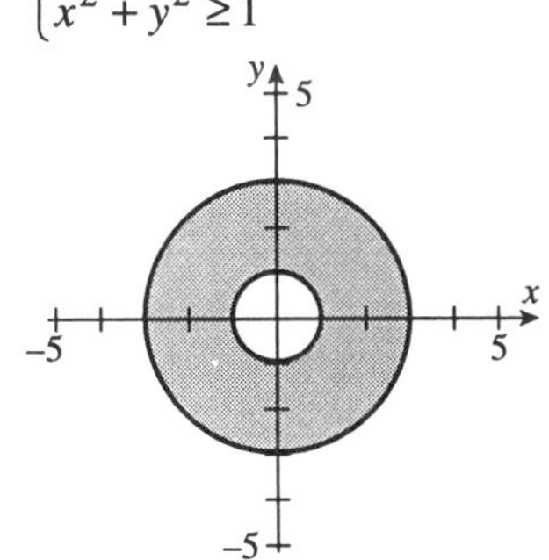

29. $\begin{cases} y > x^2 \\ y \ge 2x + 1 \end{cases}$

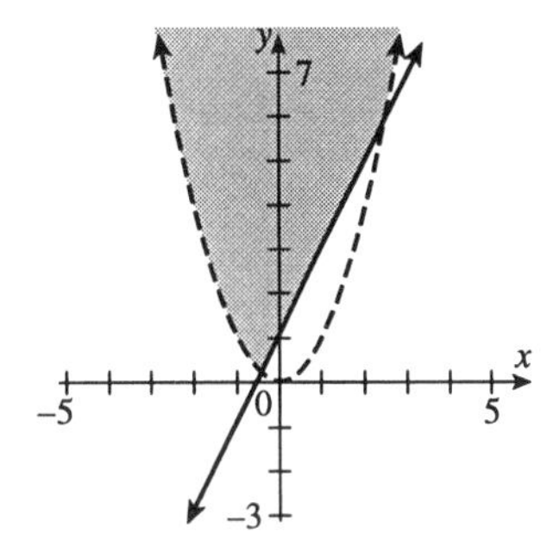

31. $\begin{cases} x > y^2 \\ y > 0 \end{cases}$

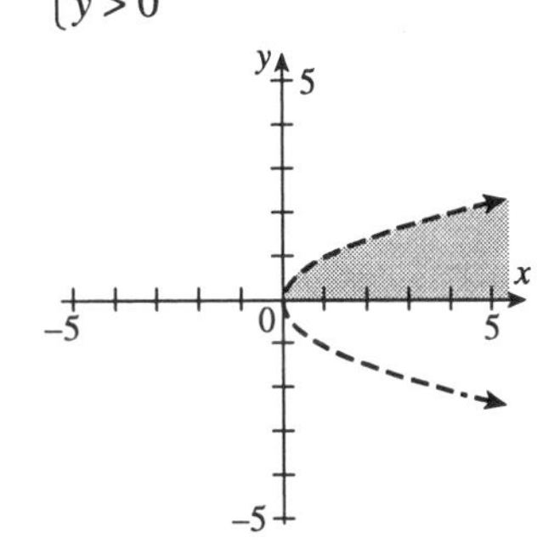

33. $\begin{cases} x^2 + y^2 > 9 \\ y > x^2 \end{cases}$

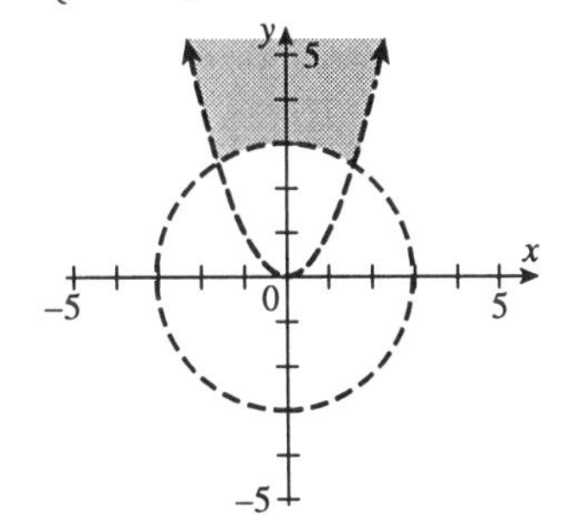

35. $\begin{cases} \dfrac{x^2}{4}+\dfrac{y^2}{9} \geq 1 \\ x^2+y^2 \geq 4 \end{cases}$

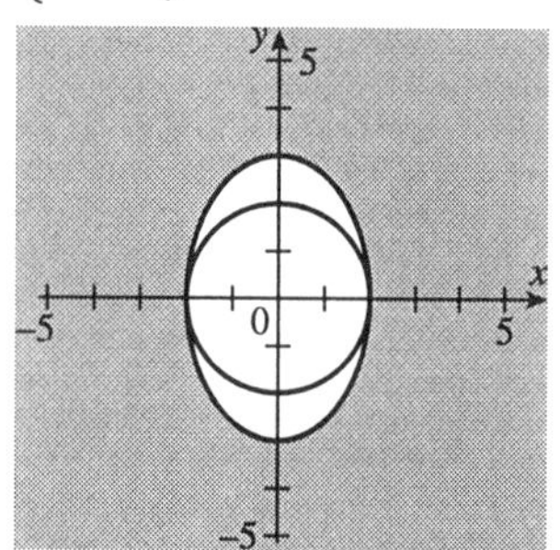

37. $\begin{cases} x^2-y^2 \geq 1 \\ y \geq 0 \end{cases}$

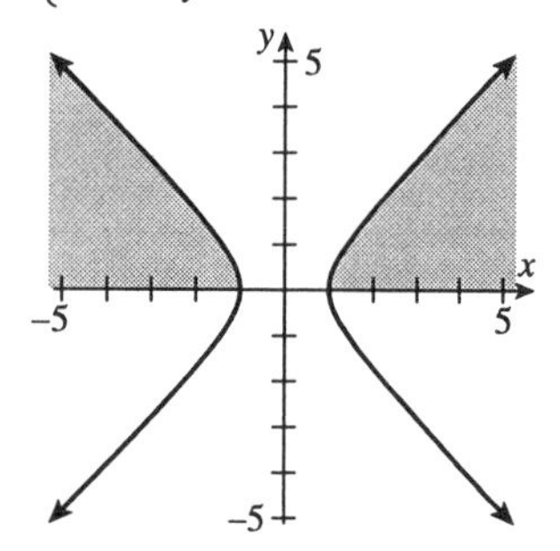

39. $\begin{cases} x+y \geq 1 \\ 2x+3y < 1 \\ x > -3 \end{cases}$

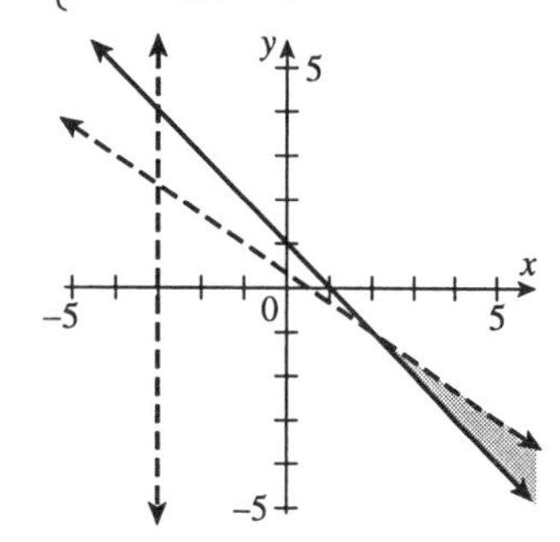

41. $\begin{cases} x^2-y^2 < 1 \\ \dfrac{x^2}{16}+y^2 \leq 1 \\ x \geq -2 \end{cases}$

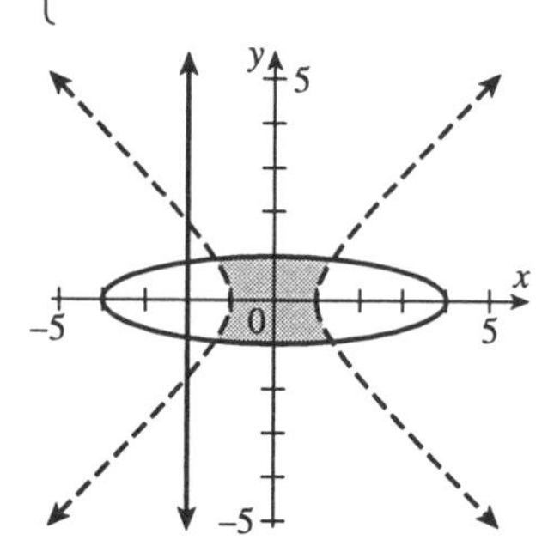

43. This is not a function because a vertical line can cross the graph in two places.

45. This is a function because no vertical line can cross the graph in more than one place.

47. $f(x)=3x^2-2$
$f(-1)=3(-1)^2-2$
$f(-1)=1$

49. $f(x)=3x^2-2$
$f(a)=3a^2-2$

Chapter 9 - Review

1. (–6, 3) and (8, 4)
$d=\sqrt{[8-(-6)]^2+(4-3)^2}$
$d=\sqrt{(14)^2+1^2}$
$d=\sqrt{197}$ units

2. (3, 5) and (8, 9)
$d=\sqrt{(8-3)^2+(9-5)^2}$
$d=\sqrt{5^2+4^2}$
$d=\sqrt{25+16}$
$d=\sqrt{41}$ units

3. $(-4, -6)$ and $(-1, 5)$
$d = \sqrt{[-1-(-4)]^2 + [5-(-6)]^2}$
$d = \sqrt{3^2 + 11^2}$
$d = \sqrt{9+121} = \sqrt{130}$ units

4. $(-1, 5)$ and $(2, -3)$
$d = \sqrt{[2-(-1)]^2 + (-3-5)^2}$
$d = \sqrt{3^2 + (-8)^2}$
$d = \sqrt{9+64} = \sqrt{73}$ units

5. $(-\sqrt{2},\ 0)$ and $(0,\ -4\sqrt{6})$
$d = \sqrt{(0+\sqrt{2})^2 + (-4\sqrt{6}-0)^2}$
$d = \sqrt{(\sqrt{2})^2 + (-4\sqrt{6})^2}$
$d = \sqrt{2+96}$
$d = \sqrt{98}$
$d = 7\sqrt{2}$ units

6. $(-\sqrt{5},\ -\sqrt{11})$ and $(-\sqrt{5},\ -3\sqrt{11})$
$d = \sqrt{(-\sqrt{5}+\sqrt{5})^2 + (-3\sqrt{11}+\sqrt{11})^2}$
$d = \sqrt{0 + (-2\sqrt{11})^2}$
$d = 2\sqrt{11}$ units

7. $(7.4, -8.6)$ and $(-1.2, 5.6)$
$d = \sqrt{(-1.2-7.4)^2 + (5.6+8.6)^2}$
$d = \sqrt{(-8.6)^2 + (14.2)^2}$
$d = \sqrt{275.6}$
$d \approx 16.60$ units

8. $(2.3, 1.8)$ and $(10.7, -9.2)$
$d = \sqrt{(10.7-2.3)^2 + (-9.2-1.8)^2}$
$d = \sqrt{(8.4)^2 + (11)^2}$
$d = \sqrt{191.56} \approx 13.84$ units

9. $(2, 6)$ and $(-12, 4)$
$\left(\frac{2-12}{2},\ \frac{6+4}{2}\right)$
$(-5, 5)$

10. $(-3, 8)$ and $(11, 24)$
$\left(\frac{-3+11}{2},\ \frac{8+24}{2}\right)$
$(4, 16)$

11. $(-6, -5)$ and $(-9, 7)$
$\left(\frac{-6-9}{2},\ \frac{-5+7}{2}\right)$
$\left(-\frac{15}{2},\ 1\right)$

12. $(4, -6)$ and $(-15, 2)$
$\left(\frac{4-15}{2},\ \frac{-6+2}{2}\right)$
$\left(-\frac{11}{2},\ -2\right)$

13. $\left(0,\ -\frac{3}{8}\right)$ and $\left(\frac{1}{10},\ 0\right)$
$\left(\frac{0+\frac{1}{10}}{2},\ \frac{-\frac{3}{8}+0}{2}\right)$
$\left(\frac{1}{20},\ -\frac{3}{16}\right)$

14. $\left(\frac{3}{4},\ -\frac{1}{7}\right)$ and $\left(-\frac{1}{4},\ -\frac{3}{7}\right)$
$\left(\frac{\frac{3}{4}-\frac{1}{4}}{2},\ \frac{-\frac{1}{7}-\frac{3}{7}}{2}\right)$
$\left(\frac{1}{4},\ -\frac{2}{7}\right)$

15. $(\sqrt{3},\ -2\sqrt{6})$ and $(\sqrt{3},\ -4\sqrt{6})$
$\left(\frac{\sqrt{3}+\sqrt{3}}{2},\ \frac{-2\sqrt{6}-4\sqrt{6}}{2}\right)$
$(\sqrt{3},\ -3\sqrt{6})$

16. $(-5\sqrt{3},\ 2\sqrt{7})$ and $(-3\sqrt{3},\ 10\sqrt{7})$
$\left(\frac{-5\sqrt{3}-3\sqrt{3}}{2},\ \frac{2\sqrt{7}+10\sqrt{7}}{2}\right)$
$(-4\sqrt{3},\ 6\sqrt{7})$

17. $C(-4, 4)$, $r = 3$
$(x-(-4))^2 + (y-4)^2 = 3^2$
or $(x+4)^2 + (y-4)^2 = 9$

18. $C(5, 0)$, diameter 10, $r = 5$
$(x-h)^2 + (y-k)^2 = r^2$
$(x-5)^2 + y^2 = 25$

19. $C(-7, -9)$, $r = \sqrt{11}$
$(x-(-7))^2 + (y-(-9))^2 = \sqrt{11}^2$
$(x+7)^2 + (y+9)^2 = 11$

20. $C(0, 0)$, diameter 7, $r = \frac{7}{2}$
$(x-h)^2 + (y-k)^2 = r^2$
$x^2 + y^2 = \frac{49}{4}$

21. $x^2 + y^2 = 7$
or $(x-0)^2 + (y-0)^2 = \sqrt{7}^2$
$C(0, 0)$

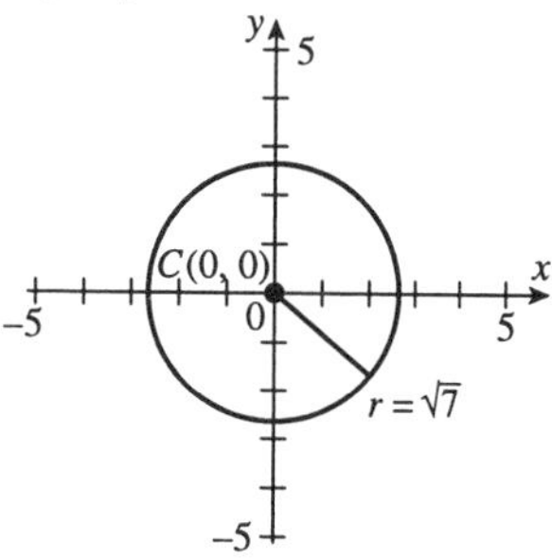

22. $x = 2(y-5)^2 + 4$
vertex (4, 5)

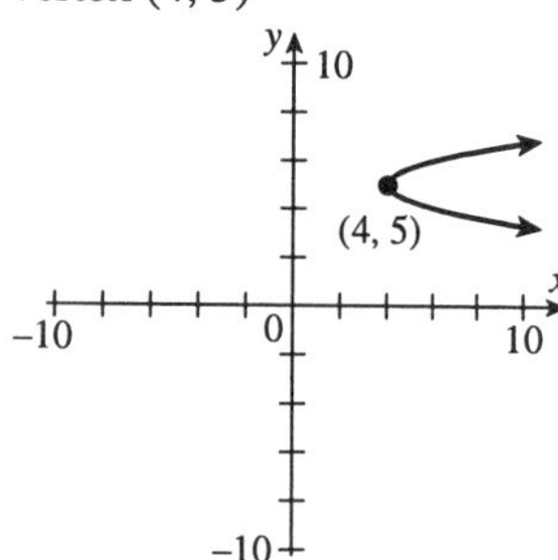

23. $x = -(y+2)^2 + 3$
$V(3, -2)$

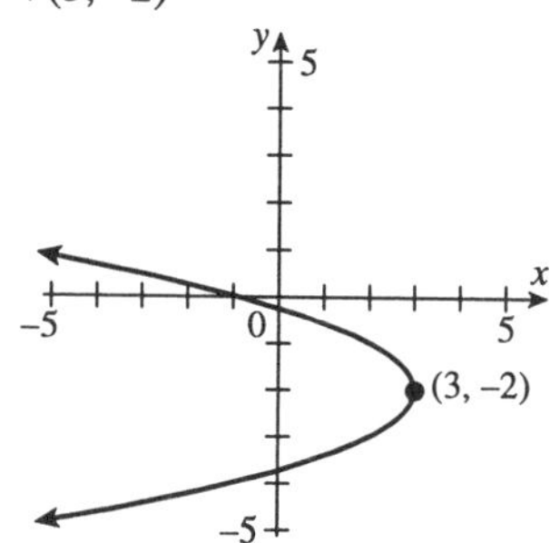

24. $(x-1)^2 + (y-2)^2 = 4$
$C(1, 2)$, $r = 2$

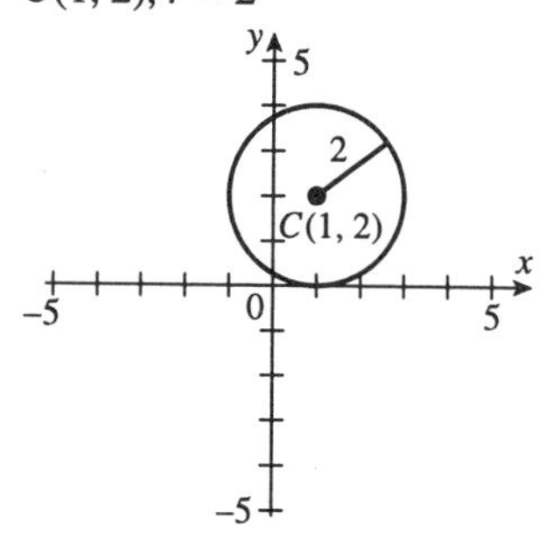

25. $y = -x^2 + 4x + 10$
$y = -(x^2 - 4x) + 10$
$y = -\left(x^2 - 4x + \left(\frac{4}{2}\right)^2\right) + 10 + 4$
$y = -(x-2)^2 + 14$
$V(2, 14)$

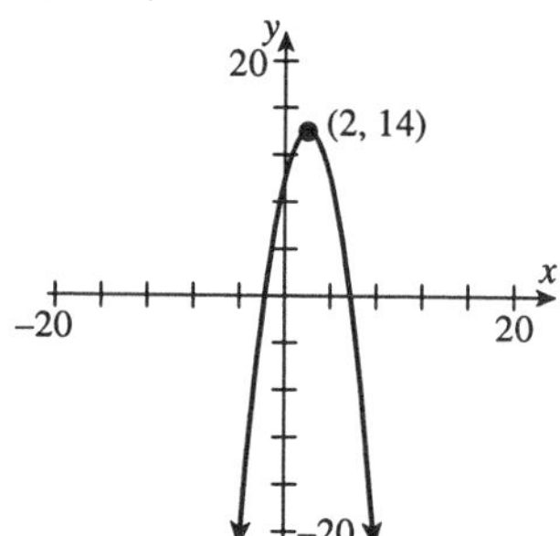

26. $x = -y^2 - 4y + 6$
$\frac{-b}{2a} = \frac{4}{2(-1)} = -2$

$x = -(-2)^2 - 4(-2) + 6 = 10$

$V(10, -2)$

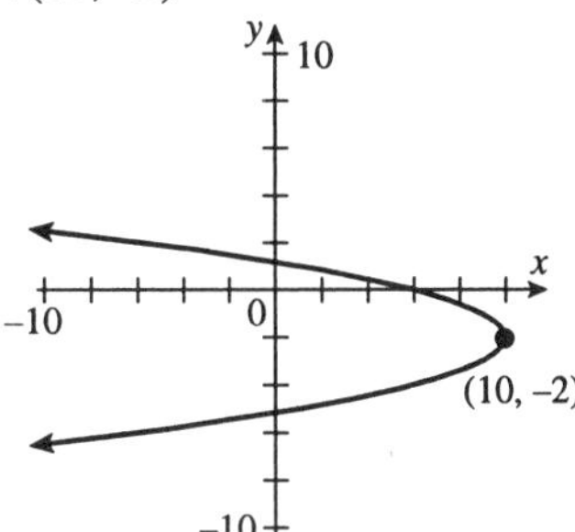

27. $x = \frac{1}{2}y^2 + 2y + 1$

$x = \frac{1}{2}(y^2 + 4y) + 1$

$x = \frac{1}{2}\left(y^2 + 4y + \left(\frac{4}{2}\right)^2\right) + 1 - 2$

$x = \frac{1}{2}(y+2)^2 - 1$

$V(-1, -2)$

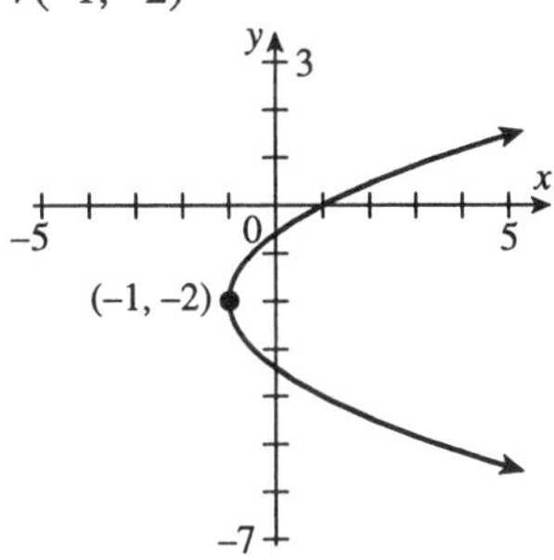

28. $y = -3x^2 + \frac{1}{2}x + 4$

$\frac{-b}{2a} = \frac{-\frac{1}{2}}{2(-3)} = \frac{1}{12}$

$y = -3\left(\frac{1}{12}\right)^2 + \frac{1}{2}\left(\frac{1}{12}\right) + 4 = \frac{193}{48}$

$V\left(\frac{1}{12}, \frac{193}{48}\right)$

29. $x^2 + y^2 + 2x + y = \frac{3}{4}$

$x^2 + 2x + \left(\frac{2}{2}\right)^2 + y^2 + y + \left(\frac{1}{2}\right)^2$

$= \frac{3}{4} + 1 + \frac{1}{4}$

$(x+1)^2 + \left(y + \frac{1}{2}\right)^2 = 2$

or $(x+1)^2 + \left(y + \frac{1}{2}\right)^2 = \sqrt{2}^2$

$C\left(-1, -\frac{1}{2}\right)$

30. $x^2 + y^2 + 3y = \frac{7}{4}$

$x^2 + y^2 + 3y + \frac{9}{4} = \frac{7}{4} + \frac{9}{4}$

$x^2 + \left(y + \frac{3}{2}\right)^2 = 4$

$C\left(0, -\frac{3}{2}\right), r = 2$

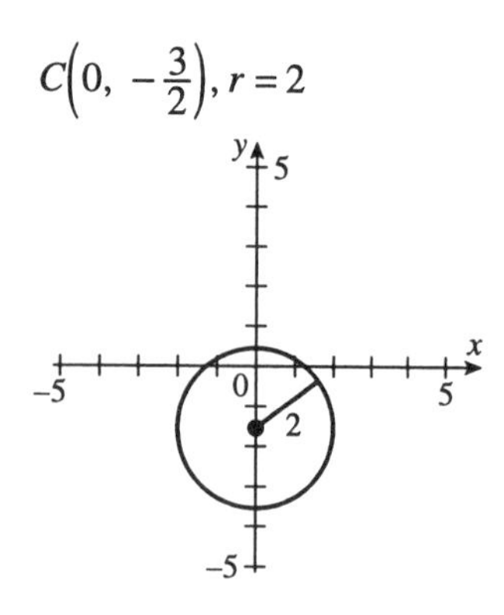

31. $4x^2 + 4y^2 + 16x + 8y = 1$

$4x^2 + 16x + 4y^2 + 8y = 1$

$4(x^2 + 4x) + 4(y^2 + 2y) = 1$

$4\left(x^2 + 4x + \left(\frac{4}{2}\right)^2\right) + 4\left(y^2 + 2y + \left(\frac{2}{2}\right)^2\right)$

$= 1 + 16 + 4$

$4(x+2)^2 + 4(y+1)^2 = 21$

$(x+2)^2 + (y+1)^2 = \frac{21}{4}$

or $(x+2)^2 + (y+1)^2 = \left(\frac{\sqrt{21}}{2}\right)^2$

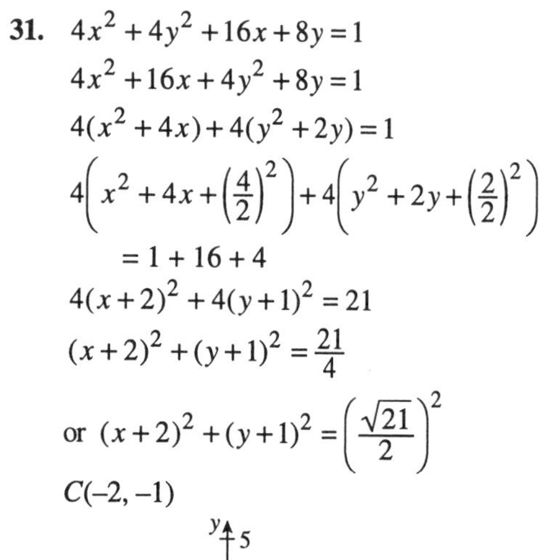

$C(-2, -1)$

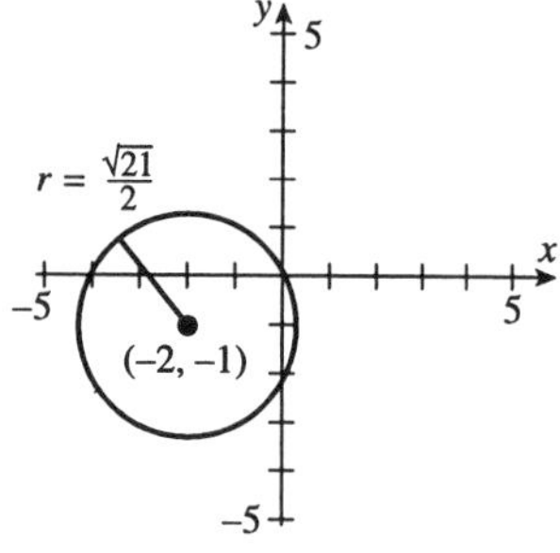

32. $3x^2 + 6x + 3y^2 = 9$

$x^2 + 2x + y^2 = 3$

$x^2 + 2x + 1 + y^2 = 3 + 1$

$(x+1)^2 + y^2 = 4$

$C(-1, 0), r = 2$

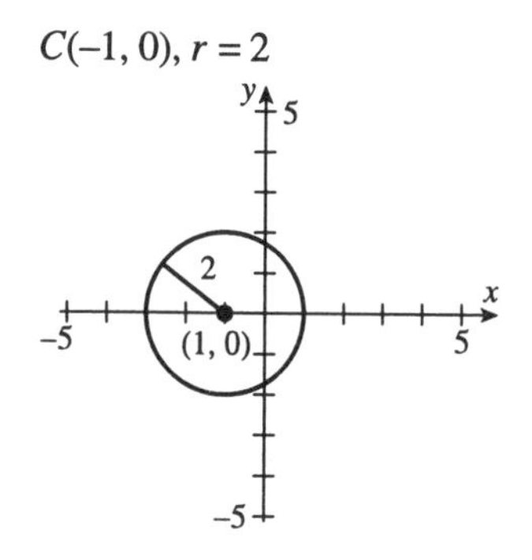

33. $y = x^2 + 6x + 9$

$y = (x+3)^2 + 0$

$V(-3, 0)$

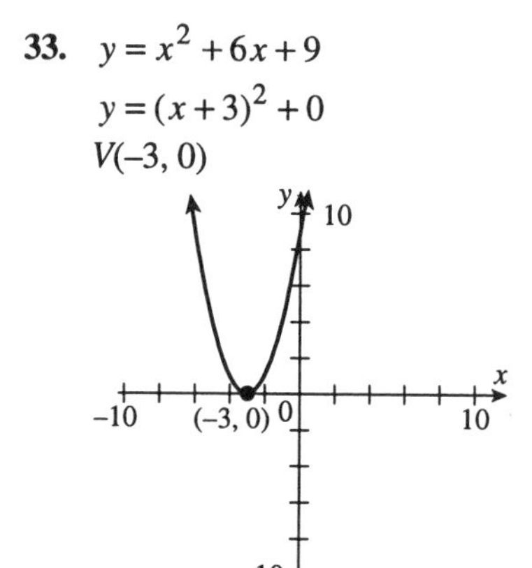

34. $x = y^2 + 6y + 9$

$x = (y+3)^2$

$V(0, -3)$

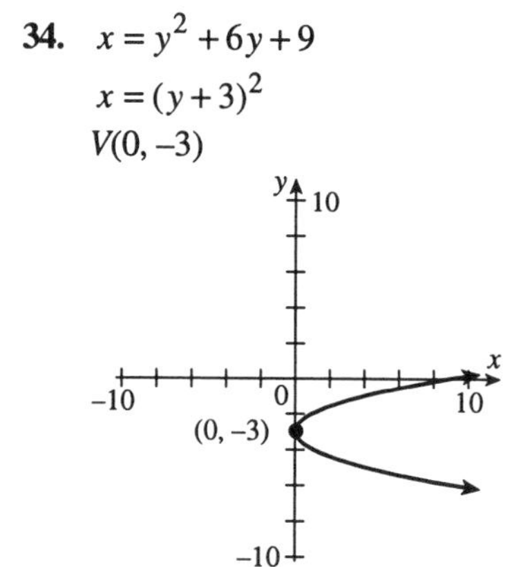

35. $C(5.6, -2.4), \ r = \frac{6.2}{2} = 3.1$

$(x - 5.6)^2 + (y - (-2.4))^2 = 3.1^2$

or $(x - 5.6)^2 + (y + 2.4)^2 = 9.61$

36. $x^2 + \frac{y^2}{4} = 1$

$a = 2,\ b = 1,\ C(0, 0)$

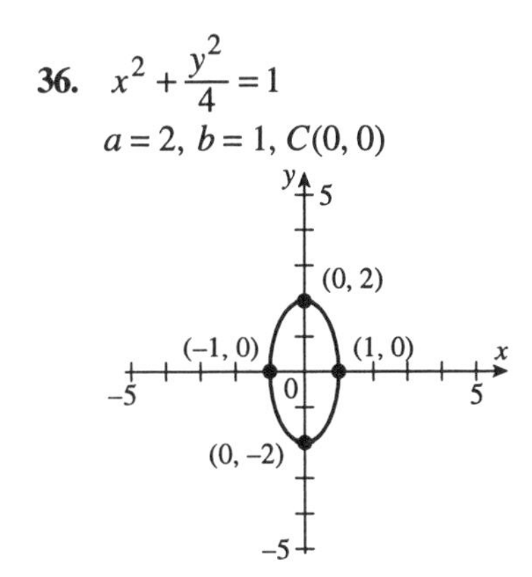

37. $x^2 - \frac{y^2}{4} = 1$

or $\frac{(x-0)^2}{1^2} - \frac{(y-0)^2}{2^2} = 1$

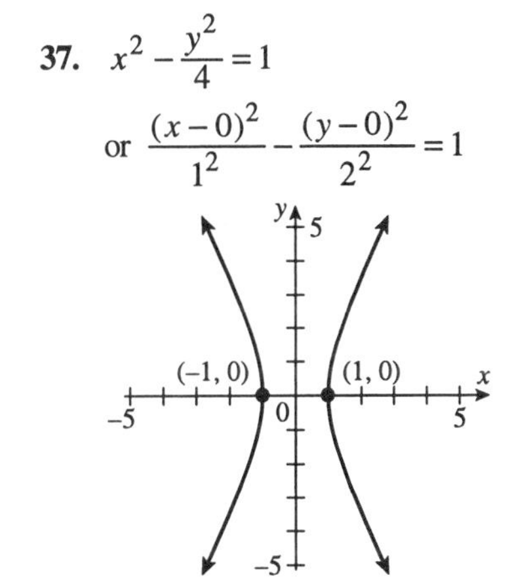

38. $\frac{y^2}{4} - \frac{x^2}{16} = 1$

$a = 2,\ b = 4,\ C(0, 0)$

39. $\frac{y^2}{4} + \frac{x^2}{16} = 1$

or $\frac{(x-0)^2}{4^2} + \frac{(y-0)^2}{2^2} = 1$

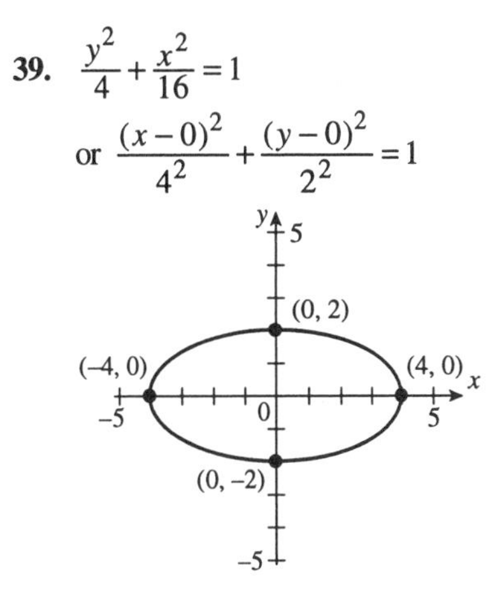

40. $\frac{x^2}{5} + \frac{y^2}{5} = 1$

$C(0, 0),\ r = \sqrt{5}$

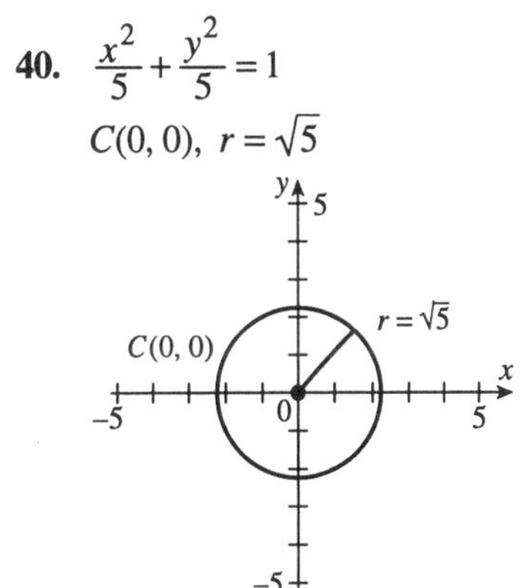

41. $\frac{x^2}{5} - \frac{y^2}{5} = 1$

or $\frac{(x-0)^2}{\sqrt{5}^2} - \frac{(y-0)^2}{\sqrt{5}^2} = 1$

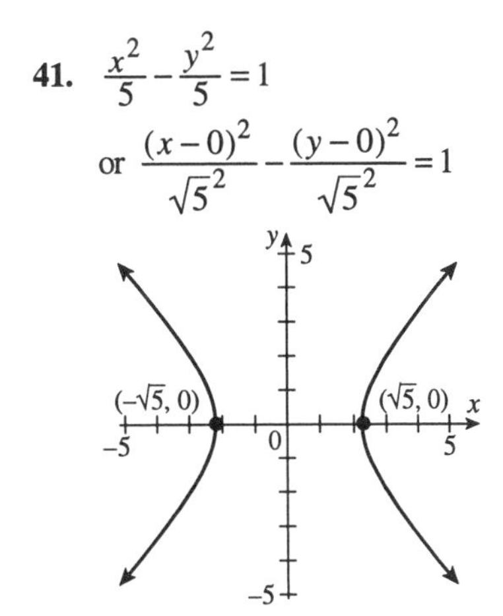

42. $-5x^2 + 25y^2 = 125$

$\frac{y^2}{5} - \frac{x^2}{25} = 1$

$a=\sqrt{5},\ b=5, C(0, 0)$

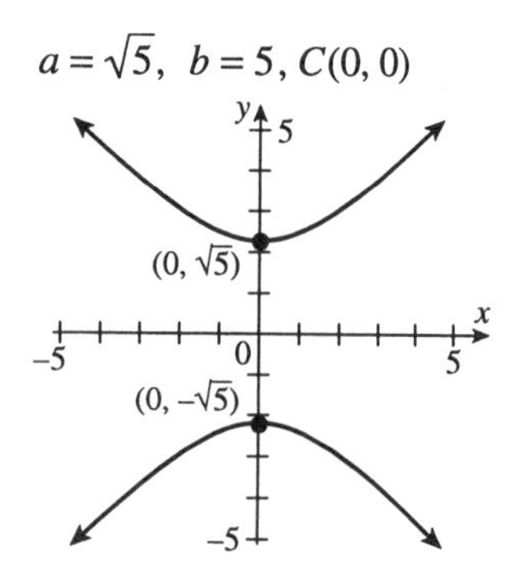

43. $4y^2+9x^2=36$

$\frac{y^2}{9}+\frac{x^2}{4}=1$

$\frac{(x-0)^2}{2^2}+\frac{(y-0)^2}{3^2}=1$

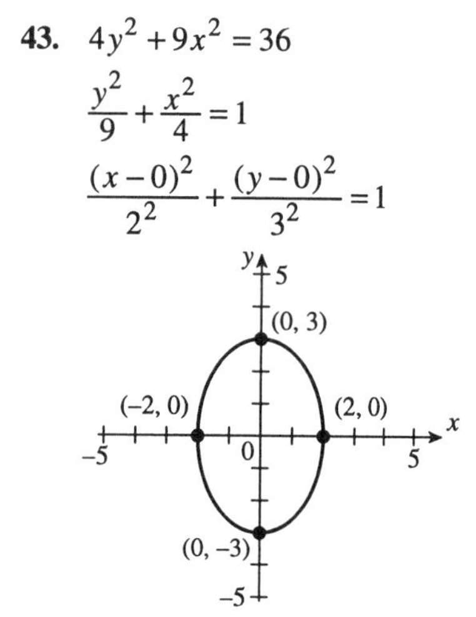

44. $\frac{(x-2)^2}{4}+(y-1)^2=1$

$a=2,\ b=1,\ C(2, 1)$

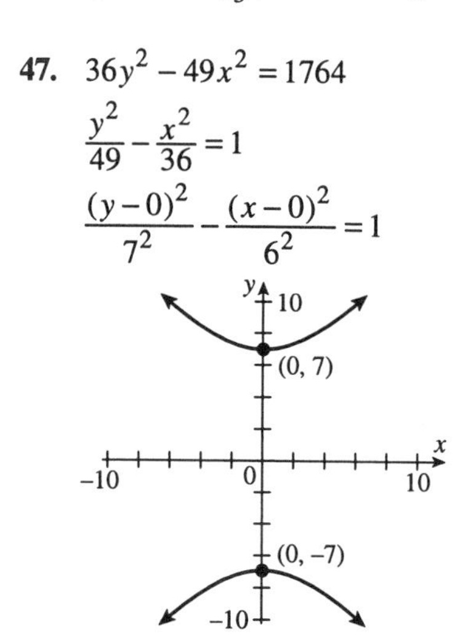

45. $\frac{(x+3)^2}{9}+\frac{(y-4)^2}{25}=1$

$\frac{(x+3)^2}{3^2}+\frac{(y-4)^2}{5^2}=1$

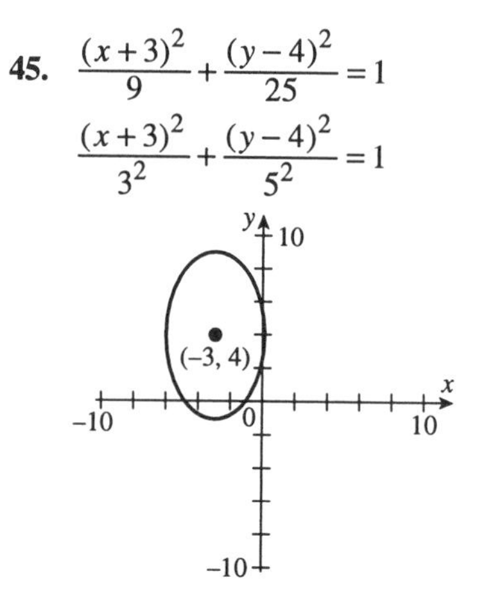

46. $x^2-y^2=1$

$a=1,\ b=1,\ C(0, 0)$

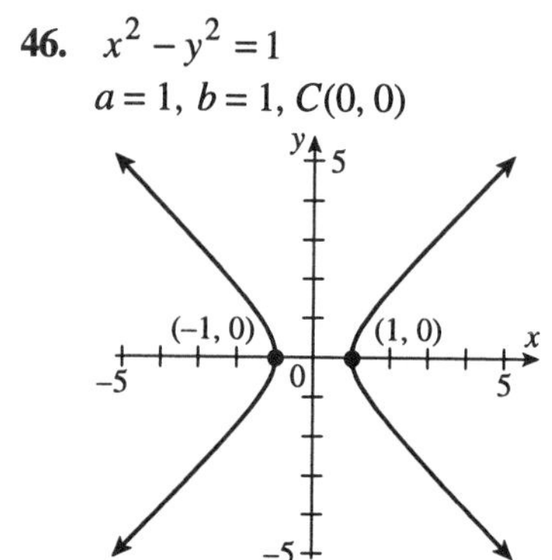

47. $36y^2-49x^2=1764$

$\frac{y^2}{49}-\frac{x^2}{36}=1$

$\frac{(y-0)^2}{7^2}-\frac{(x-0)^2}{6^2}=1$

y
10
(0, 7)
x
−10
0
10
(0, −7)
−10

48. $y^2=x^2+9$

$\frac{y^2}{9}-\frac{x^2}{9}=1$

$a = 3,\ b = 3,\ C(0, 0)$

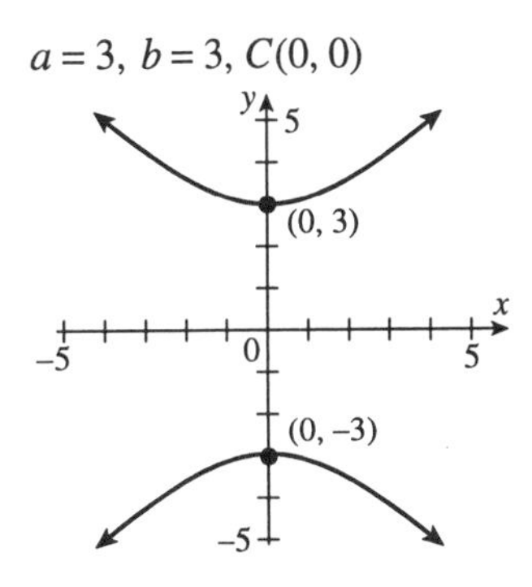

49. $x^2 = 4y^2 - 16$

$16 = 4y^2 - x^2$

$1 = \frac{y^2}{4} - \frac{x^2}{16}$

$\frac{(y-0)^2}{2^2} - \frac{(x-0)^2}{4^2} = 1$

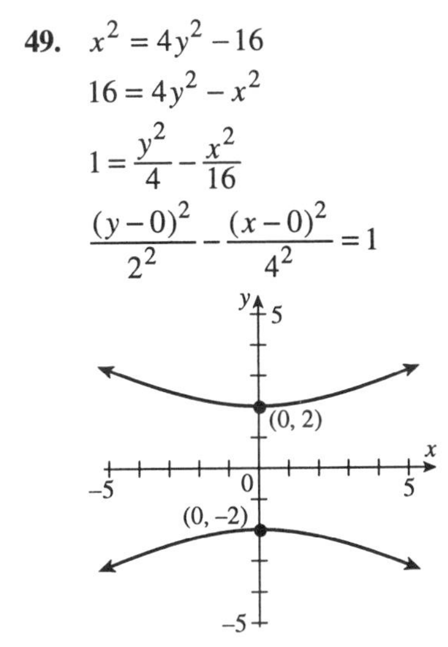

50. $100 - 25x^2 = 4y^2$

$\frac{x^2}{4} + \frac{y^2}{25} = 1$

$a = 5,\ b = 2,\ C(0, 0)$

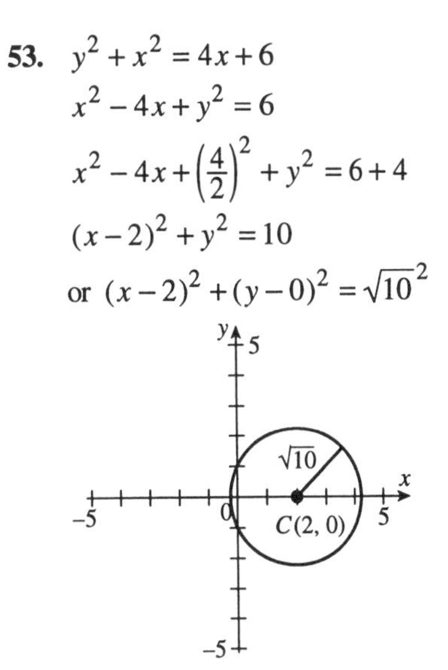

51. $y = x^2 + 4x + 6$

$= x^2 + 4x + \left(\frac{4}{2}\right)^2 + 6 - 4$

$y = (x+2)^2 + 2$

$V(-2, 2)$

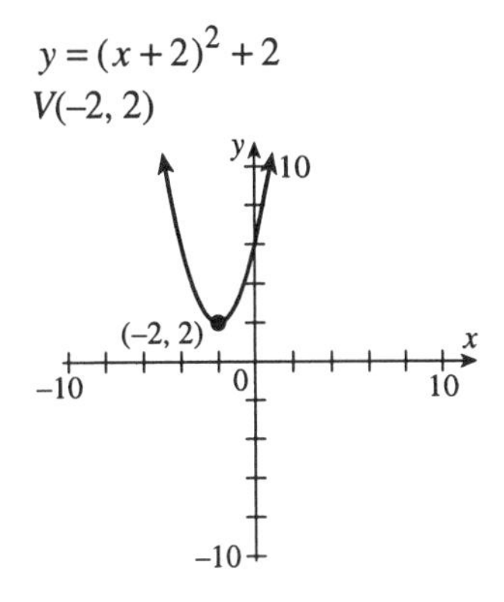

52. $y^2 = x^2 + 6$

$\frac{y^2}{6} - \frac{x^2}{6} = 1$

$a = \sqrt{6},\ b = \sqrt{6},\ C(0, 0)$

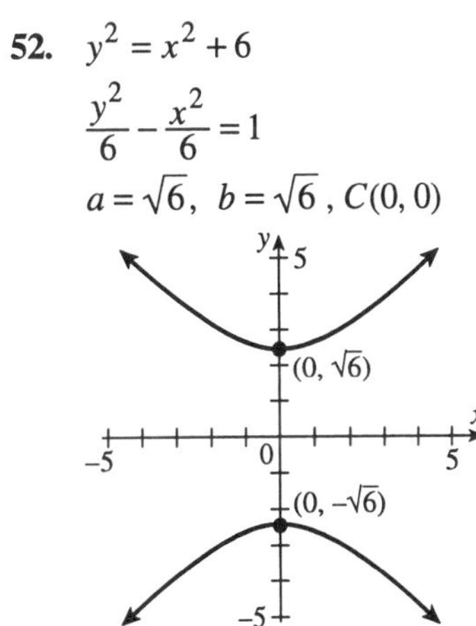

53. $y^2 + x^2 = 4x + 6$

$x^2 - 4x + y^2 = 6$

$x^2 - 4x + \left(\frac{4}{2}\right)^2 + y^2 = 6 + 4$

$(x-2)^2 + y^2 = 10$

or $(x-2)^2 + (y-0)^2 = \sqrt{10}^2$

54. $y^2 + 2x^2 = 4x + 6$

$y^2 + 2(x^2 - 2x + 1) = 6 + 2$

$y^2 + 2(x-1)^2 = 8$

$\frac{y^2}{8} + \frac{(x-1)^2}{4} = 1$

$a = 2\sqrt{2},\ b = 2,\ C(1, 0)$

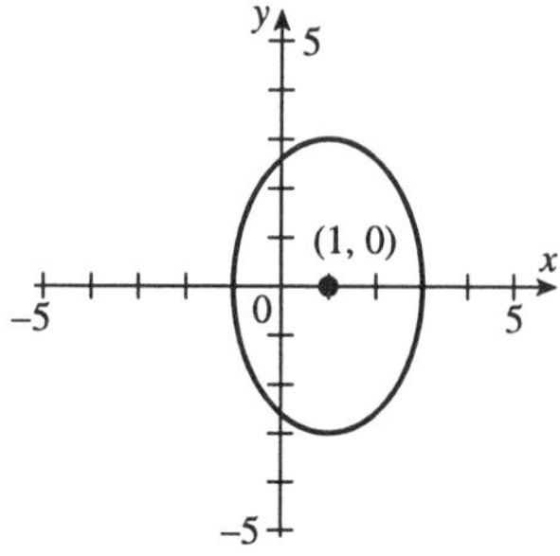

55. $x^2 + y^2 - 8y = 0$

$x^2 + y^2 - 8y + \left(\frac{8}{2}\right)^2 = 0 + 16$

$x^2 + (y-4)^2 = 16$

or $(x-0)^2 + (y-4)^2 = 4^2$

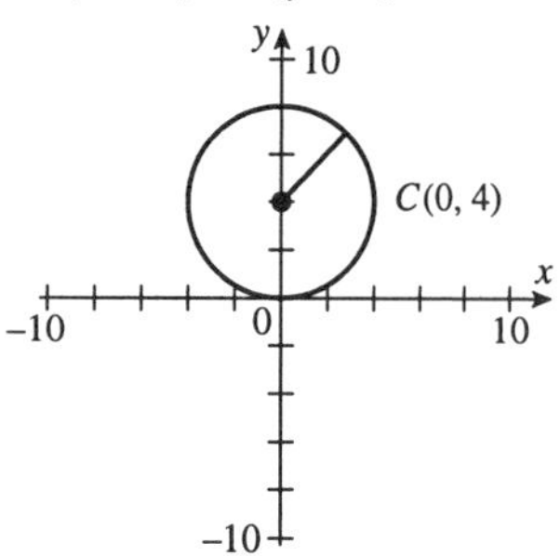

56. $x - 4y = y^2$

$x = y^2 + 4y + 4 - 4$

$x = (y+2)^2 - 4$

$V(-4, -2)$

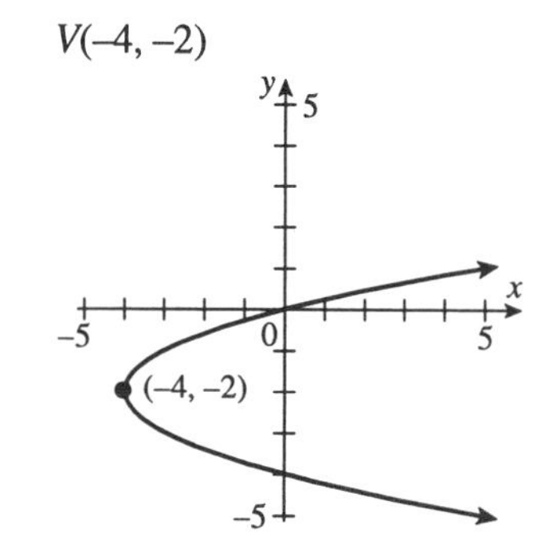

57. $x^2 - 4 = y^2$

$x^2 - y^2 = 4$

$\frac{x^2}{4} - \frac{y^2}{4} = 1$

or $\frac{(x-0)^2}{2^2} - \frac{(y-0)^2}{2^2} = 1$

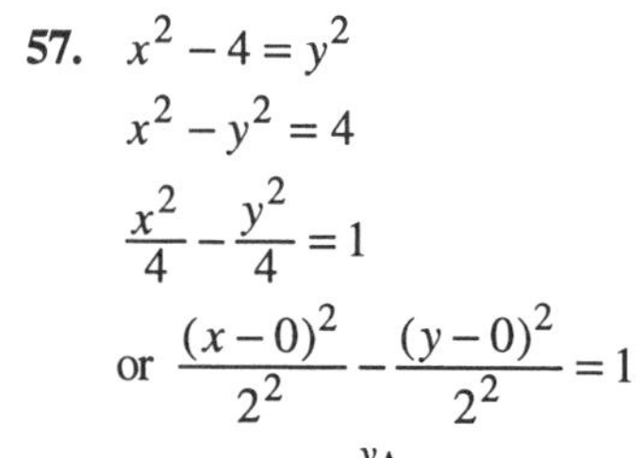

58. $x^2 = 4 - y^2$

$x^2 + y^2 = 4$

$C(0, 0),\ r = 2$

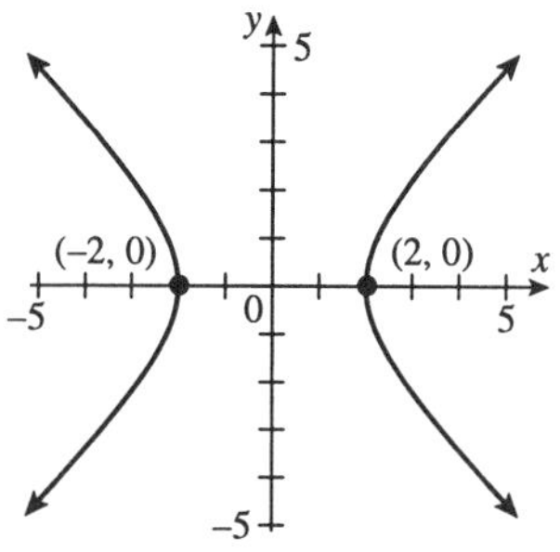

59. $6(x-2)^2 + 9(y+5)^2 = 36$

$\frac{(x-2)^2}{6} + \frac{(y+5)^2}{4} = 1$

$$\frac{(x-2)^2}{\sqrt{6}^2}+\frac{(y+5)^2}{2^2}=1$$

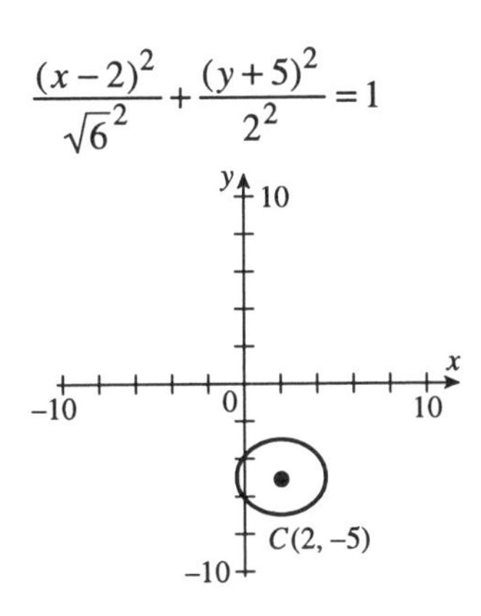

60. $36y^2 = 576 + 16x^2$

$\frac{y^2}{16}-\frac{x^2}{36}=1$

$a = 4,\ b = 6,\ C(0, 0)$

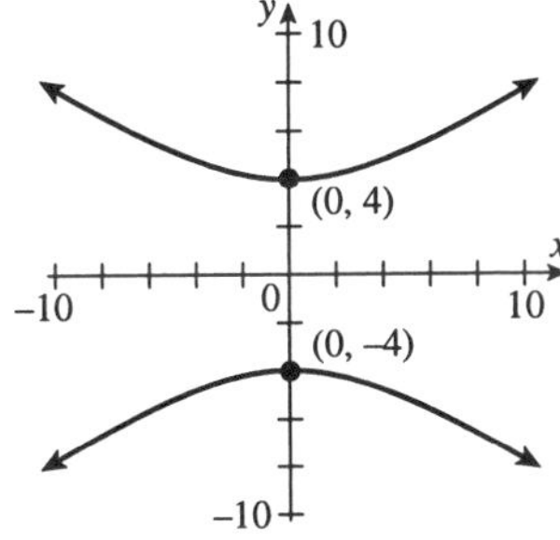

61. $\frac{x^2}{16}-\frac{y^2}{25}=1$

or $\frac{(x-0)^2}{4^2}-\frac{(y-0)^2}{5^2}=1$

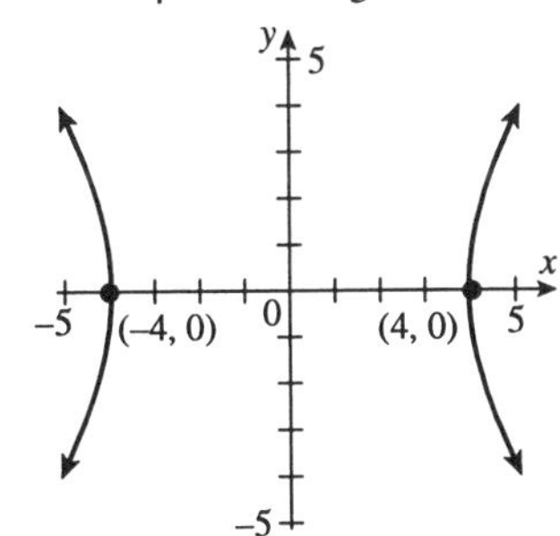

62. $3(x-7)^2 + 3(y+4)^2 = 1$

$(x-7)^2 + (y+4)^2 = \frac{1}{3}$

$C(7, -4),\ r = \frac{\sqrt{3}}{3}$

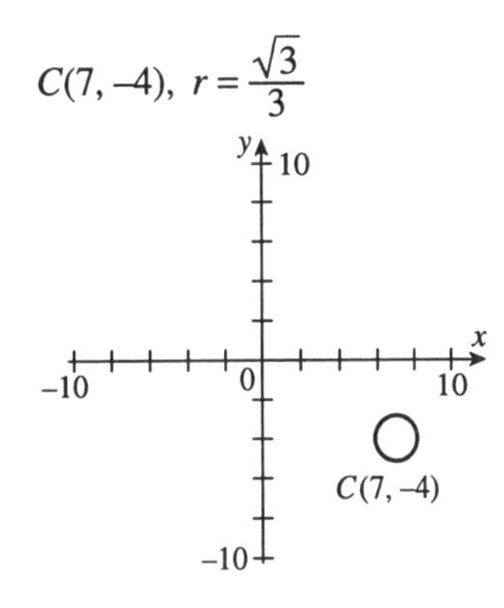

63. The answers are the same.

64. The answers are the same.

65. The answers are the same.

66. The answers are the same.

67. $\begin{cases} y = 2x - 4 & (1) \\ y^2 = 4x & (2) \end{cases}$

Substituting (1) in (2) gives

$(2x-4)^2 = 4x$

$4x^2 - 16x + 16 = 4x$

$4x^2 - 20x + 16 = 0$

$x^2 - 5x + 4 = 0$

$(x-1)(x-4) = 0$

$x - 1 = 0$ or $x - 4 = 0$

$x = 1$ or $x = 4$

$y = -2$ or $y = 4$

68. $\begin{cases} x^2 + y^2 = 4 \\ x - y = 4 \text{ or } x = y + 4 \end{cases}$

Substitute.

$(y+4)^2 + y^2 = 4$

$y^2 + 8y + 16 + y^2 = 4$

$2y^2 + 8y + 16 = 4$

$2y^2 + 8y + 12 = 0$

$y^2 + 4y + 6 = 0$

$b^2 - 4ac = 4^2 - 4\cdot1\cdot6 = 16 - 24 = -8$

Therefore, no real solutions exist.

{ }

69. $\begin{cases} y = x + 2 & (1) \\ y = x^2 & (2) \end{cases}$

Substituting (1) in (2) gives

$x + 2 = x^2$

$0 = x^2 - x - 2$

$0 = (x + 1)(x - 2)$

$x + 1 = 0$ or $x - 2 = 0$

$x = -1$ or $x = 2$

$y = 1$ or $y = 4$

70. $\begin{cases} y = x^2 - 5x + 1 \\ y = -x + 6 \end{cases}$

Substitute.

$-x + 6 = x^2 - 5x + 1$

$0 = x^2 - 4x - 5$

$0 = (x - 5)(x + 1)$

$x - 5 = 0$ or $x + 1 = 0$

$x = 5$ or $x = -1$

$x = 5$: $y = -5 + 6 = 1$

$x = -1$: $y = -(-1) + 6 = 1 + 6 = 7$

$\{(5, 1), (-1, 7)\}$

71. $\begin{cases} 4x - y^2 = 0 & (1) \\ 2x^2 + y^2 = 16 & (2) \end{cases}$

From (1) we have $y^2 = 4x$.

Substituting in (2) gives

$2x^2 + 4x = 16$

$2x^2 + 4x - 16 = 0$

$x^2 + 2x - 8 = 0$

$(x - 2)(x + 4) = 0$

$x - 2 = 0$ or $x + 4 = 0$

$x = 2$ or $x = -4$

$y = \pm 2\sqrt{2}$ reject

72. $\begin{cases} x^2 + 4y^2 = 16 \\ x^2 + y^2 = 4 \end{cases}$

Subtract.

$3y^2 = 12$

$y^2 = 4$ so $y = \pm 2$

Substitute back.

$x^2 + 4 = 4$

$x^2 = 0$ so $x = 0$

$\{(0, 2), (0, -2)\}$

73. $\begin{cases} x^2 + y^2 = 10 & (1) \\ 9x^2 + y^2 = 18 & (2) \end{cases}$

Subtracting (1) from (2) we have

$8x^2 = 8$

$x^2 = 1$

$x = \pm 1$

Substitution gives

$1 + y^2 = 10$

$y^2 = 9$

$y = \pm 3$

74. $\begin{cases} x^2 + 2y = 9 \\ 5x - 2y = 5 \end{cases}$

Add.

$x^2 + 5x = 14$

$x^2 + 5x - 14 = 0$

$(x + 7)(x - 2) = 0$

$x + 7 = 0$ or $x - 2 = 0$

$x = -7$ or $x = 2$

$x = -7$: $(-7)^2 + 2y = 9$

$49 + 2y = 9$

$2y = -40$

so $y = -20$

$x = 2$: $2^2 + 2y = 9$

$4 + 2y = 9$

$2y = 5$ so $y = \frac{5}{2}$

$\left\{(-7, -20), \left(2, \frac{5}{2}\right)\right\}$

75. $\begin{cases} y = 3x^2 + 5x - 4 & (1) \\ y = 3x^2 - x + 2 & (2) \end{cases}$

Subtracting (2) from (1) gives $0 = 6x - 6$ or $x = 1$

Substitution gives $y = 3(1) - 1 + 2$ or $y = 4$

76. $\begin{cases} x^2 - 3y^2 = 1 \text{ or } x^2 = 3y^2 + 1 \\ 4x^2 + 5y^2 = 21 \end{cases}$

Substitute.

$4(3y^2 + 1) + 5y^2 = 21$

$12y^2 + 4 + 5y^2 = 21$

$17y^2 + 4 = 21$

$17y^2 = 17$

$y^2 = 1$, so $y = \pm 1$

Substitute back.

$x^2 - 3(1) = 1$

$x^2 - 3 = 1$

$x^2 = 4$ so $x = \pm 2$

$\{(2, 1), (2, -1), (-2, 1), (-2, -1)\}$

77. area: $xy = 150$

perimeter: $2x + 2y = 50$

Solve the equation for the perimeter for x and substitute in equation for area

$2x + 2y = 50$

$x + y = 25$

$x = 25 - y$

$(25 - y)y = 150$

$25y - y^2 = 150$

$0 = y^2 - 25y + 150$

$0 = (y - 15)(y - 10)$

$y - 15 = 0$ or $y - 10 = 0$

$y = 15$ or $y = 10$

$(15)(10) = 150$

The room is 15 feet by 10 feet.

78. Four real solutions

79.

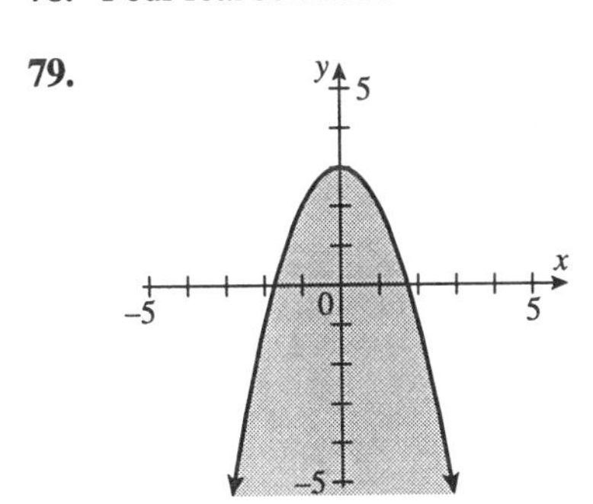

80. $x^2 + y^2 < 9$ First graph.

$x^2 + y^2 = 9$

or $(x - 0)^2 + (y - 0)^2 = 3^2$

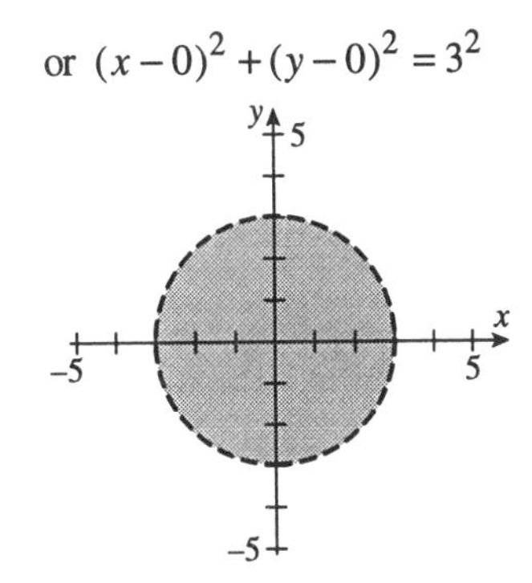

81.

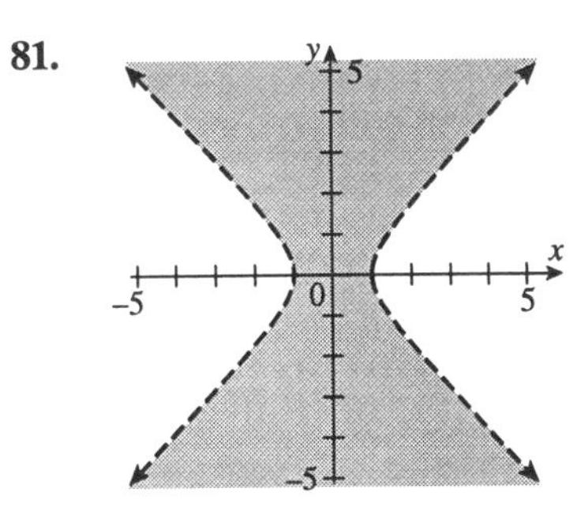

82. $\frac{x^2}{4} + \frac{y^2}{9} \geq 1$ First graph.

$\frac{x^2}{4} + \frac{y^2}{9} = 1$ or

$\frac{(x-0)^2}{2^2} + \frac{(y-0)^2}{3^2} = 1$

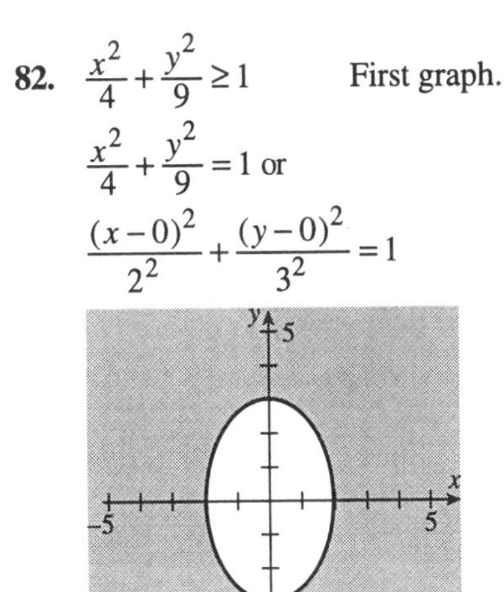

83.

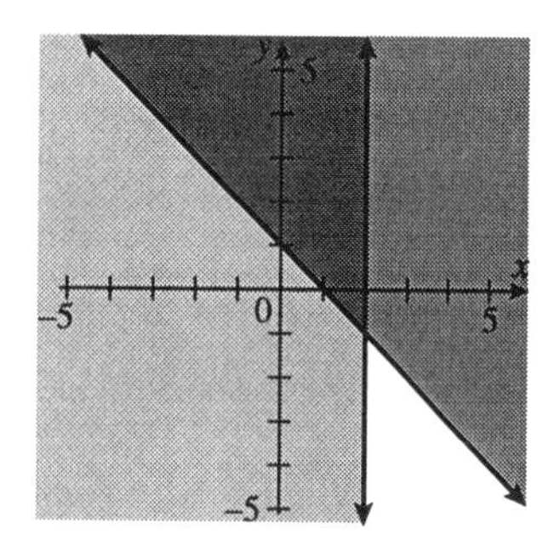

84. $\begin{cases} 3x+4y \le 12 \\ x-2y > 6 \end{cases}$ First graph.

$\begin{cases} 3x+4y = 12 \\ x-2y = 6 \end{cases}$ or $\begin{cases} y = -\frac{3}{4}x+3 \\ y = \frac{x}{2}-3 \end{cases}$

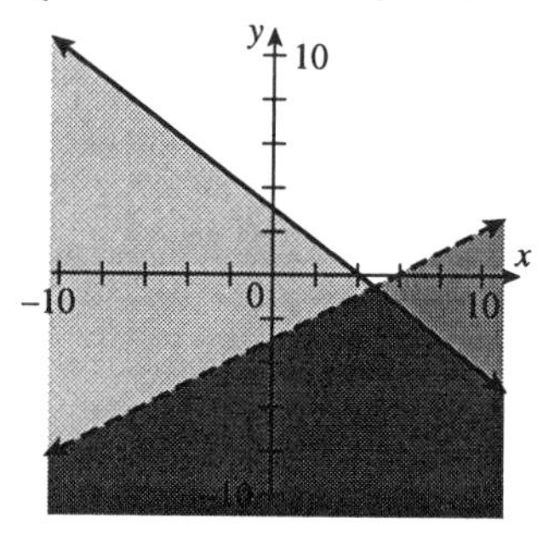

85.

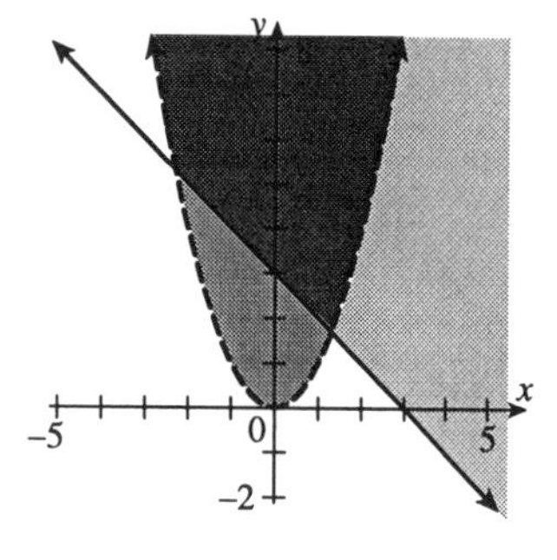

86. $\begin{cases} x^2+y^2 \le 16 \\ x^2+y^2 \ge 4 \end{cases}$ First graph.

$\begin{cases} x^2+y^2 = 16 \\ x^2+y^2 = 4 \end{cases}$ or

$\begin{cases} (x-0)^2+(y-0)^2 = 4^2 \\ (x-0)^2+(y-0)^2 = 2^2 \end{cases}$

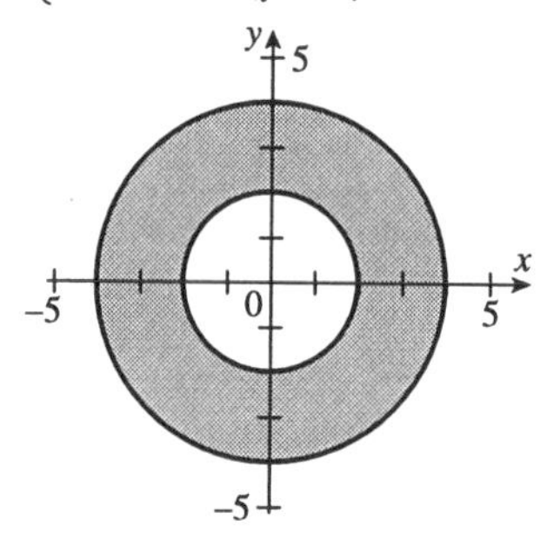

87.

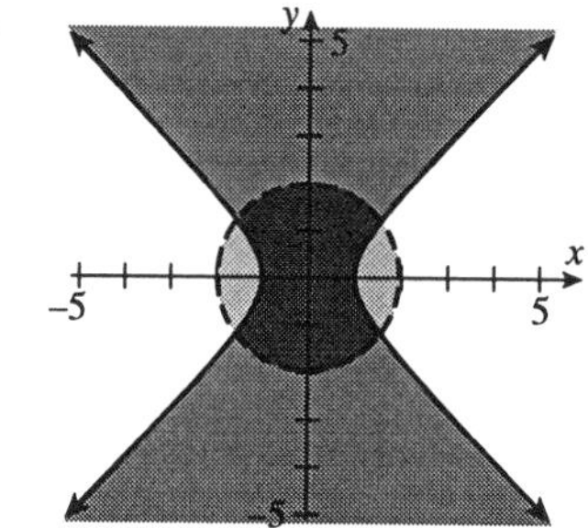

88. $\begin{cases} x^2+y^2 < 4 \\ y \ge x^2-1 \\ x \ge 0 \end{cases}$ First graph.

$\begin{cases} x^2+y^2 = 4 \\ y = x^2-1 \\ x = 0 \end{cases}$ or

$\begin{cases} (x-0)^2+(y-0)^2 = 2^2 \\ y = 1\cdot(x-0)^2-1 \\ x = 0 \end{cases}$

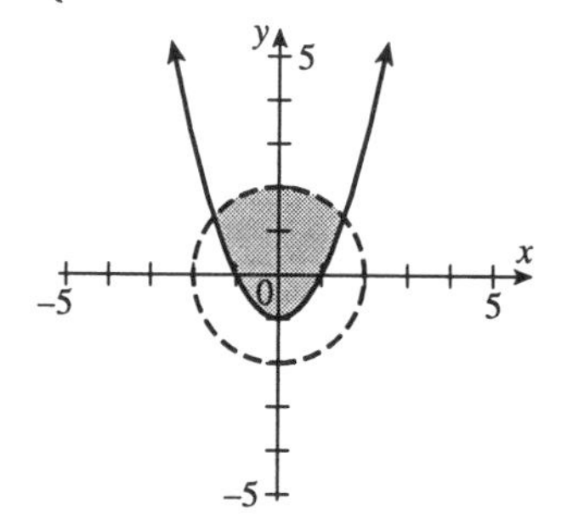

Chapter 9 - Test

1. (–6, 3) and (–8, –7)
$d=\sqrt{(-8+6)^2+(-7-3)^2}$
$d=\sqrt{(-2)^2+(-10)^2}$
$d=\sqrt{4+100}$
$d=\sqrt{104}$
$d=2\sqrt{26}$ units

2. $(-2\sqrt{5},\ \sqrt{10})$ and $(-\sqrt{5},\ 4\sqrt{10})$
$d=\sqrt{(-\sqrt{5}+2\sqrt{5})^2+(4\sqrt{10}-\sqrt{10})^2}$
$d=\sqrt{(\sqrt{5})^2+(3\sqrt{10})^2}$
$d=\sqrt{5+90}$
$d=\sqrt{95}$ units

3. (–2, –5) and (–6, 12)
$\left(\frac{-2-6}{2},\ \frac{-5+12}{2}\right)$
$\left(-4,\ \frac{7}{2}\right)$

4. $\left(-\frac{2}{3},\ -\frac{1}{5}\right)$ and $\left(-\frac{1}{3},\ \frac{4}{5}\right)$
$\left(\frac{-\frac{2}{3}-\frac{1}{3}}{2},\ \frac{-\frac{1}{5}+\frac{4}{5}}{2}\right)$
$\left(-\frac{1}{2},\ \frac{3}{10}\right)$

5. $x^2+y^2=36$ or
$(x-0)^2+(y-0)^2=6^2$
$C(0, 0)$.
$x=0$: $\quad 0^2+y^2=36,\ y^2=36,$
$y=\pm 6$
are the y-intercepts.
$y=0$: $\quad x^2+0^2=36,\ x^2=36,$
$x=\pm 6$
are the x-intercepts.

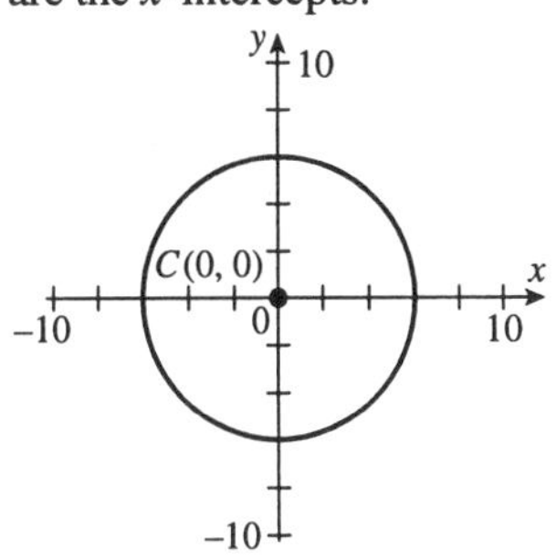

6. $x^2-y^2=36$
$\frac{x^2}{36}-\frac{y^2}{36}=1$
or $\frac{(x-0)^2}{6^2}-\frac{(y-0)^2}{6^2}=1$
$C(0, 0),\ V_1(-6,\ 0)$ and $V_2(6,\ 0)$
$x=0$: $\quad 0^2-y^2=36,\ -y^2=36,$
$y^2=-36$
So there are no y-intercepts.
$y=0$: $\quad x^2-0^2=36,\ x^2=36,$
$x=\pm 6$
are the x-intercepts. The asymptotes are given by $y-0=\pm\frac{6}{6}(x-0)$ or $y=\pm x$.

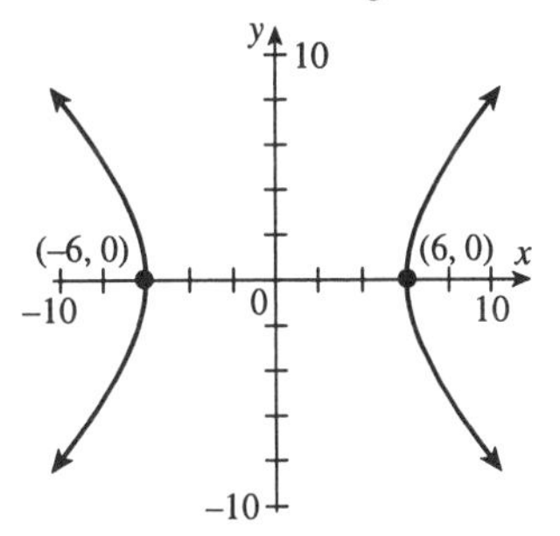

7. $16x^2+9y^2=144$
$\frac{x^2}{9}+\frac{y^2}{16}=1$ or
$\frac{(x-0)^2}{3^2}+\frac{(y-0)^2}{4^2}=1$
$C(0, 0),\ V_1(0,\ -4)$ and $V_2(0,\ 4)$
$x=0$: $\quad 16(0)^2+9y^2=144$
$9y^2=144$

$y^2 = 16$, $y = \pm 4$ are the y-intercepts.

$y = 0$: $\quad 16x^2 + 9(0)^2 = 144$

$16x^2 = 144$

$x^2 = 9$, $x = \pm 3$ are the x-intercepts.

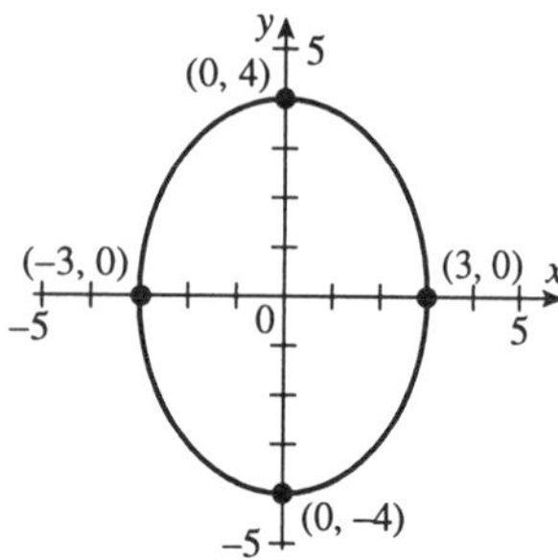

$x = 0$: $y^2 = 16$, $y = \pm 4$ are the y-intercepts

$y = 0$: $\quad (x+3)^2 = 25$

$x + 3 = \pm 5$

$x = -3 \pm 5 = 2$ and -8 are the x-intercepts.

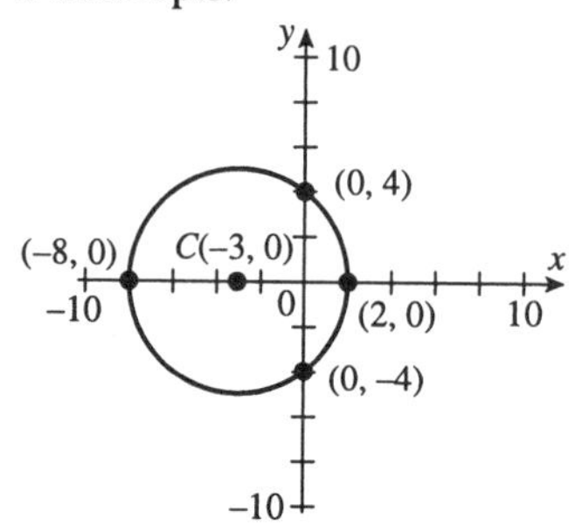

8. $y = x^2 - 8x + 16$

$y = (x-4)^2 + 0$

$V(4, 0)$

$y = 0$: $\quad (x-4)^2 = 0$

$x - 4 = 0$

$x = 4$

is the x-intercept.

$x = 0$: $\quad y = 0^2 - 8(0) + 16 = 16$

is the y-intercept.

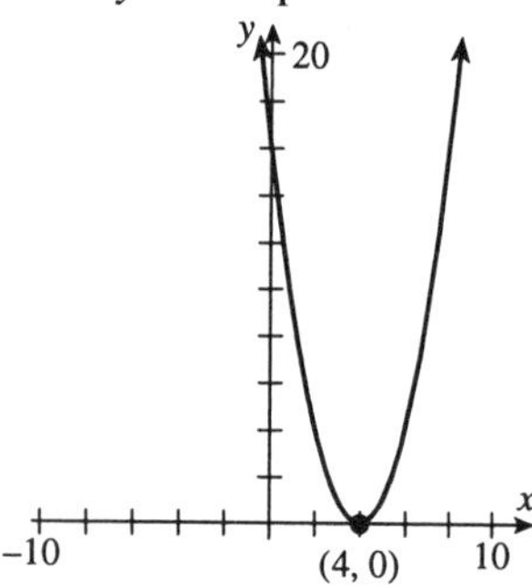

9. $x^2 + y^2 + 6x = 16$

$x^2 + 6x + \left(\frac{6}{2}\right)^2 + y^2 = 16 + 9$

$(x+3)^2 + y^2 = 25$

$(x+3)^2 + (y-0)^2 = 5^2$

$C(-3, 0)$, and $r = 5$

10. $x = y^2 + 8y - 3$

$= y^2 + 8y + \left(\frac{8}{2}\right)^2 - 3 + 16$

$x = (y+4)^2 + 13$

$V(13, -4)$

$x = 0$: $\quad 0 = (y+4)^2 + 13$

$(y+4)^2 = -13$

So there are no y-intercepts.

$y = 0$: $\quad x = 0^2 + 8(0) - 3 = -3$ is the x-intercept.

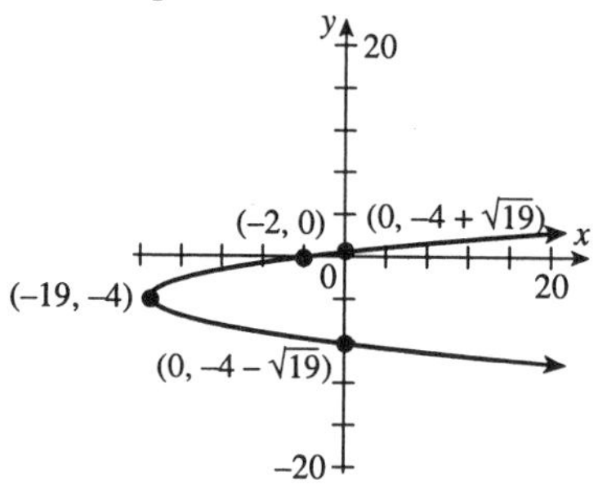

11. $\dfrac{(x-4)^2}{16} + \dfrac{(y-3)^2}{9} = 1$

or $\dfrac{(x-4)^2}{4^2} + \dfrac{(y-3)^2}{3^2} = 1$

$C(4, 3)$, $V_1(0, 3)$, and $V_2(8, 3)$

$x = 0$: $\quad \dfrac{(0-4)^2}{16} + \dfrac{(y-3)^2}{9} = 1$

$1 + \dfrac{(y-3)^2}{9} = 1$

$$\frac{(y-3)^2}{9}=0$$
$$(y-3)^2=0$$

$y-3=0$ or $y=3$ is the y-intercept.

$y=0$: $\frac{(x-4)^2}{16}+\frac{(0-3)^2}{9}=1$

$$\frac{(x-4)^2}{16}+1=1$$
$$\frac{(x-4)^2}{16}=0$$
$$(x-4)^2=0$$
$$x-4=0$$

$x=4$ is the x-intercept.

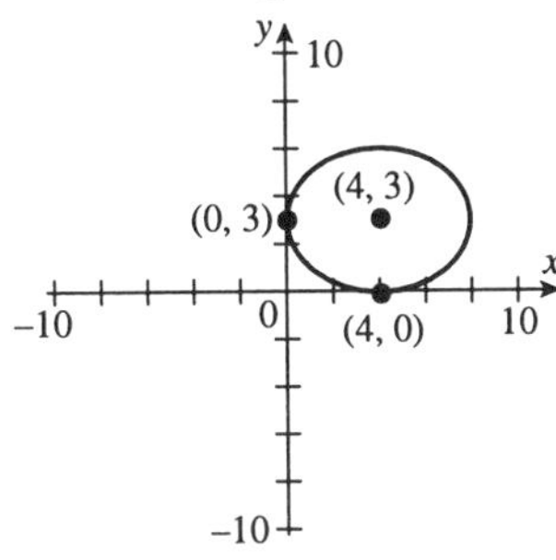

12. $y^2-x^2=0$

$y^2=x^2$

$y=\pm x$

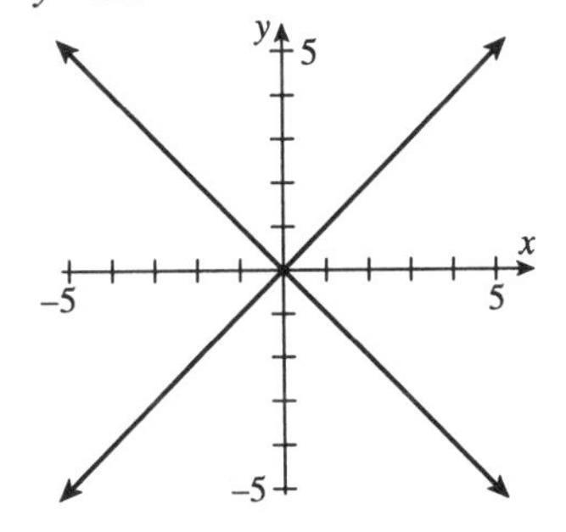

13. $\begin{cases} x^2+y^2=169 \\ 5x+12y=0 \end{cases}$

$12y=-5x$

$y=-\frac{5x}{12}$

Substitute.

$x^2+\left(-\frac{5x}{12}\right)^2=169$

$x^2+\frac{25x^2}{144}=169$

$\frac{169x^2}{144}=169$

$x^2=144$ so $x=\pm 12$

Substitute back.

$x=12$: $y=-\frac{5}{12}(12)=-5$

$x=-12$: $y=-\frac{5}{12}(-12)=5$

$\{(12, -5), (-12, 5)\}$

14. $\begin{cases} x^2+y^2=26 \\ x^2-y^2=24 \end{cases}$

Add.

$2x^2=50$

$x^2=25$ so $x=\pm 5$

Substitute back.

$25+y^2=26$

$y^2=1$ so $y=\pm 1$

$\{(5, 1), (5, -1), (-5, 1), (-5, -1)\}$

15. $\begin{cases} y=x^2-5x+6 \\ y=2x \end{cases}$

Substitute.

$2x=x^2-5x+6$

$0=x^2-7x+6$

$0=(x-1)(x-6)$

$x-1=0$ or $x-6=0$

$x=1$ or $x=6$

$x=1$: $y=2(1)=2$

$x=6$: $y=2(6)=12$

$\{(1, 2), (6, 12)\}$

16. $\begin{cases} x^2+4y^2=5 \\ y=x \end{cases}$

Substitute.

$x^2+4x^2=5$

$5x^2=5$

$x^2=1$ so $x=\pm 1$

$x=1$: $y=1$

$x=-1$: $y=-1$

$\{(1, 1), (-1, -1)\}$

17. $\begin{cases} 2x+5y \geq 10 \\ y \geq x^2+1 \end{cases}$ First graph.

$\begin{cases} 2x+5y=10 \\ y=x^2+1 \end{cases}$ or

$\begin{cases} y=-\frac{2}{5}x+2 \\ y=1\cdot(x-0)^2+1 \end{cases}$

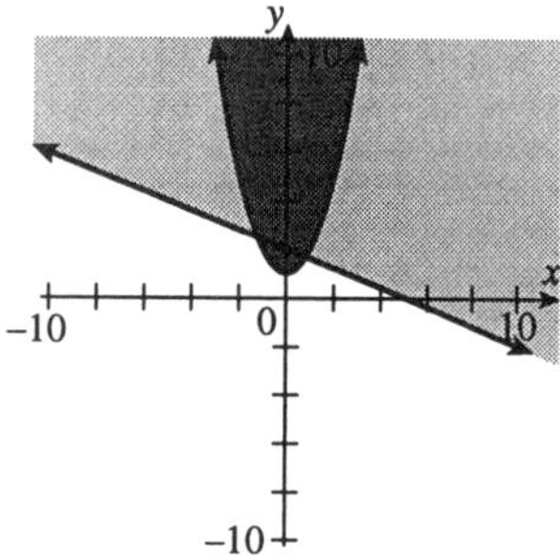

18. $\begin{cases} \frac{x^2}{4}+y^2 \leq 1 \\ x+y>1 \end{cases}$ First graph.

$\begin{cases} \frac{x^2}{4}+y^2=1 \\ x+y=1 \end{cases}$ or

$\begin{cases} \frac{(x-0)^2}{2^2}+\frac{(y-0)^2}{1^2}=1 \\ y=-x+1 \end{cases}$

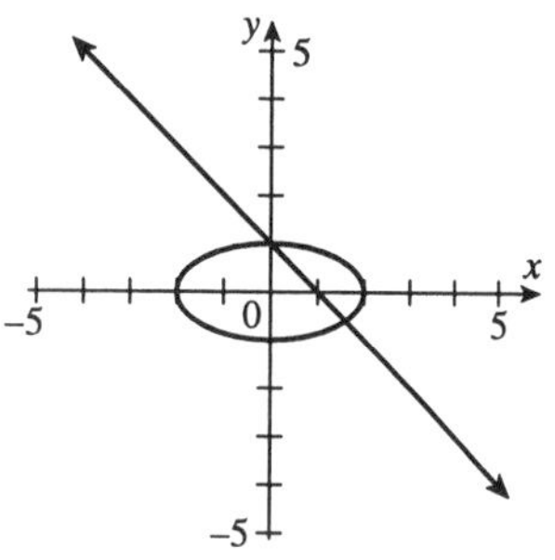

19. $\begin{cases} x^2+y^2>1 \\ \frac{x^2}{4}-y^2 \geq 1 \end{cases}$ First graph.

$\begin{cases} x^2+y^2=1 \\ \frac{x^2}{4}-y^2=1 \end{cases}$ or

$\begin{cases} (x-0)^2+(y-0)^2=1^2 \\ \frac{(x-0)^2}{2^2}-\frac{(y-0)^2}{1^2}=1 \end{cases}$

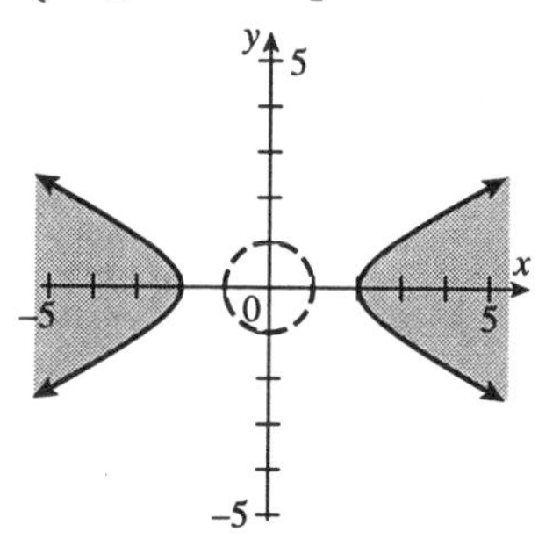

20. $\begin{cases} x^2+y^2 \geq 4 \\ x^2+y^2<16 \\ y \geq 0 \end{cases}$ First graph.

$\begin{cases} x^2+y^2=4 \\ x^2+y^2=16 \\ y=0 \end{cases}$ or

$\begin{cases} (x-0)^2+(y-0)^2=2^2 \\ (x-0)^2+(y-0)^2=4^2 \\ y=0 \end{cases}$

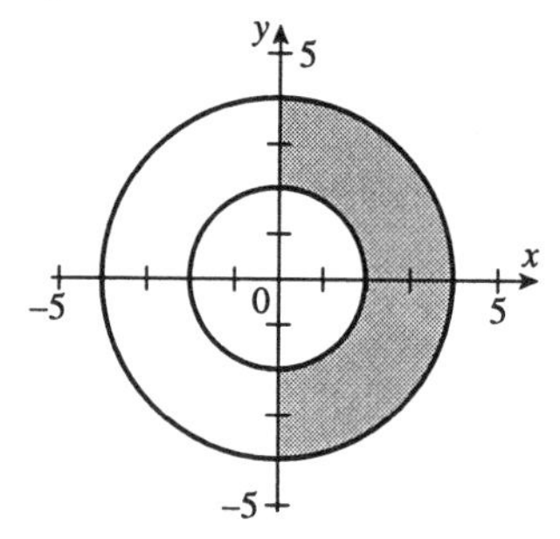

21. Graph B; vertex in third quadrant, opens to the right.

22. $100x^2+225y^2=22,500$

$\frac{x^2}{225}+\frac{y^2}{100}=0$

$a=\sqrt{225}=15$

$b=\sqrt{100}=10$

Width = 15 + 15 = 30 feet
Height = 10 feet

Chapter 9 - Cumulative Review

1. a. $3^2 = 9$

   b. $\left(\frac{1}{2}\right)^4 = \frac{(1)^4}{(2)^4} = \frac{1}{16}$

   c. $-5^2 = -25$

   d. $(-5)^2 = 25$

   e. $-5^3 = -125$

   f. $(-5)^3 = -125$

2. $\frac{x+5}{2} + \frac{1}{2} = 2x - \frac{x-3}{8}$
   $4(x+5) + 4 = 16x - (x-3)$
   $4x + 20 + 4 = 16x - x + 3$
   $4x + 24 = 15x + 3$
   $21 = 11x$
   $\frac{21}{11} = x$
   $\left\{\frac{21}{11}\right\}$

3. $A = P\left(1 + \frac{r}{n}\right)^{nt}$
   $A = 10,000\left(1 + \frac{0.05}{4}\right)^{4 \cdot 3}$
   $A = 10,000(1.0125)^{12}$
   $A = \$11,607.55$

4. $2|x| + 25 = 23$
   $2|x| = -2$
   $|x| = -1$
   An absolute value cannot be negative.
   No solution
   { }

5. $\left|2x - \frac{1}{10}\right| < -13$
   The absolute value of a number is always nonnegative and can never be less than −13. Thus there is no solution.
   { }

6. $y = |x|$

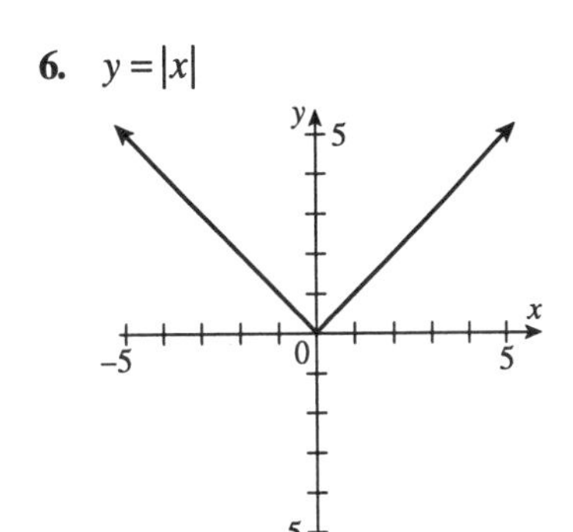

7. a. \$11.5 billion

   b. \$13,800 − \$200 = \$13,600

8. $y = \frac{1}{4}x - 3$

9. $\begin{cases} -\frac{x}{6} + \frac{y}{2} = \frac{1}{2} \\ \frac{x}{3} - \frac{y}{6} = -\frac{3}{4} \end{cases}$
   Solve the first equation for x.
   $x - 3y = -3$
   $x = 3y - 3$
   Substitute into second equation.
   $\frac{3y-3}{3} - \frac{y}{6} = -\frac{3}{4}$
   $12y - 12 - 2y = -9$
   $10y = 3$
   $y = \frac{3}{10}$
   $x = 3y - 3$
   $x = 3\left(\frac{3}{10}\right) - 3$
   $x = \frac{9}{10} - \frac{30}{10}$
   $x = -\frac{21}{10}$
   $\left\{\left(-\frac{21}{10}, \frac{3}{10}\right)\right\}$

10. Let x = the smallest angle.
Then $x + 80$ = largest angle and
$x + 10$ = third angle
$x + x + 80 + x + 10 = 180$
$3x + 90 = 180$
$3x = 90$
$x = 30$
$x + 80 = 110$
$x + 10 = 40$
30°, 40°, 110°

11. **a.** $7^0 = 1$

b. $-7^0 = -1$

c. $(2x+5)^0 = 1$

d. $2x^0 = 2(1) = 2$

12. **a.** $(2x^0y^{-3})^{-2} = 2^{-2}x^0y^6$
$= \frac{(1)y^6}{2^2} = \frac{y^6}{4}$

b. $\left(\frac{x^{-5}}{x^{-2}}\right)^{-3} = \frac{x^{15}}{x^6} = x^{15-6} = x^9$

c. $\left(\frac{2}{7}\right)^{-2} = \frac{2^{-2}}{7^{-2}} = \frac{7^2}{2^2} = \frac{49}{4}$

d. $\frac{5^{-2}x^{-3}y^{11}}{x^2y^{-5}} = \frac{y^{11+5}}{5^2x^{2+3}} = \frac{y^{16}}{25x^5}$

13.
$$\begin{array}{rrrr} & 11x^3 - 12x^2 & + x - 3 \\ + & x^3 & - 10x + 5 \\ \hline & 12x^3 - 12x^2 & - 9x + 2 \end{array}$$

14. $x^2 - 12x + 35 = (x-5)(x-7)$

15. $\frac{8x^3+125}{x^4+5x^2+4} \div \frac{2x+5}{2x^2+8}$
$= \frac{(2x+5)(4x^2-10x+25)}{(x^2+4)(x^2+1)} \cdot \frac{2(x^2+4)}{2x+5}$
$= \frac{2(4x^2-10x+25)}{x^2+1}$

16. $\frac{10x^2-5x+20}{5}$
$= \frac{10x^2}{5} - \frac{5x}{5} + \frac{20}{5}$
$= 2x^2 - x + 4$

17. **a.** $\sqrt[3]{1} = 1$

b. $\sqrt[3]{-64} = -4$

c. $\sqrt[3]{\frac{8}{125}} = \frac{2}{5}$

d. $\sqrt[3]{x^6} = x^2$

e. $\sqrt[3]{-8x^9} = -2x^3$

18. **a.** $\frac{\sqrt{28}}{\sqrt{45}} \cdot \frac{\sqrt{28}}{\sqrt{28}} = \frac{\sqrt{(28)^2}}{\sqrt{1260}}$
$= \frac{28}{6\sqrt{35}} = \frac{14}{3\sqrt{35}}$

b. $\frac{\sqrt[3]{2x^2}}{\sqrt[3]{5y}} \cdot \frac{\sqrt[3]{4x}}{\sqrt[3]{4x}} = \frac{\sqrt[3]{8x^3}}{\sqrt[3]{20xy}} = \frac{2x}{\sqrt[3]{20xy}}$

19. $x^2 = 50$
$x = \pm\sqrt{50}$
$x = \pm 5\sqrt{2}$

20. $(x+3)(x-3) \ge 0$
$x + 3 = 0$ or $x - 3 = 0$
$x = -3$ $\quad$ $x = 3$

Region	Test Point	$(x+3)(x-3) \ge 0$
$(-\infty, -3]$	-4	$(-1)(-7) \ge 0$
$[-3, 3]$	0	$(3)(-3) < 0$
$[3, \infty)$	4	$(7)(1) \ge 0$

Therefore, the solution is
$(-\infty, -3] \cup [3, \infty)$

21. **a.** $F(x) = x^2 + 2$

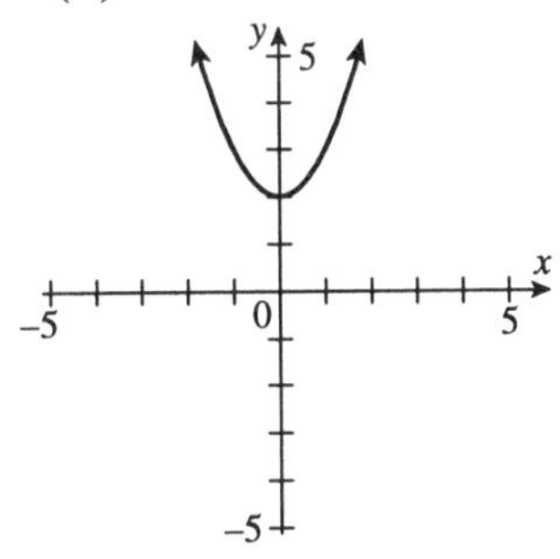

b. $g(x) = x^2 - 3$

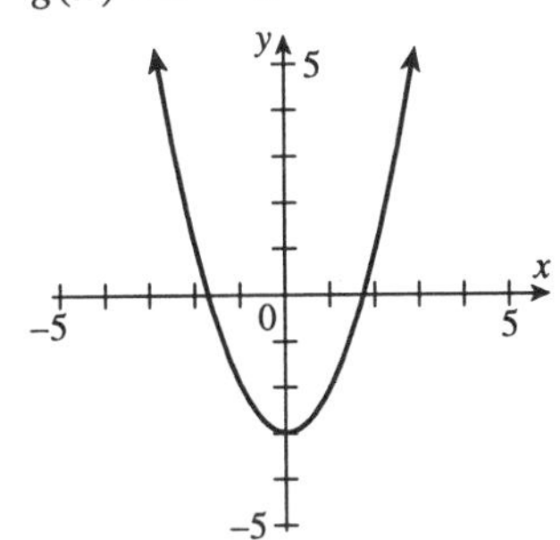

22. (2, –5) and (1, –4)

$d = \sqrt{(1-2)^2 + (-4+5)^2}$

$d = \sqrt{(-1)^2 + (1)^2}$

$d = \sqrt{1+1}$

$d = \sqrt{2}$

23. $\begin{cases} y = \sqrt{x} \\ x^2 + y^2 = 6 \end{cases}$

Substitute $y = \sqrt{x}$ into the second equation.

$x^2 + (\sqrt{x})^2 = 6$

$x^2 + x = 6$

$x^2 + x - 6 = 0$

$(x + 3)(x - 2) = 0$

$x = -3$ or $x = 2$

Disregard the negative because $\sqrt{-3}$ is not a real number.

$y = \sqrt{2}$

Solution $\{(2, \sqrt{2})\}$

Chapter 10

Exercise Set 10.1

1. $(f \circ g)(2) = f(g(2))$
$= f(-4) = (-4)^2 - 6(-4) + 2$
$= 16 + 24 + 2 = 42$

3. $(g \circ f)(-1) = g(f(-1))$
$= g(9) = -2(9) = -18$

5. $(g \circ h)(0) = g(h(0))$
$= g(0) = -2(0) = 0$

7. $(f \circ g)(x) = f(g(x)) = f(5x)$
$= (5x)^2 + 1 = 25x^2 + 1$
$(g \circ f)(x) = g(f(x)) = g(x^2 + 1)$
$= 5(x^2 + 1) = 5x^2 + 5$
$(f \circ f)(x) = f(f(x)) = f(x^2 + 1)$
$= (x^2 + 1)^2 + 1 = x^4 + 2x^2 + 2$

9. $(f \circ g)(x) = f(g(x)) = f(x + 7)$
$= 2(x + 7) - 3 = 2x + 14 - 3 = 2x + 11$
$(g \circ f)(x) = g(f(x)) = g(2x - 3)$
$= (2x - 3) + 7 = 2x + 4$
$(f \circ f)(x) = f(f(x)) = f(2x - 3)$
$= 2(2x - 3) - 3 = 4x - 9$

11. $(f \circ g)(x) = f(g(x)) = f(-2x)$
$= (-2x)^3 + (-2x) - 2$
$= -8x^3 - 2x - 2$
$(g \circ f)(x) = g(f(x)) = g(x^3 + x - 2)$
$= -2(x^3 + x - 2) = -2x^3 - 2x + 4$

13. $(f \circ g)(x) = f(g(x)) = f(-5x + 2)$
$= \sqrt{-5x + 2}$
$(g \circ f)(x) = g(f(x)) = g(\sqrt{x})$
$= -5\sqrt{x} + 2$

15. $H(x) = (g \circ h)(x)$
$= g(h(x)) = g(x^2 + 2) = \sqrt{x^2 + 2}$

17. $F(x) = (h \circ f)(x)$
$= h(f(x)) = h(3x)$
$= (3x)^2 + 2 = 9x^2 + 2$

19. $G(x) = (f \circ g)(x)$
$= f(g(x)) = f(\sqrt{x}) = 3\sqrt{x}$

21. $f = \{(-1, -1), (1, 1), (0, 2), (2, 0)\}$ is a one-to-one function.
$f^{-1} = \{(-1, -1), (1, 1), (2, 0), (0, 2)\}$

23. $h = \{(10, 10)\}$ is a one-to-one function.
$h^{-1} = \{(10, 10)\}$

25. $f = \{(11, 12), (4, 3), (3, 4), (6, 6)\}$ is a one-to-one function.
$f^{-1} = \{(12, 11), (3, 4), (4, 3), (6, 6)\}$

27. This function is not one-to-one because there are two years with the same score: (1975, 18.6) and (1985, 18.6)

29. This function is one-to-one.

Rank (input)	1	49	12	2	45
State (output)	CA	VT	VA	TX	SD

31. $f(x) = x^3 + 2$

a. $f(1) = 1^3 + 2 = 3$

b. $f^{-1}(3) = 1$

33. $f(x) = x^3 + 2$

a. $f(-1) = (-1)^3 + 2 = 1$

b. $f^{-1}(1) = -1$

35. The graph represents a one-to-one function.

37. The graph does not represent a one-to-one function.

39. The graph represents a one-to-one function.

41. The graph does not represent a one-to-one function.

43. $f(x) = x + 4$

$f^{-1}(x) = x - 4$

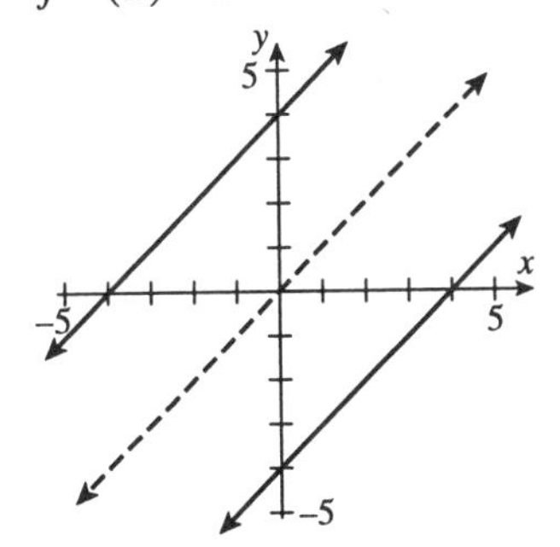

45. $f(x) = 2x - 3$

$f^{-1}(x) = \frac{x+3}{2}$

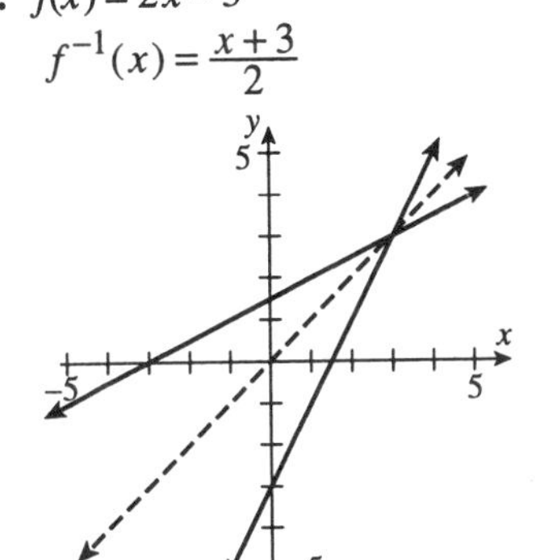

47. $f(x) = \frac{12x - 4}{3}$

$f^{-1}(x) = \frac{3x+4}{12}$

49. $f(x) = x^3$

$f^{-1}(x) = \sqrt[3]{x}$

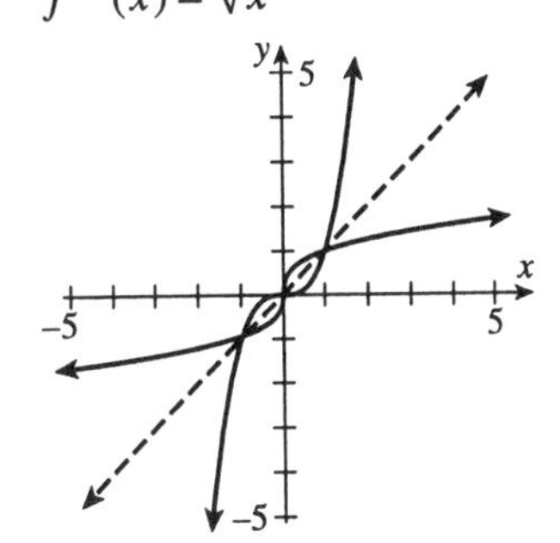

51. $f(x) = \frac{x-2}{5}$

$f^{-1}(x) = 5x + 2$

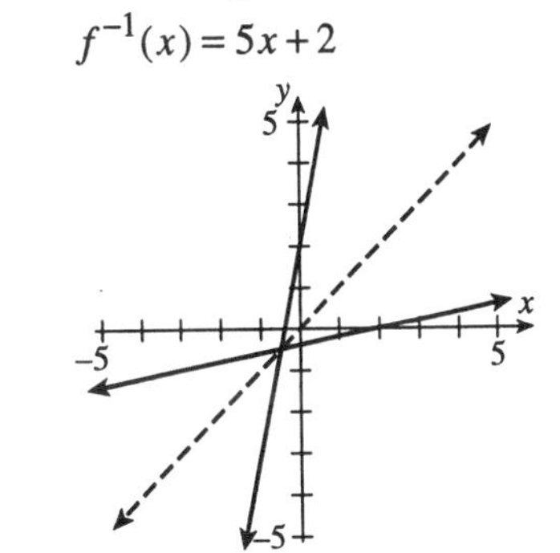

53. $g(x) = x^2 + 5, \ x \geq 0$

$g^{-1}(x) = \sqrt{x - 5}, \ x \geq 5$

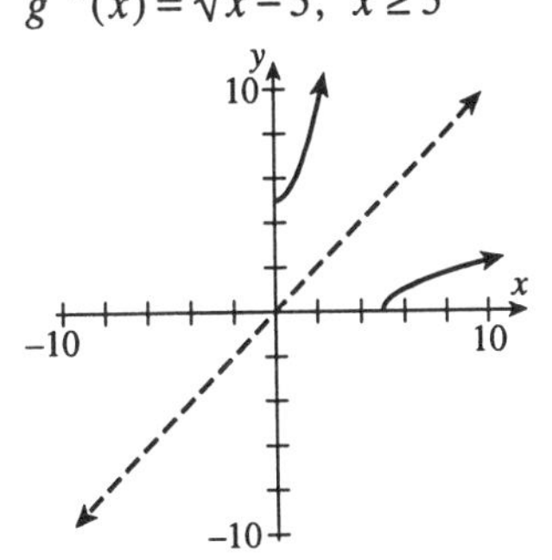

55. $(f \circ f^{-1})(x)$

$= f(f^{-1}(x)) = f\left(\frac{x-1}{2}\right)$

$= 2\left(\frac{x-1}{2}\right) + 1 = x - 1 + 1 = x$

$(f^{-1} \circ f)(x)$

$= f^{-1}(f(x)) = f^{-1}(2x+1)$
$= \frac{(2x+1)-1}{2} = \frac{2x}{2} = x$

57. $(f \circ f^{-1})(x) = f(f^{-1}(x)) = f(\sqrt[3]{x-6})$
$= (\sqrt[3]{x-6})^3 + 6 = x - 6 + 6 = x$
$(f^{-1} \circ f)(x) = f^{-1}(f(x)) = f^{-1}(x^3+6)$
$= \sqrt[3]{(x^3+6)-6} = \sqrt[3]{x^3} = x$

59. a. (0, 1), (1, 2), (2, 5)

b. $\left(\frac{1}{4}, -2\right), \left(\frac{1}{2}, -1\right)$, (1, 0), (2, 1), (5, 2)

c.

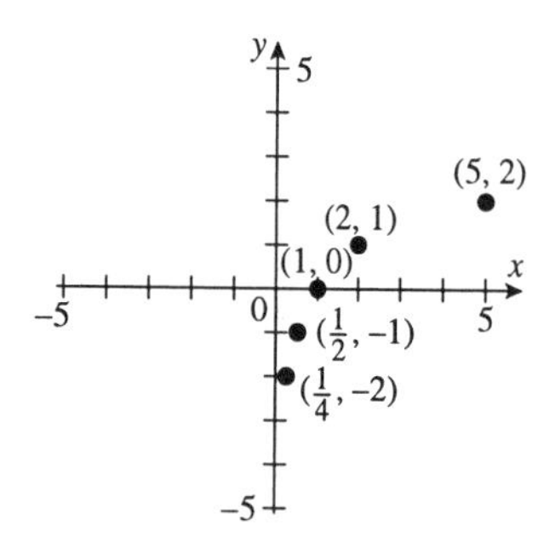

d.

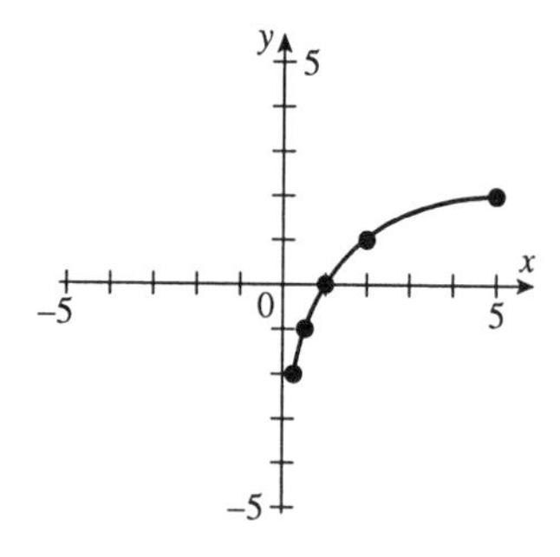

61. $f(x) = 3x + 1$ or $y = 3x + 1$
Find inverse
$x = 3y + 1$
$3y = x - 1$
$y = \frac{x-1}{3}$

$f^{-1}(x) = \frac{x-1}{3}$

63. $f(x) = \sqrt[3]{x+1}$ or $y = \sqrt[3]{x+1}$
Find inverse
$x = \sqrt[3]{y+1}$
$x^3 = y + 1$
$y = x^3 - 1$

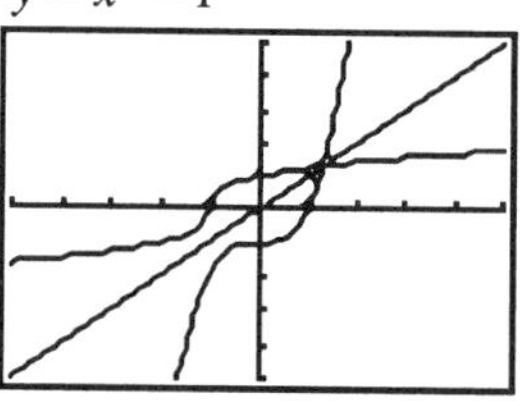

65. $25^{1/2} = \sqrt{25} = 5$

67. $16^{3/4} = (16^3)^{1/4} = (4096)^{1/4} = 8$

69. $9^{-3/2} = \frac{1}{9^{3/2}} = \frac{1}{\sqrt{9^3}} = \frac{1}{\sqrt{729}} = \frac{1}{27}$

71. $f(x) = 3^x$
$f(2) = 3^2 = 9$

73. $f(x) = 3^x$
$f\left(\frac{1}{2}\right) = 3^{1/2} = \sqrt{3} \approx 1.73$

Section 10.2

Graphing Calculator Explorations

1. 81.98%

3. 22.54%

Exercise Set 10.2

1. $y = 4^x$

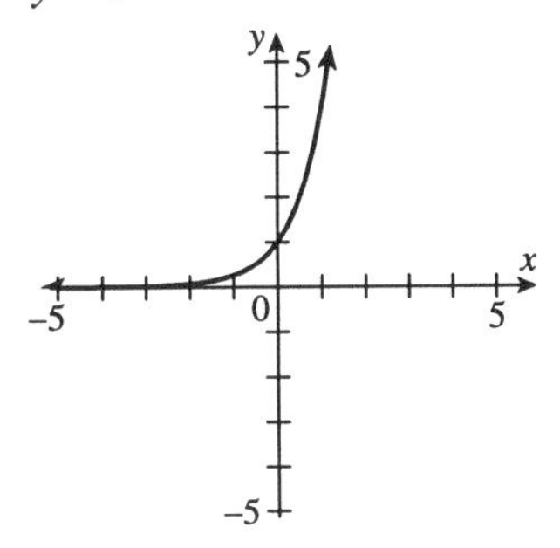

3. $y = 1 + 2^x$ or $y = 2^x + 1$

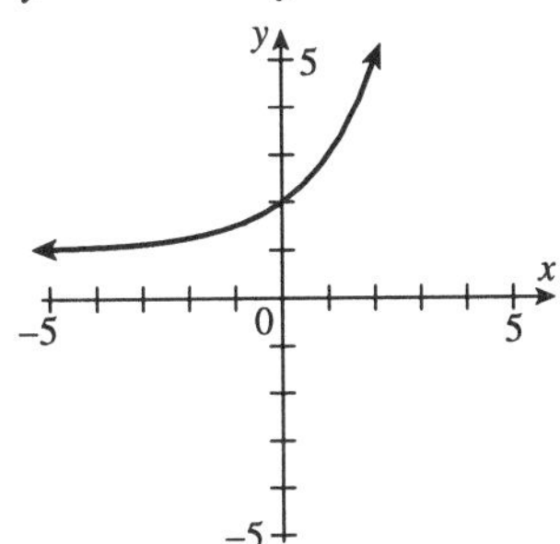

5. $y = \left(\frac{1}{4}\right)^x$

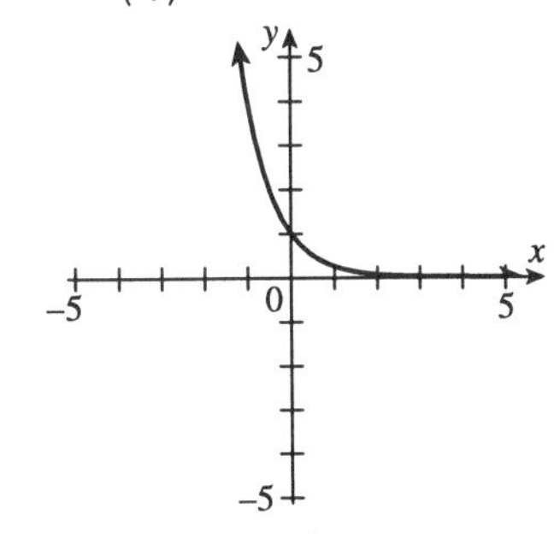

7. $y = \left(\frac{1}{2}\right)^x - 2$

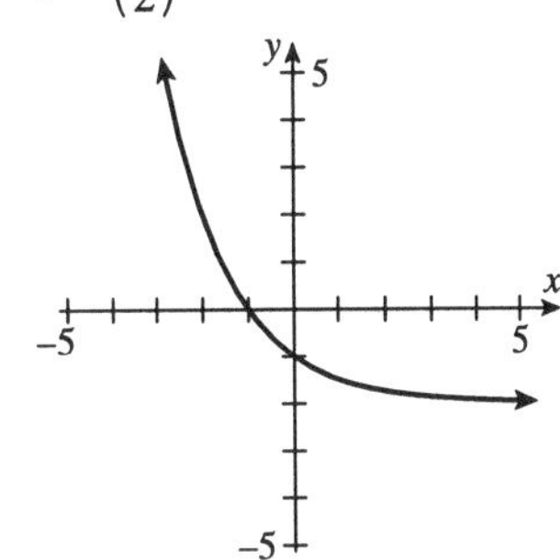

9. $y = -2^x$

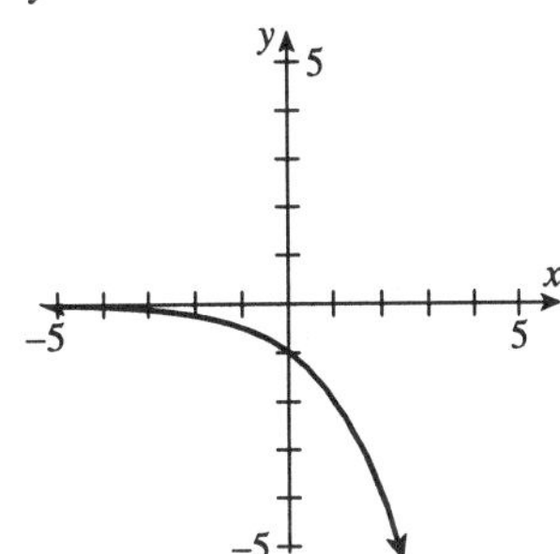

11. $y = 3^x - 2$

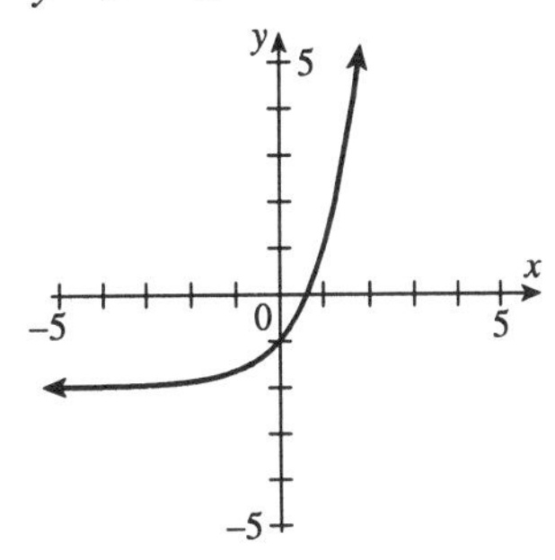

13. $y = -\left(\frac{1}{4}\right)^x$

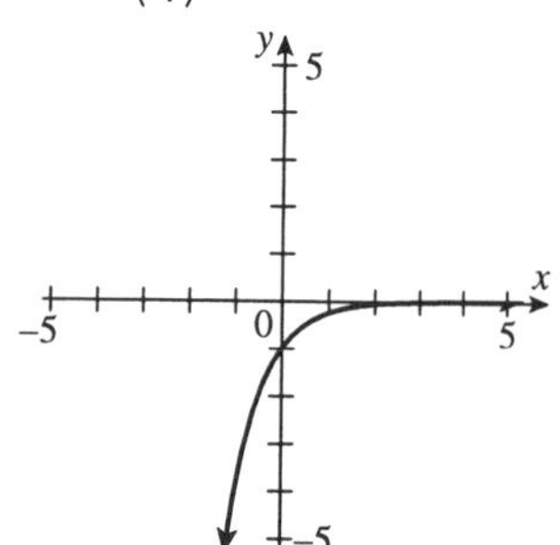

15. $y = \left(\frac{1}{3}\right)^x + 1$

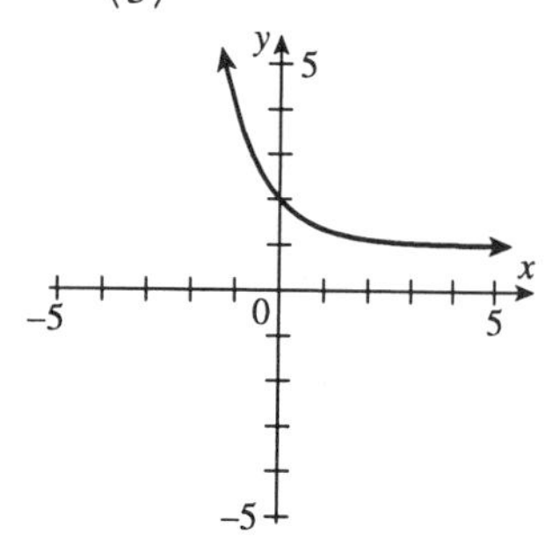

17. Answers may vary.

19. $3^x = 27$
$3^x = 3^3$
$x = 3$
$\{3\}$

21. $16^x = 8$
$(2^4)^x = 2^3$
$2^{4x} = 2^3$
$4x = 3$
$x = \frac{3}{4}$
$\left\{\frac{3}{4}\right\}$

23. $32^{2x-3} = 2$
$(2^5)^{2x-3} = 2$
$2^{10x-15} = 2^1$
$10x - 15 = 1$
$10x = 16$
$x = \frac{16}{10} = \frac{8}{5}$
$\left\{\frac{8}{5}\right\}$

25. $\frac{1}{4} = 2^{3x}$
$2^{-2} = 2^{3x}$
$3x = -2$
$x = -\frac{2}{3}$
$\left\{-\frac{2}{3}\right\}$

27. $5^x = 625$
$5^x = 5^4$
$x = 4$
$\{4\}$

29. $4^x = 8$
$(2^2)^x = 2^3$
$2^{2x} = 2^3$
$2x = 3$
$x = \frac{3}{2}$
$\left\{\frac{3}{2}\right\}$

31. $27^{x+1} = 9$
$(3^3)^{x+1} = 3^2$
$3^{3x+3} = 3^2$
$3x + 3 = 2$
$3x = -1$
$x = -\frac{1}{3}$
$\left\{-\frac{1}{3}\right\}$

33. $81^{x-1} = 27^{2x}$
$(3^4)^{x-1} = (3^3)^{2x}$
$3^{4x-4} = 3^{6x}$
$4x - 4 = 6x$
$-4 = 2x$
$x = -2$
$\{-2\}$

35. $f(x) = b^x,\ b > 1$
$f(1) = b^1 = 3$ or $b = 3$ so $f(x) = 3^x$

37. $f(x) = b^x,\ 0 < b < 1$
$f(1) = b^1 = \frac{1}{2}$ or $b = \frac{1}{2}$ so $f(x) = \left(\frac{1}{2}\right)^x$.

39. $y = 30(2.7)^{-.004t},\ t = 50$
$= 30(2.7)^{-(.004)(50)}$
$= 30(2.7)^{-.2} \approx 24.6$
Therefore, approximately 24.6 lbs of uranium will remain after 50 days.

41. $y = 200(2.7)^{.08t},\ t = 12$
$= 200 \cdot (2.7)^{.08(12)}$
$= 200(2.7)^{.96} \approx 519$
There should be approximately 519 rats by next January.

43. $y = 5(2.7)^{-.15t},\ t = 10$
$= 5(2.7)^{-.15(10)}$
$= 5(2.7)^{-1.5} \approx 1.1$ grams

45. $y = 15{,}525{,}000(2.7)^{0.007(6)}$
$y = 16{,}186{,}348$
Approximately 16,190,000 residents

47. $A = P\left(1 + \frac{r}{n}\right)^{nt}$
$t = 3$, $P = 6{,}000$, $r = .08$, and $n = 12$.
Then
$A = 6000\left(1 + \frac{.08}{12}\right)^{12(3)}$
$= 6000(1.006)^{36} \approx 7621.42$
Erica would owe $7621.42 after 3 years.

49. $A = P\left(1 + \frac{r}{n}\right)^{nt}$, where $P = 2000$, $r = .06$, $n = 2$, and $t = 12$.
Then: $A = 2000\left(1 + \frac{.06}{2}\right)^{2(12)}$
$= 2000(1.03)^{24} \approx 4065.59$
Janina has approximately $4065.59 in her savings account.

51. Results are the same.

53. ≈ 18.62 lb

55. ≈ 50.41 g

57. $5x - 2 = 18$
$5x = 20$
$x = 4$
$\{4\}$

59. $3x - 4 = 3(x + 1)$
$3x - 4 = 3x + 3$
$-4 = 3$
Inconsistent equation.
$\{\ \}$

61. $x^2 + 6 = 5x$
$x^2 - 5x + 6 = 0$
$(x - 3)(x - 2) = 0$
$x - 3 = 0$ or $x - 2 = 0$
$x = 3$ or $x = 2$
$\{3, 2\}$

63. $2^x = 8$
$2^3 = 8$
$\{3\}$

65. $5^x = \frac{1}{5}$
$5^{-1} = \frac{1}{5}$
$\{-1\}$

Exercise Set 10.3

1. $\log_2 8 = 3$

3. $\log_3 \frac{1}{9} = -2$

5. $\log_{25} 5 = \frac{1}{2}$

7. $\log_{1/2} 2 = -1$

9. $\log_7 1 = 0$

11. $\log_2 2^4 = 4$

13. $\log_{10} 100 = 2$

15. $3^{\log_3 5} = 5$

17. $\log_3 81 = 4$

19. $\log_4\left(\frac{1}{64}\right) = -3$

21. Answers may vary.

23. $\log_3 9 = x$
$2 = x$
$\{2\}$

25. $\log_3 x = 4$
$x = 3^4 = 81$
$\{81\}$

27. $\log_x 49 = 2$
$x^2 = 49$
$x = \pm 7$
We discard the negative base.
$\{7\}$

29. $\log_2 \frac{1}{8} = x$
$2^x = \frac{1}{8}$
$2^x = 2^{-3}$
$x = -3$
$\{-3\}$

31. $\log_3\left(\frac{1}{27}\right) = x$
$\frac{1}{27} = 3^x$
$3^{-3} = 3^x$
$-3 = x$
$\{-3\}$

33. $\log_8 x = \frac{1}{3}$
$x = 8^{1/3} = 2$
$\{2\}$

35. $\log_4 16 = x$
$4^x = 16$
$4^x = 4^2$
$x = 2$
$\{2\}$

37. $\log_{3/4} x = 3$
$\left(\frac{3}{4}\right)^3 = x$
$x = \frac{3^3}{4^3} = \frac{27}{64}$
$\left\{\frac{27}{64}\right\}$

39. $\log_x 100 = 2$ or $x^2 = 100$
$x = \pm 10$ and we discard the negative base.
$\{10\}$

41. $\log_5 5^3 = 3$

43. $2^{\log_2 3} = 3$

45. $\log_9 9 = 1$

47. $y = \log_3 x$
$y = 0$:
$\log_3 x = 0$
$x = 3^0 = 1$ is the only x-intercept
$x = 0$:
$y = \log 0$ which is not defined.
No y-intercept exists.

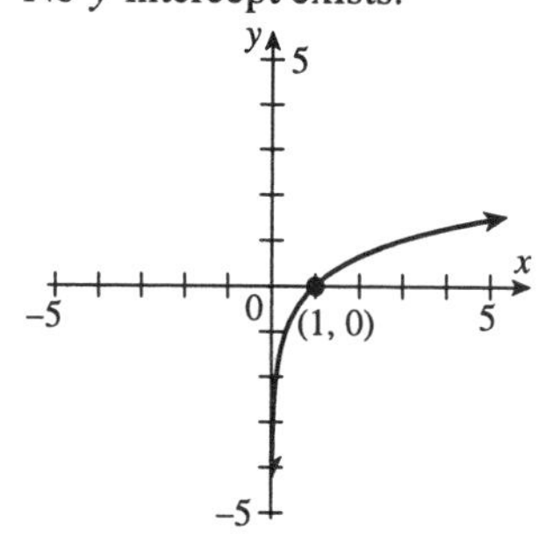

49. $f(x) = \log_{1/4} x$ or $y = \log_{1/4} x$
$y = 0$:
$0 = \log_{1/4} x$

$x = \left(\frac{1}{4}\right)^0 = 1$ is the x-intercept
$x = 0$:
$y = \log_{1/4} 0$ which is not defined.
No y-intercept exists.

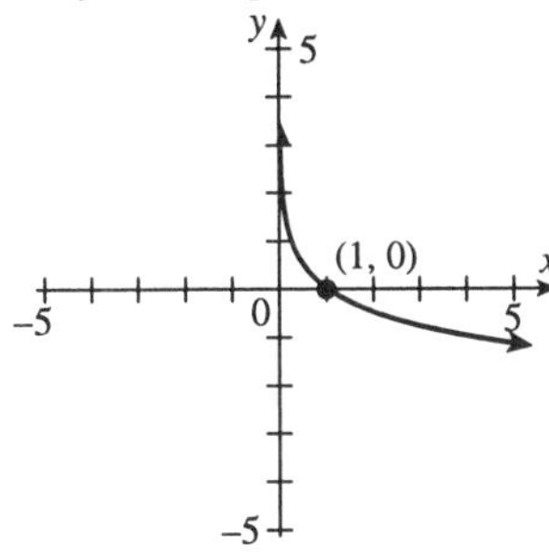

51. $f(x) = \log_5 x$ or $y = \log_5 x$
$x = 0$:
$y = \log_5 0$ is not defined so there is no y-intercept.
$y = 0$:
$0 = \log_5 x$
$x = 5^0 = 1$ is the x-intercept.

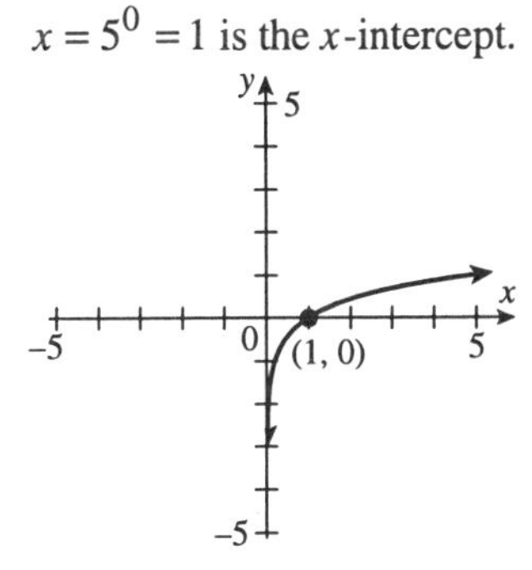

53. $f(x) = \log_{1/6} x$ or $y = \log_{1/6} x$
$x = 0$:
$y = \log_{1/6} 0$ is not defined so there is no y-intercept.
$y = 0$:
$0 = \log_{1/6} x$
$x = \left(\frac{1}{6}\right)^0 = 1$ is the x-intercept.

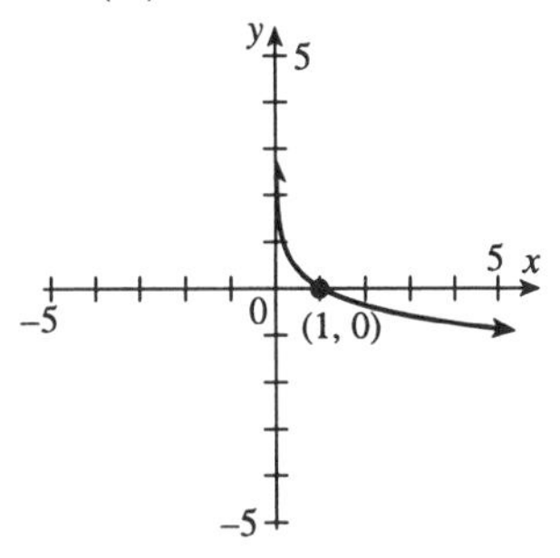

55. $y = 4^x$; $y = \log_4 x$
$x = 0$: $y = 4^0 = 1$ is the y-intercept of $y = 4^x$; hence the x-intercept of $y = \log_4 x$.
$y = 0$: $4^x = 0$ has no solution so $y = 4^x$ has no x-intercept, hence $y = \log_4 x$ has no y-intercept.

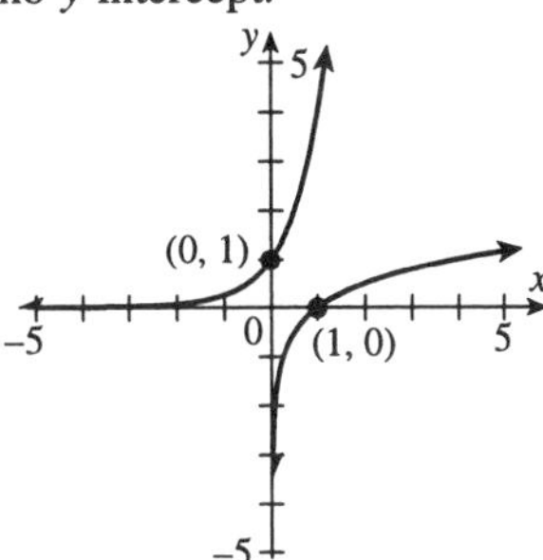

57. $y = \left(\frac{1}{3}\right)^x$; $y = \log_{1/3} x$
$x = 0$: $y = \left(\frac{1}{3}\right)^0 = 1$ is the y-intercept of $y = \left(\frac{1}{3}\right)^x$; hence the x-intercept of $y = \log_{1/3} x$.
$y = 0$: $0 = \left(\frac{1}{3}\right)^x$ has no solution so $y = \left(\frac{1}{3}\right)^x$ has no x-intercept; hence

$y = \log_{1/3} x$ has no y-intercept.

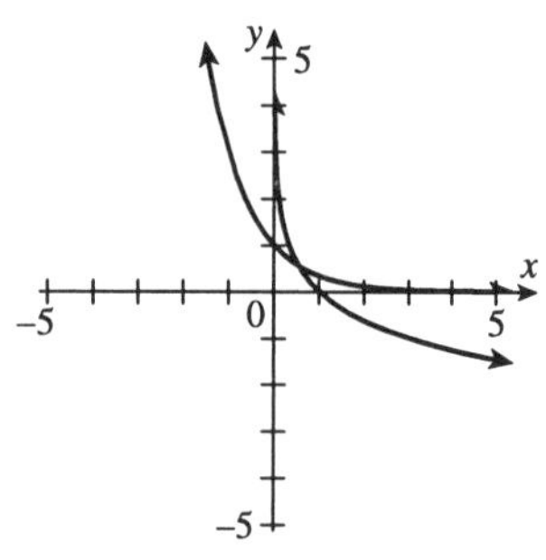

59. $\log_{10}(1-k) = \frac{-0.3}{H}$, $H = 8$

$\log_{10}(1-k) = \frac{-0.3}{8} = -.0375$

$1-k = 10^{-.0375}$

$1-10^{-.0375} = k$, so $k \approx .0827$

61. $\log_3 10$ is between 2 and 3 because $3^2 = 9$ and $3^3 = 27$

63. $\frac{x-5}{5-x} = \frac{x-5}{-(x-5)} = \frac{1}{-1} = -1$

65. $\frac{x^2-3x-10}{2+x} = \frac{(x+2)(x-5)}{x+2} = x-5$

67. $\frac{3x}{x+3} + \frac{9}{x+3} = \frac{3x+9}{x+3} = \frac{3(x+3)}{x+3} = 3$

69. $\frac{5}{y+1} - \frac{4}{y-1}$

$= \frac{5(y-1)-4(y+1)}{(y+1)(y-1)}$

$= \frac{5y-5-4y-4}{y^2-1}$

$= \frac{y-9}{y^2-1}$

Exercise Set 10.4

1. $\log_5 2 + \log_5 7 = \log_5(2\cdot 7) = \log_5 14$

3. $\log_4 9 + \log_4 x = \log_4(9x)$

5. $\log_{10} 5 + \log_{10} 2 + \log_{10}(x^2+2)$

$= \log_{10}[5\cdot 2(x^2+2)]$

$= \log_{10}(10x^2+20)$

7. $\log_5 12 - \log_5 4 = \log_5\left(\frac{12}{4}\right) = \log_5 3$

9. $\log_2 x - \log_2 y = \log_2\left(\frac{x}{y}\right)$

11. $\log_4 2 + \log_4 10 - \log_4 5$

$= \log_4 2\cdot 10 - \log_4 5 = \log_4\left(\frac{20}{5}\right)$

$= \log_4 4 = 1$

13. $2\log_2 5 = \log_2 5^2 = \log_2 25$

15. $3\log_5 x + 6\log_5 z$

$= \log_5 x^3 + \log_5 z^6 = \log_5(x^3z^6)$

17. $\log_{10} x - \log_{10}(x+1) + \log_{10}(x^2-2)$

$= \log_{10}\frac{x}{x+1} + \log_{10}(x^2-2)$

$= \log_{10}\frac{x\left(x^2-2\right)}{x+1}$

$= \log_{10}\left(\frac{x^3-2x}{x+1}\right)$

19. $\log_3\left(\frac{4y}{5}\right) = \log_3 4y - \log_3 5$

$= \log_3 4 + \log_3 y - \log_3 5$

21. $\log_2\left(\frac{x^3}{y}\right) = \log_2 x^3 - \log_2 y$

$= 3\log_2 x - \log_2 y$

23. $\log_b \sqrt{7x} = \log_b(7x)^{1/2}$

$= \frac{1}{3}\log_b(7x)$

$= \frac{1}{2}[\log_b 7 + \log_b x]$

$= \frac{1}{2}\log_b 7 + \frac{1}{2}\log_b x$

25. $\log_b\left(\frac{5}{3}\right) = \log_b 5 - \log_b 3$
$\approx .7 - .5 = .2$

27. $\log_b 15 = \log_b(5 \cdot 3)$
$= \log_b 5 + \log_b 3 \approx .7 + .5 = 1.2$

29. $\log_b \sqrt[3]{5} = \log_b 5^{1/3}$
$= \frac{1}{3}\log_b 5 \approx \frac{1}{3}(.7) \approx .23$

31. $\log_4 5 + \log_4 7 = \log_4(5 \cdot 7) = \log_4 35$

33. $\log_3 8 - \log_3 2 = \log_3\left(\frac{8}{2}\right) = \log_3 4$

35. $\log_7 6 + \log_7 3 - \log_7 4$
$= \log_7(6 \cdot 3) - \log_7 4$
$= \log_7\left(\frac{18}{4}\right) = \log_7 \frac{9}{2}$

37. $3\log_4 2 + \log_4 6 = \log_4 2^3 + \log_4 6$
$= \log_4 8 + \log_4 6 = \log_4(8 \cdot 6) = \log_4 48$

39. $3\log_2 x + \frac{1}{2}\log_2 x - 2\log_2(x+1)$
$= \log_2 x^3 + \log_2 x^{1/2} - \log_2(x+1)^2$
$= \log_2(x^3 \cdot x^{1/2}) - \log_2(x+1)^2$
$= \log_2 x^{7/2} - \log_2(x+1)^2$
$= \log_2 \frac{x^{7/2}}{(x+1)^2}$

41. $2\log_8 x - \frac{2}{3}\log_8 x + 4\log_8 x$
$= \left(2 - \frac{2}{3} + 4\right)\log_8 x = \frac{16}{3}\log_8 x$
$= \log_8 x^{16/3}$

43. $\log_7\left(\frac{5x}{4}\right) = \log_7 5x - \log_7 4$
$= \log_7 5 + \log_7 x - \log_7 4$

45. $\log_5 x^3(x+1)$
$= \log_5 x^3 + \log(x+1)$
$= 3\log_5 x + \log_5(x+1)$

47. $\log_6 \frac{x^2}{x+3} = \log_6 x^2 - \log_6(x+3)$
$= 2\log_6 x - \log_6(x+3)$

49. $\log_b 8 = \log_b 2^3 = 3\log_b 2$
$= 3(.43) = 1.29$

51. $\log_b\left(\frac{3}{9}\right) = \log_b\left(\frac{1}{3}\right)$
$= \log_b 3^{-1} = (-1)\log_b 3 = -(.68) = -.68$

53. $\log_b \sqrt{\frac{2}{3}} = \log_b\left(\frac{2}{3}\right)^{1/2} = \frac{1}{2}\log_b \frac{2}{3}$
$= \frac{1}{2}(\log_b 2 - \log_b 3)$
$= \frac{1}{2}(.43 - .68)$
$= \frac{1}{2}(-.25) = -.125$

55. $\log_2 x^3 = 3\log_2 x$
True

57. $\frac{\log_7 10}{\log_7 5} = \log_7 2$
False

59. $\frac{\log_7 x}{\log_7 y} = (\log_7 x) - (\log_7 y)$
False

61. $y = 10^x$ and $y = \log_{10} x$

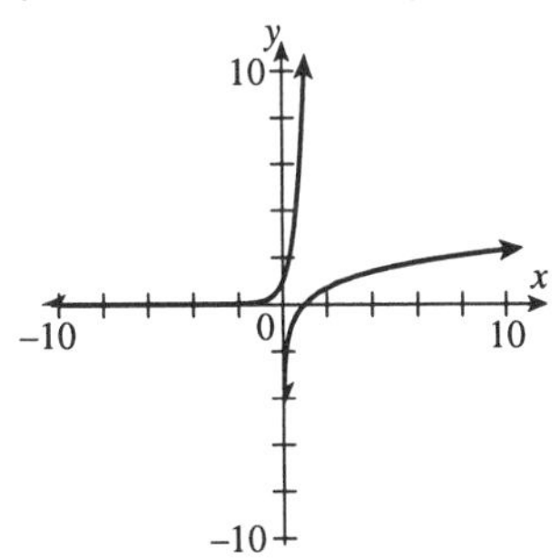

63. $\log_{10} \frac{1}{10} = x$
$10^x = \frac{1}{10}$

$10^{-1} = \frac{1}{10}$
$\{-1\}$

65. $\log_7 \sqrt{7} = x$
$7^x = \sqrt{7}$
$7^x = 7^{1/2}$
$x = \frac{1}{2}$
$\left\{\frac{1}{2}\right\}$

Exercise Set 10.5

1. $\log 8 \approx .9031$

3. $\log 2.31 \approx .3636$

5. $\ln 2 \approx .6931$

7. $\ln 0.0716 \approx -2.6367$

9. $\log 12.6 \approx 1.1004$

11. $\ln 5 \approx 1.6094$

13. $\log 41.5 \approx 1.6180$

15. Answers may vary.

17. $\log 100 = \log 10^2 = 2$

19. $\log \frac{1}{1000} = \log 10^{-3} = -3$

21. $\ln e^2 = 2$

23. $\ln \sqrt[4]{e} = \ln e^{1/4} = \frac{1}{4}$

25. $\log 10^3 = 3$

27. $\ln e^2 = 2$

29. $\log 0.0001 = \log 10^{-4} = -4$

31. $\ln \sqrt{e} = \ln e^{1/2} = \frac{1}{2}$

33. ln 50 is larger
Answers may vary.

35. $\log x = 1.3$
$x = 10^{1.3} \approx 19.9526$

37. $\log 2x = 1.1$
$2x = 10^{1.1}$
$x = \frac{1}{2}10^{1.1} \approx 6.2946$

39. $\ln x = 1.4$
$x = e^{1.4} \approx 4.0552$

41. $\ln(3x - 4) = 2.3$
$3x - 4 = e^{2.3}$
$3x = 4 + e^{2.3}$
$x = \frac{4 + e^{2.3}}{3} \approx 4.6581$

43. $\log x = 2.3$
$x = 10^{2.3} \approx 199.5262$

45. $\ln x = -2.3$
$x = e^{-2.3} \approx .1003$

47. $\log(2x + 1) = -0.5$
$2x + 1 = 10^{-.5}$
$2x = -1 + 10^{-.5}$
$x = \frac{-1 + 10^{-.5}}{2} \approx -.3419$

49. $\ln 4x = 0.18$
$4x = e^{.18}$
$x = \frac{e^{.18}}{4} \approx .2993$

51. $\log_2 3 = \frac{\ln 3}{\ln 2} \approx 1.5850$

53. $\log_{1/2} 5 = \frac{\ln 5}{\ln\left(\frac{1}{2}\right)} \approx -2.3219$

55. $\log_4 9 = \frac{\ln 9}{\ln 4} \approx 1.5850$

57. $\log_3\left(\frac{1}{6}\right) = \log_3 6^{-1}$
$= (-1)\log_3 6$
$= -\frac{\ln 6}{\ln 3} \approx -1.6309$

59. $\log_8 6 = \frac{\ln 6}{\ln 8} \approx .8617$

61. $R = \log\left(\frac{a}{T}\right) + B,\ a = 200,$
$T = 1.6$, and $B = 2.1$
$R = \log\left(\frac{200}{1.6}\right) + 2.1 \approx 4.2$
The earthquake measures 4.2 on the Richter scale.

63. $R = \log\left(\frac{a}{T}\right) + B,\ a = 400,$
$T = 2.6$, and $B = 3.1$
$R = \log\left(\frac{400}{2.6}\right) + 3.1 \approx 5.3$
The earthquake measures 5.3 on the Richter scale.

65. $A = Pe^{rt}$, $t = 12$, $P = 1400$, and $r = .08$
$A = 1400e^{(.08)12} = 1400e^{.96} \approx 3656.38$
Dana has \$3656.38 after 12 years.

67. $A = Pe^{rt}$, $t = 4$, $r = .06$, and $P = 2000$
$A = 2000e^{(.06)4} = 2000e^{.24} \approx 2542.50$
Paul owes \$2542.50 at the end of 4 years.

69. $f(x) = e^x$ or $y = e^x$

71. $f(x) = e^{-3x}$ or $y = e^{-3x}$

73. $f(x) = e^x + 2$ or $y = e^x + 2$

75. $f(x) = e^{x-1}$ or $y = e^{x-1}$

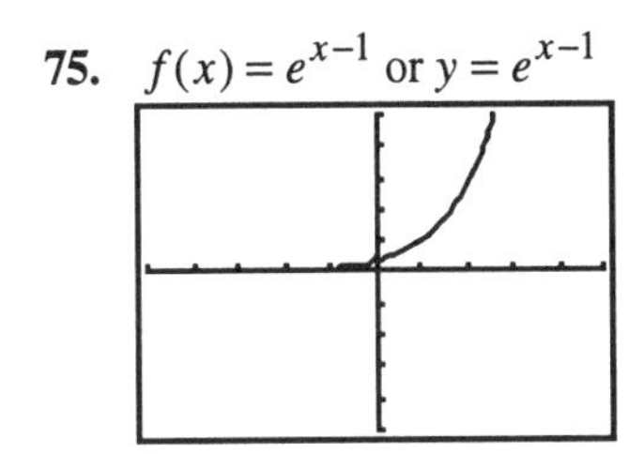

77. $f(x) = 3e^x$ or $y = 3e^x$

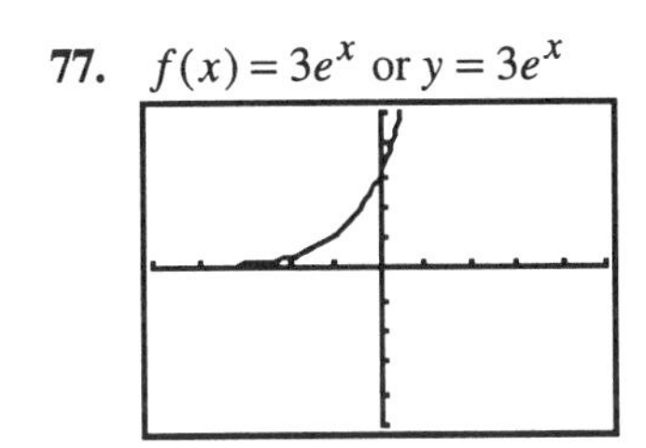

79. $f(x) = \ln x$ or $y = \ln x$

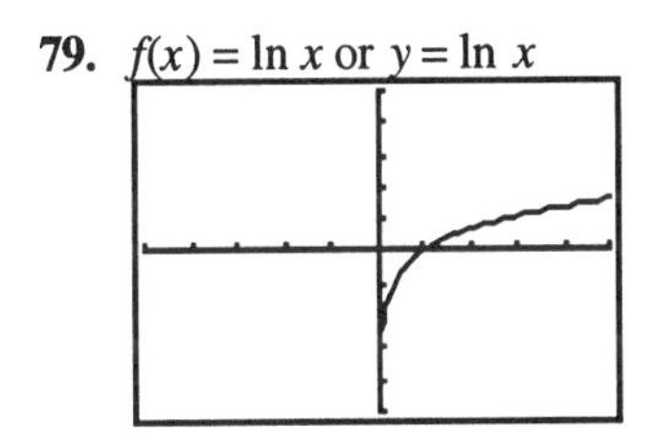

81. $f(x) = -2\log x$ or $y = -2\log x$

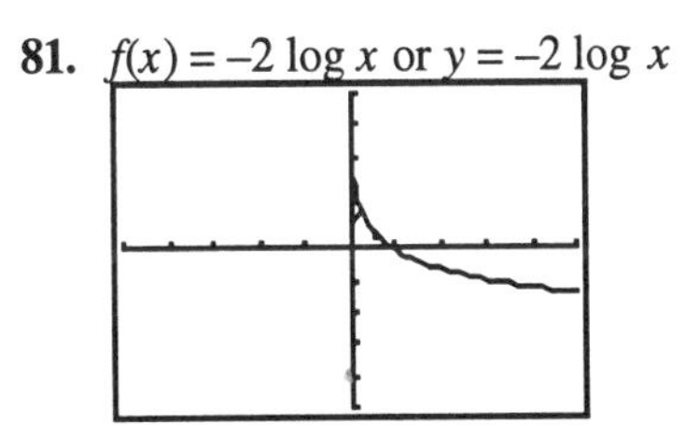

83. $f(x) = \ln(x + 2)$ or $y = \ln(x + 2)$

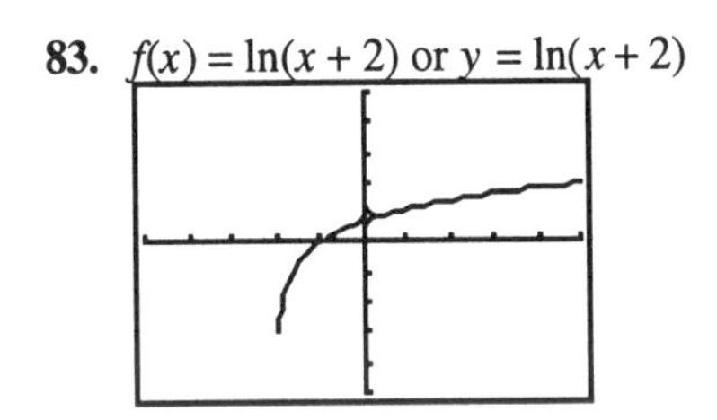

85. $f(x) = \log x - 3$ or $y = \log x - 3$

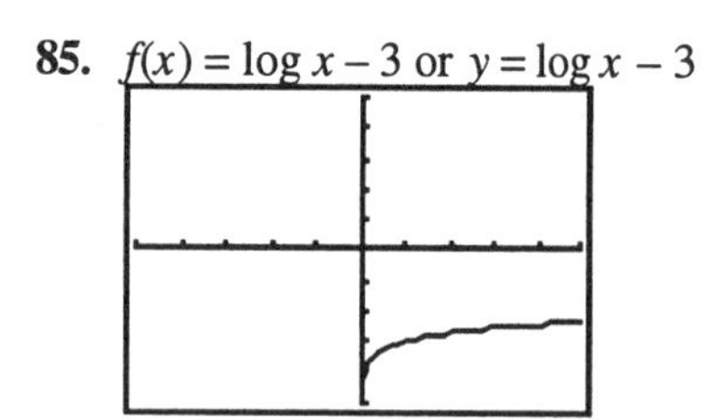

87. $f(x) = e^x$

$f(x) = e^x + 2$

$f(x) = e^x - 3$

89. $6x - 3(2 - 5x) = 6$
$6x - 6 + 15x = 6$
$21x - 6 = 6$
$21x = 12$
$x = \frac{12}{21} = \frac{4}{7}$
$\left\{\frac{4}{7}\right\}$

91. $2x + 3y = 6x$
$3y = 4x$
$x = \frac{3y}{4}$

93. $x^2 + 7x = -6$
$x^2 + 7x + 6 = 0$
$(x + 6)(x + 1) = 0$
$x + 6 = 0$ or $x + 1 = 0$
$x = -6$ or $x = -1$
$\{-6, -1\}$

95. $\begin{cases} x + 2y = -4 \\ 3x - y = 9 \end{cases}$

$\rightarrow \begin{cases} x + 2y = -4 \\ 6x - 2y = 18 \end{cases}$

Add.
$7x = 14$
$x = 2$
Substitute back.
$2 + 2y = -4$
$2y = -6$
$y = -3$
$\{(2, -3)\}$

Exercise Set 10.6

1. $3^x = 6$
$x = \log_3 6 = \frac{\ln 6}{\ln 3} \approx 1.6309$

3. $3^{2x} = 3.8$
$2x = \log_3 3.8 = \frac{\ln 3.8}{\ln 3}$ so
$x = \frac{\ln 3.8}{2 \ln 3} \approx .6076$
$\left\{\frac{\ln 3.8}{2 \ln 3}\right\}$; $\{.6076\}$

5. $2^{x-3} = 5$
$x - 3 = \log_2 5$
$x = 3 + \log_2 5 = 3 + \frac{\ln 5}{\ln 2} = 5.3219$

7. $9^x = 5$
$x = \log_9 5 = \frac{\ln 5}{\ln 9} \approx .7325$

9. $4^{x+7} = 3$
$x + 7 = \log_4 3$
$x = -7 + \log_4 3 = -7 + \frac{\ln 3}{\ln 4} \approx -6.2075$

11. $7^{3x-4} = 11$
$3x - 4 = \log_7 11$
$3x = 4 + \log_7 11$
$x = \frac{4 + \log_7 11}{3} = 4 + \frac{\frac{\ln 11}{\ln 7}}{3} \approx 1.7441$

13. $e^{6x} = 5$
$6x = \ln 5$
$x = \frac{1}{6} \ln 5 \approx .2682$

15. $\log_2(x + 5) = 4$
$x + 5 = 2^4$
$x + 5 = 16$
$x = 11$
$\{11\}$

17. $\log_3 x^2 = 4$
$x^2 = 3^4$
$x^2 = 81$
$x = \pm 9$
$\{9, -9\}$

19. $\log_4 2 + \log_4 x = 0$
$\log_4(2x) = 0$
$2x = 4^0$
$2x = 1$
$x = \frac{1}{2}$
$\left\{\frac{1}{2}\right\}$

21. $\log_2 6 - \log_2 x = 3$
$\log_2\left(\frac{6}{x}\right) = 3$
$\frac{6}{x} = 2^3$
$\frac{6}{x} = 8$
$8x = 6$
$x = \frac{6}{8} = \frac{3}{4}$
$\left\{\frac{3}{4}\right\}$

23. $\log_4 x + \log_4(x+6) = 2$
$\log_4 x(x+6) = 2$
$x(x+6) = 4^2$
$x^2 + 6x = 16$
$x^2 + 6x - 16 = 0$
$(x+8)(x-2) = 0$
$x + 8 = 0$ or $x - 2 = 0$
$x = -8$ or $x = 2$
We discard –8 as extraneous.
$\{2\}$

25. $\log_5(x+3) - \log_5 x = 2$
$\log_5\left(\frac{x+3}{x}\right) = 2$
$\frac{x+3}{x} = 5^2$
$\frac{x+3}{x} = 25$
$x + 3 = 25x$
$3 = 24x$
$x = \frac{3}{24} = \frac{1}{8}$
$\left\{\frac{1}{8}\right\}$

27. $\log_3(x-2) = 2$
$x - 2 = 3^2$
$x - 2 = 9$
$x = 11$
$\{11\}$

29. $\log_4(x^2 - 3x) = 1$
$x^2 - 3x = 4^1$
$x^2 - 3x = 4$
$x^2 - 3x - 4 = 0$
$(x-4)(x+1) = 0$
$x - 4 = 0$ or $x + 1 = 0$
$x = 4$ or $x = -1$
$\{4, -1\}$

31. $\ln 5 + \ln x = 0$
$\ln(5x) = 0$
$e^0 = 5x$
$1 = 5x$
$\frac{1}{5} = x$
$\left\{\frac{1}{5}\right\}$

33. $3\log x - \log x^2 = 2$
$3\log x - 2\log x = 2$
$\log x = 2$
$10^2 = 100$
$\{100\}$

35. $\log_2 x + \log_2(x+5) = 1$
$\log_2 x(x+5) = 1$
$x(x+5) = 2^1$
$x^2 + 5x = 2$
$x^2 + 5x - 2 = 0$
$a = 1$, $b = 5$, and $c = -2$
$x = \frac{-5 \pm \sqrt{5^2 - 4(1)(-2)}}{2(1)}$
$= \frac{-5 \pm \sqrt{25+8}}{2} = \frac{-5 \pm \sqrt{33}}{2}$

$= -\frac{5}{2} \pm \frac{\sqrt{33}}{2}$

Discard $-\frac{5}{2} - \frac{\sqrt{33}}{2}$

$\left\{-\frac{5}{2} + \frac{\sqrt{33}}{2}\right\}$

37. $\log_4 x - \log_4(2x-3) = 3$

$\log_4\left(\frac{x}{2x-3}\right) = 3$

$\frac{x}{2x-3} = 4^3$

$\frac{x}{2x-3} = 64$

$x = 64(2x-3)$

$x = 128x - 192$

$192 = 127x$

$x = \frac{192}{127}$

$\left\{\frac{192}{127}\right\}$

39. $\log_2 x + \log_2(3x+1) = 1$

$\log_2 x(3x+1) = 1$

$x(3x+1) = 2^1$

$3x^2 + x = 2$

$3x^2 + x - 2 = 0$

$(3x-2)(x+1) = 0$

$3x - 2 = 0$ or $x + 1 = 0$

$3x = 2$ or $x = -1$

$x = \frac{2}{3}$

Discard –1 as extraneous.

$\left\{\frac{2}{3}\right\}$

41. $y = y_0 e^{.043t}$, $y_0 = 83$ and $t = 5$

$y = 83e^{.043(5)} = 83e^{.215} \approx 103$

There should be 103 wolves in 5 years.

43. $y = y_0 e^{.026t}$, $y_0 = 9,000,000$ and $t = 5$

$y = 9,000,000e^{.026(5)} = 9,000,000e^{.13}$

= 10,249,456

There will be approximately 10,250,000 inhabitants by 1995.

45. $A = P\left(1+\frac{r}{n}\right)^{nt}$, $P = 600$,

$A = 2(600) = 1200$, $r = .07$, and $n = 12$

$1200 = 600\left(1+\frac{.07}{12}\right)^{12t}$

$2 = (1.0058\bar{3})^{12t}$

$12t = \log_{1.0058\bar{3}}(2)$

$t = \frac{1}{12}\log_{1.0058\bar{3}}(2)$

$= \frac{1}{12}\frac{\ln 2}{\ln(1.0058\bar{3})} \approx 10$

It would take approximately 10 years for the \$600 to double.

47. $A = P\left(1+\frac{r}{n}\right)^{nt}$, $P = 1200$,

$A = P + I = 1200 + 200 = 1400$,

$r = .09$, and $n = 4$

$1400 = 1200\left(1+\frac{.09}{4}\right)^{4t}$

$\frac{7}{6} = (1.0225)^{4t}$

$4t = \log_{1.0225}\left(\frac{7}{6}\right)$

$t = \frac{1}{4}\log_{1.0225}\left(\frac{7}{6}\right)$

$= \frac{1}{4}\frac{\ln\left(\frac{7}{6}\right)}{\ln 1.0225} \approx 1.73$

It would take the investment approximately 1.7 years to earn \$200.

49. $A = P\left(1+\frac{r}{n}\right)^{nt}$, $P = 1000$,

$A = 2(1000) = 2000$, $r = .08$, and $n = 2$.

$2000 = 1000\left(1+\frac{.08}{2}\right)^{2t}$

$2 = (1.04)^{2t}$

$2t = \log_{1.04} 2 = \frac{\ln 2}{\ln 1.04}$

$t = \frac{1}{2}\frac{\ln 2}{\ln 1.04} \approx 8.8$

It would take approximately 8.8 years for the \$1000 to double.

51. $W = .00185h^{2.67}$, $w = 85$

$85 = .00185h^{2.67}$

$\frac{85}{.00185} = h^{2.67}$

$h = \left(\frac{85}{.00185}\right)^{1/2.67} \approx 55.7$

The expected height of the boy is approximately 55.7 inches.

53. $P = 14.7e^{-.21x}$, $x = 1$

$= 14.7e^{-.21(1)} = 14.7e^{-.21} \approx 11.9$

The average atmospheric pressure of Denver is approximately 11.9 lbs/in.^2.

55. $P = 14.7e^{-.21x}$, $P = 7.5$

$7.5 = 14.7e^{-.21x}$

$\frac{7.5}{14.7} = e^{-.21x}$

$-.21x = \ln\left(\frac{7.5}{14.7}\right)$

$x = -\ln\left(\frac{7.5}{14.7}\right) \approx 3.2$

The elevation of the jet is approximately 3.2 miles.

57. $t = \frac{1}{c}\ln\left(\frac{A}{A-N}\right)$

$t = \frac{1}{0.09}\ln\left(\frac{75}{75-50}\right)$

$t = \frac{1}{0.09}\ln(3)$

$t \approx 12.21$

It will take 13 weeks.

59. $t = \frac{1}{c}\ln\left(\frac{A}{A-N}\right)$

$t = \frac{1}{0.07}\ln\left(\frac{210}{210-150}\right)$

$t = \frac{1}{0.07}\ln(3.5)$

$t \approx 17.9$

It will take 18 weeks.

61. $y_1 = e^{0.3x}$ and $y_2 = 8$

$\{6.93\}$

63. $y_1 = 2\log(-5.6x + 1.3)$

$y_2 = -x - 1$

$\{-3.68\}$

65. $y_1 = 7^{3x-4}$

$y_2 = 11$

$\{1.74\}$

67. $y_1 = \ln 5$

$y_2 = -\ln x$

$\{0.2\}$

69. $\frac{x^2 - y + 2z}{3x} = \frac{(-2)^2 - (0) + 2(3)}{3(-2)}$

$= \frac{4+6}{-6} = \frac{-10}{6} = -\frac{5}{3}$

71. $\frac{3z - 4x + y}{x + 2z} = \frac{3(3) - 4(-2) + 0}{-2 + 2(3)}$

$= \frac{9+8}{-2+6} = \frac{17}{4}$

73. $f(x) = 5x + 2$ so $f^{-1}(x) = \frac{x-2}{5}$

Chapter 10 - Review

1. $(f \circ g)(x) = f(g(x))$

$= f(x+1) = (x+1)^2 - 2$

$= x^2 + 2x + 1 - 2 = x^2 + 2x - 1$

2. $(g \circ f)(x) = g(f(x))$

$= g(x^2 - 2) = x^2 - 2 + 1 = x^2 - 1$

3. $(h \circ g)(2) = h(g(2))$

$= h(3) = 3^3 - 3^2 = 18$

4. $(f \circ f)(x) = f(f(x))$

$= f(x^2 - 2) = (x^2 - 2)^2 - 2$

$= x^4 - 4x^2 + 4 - 2 = x^4 - 4x^2 + 2$

5. $(f \circ g)(-1) = f(g(-1))$

$= f(0) = 0^2 - 2 = -2$

6. $(h \circ h)(2) = h(h(2))$

$= h(4) = 4^3 - 4^2 = 48$

7. The function is one-to-one.

$h^{-1} = \{(14, -9), (8, 6), (12, -11), (15, 15)\}$

8. The function is not one-to-one.

9. The function is one-to-one.

Rank (input)	2	4	1	3
Region (output)	West	Midwest	South	Northeast

10. The function is not one-to-one.

11. $f(x)=\sqrt{x+2}$

a. $f(7)=\sqrt{7+2}=\sqrt{9}=3$

b. $f^{-1}(3)=7$

12. $f(x)=\sqrt{x+2}$

a. $f(-1)=\sqrt{-1+2}=\sqrt{1}=1$

b. $f^{-1}(1)=-1$

13. The graph does not represent a one-to-one function.

14. The graph is not a one-to-one function since it fails the horizontal line test.

15. The graph does not represent a one-to-one function.

16. The graph is a one-to-one function since it passes the horizontal line test.

17. $f(x)=6x+11$ so $f^{-1}(x)=\frac{x-11}{6}$

18. $f(x)=12x$

$y=12x$

$x=\frac{y}{12}$

$y=\frac{x}{12}$

$f^{-1}(x)=\frac{x}{12}$

19. $q(x)=mx+b$ so $q^{-1}(x)=\frac{x-b}{m}$

20. $g(x)=\frac{12x-7}{6}$

$y=\frac{12x-7}{6}$

$\frac{6y+7}{12}=x$

$\frac{6x+7}{12}=y$

$g^{-1}(x)=\frac{6x+7}{12}$

21. $r(x)=\frac{13}{2}x-4$ so $r^{-1}(x)=\frac{2}{13}(x+4)$

22. $y=g(x)=\sqrt{x}$

$y^2=x$

$x^2=y=f^{-1}(x),\ x\geq 0$

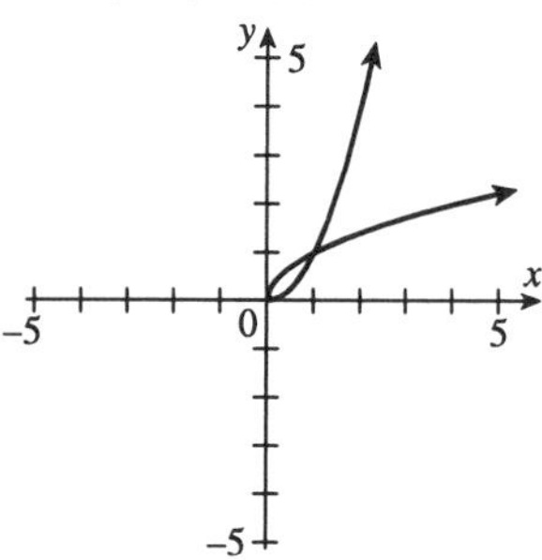

23. $h(x)=5x-5$

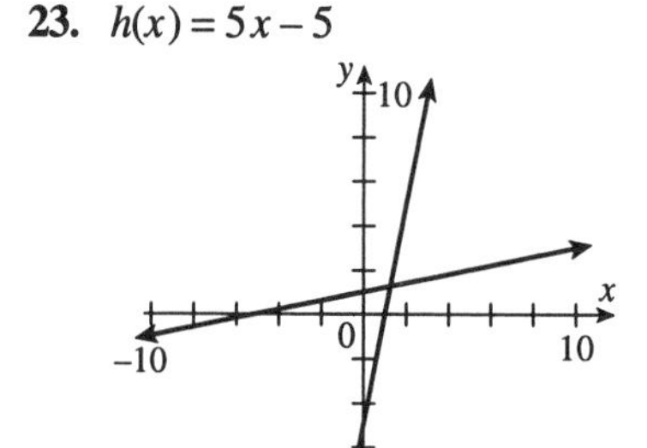

24. $f(x)=2x-3$ or $y=2x-3$

Find inverse

$x=2y-3$

$y = \frac{x+3}{2}$

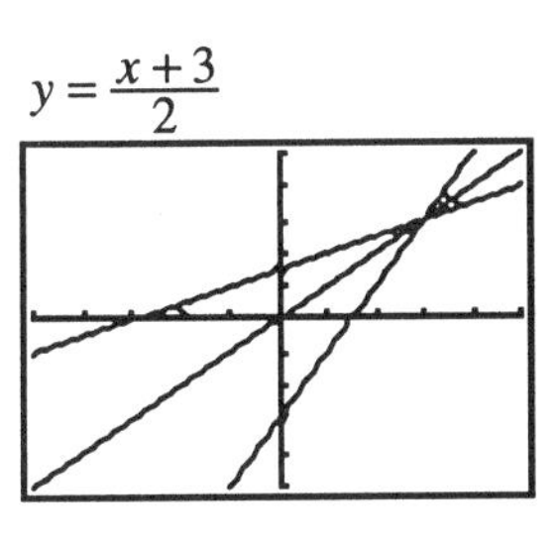

25. $4^x = 64$
$4^x = 4^3$
$x = 3$

26. $3^x = \frac{1}{9}$
$3^x = 3^{-2}$ so $x = -2$
$\{-2\}$

27. $2^{3x} = \frac{1}{16}$
$2^{3x} = 2^{-4}$
$3x = -4$
$x = -\frac{4}{3}$

28. $5^{2x} = 125$
$5^{2x} = 5^3$
$2x = 3$
$x = \frac{3}{2}$
$\left\{\frac{3}{2}\right\}$

29. $9^{x+1} = 243$
$(3^2)^{x+1} = 3^5$
$3^{2x+2} = 3^5$
$2x + 2 = 5$
$2x = 3$
$x = \frac{3}{2}$

30. $8^{3x-2} = 4$
$(2^3)^{3x-2} = 2^2$
$2^{9x-6} = 2^2$
$9x - 6 = 2$
$9x = 8$
$x = \frac{8}{9}$
$\left\{\frac{8}{9}\right\}$

31.

x	$y = 3^x$
–2	$\frac{1}{9}$
–1	$\frac{1}{3}$
0	1
1	3
2	9

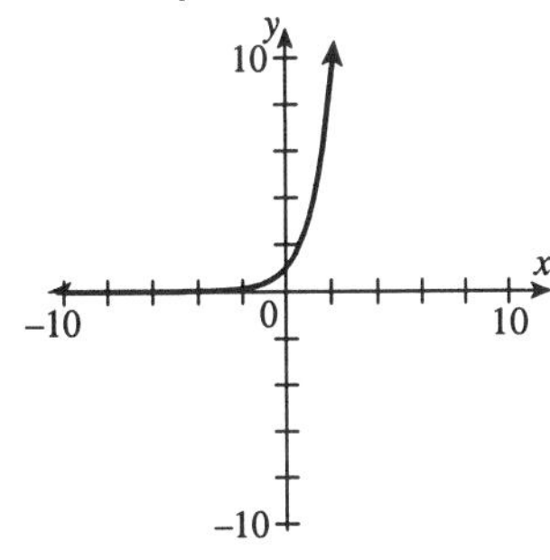

32. $y = \left(\frac{1}{3}\right)^x$

33.

x	$y = 4 \cdot 2^x$
–2	1
–1	2
0	4
1	8

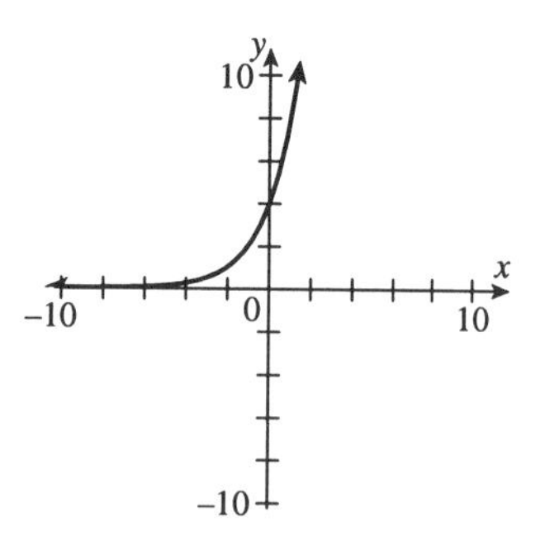

34. $y = 2^x + 4$

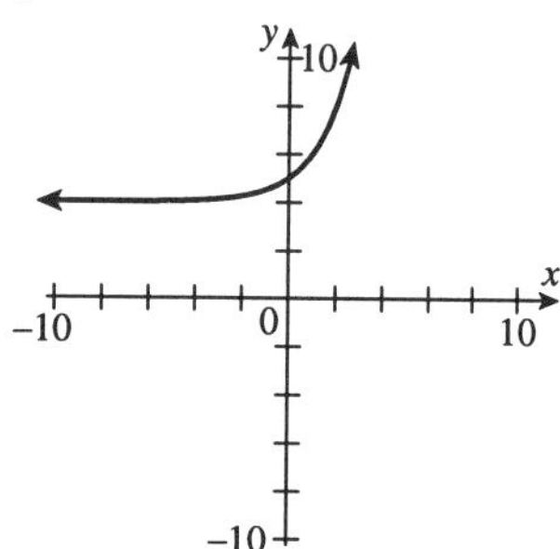

35. $A = P\left(1+\frac{r}{n}\right)^{nt}$

$A = 1600\left(1+\frac{.09}{2}\right)^{(2)(7)}$

$A = \$2963.10$

36. $A = P\left(1+\frac{r}{n}\right)^{nt}$, $P = 800$, $r = .07$, $n = 4$, and $t = 5$

$A = 800\left(1+\frac{.07}{4}\right)^{4(5)} = 800(1.0175)^{20}$

≈ 1131.82

The certificate is worth \$1131.82 at the end of 5 years.

37. The results are the same.

38. $7^2 = 49$

$\log_7 49 = 2$

39. $2^{-4} = \frac{1}{16}$

$-4 = \log_2\left(\frac{1}{16}\right)$

40. $\log_{1/2} 16 = -4$

$\left(\frac{1}{2}\right)^{-4} = 16$

41. $\log_{.4} .064 = 3$

$.4^3 = .064$

42. $\log_4 x = -3$

$x = 4^{-3} = \frac{1}{64}$

43. $\log_3 x = 2$

$x = 3^2 = 9$

$\{9\}$

44. $\log_3 1 = x$

$3^x = 1$

$3^x = 3^0$

$x = 0$

45. $\log_4 64 = x$

$3 = x$

$\{3\}$

46. $\log_x 64 = 2$

$x^2 = 64$

$x = \pm\sqrt{64} = \pm 8$

$x = 8$ since base > 0

47. $\log_x 81 = 4$

$x^4 = 81$

$x = \pm 3$

We discard the negative base −3.

$\{3\}$

48. $\log_4 4^5 = x$

$x = 5$ from property 2

49. $\log_7 7^{-2} = x$

$-2 = x$

$\{-2\}$

50. $5^{\log_5 4} = x$

$x = 4$ from property 3

51. $2^{\log_2 9} = x$
$9 = x$
$\{9\}$

52. $\log_2(3x-1) = 4$
$3x - 1 = 2^4 = 16$
$3x = 17$
$x = \frac{17}{3}$

53. $\log_3(2x+5) = 2$
$2x + 5 = 3^2$
$2x + 5 = 9$
$2x = 4$
$x = 2$
$\{2\}$

54. $\log_4(x^2 - 3x) = 1$
$x^2 - 3x = 4^1 = 4$
$x^2 - 3x - 4 = 0$
$(x+1)(x-4) = 0$
$x + 1 = 0$ or $x - 4 = 0$
$x = -1$ or $x = 4$

55. $\log_8(x^2 + 7x) = 1$
$x^2 + 7x = 8^1$
$x^2 + 7x = 8$
$x^2 + 7x - 8 = 0$
$(x-8)(x-1) = 0$
$x + 8 = 0$ or $x - 1 = 0$
$x = -8$ or $x = 1$
$\{-8, 1\}$

56.

x	$y = 2^x$
−2	$\frac{1}{4}$
−1	$\frac{1}{2}$
0	1
1	2
2	4

x	$y = \log_2 x$
$\frac{1}{4}$	−2
$\frac{1}{2}$	−1
1	0
2	1
4	2

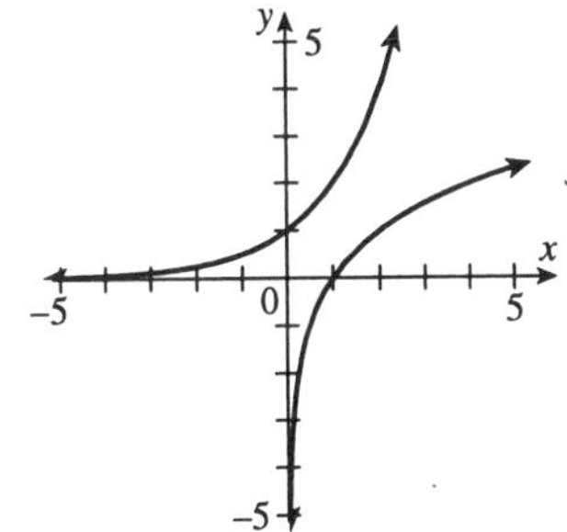

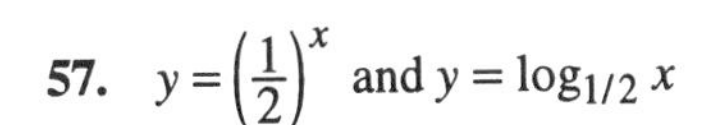
57. $y = \left(\frac{1}{2}\right)^x$ and $y = \log_{1/2} x$

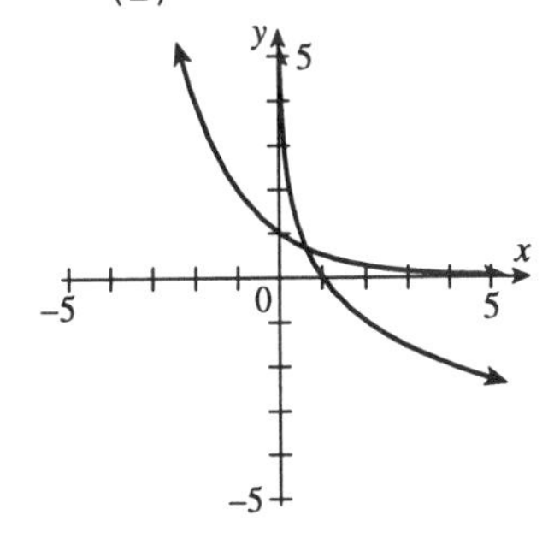

58. $\log_3 8 + \log_3 4 = \log_3(8)(4) = \log_3 32$

59. $\log_2 6 + \log_2 3 = \log_2(6 \cdot 3) = \log_2 18$

60. $\log_7 15 - \log_7 20 = \log_7 \frac{15}{20} = \log_7 \frac{3}{4}$

61. $\log 18 - \log 12 = \log \frac{18}{12} = \log\left(\frac{3}{2}\right)$

62. $\log_{11} 8 + \log_{11} 3 - \log_{11} 6$
$= \log_{11} \frac{(8)(3)}{6} = \log_{11} 4$

63. $\log_5 14 + \log_5 3 - \log_5 21$
$= \log_5(14 \cdot 3) - \log_5 21$
$= \log_5\left(\frac{42}{21}\right) = \log_5 2$

64. $2\log_5 x - 2\log_5(x+1) + \log_5 x$
$= \log_5 x^2 - \log_5(x+1)^2 + \log_5 x$
$= \log_5 \frac{(x^2)(x)}{(x+1)^2} = \log_5 \frac{x^3}{(x+1)^2}$

65. $4\log_3 x - \log_3 x + \log_3(x+2)$
$= 3\log_3 x + \log_3(x+2)$
$= \log_3 x^3 + \log_3(x+2)$
$= \log_3[x^3(x+2)]$
$= \log_3(x^4 + 2x^3)$

66. $\log_3 \frac{x^3}{x+2} = \log_3 x^3 - \log_3(x+2)$
$= 3\log_3 x - \log_3(x+2)$

67. $\log_4 \frac{x+5}{x^2} = \log_4(x+5) - \log_4 x^2$
$= \log_4(x+5) - 2\log_4 x$

68. $\log_2 \frac{3x^2 y}{z}$
$= \log_2 3 + \log_2 x^2 + \log_2 y - \log_2 z$
$= \log_2 3 + 2\log_2 x + \log_2 y - \log_2 z$

69. $\log_7 \frac{yz^3}{x} = \log_7(yz^3) - \log_7 x$
$= \log_7 y + \log_7 z^3 - \log_7 x$
$= \log_7 y + 3\log_7 z - \log_7 x$

70. $\log_6 50$
$= \log_6(5)(5)(2)$
$= \log_6(5) + \log_6(5) + \log_6(2)$
$= 0.83 + 0.83 + 0.36 = 2.02$

71. $\log_b \frac{4}{5} = \log_b 4 - \log_b 5$
$= \log_b 2^2 - \log_b 5$
$= 2\log_b 2 - \log_b 5$
$= 2(.36) - .83 = .72 - .83 = -.11$

72. $\log 3.6 \approx 0.5563$

73. $\log .15 \approx -.8239$

74. $\ln 1.25 \approx 0.2231$

75. $\ln 4.63 \approx 1.5326$

76. $\log 1000 = 3$

77. $\log \frac{1}{10} = \log 10^{-1} = -1$

78. $\ln \frac{1}{e} = \ln 1 - \ln e = 0 - 1 = -1$

79. $\ln(e^4) = 4$

80. $\ln(2x) = 2$
$2x = e^2$
$x = \frac{e^2}{2}$

81. $\ln(3x) = 1.6$
$3x = e^{1.6}$
$x = \frac{1}{3}e^{1.6}$
$\left\{\frac{1}{3}e^{1.6}\right\}$

82. $\ln(2x - 3) = -1$
$2x - 3 = e^{-1}$
$x = \frac{e^{-1} + 3}{2}$

83. $\ln(3x + 1) = 2$
$3x + 1 = e^2$
$3x = -1 + e^2$
$x = \frac{-1 + e^2}{3}$
$\left\{\frac{-1 + e^2}{3}\right\}$

84. $\frac{I}{I_0} = e^{-kx}$
$\frac{0.3I_0}{I_0} = e^{-2.1x}$
$0.3 = e^{-2.1x}$

$\frac{\ln .03}{-2.1} = x$
$x = 1.67$ mm

85. $\ln\left(\frac{I}{I_0}\right) = -kx,\ k = 3.2$

$\frac{I}{I_0} = .02$
$\ln .02 = -3.2x$
$x = \frac{-\ln .02}{3.2} \approx 1.22$
2% of the original radioactivity will penetrate at a depth of approximately 1.22 mm.

86. $\log_5 1.6 = \frac{\log 1.6}{\log 5} = 0.2920$

87. $\log_3 4 = \frac{\log 4}{\log 3} \approx 1.2619$

88. $A = Pe^{rt}$
$A = 1450e^{(0.06)(5)}$
$A = \$1957.30$

89. $A = Pe^{rt}$, $P = 940$, $r = .11$, and $t = 3$
$A = 940e^{.11(3)} = 940e^{.33} \approx 1307.51$
The \$940 investment grows to \$1307.51 in 3 years.

90. $3^{2x} = 7$
$2x \log 3 = \log 7$
$x = \frac{\log 7}{2 \log 3} \approx 0.8856$

91. $6^{3x} = 5$
$3x = \log_6 5$
$x = \frac{1}{3}\log_6 5 = \frac{1}{3}\frac{\ln 5}{\ln 6} = .2994$

92. $3^{2x+1} = 6$
$(2x+1)\log 3 = \log 6$
$2x = \frac{\log 6}{\log 3} - 1$
$x = \frac{1}{2}\left(\frac{\log 6}{\log 3} - 1\right) \approx 0.3155$

93. $4^{3x+2} = 9$
$3x + 2 = \log_4 9$
$3x = \log_4 9 - 2$
$x = \frac{\log_4 9 - 2}{3} = \frac{\frac{\ln 9}{\ln 4} - 2}{3} \approx -.1383$

94. $5^{3x-5} = 4$
$(3x-5)\log 5 = \log 4$
$3x = \frac{\log 4}{\log 5} + 5$
$x = \frac{1}{3}\left[\frac{\log 4}{\log 5} + 5\right] \approx 1.9538$

95. $8^{4x-2} = 3$
$4x - 2 = \log_8 3$
$4x = \log_8 3 + 2$
$x = \frac{\log_8 3 + 2}{4}$
$= \frac{\frac{\ln 3}{\ln 8} + 2}{4} \approx .6321$

96. $2 \cdot 5^{x-1} = 1$
$\log 2 + (x-1)\log 5 = \log 1$
$(x-1)\log 5 = \log 2$
$x = \left[\frac{-\log 2}{\log 5} + 1\right] \approx 0.5693$

97. $3 \cdot 4^{x+5} = 2$
$4^{x+5} = \frac{2}{3}$
$x + 5 = \log_4\left(\frac{2}{3}\right)$
$x = -5 \log_4\left(\frac{2}{3}\right)$
$= -5 + \frac{\ln\left(\frac{2}{3}\right)}{\ln 4} \approx -5.2925$

98. $\log_5 2 + \log_5 x = 2$
$\log_5 2x = 2$
$2x = 5^2 = 25$
$x = \frac{25}{2}$

99. $\log_3 x + \log_3 10 = 2$
$\log_3(10x) = 2$

$10x = 3^2$
$10x = 9$
$x = \frac{9}{10} = .9$
$\{.9\}$

100. $\log(5x) - \log(x + 1) = 4$
$\log \frac{5x}{x+1} = 4$
$\frac{5x}{x+1} = 10^4 = 10{,}000$
$5x = 10{,}000x + 10{,}000$
$-10{,}000 = 9995x$
$-1.0005 = x$
No solution since $x > 0$.

101. $\ln(3x) - \ln(x - 3) = 2$
$\ln\left(\frac{3x}{x-3}\right) = 2$
$\frac{3x}{x-3} = e^2$
$3x = e^2(x-3)$
$3x = e^2x - 3e^2$
$3x - e^2x = -3e^2$
$(3 - e^2)x = -3e^2$
$x = \frac{-3e^2}{3-e^2}$
$\left\{\frac{-3e^2}{3-e^2}\right\}$

102. $\log_2 x + \log_2 2x - 3 = 1$
$\log_2 (x)(2x) = 4$
$2x^2 = 2^4 = 16$
$x^2 = 8$
$x = \pm\sqrt{8} = \pm 2\sqrt{2}$
$x = 2\sqrt{2}$ since $x > 0$

103. $-\log_6(4x + 7) + \log_6 x = 1$
$\log_6 \frac{x}{4x+7} = 1$
$\frac{x}{4x+7} = 6^1$
$\frac{x}{4x+7} = 6$
$x = 6(4x + 7)$
$x = 24x + 42$
$-42 = 23x$
$x = -\frac{42}{23}$
We discard $x = -\frac{42}{23}$ as extraneous leaving no solution.
$\{\ \}$

104. $y = y_0 e^{kt}$
$y = 155{,}000e^{(0.06)(4)}$
$y = 197{,}044$ ducks

105. $y = y_0 e^{kt}$, $k = .017$,
$y_0 = 203{,}583{,}886$ and $t = 10$
$y = 203{,}583{,}886e^{.017(10)}$
$= 203{,}583{,}886e^{.17} \approx 241{,}308{,}968$
The expected population of Indonesia by the year 2005 is approximately 241,308,968.

106. $500{,}000 = 230{,}000e^{0.0316t}$
$\ln \frac{50}{23} = 0.0316t$
$\frac{1}{0.0316} \ln \frac{50}{23} = t$
$t = 24.6$ years ≈ 25 years

107. $y = y_0 e^{kt}$, $k = .0036$,
$y_0 = 650{,}000$, and $y = 700{,}000$
$700{,}000 = 650{,}000e^{.0036t}$
$\frac{14}{3} = e^{.0036t}$
$.0036t = \ln\left(\frac{14}{13}\right)$
$t = \frac{\ln\left(\frac{14}{13}\right)}{.0036} \approx 20.6$
It will take approximately 21 years for the population of Memphis to reach 700,000.

108. $2(50{,}500{,}000) = 50{,}500{,}000e^{(0.021)t}$
$2 = e^{0.021t}$
$t = \frac{\ln 2}{0.021} = 33$ years

109. $y = y_0 e^{kt}$, $y_0 = 15.5$, $y = 3(15.5) = 46.5$, and $k = .007$
$46.5 = 15.5e^{.007t}$

$3 = e^{.007t}$
$.007t = \ln 3$
$t = \frac{\ln 3}{.007} \approx 157$
It will take Mexico City approximately 157 years to triple in population.

110. $A = P\left(1+\frac{r}{n}\right)^{nt}$

$10,000 = 5000\left(1+\frac{.08}{4}\right)^{4t}$

$2 = (1.02)^{4t}$
$\log 2 = 4t \log 1.02$
$t = \frac{\log 2}{4\log 1.02} = 8.8$ years

111. $A = P\left(1+\frac{r}{n}\right)^{nt}$, $P = 6000$, $A = 10,000$, $r = .06$ and $n = 12$

$10,000 = 6,000\left(1+\frac{.06}{12}\right)^{12t}$

$\frac{5}{3} = (1.005)^{12t}$

$12t = \log_{1.005}\left(\frac{5}{3}\right)$

$t = \frac{1}{12}\log_{1.005}\left(\frac{5}{3}\right)$

$= \frac{1}{12}\frac{\ln\left(\frac{5}{3}\right)}{\ln(1.005)} \approx 8.54$

It was invested for approximately 8.5 years.

112. $\{0.69\}$

113. $\{2.82\}$

Chapter 10 - Test

1. $(f \circ h)(0) = f(h(0)) = f(5) = 5$

2. $(g \circ f)(x) = g(f(x)) = g(x) = x - 7$

3. $(g \circ h)(x) = g(h(x))$
$= g(x^2 - 6x + 5)$
$= x^2 - 6x + 5 - 7 = x^2 - 6x - 2$

4. $f(x) = 7x - 14$ so $f^{-1}(x) = \frac{x+14}{7}$

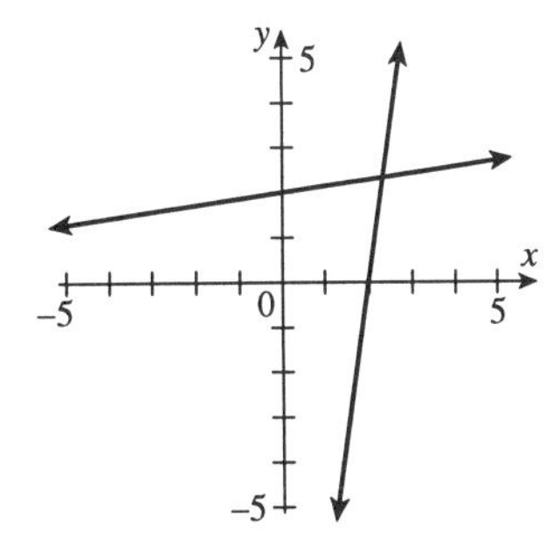

5. The graph does represent a one-to-one function.

6. The graph does not represent a function, hence does not represent a one-to-one function.

7. $y = 6 - 2x$ or $f(x) = -2x + 6$ so
$f^{-1}(x) = \frac{x-6}{-2}$ or $f^{-1}(x) = -\frac{x-6}{2}$

8. $f = \{(0, 0), (2, 3), (-1, 5)\}$ so
$f^{-1} = \{(0, 0), (3, 2), (5, -1)\}$

9. This function is not one-to-one.

10. $\log_3 6 + \log_3 4 = \log_3(6 \cdot 4) = \log_3(24)$

11. $\log_5 x + 3\log_5 x - \log_5(x+1)$
$= 4\log_5 x - \log_5(x+1)$
$= \log_5 x^4 - \log_5(x+1)$
$= \log_5 \frac{x^4}{x+1}$

12. $\log_6 \frac{2x}{y^3} = \log_6 2x - \log_6 y^3$
$= \log_6 2 + \log_6 x - 3\log_6 y$

13. $\log_b\left(\frac{3}{25}\right) - \log_b 3 = \log_b 25$
$= \log_b 3 - \log_b 5^2$
$= \log_b 3 - 2\log_b 5$
$= .79 - 2(1.16) = -1.53$

14. $\log_7 8 = \frac{\ln 8}{\ln 7} \approx 1.0686$

15. $8^{x-1}=\frac{1}{64}$
$8^{x-1}=8^{-2}$
$x-1=-2$
$x=-1$
$\{-1\}$

16. $3^{2x+5}=4$
$2x+5=\log_3 4$
$2x=-5+\log_3 4$
$x=\frac{-5+\log_3 4}{2}=\frac{-5+\frac{\ln 4}{\ln 3}}{2}\approx -1.8691$

17. $\log_3 x=-2$
$x=3^{-2}=\frac{1}{9}$
$\left\{\frac{1}{9}\right\}$

18. $\ln\sqrt{e}=x$
$\ln e^{1/2}=x$
$\frac{1}{2}=x$
$\left\{\frac{1}{2}\right\}$

19. $\log_8(3x-2)=2$
$3x-2=8^2$
$3x-2=64$
$3x=66$
$x=\frac{66}{3}=22$
$\{22\}$

20. $\log_5 x+\log_5 3=2$
$\log_5(3x)=2$
$3x=5^2$
$3x=25$
$x=\frac{25}{3}$
$\left\{\frac{25}{3}\right\}$

21. $\log_4(x+1)-\log_4(x-2)=3$
$\log_4\frac{x+1}{x-2}=3$
$\frac{x+1}{x-2}=4^3$
$x+1=64(x-2)$
$x+1=64x-128$
$129=63x$
$x=\frac{43}{21}$
$\left\{\frac{43}{21}\right\}$

22. $\ln(3x+7)=1.31$
$3x+7=e^{1.31}$
$3x=e^{1.31}-7$
$x=\frac{e^{1.31}-7}{3}\approx -1.0979$

23. $y=\left(\frac{1}{2}\right)^x+1$

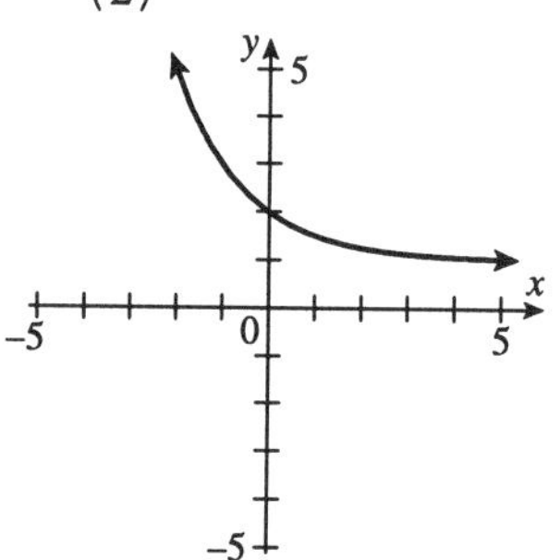

24. $y=3^x$ and $y=\log_3 x$

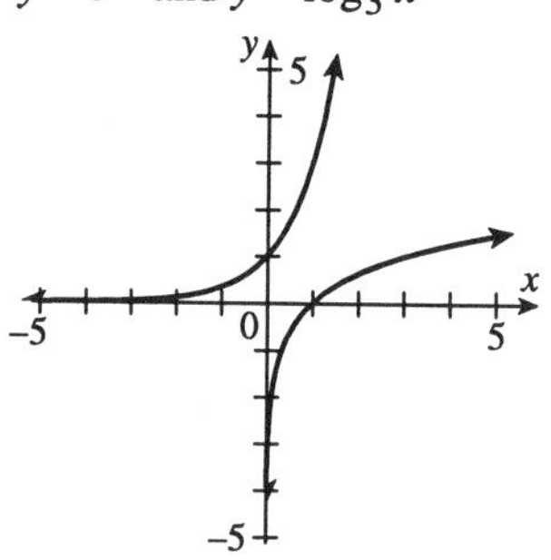

25. $A=P\left(1+\frac{r}{n}\right)^{nt}$, $P=4000$, $t=3$, $r=.09$, and $n=12$
$A=4000\left(1+\frac{.09}{12}\right)^{12(3)}$
$=4000(1.0075)^{36}\approx 5234.58$
\$5234.58 will be in the account.

26. $A = P\left(1+\frac{r}{n}\right)^{nt}$, $P = 2000$, $A = 3000$, $r = .07$, and $n = 2$

$3000 = 2000\left(1+\frac{.07}{2}\right)^{2t}$

$1.5 = (1.035)^{2t}$

$2t = \log_{1.035} 1.5$

$t = \frac{1}{2}\log_{1.035} 1.5$

$= \frac{1}{2}\frac{\ln 1.5}{\ln 1.035} \approx 5.9$

It would take 6 years for the investment to reach $3000.

27. $y = y_0 e^{kt}$, $y_0 = 57{,}000$, $k = .026$, and $t = 5$

$y = 57{,}000e^{.026(5)}$

$= 57{,}000e^{.13} \approx 64{,}913$

There will be approximately 64,913 prairie dogs 5 years from now.

28. $y = y_0 e^{kt}$, $y_0 = 400$, $y = 1000$, and $k = .062$

$1000 = 400e^{.062t}$

$2.5 = e^{.062t}$

$.062t = \ln 2.5$

$t = \frac{\ln 2.5}{.062} \approx 14.8$

It will take the naturalists approximately 15 years to reach their goal.

29. $\log(1+k) = \frac{.3}{D} = D = 56$

$\log(1+k) = \frac{.3}{56}$

$1 + k = 10^{.3/56}$

$k = -1 + 10^{.3/56} \approx .012$

The rate of population increase is approximately 1.2%.

30. $y_1 = e^{0.2x}$

$y_2 = e^{-0.4x} + 2$

$\{3.95\}$

Chapter 10 - Cumulative Review

1. **a.** $20 - x$

b. $x + 8$

c. $90 - x$

d. $x + 1$

2. $\frac{2}{5}(x-6) \ge x - 1$

$2(x-6) \ge 5(x-1)$

$2x - 12 \ge 5x - 5$

$0 \ge 3x + 7$

$x = -\frac{7}{3}$ is a critical point

Region	Test point	$\frac{2}{5}(x-6) \ge x-1$
$\left(-\infty, -\frac{7}{3}\right]$	-4	$-4 \ge -5$
$\left[-\frac{7}{3}, \infty\right)$	1	$-2 < 0$

The solution is $\left(-\infty, -\frac{7}{3}\right]$

3. **a.** $1580

b. greater than $1000

4. $x = 2$

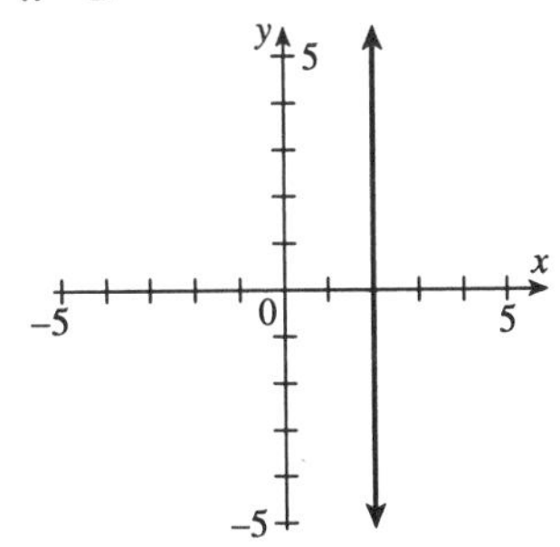

5. $\begin{cases} -5x - 3y = 9 \\ 10x + 6y = -18 \end{cases}$

Solve the first equation for x and substitute into second equation

$-5x - 3y = 9$

$5x = -3y - 9$

$x = \frac{-3y-9}{5}$

$10\left(\frac{-3y-9}{5}\right) + 6y = -18$

$2(-3y-9) + 6y = -18$

$-6y - 18 + 6y = -18$

$-6y - 18 = -6y - 18$

Identity

$\{(x, y) | -5x - 3y = 9\}$

6. $\begin{cases} 2x + 4y = 1 & (1) \\ 4x - 4z = -1 & (2) \\ y - 4z = -3 & (3) \end{cases}$

Multiply equation (3) by –4 and add the result to equation (1).

$$\begin{array}{rl} 2x + 4y = 1 & \\ -4y + 16z = 12 & \\ \hline 2x + 16z = 13 & (4) \end{array}$$

Solve for z using equations (4) and (2).

$2x + 16z = 13$

$4x - 4z = -1$

Multiply equation (4) by –2 and add the result to equation (2).

$$\begin{array}{r} -4x - 32z = -26 \\ 4x - 4z = -1 \\ \hline -36z = -27 \\ z = \frac{3}{4} \end{array}$$

Substitute back.

$2x + 16\left(\frac{3}{4}\right) = 13$

$2x + 12 = 13$

$2x = 1$

$x = \frac{1}{2}$

Substitute back.

$2\left(\frac{1}{2}\right) + 4y = 1$

$1 + 4y = 1$

$4y = 0$

$y = 0$

$\left\{\left(\frac{1}{2}, 0, \frac{3}{4}\right)\right\}$

7. a. $\frac{x^7}{x^4} = x^{7-4} = x^3$

b. $\frac{5^8}{5^2} = 5^{8-2} = 5^6$

c. $\frac{20x^6}{4x^5} = \frac{20}{4}x^{6-5} = 5x$

d. $\frac{12y^{10}z^7}{14y^8z^7} = \frac{6y^{10-8}}{7z^{7-7}} = \frac{6y^2}{7}$

8.
$$\begin{array}{r} 11x^3 - 12x^2 + x - 3 \\ +\quad x^3 - 10x + 5 \\ \hline 12x^3 - 12x^2 - 9x + 2 \end{array}$$

9. a. $(2x^3)(5x^6) = (2)(5)x^{3+6} = 10x^9$

b. $(7y^4z^4)(-xy^{11}z^5)$

$= 7(-1)xy^{4+11}z^{4+5} = -7xy^{15}z^9$

10.
$$\begin{array}{r} x^2 + 2x + 8 \\ 4x^2 + 7 \\ \hline 7x^2 + 14x + 56 \\ 4x^4 + 8x^3 + 32x^2 \\ \hline 4x^4 + 8x^3 + 39x^2 + 14x + 56 \end{array}$$

11. $2x^2 + 9x - 5 = 0$

$(2x - 1)(x + 5) = 0$

$2x - 1 = 0$ or $x + 5 = 0$

$x = \frac{1}{2}$ or $x = -5$

$\left\{-5, \frac{1}{2}\right\}$

12. $f(x) = x^3 - 4x$

$f(0) = 0$ is the y-intercept.

$0 = x^3 - 4x$

$0 = x(x^2 - 4)$

$0 = x(x - 2)(x + 2)$

$x = 0$, $x = 2$, $x = -2$ are the x-intercepts.

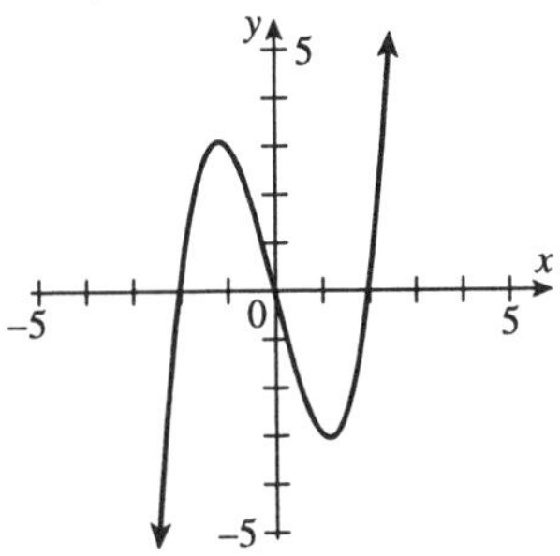

13. $\frac{3}{x} - \frac{x+21}{3x} = \frac{5}{3}$
$9 - (x + 21) = 5x$
$9 - x - 21 = 5x$
$-12 = 6x$
$-2 = x$
$\{-2\}$

14. a. $4^{3/2} = \sqrt{4^3} = \sqrt{64} = 8$

b. $-16^{3/4} = -\sqrt[4]{16^3} = -\sqrt[4]{4096} = -8$

c. $(-27)^{2/3} = \sqrt[3]{(-27)^2} = \sqrt[3]{729} = 9$

d. $\left(\frac{1}{9}\right)^{3/2} = \sqrt{\left(\frac{1}{9}\right)^3} = \sqrt{\frac{1}{729}} = \frac{1}{27}$

e. $(4x-1)^{3/5} = \sqrt[5]{(4x-1)^3}$

15. a. $(2 - 5i)(4 + i)$
$= (2)(4) + 2i - 20i - 5i^2$
$= 8 - 18i + 5 = 13 - 18i$

b. $(2 - i)^2 = (2 - i)(2 - i)$
$= 4 - 4i + i^2$
$= 4 - 4i - 1 = 3 - 4i$

c. $(7 + 3i)(7 - 3i)$
$= 49 - 9i^2 = 49 + 9 = 58$

16. $(x + 2)(x - 1)(x - 5) \le 0$
$x = -2$, $x = 1$, $x = 5$ are critical points

Region	Test point	$(x + 2)(x - 1)(x - 5)$
$(-\infty, -2]$	-3	$(-1)(-4)(-8) < 0$
$[-2, 1]$	0	$(2)(-1)(-5) > 0$
$[1, 5]$	2	$(4)(1)(0) = 0$
$[5, \infty)$	6	$(8)(5)(1) > 0$

The solution is $(-\infty, -2] \cup [1, 5]$

17. $f(x) = -2x^2$

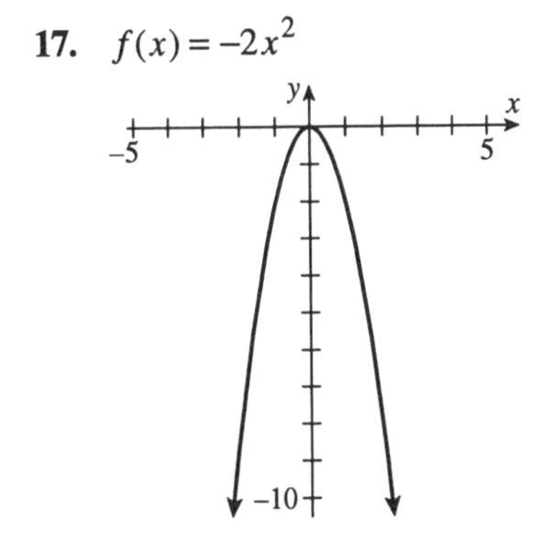

18. $x^2 + y^2 = 4$

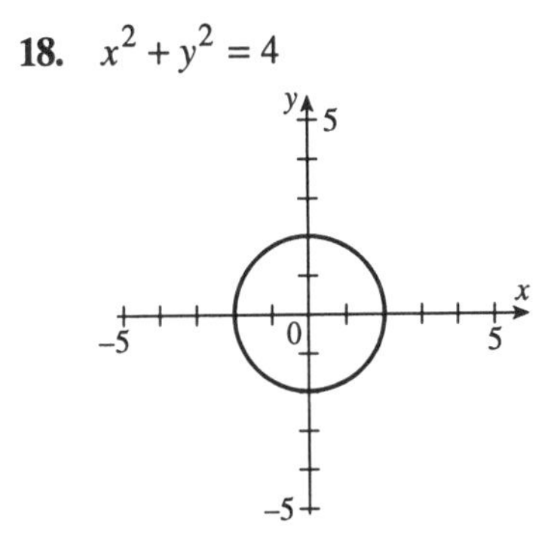

19. a. $(f \circ g)(2)$
$= f(g(2)) = f(5) = 5^2 = 25$
$(g \circ f)(2)$
$= g(f(2)) = g(4) = 4 + 3 = 7$

b. $(f \circ g)(x) = f(g(x))$
$= f(x+3) = (x+3)^2 = x^2 + 6x + 9$
$(g \circ f)(x) = g(f(x))$
$= g(x^2) = x^2 + 3$

20. a. $2^x = 16$
$2^x = 2^4$
$x = 4$
$\{4\}$

b. $9^x = 27$
$3^{2x} = 3^3$
$2x = 3$
$x = \frac{3}{2}$
$\left\{\frac{3}{2}\right\}$

c. $4^{x+3} = 8^x$
$(2^2)^{x+3} = (2^3)^x$
$2(x + 3) = 3x$
$2x + 6 = 3x$
$6 = x$
$\{6\}$

21. a. $\log_b 6 = \log_b (2 \cdot 3)$
$= \log_b 2 + \log_b 3 = 0.43 + 0.68 = 1.11$

b. $\log_b 9 = \log_b 3^2 = 2 \log_b 3$
$= 2(0.68) = 1.36$

c. $\log_b \sqrt{2} = \log_b 2^{1/2} = \frac{1}{2} \log_b 2$
$= \frac{1}{2}(0.43) = 0.215$

22. $\ln 3x = 5$
$e^5 = 3x$
$x = \frac{e^5}{3} \approx 49.4711$

23. $\log_2 x + \log_2 (x - 1) = 1$
$\log_2 x(x - 1) = 1$
$\log_2 (x^2 - x) = 1$
$2^1 = x^2 - x$
$0 = x^2 - x - 2$
$0 = (x - 2)(x + 1)$
$x - 2 = 0$ or $x + 1 = 0$
$x = 2$ or $x = -1$
-1 is an extraneous solution
$\{2\}$

Chapter 11

Exercise Set 11.1

1. $a_n = n+4$
$a_1 = 1+5 = 5$
$a_2 = 2+4 = 6$
$a_3 = 3+4 = 7$
$a_4 = 4+4 = 8$
$a_5 = 5+4 = 9$
or 5, 6, 7, 8, 9

3. $a_n = (-1)^n$
$a_1 = (-1)^1 = -1$
$a_2 = (-1)^2 = 1$
$a_3 = (-1)^3 = -1$
$a_4 = (-1)^4 = 1$
$a_5 = (-1)^5 = -1$
or −1, 1, −1, 1, −1

5. $a_n = \frac{1}{n+3}$
$a_1 = \frac{1}{1+3} = \frac{1}{4}$
$a_2 = \frac{1}{2+3} = \frac{1}{5}$
$a_3 = \frac{1}{3+3} = \frac{1}{6}$
$a_4 = \frac{1}{4+3} = \frac{1}{7}$
$a_5 = \frac{1}{5+3} = \frac{1}{8}$
or $\frac{1}{4}, \frac{1}{5}, \frac{1}{6}, \frac{1}{7}, \frac{1}{8}$

7. $a_n = 2n$
$a_1 = 2(1) = 2$
$a_2 = 2(2) = 4$
$a_3 = 2(3) = 6$
$a_4 = 2(4) = 8$
$a_5 = 2(5) = 10$
or 2, 4, 6, 8, 10

9. $a_n = -n^2$
$a_1 = -1^2 = -1$
$a_2 = -2^2 = -4$
$a_3 = -3^2 = -9$
$a_4 = -4^2 = -16$
$a_5 = -5^2 = -25$
or −1, −4, −9, −16, −25

11. $a_n = 2^n$
$a_1 = 2^1 = 2$
$a_2 = 2^2 = 4$
$a_3 = 2^3 = 8$
$a_4 = 2^4 = 16$
$a_5 = 2^5 = 32$
or 2, 4, 8, 16, 32

13. $a_n = 2n+5$
$a_1 = 2(1)+5 = 2+5 = 7$
$a_2 = 2(2)+5 = 4+5 = 9$
$a_3 = 2(3)+5 = 6+5 = 11$
$a_4 = 2(4)+5 = 8+5 = 13$
$a_5 = 2(5)+5 = 10+5 = 15$
or 7, 9, 11, 13, 15

15. $a_n = (-1)^n n^2$
$a_1 = (-1)^1(1^2) = -1(1) = -1$
$a_2 = (-1)^2(2^2) = 1(4) = 4$
$a_3 = (-1)^3(3^2) = -1(9) = -9$
$a_4 = (-1)^4(4^2) = 1(16) = 16$
$a_5 = (-1)^5(5^2) = -1(25) = -25$
or −1, 4, −9, 16, −25

17. $a_n = 3n^2$
$a_5 = 3(5)^2 = 3(25) = 75$

19. $a_n = 6n-2$
$a_{20} = 6(20)-2 = 120-2 = 118$

21. $a_n = \frac{n+3}{n}$

$a_{15} = \frac{15+3}{15} = \frac{18}{15} = \frac{6}{5}$

23. $a_n = (-3)^n$

$a_6 = (-3)^6 = 729$

25. $a_n = \frac{n-2}{n+1}$

$a_6 = \frac{6-2}{6+1} = \frac{4}{7}$

27. $a_n = \frac{(-1)^n}{n}$

$a_8 = \frac{(-1)^8}{8} = \frac{1}{8}$

29. $a_n = -n^2 + 5$

$a_{10} = -10^2 + 5 = -100 + 5 = -95$

31. $a_n = \frac{(-1)^n}{n+6}$

$a_{19} = \frac{(-1)^{19}}{19+6} = -\frac{1}{25}$

33. 3, 7, 11, 15 or
4(1) – 1, 4(2) – 1, 4(3) – 1, 4(4) – 1
In general, $a_n = 4n - 1$

35. –2, –4, –8, –16 or
$-2,\ -2^2,\ -2^3,\ -2^4$
In general, $a_n = -2^n$

37. $\frac{1}{3}, \frac{1}{9}, \frac{1}{27}, \frac{1}{81}$ or
$\frac{1}{3}, \frac{1}{3^2}, \frac{1}{3^3}, \frac{1}{3^4}$
In general, $a_n = \frac{1}{3^n}$

39. $a_n = 32n - 16$
$a_2 = 32(2) - 16 = 64 - 16 = 48$ ft
$a_3 = 32(3) - 16 = 96 - 16 = 80$ ft
$a_4 = 32(4) - 16 = 128 - 16 = 112$ ft

41. 0.10, 0.20, 0.40 or
$0.10, 0.10(2),\ 0.10(2^2)$
In general, $a_n = 0.10(2^{n-1})$
$a_{14} = 0.10(2^{13}) \approx \$819.20$

43. $a_n = 75(2)^{n-1}$
$a_6 = 75(2)^5 = 75(32) = 2400$ cases
$a_1 = 75(2)^0 = 75(1) = 75$ cases

45. 800, 400, 200, 100, 50
The population estimate for 2000 is 50 sparrows. Continuing the sequence: 25, 12, 6, 3, 1, 0. We estimate the sparrows will be extinct in 2006.

47. $a_n = \frac{1}{\sqrt{n}}$

$a_1 = \frac{1}{\sqrt{1}} = \frac{1}{1} = 1$

$a_2 = \frac{1}{\sqrt{2}} \approx 0.7071$

$a_3 = \frac{1}{\sqrt{3}} \approx 0.5774$

$a_4 = \frac{1}{\sqrt{4}} = \frac{1}{2} = 0.5$

$a_5 = \frac{1}{\sqrt{5}} \approx 0.4472$

49. $a_n = \left(1 + \frac{1}{n}\right)^n$

$a_1 = \left(1 + \frac{1}{1}\right)^1 = (2)^1 = 2$

$a_2 = \left(1 + \frac{1}{2}\right)^2 = \left(\frac{3}{2}\right)^2 = 2.25$

$a_3 = \left(1 + \frac{1}{3}\right)^3 = \left(\frac{4}{3}\right)^3 \approx 2.3704$

$a_4 = \left(1 + \frac{1}{4}\right)^4 = \left(\frac{5}{4}\right)^4 \approx 2.4414$

$a_5 = \left(1 + \frac{1}{5}\right)^5 = \left(\frac{6}{5}\right)^5 \approx 2.4883$

51. $f(x) = (x-1)^2 + 3$

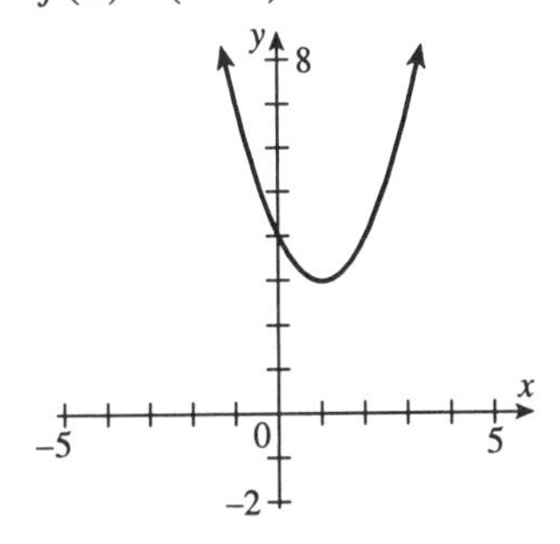

53. $f(x) = 2(x+4)^2 + 2$

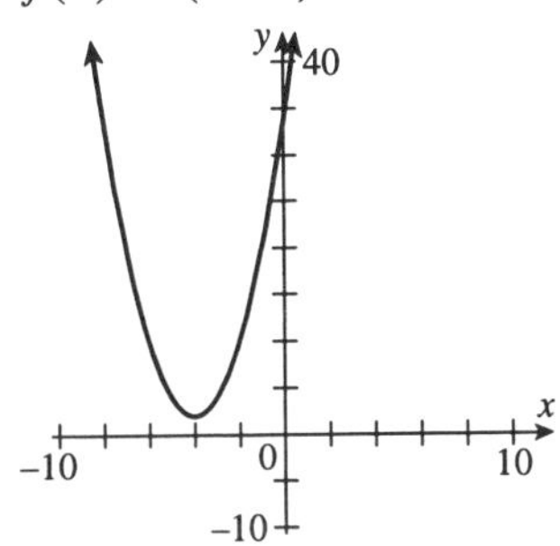

55. $d = \sqrt{[-7-(-4)]^2 + [-3-(-1)]^2}$
$d = \sqrt{(-7+4)^2 + (-3+1)^2}$
$d = \sqrt{(-3)^2 + (-2)^2}$
$d = \sqrt{9+4} = \sqrt{13}$

57. $d = \sqrt{(-3-2)^2 + [-3-(-7)]^2}$
$d = \sqrt{(-5)^2 + (-3+7)^2}$
$d = \sqrt{(-5)^2 + (4)^2}$
$d = \sqrt{25+16} = \sqrt{41}$

Exercise Set 11.2

1. $a_1 = 4;\ d = 2$
4, 6, 8, 10, 12

3. $a_1 = 6,\ d = -2$
6, 4, 2, 0, −2

5. $a_1 = 1,\ r = 3$
1, 3, 9, 27, 81

7. $a_1 = 48,\ r = \frac{1}{2}$
48, 24, 12, 6, 3

9. $a_1 = 12,\ d = 3$
$a_n = 12 + (n-1)3$
$a_8 = 12 + 7(3) = 12 + 21 = 33$

11. $a_1 = 7,\ r = -5$
$a_n = a_1 r^{n-1}$
$a_4 = 7(-5)^3 = 7(-125) = -875$

13. $a_1 = -4,\ d = -4$
$a_n = -4 + (n-1)(-4)$
$a_{15} = -4 + 14(-4) = -4 - 56 = -60$

15. 0, 12, 24
$a_1 = 0$ and $d = 12$
$a_n = 0 + (n-1)12$
$a_9 = 8(12) = 96$

17. 20, 18, 16
$a_1 = 20$ and $d = -2$
$a_n = 20 + (n-1)(-2)$
$a_{25} = 20 + 24(-2) = 20 - 48 = -28$

19. 2, −10, 50
$a_1 = 2$ and $r = -5$
$a_n = 2(-5)^{n-1}$
$a_5 = 2(-5)^4 = 2(625) = 1250$

21. $a_4 = 19,\ a_{15} = 52$
Use the relationship:
$a_4 + 11d = a_{15}$
$19 + 11d = 52$
$11d = 33$
$d = 3$
Now use the relationship:
$a_8 = a_4 + 4d$
$a_8 = 19 + 4(3) = 19 + 12 = 31$

23. $a_2 = -1$ and $a_4 = 5$
Use the relationship:
$a_2 + 2d = a_4$
$-1 + 2d = 5$
$2d = 6$

$d=3$
Now use the relationship:
$a_9 = a_4 + 5d$
$a_9 = 5+5(3)$
$a_9 = 5+15 = 20$

25. $a_2 = -\frac{4}{3}$ and $a_3 = \frac{8}{3}$
Use the relationship:
$a_2 r = a_3$
$-\frac{4}{3}r = \frac{8}{3}$
$r = -2$
Now use the relationship:
$a_1 r = a^2$
$a_1(-2) = \frac{-4}{3}$
$a_1 = \frac{2}{3}$

27. Answers may vary.

29. 2, 4, 6
$a_1 = 2$ and $d = 2$

31. 5, 10, 20
$a_1 = 5$ and $r = 2$

33. $\frac{1}{2}, \frac{1}{10}, \frac{1}{50}$
$a_1 = \frac{1}{2}$ and $r = \frac{1}{5}$

35. $x, 5x, 25x$
$a_1 = x$ and $r = 5$

37. $p, p+4, p+8$
$a_1 = p$ and $d = 4$

39. $a_1 = 14$ and $d = \frac{1}{4}$
$a_n = 14 + (n-1)\frac{1}{4}$
$a_{21} = 14 + 20\left(\frac{1}{4}\right) = 14+5 = 19$

41. $a_1 = 3$ and $r = -\frac{2}{3}$
$a_n = 3\left(-\frac{2}{3}\right)^{n-1}$
$a_4 = 3\left(-\frac{2}{3}\right)^3 = 3\left(-\frac{8}{27}\right) = -\frac{8}{9}$

43. $\frac{3}{2}, 2, \frac{5}{2}, \ldots$
$a_1 = \frac{3}{2}$ and $d = \frac{1}{2}$
$a_n = \frac{3}{2} + (n-1)\frac{1}{2}$
$a_{15} = \frac{3}{2} + 14\left(\frac{1}{2}\right) = \frac{17}{2}$

45. $24, 8, \frac{8}{3}, \ldots$
$a_1 = 24$ and $r = \frac{1}{3}$
$a_n = 24\left(\frac{1}{3}\right)^{n-1}$
$a_6 = 24\left(\frac{1}{3}\right)^5 = 24\left(\frac{1}{243}\right) = \frac{8}{81}$

47. $a_3 = 2$ and $a_{17} = -40$
Use the relationship:
$a_3 + 14d = a_{17}$
$2 + 14d = -40$
$14d = -42$
$d = -3$
Now use the relationship:
$a_{10} = a_3 + 7d$
$a_{10} = 2 + 7(-3) = 2 - 21 = -19$

49. 54, 58, 62
$a_1 = 54$ and $d = 4$
$a_n = 54 + (n-1)4$
$a_{20} = 54 + 19(4) = 54 + 76 = 130$
There are 130 seats in the twentieth row.

51. $a_1 = 6$ and $r = 3$
$a_n = 6 \cdot 3^{n-1} = 2 \cdot 3 \cdot 3^{n-1} = 2 \cdot 3^n$

53. 486, 162, 54, 18, 6
$a_1 = 486$ and $r = \frac{1}{3}$
$a_n = 486\left(\frac{1}{3}\right)^{n-1}$

Solve $486\left(\frac{1}{3}\right)^{n-1}=1$

$\left(\frac{1}{3}\right)^{n-1}=\frac{1}{486}$

$\frac{1}{3^{n-1}}=\frac{1}{486}$

$3^{n-1}=486$

$3^n=1458$

$n=\log_3 1458$

$n=\frac{\ln 1458}{\ln 3}\approx 6.6$

The ball will rebound less than a foot on the 7th bounce.

55. $a_1=2000$ and $d=125$

$a_n=2000+(n-1)125$

$a_{12}=2000+11(125)$

$a_{12}=2000+1375=3375$

His salary at the end of his training is \$3375.

57. $a_1=400$ and $r=\frac{1}{2}$

12 hrs = 4(3 hrs), so we seek the fourth term after a_1, namely a_5.

$a_n=a_1r^{n-1}$

$a_5=400\left(\frac{1}{2}\right)^4=\frac{400}{16}=25$

25 grams of the radioactive material remains after 12 hours.

59. $a_1=11,782.40$

$r=0.5$

$a_2=(11,782.40)(0.5)=\$5891.20$

$a_3=(5891.20)(0.5)=\$2945.60$

$a_4=(2945.60)(0.5)=\$1472.80$

61. $a_1=19.652$ and $d=-0.034$

$a_2=19.652-0.034=19.618$

$a_3=19.618-0.034=19.584$

$a_4=19.584-0.034=19.550$

63. Answers may vary.

65. $\frac{1}{3(1)}+\frac{1}{3(2)}+\frac{1}{3(3)}$

$=\frac{1}{3}\left(\frac{1}{1}+\frac{1}{2}+\frac{1}{3}\right)$

$=\frac{1}{3}\left(\frac{6}{6}+\frac{3}{6}+\frac{2}{6}\right)$

$=\frac{1}{3}\left(\frac{11}{6}\right)=\frac{11}{18}$

67. $3^0+3^1+3^2+3^3=1+3+9+27=40$

69. $\frac{8-1}{8+1}+\frac{8-2}{8+2}+\frac{8-3}{8+3}$

$=\frac{7}{9}+\frac{6}{10}+\frac{5}{11}$

$=\frac{770}{990}+\frac{594}{990}+\frac{450}{990}$

$=\frac{1814}{990}=\frac{907}{495}$

Exercise Set 11.3

1. $\sum_{i=1}^{4}(i-3)$

$=(1-3)+(2-3)+(3-3)+(4-3)$

$=-2+(-1)+0+1=-2$

3. $\sum_{i=4}^{7}(2i+4)$

$=[2(4)+4]+[2(5)+4]+[2(6)+4]+[2(7)+4]$

$=12+14+16+18=60$

5. $\sum_{i=2}^{4}(i^2-3)$

$=(2^2-3)+(3^2-3)+(4^2-3)$

$=1+6+13=20$

7. $\sum_{i=1}^{3}\frac{1}{i+5}$

$=\frac{1}{1+5}+\frac{1}{2+5}+\frac{1}{3+5}$

$=\frac{1}{6}+\frac{1}{7}+\frac{1}{8}=\frac{28}{168}+\frac{24}{168}+\frac{21}{168}=\frac{73}{168}$

9. $\sum_{i=1}^{3} \frac{1}{6i} = \frac{1}{6(1)} + \frac{1}{6(2)} + \frac{1}{6(3)}$
$= \frac{1}{6} + \frac{1}{12} + \frac{1}{18} = \frac{6+3+2}{36} = \frac{11}{36}$

11. $\sum_{i=2}^{6} 3i$
$= 3(2) + 3(3) + 3(4) + 3(5) + 3(6)$
$= 6 + 9 + 12 + 15 + 18 = 60$

13. $\sum_{i=3}^{5} i(i+2)$
$= 3(3 + 2) + 4(4 + 2) + 5(5 + 2)$
$= 15 + 24 + 35 = 74$

15. $\sum_{i=1}^{5} 2^i = 2^1 + 2^2 + 2^3 + 2^4 + 2^5$
$= 2 + 4 + 8 + 16 + 32 = 62$

17. $\sum_{i=1}^{4} \frac{4i}{i+3} = \frac{4(1)}{1+3} + \frac{4(2)}{2+3} + \frac{4(3)}{3+3} + \frac{4(4)}{4+3}$
$= 1 + \frac{8}{5} + 2 + \frac{16}{7} = \frac{105}{35} + \frac{56}{35} + \frac{80}{35} = \frac{241}{35}$

19. $1 + 3 + 5 + 7 + 9$
$= [(2) - 1] + [2(2) - 1] + [2(3) - 1]$
$+ [2(4) - 1] + [2(5) - 1]$
$= \sum_{i=1}^{5} (2i - 1)$

21. $4 + 12 + 36 + 108$
$= 4 + 4(3) + 4(3^2) + 4(3^3)$
$= \sum_{i=1}^{4} 4 \cdot 3^{i-1}$

23. $12 + 9 + 6 + 3 + 0 + (-3)$
$= [-3(1) + 15] + [-3(2) + 15]$
$+ [-3(3) + 15] + [-3(4) + 15]$
$+ [-3(5) + 15] + [-3(6) + 15]$
$= \sum_{i=1}^{6} (-3i + 15)$

25. $12 + 4 + \frac{4}{3} + \frac{4}{9} = \frac{4}{3^{-1}} + \frac{4}{3^0} + \frac{4}{3^1} + \frac{4}{3^2}$
$= \sum_{i=1}^{4} \frac{4}{3^{i-2}}$

27. $1 + 4 + 9 + 16 + 25 + 36 + 49$
$= 1^2 + 2^2 + 3^2 + 4^2 + 5^2 + 6^2 + 7^2$
$= \sum_{i=1}^{7} i^2$

29. $a_n = (n+2)(n-5)$
$a_1 = (1+2)(1-5) = 3(-4) = -12$
$a_2 = (2+2)(2-5) = 4(-3) = -12$
$a_1 + a_2 = -12 + (-12) = -24$

31. $a_n = n(n-6)$
$= a_1 + a_2 = 1(1-6) + 2(2-6)$
$= 1(-5) + 2(-4) = -13$

33. $a_n = (n+3)(n+1)$
$a_1 = (1+3)(1+1) = 4(2) = 8$
$a_2 = (2+3)(2+1) = 5(3) = 15$
$a_3 = (3+3)(3+1) = 6(4) = 24$
$a_4 = (4+3)(4+1) = 7(5) = 35$
$\sum_{i=1}^{4} a_i = 8 + 15 + 24 + 35 = 82$

35. $a_n = -2n$
$\sum_{i=1}^{4} (-2i)$
$= -2(1) + (-2)(2) + (-2)(3) + (-2)(4)$
$= -2 - 4 - 6 - 8 = -20$

37. $a_n = \frac{n}{-3}$
$a_1 + a_2 + a_3 = -\frac{1}{3} - \frac{2}{3} - \frac{3}{3} = -2$

39. $1, 2, 3, \ldots, 10$
$a_n = n$
$\sum_{i=1}^{10} i = 1 + 2 + 3 + \cdots + 10$

$= \frac{10(11)}{2} = 55$

A total of 55 trees were planted.

41. $a_1 = 6$ and $r = 2$

$a_n = 6 \cdot 2^{n-1}$

$a_5 = 6 \cdot 2^4 = 6 \cdot 16 = 96$

There will be 96 fungus units at the beginning of the 5th day.

43. $a_1 = 50$ and $r = 2$

Since 48 = 4(12), we seek the fourth term after a_1, namely a_5.

$a_n = 50(2)^{n-1}$

$a_5 = 50(2)^4 = 50(16) = 800$

There are 800 bacteria after 48 hours.

45. $a_n = (n+1)(n+2)$

$a_4 = (4+1)(4+2) = 5(6) = 30$ opossums

$a_1 = (1+1)(1+2) = 2(3) = 6$

$a_2 = (2+1)(2+2) = 3(4) = 12$

$a_3 = (3+1)(3+2) = 4(5) = 20$

$\sum_{i=1}^{4} a_i = 6 + 12 + 20 + 30 = 68$ opossums

47. $a_n = 100(0.5)^n$

$a_4 = 100(0.5)^4 = 6.25$ lbs of decay.

$a_1 = 100(0.5)^1 = 50$

$a_2 = 100(0.5)^2 = 25$

$a_3 = 100(0.5)^3 = 12.5$

$\sum_{i=1}^{4} a_i = 50 + 25 + 12.5 + 6.25$

$= 93.75$ lbs of decay

49. $a_1 = 40$ and $r = \frac{4}{5}$

$a_5 = 40\left(\frac{4}{5}\right)^4 = 16.384$ or 16.4 in.

$a_2 = 40\left(\frac{4}{5}\right)^1 = 32$

$a_3 = 40\left(\frac{4}{5}\right)^2 = 25.6$

$a_4 = 40\left(\frac{4}{5}\right)^3 = 20.48$

$\sum_{i=1}^{5} a_i = 40 + 32 + 25.6 + 20.48 + 16.384$

$= 134.464$ or 134.5 in.

51. a. $\sum_{i=1}^{7} i + i^2$

$= (1+1^2) + (2+2^2) + (3+3^2)$
$+ (4+4^2) + (5+5^2) + (6+6^2)$
$+ (7+7^2)$

$= 2 + 6 + 12 + 20 + 30 + 42 + 56$

b. $\sum_{i=1}^{7} i + \sum_{i=1}^{7} i^2$

$= 1 + 2 + 3 + 4 + 5 + 6 + 7 + 1 + 4$
$+ 9 + 16 + 25 + 36 + 49$

c. They are equal; 168

d. True

53. $\frac{5}{1-\frac{1}{2}} = \frac{5}{\frac{1}{2}} = 5 \cdot \frac{2}{1} = 10$

55. $\frac{\frac{1}{3}}{1-\frac{1}{10}} = \frac{\frac{1}{3}}{\frac{9}{10}} = \frac{1}{3} \cdot \frac{10}{9} = \frac{10}{27}$

57. $\frac{3(1-2^4)}{1-2} = \frac{3(1-16)}{-1}$

$= \frac{3(-15)}{-1} = \frac{-45}{-1} = 45$

59. $\frac{10}{2}(3+15) = \frac{10}{2}(18) = \frac{180}{2} = 90$

Exercise Set 11.4

1. 1, 3, 5, 7, ...

$a_1 = 1,\ d = 2,\ n = 6$

$S_6 = \frac{6}{2}[2(1) + (6-1)2]$

$S_6 = 3[2 + 10] = 3(12) = 36$

3. $4,\ 12,\ 36,\ \cdots$
$a_1 = 4,\ r = 3,\ n = 5$
$S_5 = \frac{4(1-3^5)}{1-3} = -2(1-243)$
$S_5 = -2(-242) = 484$

5. $3,\ 6,\ 9,\ \cdots$
$a_1 = 3,\ d = 3,\ n = 6$
$S_6 = \frac{6}{2}[2(3)+(6-1)(3)]$
$S_6 = 3[6+15] = 3(21) = 63$

7. $2,\ \frac{2}{5},\ \frac{2}{25},\ \cdots$
$a_1 = 2,\ r = \frac{1}{5},\ n = 4$
$S_4 = \frac{2\left[1-\left(\frac{1}{5}\right)^4\right]}{1-\frac{1}{5}} = \frac{5}{2}\left(1-\frac{1}{625}\right)$
$S_4 = \frac{5}{2}\left(\frac{624}{625}\right) = \frac{312}{125} = 2.496$

9. $1,\ 2,\ 3,\ \cdots,\ 10$
$a_1 = 1,\ d = 1,\ n = 10$
$S_{10} = \frac{10}{2}[2(1)+(10-1)(1)]$
$S_{10} = 5[2+9] = 5(11) = 55$

11. $1, 3, 5, 7$
$a_1 = 1,\ d = 2,\ n = 4$
$S_4 = \frac{4}{2}[2(1)+(4-1)2]$
$S_4 = 2[2+6] = 2(8) = 16$

13. $12,\ 6,\ 3,\ \cdots$
$a_1 = 12$ and $r = \frac{1}{2}$
$S_\infty = \frac{12}{1-\frac{1}{2}} = \frac{12}{\frac{1}{2}} = 24$

15. $\frac{1}{10},\ \frac{1}{100},\ \frac{1}{1000},\ \cdots$
$a_1 = \frac{1}{10}$ and $r = \frac{1}{10}$
$S_\infty = \frac{\frac{1}{10}}{1-\frac{1}{10}} = \frac{\frac{1}{10}}{\frac{9}{10}} = \frac{1}{9}$

17. $-10,\ -5,\ -\frac{5}{2},\ \cdots$
$a_1 = -10$ and $r = \frac{1}{2}$
$S_\infty = \frac{-10}{1-\frac{1}{2}} = \frac{-10}{\frac{1}{2}} = -20$

19. $2,\ -\frac{1}{4},\ \frac{1}{32},\ \cdots$
$a_1 = 2$ and $r = -\frac{1}{8}$
$S_\infty = \frac{2}{1-\left(-\frac{1}{8}\right)} = \frac{2}{\frac{9}{8}} = \frac{16}{9}$

21. $\frac{2}{3},\ -\frac{1}{3},\ \frac{1}{6},\ \cdots$
$a_1 = \frac{2}{3}$ and $r = -\frac{1}{2}$
$S_\infty = \frac{\frac{2}{3}}{1-\left(-\frac{1}{2}\right)} = \frac{\frac{2}{3}}{\frac{3}{2}} = \frac{4}{9}$

23. $-4,\ 1,\ 6,\ \cdots$
$a_1 = -4,\ d = 5$, and $n = 10$
$S_{10} = \frac{10}{2}[2(-4)+(10-1)5]$
$S_{10} = 5[-8+45] = 5(37) = 185$

25. $3,\ \frac{3}{2},\ \frac{3}{4},\ \cdots$
$a_1 = 3,\ r = \frac{1}{2},\ n = 7$
$S_7 = \frac{3\left[1-\left(\frac{1}{2}\right)^7\right]}{1-\frac{1}{2}} = 6\left(1-\frac{1}{128}\right)$
$S_7 = 6\left(\frac{127}{128}\right) = \frac{381}{64}$

27. $-12,\ 6,\ -3,\ \cdots$
$a_1 = -12,\ r = -\frac{1}{2},\ n = 5$
$S_5 = \frac{-12\left[1-\left(-\frac{1}{2}\right)^5\right]}{1-\left(-\frac{1}{2}\right)}$
$S_5 = -8\left(1+\frac{1}{32}\right) = -8\left(\frac{33}{32}\right)$
$S_5 = -\frac{33}{4}$ or -8.25

29. $\frac{1}{2}, \frac{1}{4}, 0, \cdots$
$a_1 = \frac{1}{2},\ d = -\frac{1}{4}, n = 20$
$S_{20} = \frac{20}{2}\left[2\left(\frac{1}{2}\right) + (20-1)\left(-\frac{1}{4}\right)\right]$
$S_{20} = 10\left(1 - \frac{19}{4}\right) = 10\left(\frac{-15}{4}\right) = -\frac{75}{2}$

31. $a_1 = 8,\ r = -\frac{2}{3},\ n = 3$
$S_3 = \frac{8\left[1-\left(-\frac{2}{3}\right)^3\right]}{1-\left(-\frac{2}{3}\right)} = \frac{24}{5}\left(1+\frac{8}{27}\right)$
$S_3 = \frac{24}{5}\left(\frac{35}{27}\right) = \frac{8}{1}\left(\frac{7}{9}\right) = \frac{56}{9}$

33. 4000, 3950, 3900, 3850, 3800
$a_1 = 4000,\ d = -50, n = 12$
$a_{12} = 4000 + 11(-50)$
$a_{12} = 4000 - 550$
$a_{12} = 3450$ cars sold in month 12.
$S_{12} = \frac{12}{2}[4000 + 3450]$
$S_{12} = 6(7450)$
$S_{12} = 44,700$ cars sold in the first 12 months.

35. Firm A:
$a_1 = 22,000,\ d = 1000, n = 10$
$S_{10} = \frac{10}{2}[2(22,000) + (10-1)1000]$
$S_{10} = 5[44,000 + 9000]$
$S_{10} = 5[53,000] = \$265,000$
Firm B:
$a_1 = 20,000,\ d = 1200, n = 10$
$S_{10} = \frac{10}{2}[2(20,000) + (10-1)1200]$
$S_{10} = 5[40,000 + 10,800]$
$S_{10} = 5[50,800] = \$254,000$
Thus, Firm A is making the more profitable offer.

37. $a_1 = 30,000,\ r = 1.10,\ n = 4$
$a_4 = 30,000(1.10)^{4-1}$
$a_4 = \$39,930$ made during her fourth year of business.
$S_4 = \frac{30,000(1-1.10^4)}{1-1.10}$
$S_4 = \$139,230$ made during the first four years.

39. $a_1 = 30,\ r = 0.9,\ n = 5$
$a_5 = 30(0.9)^{5-1} = 19.683$ or approximately 20 minutes to assemble the first computer.
$S_5 = \frac{30(1-0.9^5)}{1-0.9} = 122.853$, or approximately 123 minutes to assemble the first 5 computers.

41. $a_1 = 20$ and $r = \frac{4}{5}$
$S_\infty = \frac{20}{1-\frac{4}{5}} = \frac{20}{\frac{1}{5}} = 100$
We double the number (to account for the flight up as well as down) and subtract 20 (since the first bounce was preceded by only a downward flight). Thus, the ball traveled a total distance of $2(100) - 20 = 180$ feet.

43. Player A:
$a_1 = 1,\ d = 1, n = 9$
$S_9 = \frac{9}{2}[2(1) + (9-1)1]$
$S_9 = \frac{9}{2}[2+8] = \frac{9}{2}(10) = 45$ points
Player B:
$a_1 = 10,\ d = 1$, and $n = 6$
$S_6 = \frac{6}{2}[2(10) + (6-1)1]$
$S_6 = 3[20+5] = 3(25) = 75$ points

45. $a_1 = 200,\ d = -5, n = 20$
$S_{20} = \frac{20}{2}[2(200) + (20-1)(-5)]$
$S_{20} = 10[400 - 95] = 10[305] = 3050$
Thus, \$3050 rent is paid for 20 days during the holiday rush.

47. $a_1 = 0.01,\ r = 2,\ n = 30$
$S_{30} = \frac{0.01(1-2^{30})}{1-2} = 10,737,418.23$

He would pay $10,737,418.23 in room and board for the 30 days.

49. $0.88\overline{8} = 0.8 + 0.08 + 0.008 + \cdots$
$a_1 = 0.8$ and $r = 0.1$
$S_\infty = \frac{0.8}{1-0.1} = \frac{0.8}{0.9} = \frac{8}{9}$

51. Answers may vary.

53. $6 \cdot 5 \cdot 4 \cdot 3 \cdot 2 \cdot 1 = (30)(12)(2) = 720$

55. $\frac{3 \cdot 2 \cdot 1}{2 \cdot 1} = 3$

57. $(x+5)^2 = x^2 + 2(5x) + 25$
$= x^2 + 10x + 25$

59. $(2x-1)^3$
$= (2x-1)(2x-1)^2$
$= (2x-1)(4x^2 - 4x + 1)$
$= 8x^3 - 8x^2 + 2x - 4x^2 + 4x - 1$
$= 8x^3 - 12x^2 + 6x - 1$

Exercise Set 11.5

1. $(m+n)^3$
$= 1 \cdot m^3 + 3 \cdot m^2n + 3 \cdot mn^2 + 1 \cdot n^3$
$= m^3 + 3m^2n + 3mn^2 + n^3$

3. $(c+d)^5$
$= 1 \cdot c^5 + 5 \cdot c^4d + 10 \cdot c^3d^2 + 10 \cdot c^2d^3 + 5 \cdot cd^4 + 1 \cdot d^5$
$= c^5 + 5c^4d + 10c^3d^2 + 10c^2d^3 + 5cd^4 + d^5$

5. $(y-x)^5 = [y+(-x)]^5$
$= 1y^5 + 5y^4(-x) + 10y^3(-x)^2 + 10y^2(-x)^3 + 5y(-x)^4 + (-x)^5$
$= y^5 - 5y^4x + 10y^3x^2 - 10y^2x^3 + 5yx^4 - x^5$

7. Answers may vary.

9. $\frac{8!}{7!} = \frac{8 \cdot 7!}{7!} = 8$

11. $\frac{7!}{5!} = \frac{7 \cdot 6 \cdot 5!}{5!} = 7 \cdot 6 = 42$

13. $\frac{10!}{7!2!} = \frac{10 \cdot 9 \cdot 8 \cdot 7!}{7!2!} = \frac{10 \cdot 9 \cdot 8}{2 \cdot 1} = \frac{720}{2} = 360$

15. $\frac{8!}{6!0!} = \frac{8 \cdot 7 \cdot 6!}{6! \cdot 1} = 8 \cdot 7 = 56$

17. $(a+b)^7$
$= a^7 + 7a^6b + \frac{7 \cdot 6}{2!}a^5b^2 + \frac{7 \cdot 6 \cdot 5}{3!}a^4b^3 + \frac{7 \cdot 6 \cdot 5 \cdot 4}{4!}a^3b^4 + \frac{7 \cdot 6 \cdot 5 \cdot 4 \cdot 3}{5!}a^2b^5$
$+ \frac{7 \cdot 6 \cdot 5 \cdot 4 \cdot 3 \cdot 2}{6!}ab^6 + b^7$
$a^7 + 7a^6b + 21a^5b^2 + 35a^4b^3 + 35a^3b^4 + 21a^2b^5 + 7ab^6 + b^7$

19. $(a+2b)^5$
$= a^5 + 5a^4(2b) + \frac{5\cdot4}{2!}a^3(2b)^2 + \frac{5\cdot4\cdot3}{3!}a^2(2b)^3 + \frac{5\cdot4\cdot3\cdot2}{4!}a(2b)^4 + (2b)^5$
$= a^5 + 10a^4b + 40a^3b^2 + 80a^2b^3 + 80ab^4 + 32b^5$

21. $(q+r)^9$
$= q^9 + 9q^8r + \frac{9\cdot8}{2!}q^7r^2 + \frac{9\cdot8\cdot7}{3!}q^6r^3 + \frac{9\cdot8\cdot7\cdot6}{4!}q^5r^4 + \frac{9\cdot8\cdot7\cdot6\cdot5}{5!}q^4r^5 + \frac{9\cdot8\cdot7\cdot6\cdot5\cdot4}{6!}q^3r^6$
$+ \frac{9\cdot8\cdot7\cdot6\cdot5\cdot4\cdot3}{7!}q^2r^7 + \frac{9\cdot8\cdot7\cdot6\cdot5\cdot4\cdot3\cdot2}{8!}qr^8 + r^9$
$= q^9 + 9q^8r + 36q^7r^2 + 84q^6r^3 + 126q^5r^4 + 126q^4r^5 + 84q^3r^6 + 36q^2r^7 + 9qr^8 + r^9$

23. $(4a+b)^5$
$= (4a)^5 + 5(4a)^4b + \frac{5\cdot4}{2!}(4a)^3b^2 + \frac{5\cdot4\cdot3}{3!}(4a)^2b^3 + \frac{5\cdot4\cdot3\cdot2}{4!}(4a)b^4 + b^5$
$= 1024a^5 + 1280a^4b + 640a^3b^2 + 160a^2b^3 + 20ab^4 + b^5$

25. $(5a-2b)^4$
$= (5a)^4 + 4(5a)^3(-2b) + \frac{4\cdot3}{2!}(5a)^2(-2b)^2 + \frac{4\cdot3\cdot2}{3!}(5a)(-2b)^3 + (-2b)^4$
$= 625a^4 - 1000a^3b + 600a^2b^2 - 160ab^3 + 16b^4$

27. $(2a+3b)^3$
$= (2a)^3 + 3(2a)^2(3b) + \frac{3\cdot2}{2!}(2a)(3b)^2 + (3b)^3$
$= 8a^3 + 36a^2b + 54ab^2 + 27b^3$

29. $(x+2)^5$
$= x^5 + 5x^4(2) + \frac{5\cdot4}{2!}x^3(2^2) + \frac{5\cdot4\cdot3}{3!}x^2(2^3) + \frac{5\cdot4\cdot3\cdot2}{4!}x(2^4) + 2^5$
$= x^5 + 10x^4 + 40x^3 + 80x^2 + 80x + 32$

31. 5th term of $(c-d)^5$ corresponds to $r=4$:
$\frac{5!}{4!(5-4)!}c^{5-4}(-d)^4 = 5cd^4$

33. 8th term of $(2c+d)^7$ corresponds to $r=7$:
$\frac{7!}{7!(7-7)!}(2c)^{7-7}d^7 = d^7$

35. 4th term of $(2r-s)^5$ corresponds to $r=3$:
$\frac{5!}{3!(5-3)!}(2r)^{5-3}(-s)^3 = -40r^2s^3$

37. 3rd term of $(x+y)^4$ corresponds to $r=2$:
$\frac{4!}{2!(4-2)!}x^{4-2}y^2 = 6x^2y^2$

39. 2nd term of $(a+3b)^{10}$ corresponds to $r=1$:
$\frac{10!}{1!(10-1)!}a^{10-1}(3b)^1 = 30a^9b$

41. $f(x)=|x|$

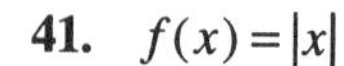

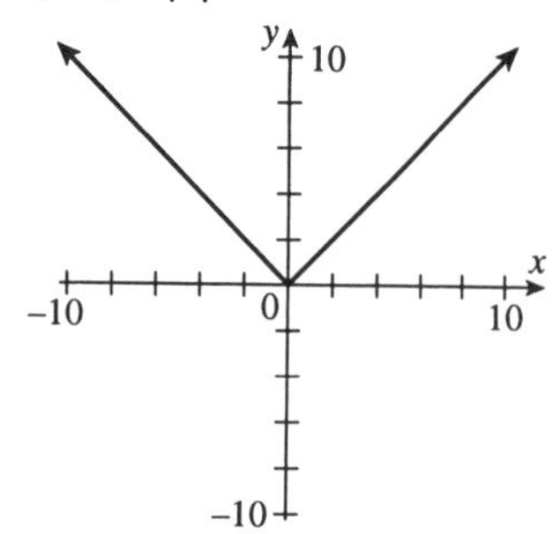

Not one-to-one

43. $H(x)=2x+3$

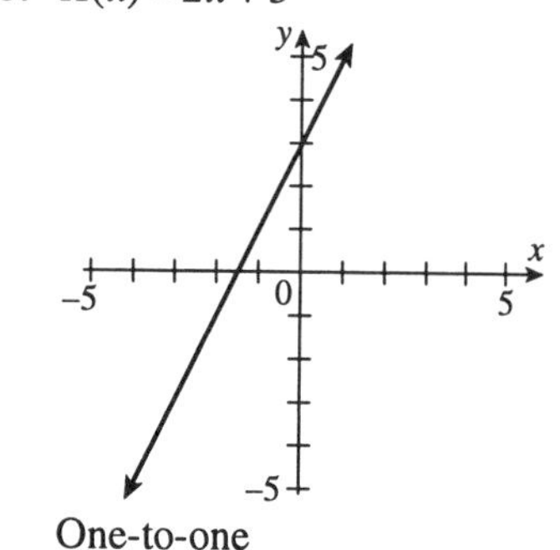

One-to-one

45. $f(x)=x^2+3$

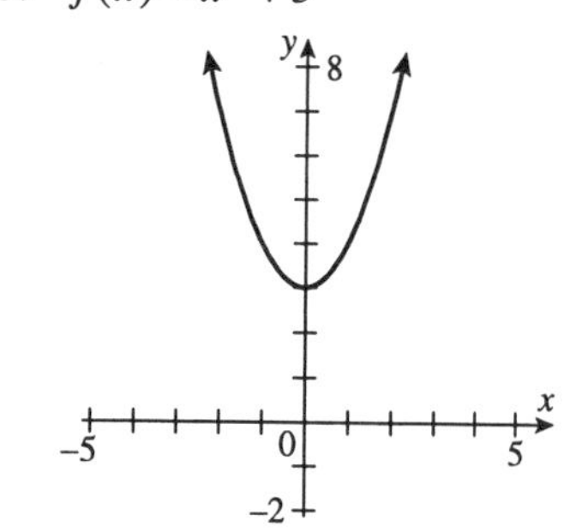

Chapter 11 - Review

1. $a_n=-3n^2$

$a_1=-3(1)^2=-3$

$a_2=-3(2)^2=-12$

$a_3=-3(3)^2=-27$

$a_4=-3(4)^2=-48$

$a_5=-3(5)^2=-75$

or −3, −12, −27, −48, −75

2. $a_n=n^2+2n$

$a_1=1^2+2(1)=1+2=3$

$a_2=2^2+2(2)=4+4=8$

$a_3=3^2+2(3)=9+6=15$

$a_4=4^2+2(4)=16+8=24$

$a_5=5^2+2(5)=25+10=35$

3. $a_n=\dfrac{(-1)^n}{100}$

$a_{100}=\dfrac{(-1)^{100}}{100}=\dfrac{1}{100}$

4. $a_n=\dfrac{2n}{(-1)^2}$

$a_{50}=\dfrac{2(50)}{(-1)^2}=100$

5. $\frac{1}{6}, \frac{1}{12}, \frac{1}{18}, \ldots$

or $\frac{1}{6\cdot 1}, \frac{1}{6\cdot 2}, \frac{1}{6\cdot 3}, \ldots$

In general, $a_n=\dfrac{1}{6n}$

6. −1, 4, −9, 16, …

$a_n=(-1)^n n^2$

7. $a_n=32n-16$

$a_5=32(5)-16=160-16=144$ ft

$a_6=32(6)-16=192-16=176$ ft

$a_7=32(7)-16=224-16=208$ ft

8. $a_n=100(2)^{n-1}$

$10,000=100(2)^{n-1}$

$100=2^{n-1}$

$\log 100=(n-1)\log 2$

$n=\dfrac{\log 100}{\log 2}+1=7.6$

Eighth day culture will be at least 10,000.

Originally, $a_1=100(2)^0=100.$

9. 450, 1350, 4050, 12,150; 36,450

We predict the number of infected people to be 36,450 in 1992.

10. $a_n = 50 + (n-1)8$
$a_1 = 50$
$a_2 = 50 + 8 = 58$
$a_3 = 50 + 2(8) = 66$
$a_4 = 50 + 3(8) = 74$
$a_5 = 50 + 4(8) = 82$
$a_6 = 50 + 5(8) = 90$
$a_7 = 50 + 6(8) = 98$
$a_8 = 50 + 7(8) = 106$
$a_9 = 50 + 8(8) = 114$
$a_{10} = 50 + 9(8) = 122$

11. $a_1 = -2$ and $r = \frac{2}{3}$
$-2, -\frac{4}{3}, -\frac{8}{9}, -\frac{16}{27}, -\frac{32}{81}$

12. $a_n = 12 + (n-1)(-1.5)$
$a_1 = 12$
$a_2 = 12 + (1)(-1.5) = 10.5$
$a_3 = 12 + 2(-1.5) = 9$
$a_4 = 12 + 3(-1.5) = 7.5$
$a_5 = 12 + 4(-1.5) = 6$

13. $a_1 = -5,\ d = 4,$ and $n = 30$
$a_{30} = 5 + (30-1)4 = -5 + 116 = 111$

14. $a_n = 2 + (n-1)\frac{3}{4}$
$a_{11} = 2 + 10\left(\frac{3}{4}\right) = \frac{38}{4} = \frac{19}{2}$

15. 12, 7, 2, ⋯
$a_1 = 12,\ d = -5,$ and $n = 20$
$a_{20} = 12 + (20-1)(-5) = 12 - 95 = -83$

16. $a_n = a_1 r^{n-1}$
$a_n = 4\left(\frac{3}{2}\right)^{n-1}$
$a_6 = 4\left(\frac{3}{2}\right)^{6-1} = 4\left(\frac{3}{2}\right)^5 = \frac{4(243)}{32} = \frac{243}{8}$

17. $a_4 = 18$ and $a_{20} = 98$
Use the relationship:
$a_4 + 16d = a_{20}$
$18 + 16d = 98$
$16d = 80$
$d = 5$
Now use the relationship:
$a_4 = a_1 + 3d$
$18 = a_1 + 3(5)$
$18 = a_1 + 15$
$a_1 = 3$

18. $-48 = a_3 = a_1 r^{3-1}$
$192 = a_4 = a_1 r^{4-1}$
$-48 = a_1 r^2$
$192 = a_1 r^3$
Dividing we have $r = -4$.
Substitution gives $-48 = 16a_1$ or $a_1 = -3$

19. $\frac{3}{10}, \frac{3}{100}, \frac{3}{1000}, \cdots$ or
$\frac{3}{10}, \frac{3}{10^2}, \frac{3}{10^3}, \cdots$
In general, $a_n = \frac{3}{10^n}$

20. 50, 58, 66, ⋯
$a_n = 50 + (n-1)8$

21. $\frac{8}{3}, 4, 6, \cdots$
Geometric, since $\frac{4}{\frac{8}{3}} = \frac{6}{4} = \frac{3}{2}$
$a_1 = \frac{8}{3}$ and $r = \frac{3}{2}$

22. arithmetic $a_1 = -10.5,\ d = 4.4$

23. $7x, -14x, 28x$
Geometric, since
$\frac{-14x}{7x} = \frac{28x}{-14x} = -2$
$a_1 = 7x$ and $r = -2$

24. neither

25. $a_1 = 8$ and $r = 0.75$
8, 6, 4.5, 3.4, 2.5, 1.9
Yes, a ball that rebounds to a height of 2.5 feet after the fifth bounce is good, since $2.5 \geq 1.9$.

26. $a_n = 25 + (n-1)(-4)$
$a_n = 25 + 6(-4) = 1$

27. $a_1 = 1$ and $r = 2$
$a_n = 1 \cdot 2^{n-1}$ or
$a_n = 2^{n-1}$
$a_{10} = 2^{10-1} = 2^9 = \$512$
$a_{30} = 2^{30-1} = 2^{29} = \$536,870,912$

28. $a_n = a_1 r^{n-1}$
$a_5 = 30(.7)^4 = 7.203$ in.

29. $a_1 = 900$ and $d = 150$
$a_n = 900 + (n-1)150$
$a_6 = 900 + (6-1)150$
$= 900 + 750 = \$1650$/month

30. $\frac{1}{512}, \frac{1}{256}, \frac{1}{128}, \dots$
first fold $a_1 = \frac{1}{256}$, $r = 2$
$a_n = a_1 r^{n-1}$
$a_{15} = \frac{1}{256}(2)^{15-1} = 64$ inches

31. $\sum_{i=1}^{5}(2i-1)$
$= [2(1) - 1] + [2(2) - 1] + [2(3) - 1] + [2(4) - 1] + [2(5) - 1]$
$= 1 + 3 + 5 + 7 + 9 = 25$

32. $\sum_{i=1}^{5} i(i+2)$
$= 1(1 + 2) + 2(2 + 2) + 3(3 + 2) + 4(4 + 2) + 5(5 + 2)$
$= 3 + 8 + 15 + 24 + 35 = 85$

33. $\sum_{i=2}^{4} \frac{(-1)^i}{2i}$
$= \frac{(-1)^2}{2(2)} + \frac{(-1)^3}{2(3)} + \frac{(-1)^4}{2(4)}$
$= \frac{1}{4} - \frac{1}{6} + \frac{1}{8}$
$= \frac{6-4+3}{24} = \frac{5}{24}$

34. $\sum_{i=3}^{5} 5(-1)^{i-1}$
$= 5(-1)^{3-1} + 5(-1)^{4-1} + 5(-1)^{5-1}$
$= 5(1) + 5(-1) + 5(1) = 5$

35. $a_n = (n-3)(n+2)$
$S_4 = (1-3)(1+2) + (2-3)(2+2) + (3 - 3)(3 + 2) + (4 - 3)(4 + 2)$
$= -6 - 4 + 0 + 6 = -4$

36. $a_n = n^2$
$S_6 = (1)^2 + (2)^2 + (3)^2 + (4)^2 + (5)^2 + (6)^2$
$= 1 + 4 + 9 + 16 + 25 + 36 = 91$

37. $a_n = -8 + (n-1)3$
$a_1 = -8 + (1-1)3 = -8 + 0 = -8$
$a_2 = -8 + (2-1)3 = -8 + 3 = -5$
$a_3 = -8 + (3-1)3 = -8 + 6 = -2$
$a_4 = -8 + (4-1)3 = -8 + 9 = 1$
$a_5 = -8 + (5-1)3 = -8 + 12 = 4$
So $S_5 = -8 + (-5) + (-2) + 1 + 4 = -10$.

38. $a_n = 5(4)^{n-1}$
$S_3 = 5(4)^0 + 5(4)^1 + 5(4)^2$
$= 5 + 20 + 8 = 105$

39. $1 + 3 + 9 + 27 + 81 + 243$
$= 3^0 + 3^1 + 3^2 + 3^3 + 3^4 + 3^5$
$= \sum_{i=1}^{6} 3^{i-1}$

40. $6 + 2 + (-2) + (-6) + (-10) + (-14) + (-18)$
$a_n = 6 + (n-1)(-4)$
$\sum_{i=1}^{7} 6 + (i-1)(-4)$

41. $\frac{1}{4}+\frac{1}{16}+\frac{1}{64}+\frac{1}{256}=\frac{1}{4^1}+\frac{1}{4^2}+\frac{1}{4^3}+\frac{1}{4^4}$

$\sum_{i=1}^{4}\frac{1}{4^i}$

42. $1+\left(-\frac{3}{2}\right)+\frac{9}{4}$

$a_n=1\left(-\frac{3}{2}\right)^{n-1}$

$\sum_{i=1}^{3}\left(-\frac{3}{2}\right)^{i-1}$

43. $a_1=20$ and $r=2$

$a_n=20(2)^n$ represents the number of yeast where n represents the number of 8-hr periods.. Since 48 = 6(8) here, $n=6$

Now, $a_6=20(2)^6=1280$ yeast

44. $a_n=n^2+2n-1$

$a_4=(4)^2+2(4)-1=23$ cranes

$\sum_{i=1}^{4}i^2+2i-1$

$=(1+2-1)+(4+4-1)$
$+(9+6-1)+(16+8-1)$
$=2+7+14+23=46$ cranes

45. First, Job A:

$a_1=19,500$ and $d=1100$

$a_5=19,500+(5-1)1100$

$=19,500+4400=\$23,900$

Now, Job B:

$a_1=21,000$ and $d=700$

$a_5=21,000+(5-1)700$

$=21,000+2800=\$23,800$

46. $a_n=200(0.5)^n$

$a_3=200(0.5)^3=25$ kg

$\sum_{i=1}^{3}200(0.5)^i$

$=200(0.5)+200(0.5)^2+200(0.5)^3$

$=100+50+25=175$ kg

47. 15, 19, 23, ⋯

$a_1=15$ and $d=4$

$S_6=\frac{6}{2}[2(15)+(6-1)4]$

$=3[30+20]=3(50)=150$

48. 5, −10, 20, ⋯

$a_1=5,\ r=-2$

$S_n=\frac{a_1(1-r^n)}{1-r}$

$S_9=\frac{5(1-(-2)^9)}{1-(-2)}=855$

49. $a_1=1,\ d=2$, and $n=30$

$S_{30}=\frac{30}{2}[2(1)+(30-1)2]$

$=15[2+58]=15(60)=900$

50. 7, 14, 21, 28, ⋯

$a_n=7+(n-1)7$

$a_{20}=7+(20-1)7=140$

$S_n=\frac{n}{2}(a_1+a_n)$

$S_{20}=\frac{20}{2}(7+140)=1470$

51. 8, 5, 2, ⋯

$a_1=8,\ d=-3$, and $n=20$

$S_{20}=\frac{20}{2}[2(8)+(20-1)(-3)]$

$=10[16-57]=10(-41)=-410$

52. $\frac{3}{4},\ \frac{9}{4},\ \frac{27}{4},\ \cdots$

$a_1=\frac{3}{4},\ r=3$

$S_n=\frac{a_1(1-r^n)}{1-r}$

$S_8=\frac{\frac{3}{4}(1-3^8)}{1-3}=2460$

53. $a_1=6$ and $r=5$

$S_4=\frac{6(1-5^4)}{1-5}=\frac{-3}{2}(1-625)$

$=\frac{3}{2}(624)=936$

54. $a_1 = -3,\ d = -6$

$a_n = -3 + (n-1)(-6)$

$a_{100} = -3 + (100-1)(-6) = -597$

$S_n = \frac{n}{2}(a_1 + a_n)$

$S_{100} = \frac{100}{2}(-3 + (-597))$

$S_{100} = -30,000$

55. $5,\ \frac{5}{2},\ \frac{5}{4},\ \cdots$

$a_1 = 5$ and $r = \frac{1}{2}$

$S_\infty = \frac{5}{1-\frac{1}{2}} = \frac{5}{\frac{1}{2}} = 10$

56. $18,\ -2,\ \frac{2}{9},\ \cdots$

$a_1 = 18,\ r = -\frac{1}{9}$

$S_\infty = \frac{a_1}{1-r} = \frac{18}{1+\frac{1}{9}} = \frac{81}{5}$

57. $-20,\ -4,\ -\frac{4}{5},\ \cdots$

$a_1 = -20$ and $r = \frac{1}{5}$

$S_\infty = \frac{-20}{1-\frac{1}{5}} = \frac{-20}{\frac{4}{5}} = -25$

58. $0.2,\ 0.02,\ 0.002,\ \cdots$

$a_1 = 0.2,\ r = \frac{1}{10}$

$S_\infty = \frac{a_1}{1-r} = \frac{.2}{1-\frac{1}{10}} = \frac{2}{9}$

59. $a_1 = 20,000,\ r = 1.15$, and $n = 4$.

$a_4 = 20,000(1.15)^{4-1} = \$30,418$

earned in his fourth year.

$S_4 = \frac{20,000(1-1.15^4)}{1-1.15} = \$99,868$

earned in his first four years.

60. $a_n = 40(.8)^{n-1}$

$a_4 = 40(.8)^{4-1} = 20.48$ min.

$S_n = \frac{a_1(1-r^n)}{1-r} = \frac{40(1-.8^4)}{1-.8} = 118$ min.

61. $a_1 = 100,\ d = -7$, and $n = 7$

$a_7 = 100 + (7-1)(-7) = 100 - 42$

= \$58 rent paid for the seventh day.

$S_7 = \frac{7}{2}[2(100) + (7-1)(-7)]$

$= \frac{7}{2}[200 - 42] = \frac{7}{2}(158)$

= \$553 rent paid for the first seven days.

62. $a_1 = 15,\ r = .8$

$S_\infty = \frac{a_1}{1-r} = \frac{15}{1-.8} = 75$ feet downward

$a_1 = 12,\ r = .8$

$S_\infty = \frac{a_1}{1-r} = \frac{12}{1-.8} = 60$ feet upward

The total is 135 feet.

63. $1800,\ 600,\ 200,\ \cdots$

$a_1 = 1800,\ r = \frac{1}{3}$, and $n = 6$

$S_6 = 1800\frac{\left(1-\left(\frac{1}{3}\right)^6\right)}{1-\frac{1}{3}}$

$= 2700\left(1 - \frac{1}{729}\right)$

$= 2700\left(\frac{728}{729}\right) = \frac{72800}{27}$

≈ 2696 mosquitoes killed during the first six days after the spraying

64. $1800,\ 600,\ 200,\ \cdots$

For which n is $a_n > 1$?

$a_n = a_1 r^{n-1} = 1800\left(\frac{1}{3}\right)^{n-1} > 1$

$(n-1)\log\left(\frac{1}{3}\right) > \log\frac{1}{1800}$

$(n-1)(-0.4771213) > (-3.2552725)$

$n - 1 < 6.8$

$n < 7.8$

No longer effective on the 8th day.

65. $0.55\overline{5} = .5 + .05 + .005 + \cdots$
$a_1 = .5$ and $r = .1$
$S_\infty = \frac{.5}{1-.1} = \frac{.5}{.9} = \frac{5}{9}$
Thus, $0.55\overline{5} = \frac{5}{9}$

66. 27, 30, 33, ⋯
$a_n = 27 + (n-1)(3)$
$a_{20} = 27 + (20-1)(3) = 84$
$S_n = \frac{n}{2}(a_1 + a_n)$
$S_{20} = \frac{20}{2}(27+84) = 1110$

67. $(x+z)^5 = 1 \cdot x^5 + 5 \cdot x^4 z + 10 \cdot x^3 z^2 + 10 \cdot x^2 z^3 + 5 \cdot xz^4 + 1 \cdot z^5$
$= x^5 + 5x^4 z + 10x^3 z^2 + 10x^2 z^3 + 5xz^4 + z^5$

68. $(y-r)^6 = y^6 - 6y^5 r + 15y^4 r^2 - 20y^3 r^3 + 15y^2 r^4 - 6yr^5 + r^6$

69. $(2x+y)^4 = 1 \cdot (2x)^4 + 4 \cdot (2x)^3 y + 6 \cdot (2x)^2 y^2 + 4 \cdot (2x) y^3 + 1 \cdot y^4$
$= 16x^4 + 32x^3 y + 24x^2 y^2 + 8xy^3 + y^4$

70. $(3y-z)^4 = (3y)^4 + 4(3y)^3(-z) + 6(3y)^2(-z)^2 + 4(3y)(-z)^3 + (-z)^4$
$= 81y^4 - 108y^3 z + 54y^2 z^2 - 12yz^3 + z^4$

71. $(b+c)^8$
$= b^8 + 8b^7 c + \frac{8 \cdot 7}{2!} b^6 c^2 + \frac{8 \cdot 7 \cdot 6}{3!} b^5 c^3 + \frac{8 \cdot 7 \cdot 6 \cdot 5}{4!} b^4 c^4 + \frac{8 \cdot 7 \cdot 6 \cdot 5 \cdot 4}{5!} b^3 c^5 + \frac{8 \cdot 7 \cdot 6 \cdot 5 \cdot 4 \cdot 3}{6!} b^2 c^6$
$+ \frac{8 \cdot 7 \cdot 6 \cdot 5 \cdot 4 \cdot 3 \cdot 2}{7!} bc^7 + c^8$
$= b^8 + 8b^7 c + 28b^6 c^2 + 56b^5 c^3 + 70b^4 c^4 + 56b^3 c^5 + 28b^2 c^6 + 8bc^7 + c^8$

72. $(x-w)^7 = x^7 - 7x^6 w + 21x^5 w^2 - 35x^4 w^3 + 35x^3 w^4 - 21x^2 w^5 + 7xw^6 - w^7$

73. $(4m-n)^4 = (4m + (-n))^4$
$= (4m)^4 + 4(4m)^3(-n) + \frac{4 \cdot 3}{2!}(4m)^2(-n)^2 + \frac{4 \cdot 3 \cdot 2}{3!}(4m)(-n)^3$
$= 256m^4 - 256m^3 n + 96m^2 n^2 - 16mn^3 + n^4$

74. $(p-2r)^5 = p^5 + 5p^4(-2r) + 10p^3(-2r)^2 + 10p^2(-2r)^3 + 5p(-2r)^4 + (-2r)^5$
$= p^5 - 10p^4 r + 40p^3 r^2 - 80p^2 r^3 + 80pr^4 - 32r^5$

75. 4th term of $(a+b)^7$ corresponds to $r = 3$.
$\frac{7!}{3!(7-3)!} a^{7-3} b^3 = 35a^4 b^3$

76. Given $(y+2z)^{10}$. The 11th term is $\frac{n!}{r!(n-r)!} a^{n-r} b^r$ where $n = 10$, $r = 10$, $a = y$, $b = 2z$.
$\frac{10!}{10!0!} y^{10-10} (2z)^{10} = 1024z^{10}$

Chapter 11 - Test

1. $a_n = \frac{(-1)^n}{n+4}$

$a_1 = \frac{(-1)^1}{1+4} = -\frac{1}{5}$

$a_2 = \frac{(-1)^2}{2+4} = \frac{1}{6}$

$a_3 = \frac{(-1)^3}{3+4} = -\frac{1}{7}$

$a_4 = \frac{(-1)^4}{4+4} = \frac{1}{8}$

$a_5 = \frac{(-1)^5}{5+4} = -\frac{1}{9}$

or $-\frac{1}{5}, \frac{1}{6}, -\frac{1}{7}, \frac{1}{8}, -\frac{1}{9}$

2. $a_n = \frac{3}{(-1)^n}$

$a_1 = \frac{3}{(-1)^1} = -3$

$a_2 = \frac{3}{(-1)^2} = 3$

$a_3 = \frac{3}{(-1)^3} = -3$

$a_4 = \frac{3}{(-1)^4} = 3$

$a_5 = \frac{3}{(-1)^5} = -3$

or –3, 3, –3, 3, –3

3. $a_n = 10 + 3(n-1)$

$a_{80} = 10 + 3(80-1) = 10 + 237 = 247$

4. $a_n = (n+1)(n-1)(-1)^n$

$a_{200} = (200+1)(200-1)(-1)^{200}$

$= 200^2 - 1^2 = 40000 - 1 = 39{,}999$

5. $\frac{2}{5}, \frac{2}{25}, \frac{2}{125}, \cdots$ or $\frac{2}{5}, \frac{2}{5^2}, \frac{2}{5^3}, \cdots$

In general, $a_n = \frac{2}{5^n}$

6. $-9, 18, -27, 36, \cdots$

or $(-1)^1 9\cdot 1, (-1)^2 9\cdot 2,$

$(-1)^3 9\cdot 3, (-1)^4 9\cdot 4, \cdots$

In general, $a_n = (-1)^n 9n$

7. $a_n = 5(2)^{n-1}$

Geometric sequence with $a_1 = 5$ and $r = 2$.

$S_5 = \frac{5(1-2^5)}{1-2} = -5(1-32) = 5(31) = 155$

8. $a_n = 18 + (n-1)(-2)$

Arithmetic sequence with $a_1 = 18$ and $d = -2$.

$S_{30} = \frac{30}{2}[2(18) + (30-1)(-2)]$

$= 15[36 - 58] = 15(-22) = -330$

9. $a_1 = 24$ and $r = \frac{1}{6}$

$S_\infty = \frac{24}{1-\frac{1}{6}} = \frac{24}{\frac{5}{6}} = \frac{144}{5}$

10. $\frac{3}{2}, -\frac{3}{4}, \frac{3}{8}, \cdots$

$a_1 = \frac{3}{2}$ and $r = -\frac{1}{2}$

$S_\infty = \frac{\frac{3}{2}}{1-\left(-\frac{1}{2}\right)} = \frac{\frac{3}{2}}{\frac{3}{2}} = 1$

11. $\sum_{i=1}^{4} i(i-2)$

$= 1(1-2) + 2(2-2) + 3(3-2) + 4(4-2)$

$= -1 + 0 + 3 + 8 = 10$

12. $\sum_{i=2}^{4} 5(2)^i(-1)^{i-1}$

$= 5(2)^2(-1)^{2-1} + 5(2)^3(-1)^{3-1} + 5(2)^4(-1)^{4-1} = -20 + 40 - 80 = -60$

13. $(a-b)^6 = (a+(-b))^6$
$= 1\cdot a^6 + 6\cdot a^5(-b) + 15\cdot a^4(-b)^2 + 200^3(-b)^3 + 15\cdot a^2(-b)^4 + 6\cdot a(-b)^5 + 1\cdot(-b)^6$
$= a^6 - 6a^5b + 15a^4b^2 - 20a^3b^3 + 15a^2b^4 - 6ab^5 + b^6$

14. $(2x+y)^5 = 1\cdot(2x)^5 + 5\cdot(2x)^4y + 10\cdot(2x)^3y^2 + 10\cdot(2x)^2y^3 + 5\cdot(2x)y^4 + y^5$
$= 32x^5 + 80x^4y + 80x^3y^2 + 40x^2y^3 + 10xy^4 + y^5$

15. $(y+z)^8 = y^8 + 8y^7z + \frac{8\cdot7}{2!}y^6z^2 + \frac{8\cdot7\cdot6}{3!}y^5z^3 + \frac{8\cdot7\cdot6\cdot5}{4!}y^4z^4 + \frac{8\cdot7\cdot6\cdot5\cdot4}{5!}y^3z^5$
$+ \frac{8\cdot7\cdot6\cdot5\cdot4\cdot3}{6!}y^2z^6 + \frac{8\cdot7\cdot6\cdot5\cdot4\cdot3\cdot2}{7!}yz^7 + z^8$
$= y^8 + 8y^7z + 28y^6z^2 + 56y^5z^3 + 70y^4z^4 + 56y^3z^5 + 28y^2z^6 + 8yz^7 + z^8$

16. $(2p+r)^7 = (2p)^7 + 7(2p)^6r + \frac{7\cdot6}{2!}(2p)^5r^2 + \frac{7\cdot6\cdot5}{3!}(2p)^4r^3 + \frac{7\cdot6\cdot5\cdot4}{4!}(2p)^3r^4$
$+ \frac{7\cdot6\cdot5\cdot4\cdot3}{5!}(2p)^2r^5 + \frac{7\cdot6\cdot5\cdot4\cdot3\cdot2}{6!}(2p)r^6 + r^7$
$= 128p^7 + 448p^6r + 672p^5r^2 + 560p^4r^3 + 280p^3r^4 + 84p^2r^5 + 14pr^6 + r^7$

17. $a_n = 250 + 75(n-1)$
$a_{10} = 250 + 75(10-1)$
$= 250 + 675 = 925$ people in the town at the beginning of the tenth year.
$a_1 = 250 + 75(1-1) = 250$ people in the town at the beginning of the first year.

18. 1, 3, 5, ...
$a_1 = 1,\ \ d = 2,$ and $n = 8$
$a_8 = 1 + (8-1)2 = 1 + 14 = 15$
We want 1+ 3 + 5 + ... + 15
$s_8 = \frac{8}{2}[1+15] = 4(16) = 64$ shrubs planted in the 8 rows.

19. $a_1 = 80,\ \ r = \frac{3}{4},$ and $n = 4$
$a_4 = 80\left(\frac{3}{4}\right)^{4-1} = 80\left(\frac{27}{64}\right) = \frac{135}{4}$ or
33.75 cm on the 4th swing.
$S_4 = \frac{80\left(1-\left(\frac{3}{4}\right)^4\right)}{1-\frac{3}{4}} = 320\left(1-\frac{81}{256}\right)$
$= 320\left(\frac{175}{256}\right) = \frac{875}{4} = 218.75$ cm on the first 4 swings.

20. $a_1 = 80$ and $r = \frac{3}{4}$
$S_\infty = \frac{80}{1-\frac{3}{4}} = \frac{80}{\frac{1}{4}} = 320$ cm before the pendulum comes to rest.

21. 16, 48, 80, ...
$a_1 = 16,\ \ d = 32,$ and $n = 10$
$a_{10} = 16 + (10-1)32 = 16 + 288$
$= 304$ feet fallen during the 10th second.
$S_{10} = \frac{10}{2}[16+304] = 5(320) = 1600$ ft fallen during the first 10 seconds.

22. $0.42\overline{42} = .42 + .0042 + .000042 + \cdots$
$a_1 = .42$ and $r = .01$
$S_\infty = \frac{.42}{1-.01} = \frac{.42}{.99} = \frac{42}{99} = \frac{14}{33}$
Thus, $0.42\overline{42} = \frac{14}{33}$.

Chapter 11 - Cumulative Review

1. $25 - 3.5c = 6$
$25 - 6 = 3.5c$
$19 = 3.5c$
$\frac{19}{3.5} = c$

$\frac{38}{7} = c$
$\left\{\frac{38}{7}\right\}$

2. $|x| \le 3$
$x \le 3$ or $x \ge -3$
$[-3, 3]$

3. a. Function

b. Not a function

c. Function

4. $2x + 3y = -6$
$3y = -2x - 6$
$y = -\frac{2}{3}x - 2$
$m = -\frac{2}{3}$
perpendicular line has $m = \frac{3}{2}$
$y - 4 = \frac{3}{2}(x - 4)$
$y = \frac{3}{2}x - 6 + 4$
$y = \frac{3}{2}x - 2$
$f(x) = \frac{3}{2}x - 2$

5. $y = 50x$
$y = 30x + 10{,}000$
$50x = 30x + 10{,}000$
$20x = 10{,}000$
$x = 500$
500 units to break even

6. Let x be the second number.
Then $x - 4$ is the first number.
$4(x - 4) = 6 + 2x$
$4x - 16 = 6 + 2x$
$4x - 2x = 16 + 6$
$2x = 22$
$x = 11$
$x - 4 = 7$
The numbers are 7 and 11.

7. a. $5^{-2} = \frac{1}{5^2} = \frac{1}{25}$

b. $2x^{-3} = \frac{2}{x^3}$

c. $(3x)^{-1} = \frac{1}{(3x)^1} = \frac{1}{3x}$

d. $\frac{m^5}{m^{15}} = \frac{1}{m^{15-5}} = \frac{1}{m^{10}}$

e. $\frac{3^3}{3^6} = \frac{1}{3^{6-3}} = \frac{1}{3^3} = \frac{1}{27}$

f. $2^{-1} + 3^{-2} = \frac{1}{2^1} + \frac{1}{3^2} = \frac{1}{2} + \frac{1}{9}$
$= \frac{9}{18} + \frac{2}{18} = \frac{11}{18}$

g. $\frac{1}{t^{-5}} = t^5$

8.
$$\begin{array}{rrrrr} & 12z^5 & & -12z^3 & +z \\ - & & (-3z^4 & +z^3 & +12z) \\ \hline & 12z^5 & +3z^4 & -13z^3 & -11z \end{array}$$

9. $y^3 - 64 = y^3 - 4^3 = (y-4)(y^2 + 4y + 16)$

10. $\frac{2x-1}{2x^2 - 9x - 5} + \frac{x+3}{6x^2 - x - 2}$
$= \frac{2x-1}{(2x+1)(x-5)} + \frac{x+3}{(2x+1)(3x-2)}$
$= \frac{(2x-1)(3x-2) + (x+3)(x-5)}{(2x+1)(x-5)(3x-2)}$
$= \frac{6x^2 - 7x + 2 + x^2 - 2x - 15}{(2x+1)(x-5)(3x-2)}$
$= \frac{7x^2 - 9x - 13}{(2x+1)(x-5)(3x-2)}$

11.
$$\begin{array}{r|l} & 3x^2+2x+3 \\ x^2-1 & 3x^4+2x^3 \qquad -8x+6 \\ & \underline{3x^4 \qquad -3x^2} \\ & 2x^3+3x^2-8x \\ & \underline{2x^3 \qquad -2x} \\ & 3x^2-6x+6 \\ & \underline{3x^2 \qquad -3} \\ & -6x+9 \end{array}$$

Answer: $3x^2+2x+3-\frac{6x+9}{x^2-1}$

12. a. $\sqrt{\frac{25}{49}}=\frac{\sqrt{25}}{\sqrt{49}}=\frac{5}{7}$

b. $\sqrt[3]{\frac{8}{27}}=\frac{\sqrt[3]{8}}{\sqrt[3]{27}}=\frac{2}{3}$

c. $\sqrt{\frac{x}{9}}=\frac{\sqrt{x}}{\sqrt{9}}=\frac{\sqrt{x}}{3}$

d. $\sqrt[4]{\frac{3}{16y^4}}=\frac{\sqrt[4]{3}}{\sqrt[4]{16y^4}}=\frac{\sqrt[4]{3}}{2y}$

13. $\sqrt{-10x-1}+3x=0$
$\sqrt{-10x-1}=-3x$
$-10x-1=9x^2$
$0=9x^2+10x+1$
$0=(9x+1)(x+1)$
$9x+1=0$ or $x+1=0$
$x=-\frac{1}{9}$ or $x=-1$
$\left\{-\frac{1}{9}, -1\right\}$

14. $3x^2+16x+5=0$
$(3x+1)(x+5)=0$
$3x+1=0$ or $x+5=0$
$x=-\frac{1}{3}$ or $x=-5$
$\left\{-\frac{1}{3}, -5\right\}$

15. $f(x)=3x^2+3x+1$

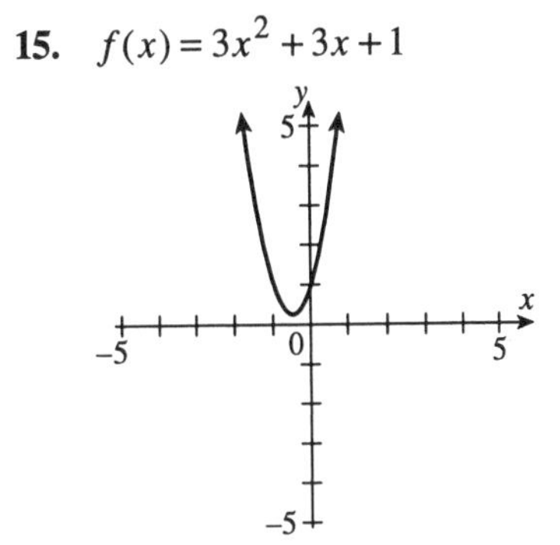

16. $\frac{x^2}{9}+\frac{y^2}{16}=1$

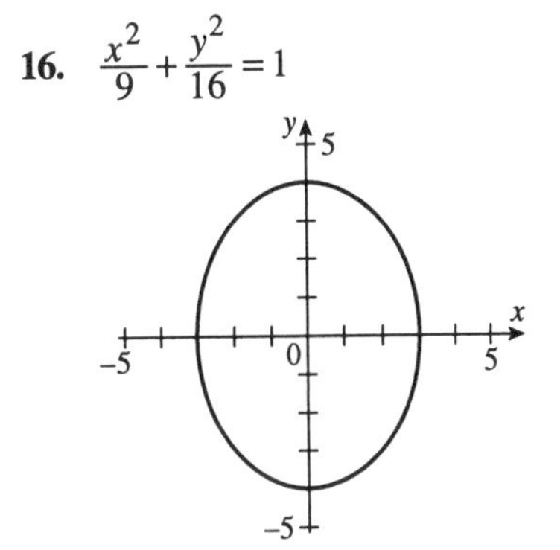

17. $\begin{cases} x^2+2y^2=10 \\ x^2-y^2=1 \end{cases}$

$$\begin{array}{r} x^2+2y^2=10 \\ \underline{-x^2+y^2=-1} \\ 3y^2=9 \\ y^2=3 \\ y=\pm\sqrt{3} \end{array}$$

$y=\sqrt{3}$
$x^2=y^2+1$
$x^2=(\sqrt{3})^2+1$
$x^2=4$
$x=\pm 2$

$y=-\sqrt{3}$
$x^2=y^2+1$
$x^2=(-\sqrt{3})^2+1$
$x^2=4$
$x=\pm 2$

$\{(2, \sqrt{3}), (2, -\sqrt{3}), (-2, \sqrt{3}), (-2, -\sqrt{3})\}$

18. a. one-to-one

b. not one-to-one

c. one-to-one

d. not one-to-one

e. not one-to-one

19. $A = P\left(1+\frac{r}{n}\right)^{nt}$

$A = 1600\left(1+\frac{0.09}{12}\right)^{12\cdot 5}$
$A = \$2505.09$

20. a. $\log_4 \frac{1}{4} = x$

$4^x = \frac{1}{4}$
$x = -1$
$\{-1\}$

b. $\log_5 x = 3$

$5^3 = x$
$125 = x$
$\{125\}$

c. $\log_x 25 = 2$

$x^2 = 25$
$x = 5$
$\{5\}$

d. $\log_3 1 = x$

$3^x = 1$
$x = 0$
$\{0\}$

e. $\log_b 1 = x$

$b^x = 1$
$x = 0$
$\{0\}$

21. $\log(x+2) - \log x = 2$

$\log \frac{x+2}{x} = 2$

$10^2 = \frac{x+2}{x}$

$100 = \frac{x+2}{x}$

$100x = x + 2$
$99x = 2$

$x = \frac{2}{99}$

$\left\{\frac{2}{99}\right\}$

22. $a_n = \frac{(-1)^n}{3n}$

a. $a_1 = \frac{(-1)^1}{3(1)} = -\frac{1}{3}$

b. $a_8 = \frac{(-1)^8}{3(8)} = \frac{1}{24}$

c. $a_{100} = \frac{(-1)^{100}}{3(100)} = \frac{1}{300}$

d. $a_{15} = \frac{(-1)^{15}}{3(15)} = -\frac{1}{45}$

23. $a_1 = 20,000,\ d = 800$
$a_n = 20,000 + (n-1)(800)$
$= 19,200 + 800n$
$a_4 = 19,200 + 800(4) = 22,400$
$22,400

24. Let $a = x, b = 2y$
$(x+2y)^5$
$= x^5 + \frac{5}{1!}x^4(2y) + \frac{5\cdot 4}{2!}x^3(2y)^2$
$+ \frac{5\cdot 4\cdot 3}{3!}x^2(2y)^3 + \frac{5\cdot 4\cdot 3\cdot 2}{4!}x(2y)^4$
$+(2y)^5$
$= x^5 + 10x^4y + 40x^3y^2 + 80x^2y^3$
$+ 80xy^4 + 32y^5$